The Mechanism of Induced Seismicity

Edited by
Cezar I. Trifu

2002

Springer Basel AG

Reprint from Pure and Applied Geophysics
(PAGEOPH), Volume 159 (2002), No. 1-3

Editors:

Cezar I. Trifu
Engineering Seismology Group Canada Inc.
1, Hyperion Court
Kingston, Ontario K7K 7G3
Canada
e-mail: trifu@esg.ca

A CIP catalogue record for this book is available from the Library of Congress, Washington D.C., USA

Deutsche Bibliothek Cataloging-in-Publication Data

The mechanism of induced seismicity / ed. by Cezar I. Trifu. - Springer Basel AG, 2002

(Pageoph topical volumes)
ISBN 3-7643-6653-2

ISBN 978-3-7643-6653-7 ISBN 978-3-0348-8179-1 (eBook)
DOI 10.1007/978-3-0348-8179-1

Originally published by Birkhäuser Verlag 2002
Softcover reprint of the hardcover 1st edition 2002
Member of the BertelsmannSpringer Publishing Group
Printed on acid-free paper produced from chlorine-free pulp

9 8 7 6 5 4 3 2 1

Contents

Pure appl. geophys. 159 (2002) 001–005
0033 – 4553/02/030001 – 05 $ 1.50 + 0.20/0

Pure and Applied Geophysics

Introduction

Over the last 12 years, Pure and Applied Geophysics has published six volumes on studies related to Induced Seismicity, raising its profile considerably. The journal's international level of peer-review played a significant role in establishing a high level of scientific credibility to the topic. Additionally, the relatively large amount of contributions published in these special volumes has increased the awareness to the diversity of problems that are not only of interest to the theoretical and applied seismology, but also to the rock and soil mechanics, mining, geotechnical, petroleum, and geothermal engineering.

It is perhaps symbolic that the journal has decided to publish another volume on Induced Seismicity in the first year of the new millennium. This is a time when the world's need for natural resources must be met by the use of more efficient, safer, and therefore implicitly smarter technologies for the extraction of minerals and hydrocarbons, the injection of fluids into the ground as related to the exploitation of geothermal energy and waste disposal, as well as geotechnical projects such as underground nuclear waste repositories, hydrocarbon storage caverns, tunnels, underground water purification plants, and water dams. These activities will demand a wide range of processes to be performed at increasingly greater depths, under constantly more difficult conditions, and will consequently require a superior level of control to ensure they are economically efficient and safe for the operators, their tools, and the environment.

It is well known that microseismic monitoring can provide direct, real-time observations of the occurrence of local instabilities in the ground. However, in order to achieve the desired level of control, the mechanism of induced seismicity must be fully understood. It is the characterization of the process of stress transfer through the adjacent rock mass that is the focus of this special volume. This extremely challenging task requires additional scientific advancements to be made in order to allow possible scenarios to be reliably considered in assessing the optimum strategy to implement at a site.

Different from all the other special volumes, the present collection of papers is not the result of a symposium, rather the result of an impressive response to a Call-for-Papers published in EOS (the newsletter of the American Geophysical Union). This volume contains 24 original contributions, authored by 70 different specialists from 12 countries in five continents. These studies cover an unprecedented wide range of applications, from mining-induced seismicity to geotechnical applications, from the monitoring of petroleum reservoirs to that of fluid injections in geothermal areas, and to seismicity associated with water reservoirs. Exploring current advances in the field of induced seismicity, the volume is a valuable source of information for specialists and provides reference material for graduate students.

The first group of papers includes seven contributions related to seismicity induced by mining or associated with the presence of mines. A study by Westaway concludes that an increase in seismicity of coastal northern California occurred during Spring several decades before the M 7.8 San Francisco earthquake of 1906. His analysis further suggested that this effect was caused by stress changes due to seasonal sediment loading near active strike-slip faults, transported during the process of hydraulic mining of gold. Ogasawara *et al.* examined the observations from a multidisciplinary monitoring array within the immediate proximity of the abandoned and flooded Ikuno mine before, during, and after the M 7.2 Kobe earthquake of 1995. Their results support the conclusion that the post-seismic increase in the microseismicity at the mine can be explained by the increase in the pore pressure under the dynamic stress changes associated with large earthquakes. In a subsequent study carried out at the same mine site, Ogasawara *et al.* analyzed the potential of heavy rainfall to inducing microseismicity. They found that the localized and delayed diffusion of water played a major role in this process, in spite of the slow change in the pore pressure, which was smaller than the direct stress change due to water load increase. Three contributions evaluate failure components of mining-induced events. Teyssoneyre *et al.* studied the mechanism and the source time function of a M 4.8 mine collapse based on regional body wave data. They concluded that the collapse was actually composed of two subevents, the first with a large isotropic component, and the second with an almost pure-shear mechanism. Domanski *et al.* apply the empirical Green's function deconvolution technique to analyze *P* waves from seven pure-shear events with M 2.1 to 2.9 from the Rudna copper mine, Poland. Their investigation indicates that most of the events are characterized by unilateral ruptures that propagated with relatively low velocity, between 0.3 and 0.6β. The use of an original seismic moment tensor inversion method for retrieving failure components is examined by Trifu and Shumila on microseismic events (M-1.2 to 0) which occurred at Kidd mine, Canada. Reliable results are derived for 21 events, of which seven are found to exhibit large positive isotropic components (72–76%), associated with incipient pillar bursting. The section on mining-induced seismicity concludes with a study by Milev and Spottiswoode, who investigate the effect of the rock properties on microseismicity recorded at several mines in the Witwatersrand Basin, South Africa. They suggested that the ratio between the cumulative seismic moment and the volume of elastic convergence is controlled by the rock type.

A second group of six papers comprises studies of induced microfracturing, relevant to geotechnical applications. Pettitt *et al.* analyze the excavation damaged zone around two large-diameter boreholes employed to test the viability of retrieving simulated nuclear waste canisters at the Äspö Hard Rock Laboratory, Sweden. Combining high-resolution *P*-wave velocity measurements and acoustic emission source locations, they attempt to provide a quantitative estimate of the effect of the damaged zone on the rock mass. Collins *et al.* examine the benefits of employing small-scale microseismic and acoustic emission systems to investigate the temporal fracture mechanics of microcrack formation associated with a tunnel sealing experiment at the Underground Research Laboratory nuclear waste test site in Canada. They associate microseismic events with clusters of acoustic emissions and outline that both types of sources are generally characterized by deviatoric failure components. Using the same experimental setup, Hazzard *et al.* employed a bonded-particle model to simulate shear microfractures induced by the tunnel excavation. Comparing the modeling results with information provided by the monitoring of microseismicity and acoustic emissions, the authors identify similarities in both the presence of foreshocks associated with macro-slip events, and the patterns of energy release during loading. Hildyard and Young attempt to model the seismic wave interaction with fractured rock surrounding underground openings, through experiments such as a rockburst simulation, *in situ* events generating acoustic emissions, and laboratory fractures. Their results highlight that realistic wave modeling around openings requires the presence of a stress-dependent fracture stiffness coupling the surfaces of the fracture. Ultrasonic attenuation tomography and enhanced velocity tomography were studied by Dębski and Young for an earlier laboratory experiment of thermally induced fractures in granite. Using a Bayesian inversion, the authors conclude that attenuation tomography provides only a rough structure of the sample, whereas enhanced velocity tomography offers an accurate imagining of the micro-structural rock changes. This section concludes with a paper by Ogasawara, who studies the decays in earthquake swarms and laboratory rock fracture trials by examining acoustic emissions generated during thermal stress relaxation. He proposes simple and original experiments using a rice cooker, and provides additional support for his findings by comparing them with numerical modeling results.

A third group of papers, including five contributions, is dedicated to the analysis of seismicity induced in hydrocarbon fields. Phillips *et al.* provide a review of their results on spatial and temporal clustering of microseismicity associated with both hydrocarbon and geothermal reservoirs. They conclude that induced seismicity patterns are dominated by horizontal, linear features that migrate in time towards the well in sedimentary environments, whereas an interacting network of fractures is apparent for crystalline rock and brittle sediments. Microseismicity induced by fluid injection was analyzed by House and Flores to study the fracture system, fluid pathways, and *in situ* stresses. The authors found that hypocenters farthest from the injection define a nearly horizontal tube of seismicity towards the injection well. Both *P*- and *S*-wave velocities along vertical paths are higher than those along horizontal paths, and the orientation of seismically derived minimum stress is consistent with regional data. Zoback and Zinke examine the behavior of the Valhall and Ekofisk oil reservoirs in Norway. Based on *in situ* stress, pore pressure, and microseismic data, they demonstrate how, as a result of production-induced pore-pressure and stress changes, normal faulting has spread out from the crests of the anticlinal structures to the flanks, now observed in and above the productive horizons. An analysis of the deformation processes associated with the Romashkino oil field reservoir in Russia was carried out by Turuntaev and Razumnaya using induced seismicity and production data. The study finds that the efficiency of water injection for oil recovery decreases with increasing seismicity, and recommends the installation of permanent microseismic networks for the monitoring of hydrocarbon fields. To increase the accuracy of real-time event locations, Baziw and Weir-Jones attempt to improve the automatic picking of first wave arrivals. As such, they employ a real-time Kalman filter, modeling the background noise as a Gauss-Markov process and the seismic wavelet as a periodic process with a random walk initial amplitude, to seismic data collected in Northern Alberta, Canada.

Applications of induced seismicity associated with geothermal fields count for five contributions and constitute the fourth group of studies included in this special volume. Rutledge *et al.* investigated the near-surface, above reservoir microseismicity recorded at the Geysers field, California. Their analysis shows that this seismicity is strongly correlated in time with the reservoir injection and the associated induced seismicity occurring in the reservoir, immediately below, suggesting the presence of a remote triggering mechanism that was previously observed at a larger scale. The next study, authored by Sasaki and Kaieda, evaluates the stress state from the inversion of focal mechanisms of microseismic events induced at Hijiori, Japan, where the geothermal reservoir was developed through hydraulic fracturing. The failure mechanisms indicate that no tensile fractures were created in intact rock and that the microseismicity was caused by shear failures. The following three studies are carried out at the geothermal site Soultz-sous-Forêts, France. Moriya *et al.* analyzed the microseismic multiplets induced during hydraulic fracturing. The correlation found between the orientation of multiplet planes, fracture distribution detected by borehole imagers, and the stress field, indicates that the multiplet fractures represent seismically activated pre-existing planes that interlink fluid flow structures in a mesh fracture-reservoir system. The evaluation of the deep, subsurface structure has been attempted by Soma *et al.* using the acoustic emission reflectivity method in the time-frequency domain. The results obtained are compared with independent borehole observations and the frequency dependence of identified reflectors from hodogram linearity is further investigated. Rowe *et al.* are proposing an automatic, correlation-based algorithm for adjusting phase picks in large seismic data sets to improve the accuracy of event locations. Their methodology, including innovative signal processing and clustering and cluster stacking techniques, allow for a significant improvement in both *P*- and *S*-wave arrival consistency, and an increased resolution of hypocenter relocations.

The last type of application included in this collection of articles involving induced seismicity is related to water dam reservoirs. Although represented here by only one study, this area of research is no less significant than other areas previously discussed. Here, a detailed review of 16 case studies of reservoir-induced seismicity of moderate magnitude (M 1.6 to 4.2) in Brazil is presented by Assumpção *et al.* Their analysis indicates that the associated induced seismicity hazard varies throughout the country, with some regions more prone than others to the occurrence of such phenomena. However, no correlation was found between the induced seismic hazard and the level of natural seismicity.

The wide variety of research topics involving induced seismicity included in this present volume is indicative of the increased interest of the scientific community to address some of the critical problems that lie ahead. The demand for more resources will push related industries to undertake new technical challenges that will require rapid decisions based on educated and objective assessment. The availability of microseismic monitoring technology at relatively lower costs makes this tool not only attractive for the gathering of data remotely, but for making more real-time assessments of rapidly changing conditions due to human intervention with the environment. Therefore, induced seismology will continue to be called upon by the industry to address some of the practical problems encountered, and contribute, along with other disciplines to offering a better understanding of how their operations can be controlled.

Acknowledgments

This special volume was made possible due to numerous authors who expressed interest in contributing to it, and who are acknowledged for the quality of their work, their commitment and resourcefulness during the revision process. Additionally, this volume benefited enormously from the dedicated work of numerous reviewers whose time, expertise, and objectivity have been oriented to ensuring that the scientific community will circulate a sound and efficient message. As such, I would like to express my most sincere gratitude to Douglas Alde, Hiroshi Asanuma, Igor Beresnev, Artur Cichowicz, Francois Cornet, Torsten Dahm, Diane Doser, Bernard Dost, Mariana Eneva, Steve Falls, Bruno Feignier, Joan Gomberg, Jean-Robert Grasso, Andre Herrero, Sue Hough, Leigh House, Hisao Ito, Andrzej Kijko, David Lockner, Shawn Maxwell, Art McGarr, Steve McKinnon, Andrew Michael, Koji Nagano, Masao Nakatani, Stephane Nechtschein, Stefan Nielsen, David Oppenheimer, Scott Phillips, Gerhard Pratt, Bob Simpson, David Simpson, Shahriar Talebi, Pradeep Talwani, John Tinucci, Ted Urbancic, Bill Walter, Stefan Wiemer, Uli Zimmer, Mark Zoback, and Waclaw Zuberek for their insightful comments and generous suggestions made throughout the review process. Special thanks go to Renata Dmowska, the Editor-in-Chief for Topical Issues at Pure and Applied Geophysics, whose interest in having this volume organized with the journal was instrumental throughout every stage of this project. Last, but not least, I would like to thank my colleagues at ESG for their continuous encouragement and support.

Cezar I. Trifu
Editor

Pure appl. geophys. 159 (2002) 007–062
0033–4553/02/030007–56 $ 1.50 + 0.20/0

Pure and Applied Geophysics

Seasonal Seismicity of Northern California Before the Great 1906 Earthquake

ROB WESTAWAY[1]

Abstract—During the late nineteenth century, coastal parts of northern California experienced seismicity which was more intense than at any other time in its historical record and was concentrated in Spring. The seasonal character of this anomalous seismicity was statistically significant from the 1880s, if not earlier, until the time of the great 1906 earthquake. It can be explained as a result of seasonal stress transients caused by the large seasonal sediment loads which accumulated in coastal bays that overlie active strike-slip faults, as a result of hydraulic mining of gold in inland areas. Such localised loading will increase the absolute vertical stress and reduce the absolute horizontal stress, thus facilitating shear failure by unclamping vertical fault planes. However, rather than simply influencing the timing of earthquakes which would otherwise have occurred anyway sometime later, this process appears to have affected the character of the region's seismicity, causing a sequence of magnitude 5–6 earthquakes which propagated outward from these coastal depocentres. These results have three important implications. First, the proposed physical mechanism is only feasible provided the deepest continental crust in this region has a viscosity no greater than $\sim 10^{17}$ Pa s, giving it a Maxwell time of the order of weeks. These observations of seasonal seismicity thus suggest that such low viscosities are reasonable, a conclusion which has been difficult to establish using other evidence. They indicate that future calculations of the stress field in this region, for instance for investigating the role of earthquake triggering, should incorporate this response by lower-crustal flow, not just the immediate elastic effect. Second, the conclusion that seasonal stress transients caused by seasonal changes in surface processes can affect seismicity, has important implications for hazard assessment. It raises the possibility that surface processes may influence the seasonal timing and character of seismicity in other regions also. Finally, it indicates that the decades of intense seismicity, widely interpreted as precursory activity to the 1906 earthquake, were atypical of the longer-term record. Analogous sequences of moderate-sized events cannot necessarily be expected to provide advance warning of future large earthquakes in this region.

Key words: California, earthquake hazard, seismicity, San Andreas fault.

Introduction

MCCLELLAN (1984) identified an apparent seasonal pattern to the seismicity of coastal parts of northern California (Fig. 1) before the great San Francisco earthquake of 18 April 1906. He established that during 1855 to 1906 earthquakes with magnitude $(M) \geq 5.5$ were concentrated in Spring, with a probability that his observed distribution arose by chance of only $\sim 3\%$. In contrast, the seismicity

[1] 16 Neville Square, Durham DH1 3PY, England.

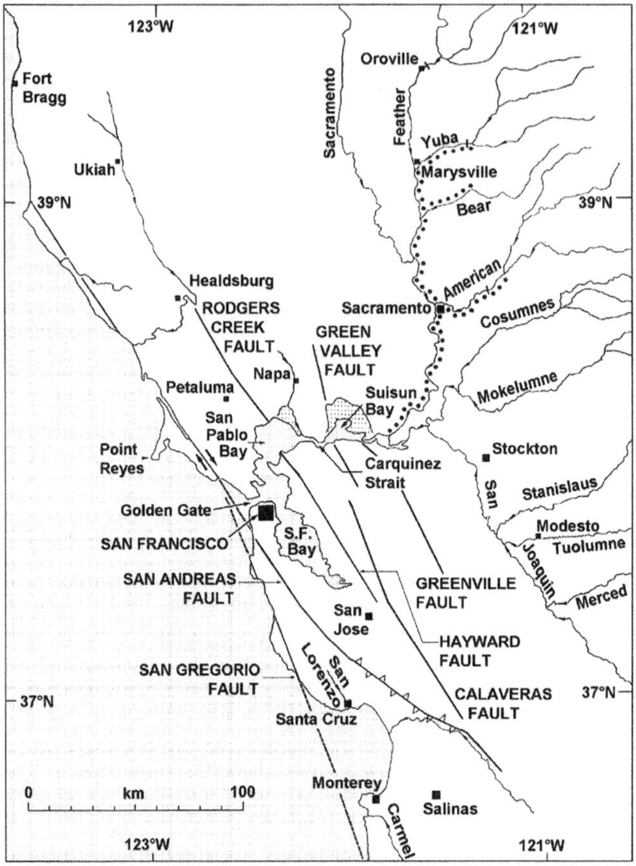

Figure 1

Map of northern California showing the geometry of drainage into San Francisco Bay and the adjacent active strike-slip faults. Chevron ornament marks the stepover on the San Andreas faults in the Santa Cruz Mountains. Fine dots mark tidal flats adjoining San Pablo and Suisun Bays. Large dots mark the most important paths of hydraulic mining debris, from GILBERT (1917). Positions of gauging stations on the Feather, Yuba, and American Rivers are from CLAPP (1907).

between 1907 and the time of MCCLELLAN's (1984) study showed no significant seasonal pattern.

Most earthquakes in this region occur on northwest-trending right-lateral strike-slip faults, which take up shearing motion between the Pacific and North American plates (Fig. 1). MCCLELLAN (1984) noted that such faults can be "unclamped", or brought nearer the condition for shear failure, by a decrease in the southwest component of horizontal compressive stress or by an increase in pore fluid pressure. He thus suggested several possible explanations for the observed seasonal seismicity. First, noting that the study region has strongly seasonal rainfall (Fig. 2), he postulated that seepage of Winter rainfall underground may increase the pore fluid

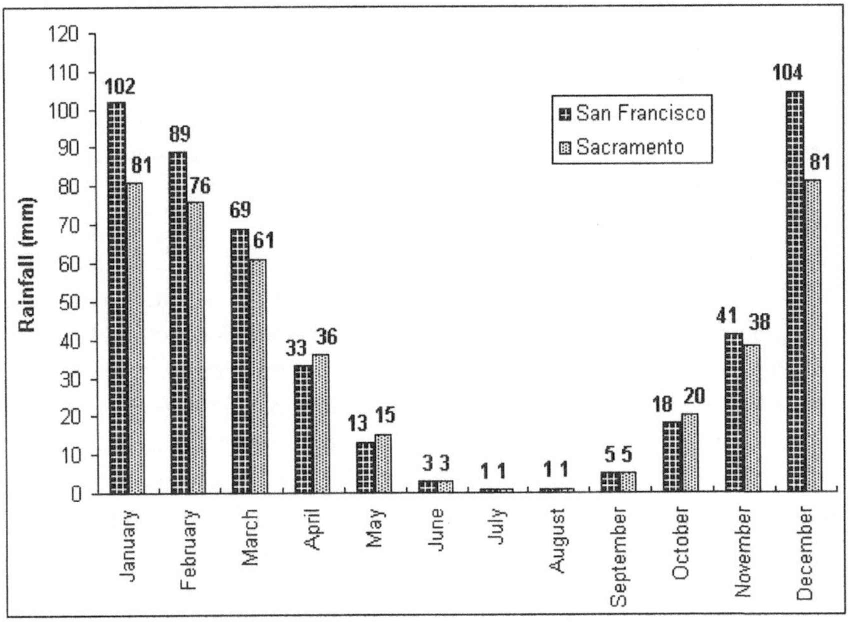

Figure 2
Mean monthly rainfall at San Francisco and Sacramento. Data are from METEOROLOGICAL OFFICE (1980).

pressure across faults. Second, he suggested that seasonal variations in tidal stresses caused by the changing inclination of the Earth's axis relative to the Sun, seasonal variations in sea level, or seasonal variations in the rate of the Earth's axial rotation, may also reduce the normal stress or increase the shear stress across faults. However, MCCLELLAN (1984) estimated neither the magnitude of the stress transient required to cause seasonal seismicity in this region, nor the magnitude of the stress transient which can be caused by any of these processes. These processes are discussed in detail below, where it is established that none of them can sustain either the required magnitude of stress transient or the observed seasonal pattern. Other processes thus have to be identified if this region's seasonal seismicity is to be explained.

MCCLELLAN (1984) further suggested that the seasonal seismicity before 1906 typifies time ahead of a great earthquake, with northern California already near the condition for shear failure due to the earlier build-up of shear stress. As no equivalent seasonal seismicity was evident to him up to 1982, he concluded that no large earthquake was imminent. The Loma Prieta earthquake of 18 October 1989 was the second largest in historical time in northern California. Although it was much smaller than the 1906 event, if MCCLELLAN (1984) had correctly identified the cause of the seasonal seismicity before 1906, then a similar pattern should have occurred before 1989. Nonetheless, even if one restricts study to the immediate epicentral area of the Loma Prieta event and looks at events below magnitude 5.5, no

Spring concentration is evident. Microseismicity which was judged precursory to the Loma Prieta event was noted by SHAW *et al.* (1992) on the southern Calaveras fault, up to ~30 km northeast of its epicentral area. OLSON and HILL (1993) suggested instead that the Lake Elsman earthquakes of 27 June 1988 and 9 August 1989 (Table 1) were the main precursors of the Loma Prieta event. However, neither this seismicity, nor the smaller local events also studied by OLSON and HILL (1993), showed any Spring concentration.

Other studies (e.g., BUFE and VARNES, 1993; BOWMAN *et al.*, 1998; JAUMÉ and SYKES, 1999) have proposed that the *overall* seismicity of northern California has progressively increased in the decades preceding both the 1906 and 1989 earthquakes (although this characteristic was much less clear before 1989 than before 1906; see below). They also suggested that this effect indicates that the region was evolving at these times towards a "critical state": moderate-sized earthquakes progressively redistributed elastic strain energy until a state of stress was reached which allowed shear failure to occur over many hundreds or thousands of square kilometres of fault plane. Numerical models of seismicity, consisting of patches of fault with slight variations in properties, where failure of one patch can change the state of stress of its neighbours (e.g., RYDELEK and SACKS, 1999), can generate analogous behaviour.

The aim of this study is to suggest a related explanation for the seasonal character of the seismicity before 1906. At this time, many faults were already close to their condition for shear failure, and so could be made to slip by small changes to the state of stress. It is proposed that the seasonal character to the seismicity originated as a result of fault unclamping due to stress transients caused by seasonal variations in sediment loading in coastal bays near major active strike-slip faults (Fig. 1). Seasonal sediment transport can be expected in this region given its seasonal rainfall (Fig. 2). Moreover, the quantity of sediment transported in the late nineteenth and earliest twentieth centuries was vast, due to large-scale hydraulic mining of gold in the Sierra Nevada mountains (see below). This region's major rivers are now managed using flood control reservoirs and overflow channels (e.g., FISCHER, 1994), which even out discharge and reduce sediment transport, and hydraulic mining has long been forbidden (see below), so the sediment supply is now reduced. As a result, seasonal stress transients are smaller now than in the late nineteenth century, and no equivalent Spring concentration of seismicity would be expected ahead of the Loma Prieta event or when the region is near shear failure ahead of any future great earthquake.

Seismicity of Coastal Northern California

The major active right-lateral faults in the study region (Fig. 1) are well documented (e.g., BROWN, 1990): several having ruptured in historical time (e.g., ELLSWORTH, 1990). The 18 April 1906 San Francisco earthquake ($M \sim 7.8$) ruptured

Table 1

Seismicity of northern California

Date	Time	Locality	Lat.	Long.	M	as	M_E	M_M	I_o	M_B	Population		
1851–1878													
15 May 1851	16:10	San Francisco	37.8	122.4		–			6	5.1			
23 Nov 1852	07:00	Redwood City	37.5	122.4		–			felt				
27 Aug 1855	11:00	San Francisco	38.1	122.5	4.9	M_I			6	5.5			
2 Jan 1856	18:15	San Mateo County	37.5	122.5	5.3	M_I			6	5.5			B
15 Feb 1856	13:25	San Mateo County	37.5	122.3	5.5	M_I	5 3/4	5.9	7	5.8	*M*	I	A
26 Nov 1858	10:00	Alameda County	37.5	121.9	6.1	M_I	6 1/4	5.9	7	6.1	*M*	I	A
5 Oct 1859	20:16	San Francisco	37.8	122.4		–			6				
17 April 1860	03:30	Monterey Bay	36.9	122.1						5.5			
4 July 1861	00:11	San Ramon Valley	37.8	122.0	5.6	M_I	5 3/4		8	5.8	*N*		A
19 Dec 1863	22:38	Menlo Park	37.5	122.2	4.8	M_I			5	5.1			
26 Feb 1864	13:47	Santa Clara County	37.1	121.7	5.9	M_I	6	5.8	6	5.9	*M*	I	A
5 March 1864	16:49	Alameda County	37.7	122.0	5.7	M_I	5 3/4		6	5.9	*N*		A
21 May 1864	02:01	Alameda County	37.6	122.1	5.3	M_I			6	5.6			(B)
22 July 1864	06:40	Fremont	37.6	122.0	4.7	M_I			5	5.0			
8 March 1865	14:30	Sonoma County	38.4	122.6	4.7	M_I			7	5.2			
24 May 1865	11:21	Santa Clara County	37.1	121.8	5.5	M_I			7	5.9	*N*		A
8 Oct 1865	20:46	Santa Cruz Mountains	37.2	121.9	6.3	M_I	6 1/2	6.2	8	6.5	*M*	I	A
26 March 1866	20:12	Gilroy	37.1	121.6	5.4	M_I		5.8	7	5.6			B
15 July 1866	06:30	Stanislaus County	37.5	121.3	5.8	M_I	6	5.7	5	5.9	*M*		A
21 Oct 1868	15:53	Hayward	37.7	122.1	6.8	M_I	7	6.7	9	6.8	*M*	I	A
2 Apr 1869	01:50	Hayward	37.7	122.1						5.0			
8 Oct 1869	09:30	Ukiah	39.1	123.1	5.0	M_I			7				B
17 Feb 1870	20:12	Los Gatos	37.2	122.1	5.8	M_I	6	5.5	7	5.8	*M*	I	A
2 April 1870	19:48	Oakland	37.9	122.3	5.3	M_I			6	5.6			B
29 May 1876	18:55	Freestone	38.4	122.9	4.2	M_I			6	4.7			
1879–1906													
9 Jan 1880	13:45	Paicines	36.7	121.3		–			felt				
10 April 1881	10:00	Stanislaus County	37.4	121.4	5.9	M_I	6	6.0	6	6.1	*M*	I	A
6 March 1882	21:45	San Benito County	36.9	121.2	5.7	M_I	5 3/4	5.5	6		*M*		A
27 June 1882	13:32	Santa Cruz	37.2	122.0		–			7	5.9			
30 March 1883	15:45	Gilroy	36.9	121.6	5.6	M_I		5.7	7	5.9	*M*	I	A

Table 1

Continued

Date	Time	Locality	Lat.	Long.	M	as	M_E	M_M	I_o	M_B	Population		
26 March 1884	00:40	Santa Cruz	37.1	122.2	5.9	M_I	6		6	6.1	*N*		A
31 March 1885	07:56	San Benito County	36.7	121.3	5.5	M_I	5 ¾	5.6	7		*M*	I	A
2 April 1885	15:25	San Benito County	36.8	121.4	5.4	M_I		5.5	5		*M*		
1 Aug 1885	00:10	Cloverdale	38.8	123.0		–			6				
3 Dec 1887	18:55	Mendocino County	39.3	123.6	4.9	M_I			7				
18 Feb 1888	10:50	Mendocino County	39.2	123.6	4.9	M_I			5				
29 Feb 1888	22:50	Petaluma	38.3	122.7	4.2	M_I			6	4.6			
17 Sept 1888	11:51	Morgan Hill	37.1	121.8	4.5	M_I			5	5.0			
18 Nov 1888	22:28	Oakland	37.9	122.3	4.3	M_I			7	5.1			
15 April 1889	03:28	Santa Cruz	37.1	121.9	4.8	M_I			6	5.2			
19 May 1889	11:10	Antioch	38.0	121.9	6.0	M_I	6 ¼	5.6	7	6.0	*M*	I	A
31 July 1889	12:47	Oakland	37.8	122.2	5.2	M_I			7	5.6			B
24 April 1890	11:36	Pajaro Gap	36.9	121.6	6.0	M_I	6 ¼	5.9	8	6.3	*M*	I	A
2 Jan 1891	20:00	San Jose	37.3	121.8	5.5	M_I			6	5.7	*N*		A
12 Oct 1891	06:28	Napa	38.3	122.4	5.5	M_I		5.6	8	5.8	*M*	I	A
19 April 1892	10:50	Vacaville	38.4	122.0	6.4	M_I	6 ½	6.8	9	6.4	*M*	I	A
21 April 1892	17:43	Winters	38.5	121.9	6.2	M_I	6 ¼	6.3	9	6.4	*M*		
30 April 1892	00:09	Davis	38.4	121.8	5.5	M_I			6	5.5	*N*		
13 Nov 1892	12:45	Hollister	36.8	121.5	5.6	M_I	5 ¾		6	5.5	*N*		A
30 June 1893	13:30	San Pablo Bay	38.0	122.4	4.6	M_I			5	5.2			
9 Aug 1893	09:15	Santa Rosa	38.4	122.6	5.1	M_I			7	5.5			B
20 June 1897	20:14	Gilroy	37.0	121.5	6.2	M_I	6 ¼	6.0	8	6.3	*M*	I	A
31 March 1898	07:43	Mare Island	38.2	122.4	6.2	M_I	6 ½	6.3	8	6.3	*M*	I	A
15 April 1898	07:07	Greenwood	39.2	123.8	6.9	M_S	6 ½	6.7	8		*M*	I	A
30 April 1899	22:41	Watsonville	36.9	121.7	5.6	M_I		5.7	7	5.9	*M*	I	A
2 June 1899	07:19	San Mateo County	37.7	122.5	5.4	M_I		5.5	7	5.5	*X*	I	B
6 July 1899	20:10	Santa Clara County	37.2	121.5	5.8	M_I	5 ¾	5.5	7	5.9	*M*		A
13 Oct 1899	05:00	Santa Rosa	38.4	122.7		–			8	4.9			
30 April 1900	22:41	Watsonville	36.9	121.6	4.5	M_I			5				
19 May 1902	18:31	Solano County	38.3	121.9	5.4	M_I		5.5	7	5.9	*X*	I	B
11 June 1903	13:12	San Jose	37.4	121.9	5.8	M_I	5 ½	5.6	7	6.1	*M*	I	A
24 July 1903	20:26	Willows	39.5	122.0	4.5	M_I			7				
3 Aug 1903	06:49	San Jose	37.3	121.8	5.8	M_I	5 ½	5.6	7	6.1	*M*	I	A

30 July 1904	10:26	Winters	38.5	122.0	4.5	M_I			5					
18 April 1906	13:12	San Francisco	37.7	122.5	7.8	M_S	8¼	8.3	11	7.8	*M*	I	A	
1907–1934														
11 March 1910	06:52	Watsonville	36.9	121.8	5.8	M_S	5.8	5.7	6	5.8	*M*	I	A	
31 Dec 1910	12:11	Salinas	36.8	121.4	5.0	M_I			6				B	
1 July 1911	22:00	San Jose	37.3	121.8	6.6	M_S	6.5	6.0	8	6.2	*M*	I	A	
12 Sep 1912	17:27	Saratoga	37.3	122.2	4.5	M_I			5					S
9 Nov 1914	02:31	Santa Clara County	37.2	122.0	4.7	M_I		5.5	7	5.5	*X*	I		S
6 August 1916	19:38	Paicines	36.7	121.3	5.3	M_I			7				B	
8 August 1916	16:52	Santa Cruz	37.0	122.0	5.0	*U*			5					S
25 Nov 1919	11:03	Santa Cruz County	37.1	121.8	4.5	M_I			5					S
5 Oct 1920	19:03	Monterey County	36.6	121.7	5.2	*U*			5				B	
22 Oct 1926	12:35	Monterey Bay	36.6	122.4	6.1	M_S	6.1	6.2	7	6.1	*M*	I	A	
22 Oct 1926	13:35	Monterey Bay	36.6	122.2	6.1	M_S	6.1	6.2	7	6.1	*M*			
24 Oct 1926	22:51	Davenport	37.0	122.2	5.5	M_I		5.6	4	5.3	*M*			S
15 Feb 1927	23:54	Davenport	37.0	122.3	5.0	M_I		5.6	5	5.9	*X*		B	
4 Dec 1931	00:53	Carmel Valley	36.5	121.7	5.1	*U*			4				B	
14 June 1932	09:44	Saratoga	37.3	122.1	4.5	M_L			5					S
25 Oct 1932	01:07	Sausalito	37.8	122.5	5.1	M_L			4				B	
3 Nov 1932	18:54	Saratoga	37.2	122.2	4.9	*U*			5					S
13 Dec 1933	15:34	Los Gatos	37.2	122.0	5.0	*U*			4				B	S
1935–1962														
12 Feb 1938	20:00	Santa Cruz	37.0	122.0	4.5	M_L			6					S
22 June 1947	23:29	Santa Cruz County	37.0	121.8	4.7	M_L			6					S
29 July 1951	10:53	Pinnacles	36.6	121.2	5.0	M_L			6				B	
25 April 1954	20:33	Santa Cruz County	36.9	121.7	5.3	M_L			8				B	
5 Sept 1955	02:01	Santa Clara County	37.4	121.8	5.5	M_L		5.6	7	5.5	*M*	I	A	
24 Oct 1955	04:10	Walnut Creek	38.0	122.1	5.4	M_L			7				B	
22 March 1957	19:44	Daly City	37.7	122.5	5.3	M_L			7	5.3			B	
2 March 1959	23:27	Gilroy	37.0	121.6	5.3	M_L			6				B	
20 Jan 1960	03:25	San Benito County	36.8	121.4	5.0	M_L			6				B	
9 April 1961	07:23	San Benito County	36.7	121.3	5.6	M_L			7		*N*		A	
9 April 1961	07:25	San Benito County	36.7	121.3	5.5	M_L			7		*N*			
6 June 1962	17:50	Lakeport	39.1	123.1	5.2	M_L			7				B	
1963–1990														
22 May 1963	22:41	La Honda	37.3	122.3	4.6	M_L			6					S

Table 1

Continued

Date	Time	Locality	Lat.	Long.	M	as	M_E	M_M	I_o	M_B	Population			
14 Sept 1963	19:46	Watsonville	36.9	121.6	5.4	M_L			7				B	
16 Nov 1964	02:46	Corralitos	37.1	121.7	5.0	M_L			7				B	S
7 Sep 1967	12:39	Santa Cruz County	37.0	121.8	4.7	M_L			6					S
18 Dec 1967	17:24	Corralitos	37.0	121.8	5.3	M_L			6				B	S
2 Oct 1969	04:56	Santa Rosa	38.5	122.7	5.6	M_L		5.5	8	5.6	M	I	A	
2 Oct 1969	06:19	Santa Rosa	38.5	122.7	5.7	M_L			8	5.7	N			
24 Feb 1972	15:56	San Benito County	36.6	121.2	5.1	M_L			6				B	
12 Nov 1973	18:17	Los Gatos	37.2	122.0	4.5	M_L			4				S	
28 Nov 1974	23:01	Thanksgiving Day	36.9	121.5	5.2	M_L			6				B	
6 Aug 1979	17:05	Coyote Lake	37.1	121.5	5.9	M_L	5.7	5.9	7	5.7	M	I	A	
24 Jan 1980	19:00	Livermore	37.9	121.8	5.8	M_L	5.8	5.8	7	5.8	M		A	
24 Jan 1980	19:01	Livermore	37.9	121.8	5.1	M_L			Felt					
27 Jan 1980	02:33	Livermore	37.7	121.7	5.4	M_L		5.5	7	5.4	X			
18 Aug 1982	08:43	Santa Cruz County	37.0	121.7	4.5	M_L			5					S
24 April 1984	21:15	Morgan Hill	37.3	121.7	6.2	M_L	6.1		8	6.2	N		A	
26 Jan 1986	19:20	Paicines	36.8	121.3	5.5	M_L			7		N		A	
31 March 1986	11:55	Mount Lewis	37.5	121.7	5.7	M_L			6	5.6	N		A	
20 Feb 1988	08:39	Tres Pinos	36.8	121.3	5.1	M_L			5				B	
13 June 1988	01:45	San Jose	37.4	121.8	5.3	M_L			6				B	
27 June 1988	18:43	Lake Elsman	37.1	121.9	5.3	M_L			6					S
8 Aug 1989	08:13	Lake Elsman	37.1	122.0	5.4	M_L			7				B	S
18 Oct 1989	00:04	Loma Prieta	37.0	121.9	7.0	M_L	7.1		9	6.9	N		A	S

Earthquakes listed by STOVER and COFFMAN (1993) for the time scale 1851 to 1990 and for the region bounded by the coordinates 36.5°N 121.0°W, 38.0°N 121.0°W, 39.5°N 122.0°W, 39.5°N 125.0°W, 36.5°N 123°W, and 36.5°N 121.0°W (Fig. 1). Events of 17 April 1860 and 2 April 1869, from BAKUN (1999), were omitted by STOVER and COFFMAN (1993). M is the estimated magnitude, followed by the method used. I_o is the Modified Mercalli Intensity in the epicentral area, and M_B is the magnitude assigned by BAKUN (1999). Earthquake populations are identified in the final four columns. In the first column, M designates events considered by MCCLELLAN (1984) as having $M \geq 5.5$, X designates events considered by MCCLELLAN (1984) as having $M \geq 5.5$ but which STOVER and COFFMAN (1993) list as having $M < 5.5$, and N designates events omitted by MCCLELLAN (1984) but which STOVER and COFFMAN (1993) list as having $M \geq 5.5$. In the second column, I designates events which MCCLELLAN (1984) regards as independent. In the third column, A and B designate events regarded as independent in this study, with $M \geq 5.5$ and $M \geq 5.0$. In the fourth column, S denotes earthquakes since 1906 beneath the Santa Cruz Mountains (within an area bounded by 37.5°N 122.4°W, 37.2°N 122.4°W, 37.0°N 122.2°W, 37.0°N 121.7°W, 37.1°N 121.7°W, 37.5°N 122.3°W, and 37.5°N 122.4°W).

the San Andreas fault for ~400 km distance northward from near Santa Cruz, with up to ~6 m of slip. The Loma Prieta earthquake of 18 October 1989 (M 7.0) ruptured a subparallel oblique reverse fault segment along the leftward step in the San Andreas fault near Santa Cruz (e.g., LISOWSKI *et al.*, 1990). The Santa Cruz Mountains earthquake on 8 October 1865 (with estimated $M \sim 6.3$) may have ruptured the same fault. The Hayward earthquake of 21 October 1868 (estimated $M \sim 6.8$) ruptured a ~40 km length of the Hayward fault between the latitude of San Francisco and the southern end of its Bay (e.g., TOWNLEY and ALLEN, 1939). A reassessment by TOPPOZADA and BORCHERDT (1998) indicates that the earlier event of 10 June 1836 (estimated $M \sim 6.3$) probably ruptured a ~20 km length of the San Andreas fault near Salinas (Fig. 1). This event was previously thought to have ruptured the northern ~40 km of the Hayward fault, and to have had $M \sim 6.8$. Conversely, TOPPOZADA and BORCHERDT (1998) deduced that the June 1838 earthquake had $M \sim 7.5$ and ruptured the ~140 km length of the San Andreas fault from San Francisco to Salinas (Fig. 1). It was previously thought to have only ruptured ~60 km of fault ($M \sim 7$) (e.g., ELLSWORTH, 1990).

Trenching data, geodetic data, and other evidence, reveals how the ~50 mm a^{-1} of relative motion between the Pacific and North American plates (e.g., DEMETS *et al.*, 1990) is partitioned between these faults, and in most cases has resolved the timing of the last major prehistoric earthquake. The San Andreas fault (Fig. 1) is slipping at 24 ± 3 mm a^{-1} north of the Golden Gate and 17 ± 4 mm a^{-1} on the San Francisco peninsula (e.g., BROWN, 1990; SCHWARTZ *et al.*, 1998; HALL *et al.*, 1999). Radiocarbon dating (e.g., SCHWARTZ *et al.*, 1998; HALL *et al.*, 1999) indicates that earthquake activity on it comparable in scale to 1906 probably occurred in the early to mid 17th century. However, these data cannot resolve a single great earthquake, equivalent to 1906, from a sequence of smaller events spanning several decades: possibly analogous to the activity on the North Anatolian fault zone in Turkey during 1939–1967 (e.g., AMBRASEYS and FINKEL, 1995) and on the Dead Sea fault zone in the Middle East during 1759–1796, 1404–1407, 1157–1202, and earlier (e.g., AMBRASEYS and BARAZANGI, 1989; BEN-MENAHEM, 1991). SCHWARTZ *et al.* (1998) preferred it to have been a single event, giving a recurrence interval of ~250 to ~300 years, although this was more a value judgment than a solid conclusion.

WEBER and NOLAN (1995) estimated the slip rate on the San Gregorio fault as 4 to 11 mm a^{-1} where it comes ashore west of Santa Cruz (Fig. 1), consistent with the estimated 7 mm a^{-1} southward decrease on the San Andreas fault where the San Gregorio fault splays from it. The Hayward and Rodgers Creek faults slip at 9 ± 2 mm a^{-1} (e.g., SCHWARTZ *et al.*, 1993; LIENKAMPER and BORCHERDT, 1996). As the northern Calveras fault slips at 5 ± 2 mm a^{-1} according to KELSON *et al.* (1996), its slip rate south of where the Hayward fault splays from it can be estimated as ~15 mm a^{-1}. Unpublished data compiled by WORKING GROUP ON CALIFORNIA EARTHQUAKE PROBABILITIES (1999) indicates that the Green Valley and Greenville faults (Fig. 1) are slipping at ~5 and ~2 mm a^{-1}.

South of $\sim$ 36.9°N, the seismicity has a different character from farther north. Here, the San Andreas fault is creeping aseismically (e.g., JOHNSON *et al.*, 1974), and also slips in frequent microearthquakes (e.g., OLSON and HILL, 1993; RUBIN *et al.*, 1999) and moderate-sized events up to magnitude $\sim$5.5.

Seasonal Character of Seismicity

MC CLELLAN (1984) studied earthquakes during 1855 to 1982, with magnitudes $\geq$5.5, in the region demarcated in the caption to Table 1. To avoid bias due to aftershocks, he selected independent earthquakes by rejecting events within one year or 1°($\sim$ 110 km) distance of an earlier event. The 27 such events comprised 19 before and 7 since 18 April 1906. MCCLELLAN (1984) looked for evidence of seasonality using the Schuster test (e.g., SCHUSTER, 1897), where each earthquake is represented as a unit vector whose phase is determined by its date within the year, and these vectors are added. The probability P of obtaining a resultant vector longer than a given value of R by adding n unit vectors with random phase is given by

$$P = \exp(-R^2/n) \ . \tag{1}$$

The method is formally valid for $n \geq 10$ but gives an approximate result for smaller n. I regard as seasonally significant, distributions of earthquakes which have a probability, from equation (1), of $\leq 5\%$ of arising through chance, equivalent to a 95% confidence limit.

Estimation of Earthquake Magnitudes

Magnitude estimation is important in this study, because previous work (e.g., SYKES and JAUMÉ, 1990; BUFE and VARNES, 1993) has tentatively suggested a "rule of thumb" that precursory seismicity to an earthquake of a given magnitude is clearest for earthquakes down to two magnitude units smaller; the rate of even smaller events showing no such clear time-dependence. In the light of these observations, it seemed reasonable to test whether the seasonal character of such precursory seismicity varies with magnitude in an analogous manner with magnitude. The first requirement is thus to establish an earthquake population which can be tested for seasonality, including all known events above a specified magnitude threshold. MCCLELLAN (1984) used data from ELLSWORTH *et al.* (1981). However, later compilations (e.g., ELLSWORTH, 1990; STOVER and COFFMAN, 1993; BAKUN, 1999) report not only different magnitudes but also a different rank order for many of the earthquakes studied by MCCLELLAN (1984) (Table 1).

MCCLELLAN (1984) regarded magnitude 5.5 as the limit of completeness of the historical seismicity record. The fact (evident from Table 1) that he omitted some larger events indicates that his catalogue was *not* complete at this level; indeed,

TOPPOZADA *et al.* (1988) considered the region's catalogued historical seismicity since 1850 to be complete only above magnitude 6. The present study will adopt a threshold of 5, to incorporate more earthquakes to allow additional analysis. There is no guarantee that the cataloguing of events of this size was complete before the regional seismograph network was deployed in 1932 (see below). However, as there is no known reason why earthquakes should be preferentially missed in any particular season, the statistical analysis is unlikely to be adversely affected. Because events below magnitude 5 *were* documented in the late 19th century (Table 1), it is unlikely that many larger ones have been missed.

The historical record divides into four eras by the evidence available on earthquake sizes. Before the late 1890s, the only evidence is from macroseismic effects. Before California joined the United States in 1848 this evidence is extremely limited. As the population increased rapidly in later decades, newspaper reports of earthquake effects form an important source of information, and have been used (e.g., by TOPPOZADA *et al.*, 1981) to estimate magnitudes. AGNEW and SIEH (1978) compiled a population map of California using data from the 1860 census. The population of coastal northern California south of $\sim 38.5°$N was already sufficient at this time – and sufficiently uniformly distributed – to make it unlikely that any large earthquakes have been missed. The method for determining macroseismic or intensity magnitudes, M_I, uses the areas over which effects of a given intensity or greater were sustained (e.g., TOPPOZADA, 1975). The scale is calibrated using later earthquakes for which seismological magnitudes are also available.

From the late 1890s to the early 1930s, information is also available from global seismograph stations. The low-amplification instruments of this era were suited to recording surface waves, enabling surface-wave magnitudes (M_S) to be estimated. The local seismograph network, deployed across California from 1932 onward, led to estimates of local magnitude M_L from amplitudes of short-period seismograms (RICHTER, 1935). M_L values have been determined, using a consistent definition, from this time to the present day. Because of this continuity, studies of macroseismic effects have calibrated M_I to be consistent with M_L. For instance, TOPPOZADA (1975) calibrated an intensity magnitude, consistent with M_L, as

$$M_I = 0.85 \log_{10}(A_6/\mathrm{km}^2) + 2.56 \ , \tag{2}$$

where A_6 is the area which experienced macroseismic effects of Modified Mercalli Intensity 6 or greater, the threshold for damage to property. However, M_L saturates at $\sim$7, as larger earthquakes (measured by the dimensions of the fault plane) radiate seismic waves no more strongly at periods of $\sim$1s. Equation (2) predicts that A_6 is $\sim$740 km^2 for M_I5, equivalent to a circular area with a diameter of $\sim$30 km. Given the density of settlement in the region during the late nineteenth century, from AGNEW and SIEH (1978), it is indeed difficult to believe that many earthquakes of this size went unreported. Finally, in recent decades attempts have been made to relate the logarithm of seismic moment to a moment magnitude scale (e.g., HANKS and KANAMORI, 1979), here designated as M_W.

ELLSWORTH (1990) advocated a definitive magnitude measure for each earthquake throughout the historical record, which he called summary magnitude and which I designate as M_E. He used M_I up to 1898 and mainly relied on M_S thereafter, using M_L or M_W instead for a few events. He preferred M_S to M_L for events since the 1930s because of the saturation problem for M_L. ELLSWORTH (1990) also reported that M_I typically underestimates M_L by 0.25 and M_S by 0.3 for earthquakes for which these measures had been estimated independently. He thus proposed a correction for M_I values for 19th century events, by adding 0.15 and then rounding to the nearest quarter, with the aim of bringing them into line with other measures of earthquake size. Later studies (e.g., JACKSON *et al.*, 1995) have quoted these corrected M_I values, many of which thus exceed the original reported estimates.

To investigate this possible miscalibration, I examine how well Toppozada's (1975) equation for deriving M_I from A_6 (equation (2)) matches M_L. Figure 3 shows $\Delta M = M_L - M_I$ against M_L for the 40 earthquakes for which TOPPOZADA (1975) measured A_6. For the largest earthquakes, substantially above M_L 6, ΔM is > 0, and M_I does underestimate M_L, by typically ~ 0.2, as ELLSWORTH (1990) noted. However,

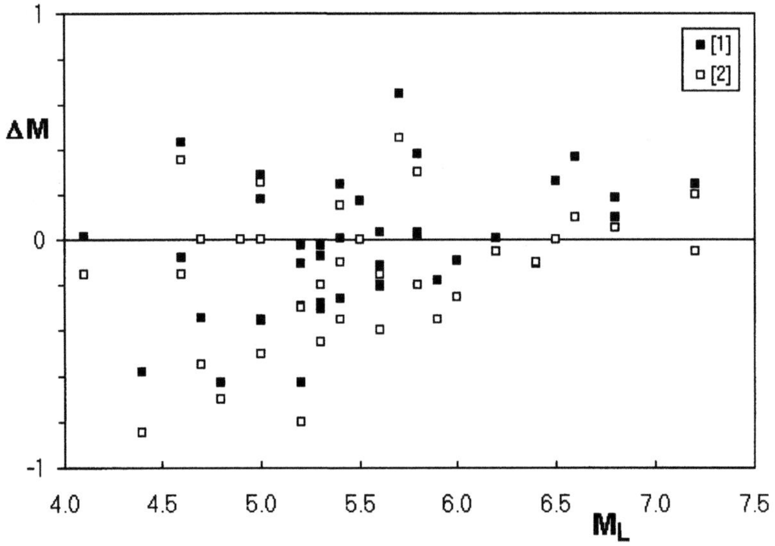

Figure 3

Solid symbols [1] plot $\Delta M = M_L - M_I$ against M_L for 40 earthquakes from 1950 to 1971 used by TOPPOZADA (1975) to calibrate M_I. M_L values are from TOPPOZADA (1975), except for the 1952 Kern County earthquake for which the correct M_L value of 7.2 has been used instead. M_I values have been calculated from A_6, the area enclosed by the intensity 6 isoseismal, using equation (2). These data indicate no overall bias between M_I and M_L, but suggest that ΔM is systematically slightly above 0 for $M_L > 6$ and systematically slightly below 0 for $M_L < 5.5$. Open symbols [2] plot ΔM against M_L using M_I values adjusted as proposed by ELLSWORTH (1990). These adjusted M_I values agree with M_L for $M_I > 6$, but significantly overestimate M_L for $M_L < 5.5$.

for the smaller events this mismatch typically decreases, and around M_L 5 the opposite mismatch develops. Figure 3 also shows the alternative estimates of ΔM, using the modified M_I values from ELLSWORTH (1990), also plotted against M_L . These are consistent with M_L for $M_L > 6$, but substantially overestimate M_L for events around $M_L 5$. The earthquakes which provide the greatest weight of evidence for seasonal seismicity (see Table 1 and below) fall in the magnitude range from ~5.3 to ~6.0., where M_I and M_L agree closely (Fig. 3). Pending any overall recalibration of the regression analysis for M_I, I thus regard M_I and M_L as equivalent.

As a further test of any possible effect of magnitude miscalibration on earthquake seasonality, significance testing is also carried out using the earthquake catalogue of BAKUN (1999). BAKUN (1999) believed this to be complete for the late 19th century only above magnitude ~5.5, but comparison of his magnitude estimates with mine (Table 1) indicates that they diverge below magnitude 6, with his values typically larger, such that both his catalogue and mine cover essentially the same earthquakes. However, these two earthquake populations differ because his study area was smaller than mine and because the different magnitude calibrations lead to a different rank order of events. As a result, comparison of their analyses provides an indication of the effect of different magnitude calibration procedures and study region bounds on whether the late 19th century earthquake population is judged to have been seasonal: these effects are shown to make no significant difference to this conclusion.

Sampling Procedure for Seismicity

MCCLELLAN (1984) found that his population of 27 independent events was not seasonally significant, nor were the 7 events since 1907, but the seasonal pattern of the 19 events during 1855–1905 had a probability of arising by chance of only ~3%, being concentrated around 15 April. However, several problems are evident with MCCLELLAN's (1984) method and results. First, as already discussed, his intensity magnitude data were from old calibrations. Following the calibration checking described earlier, I use M_L since 1932 and M_I beforehand, using M_S instead for a few events (Table 1). Table 1 also shows the magnitude used for each earthquake by MCCLELLAN (1984), plus the magnitudes determined by ELLSWORTH (1990) and BAKUN (1999). Substantial discrepancies are evident between these different determinations. For instance, several intensity magnitudes of 5.5 or above are for events which MCCLELLAN (1984) regarded as smaller, and several events included in MCCLELLAN's (1984) study have since been assigned magnitudes below 5.5.

A second problem concerns the post-1906 earthquakes used in MCCLELLAN's (1984) study, which were formally too few for the Schuster test. I investigate more earthquakes, by expanding the time limit to 1990, reducing the magnitude threshold to 5.0, and adopting a different criterion for independent events. I regard earthquakes more than six weeks apart as independent regardless of their separation in distance, to approximate the duration of the bulk of activity in many aftershock sequences, rather than the ~110 km and 12 month separation criteria used by McClellan (1984). McClellan's (1984) criteria have the undesirable property that if earthquakes are very frequent, then all events occurring within 12 months of the first are discounted. The earliest event after this 12-month period will be included as independent, then all other events will be disregarded for 12 further months. As a result, an apparently seasonal distribution may arise from random seismicity as an artefact of the sampling procedure. McClellan's (1984) procedure also excludes some earthquakes which occurred just less than 12 months after others. For instance, the 10 April 1881 event excluded the 6 March 1882 event, and the 30 March 1883 event excluded the 26 March 1884 event (Table 1). A sequence of $M \geq 5.5$ earthquakes in Spring in four consecutive years was thus masked by a sampling procedure which excluded two of them.

My procedure (Table 1) identifies 72 independent earthquakes with $M \geq 5.0$ during 1851 to 1990 inclusive, similar to the 62 considered independent by Bakun (1999) over the same time scale but in a smaller region. I have divided this interval into five subsets of 28 years, demarcated at the ends of 1878, 1906, 1934, and 1962. These contain, respectively, 15, 23, 10, 9, and 15, events. As a result, the Schuster test is formally valid for all bar one of these populations individually, as well as when they are grouped in pairs or other combinations. The number of earthquakes documented since 1906 also now exceeds those earlier, allowing a fairer test for any post-1906 earthquake seasonality.

Figure 4a shows frequency of occurrence against magnitude for the events identified during 1851 to 1906. The distribution can be fitted with a *b*-value of 1 down to magnitude 5.5. Although this test does not prove that cataloguing is complete for this magnitude range, it provides no basis for the view that earthquakes in this magnitude range have been systematically missed. Figure 4b shows the corresponding graph for earthquakes for 1962–1969, using equivalent magnitude data from Stover and Coffman (1993). This interval of time was chosen as it was previously analysed by Bolt and Miller (1971), who showed that its earthquake population for the present study region fits a *b*-value of 1 down to M_L 3. Figure 4b and their Figure 6c agree closely for $M_L > 5$ but diverge at smaller magnitudes due to the flattening of the curve in Figure 4b, which is caused by smaller events having not been completely catalogued by Stover and Coffman (1993). This evidence indicates that ~1 is the correct *b*-value for this region, thus supporting the deduction from Figure 4a that no late 19th century earthquakes above magnitude 5.5 appear to have been missed.

Summary of Seismicity Patterns

Figure 5 illustrates how the seismicity pattern has varied over time, with respect to season and latitude. Several features are apparent. First, in 1880–1906, all

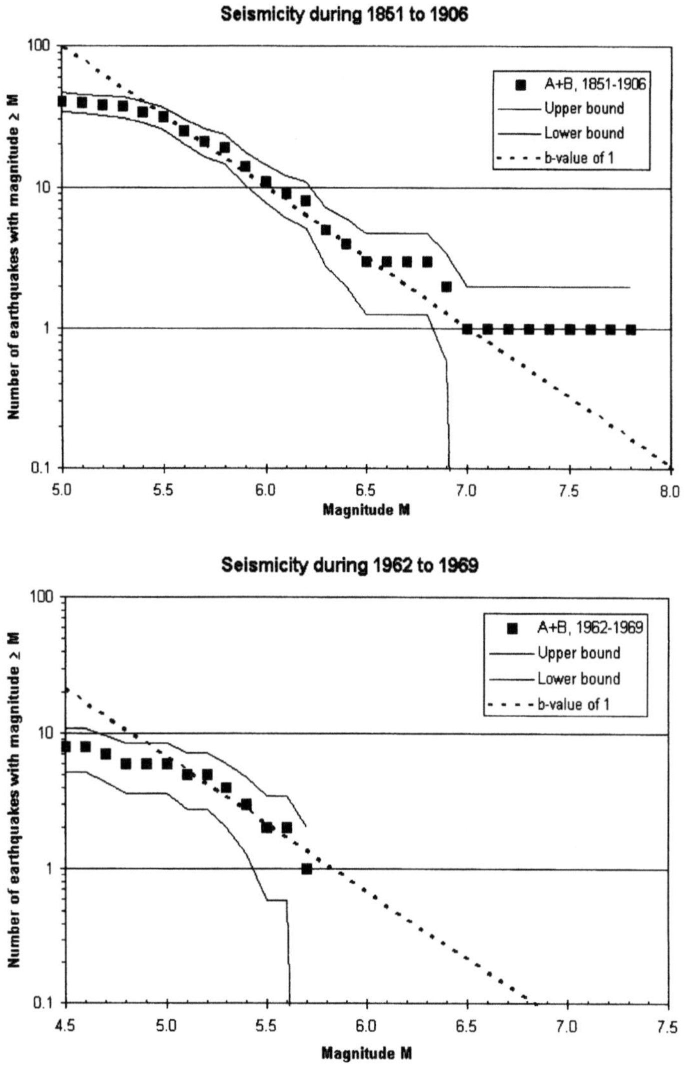

Figure 4

Graphs of frequency of occurrence against magnitude M for earthquake populations A and B in Table 1 for 1851–1906 (a) and 1962–1969 (b). Solid symbols show numbers of earthquakes with magnitudes less than or equal to a given value M. Dashed lines indicate approximate error bounds expected from statistical fluctuations in rates of occurrence, estimated as plus or minus the square root of each number of earthquakes. Dashed straight lines are fitted by eye with "b" values of 1. These fitted lines indicate that no significant numbers of earthquakes were missed down to magnitude 5.5 in (a) and 5.0 in (b).

earthquakes with $M \geq 6.0$ occurred during April, May or June. Most, plus many smaller events, indeed occurred during April. Second, since about 1950 an apparently random pattern has occurred. Third, at first sight the distribution before 1880 also appears random. However, about half of its earthquakes occurred during early Spring, raising the possibility that they were a precursor to the later seismicity

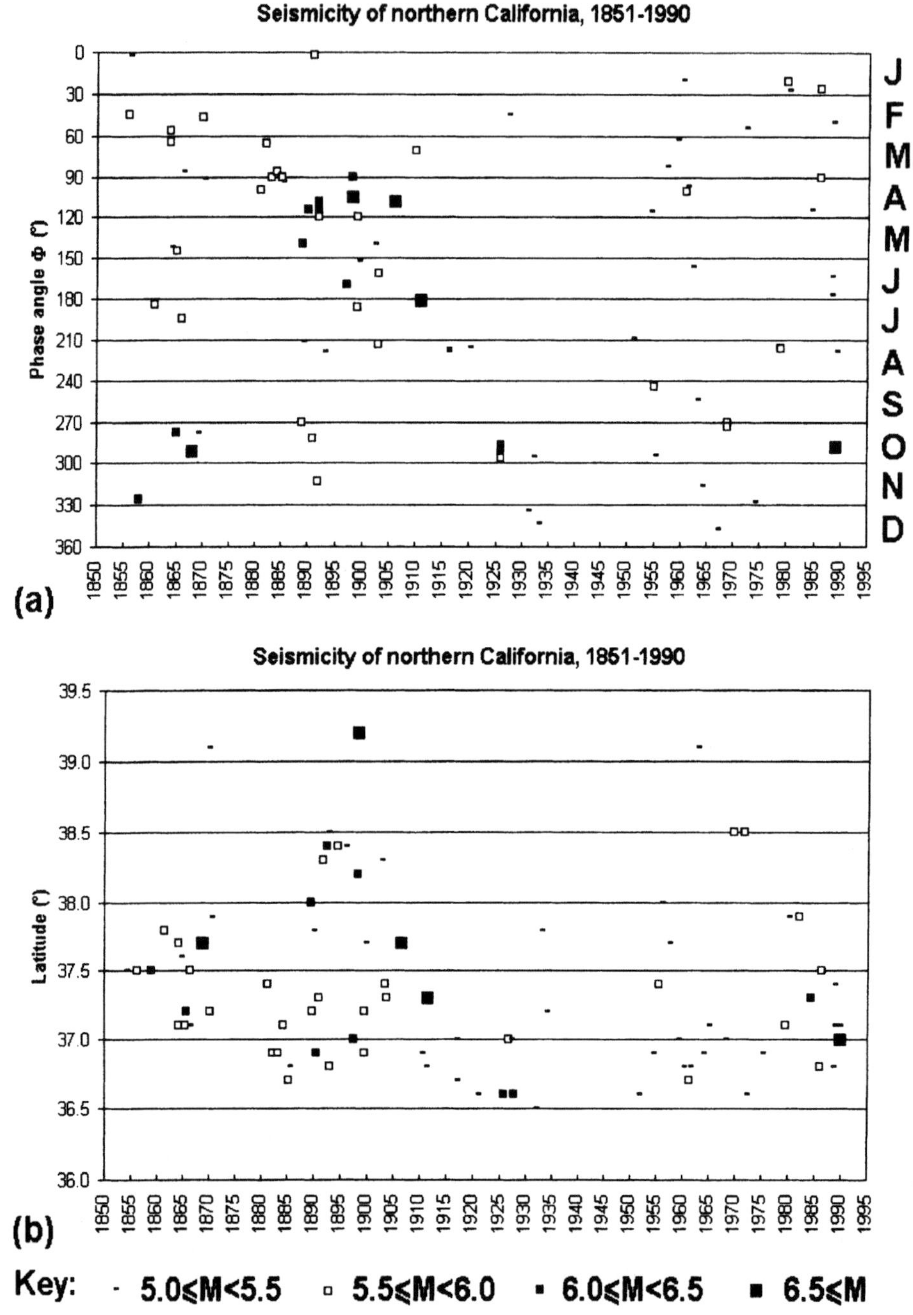

Figure 5

Distributions with respect to time of earthquakes in the study region as a function of seasonal phase angle (a) and latitude (b). Phase angle ϕ varies from 0° on 1 January to 360° on 31 December. All earthquakes in Table 1 with magnitudes $\geq$5.0 are included in this figure, regardless of whether they are judged to be independent using criteria discussed in the text.

concentration. If so, their concentration during March was somewhat earlier than the April concentration after 1880. Finally, in 1906–1950 the seismicity was so sparse and irregular that no overall pattern is evident. The seismicity maps in Figure 6 cover the most interesting part of the record, during 1860 to 1906, with events labelled by season. Maps showing this region's seismicity during other time steps have been published before (e.g., ELLSWORTH, 1990; JAUMÉ and SYKES, 1999; SYKES and JAUMÉ, 1990), and so are not repeated here.

Almost all the study region's seismicity occurred south of 38.5°N (Table 1). Since 1850, only three earthquakes with $M \geq 5.0$ have occurred between 38.5°N and 39.5°N: the Ukiah and Lakeport events of 8 October 1869 ($M = 5.0$) and 6 June 1962 ($M = 5.2$), and the much larger Greenwood event of 15 April 1898 ($M = 6.9$). The Greenwood earthquake thus shares the same seasonal timing as the other $M \geq 6.0$ events during 1880 to 1906. However, its epicentre is so far from the other events that it is debatable whether this timing can reflect the same causal process Nonetheless, it did occur on or adjacent to part of the San Andreas fault that later ruptured in 1906 (Fig. 6b), and has been considered a precursor of the 1906 event in several previous studies (e.g., SYKES and JAUMÉ, 1990; BUFE and VARNES, 1993; JAUMÉ and SYKES, 1996; JAUMÉ and SYKES 1999).

Figures 5b and 6 also indicate another pattern, noted by ELLSWORTH (1990). During the 1850s to 1870s most seismicity was concentrated between 37.5°N and 38.0°N, around San Francisco Bay. However, from 1880 until 1906 this latitude range was relatively aseismic: activity instead spread gradually northward to 38.5°N and southward to 36.5°N.

Outside 1879 to 1906, seven earthquakes with $M > 6.0$ have occurred: on 26 November 1858 (Alameda County; $M = 6.1$), 8 October 1865 (Santa Cruz Mountains; $M = 6.3$), 21 October 1868 (Hayward; $M = 6.8$), 1 July 1911 (San Jose; $M = 6.6$), 22 October 1926 (Monterey Bay; $M = 6.1$), 24 April 1984 (Morgan Hill; $M = 6.2$), and 18 October 1989 (Loma Prieta; $M = 7.0$). Their concentration in autumn raises the possibility of a further seasonal pattern, which was masked during the late 19th century. As both large earthquakes in the Santa Cruz Mountains (in 1865 and 1989) occurred at virtually the same time of year, I consider whether smaller events in this area also share the same seasonal pattern. Earthquakes in this locality have thus been compiled down to magnitude 4.5 for 1907 to 1990 (population S in Table 1).

Statistical Tests of Seasonal Seismicity

Table 2 summarises results of applying the Schuster test to the earthquake populations identified above, varying the magnitude threshold. Although the 1851–1878 population shows no significant seasonality, the 1879–1906 population is demonstrably seasonal regardless of the threshold, being concentrated around mid-April to early May. Indeed, its three largest earthquakes, on 19 April 1892 (Vacaville; $M = 6.4$), 15 April 1898 (Greenwood; $M = 6.9$), and 18 April 1906 (San Francisco;

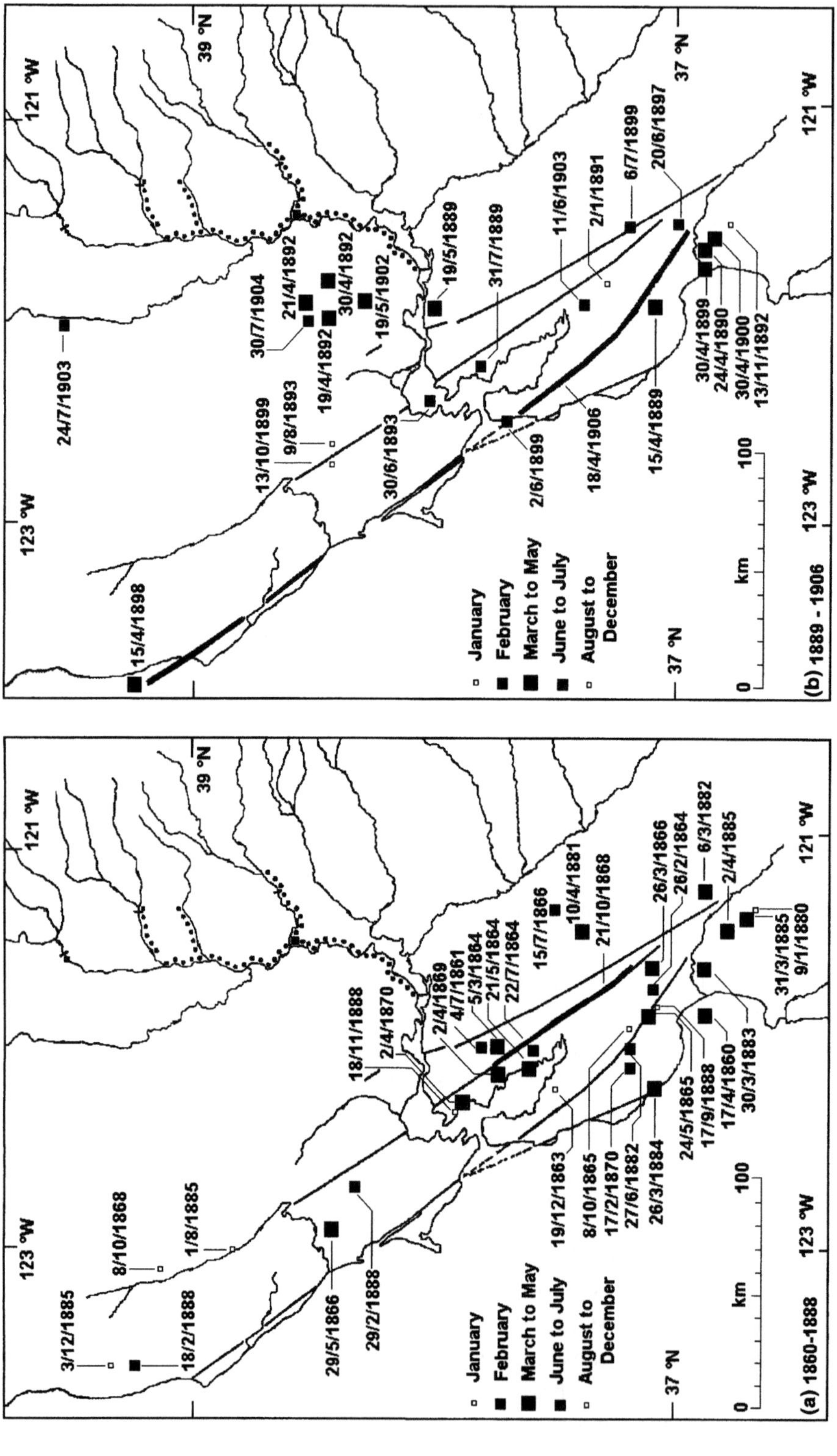

Figure 6
Maps of the seismicity of the study region (Table 1), for (a) 1860–1888 and (b) 1889–1906. Each event is labelled with its date and ornamented to highlight the seasonal pattern which existed at the time.

Table 2

Schuster test results

Criterion	N	θ	Date	R	P
1879–1906					
$M \geq 5.0$	23	128°	8 May	11.94	0.00202
$M \geq 5.3$	21	119°	29 April	11.99	0.00106
$M \geq 5.5$	19	113°	23 April	10.25	0.00396
$M \geq 5.7$	13	124°	4 May	9.87	0.000553
$M \geq 5.8$	12	129°	9 May	9.38	0.000651
$M \geq 5.9$	9	113°	23 April	8.22	0.000549
$M \geq 6.0$	7	119°	29 April	6.39	0.00296
$M \geq 6.2$	5	116°	26 April	4.46	0.01875
$M \geq 6.4$	3	107°	17 April	2.999	0.04992
1879–1906; Lat. ≤38.5°N					
$M \geq 5.0$	22	131°	11 May	11.035	0.00395
$M \geq 5.3$	20	120°	30 April	11.022	0.00230
$M \geq 5.5$	18	115°	25 April	9.265	0.00849
$M \geq 5.7$	12	126°	6 May	8.935	0.00129
$M \geq 5.8$	11	132°	12 May	8.482	0.00144
$M \geq 5.9$	8	114°	24 April	7.230	0.00145
$M \geq 6.0$	6	121°	1 May	5.418	0.00750
$M \geq 6.2$	4	118°	28 April	3.479	0.04850
1879–1906; 36.9°N ≤ Lat. ≤ 38.5°N					
$M \geq 5.0$	21	134°	14 May	10.284	0.00650
$M \geq 5.3$	19	123°	3 May	10.161	0.00437
$M \geq 5.5$	17	118°	28 April	8.361	0.01638
1851–1878					
$M \geq 5.0$	15	63°	3 Mar	2.935	0.5631
$M \geq 5.3$	14	72°	12 Mar	3.800	0.3566
$M \geq 5.5$	11	63°	3 Mar	2.187	0.6474
1865–1892					
$M \geq 5.0$	20	96°	6 March	5.397	0.2331
$M \geq 5.3$	18	89°	27 Feb	6.871	0.0726
$M \geq 5.5$	16	89°	27 Feb	4.874	0.2266
1864–1892					
$M \geq 5.0$	23	93°	3 March	7.739	0.07399
$M \geq 5.3$	21	88°	26 Feb	9.240	0.01715
$M \geq 5.5$	18	81°	19 Feb	6.695	0.08291
1851 to 1878 and 1907–1990; $M \geq 6$					
to 1878	3	298°	29 Oct	2.815	0.07124
to 1983	5	282°	13 Oct	3.499	0.08645
to 1990	7	280°	1 Oct	3.152	0.17077
1907–1990; Santa Cruz Mountains					
$M \geq 4.5$	19	280°	11 Oct	7.611	0.04740
$M \geq 4.7$	12	275°	6 Oct	6.662	0.02477
$M \geq 4.9$	9	285°	16 Oct	5.251	0.04672
$M \geq 5.0$	8	281°	12 Oct	4.312	0.09782

For each test, N is the number of earthquakes included in each population, θ is the phase of the resultant vector indicating the seasonal timing about which the earthquake population is concentrated, R is the magnitude of this resultant vector, and P is the probability, from equation (1), of a resultant vector of this magnitude arising by chance. Values of P of < 0.05 are regarded here as significant.

$M = 7.8$), all occurred during 15 to 19 April. If the Greenwood event is excluded on account of its distance from the others, the remaining population remains significantly seasonal (Table 2). Only one event in 1879–1906 which was included in these tests (on 31 March 1885; $M = 5.5$) occurred on or near the creeping part of the San Andreas fault south of 36.9°N. If this is excluded, the remaining population also remains significantly seasonal (Table 2).

Figure 7 shows the seismicity data for 1879–1906 displayed as an empirical probability density function. It indicates that the probability that an independent earthquake in the study region with $M \geq 5.0$ occurred during April within this time scale was ~0.08 per week, compared with the expected value of 1/52 or ~0.02 per week for a uniform seasonal distribution. A simple way of expressing this seasonal character is thus that earthquakes with $M \geq 5.0$ were about four times more concentrated in Spring than would be expected for a random distribution.

In principle, it would be possible to test the statistical significance of this pattern (Fig. 7) by randomising the earthquake dates from Table 1 and looking for whether concentrations in a particular season can arise by chance. Figure 8 illustrates the results of a variant of this test, in which a sequence of N random seasonal phase angles – all calculated to the nearest degree to maintain consistency with Figure 5a and Table 2 – is represented by N random integers between 0 and 360. Their vector sum is determined, then equation (1) is used to estimate the notional probability that

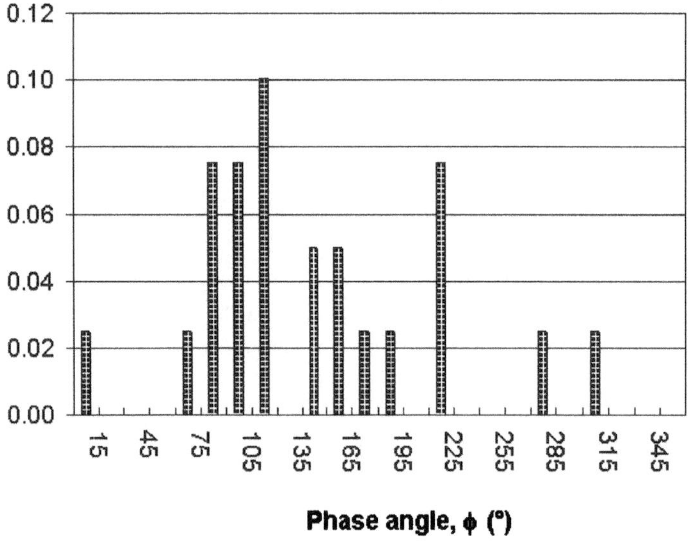

Figure 7
Empirical seasonal probability density function for earthquakes during 1879–1906, estimated using the phase angles of the 23 independent earthquakes with $M \geq 5.0$ during this time scale (Table 1). The year is broken into 30 equal intervals for which probabilities have been calculated. Each interval spans 15° of ϕ, and is labelled with its upper bound. Probabilities are normalised per week within these intervals. The peak probability density is ~0.08 per week during April, compared with the expected value of ~0.02 per week (1/52 per week) for a uniform seasonal distribution. See text for discussion.

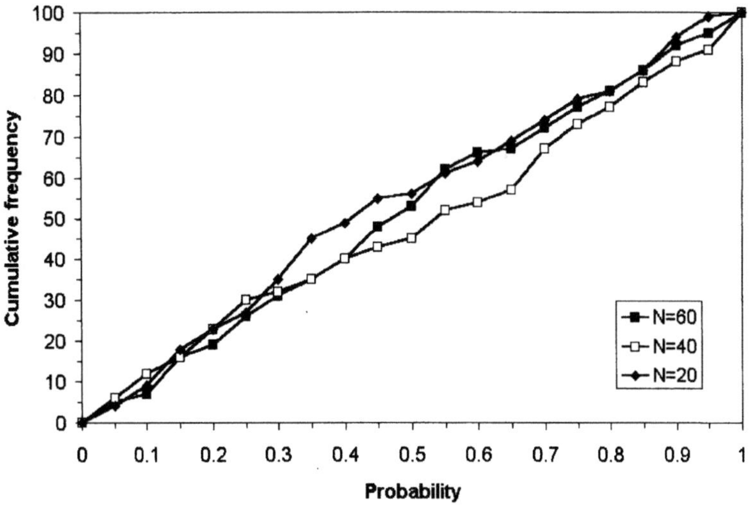

Figure 8

Tests of the likelihood of seasonal earthquake populations arising by chance from random distributions. Each graph shows, for 100 random distributions each of N earthquakes (for $N = 20$, 40, and 60), the cumulative frequency of occurrence versus the probability of each of the distributions being judged to have arisen by chance according to equation (1). See text for discussion.

this particular vector sum arose by chance. This calculation is then repeated 100 times for each value of N, each time with a different set of random numbers, and the results plotted (Fig. 8). It is evident that about 5 times out of 100, the N random phase angles will be distributed in a pattern which – the Schuster test indicates – has a $\leq 5\%$ probability of arising by chance. It is thus concluded that, although it is possible for apparently seasonal patterns to arise by chance from a random set of events, it is unlikely, and the probabilities determined using the Schuster test do provide a meaningful measure of how unlikely it is. Moreover, this conclusion is robust for values of N of 20, 40, and 60 (Fig. 8), encompassing the numbers of earthquakes in the majority of Schuster tests listed in Table 2 (as well as in Table 3 below).

Although the overall 1851–1868 population is not significantly seasonal, there is some evidence of a seasonal pattern in the later part of this period, as 6 of the 11 events in 1864–1878 occurred during February to May. A 28-year sample for 1865–1892 is not significantly seasonal. However, the three events in 1864 makes this population for $M \geq 5.3$ significantly seasonal: it is concentrated about late February, with a probability of only ~ 0.02 of originating by chance (Table 2).

In contrast, none of the three later earthquake populations, for 1907–1934, 1935–1962, and 1963–1990, shows any statistically significant evidence of seasonality. Larger groupings, comprising pairs of these populations, or all three, or all three plus the 1851–1878 population, likewise show no significant seasonal character. These negative results are expected given the data distribution in Figure 5a, and are omitted from Table 2.

It is evident from Table 1 that some pairs of earthquakes listed as independent are borderline candidates, notably those in January to February 1856, February to April 1870, September to October 1955, January to March 1986, and August to October 1989. Most of these events occurred at times when the seismicity of the overall study region showed no significant seasonal character, and this conclusion remains regardless of whether one or both events in each pair is classed as independent. However, the events in February to April 1870 contribute to the weight of evidence in support of seasonal seismicity during 1864–1892. If the event of 2 April 1870 is excluded, the probability of this earthquake population with $M \geq 5.3$ arising by chance increases from ~0.02 to ~0.04. It thus remains significant at the 95% confidence level.

Table 3 shows Schuster test results both for BAKUN's (1999) earthquake population (a) and for mine (b) (using data from Table 1 and Fig. 6) for 1860 to 1906. As already noted these populations differ due to different magnitude calibration procedures and bounds to the study region, and because BAKUN (1999) used different criteria for excluding events that he thought may be related. It is evident that these differences do not affect the main conclusion: during this time span the seismicity was concentrated in Spring in a manner which was highly statistically significant.

The significance can also be assessed of the Autumn concentration of $M > 6.0$ earthquakes outside 1879–1906. The small number of events means that this pattern has never been formally significant (Table 2). However, the Autumn concentration of earthquakes in 1907–1990 in the Santa Cruz Mountains is marginally significant (Table 2). Furthermore, many earthquakes just outside the demarcated boundaries of the Santa Cruz Mountains (see Table 1 caption) also share the same seasonality, notably the events of 5 September 1955 (Santa Clara County; M 5.5), 14 September 1963 (Watsonville; M 5.4), 28 November 1974 ("Thanksgiving Day"; M 5.2), and 6 August 1979 (Coyote Lake; M 5.9) (Table 1).

This evidence thus suggests that a pattern of seismicity concentrated in Spring began to develop in the 1860s and persisted until 1906. The precise statistical significance of the earthquakes in each population above each specified magnitude threshold is listed in Tables 2 and 3. The seismicity seems to have been concentrated around early March early in this time span, but gradually adjusted to late April. It is evident that if one wishes to propose a causal mechanism for this seasonal seismicity, a process is required which occurred at this time but not before or since. Moreover, this seasonal effect appears to have masked a different pattern, which at other times has concentrated the seismicity in part of the study region in Autumn.

Hydraulic Mining of Gold in the Sierra Nevada

Much of the western flank of the Sierra Nevada mountains is covered by Tertiary river gravels, which have been uplifted by many hundreds of metres since deposition

Table 3

Schuster test results for 1860–1906

Criterion	N	θ	Date	R	P
(a) Earthquake population from Bakun (1999)					
All (*)	46	130°	10 May	17.96	0.00090
All	43	135°	15 May	15.46	0.00385
$M \geq 5.0$	40	136°	16 May	15.10	0.00334
$M \geq 5.2$	36	132°	12 May	17.09	0.00030
$M \geq 5.5$	33	134°	14 May	15.10	0.00100
$M \geq 5.6$	30	131°	11 May	13.20	0.00301
$M \geq 5.8$	26	132°	12 May	11.96	0.00407
$M \geq 5.9$	22	131°	11 May	13.13	0.00040
$M \geq 6.0$	13	130°	10 May	7.07	0.02140
$M \geq 6.1$	12	129°	9 May	6.08	0.04577
$M \geq 6.2$	8	122°	2 May	3.55	0.20735
(b) Earthquake population from this study					
All (*)	60	135°	15 May	18.07	0.00432
$M \geq 5.0$ (*)	38	123°	3 May	17.43	0.00034
$M \geq 5.0$	34	125°	5 May	13.66	0.00415
$M \geq 5.3$	31	116°	26 April	14.74	0.00090
$M \geq 5.5$	27	115°	25 April	11.24	0.00932
$M \geq 5.7$	19	117°	27 April	9.51	0.00858
$M \geq 5.8$	17	128°	8 May	8.42	0.01552
$M \geq 5.9$	12	108°	18 April	6.83	0.02060
$M \geq 6.0$	9	125°	5 May	4.48	0.10709

These tests use (a) the magnitudes for the earthquake population defined by BAKUN (1999) (i.e., M_B listed in Table 1) and (b) the magnitudes from this study (M in Table 1). Notation is the same as in Table 2. In (a), grouping (*) includes the three events which BAKUN (1999) thought might well be aftershocks of earlier events: 2 April 1869, 2 April 1870, and 30 April 1892. BAKUN (1999) omitted the 2 April 1885 event but included those on 21 May 1864 and 21 April 1892. In (b), grouping (*) include the four events which this study considers not to be independent of other events: 21 May 1864, 2 April 1885, 21 April 1892 and 30 April 1892. The 2 April 1869 event is omitted as it was not documented by STOVER and COFFMAN (1993), but the 2 April 1870 event has been included.

(e.g., CHASE and WALLACE, 1986; HUBER, 1990). In 1848 these were found to be gold-bearing, and following the Gold Rush of 1849, mining of these placer deposits developed on an industrial scale. The main method used jets of water to wash gravel into sluices which trapped some of the gold. The waste gravel or tailings continued downslope to the Sacramento river system (Fig. 1).

Movement of this gravel caused major changes. For instance, the bed of the Yuba River, which transported the greatest volume of tailings, rose at ~0.3 m per year. Levees thus had to be built to protect towns and agricultural land. Levee breaches indeed caused Marysville (Fig. 1) to be flooded in 1875 and 1876. By 1880, ~20000 hectares of agricultural land had been damaged through inundation by flood deposits of gravel, and the Sacramento River was no longer navigable. These and other social consequences of hydraulic mining have been described in detail by KELLEY (1959) and summarised by PYNE (1980).

In 1884 a legal injunction prevented further hydraulic mining. In 1893, the U.S. Congress established the California Debris Commission to regulate this industry, which was effectively shut down by stringent licensing. Efforts also began to engineer the Sacramento river system to mitigate the effects of mining debris. However, in 1904 the California Miners' Association began lobbying for hydraulic mining to resume. They appealed to President Theodore Roosevelt, who referred the matter to the U.S. Geological Survey, where Grove K. Gilbert was assigned to investigate. He concluded that the value of the remaining gold which could be extracted was negligible compared with the costs of the loss of water resources and agricultural land and of the other forms of environmental damage which would arise if mining were to resume.

GILBERT (1917) documented the distribution of debris from the Sierra Nevada (Table 4). Its output fell from its 1884 peak by ~50% by 1887 and ~90% by 1902 (GILBERT, 1917). Contributions from natural erosion of mine sites, clandestine mining, and limited amounts of licensed mining, ensured that it did not fall immediately to zero (Fig. 9). GILBERT (1917) also estimated that between 1849 and 1914 ~1.8 km^3 of debris had entered this region's river system, of which ~1.3 km^3 was from mining; most of the rest being from soil erosion due to agriculture. Its fine component quickly reached to the coastline, contributing to the early part of the sediment accumulation in the coastal bays. The coarser component travelled more slowly: for instance, this gravel deposits reached its maximum thickness in 1900 at the point where the Yuba River leaves the Sierra Nevada mountains (near the gauging point marked in Fig. 1), but not until 1905 at Marysville, ~25 km farther west, where the Yuba joins the Feather River. Upstream of the point of peak gravel thickness at any given time, the deposits became scoured enabling some gravel to be redeposited farther downstream. Coarse debris thus continued to migrate downstream for decades after significant mining ceased, its maximum flux into the coastal bays occurring in the first decade of the century (Fig. 9).

GILBERT (1917) estimated that by 1914, ~10% of the total of debris remained within the Sierra Nevada, ~20% formed Piedmont deposits along valleys beyond the range front, a further ~20% remained in channel beds or as floodplain deposits of major rivers, and ~50% had reached coastal bays or the Pacific Ocean (Table 4). A typical thickness of ~1 m of sediment had indeed accumulated in Suisun and San Pablo bays, which straddle active strike-slip faults (Fig. 1). GILBERT's (1917) estimate that ~0.9 km^3 of debris had reached these bays by 1914 represents a time-averaged sediment flux of ~0.02 $km^3\ yr^{-1}$ (Fig. 9).

Proposed Cause of the Observed Seasonal Seismicity

It is proposed that the Spring concentration of seismicity in the late 19th century was caused by seasonal perturbations to the stress field which resulted from the

Table 4

Distributions of sediment in northern California

Locality	Subtotal (km^3)	Quantity (km^3)	
Output of debris, 1849–1914:			
Debris from mining:			
Feather River:	0.08		
Yuba River	0.52		
Bear River	0.19		
American River	0.20		
Mokelumne to Tuolumne Rivers	0.18		
Elsewhere	0.12		
		1.29	
Non-mining waste		0.54	
			1.83
Distribution of debris in 1914			
Deposits remaining within the Sierra Nevada		0.20	
Piedmont deposits		0.40	
Deposits in valley river channels		0.08	
Deposits on floodplains and tidal marshes		0.23	
Deposits in the bays:			
Suisun Bay	0.15		
Carquinez Strait	0.04		
San Pablo Bay	0.44		
San Francisco Bay	0.25		
		0.88	
Deposits in the Pacific Ocean		0.04	
			1.83

Data are from GILBERT (1917).

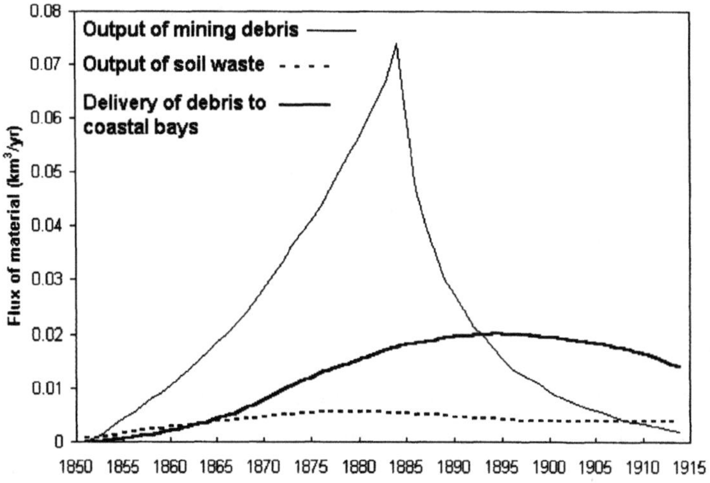

Figure 9
Movement of sediment in northern California during 1851–1914, from GILBERT (1917).

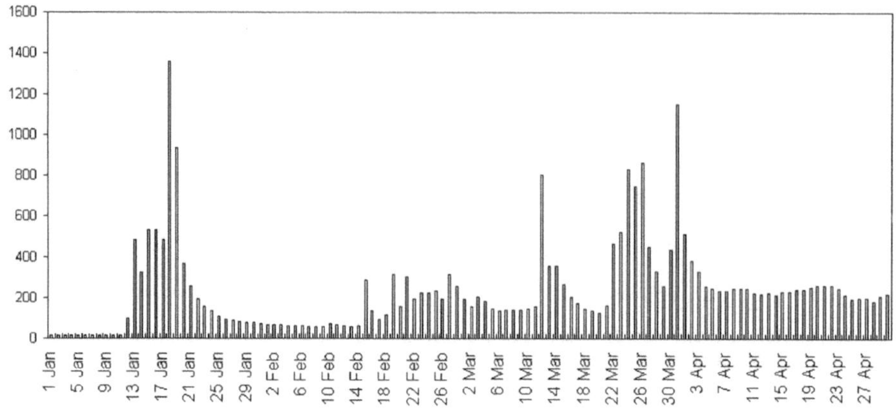

Figure 10
Discharge of water past Smartville gauging station on the Yuba River, 1 January to 30 April 1906. Data are from CLAPP (1907). Discharge of the American River at Fair Oaks and the Feather River at Oroville show similar variations. Positions of gauging stations are shown in Figure 1.

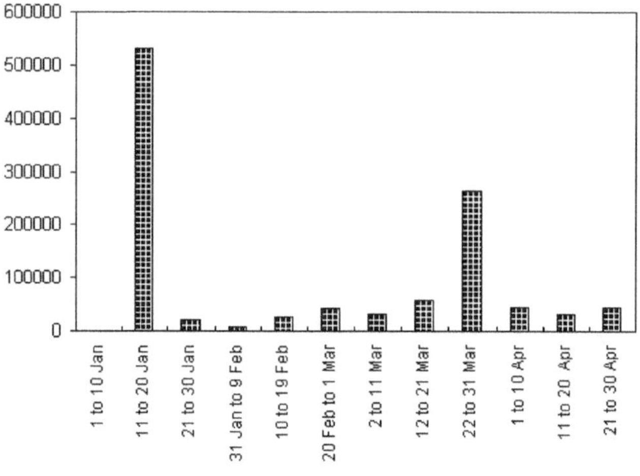

Figure 11
Estimates of the flux of suspended sediment past the gauging station at Smartville on the Yuba River during January to April 1906, from GILBERT (1917).

seasonal sediment loading in coastal bays following seasonal transport of sediment by the region's river system.

This region's strongly seasonal precipitation is most intense in December and January in inland mountain regions as well as in the coastal localities depicted in Figure 2. The greatest discharge by rivers usually occurs at this time, the largest flows being in warm Winters when rainfall melts earlier snow on high ground. In a given year, the greatest flux of sediment will be transported at times of high discharge. As examples, Figures 10 and 11 show measurements of the water discharge and sediment transport during the extreme flood in the winter of 1905–1906.

The result of these processes was the addition of a typical ~1 m thickness of sediment within the coastal bays, about two thirds of which arrived within ~25 years (1880–1905; Fig. 9). On average each year, the floor of each of these bays thus received ~30 mm of sediment, contributing a load of ~0.6 kPa (30 mm thickness $\times$ 2000 kg m^{-3} sediment density $\times$ g) with much of this load typically arriving within a few weeks at the time of peak discharge by the Sacramento River. I will model this process as the application of a sinusoidally varying load with an amplitude of 1 kPa.

Such localised loads will initially be supported by the elastic stress field within the crust. The resulting stress perturbation will increase the absolute vertical stress σ_{zz}. Considerations of equilibrium and stress compatibility (see Appendix) require this to be accompanied by a shear stress σ_{xz}, oriented to shear the base of the brittle layer away from the centre of the load, and by a change in the horizontal stress σ_{xx} beneath the load.

After a time delay of the order of the Maxwell time of the deepest continental crust, the lower crust will begin to flow in response to this sediment load. The load will cause a local excess in pressure at the base of the brittle layer compared with beneath its surroundings. As a result, lower crust will begin to flow outward from beneath it. The removal of lower crust from beneath a sediment load will require the stress field needed to support the load to be concentrated within the upper crustal brittle layer rather than being distributed throughout the crust. This will modify the perturbations to the stress tensor caused directly by loading, increasing the absolute vertical stress and decreasing the absolute horizontal stress beneath the load (see Appendix). Reduction in the northeast component of horizontal stress is equivalent to reducing the normal stress across a vertical fault striking northwest-southeast, thus unclamping the fault and so enhancing the possibility of shear failure on it.

Instantaneous and Relaxed Stress Perturbations

Analytic methods exist for calculating stress perturbations following the instantaneous application of loads (e.g., GOUGH, 1969; GOUGH and GOUGH, 1970), and have been used for instance in calculating the response to the impounding of artificial lakes. In these solutions, the Earth is treated as an elastic half space with uniform elastic properties throughout, and loads are assumed to *not* be isostatically compensated. This method has established various standard results: for instance, under a two-dimensional slab load M of uniform magnitude and half-width L, the differential stress $\Delta\sigma$ is

$$\Delta\sigma(x,z) = \frac{2Mz}{\pi}\left|\frac{L-x}{z^2+(L+x)^2}+\frac{L+x}{z^2+(L-x)^2}\right| \tag{3}$$

where the horizontal x coordinate is measured from the mid-point of the load and the vertical z coordinate is downward. The maximum differential stress $\Delta\sigma_m$ is thus

$$\Delta\sigma_m = \frac{2M}{\pi} \tag{4}$$

at $x = 0$, $z = L$. For $z \gg L$ and $x = 0$, $\Delta\sigma \propto 1/z$, being $\sim$0.2 $\Delta\sigma_m$ at $z = 10\ L$, $\sim$0.1 $\Delta\sigma_m$ at $z = 20\ L$, and so on. This means that even small loads (L say 5 km) are predicted by this theory to cause substantial stress perturbations in the mantle. The mantle is of course much stronger than the crust: its shear modulus is $\sim$70 GPa rather than $\sim$30 GPa. One would thus expect any stress transient to be more concentrated in the crust than is predicted by this analysis which treats the Earth as a uniform half space.

The abundant geological literature on calculations of the isostatic response to sediment loading concerns geological time scales, and I am unaware of any other established method for calculating the instantaneous response to sediment loads. It is thus necessary to develop a new method, which assumes more appropriate boundary conditions (see Appendix). This method allows the instantaneous stress response to be calculated, immediately after a surface load is applied. It also allows the subsequent response to be calculated once the lower crust has begun to flow in response to pressure gradients applied to it as a result of the imposition of this load.

Figure 12a shows the calculated immediate stress response beneath the centre of a sinusoidal load of amplitude $\sim$1 kPa and width $2L = 24$ km. Figure 12b shows the resulting delayed response, which is substantially greater in magnitude than the immediate response. WESTAWAY (1998) estimated the viscosity of the deepest crust in regions of normal crustal thickness and heat flow as $\sim 10^{17}$ Pa s, so its Maxwell time is of the order of 3×10^6 s or six weeks. Stress adjustment will thus be well under way in the months after a seasonal sediment load is applied, and something approaching this delayed response is thus expected within several months after each annual sediment load was applied to the coastal bays in the San Francisco area. Furthermore, at depths where the Maxwell time is long compared with time since the start of loading, stress perturbations of the form in Figure 12b will accumulate from year to year.

An alternative way to obtain the result that the perturbations to the stress tensor elements increase, from their instantaneous values, after the lower crust begins to flow, is as follows. From Figure 12a, the instantaneous annual sediment load with an assumed half-width of $\sim$12 km will increase the local pressure at the base of the brittle layer by $\sim$1 kPa, and thus establish a typical outward pressure gradient, P, of $\sim$0.08 Pa m^{-1}. From equation (A22), the typical instantaneous perturbation to the shear stress at the base of the brittle layer in the surroundings to the load is $\sim$0.5 kPa. Flow in a lower-crustal channel of thickness W in response to a pressure gradient P causes a shear traction on the upper boundary of the channel of σ_{xz} where

$$\sigma_{xz} \cong PW \tag{5}$$

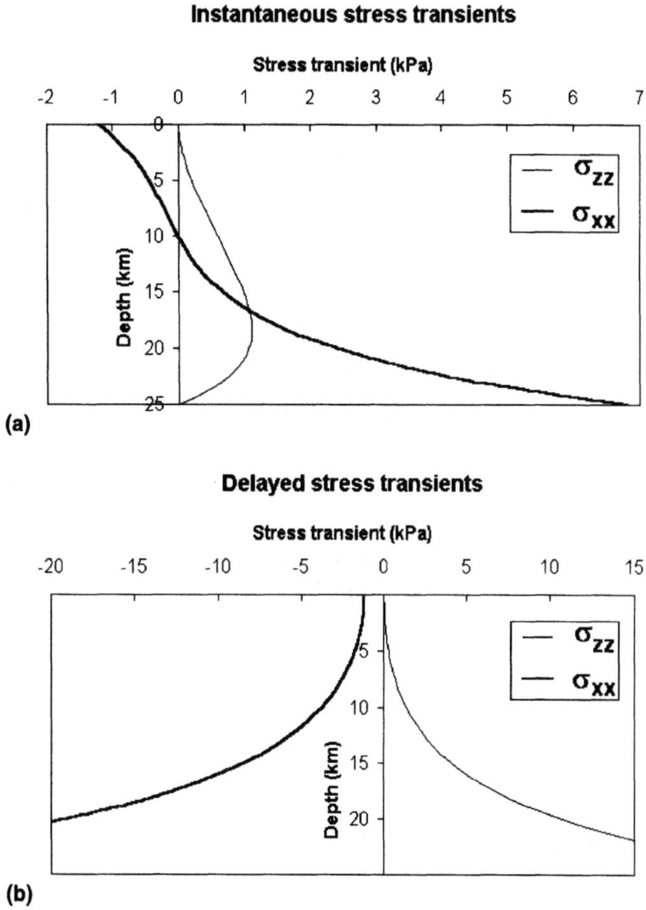

Figure 12

Graphs of perturbations to the stress tensor due to seasonal sediment loading, estimated along a vertical profile through the centre of the load between the Earth's surface and the Moho. (a) Instantaneous effect following deposition of a single year's sediment load, calculated using equation (18) with $L = 12$ km, $k = \pi/24$ km^{-1}, $D = 4.58$ MPa m^{-1} ($Dk = 600$ Pa), and $z_m = 25$ km, making $f = 1.4341$ and $u = 3.2725$. The solution is constrained to cause the magnitude of σ_{zz} to increase by up to ~1 kPa, consistent with the annual sediment load in San Pablo Bay during the late nineteenth century. (b) Subsequent relaxed stress field, calculated using equation (11), with the same values of D, k and L as before. This solution is valid at times after loading which are long compared with the Maxwell time of the deepest crust but short compared with the Maxwell time at the depth where the stress field is calculated.

(WESTAWAY, 1998). When the lower crust first begins to flow following a season's sediment loading, only the deepest crust immediately above the Moho will be affected. If W is thus, initially, say 2 km (i.e., this induced flow is confined to the lower crust between 23 and 25 km depth), then from (5) an additional contribution to the shear stress at the base of the brittle layer of ~0.08 Pa m^{-1} × ~2 km or ~0.16 kPa will develop, which will add to the shear stress already present. Considerations of equilibrium and stress compatibility (see Appendix) require corresponding adjustments to the perturbations to σ_{xx} and σ_{zz}. The resulting increase in the absolute vertical stress beneath the load increases the outward pressure gradient, which increases the shear traction caused by the resulting lower-crustal flow, which perturbs the vertical stress further, and so on. Equations (A12) to (A14) describe the resulting configuration, above the depth range affected by the induced lower crustal flow, when steady-state conditions are re-established (Fig. 13).

Implications for Conditions for Shear Failure

Some recent studies have proposed (e.g., DIETERICH, 1994) and applied (e.g., HARRIS and SIMPSON, 1998) complex failure criteria to try to explain characteristics of seismicity, or have argued that different faults in the study region obey different criteria (e.g., PARSONS *et al.*, 1999). However, in this study, whether or not the condition for shear failure is satisfied in a given locality will be determined using the conventional Coulomb–Navier criterion, using the same set of parameters in all localities. A failure parameter Φ is defined as:

$$\Phi = \tau - c(\sigma_n - p_f) - S \ , \tag{6}$$

where τ is the resolved shear stress, σ_n is the normal stress, c is the coefficient of friction, p_f is the pressure of pore fluid, and S is the cohesion. Realistic estimates are $\sim$0.5 for c and 10 MPa for S for renewed failure of a pre-existing fault plane. Formation of a new fault plane instead requires $S \sim 100$ MPa. It thus follows that once a fault plane has already formed,

$$\frac{\partial \Phi}{\partial t} = \frac{\partial \tau}{\partial t} - c\frac{\partial \sigma_n}{\partial t} + c\frac{\partial p_f}{\partial t} \ . \tag{7}$$

Several methods are available for estimating $\partial\tau/\partial t$. The first assumes that between earthquakes shear stress accumulates uniformly throughout the plate boundary zone. If the rate of relative plate motion is V and the width of this boundary zone is X, then

$$\frac{\partial \tau}{\partial t} = \frac{\mu X}{V} \tag{8}$$

and is 15 kPa yr^{-1} for $V = 50$ mm yr^{-1}, $X = 100$ km, and $\mu = 30$ GPa.

A second method considers individual faults within the plate boundary zone. Consider earthquakes which cause square patches of fault with dimension H to slip by a distance D. Let the long-term average slip rate on the fault be V. The recurrence interval of earthquakes is thus D/V and the coseismic stress drop is $\mu D/H$. The average rate of increase in shear stress between earthquakes is thus

$$\frac{\partial \tau}{\partial t} = \frac{\mu D}{H} / \frac{D}{V} = \frac{\mu V}{H} \ . \tag{9}$$

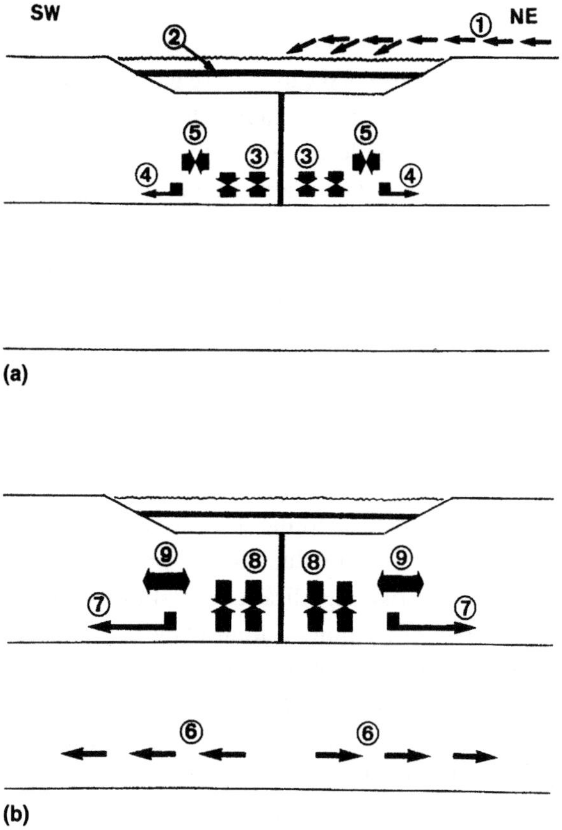

Figure 13
Cartoons illustrating the processes responsible for the Spring seismicity concentration in northern California during the late nineteenth century. (a) Immediate effects. Seasonal influxes of sediment transported in late Winter or early Spring by the Sacramento River (1) cause seasonal sedimentation in coastal bays (2). This loading causes an increase in the absolute vertical stress at depth (3). Considerations of equilibrium and stress compatibility require this to be accompanied by changes in the shear stress and horizontal stress. Near the base of the brittle layer, these changes involve an increase in the absolute horizontal stress (5) of similar magnitude to the increase in vertical stress (Fig. 12a), and an outward-directed shear stress (4). (b) Delayed effects. The local excess in pressure at the base of the brittle layer, caused by the sediment loading in (a), will induce outward flow in the lower crust from beneath the depocentre. After a time of the order of the Maxwell time of the deepest crust has elapsed, this flow (6) will begin, and will impart a shear traction on the base of the brittle layer that will add to the local shear stress (7). Considerations of equilibrium and stress compatibility require this to be accompanied by an increase in the absolute vertical stress (8) and a decrease in the absolute horizontal stress (9) beneath the depocentre (Fig. 12b). The decreasing absolute horizontal stress will unclamp vertical faults passing beneath the depocentre (thick lines in (a) and (b)), facilitating shear failure on them.

V ranges from $\sim$5 mm yr^{-1} to $\sim$30 mm yr^{-1} on the major faults in Figure 1. Taking H = 15 km, the thickness of the brittle layer, $\partial\tau/\partial t$ ranges between 10 and 60 kPa yr^{-1}.

A third method uses rates of change in interseismic shear strain from geodetic measurements, multiplied by μ to give rates of change in shear stress. Shear strain is accumulating at rates of $\sim$0.2 to $\sim$0.6 rad yr^{-1} (e.g., PRESCOTT and YU, 1986), the highest rate being near the San Andreas fault whose slip rate is the fastest. These values indicate shear stress accumulation at $\sim$6 to $\sim$18 kPa yr^{-1}.

Given this range of estimates, I adopt a nominal rate of shear stress accumulation of 10 kPa yr^{-1}.

Figure 14 illustrates how Φ may vary seasonally due to different factors. Figure 14a illustrates the combined effect of an annual increase in shear stress at 10 kPa yr^{-1} and of annual sediment loading of the magnitude which occurred in the late 19th century. The contribution $\Delta\Phi_S$ from sediment loading is depicted schematically using the function

$$\Delta\Phi_S = c\left[U_0 + (U_\infty - U_0)\tanh^3(t/t_m)\right] \;, \tag{10}$$

where t is the time since sediment loading. This function is chosen empirically to enable $\Delta\Phi_S$ to vary smoothly from a value of $c\ U_0$ immediately after loading to an ultimate value of $c\ U_\infty$, such that significant relaxation has occurred after a time $t_m = 6$ weeks. The coefficient of friction on fault planes, c, is taken as 0.5, as before. U_0 and U_∞, chosen to match the maximum perturbations to σ_{xx} in Figure 12, are -0.67 kPa and $+8.47$ kPa, respectively. The zero of time when the sediment load is regarded as deposited is taken as week 9 of the year (i.e., early March). It is evident that the increase in Φ which is predicted over the following 3 months is similar to the increase which would occur in the absence of any seasonal stress transient over the whole remaining year. As the initial value of Φ is undetermined, the time of shear failure is likewise unresolved. However, one may estimate a probability density

▶

Figure 14

Graphs estimating seasonal variations in the failure parameter Φ at the base of the brittle layer, with weeks numbered from the start of the year. If Φ reaches zero, then shear failure will occur and Φ will rapidly decrease to a value far below zero. This effect is omitted. However, the starting value of Φ is set in each case such that shear failure is expected sometime during the two-year period covered. (a) The combined effect of interseismic build-up in shear stress and late 19th century sediment loading. Solid line indicates the nominal 10 kPa yr^{-1} rate of increase in tectonic shear stress due to plate motions. Dashed line indicates the estimated response to seasonal sediment loading, from equation (10), as discussed in the text. Thick solid line shows the combined effect of both processes. The value of Φ rises steeply in Spring given to the assumed phase of the response to sediment loading. (b) The effect of seasonal water table variations within the Santa Cruz Mountains. Solid line is as in (a). Dashed line shows a hypothetical sawtooth contribution to Φ, with a peak to peak amplitude of 5 kPa, from the seasonal variation in water load. For reasons discussed in the text and summarised in Figure 16, this variation will enhance Φ when the water table is low, in Summer and early Autumn. Thick solid line shows the combined effect of both processes. The resulting value of Φ rises steeply in late Autumn. (c) The combined effect of interseismic stress build-up, late 19th century sediment loading, and seasonal water table variations. Solid line shows the combined effect of tectonic and sediment loading, is as in (a). Dashed line, illustrating groundwater loading, is as in (b). Thick solid line shows the combined effect of all processes, with Φ rising both in Spring and in Autumn.

function for the resulting seasonal seismicity as proportional to $d\Phi/dt$ at each point in time, scaled to give an overall probability of 1 throughout the year. The thick dashed line in Figure 15 shows the probability density function for the Φ variation in Figure 14a.

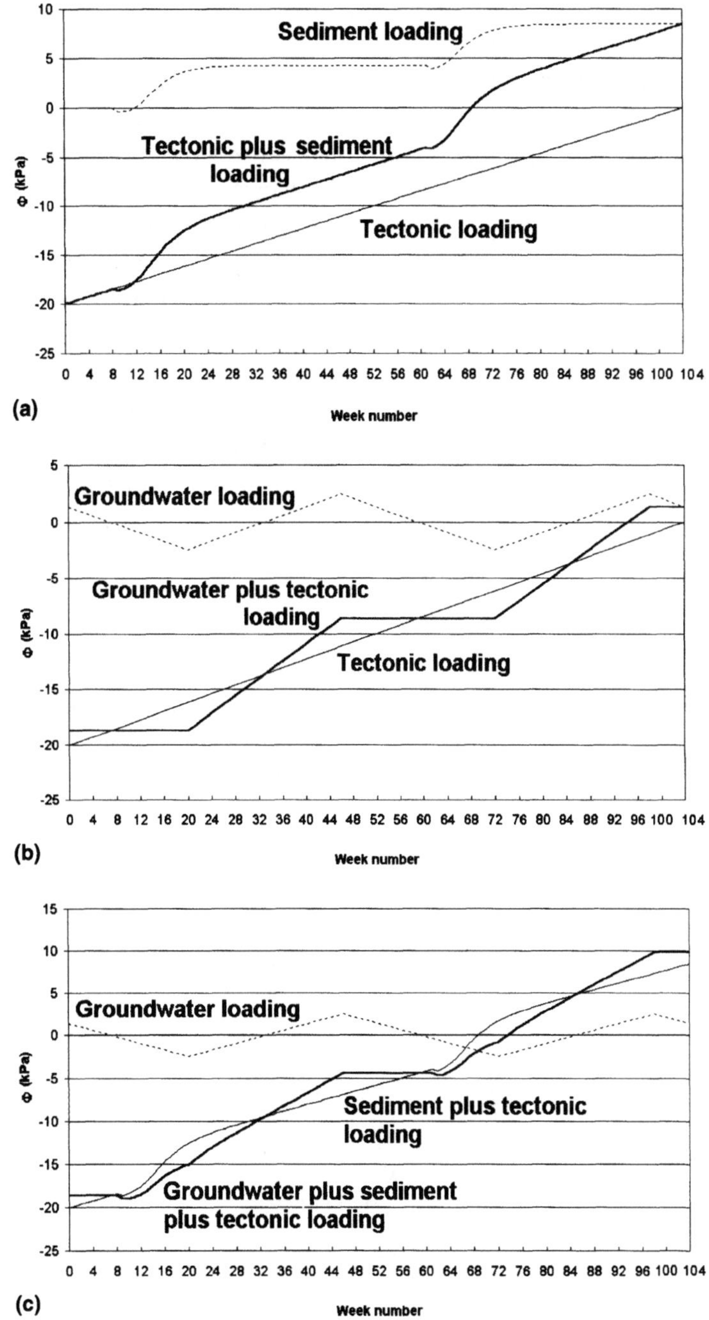

Present-day Magnitude of Stress Transients

To appreciate the scale of seasonal stress transients in the study region in the late nineteenth century, I now estimate present-day values for comparison. I first consider the four possible causes of seasonal stress transients identified by McCLELLAN (1984). However, this summary analysis suggests that these processes will either have a different effect from that inferred by McCLELLAN (1984), or will be of marginal significance. Furthermore, none of these processes can account for the difference in character of seismicity between the nineteenth and twentieth centuries. I then consider two additional processes which can cause stress transients: erosion and sediment transport, and seasonal water table changes.

No attempt is made here to calculate the stress perturbations which earthquakes cause in surrounding regions, or their effect on failure criteria: this topic has already been considered in detail by others (e.g., SIMPSON and REASENBERG, 1994; STEIN, 1999). For instance, the 1989 Loma Prieta earthquake is estimated to have affected the horizontal shear stress resolved parallel to its fault plane by between ~300 kPa in its epicentral area and ~3 kPa at distances of ~100 km (SIMPSON and REASENBERG, 1994). A pre-existing seasonal earthquake population will thus, of course, cause seasonal stress changes. However, these stress changes cannot explain how the seasonal earthquake population developed in the first place.

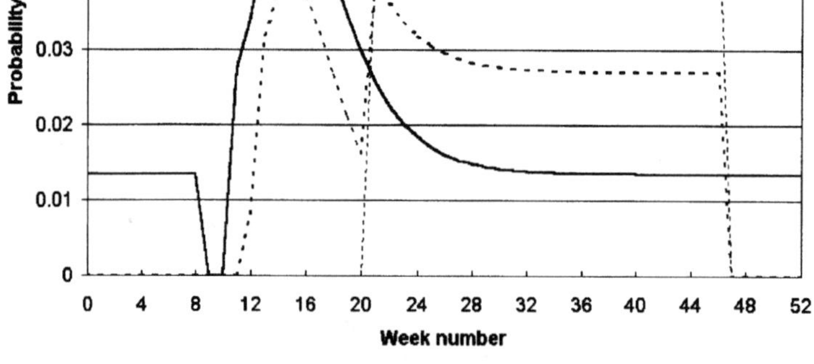

Figure 15
Probability density functions for seasonal seismicity estimated from the three stress field variations depicted in Figure 14.

Pore Fluid Pressure Increase

McClellan (1984), first, attributed the Spring seismicity concentration to an increase in pore fluid pressure at mid-crustal depths caused by rainfall in the preceding Winter. Such an explanation is contradicted by detailed modelling of hydrological changes following the 1989 Loma Prieta earthquake (e.g., Curry *et al.*, 1994; Rojstaczer and Wolf, 1994): the diffusivity of pore fluid through the crust is orders of magnitude too low for any significant response to develop at mid-crustal depths in weeks following seasonal rainfall. Further evidence of the time scale for circulation of water is provided by stable isotope measurements of pore water from < 1 km depth within the San Andreas fault zone (e.g., O'Neil, 1985). These indicate that this water fell as rain under a different climate from the present day and thus fell thousands of years ago, at least. Instead of increasing the pore pressure at mid-crustal depths, rainfall will increase the water table level near the Earth's surface, causing an increased load on the underlying crust. An investigation of the effects on Φ of seasonal water loading is thus warranted.

Seasonal Changes in Mean Sea Level

McClellan's (1984) second explanation attributed Spring seismicity to seasonal unloading of the continental shelf due to the annual reduction in mean sea level of $\sim$0.2 m between Winter and Summer. This change involves a decrease in the absolute vertical stress of $\sim$2 kPa. However, as seasonal changes in water table level of many meters occur in inland areas within this region (e.g., Rojstaczer and Wolf, 1994), it again makes sense to consider this process instead.

Tides

McClellan's (1984) third possible explanation considered that the increasing seasonal inclination of the Sun may cause the contribution of tidal stresses to Φ to be greatest in Spring. Others (e.g., Kilston and Knopoff, 1983) have also suggested that large earthquakes elsewhere in California are concentrated when tidal stresses are most favourable for shear failure. On the other hand, by studying the timing of many thousands of microearthquakes in California, Vidale *et al.* (1998a, b) have only identified a marginal increase in their frequency of occurrence at times when tidal stresses are favourably oriented compared with at other times.

The tidal contribution to variations in stress tensor elements is of the order of $\pm$0.5 kPa to $\pm$2 kPa (e.g., Shlien, 1972; Heaton, 1975; Vidale *et al.*, 1998a), a small fraction of the estimated annual change in shear stress due to the relative motion of plates in California (see above). Simpson and Reasenberg (1994) calculated that in the year following the Loma Prieta earthquake, the maximum effect of tides caused normal and shear stresses to range by $\sim$5 and $\sim$1 kPa, with maximum daily ranges of $\sim$1 and $\sim$0.2 kPa. Tidal effects thus contribute a high-frequency low-amplitude ripple onto longer-term stress variations such as in Figure 14.

Variations in Length of the Day

McCLELLAN's (1984) fourth possible cause of seasonal seismicity involved hypothetical seasonal changes in stress at the Pacific-North America plate boundary caused by variations in the torque acting on both plates, during the Spring deceleration in the rate of the Earth's axial rotation (e.g., CHALLINOR, 1971). This change in rotation rate is caused by the increase in the strength of westerly winds in the Southern Hemisphere during the transition to the southern Winter (e.g., LAMBECK and HOPGOOD, 1981, 1982), given conservation of the overall angular momentum of the solid Earth and its atmosphere. Whatever system of torques causes the solid Earth's angular momentum to change thus most likely acts mainly upon land areas in the Southern Hemisphere.

The magnitude of the torque required to cause the seasonal change in the solid Earth's angular momentum can be estimated by multiplying the Earth's moment of inertia about its rotation axis by its seasonal rate of change in angular velocity. The Earth's rotation period typically increases each Spring by ~2 ms over 3 months (e.g., CHALLINOR, 1971). The resulting change in angular velocity by about 2 parts in 10^8 (e.g., LAMBECK and HOPGOOD, 1982) requires a mean angular acceleration of $\sim 2 \times 10^{-20}$ s^{-2}. Multiplying this by the Earth's moment of inertia of $\sim 10^{38}$ kg m^2 gives a mean torque Γ of $\sim 2 \times 10^{18}$ N m. To derive an upper bound to the importance of this effect for California, suppose that this torque is concentrated not in the Southern Hemisphere but on the North American plate in the vicinity of the San Andreas fault zone, around latitude $\phi = 37°$N. The associated force F acting on this plate (which will be eastward in Autumn and westward in Spring) will be

$$F = \frac{\Gamma}{R\cos(\phi)} \tag{11}$$

where R is the radius of the Earth (~6400 km). F will thus have a magnitude of $\sim 4 \times 10^{11}$ N. If it acts across an area which is ~400 km long (north south) and ~100 km wide (east-west), it will contribute to changing the (east, down) shear stress in the crust within this area by ~10 Pa. A stress perturbation of this order will be negligible compared with other factors. Furthermore, because seasonal torques due to changes in atmospheric angular momentum will in reality be more widely distributed, and concentrated in the Southern Hemisphere, the resulting stress changes around the San Andreas fault zone will be much smaller than even this low estimate.

Effects of Erosion and Sediment Transport

Unlike in the late 19th century, the Sacramento river system is highly regulated at present (e.g., FISCHER, 1994). Much of the precipitation each winter is now stored in flood control reservoirs, and is thus unable to transport sediment to the coast. When flooding does occur, most floodwater and the associated suspended sediment is diverted into overflow channels. Much of this sediment is thus deposited over broad regions rather than accumulating at the coast.

At present, the many short rivers which drain the Coast Ranges into San Francisco and Monterey Bays are probably more important sources of sediment. Data for two such rivers, the Carmel and San Lorenzo (Table 5), indicate that their drainage basins are eroding at typical rates of $\sim$0.3 mm yr^{-1}. Assuming these rates are applicable throughout the $\sim$9000 and $\sim$12000 km^2 areas drained by small rivers flowing into San Francisco and Monterey Bays, respectively, the typical fluxes of sediment into these bays can be estimated as $\sim$0.004 and $\sim$0.005 km^3 yr^{-1}. These figures are probably overestimates, as much of the region which drains into these bays has much lower rainfall than the mountainous coastal regions traversed by the rivers listed in Table 5.

However, weather conditions occasionally cause extremes of sediment transport. An example involved the storm of 3–5 January 1982, when >500 mm of rain fell in some parts of the San Francisco Bay region, on land which was already saturated by earlier rainfall (e.g., BLODGETT and POESCHEL, 1988). As a result, debris flows carried large quantities of material into rivers within the Coast Ranges, causing abnormal fluxes of sediment to reach the coastal bays. NOLAN and MARRON (1988) estimated that $\sim 0.94 \times 10^9$ kg or $\sim$0.00047 km^3 of sediment was transported during this three-day storm by the San Lorenzo River, which drains a $\sim$300 km^2 area of the Santa Cruz Mountains (Fig. 1), more than six times the typical quantity for a complete year (Table 5). Nonetheless, even if such a high rate of erosion occurred throughout the $\sim$2100 km^2 total area of these mountains, the flux of material would be only $\sim$0.003 km^3, one sixth of the typical annual flux reaching San Francisco Bay in the late 19th century.

Although crude and approximate, these calculations do indicate that rates of sediment transport in the study region are too low at present to have a significant effect on the seasonal character of seismicity. However, BROWN (1988) noted that the sediment transport during the 1982 storm was exacerbated by urban development.

Table 5

Sediment yields from small rivers

River	Area (km^2)	Load (10^9 kg yr^{-1})	Yield (10^3 kg km^{-2} yr^{-1})	Flux (km^3 yr^{-1})	Erosion Rate (mm yr^{-1})
San Lorenzo	290	0.15	507	0.00008	0.26
Carmel	630	0.40	635	0.00020	0.32

Area is area of drainage basin. Load is average mass of sediment transported per year. Yield is Load divided by Area. Flux is volume of sediment transported per year, estimated as Load divided by a nominal density of 2000 kg m^{-3}. Erosion Rate, spatially averaged across the drainage basin, is Flux divided by Area. Data for San Lorenzo River are from NOLAN and MARRON (1988), Load being the mean value for 1973 to 1981. Data for Carmel River are from MILLIMAN and SYVITSKI (1992).

The lack of undeveloped flat land means that much of the region's recent dramatic population growth has required building in areas of substantial relief. BROWN (1988) deduced that the resulting disturbance to sites and removal of vegetation, which protects the underlying rock against erosion, is contributing to this increase in rates of erosion. One may thus predict that if quantities of sediment transported during future storms continue to increase, then the magnitude of seasonal stress transients may eventually reach the levels of the late nineteenth century, raising the possibility of a resumption of seismicity with a concentration in Spring.

Seasonal Water Table Changes

Winter rainfall will recharge shallow aquifers. In the Santa Cruz Mountains, where the mean annual rainfall exceeds 1000 mm, the water table typically rises by ~5 m each Spring and falls to a minimum in Autumn ahead of the Winter recharge, with annual variations from near zero to as much as ~18 m (e.g., CURRY *et al.*, 1994). Assuming that the pore space within rock which can be occupied by water is ~30% of the total, a 5-m rise in the water table will load the underlying crust by ~15 kPa.

The Loma Prieta earthquake of October 1989 occurred after several dry years followed intense rainfall at the start of 1986, at a time when the water table was at its lowest for 15 years (ROJSTACZER and WOLF, 1994). The previous earthquake in the same region, in October 1865, also occurred after several dry years followed the intense rainfall in the winter of 1861–1862 (data from EXNER *et al.* (1929)). If this seasonal association signifies anything, it is that the reduction in the absolute vertical stress caused by drawdown of the water table in this locality has been conducive to shear failure.

Figure 16 suggests a physical mechanism which may cause this seasonal seismicity pattern in this locality, where the San Andreas fault steps leftward (Fig. 1) (e.g., BÜRGMANN *et al.*, 1994; LANGENHEIM *et al.*, 1997). Such stepovers can be approximated as zones of oblique distributed shortening containing throughgoing strike-slip faults (e.g., WESTAWAY, 1995). Within the stepover region, the absolute horizontal stress thus exceeds the absolute vertical stress. Figure 16a shows a cross section through such a stepover with the water table at a given level. The local isostatic balance is assumed to be governed by the excess pressure at the base of the brittle layer within the stepover, compared with its surroundings, which drives outward lower-crustal flow from beneath the stepover. In this initial configuration, some gravitational potential energy is stored in the topography, and some elastic strain energy is stored by the horizontal compression of the rocks within the stepover. In Figure 16b, the water table level has dropped, representing the situation in Autumn of a typical year, thus reducing the load applied to the crust. This reduction will enable the crust to adjust by converting elastic strain energy stored in horizontal compression to gravitational potential energy. As a result, the mountain range will

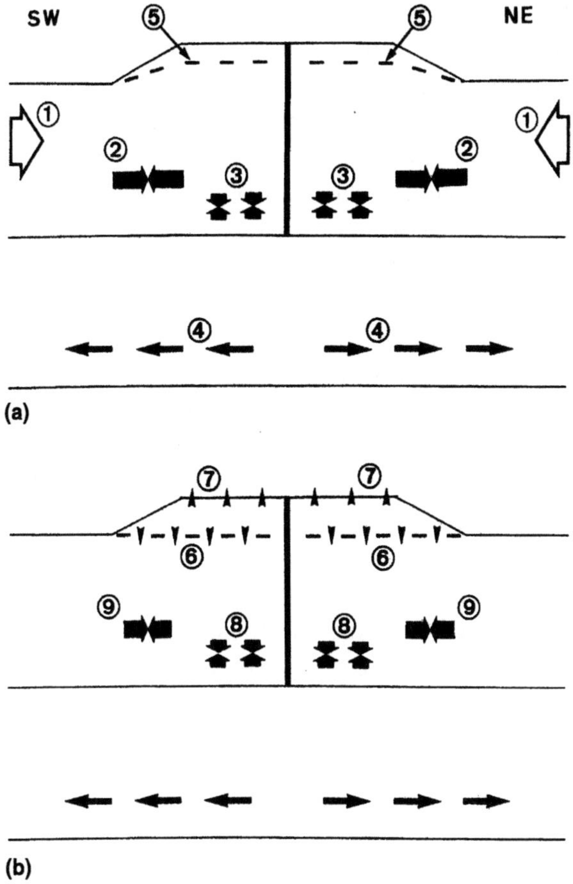

Figure 16
Cartoons illustrating the processes responsible for the Autumn seismicity concentration in the Santa Cruz Mountains during the twentieth century. (a) Initial state. The Santa Cruz Mountains form a zone of oblique convergence (1) around a leftward step in the right-lateral San Andreas fault. The resulting component of distributed shortening perpendicular to this fault segment causes the absolute horizontal stress (2) to exceed the absolute vertical stress (3). The associated component of thickening of the brittle layer causes the pressure at the base of the brittle layer to be greater beneath these mountains than beneath its surroundings. This causes a component of outward lower-crustal flow (4) which contributes to maintaining isostatic equilibrium. The initial level of the water table (5) provides a water load which contributes to the local excess pressure at the base of the brittle layer. (b) Effects following a reduction in water table level. A reduction in water table level (6) will act to reduce the pressure near the base of the brittle layer. However, it is not possible for the flow regime in the lower crust to respond immediately to such a change, because its Maxwell time is finite. The region will instead adjust by converting elastic strain energy stored by its horizontal compression into gravitational potential energy, by uplifting the Earth's surface (7) so as to keep the vertical stress at the base of the brittle layer the same as in (a) (8), causing the rate of outward lower-crustal flow to remain the same as before. The associated reduction in horizontal stress (9) will facilitate shear failure by reducing the normal stress across the active fault plane.

uplift somewhat, and the absolute horizontal stress will decrease. From (6), this change will unclamp any fault planes within the stepover, enhancing the condition for shear failure; thus facilitating seismicity at times of low water table levels.

Figure 14b indicates schematically how this process may contribute to seasonal seismicity. It shows the inferred contributions to Φ from tectonic stress build-up and seasonal water table changes, the latter represented as a sawtooth variation with peak to peak amplitude 5 kPa and which reaches its maximum in week 46 of the year (i.e., in mid-November). This amplitude chosen is the minimum which can cause significant seasonal seismicity, given the assumed rate of shear stress build-up of 10 kPa yr^{-1}. Figure 15 illustrates the probability density function for this variation in Φ. It predicts that earthquakes can only occur between weeks 20 and 46, in Summer and Autumn.

No attempt is made here to calculate stress changes or seasonal amounts of surface uplift caused by the process in Figure 16; such details will be presented elsewhere. However, the existence of this process is potentially testable, as it predicts that the uplift of this mountain range does not occur uniformly from year to year, but will instead show seasonal fluctuations. Amounts of uplift will also be greatest in years when the water table is substantially drawn down. To the best of my knowledge, no-one has yet tried to measure the uplift of this mountain range using continuous geodetic measurements. Such measurements, carried out in co-ordination with monitoring of water table fluctuations to indicate local loads on the lower crust, may provide insights into the present-day pattern of stress transients in this region. The roughly uniform uplift rates deduced over much longer time scales using geological evidence, for instance by VALENSISE and WARD (1991), VALENSISE (1994), and LANGENHEIM *et al.* (1997), average out these seasonal fluctuations.

Summary of Results and their Implications

During the late 19th century and up to the great San Francisco earthquake of 18 April 1906, northern California experienced more intense seismicity than at any other time in its historical record. This activity, at the magnitude >5 level, affected localities both within, and north and south of, the San Francisco Bay region, and – like the great 1906 event – was concentrated in Spring. It is proposed that this seasonal character was influenced by seasonal stress transients associated with seasonal sedimentation in coastal bays which straddle active strike-slip faults. Such transients will be superposed onto the effects of the steady interseismic build-up in shear stress at this plate boundary, and will thus increase the probability of earthquakes shortly after the time of sedimentation by increasing the likelihood that some point within the crust is brought to the condition for shear failure. The magnitude of these seasonal stress transients were particularly large in the late 19th century, due to the quantities of sediment released into the region's river system due to hydraulic mining of gold; they are estimated as causing seasonal increases in the Navier–Coulomb failure parameter Φ by $\sim$5 kPa (Fig. 14).

It is further suggested that outside this time of extreme sedimentation in the late 19th century, the most important influence on the seasonal character of seismicity is loading by seasonal water table changes. In the Santa Cruz Mountains, this concentrates seismicity in Autumn. The greater difficulty in resolving the significant seasonal character in the study region's seismicity in 1851–1878 compared with 1879–1906 may thus relate to the presence of two seasonal forcing mechanisms which were out of phase. The Spring sediment loading in coastal bays presumably overwhelmed any other forcing process during 1879–1906. To illustrate this masking effect, Figure 14c shows the predicted variations in Φ when seasonal water table changes in the Santa Cruz Mountains and nearby seasonal sedimentation create stress transients of similar magnitude. Compared with Figures 14a and 14b the rate of increase in Φ is more nearly uniform, and the probability of earthquakes is more uniform throughout the year (Fig. 15).

Implications of a Weak Lower Crust

To account for the seasonal seismicity in the 19th century, it is suggested that that the deepest crust in this region has a Maxwell time of the order of weeks, and thus a viscosity of the order of 10^{17} Pa s. WESTAWAY (1998) showed that lower-crustal flow can be characterised by an "effective" viscosity η_e, which is ~60 times the viscosity at the Moho, and represents the viscosity which an isoviscous lower-crustal channel would require to give the same flow rate as in real lower-crust which has a temperature-dependent viscosity that decreases downwards. The present study is thus, in effect, implying that the lower crust beneath California has $\eta_e \sim 6 \times 10^{18}$ Pa s. Such a low viscosity has not previously been considered for this region. It has other consequences also, for instance, it requires that the region's topography is being dynamically supported by lower-crustal flow, suggesting that this effect will need to be considered along with the other contributing factors modelled previously (e.g., by BÜRGMANN *et al.*, 1994). It is not possible to provide independent proof of such a low viscosity in the lower crust. However, recent studies (e.g., WESTAWAY, 1998; MITCHELL and WESTAWAY, 1999) infer similar viscosities to account for other geological processes including the growth of topography in other regions. WESTAWAY (2001) has argued that lower-crustal flow has contributed significantly to the development of relief in north-western Europe during the Quaternary, a view which requires a similar typical value of η_e to that deduced in the present study. Conversely, if it could be shown that η_e for the crust beneath California is much greater than $\sim 10^{19}$ Pa s, the explanation for this region's seasonal seismicity in the late 19th century would be shown to be incorrect, in which case this seasonal characteristic would remain a mystery.

Many people have tried calculating stress perturbations caused in the surroundings of regions affected by individual northern California earthquakes. Some such studies (e.g., HARRIS and SIMPSON, 1998; PARSONS *et al.*, 1999) have identified evidence that they cannot readily explain by this approach: for instance, the 1911 San Jose earthquake occurred in a locality which was so severely de-stressed immediately after the 1906 earthquake that the subsequent loading by five years of plate motion cannot possibly bring it back to the condition for shear failure if the Coulomb–Navier criterion is used. As a result, they have proposed other, much more complex, failure criteria, or have argued that different faults obey different failure criteria. However, these studies have worked with instantaneous elastic stresses, and have neglected the effect – considered in the present study – of the subsequent redistribution of stress as a result of flow induced in the lower crust. In contrast, the present study has illustrated that the delayed stress changes induced in the study region by a simple surface load can be quite complex. It follows that the delayed stress changes induced on other faults by the loading effect of slip on any one fault are likely to be even more complex, and may explain anomalous aspects of the region's seismicity such as that mentioned above. Future investigation of stress changes accompanying lower-crustal flow that is induced by individual earthquakes can thus be identified as a potentially promising line of enquiry for the future. However, it is likely that any solution of this general problem will be extremely complex: it may well require coupling of numerical models for the behaviour of inhomogeneous fault planes in the three-dimensional pattern observed in the study region under the influence of applied tectonic loads with numerical models for flow in the lower crust and for the response to applied surface loads.

Effects of Repeated Seasonal Loading and Seismicity on the Stress Field

Previous studies (e.g., SYKES and JAUMÉ, 1990; BUFE and VARNES, 1993; JAUMÉ and SYKES, 1996, 1999; BOWMAN *et al.*, 1998; HARRIS and SIMPSON, 1998) have tried to explain the varying rates of seismicity in this study region as combinations of precursor and aftershock activity around its largest earthquakes. However, there is little consensus between these different views. For instance, SYKES and JAUMÉ (1990) and JAUMÉ and SYKES (1996, 1999) regarded the moderate seismicity which began in the 1870s (Fig. 6) as precursory activity to the 1906 event. SYKES and JAUMÉ (1990) regarded the greatly reduced seismicity after 1906 as marking the "stress shadow" of the 1906 event, which de-stressed such a large region that few fault planes were left near the condition for shear failure. They also debated whether the increase in seismicity after 1955 (Fig. 5) was best regarded as precursory activity to the 1989 event or as recovery from the 1906 stress shadow. SYKES and JAUMÉ (1990) also interpreted the cluster of events which occurred in 1855 to 1868 (Fig. 5) as precursors of the 1868 Hayward earthquake. BUFE and VARNES (1993) claimed instead that the seismicity up until 1892 involved an overall decrease in moment rate with time and it – including the 1868 earthquake – was more likely a prolonged aftershock sequence of the 1838 event, but accepted that precursors to the 1906 earthquake did occur after 1892. They also regarded the increase in seismicity after 1955 as unequivocal precursory activity to the 1989 event, and indeed claimed that this pattern began as early as 1927 and was clear enough as early as 1969 to allow this specific earthquake to have been predicted then. In contrast, BOWMAN *et al.* (1998) called for greater objectivity in discriminating between apparent precursory activity of an event such as in 1989 and the recovery from the stress shadow of an earlier larger event such as in 1906. Although they considered the evidence for a precursory acceleration in seismicity ahead of 1989 to be strong, inspection of their data (their Fig. 5) indicates that only the 1984 Morgan Hill and 1986 Mount Lewis earthquakes separate such a precursory acceleration in seismicity from a uniform rate of recovery from the 1906 stress shadow. Thus, whether one believes that the Loma Prieta earthquake had precursory activity depends entirely on whether one believes these two earthquakes to be its precursors or random "background" seismicity. Although basing a claim about precursory activity on such scanty evidence seems risky, the evidence for precursory activity to both the 1868 and the 1906 earthquakes seems instead to be much stronger, as was originally noted by SYKES and JAUMÉ (1990).

It is proposed in this study that the precursory seismicity before 1906 was caused by the large seasonal stress perturbations in the late 19th century and, by implication, the relative lack of precursory seismicity before 1989 related to the relative absence of such perturbations at that time. However, it is important to consider whether the abnormally large seasonal stress transients in the late 19th century simply brought forward earthquakes which would have happened anyway, or whether they caused additional seismicity.

The evidence that more $M > 5$ earthquakes occurred during the decades before 1906 than at any other time (Table 1), and their spatial distribution spreading outward from the localities with most rapid sedimentation (Fig. 5b), suggest that the seasonal stress transients at this time did cause additional seismicity. Furthermore, the loading by the annual flux of sediment would do work on the crust in its vicinity which would result in additional elastic strain energy being stored, in addition to the component stored as a result of plate motions, and subsequently redistributed following each earthquake. This reasoning also suggests that this seasonal loading may have not simply brought forward earthquakes which would have happened anyway sometime later: it may well have instead led to the regional stress field evolving in a manner which was unique to this time.

After each earthquake, the stress field will adjust. The value of Φ is greatly reduced adjacent to the fault which slipped, but may increase elsewhere as elastic strain energy is redistributed. Theory (e.g., CARLSON, 1991; SHAW *et al.*, 1992) has established that the regular occurrence of moderate-sized earthquakes in a region has a systematic effect on the local stress field. It predicts that the stress field becomes "rougher" (i.e., develops sharper lateral variations) on a local scale, but "smoother" (i.e., more nearly uniform) when spatially averaged over many fault segments. Carlson's (1991) work was motivated by the need to try to explain how great earthquakes (such as the 1906 event in California) can initiate a "cascade" of fault ruptures, which require several adjacent fault segments to simultaneously approach the condition for shear failure. Using a simple rheological model for faulting, CARLSON (1991) showed that after many moderate earthquakes in a region, the conditions can become set for a very large earthquake. If this is so, it implies that stress transients caused by hydraulic mining were responsible not only for the intensified sequence of moderate earthquakes illustrated in Figure 5, but also for the great 1906 earthquake itself. At other times, independent ruptures have occurred instead on fault segments which slipped together in 1906. For instance, the Santa Cruz Mountains event of 8 October 1865 and the Loma Prieta event of 18 October 1989 ruptured faults near the southern end of the 1906 rupture, and the event of June 1838 appears to have ruptured this fault segment plus the peninsular segment of the San Andreas fault south of San Francisco (TOPPOZADA and BORCHERDT, 1998).

It is suggested here that by the 1850s, the Hayward fault was close to its condition for shear failure. Seasonal loading of the crust, which was already substantial at this time (Fig. 9), facilitated the sequence of moderate-sized earthquakes at this time, which redistributed elastic strain energy such that in the end the Hayward fault could slip in the 1868 earthquake. This slip de-stressed a wide region, and it took a decade or so before the surrounding region was again close enough to the condition for shear failure that such loading effects could once again establish a seasonal pattern of moderate-sized earthquakes. This ultimately contributed to establishing the conditions for the 1906 earthquake to occur. This event de-stressed the entire region, and subsequent stress recovery led to the 1989 earthquake.

It is, further, suggested that the repeated annual loading to the crust which occurred in the late 19th century assisted in bringing patches of fault to the condition for shear failure, and so helped to create the conditions for shear failure in the 1868 and 1906 earthquakes. It follows that the resulting sequences of precursory earthquakes, concentrated in Spring, were unique to this time in this study region and thus provide no guide to what will happen ahead of future large earthquakes. Future large earthquakes in the region, like in 1989, may indeed have little or no precursory activity, as is indeed typical for large earthquakes worldwide. Furthermore, the annual cyclic loading and the associated precursory seismicity may well mean that the manner in which the region approached criticality ahead of 1906 was different to other large earthquakes. This raises the possibility that the 1906 earthquake may have been unique, and that at other times the segments of the San Andreas fault affected by it, for instance north and south of the Golden Gate, may rupture independently. As already noted, the radiocarbon dating evidence currently available cannot resolve this point (see, e.g., SCHWARTZ *et al.*, 1998).

Using an idealised "cellular automata" model of faulting subjected to a steady build-up of tectonic stress at 3.3 kPa yr^{-1}, RYDELEK and SACKS (1999) showed that a 20 kPa *increase* in normal stress throughout a vertical model fault plane inhibited precursory microseismicity, delayed a future large earthquake by 13 years, and resulted in an eventual large earthquake that was not identical to the one in the unperturbed model run. In this case, the time delay was more than double what would be predicted for a homogenous fault plane (20 kPa/3.3 kPa $yr^{-1} \cong 6$ yr). It follows that abrupt annual sediment loading followed by stress adjustment accompanying lower-crustal flow, which will *reduce* the normal stress across vertical faults, would also have a greater effect if coupled to a model of this type than for a homogenous fault plane as in Figure 14: it would enhance precursory microseismicity and being forward a future large earthquake, which would also not be identical to what would have happened otherwise. For the homogenous model fault plane in Figure 14, tectonic loading contributes +10 kPa yr^{-1} to Φ and the superposed annual reduction in normal stress caused by sediment loading contributes an additional +5 kPa yr^{-1}. If maintained over 40 years, this sediment loading contribution would increase Φ by 5 kPa $\times$ 40 or 200 kPa and bring the following large earthquake forward by 200 kPa/10 kPa yr^{-1} or 20 years. Given the results of RYDELEK and SACKS (1999), the additional effect of fault plane inhomogeneity may have resulted in a much longer time delay. It follows that in the absence of hydraulic mining, northern California would thus have experienced no large earthquake in 1906 but instead a different – possibly smaller – earthquake some decades later. This interpretation raises the possibility of attributing legal liability for the loss of life and damage to property caused by the 1906 earthquake to the operators of hydraulic mines which had been worked in the decades beforehand.

Relationship to Induced Seismicity Beneath Reservoirs

In this regard, it is relevant to consider the analogous problem of reservoir-induced seismicity (e.g., ROTHÉ, 1970). As many examples of this phenomenon are now known, it has become standard for operators of large dams to monitor local seismicity both before and after the dam is built (e.g., GUPTA and RASTOGI, 1976), to provide evidence whether any later earthquake is "natural" or "induced". By analogy, it may in future become standard for anyone proposing any large-scale change to the hydrology or sediment transport regime of a river system in a seismically active region to analyse the changes expected in the local pattern of seasonal stress transients, to establish whether any significant effect is likely on the seismicity. Reservoir-induced seismicity may indeed itself be explicable using a variant of the process in Figure 13. Previous studies (e.g., SIMPSON *et al.*, 1988) have noticed that reservoir-induced seismicity is sometimes a prompt effect following water loading and sometimes a delayed effect. Rather than requiring separate physical processes, both prompt and delayed stress transients can potentially be explained by the process in Figure 13, depending on the viscosity of the lowermost crust in different regions. The possibility of modelling reservoir-induced seismicity in this manner, and using it to obtain estimates of lower-crustal viscosity, will be discussed in detail elsewhere.

However, the characteristic scale of reservoir loads (which typically trigger induced seismicity when water depths exceed ~100 m and water loads thus exceed

~1 MPa) relative to the ~1 kPa order-of magnitude of nineteenth-century annual sediment loads in California, does warrant comment. The stress modelling presented earlier indicates that, once the lower crust begins to flow after a load is applied, the effect on Φ depends on the combination of the magnitude of the load and its horizontal scale. As reservoirs are typically much narrower than the California coastal bays, much larger-magnitude loads are needed to achieve the same change in Φ. The main reason for this is that when a load is much narrower than the thickness of the brittle upper crust, the bulk of it will be supported within the upper-crustal brittle layer and so its effect on the pressure at the base of this layer is greatly reduced. Conversely, if a surface load is very widely distributed, the resulting horizontal gradients in pressure at the base of the brittle layer will be negligible, causing a negligible flow response in the lower crust which will not significantly affect the stress field within the brittle layer.

Other Possible Regions with Conditions Conducive to Seasonal Seismicity

The evidence of seasonal seismicity in coastal northern California during the late 19th century also raises the possibility that other regions with vigorous surface processes may also experience seasonal seismicity. To assess this possibility, Table 6 compiles sediment load data for the world's major rivers. It is evident that many of these rivers have existed in the same locations for long enough that their deltas have prograded beyond the continental shelf. They thus now discharge onto oceanic crust, which lacks a weak lower-crustal channel. Furthermore, most of the others deposit widely distributed loads on the continental shelf, rather than localised loads in enclosed bays. Based on their natural hydrological regimes, the three best candidate localities for investigation of relationships between seasonal sediment loads and any local induced seismicity are the mouths of the Colorado and Huanghe (Yellow) Rivers, and the Shatt-el-Arab, the mouth of the Tigris/Euphrates (Table 6). The Colorado was of course dammed in 1935 and, as a result, its discharge of water and sediment to the Gulf of California are now negligible. However, it previously had a strongly seasonal flow regime involving an annual nival flood caused by melting of snow in the Rocky Mountains in early Spring. The Tigris and Euphrates likewise formerly experienced nival floods due to melting of snow in the mountains of eastern Turkey, but have now also been dammed and barraged at many points. The Huanghe retains a strongly seasonal regime, with an annual flood stage caused by the summer monsoon. However, testing of the earthquake populations in any of these localities, in an attempt to confirm the predicted seasonal pattern, is beyond the scope of this study.

A potential difficulty is raised, because earlier analysis suggested that the extent of seasonal character to a region's seismicity depends on the relative magnitudes of the seasonal transient in Φ caused by surface processes compared with the annual interseismic increase caused by active tectonics. This implies that the most strongly seasonal earthquake populations are likely to exist in regions which have strongly

Table 6

Comparison of sediment loads of major rivers

River	Load ($Mt\ yr^{-1}$)	Location	Crust at mouth	Notes
Amazon	1320	Atlantic Ocean	Oceanic	(1)
Bhramaputra	1157	Bay of Bengal	Oceanic	(1)
Ganges	680	Bay of Bengal	Oceanic	(1)
Mississippi	605	Gulf of Mexico	Oceanic	(1)
Yangtse	468	South China Sea	Continental	(2)
Indus	300	Indian Ocean	Oceanic	(1)
Mekong	176	South China Sea	Continental	(2)
Colorado	167	*Gulf of California	Continental	(3)
Orinoco	165	Atlantic Ocean	Oceanic	(1)
Mackenzie	110	Beaufort Sea	Continental	(2)
Huanghe	100	Bohai Gulf	Continental	(3)
Nile	100	*Mediterranean Sea	Continental margin	(4)
La Plata	87	Rio de la Plata Gulf	Continental	(2)
Yukon	79	Bering Sea	Continental	(2)
Danube	74	Black Sea	Continental	(2)
Orange	58	Atlantic Ocean shelf	Continental	(2)
Amur	57	Sea of Okhotsk	Continental	(2)
Congo	51	Atlantic Ocean	Oceanic	(1)
Shatt-el-Arab	50	*Persian Gulf	Continental	(3)
Zambezi	48	Indian Ocean	Oceanic	(1)
Sacramento	40	$^{+}$San Francisco Bay	Continental	(3)
Niger	40	Atlantic Ocean	Oceanic	(1)
Murray	33	Southern Ocean shelf	Continental	(2)
Columbia	32	Pacific Ocean	Oceanic	(1)
Rio Grande	30	Gulf of Mexico	Oceanic	(1)
Ob	18	Kara Sea	Continental	(2)
Lena	17	Laptev Sea	Continental	(2)
Yenisei	14	Kara Sea	Continental	(2)
Saõ Francisco	7	Atlantic Ocean	Oceanic	(1)
Kolyma	6	East Siberian Sea	Continental	(2)
Saint Lawrence	2	Gulf of St. Lawrence	Continental margin	(5)
Dnepr	1	Black Sea	Continental	(2)

Sediment load data are from SUMMERFIELD and HULTON (1994): * denotes a river which is now dammed such that the modern sediment load reaching the coast is negligible; $^{+}$ denotes the estimated value for the late 19th century. Notes indicate: (1) the sediment load is mainly on oceeanic crust; (2) the sediment load is applied over a very broad region of continental shelf such that the load in any one locality is negligible; (3) the sediment load is applied in a delta on continental crust or in an enclosed bay; (4) the nature of crust beneath the delta is uncertain but probably marks a transition between continental to oceanic crust; (5) the sediment load is applied over a very broad region of continental shelf with much of it transported off the shelf by turbidity currents. The present analysis suggests that the most likely candidates for causing induced seismicity are localities labelled (3).

seasonal climates (causing strongly seasonal river discharges and sediment loads) and are deforming very slowly. Of course, the slower the deformation rate is, the fewer earthquakes of a given magnitude are expected, and the greater the difficulty will be in establishing that any seasonal variation is statistically significant. This difficulty is compounded by the possibility of multiple seasonal forcing mechanisms which are out of phase and thus mask each other's effects, as in Figure 14c. Northern California provides a unique case study for investigating seasonal seismicity, resembling a controlled experiment where the Spring forcing mechanism caused by seasonal sedimentation has been present in some decades but absent in others.

The Spectrum of Amplitudes and Time-scales of Stress Transients

This study can also be regarded as filling a data gap concerning the triggering of seismicity, between the effects of tides, which contribute changes to Φ of ~1 kPa and do not normally directly trigger seismicity, the effects of large earthquakes on adjacent fault planes, which can contribute changes to Φ of ~100 kPa or more and can trigger seismicity (e.g., STEIN, 1999), and the effects of reservoir loading, already mentioned, which can contribute changes to Φ of ~1 MPa and can trigger seismicity.

In a detailed search for a corerelation between microearthquakes and tides, VIDALE *et al.* (1998a) investigated the timing of ~13,000 microearthquakes along creeping fault segments within the San Andreas fault zone. They found an increase in seismicity rates at times when tidal stresses unclamped the faults, but which was not statistically significant. A later analysis (VIDALE *et al.*, 1998b) of ~27,500 microearthquakes found a ~1% excess in seismicity rates at times of peak tidal unclamping, which was significant at the 95% confidence level. It thus appears that changes to Φ of ~1 kPa do not normally trigger seismicity, but it can be anticipated that the threshold for a much clearer relationship occurs for changes to Φ not much greater than ~1 kPa. Conversely, as STEIN (1999) recently reviewed, many examples are now known of earthquakes triggered by coseismic stress changes of ~100 kPa or more: either as prompt effects (as in aftershock sequences) or as delayed effects. A topical example concerns the earthquakes on the North Anatolian fault zone in 1999: the İzmit event (M_w 7.4) of 17 August and the Düzce event (M_w 7.2) of 12 November, which occurred 12 weeks later at the eastern end of the earlier fault rupture. The present study suggests that changes to Φ of ~5 kPa caused by relaxation of the lower crust following sediment loading can trigger seismicity. The threshold for a significant effect thus appears to lie somewhere between Φ ~1 kPa and Φ ~5 kPa. The possibility that the delay involved in this and other examples of earthquake triggering may relate to the time scale for changes to the state of stress which accompany the lower crustal flow response following initial loading by the first earthquake is an interesting objective for the future.

The stress variations preceding seismicity are of course more complex than is suggested in Figure 14. For instance, as previously noted, the effect of tides can be regarded as a low-amplitude high-frequency ripple superposed on longer-term Φ variations such as in Figure 14. However, as VIDALE *et al.* (1998a) noted, the rate of change of stress caused by tides is high, of the order of 0.1 kPa hr^{-1}: much higher than any conceivable variation due to any other interseismic process. One might thus expect that faults ought always to be brought to the condition for shear failure at times of increasing Φ during one or other tidal stress cycle. The fact that this does not usually happen led VIDALE *et al.* (1998a) to suggest that earthquakes are preceded by rapid increases in stress, which load fault zones before failure at a rate which exceeds the rate of change of tidal stress. An alternative possibility is that when Φ approaches zero, fault zones weaken before failure at a rate which exceeds the rate of change of tidal stress. In terms of the notation in equation (6), this view implies that $d\tau/dt$ and/or $-c\,d\sigma_n/dt$ accompanying tidal stress changes is typically smaller than $-dS/dt$ caused by the reduction in cohesion (S) due to this suggested pre-seismic weakening when Φ is close to zero. This view also implies that either the typical duration of such pre-seismic weakening takes longer than half a tidal cycle, or the typical magnitude of $|\Phi|$, Φ_0, at which this weakening process begins is larger than the typical $\sim$1 kPa amplitude of tidal stress fluctuations. If so, the arguments presented earlier for the physical basis of seasonal seismicity in the late 19th century are not significantly affected: they mean that instead of raising Φ to zero, stress fluctations due to seasonal sediment loading raise Φ on a fault plane to a value of the order of $-\Phi_0$. Once this threshold is reached, the fault spontaneously weakens and eventually fails as a result, maybe a few days or weeks later. This reasoning suggests that the main factor causing the late 19th century seasonal seismicity may not have been that the stress perturbation caused by sediment loading was larger than the typical stress perturbation caused by tides: it may instead relate to the longer time scale over which this stress perturbation caused by sediment loading was built up and maintained.

Conclusions

The seismicity of coastal northern California was significantly concentrated in Spring during the decades before the great earthquake of 1906. It is suggested that this was caused by stress transients due to seasonal sediment loading near active strike-slip faults. Loading was enhanced at this time by the sediment transported into this region due to hydraulic mining of gold. The calculated magnitude of the resulting stress transients was sufficient to influence the seasonal character of seismicity, provided the deepest continental crust in this region has a viscosity no greater than $\sim 10^{17}$ Pa s.

Appendix: Instantaneous and Relaxed Stress Perturbations

I assume a two-dimensional stress field, such that no stress tensor element in the vertical x, z plane varies with position in the horizontal y direction. The stress tensor at each point within the crust is required to satisfy the equations of horizontal and vertical equilibrium:

$$\frac{\partial \sigma_{xx}}{\partial x} + \frac{\partial \sigma_{xz}}{\partial z} = 0 \tag{A1}$$

and

$$\frac{\partial \sigma_{xz}}{\partial x} + \frac{\partial \sigma_{zz}}{\partial z} = -\rho g \ . \tag{A2}$$

Because the equations governing elasticity are linear, a general solution can be found by superposing individual solutions. I thus separate the stress field caused by gravity acting on the crust from the residual stress field caused by other factors. As no internal body forces contribute to this residual stress field, the equation for equilibrium analogous to (A2) becomes

$$\frac{\partial \sigma_{xz}}{\partial x} + \frac{\partial \sigma_{zz}}{\partial z} = 0 \ . \tag{A3}$$

Assuming also that no discontinuities in elastic displacement exist within the crust, at each point the diagonal elements of the residual stress tensor also satisfy the stress compatibility equation or

$$\nabla^2(\sigma_{xx} + \sigma_{zz}) = 0 \ . \tag{A4}$$

I look for solutions in which the residual stress tensor elements vary beneath the load, and satisfy equations (A1), (A2), and (A4). If required, the overall stress field can then be determined by adding the contribution from gravitational body forces. I thus consider a scalar Airy stress function Ψ of the form

$$\Psi \equiv F(z)\cos(kx) \tag{A5}$$

where $F(z)$ is an as yet undetermined function and x is measured from the centre of the load. At each point, the residual stress tensor thus satisfies

$$\sigma_{xx} \equiv \frac{\partial^2 \Psi}{\partial z^2} \tag{A6}$$

$$\sigma_{xz} \equiv \frac{-\partial^2 \Psi}{\partial x \partial z} \tag{A7}$$

and

$$\sigma_{zz} \equiv \frac{\partial^2 \Psi}{\partial x^2} \tag{A8}$$

(e.g., HAFNER, 1951). A suitable general form for $F(z)$ is

$$F(z) \equiv A\cosh(kz) + B\sinh(kz) + Cz\cosh(kz) + Dz\sinh(kz) \tag{A9}$$

where A, B, C and D are constants whose values can be determined from boundary conditions.

At the free surface at $z = 0$, both σ_{xz} and σ_{zz} are required to be zero. This means that $A = 0$ and $C = -Bk$. The residual vertical stress σ_{zz} within the brittle layer can thus be expressed as

$$\sigma_{zz} = -k^2 \cos(kx)[B \sinh(kz) - Bkz \cosh(kz) + Dz \sinh(kz)] \ . \tag{A10}$$

In the limit of small z, this equation can be approximated as

$$\sigma_{zz} = -kD(kz)^2 \cos(kx) \ , \tag{A11}$$

indicating that kD quantifies the amplitude of the load applied to the rock surface. A set of solutions thus exists with B zero, of the form:

$$\sigma_{xx} = Dk \cos(kx)[2 \cosh(kz) + kz \sinh(kz)] \tag{A12}$$

$$\sigma_{xz} = Dk \sin(kx)[\sinh(kz) + kz \cosh(kz)] \tag{A13}$$

$$\sigma_{zz} = -Dk \cos(kx) kz \sinh(kz) \ . \tag{A14}$$

The lower crust can be regarded as a viscous fluid on long time scales, but will behave as an elastic solid at short time scales. The rate of transition from short-term to long-term behaviour is governed by the Maxwell time, t_m, where

$$t_m = \eta/\mu \ , \tag{A15}$$

η and μ being the viscosity and shear modulus at each point. Because η decreases downward due to the geothermal gradient, the deepest crust has the shortest Maxwell time.

On a time scale after loading which is short compared with the Maxwell time of the deepest crust, the entire crust will behave as an elastic body which is in welded contact with the mantle lithosphere. I assume that the instantaneous loading effect is initially supported by elastic stresses within the crust, such that σ_{zz} is unaffected at the Moho at depth z_m. This assumption is reasonable given the small spatial extent of the coastal bays in northern California (Fig. 1). It requires that $\Psi(z_m) = 0$ or

$$B \sinh(kz_m) - Bkz_m \cosh(kz_m) + Dz_m \sinh(kz_m) = 0 \ , \tag{A16}$$

so

$$B = \frac{z_m D}{[kz_m \coth(kz_m) - 1]} \ . \tag{A17}$$

or

$$B = \frac{fD}{k} \tag{A18}$$

where

$$f = \frac{kz_m}{[kz_m \coth(kz_m) - 1]} = \frac{u}{u \coth(u) - 1} \ . \tag{A19}$$

is a dimensionless parameter, with

$$u = kz_m = \frac{\pi z_m}{2L} \quad . \tag{A20}$$

The residual stress field instantaneously after loading is thus:

$$\sigma_{xx} = Dk\cos(kx)[(2 - fkz)\cosh(kz) + (kz - f)\sinh(kz)] \tag{A21}$$

$$\sigma_{xz} = Dk\sin(kx)[(1 - fkz)\sinh(kz) + kz\cosh(kz)] \tag{A22}$$

$$\sigma_{zz} = -Dk\cos(kx)[(kz + f)\sinh(kz) - fkz\cosh(kz)] \quad . \tag{A23}$$

I thus describe the immediate response to seasonal sediment loading using equations (A21) to (A23) with an appropriate value of D to represent the applied load. After several times the Maxwell time of the deepest crust, this stress field adjusts toward a form given instead by equations (A12) to (A14). Whether this relaxed response develops before the next seasonal sediment load arrives depends on the Maxwell time of the lowermost crust.

The following procedure allows D to be estimated for a given magnitude of load. The equation for vertical equilibrium (A3) relates $\partial\sigma_{xz}/\partial x$ and $\partial\sigma_{zz}/\partial z$. For the instantaneous response to loading, equations (A21) to (A23) these terms are:

$$\frac{\partial\sigma_{xz}}{\partial x} = Dk^2\cos(kx)[(1 - fkz)\sinh(kz) + kz\cosh(kz)] = \frac{\partial\sigma_{zz}}{\partial z} \quad . \tag{A24}$$

The absolute value of σ_{zz} will increase downward to a depth z_0 where $\partial\sigma_{xz}/\partial x$ is zero and σ_{zz} will equal the load applied at the Earth's surface. Equation (A24) can be solved for $w = kz_0$, which satisfies

$$\tanh(w) = \frac{w}{fw - 1} \quad . \tag{A25}$$

To model stress transients in coastal northern California, I take the crustal thickness, z_m, as 25 km, from MOONEY and WEAVER (1989), and estimate L as 12 km, the half width of San Pablo Bay. From (A18), f is thus 1.4341. By inspection, solutions to (A25) have $w > 2$ for this value of f, such that $\tanh(w)$ is very nearly 1, and equation (A25) can be approximated as

$$w = \frac{1}{f - 1} \quad , \tag{A26}$$

which has the solution $w = 2.3036$ or $z_0 = 17.6$ km. Localities around the base of the brittle layer will thus experience an increase in the absolute vertical stress which is roughly equal to the applied load of amplitude 1 kPa. The correct magnitude of the annual sediment load during the late 19th century thus requires $Dk = 0.6$ kPa (Fig. 12a).

Acknowledgments

I thank Joan Gomberg and Bob Simpson for thoughtful and constructive reviews.

REFERENCES

AGNEW, D. C. and SIEH, K. E. (1978), *A Documentary Study of the Felt Effects of the Great California Earthquake of 1857*, Bull. Seismol. Soc. Am. *68*, 1717–1729.

AMBRASEYS, N. N. and FINKEL, C., *The Seismicity of Turkey and Adjacent Areas: A Historical Review* (Eren, İstanbul, Turkey, 1995).

AMBRASEYS, N. N. and BARAZANGI, M. (1989), *The 1759 Earthquake in the Bekaa Valley: Implications for Earthquake Hazard Assessment in the Eastern Mediterranean Region*, J. Geophys. Res. *94*, 4007–4013.

BAKUN, W. H. (1999), *Seismic Activity of the San Francisco Bay Region*, Bull. Seismol. Soc. Am. *89*, 764–784.

BEN-MENAHEM, A. (1991), *Four Thousand Years of Seismicity along the Dead Sea Rift*, J. Geophys. Res. *96*, 20,195–20,216.

BLODGETT, J. C. and POESCHEL, K. R. (1988), *Peak discharge, volume, and frequency of the January 1982 flood in the Santa Cruz Mountains and vicinity*. In *Landslides, Floods, and Marine Events of the Storm of January 3–5, 1982, in the San Francisco Bay Region, California* (eds. Ellen, S. D. and Wieczorek, G. F.), U.S. Geol. Surv. Professional Paper 1434, pp. 229–243.

BOLT, B. A. and MILLER, R. D. (1971), *Seismicity of Northern and Central California, 1965–1969*, Bull. Seismol. Soc. Am. *61*, 1831–1847.

BOWMAN, D. D., OUILLON, G., SAMMIS, C. G., SORNETTE, A., and SORNETTE, D. (1998), *An Observational Test of the Critical Earthquake Concept*, J. Geophys. Res. *103*, 24,359–24,372.

BROWN, R. D., Jr. (1990), *Quaternary deformation*, In *The San Andreas Fault System* (ed. Wallace, R. E.) U.S. Geol. Surv. Professional paper 1515, pp. 83–113.

BROWN, W. M., III. (1988), *Historical setting of the storm: Perspectives on population, development, and damaging rainstorms in the San Francisco Bay region*. In *Landslides, Floods, and Marine Events of the Storm of January 3–5, 1982, in the San Francisco Bay Region, California* (eds. Ellen, S. D. and Wieczorek, G. F.), U.S. Geol. Surv. Professional Paper 1434, pp. 7–15.

BUFE, C. G. and VARNES, D. J. (1993), *Predictive Modeling of the Seismic Cycle of the Greater San Francisco Bay Region*, J. Geophys. Res. *98*, 9871–9883.

BÜRGMANN, R., ARROWSMITH, R., DUMITRU, T., and MCLAUGHLIN, R. (1994), *Rise and Fall of the Southern Santa Cruz Mountains, California, from Fission Tracks, Geomorphology, and Geodesy*, J. Geophys. Res. *99*, 20,181–20,202.

CARLSON, J. M. (1991), *Time Intervals Between Characteristic Earthquakes and Correlations with Smaller Events: An Analysis Based on a Mechanical Model of a Fault*, J. Geophys. Res. *96*, 4255–4267.

CHALLINOR, R. A. (1971), *Variations in the rate of rotation of the Earth*, Science *172*, 1022–1025.

CHASE, C. G. and WALLACE, T. C. (1986), *Uplift of the Sierra Nevada of California*, Geology *14*, 730–733.

CLAPP, W. B. (1907), *The Surface Water Supply of California, 1906*, U.S. Geological Survey Water-Supply Paper 213, 219 pp.

CURRY, R. R., EMERY, B. A., and KIDWELL, T. G. (1994). *Sources and magnitudes of increased streamflow in the Santa Cruz Mountains for the 1990 water year after the earthquake*. In *The Loma Prieta, California, Earthquake of October 17, 1989 – Hydrologic Disturbances* (ed. Rojstaczer, S.) U.S. Geol. Surv. Professional Paper 1551-E, pp. E31–E50.

DEMETS, C., GORDON, R. G., ARGUS, D. F., and STEIN, S. (1990), *Current Plate Motions*, Geophys. J. Int. *101*, 425–478.

DIETERICH, J. (1994), *A Constitutive Law for Rate of Earthquake Production and its Application to Earthquake Clustering*, J. Geophys. Res. *99*, 2601–2618.

ELLEN, S. D. and WIECZOREK, G. F. (1988), *Landslides, floods, and marine effects of the storm of January 3–5, 1982, in the San Francisco Bay region, California*, U.S. Geol. Surv. Professional Paper 1434, 281 pp.

ELLSWORTH, W. L. (1990), *Earthquake history, 1769–1989*. In *The San Andreas Fault System* (ed. Wallace, R. E.), U.S. Geol. Surv. Professional Paper 1515, pp. 153–187.

ELLSWORTH, W. L., LINDH, A. G., PRESCOTT, W. H., and HERD, D. G. (1981), *The 1906 San Francisco earthquake and the seismic cycle*. In *Earthquake Prediction: An International Review* (eds. Simpson, D.W. and Richards, P.G.) (American Geophysical Union, Washington, D.C.), pp. 126–140.

EXNER, F., WALKER, G., SIMPSON, G. C., CLAYTON, H. H., and MOSSMAN, R. C., *World Weather Records, Errata.* Smithsonian Miscellaneous Collections volume 79, publication 3019 (Smithsonian Institution, Washington, D.C., 1929).

FISCHER, K. J., *Fluvial geomorphology and flood control strategies: Sacramento River, California.* In *The Variability of Large Alluvial Rivers* (eds. Schumm, S. A. and Winkley, B. R.) (American Society of Civil Engineers Press, New York 1994), pp. 115–138.

GILBERT, G. K. (1917), *Hydraulic-Mining Debris in the Sierra Nevada*, U.S. Geol. Surv. Professional Paper 105.

GOUGH, D. I. (1969), *Incremental Stress under a Two-dimensional Artificial Lake*, Can. J. Earth Sci. *6*, 1067–1075.

GOUGH, D. I. and GOUGH, W. I. (1970), *Stress and Deflection in the Lithosphere near Lake Kariba – I*, Geophys. J. R. Astr. Soc. *21*, 65–78.

GUPTA, H. K. and RASTOGI, B. K., *Dams and Earthquakes. Developments in Geotechnical Engineering*, vol. 11 (Elsevier, Oxford, 1976).

HAFNER, W. (1951), *Stress distributions and Faulting*, Geol. Soc. Am. Bull. *62*, 373–398.

HANKS, T. C. and KANAMORI, H. (1989), *A moment-magnitude scale*, J. Geophys. Res. *84*, 2348–2350.

HALL, N. T., WRIGHT, R. H., and CLAHAN, K. B. (1999), *Paleoseismic Studies of the San Francisco Peninsula Segment of the San Andreas Fault Zone near Woodside, California*, J. Geophys. Res. *104*, 23,215–23,236.

HARRIS, R. A. and SIMPSON, R. W. (1998), *Suppression of Large Earthquakes by Stress Shadows: A Comparison of Coulomb and Rate-and-state Failure*, J. Geophys. Res. *103*, 24,439–24,451.

HEATON, T. H. (1975), *Tidal Triggering of Earthquakes*, Geophys. J. R. Astr. Soc. *43*, 307–326.

HUBER, N. K. (1990), *The Late Cenozoic Evolution of the Tuolumne River, Central Sierra Nevada, California*, Geol. Soc. Am. Bull. *102*, 102–115.

JACKSON, D., AKI, K., CORNELL, A., DIETERICH, J., HENYEY, T., MAHDYIAR, M., SCHWARTZ, D., and WARD, S. (1995), *Seismic hazards in southern California: probable earthquakes, 1994 to 2024*, Bull. Seismol. Soc. Am. *85*, 379–439.

JAUMÉ, S. C. and SYKES, L. R. (1996), *Evolution of Moderate Seismicity in the San Francisco Bay Region, 1850 to 1993: Seismicity Changes Related to the Occurrence of Large and Great Earthquakes*, J. Geophys Res. *101*, 765–789.

JAUMÉ, S. C. and SYKES, L. R. (1999), *Evolving Towards a Critical Point: A Review of Accelerating Seismic Moment/Energy Release Prior to Large and Great Earthquakes*, Pure appl. geophys. *155*, 279–306.

JOHNSON, A. G., KOVACH, R. L., and NUR, A. (1974), *Fluid-pressure Variations and Fault Creep in Central California*, Tectonophysics *23*, 257–266.

KELLEY, R., *Gold vs. Grain: The Hydraulic Mining Controversy in California's Sacramento Valley, A Chapter in the Decline of the Concept of Laissez Faire* (Arthur Clarke Publishing Co., Glendale, Calif., 1959).

KELSON, K. I., SIMPSON, G. D., LETTIS, W. R., and HARADEN, C. C. (1996), *Holocene Slip Rate and Recurrence of the Northern Calaveras Fault at Leyden Creek, Eastern San Francisco Bay Region*, J. Geophys. Res. *101*, 5961–5975.

KILSTON, S. and KNOPOFF, L. (1983), *Lunar-solar Periodicities of Large Earthquakes in Southern California*, Nature *304*, 21–25.

LAMBECK, K. and HOPGOOD, P. (1981), *The Earth's rotation and atmospheric circulation, from 1963 to 1973*, Geophys. J. R. Astr. Soc. *64*, 67–89.

LAMBECK, K. and HOPGOOD, P. (1982), *The Earth's Rotation and Atmospheric Circulation: 1958–1980*, Geophys. J. R. Astr. Soc. *71*, 581–587.

LANGENHEIM, V. E., SCHMIDT, K. M., and JACHENS, R. C. (1997), *Coseismic Deformation During the 1989 Loma Prieta Earthquake and Range-front Thrusting Along the Southwestern Margin of the Santa Clara Valley, California*, Geology *25*, 1091–1094.

LIENKAMPER, J. J. and BORCHERDT, G. (1996), *Holocene Slip Rate of the Hayward Fault at Union City, California*, J. Geophys. Res. *101*, 6099–6108.

LISOWSKI, M., PRESCOTT, W. H., SAVAGE, J. C., and JOHNSTON, M. J. (1990), *Geodetic Estimate of Coseismic Slip During the 1989 Loma Prieta, California, Earthquake*, Geophys. Res. Lett. *17*, 1437–1440.

MCCLELLAN, P. H. (1984), *Earthquake Seasonality Before the 1906 San Francisco Earthquake*, Nature *307*, 153–156.

METEOROLOGICAL OFFICE, *Tables of Temperature, Relative Humidity, Precipitation and Sunshine for the World. Part I North America and Greenland (including Hawaii and Bermuda)* (Her Majesty's Stationery Office, London, 1980).

MILLIMAN, J. D. and SYVITSKI, J. P. M. (1992), *Geomorphic/Tectonic Control of Sediment Discharge to the Ocean: The Importance of Small Mountainous Rivers*, J. Geol. *100*, 525–544.

MITCHELL, J. and WESTAWAY, R. (1999), *Chronology of Neogene and Quaternary Uplift and Magmatism in the Caucasus: Constraints from K-Ar Dating of Volcanism in Armenia*, Tectonophysics *304*, 157–186.

MOONEY, W. D. and WEAVER, C. S. (1989), *Regional crustal structure and tectonics of the Pacific Coastal States: California, Oregon, and Washington*. In *Geophysical Framework of the Continental United States* (eds. Pakiser, L. C. and Mooney, W. D.), Geol. Soc. Am. Memoir *172*, pp. 129–161.

NOLAN, K. M. and MARRON, D. C. (1988), *Stream-channel response to the storm in the Santa Cruz Mountains*. In *Landslides, Floods, and Marine Events of the Storm of January 3–5, 1982, in the San Francisco Bay Region, California* (eds. Ellen, S. D. and Wieczorek, G. F.), U.S. Geol. Surv. Professional Paper 1434, pp. 245–264.

OLSON, J. A. and HILL, D. P. (1993), *Seismicity in the southern Santa Cruz Mountains during the 20-year period before the earthquake*. In *The Loma Prieta, California, earthquake of October 17, 1989; preseismic observations* (ed. Johnston, M.J.S.), U.S. Geological Survey Professional Paper 1550-C, pp. C3-C16.

O'NEIL, J. R. (1985), *Water-rock Interactions in Fault Gouge*, Pure appl. geophys. *122*, 440–446.

PARSONS, T., STEIN, R. S., SIMPSON, R. W., and REASENBERG, P. A. (1999), *Stress Sensitivity of Fault Seismicity: A Comparison Between Limited-offset Oblique and Major Strike-slip Faults*, J. Geophys. Res. *104*, 20,183–20,202.

PRESCOTT, W. H. and SHUI-BEIH YU (1986), *Geodetic Measurement of Horizontal Deformation in the Northern San Francisco Bay Region, California*, J. Geophys. Res. *91*, 7475–7484.

PYNE, S. J., *Grove Karl Gilbert: A Great Engine of Research* (University of Texas Press, Austin, Tex., 1980).

RICHTER, C. F. (1935), *An Instrumental Earthquake Scale*, Bull. Seismol. Soc. Am. *25*, 1–32.

ROJSTACZER, S. and WOLF, S. (1991), *Hydrologic changes associated with the earthquake in the San Lorenzo and Pescadero drainage basins*, U.S. Geological Survey Open-File Report 91–567, 21 pp.

ROJSTACZER, S. and WOLF, S. (1994), *Hydrologic changes associated with the earthquake in the San Lorenzo and Pescadero drainage basins*. In *The Loma Prieta, California, Earthquake of October 17, 1989 – Hydrologic Disturbances* (ed. Rojstaczer, S.), U.S. Geol. Surv. Professional Paper 1551-E, pp. E51–E64.

ROTHÉ, J. P. (1970), *Artificial Earthquakes*, Tectonophysics *9*, 215–238.

RUBIN, A. M., GILLARD, D., and GOT, J.-L. (1999), *Streaks of Microearthquakes Along Creeping Faults*, Nature *400*, 635–641.

RYDELEK, P. A. and SACKS, I. S. (1999), *Large Earthquake Occurrence Affected by Small Stress Changes*, Bull. Seismol. Soc. Am. *89*, 822–828.

SCHUSTER, A. (1897), *On Lunar and Solar Periodicities of Earthquakes*, Proc. R. Soc. London *61*, 455–564.

SCHWARTZ, D. P., PANTOSTI, D., HECKER, S., OKUMURA, K., POWERS, T., and HAMILTON, J. (1993), *Late Holocene Behavior and Seismogenic Potential of the Rodgers Creek Fault Zone, Sonoma County, California*, California Division of Mines and Geology Spec. Publ. *113*, 393–398.

SCHWARTZ, D. P., PANTOSTI, D., OKUMURA, K., POWERS, T., and HAMILTON, J. (1998), *Paleoseismic Investigations in the Santa Cruz Mountains: Implications for the Recurrence of Large Magnitude Earthquakes on the San Andreas Fault*, J. Geophys. Res. *103*, 17,985–18,001.

SHAW, B. E., CARLSON, J. M., and LANGER, J. S. (1992), *Patterns of Seismic Activity Preceding Large Earthquakes*, J. Geophys. Res. *97*, 479–488.

SHLIEN, S. (1972), *Earthquake-tide Correlation*, Geophys. J. R. Astr. Soc. *28*, 27–34.

SIMPSON, D. W., LEITH, W. S., and SCHOLZ, C. H. (1988), *Two Types of Reservoir-induced Seismicity*, Bull. Seismol. Soc. Am. *78*, 2025–2040.

SIMPSON, R. W. and REASENBERG, P. A. (1994). *Earthquake-induced static stress changes on Central California faults*. In *The Loma Prieta, California, Earthquake of October 17, 1989 – Tectonic Processes and Models* (ed. Simpson, R. W.), U.S. Geol. Surv. Professional Paper 1550-F, pp. F55–F89.

STEIN, R. S. (1999), *The Role of Stress Transfer in Earthquake Occurrence*, Nature *402*, 605–609.

STOVER, C. W. and COFFMAN, J. L. (1993), *Seismicity of the United States, 1568–1989 (Revised)*, U.S. Geol. Surv. Professional Paper 1527, 418 pp.

SUMMERFIELD, M. A. and HULTON, N. J. (1994), *Natural Controls of Fluvial Denudation Rates in Major World Drainage Basins*, J. Geophys. Res. *99*, 13,871–13,883.

SYKES, L. R. and JAUMÉ, S. C. (1990), *Seismic Activity on Neighbouring Faults as a Long-term Precursor to Large Earthquakes in the San Francisco Bay Area*, Nature *348*, 595–599.

TOPPOZADA, T. R. (1975), *Earthquake Magnitude as a Function of Intensity Data in California and Western Nevada*, Bull. Seismol. Soc. Am. *65*, 1223–1238.

TOPPOZADA, T. R. and BORCHERDT, G. (1998), *Re-evaluation of the 1836 "Hayward Fault" and the 1838 San Andreas Fault Earthquakes*, Bull. Seismol. Soc. Am. *88*, 140–159.

TOPPOZADA, T. R., REAL, C. R., and PARKE, D. L. (1981), *Preparation of isoseismal maps and summaries of reported effects for pre-1900 California earthquakes*, California Division of Mines and Geology Open-File Rep. 81-11SAC, 182 pp.

TOPPOZADA, T. R., REAL, C. R., and PARKE, D. L., *Earthquake history of California*. In *Historical Seismograms and Earthquakes of the World* (eds. Lee, W. H. K., Meyers, H, and Shimazaki, K.) (Academic Press, London 1988), pp. 267–275.

TOWNLEY, S. D. and ALLEN, M. W. (1939), *Descriptive Catalog of Earthquakes of the Pacific Coast of the United States, 1769 to 1928*, Bull. Seismol. Soc. Am. *29*, 1–297.

VALENSISE, G. (1994), *Geologic assessment of the relative contribution of strike-slip faulting, reverse-slip faulting, and bulk squeezing in the creation of the Central Santa Cruz Mountains, California*. In *The Loma Prieta, California, Earthquake of October 17, 1989 – Tectonic Processes and Models* (ed. Simpson, R. W.), U.S. Geol. Surv. Professional Paper 1550-F, pp. F23–F47.

VALENSISE, G. and WARD, S. N. (1991), *Long-term Uplift of the Santa Cruz Coastline in Response to Repeated Earthquakes Along the San Andreas Fault*, Bull. Seismol. Soc. Am. *81*, 1694–1704.

VIDALE, J. E., AGNEW, D. C., JOHNSTON, M. S. J., and OPPENHEIMER, D. H. (1998a), *Absence of Earthquake Correlation with Earth Tides: An Indication of High Preseismic Fault Stress State*, J. Geophys. Res. *103*, 24,567–24,572.

VIDALE, J. E., AGNEW, D. C., OPPENHEIMER, D. H., RODRIQUEZ, C., and HOUSTON, H. (1998b), *A Weak Correlation Between Earthquakes and Extensional Normal Stress and Stress Rate from Lunar Tides*, EOS, Trans. AGU, Fall Meeting Suppl., 79, 641.

WEBER, G. E. and NOLAN, J. M. (1995), *Determination of Late Pleistocene – Holocene Slip Rates along the San Gregorio Fault Zone, San Mateo County, California*, U.S. Geol. Surv. Open-File Rep. 95–210, 805–807.

WESTAWAY, R. (1995), *Deformation Around Stepovers in Strike-slip Fault Zones*, J. Struct. Geol. *16*, 823–838.

WESTAWAY, R. (1998), *Dependence of Active Normal Fault Dips on Lower-crustal Flow Regimes*, J. Geol. Soc. London *155*, 233–253.

WESTAWAY, R. (1999), *The Mechanical Feasibility of Low-angle Normal Faulting*, Tectonophysics *308*, 407–443.

WESTAWAY, R. (2001), *Flow in the lower continental crust as a mechanism for the Quaternary uplift of the Rhenish Massif, north-west Europe*. In *River Basin Sediment Systems: Archives of Environmental Change* (eds. Maddy, D., Macklin, M.G., and Woodward, J.C.) (Balkema, Abingdon, England), pp. 87–167.

WORKING GROUP ON CALIFORNIA EARTHQUAKE PROBABILITIES (1999), *Earthquake Probabilities in the San Francisco Bay Region: 2000 to 2030 – A Summary of Findings*, U.S. Geol. Surv. Open-File Rep. 99–517, 36 pp.

(Received April 20, 2000, revised/accepted September 18, 2000)

Pure appl. geophys. 159 (2002) 063–090
0033 – 4553/02/030063 – 28 \$ 1.50 + 0.20/0

Pure and Applied Geophysics

Post-seismic Effects of an *M* 7.2 Earthquake and Microseismicity in an Abandoned, Flooded, Deep Mine

HIROSHI OGASAWARA,[1] YASUTOSHI KUWABARA,[1] TAKAYUMI MIWA,[1] KUNIO FUJIMORI,[2] NORIO HIRANO[3] and MAKOTO KOIZUMI[3]

Abstract—An *M* 7.2 earthquake took place 67 km southeast of an abandoned mine containing flooded, vertical ore veins at depth of 1 km. Multidisciplinary monitoring of the unweathered country rock within the mine was carried out for a distance of about one kilometer. The *M* 7.2 earthquake was followed by significant post-seismic changes in strain and tilt of $\sim 10^{-6}$, with a self-potential of ~10 mV, and an increase in the water level in the mine of about 10 cm/day. These phenomena continued for several months before returning to pre-earthquake levels. Coseismic elastic deformations were too small to account for the observed post-seismic changes, and have different senses in strain and tilt from the observed post-seismic changes. The contractions in strain and the changes in the electric self-potential and water-elevation rate strongly suggest an increase in pore pressure. An increase in microseismic activity ($M < 0$) was also observed around the flooded, vertical ore veins about 1–2 km from our multidisciplinary monitoring network, when pore-pressure reached a maximum following the *M* 7.2 earthquake.

Key words: Post-seismic effects, flooded vertical vein, induced seismicity, streaming potential, strain, water table.

1. Introduction

Immediately after the *M* 7.2 1995 Hyogo-ken Nanbu (Kobe) earthquake, Japan, the seismicity rate increased to five or six times the usual level in the Tanba area (the ellipse in Fig. 1) northeast of the aftershock zone (DISASTER PREVENTION RESEARCH INSTITUTE, KYOTO UNIVERSITY, 1995, 1996). Increased seismicity was also observed along the Yamasaki fault (Fig. 1) with a strike conjugate to the Hyogo-ken Nanbu earthquake fault (e.g., NISHIGAMI *et al.*, 1995).

One possible cause of these increased seismicity rates is the static, elastic change in stress associated with dislocation along the source fault of the 1995 Hyogo-ken Nanbu earthquake. Assuming a dislocation along the source fault consistent with

[1] Faculty of Science and Engineering, Ritsumeikan University, 1-1-1 Noji Higashi, Kusatsu, 525-8577, Japan. E-mail: ogasawar@se.ritsumei.ac.jp
[2] Graduate School of Science, Kyoto University, Sakyo-ku, Kyoto, 606-8502, Japan.
[3] Research Center of Earthquake Prediction, Disaster Prevention Research Institute, Kyoto University, Gokasho, Uji, 611-0011, Japan.

geodetic data, HASHIMOTO (1995, 1997) calculated the change in the Coulomb failure function (ΔCFF; e.g., STEIN and LISOWSKI, 1983) for faults that might slip along an E-W compressive stress field in western Japan. He found a coincidence between the relatively high ΔCFF and the epicenter of the largest aftershock. He also found that areas of increased seismicity in the Tanba area and along the Yamasaki fault were located in the higher ΔCFF areas. However the ΔCFF was usually small, at less than 0.06 MPa for most of the Tanba area, and less than 0.02 MPa for the Yamasaki fault. In addition, there are some areas where the sense of the ΔCFF does not coincide with the change in seismicity.

We should consider the effect of a passing surface wave to account for such remote seismicity, because such a wave can produce dynamic strain (e.g., OIKE and MATSUMURA, 1985) as large as several μstrain (a few 0.1 MPa) up to several hundred km from the source of the M $\sim$ 7 event (e.g., GOMBERG and BODIN, 1994).

One of the best-studied examples of such seismicity is the 1992 Landers, California, earthquake (e.g., HILL *et al.*, 1993; ANDERSON *et al.*, 1994). GOMBERG (1996) summarized a number of related studies of this earthquake, and GOMBERG and DAVIS (1996) inferred a trigger threshold dependent on strain rate, based upon responses to seismicity at The Geysers geothermal field in California. The remaining problem was to determine a mechanism to considerably delay seismicity, compared with the very short duration of the dynamic strain. Combining the rate-dependent threshold with the rate- and state-dependent frictional constitutive law (e.g., DIETERICH, 1972, 1978; RUINA, 1983), GOMBERG *et al.* (1997, 1998) successfully explained the delayed seismicity as "clock advance," or hastened seismicity as a nonlinear response to a transient stress perturbation for a hypothetical single block-spring model. Unfortunately, transient effects related to pore fluids were not used in their studies to derive a rate-dependent threshold based upon seismicity at The Geysers. There is certainly a considerable volume of water in The Geysers area (e.g., EBERHART-PHILLIPS and OPPENHEIMER, 1984; STARK and DAVIS, 1996), but these researchers had no *in situ* data for strain or pore pressure at the seismicity sources. In contrast, SLEEP and BLANPIED (1992, 1994) and HILL *et al.* (1993) suggested that the effects of pore fluids can play an important role in inducing seismicity, although there have been few *in situ* data derived from the vicinity of a seismic source area to address this issue.

In this paper we document post-seismic changes in strain, tilt, self-potential and the underground water table that strongly suggest an increase and subsequent decay in pore pressure following the 1995 Kobe earthquake (M = 7.2) at a distance of 67 km (Fig. 1). These factors were monitored using multidisciplinary sensors deployed in unweathered country rock within the Ikuno mine (OGASAWARA *et al.*, 1992; Figs. 1, 2 and 3), an abandoned, flooded rockburst mine. We also evaluate similar post-seismic changes in strain and tilt observed for a 1984 M 5.6 earthquake 30 km from the mine (TABEI, 1987), in an attempt to decipher the causes of post-seismic changes.

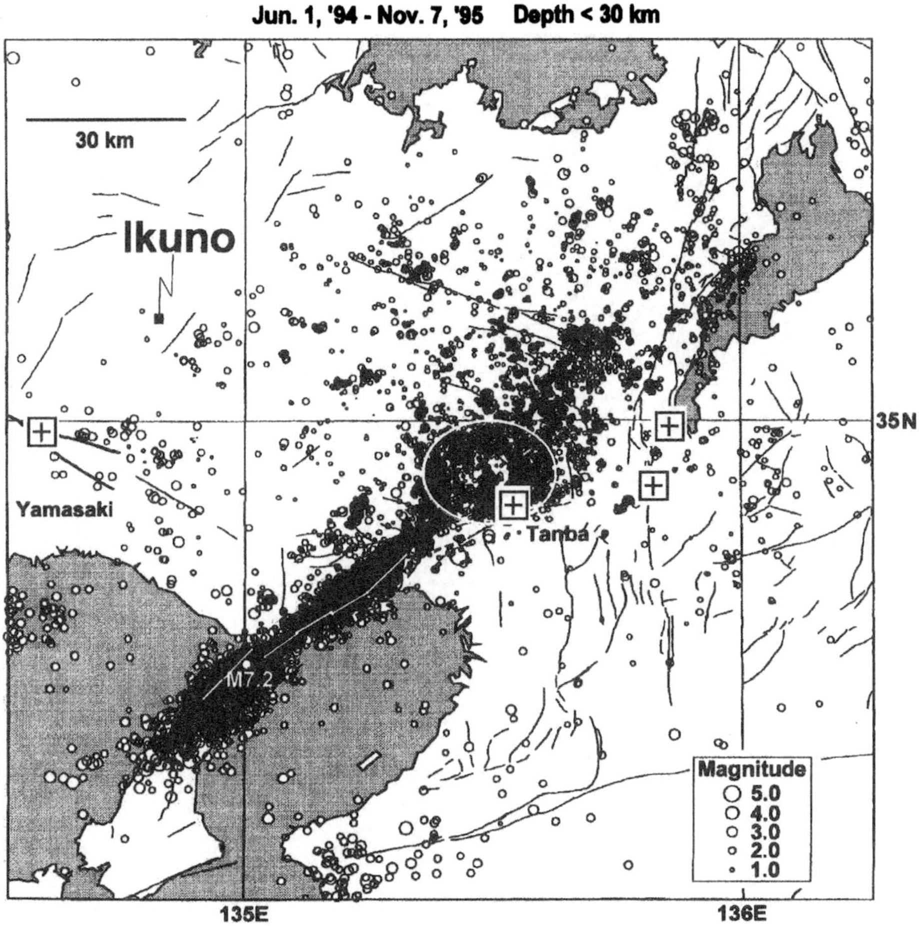

Figure 1

Location of the Ikuno mine (solid square) and the epicenter of the January 17, 1995 Hyogo-ken Nanbu earthquake, Japan (M = 7.2, Japan Meteorlogical Agency; white circle) as well as seismicity ($M > 1$) from June 1994 to November 1995 (Disaster Prevention Research Institute, Kyoto University, 1996). Fault segments (Hashimoto *et al.*, 1996) that were activated during the M 7.2 earthquake are shown by white lines. Solid lines show active faults. High seismicity is seen in the Tanba area (indicated by an ellipsoid). Squares with a cross show crustal movement stations where significant change was seen in the secular strain trend. Although no seismicity was detected, several km distant from the Ikuno mine, small and shallow earthquakes ($M < 0$) were observed in the Ikuno mine, as discussed in this paper.

After the abandoned Ikuno mine was mostly flooded, no seismicity was detected nearby the Kyoto University microearthquake observation network, which has a detection threshold of $M > 1$ (Fig. 1). However, microseismicity ($M < 0$) was detected by a small array deployed within the mine (Ogasawara *et al.*, 1992, 2001; Fig. 4c). The seismicity in the Ikuno mine is shallow (less than a few km deep) and distinct from the surrounding natural seismicity. There have been no mining

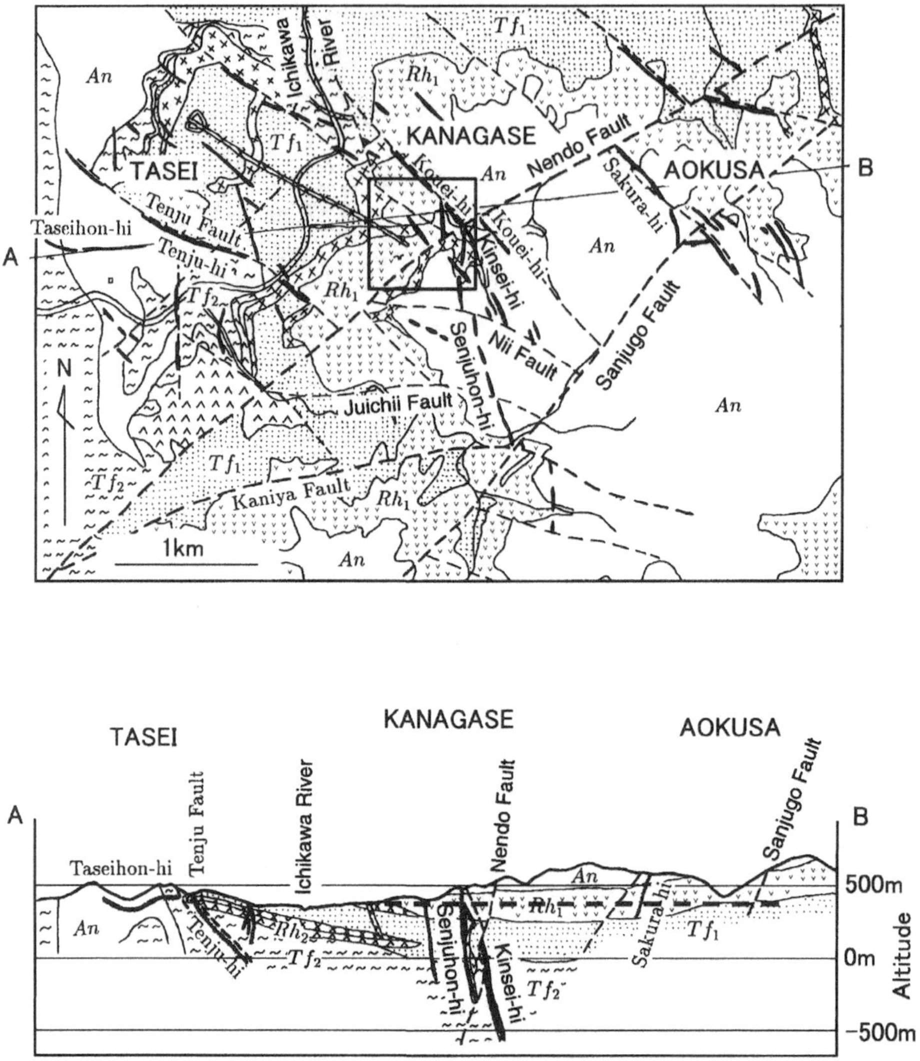

Figure 2

Geology of the Ikuno mine (simplified from the geological map supplied by the Ikuno mine). (Upper) plan view of geology around the Ikuno mine. The Ikuno mine once excavated nearly vertical veins (thick lines) intruded into Tertiary volcanic rocks. The Kanagase area was the most deeply excavated area and experienced rock bursts and water-induced earthquakes. The 750 m × 750 m square in Kanagase is the area where our multidisciplinary observation instruments are installed, as also shown in Figure 3. Dashed lines show geological faults. (Lower) vertical section A-B. A thick, horizontal, dashed line in the lower figure represents the underground level where a geophone array is deployed and a crustal movement monitoring station is located. One of the deepest ore veins was the Kinsei-hi vein where rock bursts occurred. After the Ikuno mine was closed down, the excavated area below the Ichikawa River was flooded, and earthquake swarms were induced in the Kanagase area.

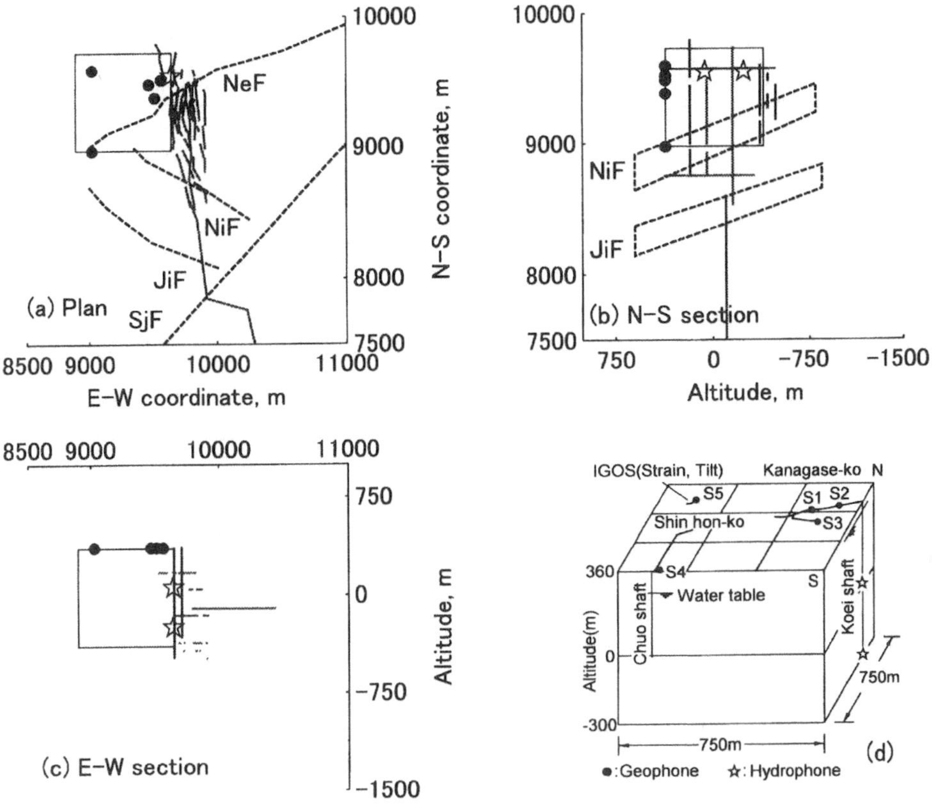

Figure 3

Configuration of the faults, the flooded tunnels at the deep vein, and the monitoring instruments. (a), (b) and (c): Solid and dashed lines are tunnels and faults, respectively. NeF, NiF, JiF and SjF: the Nendo fault, the Nii fault, the Juichii fault and Sanjugo fault, respectively. A square indicates the area where we installed the instruments. Dashed parallelograms in (b) represent fault planes of NiF and JiF projected on the N-S vertical plane. (d): The instrument configuration in the square in (a), (b), (c) and Figure 2 (after OGASAWARA *et al.*, 1992). The table of water flooding the veins was measured at the Koei shaft in the Kanagase area. Significant post-seismic changes were observed using extensometers and tiltmeters at the Ikuno Geophysical Observation Station (IGOS); self-potential was measured by electrodes in the Shinhon-ko (E3-E6) tunnel and in the water table of the Koei shaft.

operations at the Ikuno mine since its closure, and little artificial perturbation that could lead to seismicities. Any seismicity involving the mine is within a few km of our multidisciplinary monitoring network, consequently it is relatively easy to discuss triggers for such seismicity. Seismicities (Fig. 4c) have been associated with a rise of the subterranean water table during heavy rainfall episodes (Fig. 4d; OGASAWARA *et al.*, 2001). However, the Ikuno mine experienced microearthquakes following the Kobe earthquake, even though that time period had the least rainfall of any since we started multidisciplinary monitoring there (Fig. 4e). In this paper we describe the relationship between pore pressure and seismicity at the mine.

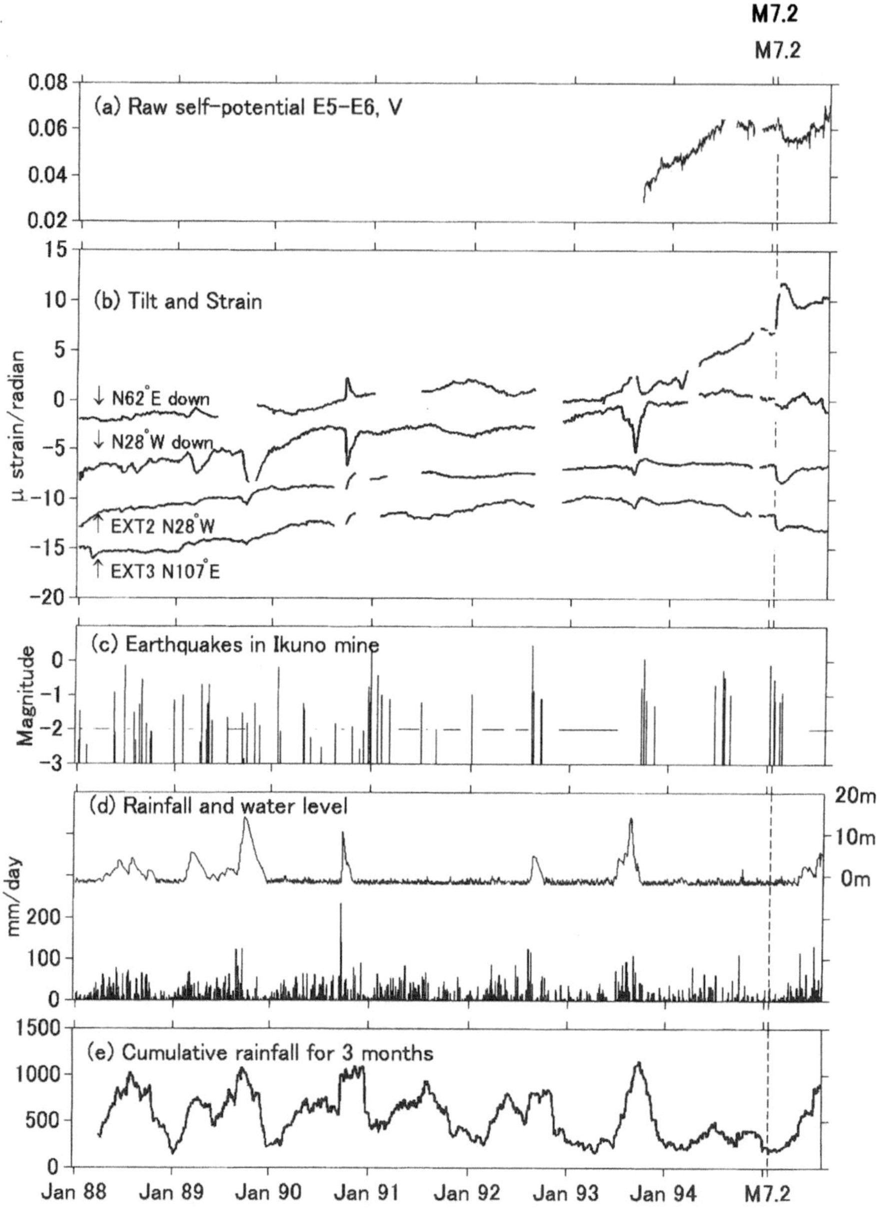

Figure 4

Records of our multidisciplinary observations encompassing an eight-year period. The onset of the *M* 7.2 earthquake is shown by the dashed vertical line. (a): Self-potential of E3-E4. (b): Tilt and strain at IGOS. Arrows indicate direction of downward tilt and extension. (c): Microearthquakes in the Kanagase area. Horizontal lines show time periods without data. (d): Daily rainfall (lower) and water level at the Koei shaft (upper). (e): Cumulative rainfall for the preceding three months.

2. Background

a) The Ikuno Mine and Mine Tremors

The ore veins in the Ikuno mine originated from Tertiary hydrothermal activity. The deepest ore veins are nearly vertical and at a depth of approximately 1 km (Figs. 2 and 3). The mine has a history of induced earthquakes, and the deeply mined, vertical ore veins had rockbursts ($M < 2$) before the mine was closed down in 1973 (TANAKA and NISHIDA, 1971; NISHIDA and TANAKA, 1972). In the late 1970s, the total amount of water flooding the mine reached several tens of millions of tons. The deep veins experienced induced earthquakes ($M < 3$) during this water impoundment (TANAKA and NISHIDA, 1977; TANAKA and OKA, 1979), although seismic activity nearly ceased once the ore veins were mostly flooded. OGASAWARA *et al.* (2001) found that a significant rise of the water table caused by heavy rainfalls activated microseismicity ($M < 0$) around the deepest, flooded, vertical ore veins and around the faults that offset them. They determined that seismicity did not immediately follow the rise in the water table, even though such a rise caused immediate changes in strain and tilt as large as 10^{-6} (Fig. 4b). They concluded that microseismicity does not directly respond elastically to the water load or to the opening of ore veins by the increase in water pressure of about 0.1 MPa. The localized diffusion of the water resulting from the increase in pore pressure may be the cause of the microseismicity.

b) Multidisciplinary Observations at the Ikuno Mine

The monitoring of crustal deformation in the Ikuno mine began in 1943 (SASSA and NISHIMURA, 1951) with horizontal-pendulum tiltmeters placed deep underground. This was followed in 1977 by the installation of extensometers with super Invar pipes, horizontal-pendulum tiltmeters and water-tube tiltmeters at the Ikuno Geophysical Observation Station at an altitude of 360 m (IGOS in Fig. 3d; TABEI *et al.*, 1985). In addition to monitoring crustal deformation, we started multidisciplinary monitoring within the Ikuno mine in 1987 (a thick, dashed, horizontal line in Figure 2 (lower) represents the level of deployment at an altitude of 360 m; OGASAWARA *et al.*, 1992). In order to monitor seismicity, geophones were installed until 1993 on the sidewall at five sites in the tunnels shown in Figure 3d. Hydrophones were temporarily installed in the Koei shaft (Fig. 3d) to check the accuracy of hypocenter locations. As no large station corrections were needed for the geophones and hydrophones, the velocity structure was approximated as homogeneous with $V_p = 4.65$ km/s. The accuracy of hypocenter determination was within several tens of meters until 1993, when a mining museum was constructed underground and we had to suspend seismic monitoring. Beyond 1993, geophone S1 was no longer available and S2 was shifted 200 m to the east, and the accuracy of hypocenter location degraded from several tens of meters to within approximately one hundred meters. The seismograms were recorded on a PC system with 12-bit, 2-kHz sampling. The detection threshold was $M > -2.5$ before 1993 and $M > -1.5$ after 1993. To monitor self-potential, Pb-$PbCl_2$ electrodes were installed in 1993 in boreholes several tens of cm deep in the footwall of the tunnel, shown as E3–E6 in Figure 3d (OGASAWARA and FUJIMORI, 1995). The Nendo fault (nendo means "clay" in Japanese) is a significant NE-SW-striking, well-fissured fault (Fig. 2) located within several tens of meters of E3 and E5 (compare E3 and E5 with NeF in Fig. 3). The strain, the tilt and the self-potential signals are recorded every five minutes. The level of water in the deep, vertical ore veins at the Koei shaft is measured once a day by an engineer with a tape measure (Fig. 3d). Further details concerning multidisciplinary monitoring and pertaining to the Ikuno mine and the history of induced seismicity are in OGASAWARA *et al.* (2001).

3. *The Effects of the M 7.2 Earthquake on the Ikuno Mine*

a) M 7.2 Earthquake that Took Place in the Least-rainfall Period

Figure 4e delineates the rainfall pattern over a seven-year period in the Ikuno mine area. It is especially noteworthy that there was little rainfall from the beginning of 1994, and that the *M* 7.2 event occurred during the lowest period of rainfall. The seismic intensity at the Ikuno mine from the 1995 Hyogo-ken Nanbu earthquake ($M = 7.2$) registered at least 4 on the Japan Meteorological Agency scale (6 or 7 on the modified Mercalli scale).

b) Significant Post-seismic Changes that Cannot be Explained by Coseismic Stress Changes

The strain began immediately after the Hyogo-ken Nanbu earthquake and reached its maximum of about 10^{-6} for every component (Figs. 4b and 5d), which corresponds to an areal contraction of about 2×10^{-6} strain.

▶

Figure 5

Significant post-seismic changes for the *M* 7.2 earthquakes. (a): Electric self-potentials E3-E4 and E5-E6 at Shinhon-ko (Fig. 3d). As unstable drift seen in E5-E6 in January 1995 was removed. E5-E6 with the drift removed and adjusted is shown in (a). (c): Tilt at IGOS. Arrows show directions of downward tilt. (d): Strain at IGOS. Arrows indicate directions of extension. (e): Earthquakes in the Kanagase area. Horizontal line represents a period without data. (f): Rainfall (lower) and the level of water flooding the deep ore veins. Note that microseismic events occurred in the Kanagase area during the period when significant changes were seen in strain and electric self-potential, whereas no significant water table rise or rainfall was seen. We have no available coseismic strain and tilt steps observed for the 1995, because the monitoring system was down for 8 days before the 1995 Hyogo-ken Nanbu earthquake. The strain data after the earthquake are shifted to adjust for this data gap in the layout of this figure.

The GPS network near the Ikuno mine showed no post-seismic displacement larger than 0.5 cm following the Hyogo-ken Nanbu earthquake (GEOGRAPHICAL SURVEY INSTITUTE, JAPAN, 1995). If we divide 0.5 cm by 10^{-6}, the order of the

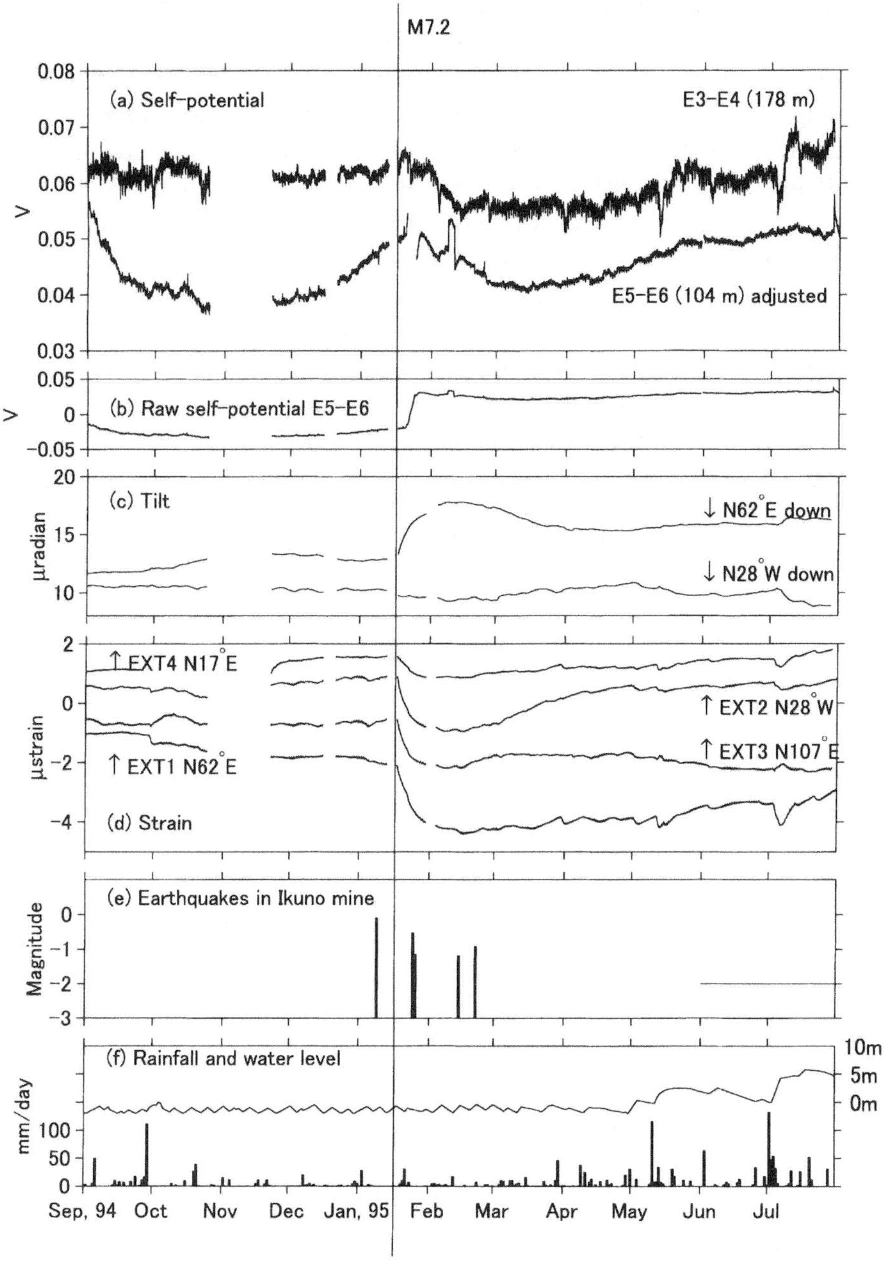

observed strain in the mine, the result is several km, which corresponds to the upper size limit of the area that experienced significant post-seismic deformation. This area is similar to that of the Ikuno mine (Fig. 2), therefore the area which sustained significant post-seismic deformation was confined within the Ikuno mine.

The recording system was not operational for eight days prior to the 1995 Hyogo-ken Nanbu earthquake, therefore we have no available coseismic strain and tilt observations that include this earthquake (Fig. 5). The coseismic strain step can be estimated by computations (e.g., OKADA, 1992) based on the dislocation theory. Assuming fault segments as shown in Figures 1 and 6 (HASHIMOTO *et al.*, 1996), we estimate the coseismic strain and tilt to be as shown in Table 1. The areal strain step at the Ikuno mine from the Hyogo-ken Nanbu earthquake was about 0.3×10^{-6}. The observed post-seismic change in strain corresponds to an areal strain of about 2×10^{-6}, thus the estimated coseismic areal strain step at the Ikuno mine is about one-seventh of the observed value. This value is too small to account for the observed post-seismic change. As shown in Table 1, the characteristics for strain and tilt differ markedly between coseismic and post-seismic changes.

Assuming a reasonable value of the elastic modulus corresponding to the V_p = 4.65 km/s used for hypocenter locations (OGASAWARA *et al.*, 2001), the observed post-seismic strain change of about 2×10^{-6} would correspond to a stress

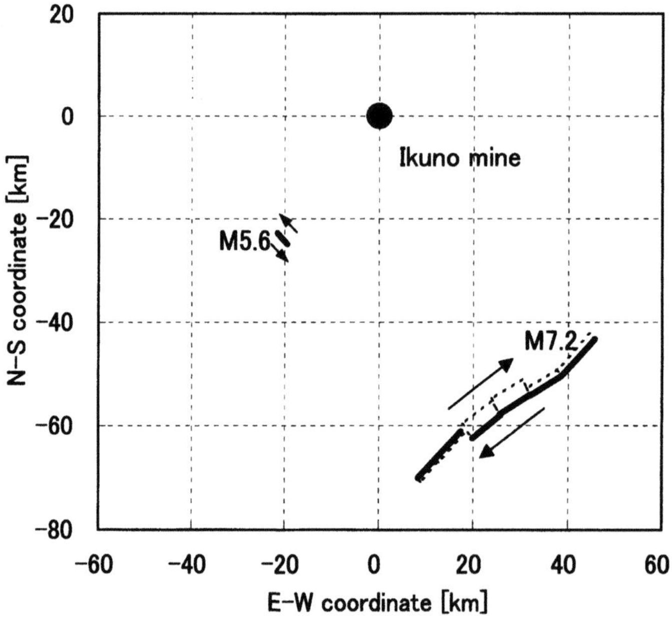

Figure 6
Relationship of the Ikuno mine to two earthquakes (*M* 5.4 in 1984 and *M* 7.2 in 1995) that produced significant post-seismic effects at the mine. The *M* 5.4 event had almost no dip-slip component (after NISHIGAMI and TSUKUDA, 1985), while the *M* 7.2 event had a significant thrust-slip component as well as a strike-slip component (data after HASHIMOTO *et al.*, 1996).

Table 1

Comparison of coseismic and post-seismic changes in strain (ε) and tilt for M 7.2 and M 5.6 events. Coseismic changes were calculated assuming fault segments shown in Figure 6. Post-seismic changes of M 5.6 are after TABEI *(1987)*

Strain	Coseismic step			Post-seismic change	
	ε_{areal}	ε_{shear}	Comp. axis	ε_{areal}	ε_{shear}
M 5.6 (1984)	-7×10^{-11}	4×10^{-9}	N93°E	-3×10^{-6}	1×10^{-7}
M 7.2 (1995)	-3×10^{-7}	2×10^{-7}	N52°W	-3×10^{-6}	1×10^{-7}

Tilt	Coseismic step		Post-seismic change	
	Radian	Direction	Radian	Direction
M 5.6 (1984)	3×10^{-9}	SE-down	2×10^{-6}	SW-down
M 7.2 (1995)	1×10^{-7}	E-down	6×10^{-6}	SW-down

change of about 0.1 MPa. This is about five times larger than the change in the Coulomb failure function shown on the map published by HASHIMOTO (1995, 1997). In addition, the observed post-seismic contraction suggests an increase in fault strength sufficient to inhibit faulting.

Similar post-seismic strain and tilt changes were observed after an $M = 5.6$ earthquake in 1984 about 30 km to the southwest of the Ikuno mine (TABEI 1987). Figure 7 indicates post-seismic changes in strain versus lapse time (in days) for the main events in 1995 and 1984. Even though the estimated coseismic strain and tilt were quite different between the 1995 and 1984 events (Table 1), the magnitude and duration of the strain changes were surprisingly similar to each other at EXT2 and EXT3. EXT1 and EXT4 were unstable in 1984, which is why we think they were dissimilar. Consequently, the elastic response to the dislocation at the source of the M 7.2 and M 5.6 events cannot account for the magnitude and the sense of the post-seismic changes in strain or tilt.

c) Self-potential, Indirect Evidence of the Perturbation in the Underground Water System

TABEI (1987) suggested that perturbation of underground water might account for the post-seismic changes in strain and tilt in 1984. However, he had no data available other than strain and tilt values to investigate further.

A clear symptom of perturbation to the underground water system is a post-seismic change in electric self-potential (hereafter SP; KUWABARA *et al.*, 1997a,b). There were gradual changes in SP from September 1993 to June 1994 (Fig. 4a), which we cannot currently explain. However, the post-seismic SP change can be easily distinguished after the onset of the M 7.2 event (Fig. 4a).

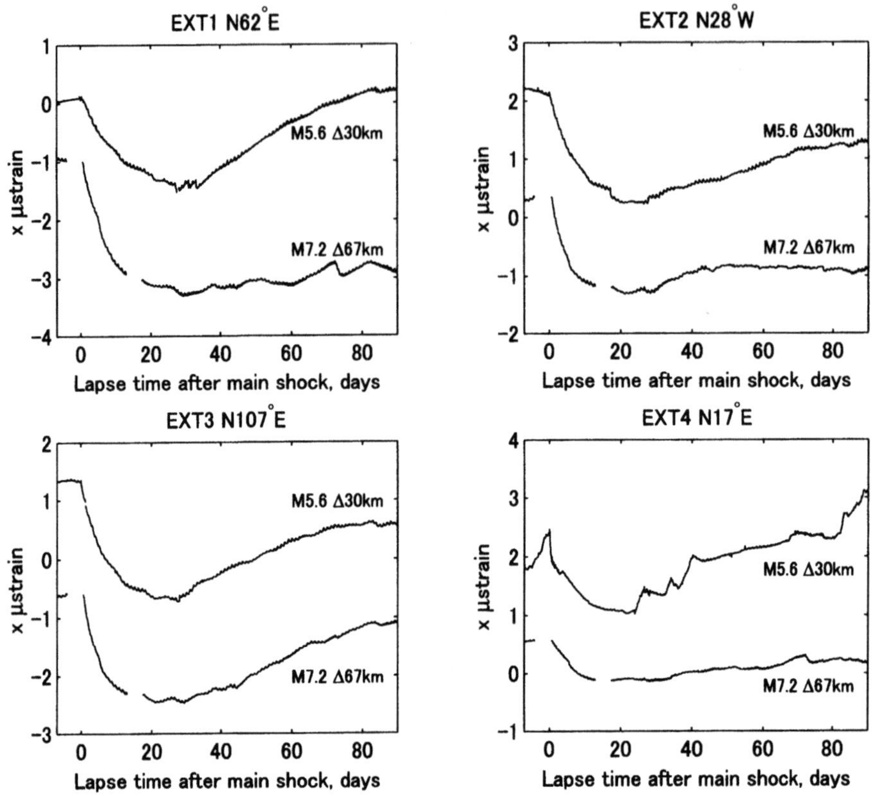

Figure 7

A comparison of post-seismic strain effects for the M 5.6 event in 1984 and the M 7.2 event in 1995. Strains are plotted versus lapse time after the onset of these events. Note that all are contracted. Regardless of the differences in magnitude and distance from the events, the EXT2 and EXT3 profiles are surprisingly similar to each other. EXT1 and EXT4 are similar, but not so similar as EXT2 and EXT3. We should take them into account that EXT1 and EXT4 were unstable in 1984.

In Figures 5a and 5b the time-axis for post-seismic SP changes is expanded. We had two dipoles, E3-E4 and E5-E6, with the same direction but different spans (Fig. 3d). Since E5-E6 was sometimes unstable (Fig. 5b), we removed the unstable change in E5-E6 in late January 1995 and shifted to the change shown in the lower part of Figure 5a. The differences in high-frequency variation (>10 cycles/day) for E3-E4 and E5-E6 (Fig. 5a) are caused by differences in the span of the dipoles and a remote artificial electric noise source, which suggests that both dipoles detected electric fields well. Independent of differences in the spans of the dipoles, there was a similar magnitude of post-seismic change observed for both dipoles, suggesting that the sources are near E3 and E5. E3 and E5 are installed within 1 m of each other and both are situated within a few tens of meters of the Nendo fault (compare Figs. 3a with 3d). The post-seismic SP change was very slow, compared with the SP changes associated with heavy rainfalls (see May or July, 1995 in Fig. 5 or 7). Rainwater

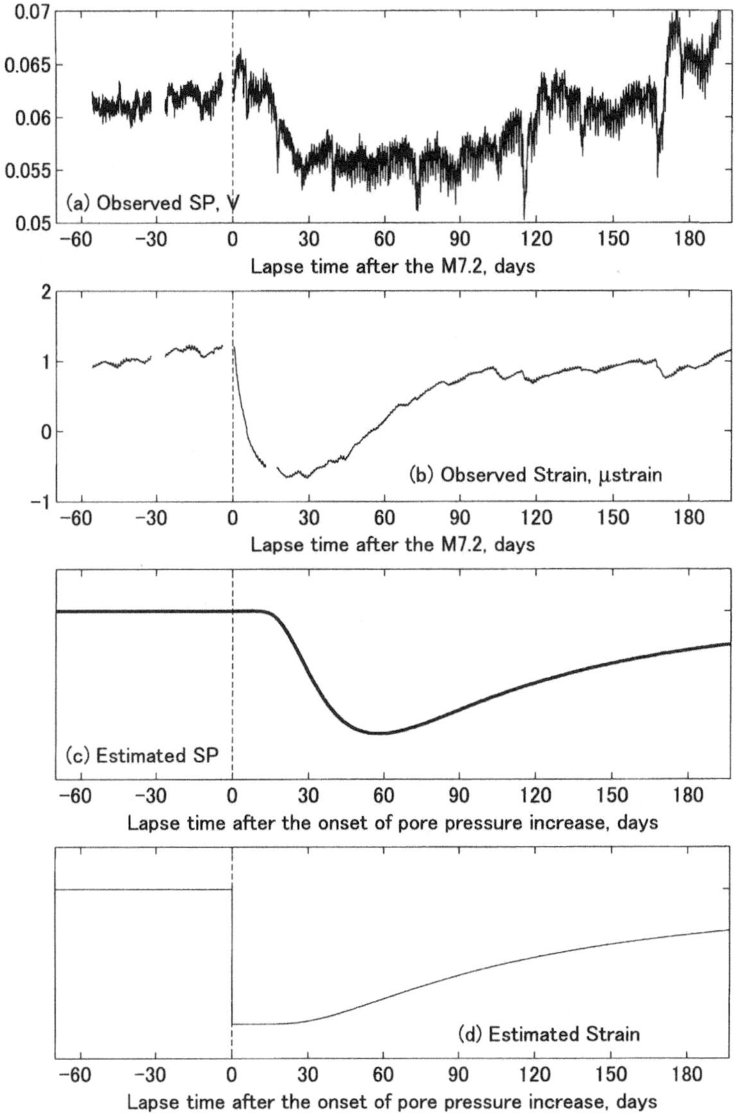

Figure 8
A comparison of post-seismic changes in self-potential and strain. (a and b): Observed SP and strain. (c and d): Theoretically estimated SP and strain, assuming an instantaneous point-source for the pore-pressure increase, a diffusivity, $a = 2 \times 10^{-4}\ \mathrm{m^2/s}$ and a distance, $r = 100$ m. Vertical dotted lines represent the onset of the *M* 7.2 earthquake and the instantaneous increase in pore pressure. Self-potential and strain are estimated according to Eqs. (1) and (2).Theoretically estimated self-potential (c) and strain (d) explain well the delayed response in self-potential to the onset and instantaneous response of strain to the onset. These also explain well the different recovery processes for self-potential and strain. Therefore, the observed self-potential and strain are strong evidence of a pore-pressure increase, although their variations seem to have different exponential changes with different time constants.

flows rapidly through open fissures near the ground surface, however post-seismic changes are not caused by such rapid water flow.

If we plot SP and strain together (Figs. 8a and b), we see the following characteristics:

(1) The post-seismic change in SP began several days after the *M* 7.2 earthquake, whereas the strain started immediately after the earthquake. This suggests that the self-potential signal is not caused by a piezo-elastic response to the coseismic stress change of the *M* 7.2 earthquake. Although the visco-elastic response of a rock mass with some time constant may be able to account for the post-seismic strain and tilt observed immediately after the *M* 7.2 event, it cannot account for the delay in post-seismic change seen only in the self-potential.

(2) Except for the delay, a change in SP is not the same as a change in strain.

However, these are important evidence that an increase in pore pressure causes both. To demonstrate this, in Figures 8c and d we show self-potential and the strain theoretically obtained by assuming an instantaneous point source for pore-pressure increase, Δp. Since SP is proportional to the gradient of pore-pressure, P (e.g., ISHIDO and MIZUTANI, 1981), for the simplest case, we can assume a point source and SP that can be expressed as:

$$SP \propto \frac{\partial P}{\partial r} = \frac{\partial}{\partial r}\left[\frac{\Delta p}{(4\pi at)^{3/2}}\exp(-r^2/4at)\right] , \tag{1}$$

where r is the distance from the source, a the diffusivity and t the lapse time after an instantaneous increase in pore pressure. Associated with Δp, radial stress σ_{rr} is given as:

$$\sigma_{rr} = -\frac{2K_1}{r^3}\left\{\mathrm{erf}\left(\frac{r}{2\sqrt{at}}\right) - \frac{r}{\sqrt{\pi at}}e^{-r^2/4at}\right\} , \tag{2}$$

where K_1 is a constant proportional to Δp and the elastic modulus. We find Eq. (2) in textbooks on thermo-elasticity as a response to a point of instantaneous heat (e.g., TAKEUCHI, 1981) and

$$K_1 = (1 + \nu)\alpha QG/2\pi(1 - \nu) , \tag{3}$$

where ν and G are Poisson's ratio and the rigidity, α the thermal expansion coefficient, Q heat instantaneously generated. However, the linearity of the pore-elastic effect is precisely analogous to the theory of linear thermo-elasticity, as RICE and CLEARY (1976) noted. Therefore, Eq. (2) can be used to express temporal change in stress associated with an instantaneous increase in pore pressure at a given point. Figures 8c and 8d show examples of theoretical SP and σ_{rr} that fit better with the observed values (Figs. 8a and 8b), assuming $a = 2 \times 10^{-4}$ m^2/s and $r = 100$ m. We learn from Eqs. (1–2) and Figures 8c and 8d as follows:

- SP is delayed with respect to the onset of pore-pressure increase, whereas strain responds instantaneously to its onset. The distance to the source and diffusivity determine the delay in SP.
- SP starts to recover only after a significant delay, compared to the quick start of recovery of strain. The rate of recovery is determined by the spatial distribution of Δp and diffusivity.

Equations (1) and (2) are Green's functions, thus we can obtain better-fitted profiles by convolving these while considering the space/time variation of Δp. Assuming an instantaneous line or plane source at a distance of ~100 m in a medium with permeability equal to $\sim 10^{-7}$ cm/s and an annual change, KUWABARA (1998) found that the profile which best fits the observed self-potential with a correlation coefficient of 0.88 is shown in Figure 9. Permeability on an order of 10^{-7} cm/s suggests diffusion along the 2nd or 3rd order geological discontinuities' zone (PUSCH, 1995), which is characterized by a fracture zone with a 1–100 m width, 10–1000 m

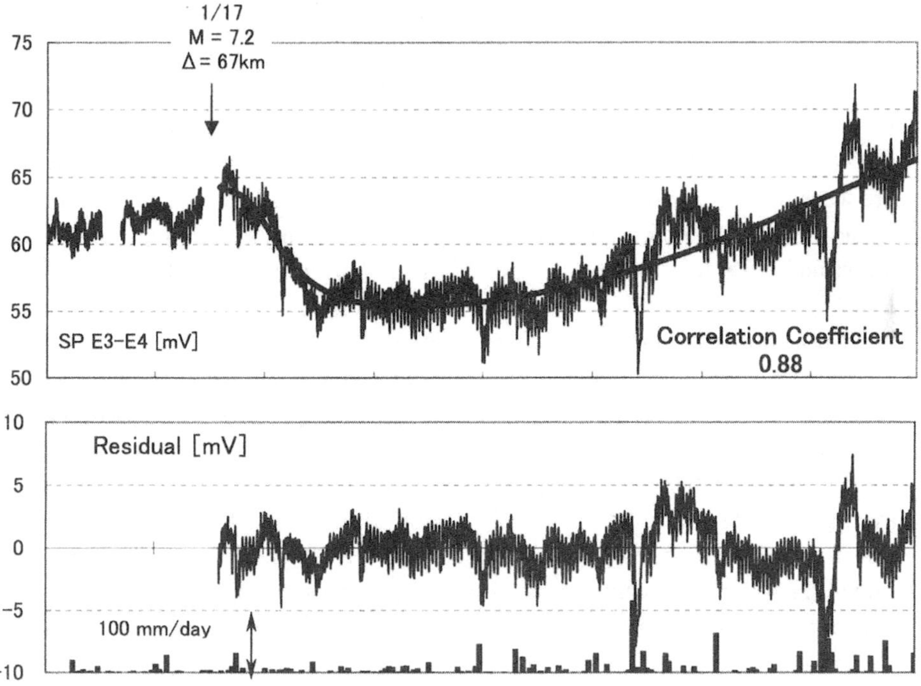

Figure 9
A model to account for the observed post-seismic change in self-potential (after KUWABARA, 1998). (Upper): The observed self-potential and a best-fit model, with a correlation coefficient of 0.88 obtained by assuming a plane source of instantaneous pore-pressure increase and annual change. (Lower): Residuals and rainfall shown by bars.

spacing and 100 m–10 km length. The Nendo faults delineated as thin dashed lines in Figures 3 and 4 are consistent with these ranges. The excavated areas are also candidates, because the 'cut and fill method' was adopted for mining at Ikuno and the underground water diffused through the excavated areas as easily as through the fault zones.

We will not attempt further to find a unique solution to account perfectly for all of the observed strain, tilt and self-potential, because data from only a single station are not enough to fully constrain the solution. However, at least there is a suggestion that Δp did not increase instantaneously, but gradually with time, and reached its maximum in early February. This happened because the observed strain shows a gradual response to the onset (Fig. 8a), regardless of the fact that the theory (Eq. (2), Fig. 8b) predicts an instantaneous strain response to the onset of an instantaneous pore-pressure increase.

d) Direct Evidence: A Significant Change in Water Flooding the Deep Vertical Ore Veins

There is another line of evidence to suggest that the underground water system was perturbed, as reflected in the post-seismic self-potential change. The water level flooding the deep ore veins is usually measured once a day by an engineer at the Koei shaft (Fig. 3d). This water level is controlled by drain pumps so that it does not overflow into the river. Therefore, the water table will rise linearly if no pumps are working, whereas it will drop linearly if either pump is working. As a result, the water table oscillates within a period of several days and a triangular shape with an amplitude of a few m (see the period from November 1994 to March 1995 in Fig. 5f).

It rained briefly before and after the Hyogo-ken Nanbu earthquake on January 17, 1995 (Fig. 4e), and no significant change in the water table at the deep veins was seen as *prima facie* evidence associated with the earthquake (Figs. 4d and 5f); we see only the normal oscillatory changes. However, if we carefully examine the rate of rise for the water table when no pumps were working (the black bars in Fig. 10a; averages are shown by open circles), a jump of about 10–15 cm/day is seen in the elevation rate when the earthquake occurred. We also find a jump in the lowering rate of the water table while the no. 1 or no. 2 pump was working (see gray bars with and without stripes in Fig. 10a, respectively), implying an interesting response. Therefore, in Figure 10b we made plots of a composite water table rate by superimposing the rising/descending rates with respect to those averaged over the period prior to the M 7.2 event. Interestingly, a moving average for seven days (the thick line in Fig. 10b) shows a gradual decrease following the jump, and a gradual recovery from March until heavy rainfall occurred in May. The rate of the water table refers to the flow rate of water that is proportional to the gradient of pore pressure according to Darcy's law, having the same physical response to pressure as the self-potential. As seen in Figure 10b, the composite water table rate (a thick line)

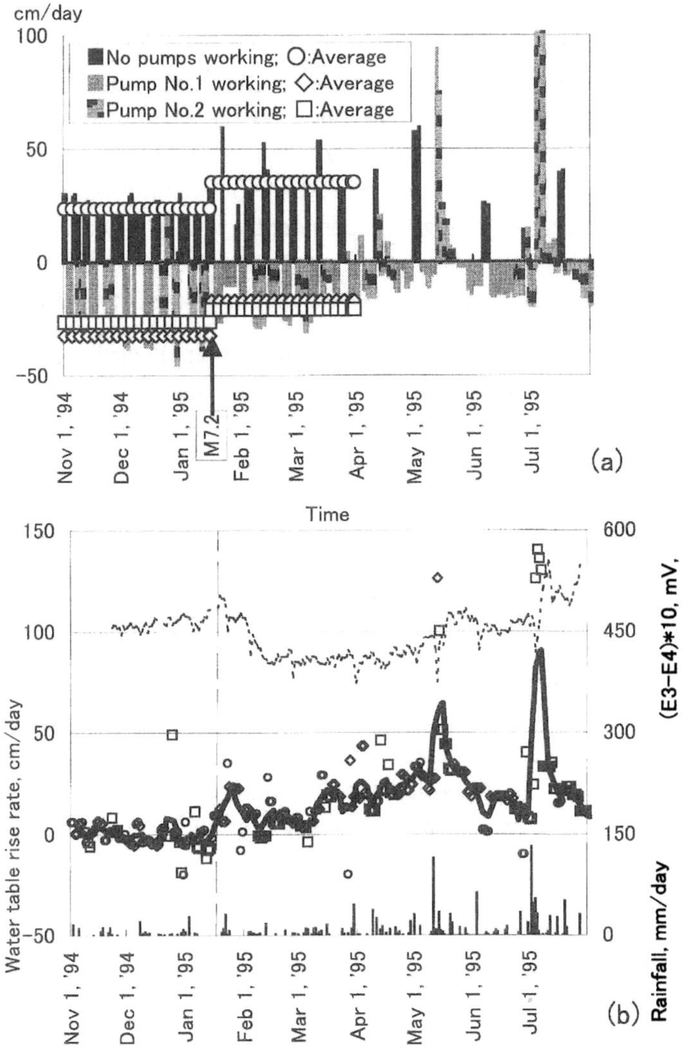

Figure 10

Changes of water level rate flooding the deep ore veins, associated with the *M* 7.2 earthquake. (a): Raw data on water level rate for periods with and without the operation of pumps are shown by bars. Average rates before and after the *M* 7.2 earthquake are represented by circles, diamonds and squares. There are jumps in rate at the onset of the *M* 7.2 event. (b): A composite water level rate. Rates compared to the average for a prior period of the *M* 7.2 are superimposed. A thick line shows the moving average for seven days. Self-potential (dotted line) and rainfall are shown for comparison. The observed self-potential and water level rate are similar from the event's onset to May with abundant rainfall, being consistent with physics that both self-potential and level rate are proportional to the pore-pressure gradient.

looks similar to self-potential (a dashed line), which strongly suggests a close physical relationship between them. Because the water discussed here floods the deep ore veins in the Ikuno mine, we suggest that pore pressure over the Ikuno mine increased along with the *M* 7.2 earthquake.

Such minute changes in the water table are seen only during periods with low rainfall. One month before and two months after the *M* 5.6 event in 1984, it rained considerably more than before and after *M* 7.2 in 1995. This gives us the rare chance to observe an *M* 7.2 event during a period of low rainfall, and we were able to distinguish the post-seismic changes independently of any rainfall effect.

e) Earthquake Activity in Ikuno Mine During the Post-M 7.2 Changes

As seen in Figure 5e, there was a swarm of four small earthquakes ($M > -1.5$) following the *M* 7.2 earthquake. It should be noted that all four small earthquakes occurred in the period of maximum contraction. As discussed in Section 3c, strain directly corresponds to pore pressure, and therefore these fours events occurred in the period with maximum pore pressure and minimum strength.

Figures 11a–11c show the hypocenters of events that occurred before 1993. Diamonds are scaled with lapse times after the onset of heavy rainfalls, and we see that larger diamonds tend to concentrate around deep ore veins (see an area with many deep tunnels in Fig. 11) and around the Nii and Juichii faults that offset the veins (see NiF and JiF in Figs. 11a and b). In Figures 11d–11f the hypocenters of microearthquakes after the *M* 7.2 event are superimposed as squares. We find that these post-seismic microearthquakes are also located in the area with post-heavy rainfall seismicity. This suggests that these events were also induced by pore-water fluctuation, since it is inferred that post-heavy rainfall seismicity is related to pore-water fluctuation.

However, an event occurred just before the *M* 7.2 event (Fig. 5e and a triangle in Figs. 11d–11f). In 1994 five events swarmed (Fig. 5e and circles in Figs. 11d–11f) in spite of the relatively low rainfall (Fig. 4e). Certain reasons are needed to explain these earthquakes; otherwise the events following the *M* 7.2 event appear to be chance phenomena.

Since ten earthquakes occurred during approximately 15 months when there was low rainfall, starting in January 1994, 1.33 ($= 2 \times 10/15$) events were expected in a two-month period, with significant post-seismic change. Similarly, by 8.67 events would be expected to occur in the following 13-month period without a post-seismic change. As the observed numbers of events were 4 and 6 for the periods with and without a post-seismic effect, the chi-square is 6.15, which corresponds to a probability of 0.023.

This probability is low but not extraordinarily low. However, here we must consider the coda of the *M* 7.2 event and the multitude of aftershocks that occurred following it (see Fig. 1). These aftershocks significantly masked small earthquakes in the proximity of the Ikuno mine. In addition, the recording system was frequently interrupted by the dead time for post-trigger processing for every aftershock

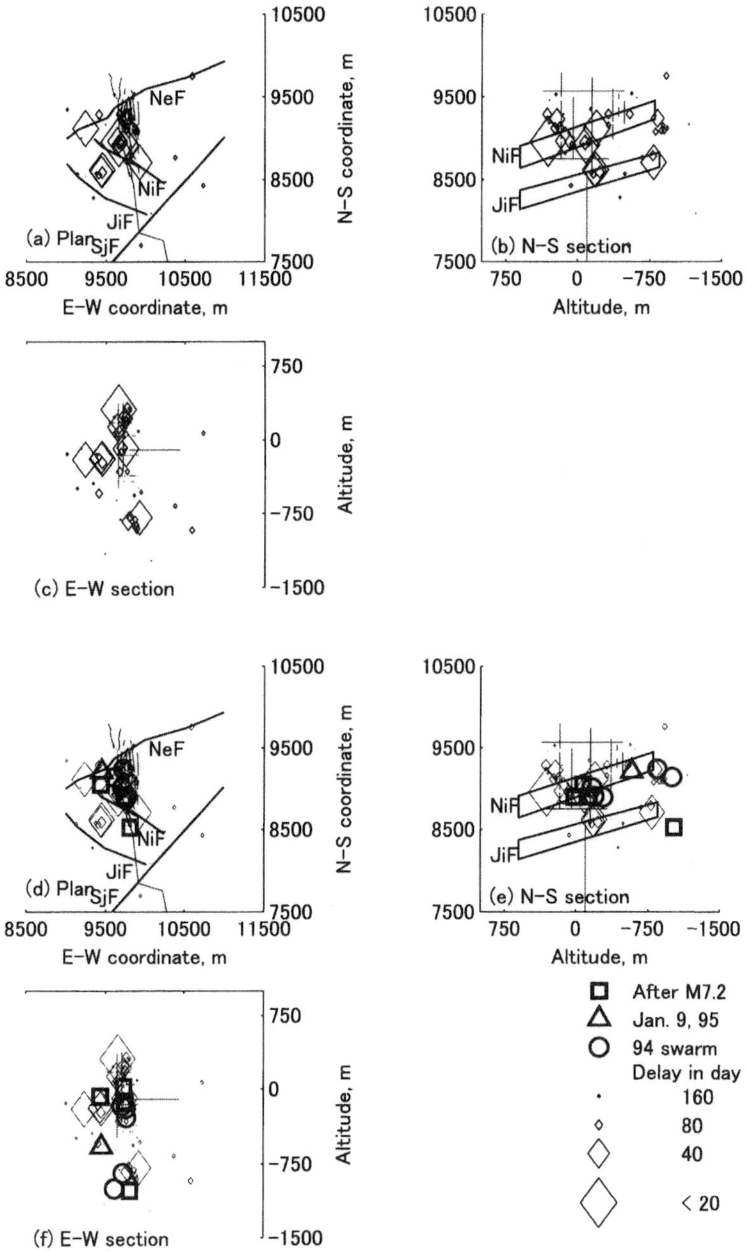

Figure 11
Comparison of hypocenter locations. (a–c): Hypocenters for 56 seismic events before 1993 (after OGASAWARA *et al.*, 2001). Diamonds are scaled with the delay after the onset of heavy rainfall (1988–1992). Thin lines: flooded tunnels in ore veins. Thick lines: faults (NeF, NiF and JiF: Nendo fault, Nii fault and Juichii fault). (d–f): Earthquakes in a low rainfall period in 1994 (circles), an earthquake 8 days before the *M* 7.2 event (a triangle) and the events after the *M* 7.2 event (squares) superimposed on Figures 11(a–c).

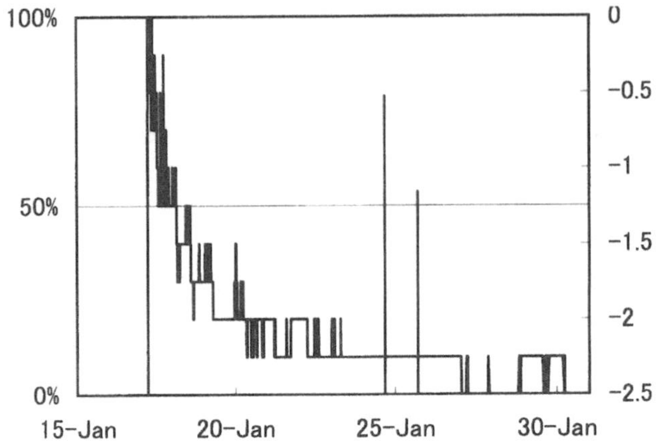

Figure 12
Effect of a coda of M 7.2 and frequent aftershocks to microearthquake monitoring in the Ikuno mine. The period when mine events could not be detected is estimated from the relationship between magnitude and the F-P time. The time for post-trigger processing is added to the F-P time to calculate the masked percentage of microearthquake monitoring in the mine. Magnitude data for aftershocks of the M 7.2 event (after the Japan Meteorological Agency).

(Fig. 12). There are known instances of microseismicities that immediately followed passage of the seismic wave of large earthquakes (e.g., HILL *et al.*, 1993). Therefore, the possibility exists that more than four seismic events occurred in the vicinity of the Ikuno mine.

The above is only our inference. Accordingly, we will endeauor next to confirm that the two swarms in 1994 and 1995 did not occur by chance. These two swarms took place in 2-month periods, spanning four months. Similar to that noted above, 2.67 events ($=4 \times 10/15$) are expected to occur on average in a 4-month period. However, all ten events swarmed in a 4-month period and no events took place in the other 11-month period. The chi-square for this case is 27.5, which corresponds to a probability of 5×10^{-7}, strongly suggesting that these two swarms did not occur by chance. In addition, the likelihood of 1995 post-seismic seismicity will increase if we can find a good reason to account for the 1994 swarm.

Over an eight-year period we have never had swarms such as those which occurred during the 1994 period of low rainfall. We also find that events in 1994 were located in an area similar to that in which heavy-rainfall seismicity took place. Therefore, we looked for any change that took place only in 1994, and found secular changes that could account for this concurrence of events. There is a clear acceleration to secular change in the S62°W tilt component shown in Figure 4b; it tilted by only a few μradians from 1988 to 1994, and then by several μradians just in 1994. At the same time, as seen in Figure 4b, there was a change in the secular trend in EXT3 with a direction nearly parallel to the E-W tectonic maximum compression;

it expanded from 1988 to 1994 by above five μstrains, whereas it reversed and started contracting by a few μstrain in 1994. Such strain and/or tilt behavior was also observed at other crustal movement stations in the area of Figure 1 (shown as squares with crosses; FURUSAWA, 1995), which depicts seismicity beginning in June 1994 (KATAO and ANDO, 1996). Consequently, we infer that the 1994 swarm at the Ikuno mine was related to a regional change in stress state. Although this is speculative, the swarm may have been caused by the stress change itself or a delayed change in strength associated with a velocity-weakening, as predicted by the rate- and state-dependent friction constitutive law (e.g., DIETERICH, 1972, 1978; RUINA, 1983).

As we could have found an explanation for the 1994 swarm, we speculate that the post-seismic swarm of four earthquakes was related to the *M* 7.2 event or to the significant post-seismic change.

f) Comparison with Post-heavy Rainfall Effects

OGASAWARA *et al.* (2001) noted significant changes in strain, tilt and water table rate following a rise of several m in the water table due to heavy rainfall exceeding several tens of mm/day (Fig. 3). They also documented that the microseismicity in flooded deep ore veins, and on faults that offset the veins, was associated with a significant elevation of underground water. We compare the characteristics of the post-heavy rainfall effect with the post-seismic effect of the *M* 7.2 event, as summarized in Table 2.

The post-*M* 7.2 effects in strain and tilt were similar in magnitude and duration to the post-heavy rainfall effect. However, the direction of the post-*M* 7.2 tilt was nearly orthogonal to the post-heavy rainfall tilt. Contraction with slight shear was observed for the post-*M* 7.2 strain, whereas there was significant shear for the post-heavy rainfall strain. The post-heavy rainfall effects in strain and tilt are well explained by the opening of vertical veins caused by a rise of the water table, but are not necessarily consistent with the faulting mechanisms of microearthquakes induced in the mine. In contrast, the post-*M* 7.2 strain and tilt are well explained by an increase and decay in pore pressure above the mine. Therefore, we infer that the post-*M* 7.2 effect in strain and tilt affected the entire mine, whereas the post-heavy rainfall effect was localized only within deep ore veins.

In contrast, the post-*M* 7.2 microseismicities were seen in a similar location to the post-heavy rainfall microseismicity, namely around the flooded deep veins and the faults that offset the veins.

As discussed above, the post-*M* 7.2 microseismicity was feasibly caused by an increase in pore pressure. However, the post-*M* 7.2 water-table rate was considerably slower than the post-heavy rainfall rate, and in fact was too small to detect during a period of such heavy rainfall. We may speculate on this situation as follows:

Pore-water perturbations that are difficult to sense during heavy rainfall periods can lead to microseismicity in the mine. A significant post-heavy rainfall rise of the

Table 2

Summary of characteristics observed in post-M 7.2 changes and other responses relevant to discussion. O: Similar or comparable, X: different. DWE: dependent upon water elevation

		M 7.2 post-seismic This study	Post-heavy rainfall	
Strain				
	Magnitude	$\sim 2 \times 10^{-6}$	$\sim 1 \times 10^{-6}$; DWE*	O?
	Duration	Several months	Several months; DWE*	O?
	Shear	Little	Significant*	X
Tilt	Magnitude	$\sim 6 \times 10^{-6}$	$<$several $\times 10^{-6}$; DWE*	O?
	Tilt direction	SW-down	SE; Pointing vein*	X
	Duration	Several months	Several months; DWE*	O?
Source of strain and tilt				
		Some broad	Within deep veins	X
Self-potential				
	Magnitude	~10 mV	Several mV	?
	Duration	Several months	≲one month	X
	Delay	Several days	Little	X
Correlation to	strain and tilt	Good but different	N/A	
Correlation to	water table rate	Good and similar	N/A	
Water level				
	Elevation	Less than change by pumping	Several m*	X
	Rate	~10 cm/day	~m/day*	X
	Duration	Several months	Depends on pumping*	X
Microseismicity				
	Hypocenters	Near deep veins	Near deep veins*	O
	Active period	A few months after the *M* 7.2 With high pore pressure	A few months after the water table elevation*	O

* after OGASAWARA *et al.* (2001).

water table as large as the post-*M* 7.2 or post-*M* 5.6 increase in pore pressure would act as a trigger to perturb pore water and lead to a decrease in strength, resulting in microseismicity. The post-heavy rainfall perturbation of pore water is localized in deep ore veins, and the perturbation itself is difficult to detect as strain and tilt at IGOS, 1–2 km from the microseismicity hypocenters. However, only significant water table elevation above veins about 1.0×1.0-km wide is detectable as associated strain and tilt for heavy rainfall periods.

5. Conclusions

The Hyogo-ken Nanbu earthquake occurred epicentrally 67 km from the Ikuno mine (Fig. 1). The mine had been closed because of rock bursts, and had vertical veins at about 1 km depth flooded after the closure. Significant post-seismic changes followed the *M* 7.2 earthquake in strain, tilt, self-potential and water table rate in unweathered, subsurface country rock (Figs. 4, 5) despite low rainfall. These changes continued for months before returning to pre-earthquake levels. Both strain and tilt changed in similar ways. The self-potential and the water level rate also changed in similar ways to each other (Fig. 10), although in a different manner from the strain and tilt changes (Fig. 8). Even though these similarities and differences are qualitative, they can be successfully explained by an increase and decay with diffusion of pore pressure, and by physical relationships between strain, self-potential, water-table rate and pore pressure (Fig. 8). In addition, four microearthquakes occurred 1–2 km from our multidisciplinary monitoring network in the mine during a period with maximum pore pressure, which corresponds to minimum strength (Fig. 5).

Post-seismic changes in strain were surprinsingly similar between the *M* 7.2 seismic event in 1995 and the *M* 5.6 event in 1984 (Fig. 7). Nonetheless, the estimated static coseismic strains were quite different and very small for the *M* 5.6 event, compared with the observed post-seismic strain (Table 1). Therefore, we infer that post-seismic change are caused not by static strain change, but probably by dynamic strain from passing seismic waves.

We observed only a small number of microearthquakes in the mine, owing to the triggering threshold of our recording equipment. We lacked continuous seismic records with which we could ascertain if there had been additional small earthquakes in the mine within the coda period of a large event or as aftershocks. We also lacked strong-motion data that would have given us insights of dynamic loading. The parts of the Ikuno mine where we were able to install instruments were very limited because we started our work 15 years after the mine had closed. We hope that these limitations can be addressed in the future. Nonetheless, we must emphasize that we did successfully monitor post-seismic effects during a period of minimal rainfall. Large earthquakes seldom occur, and this was a rare opportunity to study post-seismic effects under these conditions.

The magnitude of the events that occurred in the Ikuno mine was very small. However, they occurred in a limited area during a critical state in a mine that was easy to stimulate seismically. The wider the area over which a critical state prevails, the larger will be the expected magnitude. Although our conclusions are speculative, we still consider that they are applicable to larger natural earthquakes that are shallower than several km. For example, OGASAWARA (2001) carried out an analogous experiment and follow-up numerical analysis to demonstrate that both the power law and exponential decay in seismicity can be explained by transient pore pressure and induced stress. The conditions intrinsic to the Ikuno mine are a large volume of water in a seismogenic zone with open ore veins. Areas with subsurface hydraulic activity contain considerable water and often accompany shallow earthquake swarms. Therefore, we speculate that such areas may be stimulated in a manner similar to the post-*M* 7.2 or post-heavy rainfall effects in the Ikuno mine. A similar situation prevails where earthquake swarms are associated with upward intrusion of fluid into vertical cracks. For example, earthquake swarms are common off the Izu Peninsula, Japan. These swarms generally accompany significant crustal movement caused by the tensile opening of vertical cracks as wide across as several tens of cms. These cracks are not as narrow as the open veins in the Ikuno mine (e.g., TADA and HASHIMOTO, 1988, 1989; OKADA and YAMAMOTO, 1991). Accordingly we think that multidisciplinary monitoring in mines with flooded, deep veins and induced seismicity is very important to understanding the nature of seismogenic zones under these conditions.

In point of fact, most remote areas (> one fault length) around the *M* 7.2 event increased in water discharge independent of coseismic static strain (KOIZUMI *et al.*, 1996; Fig. 13), which suggests that pore-pressure increases were common phenomena in many places. However, it is clear that a pore-pressure increase cannot account for the triggered seismicity, if any, in water-poor areas. A number of

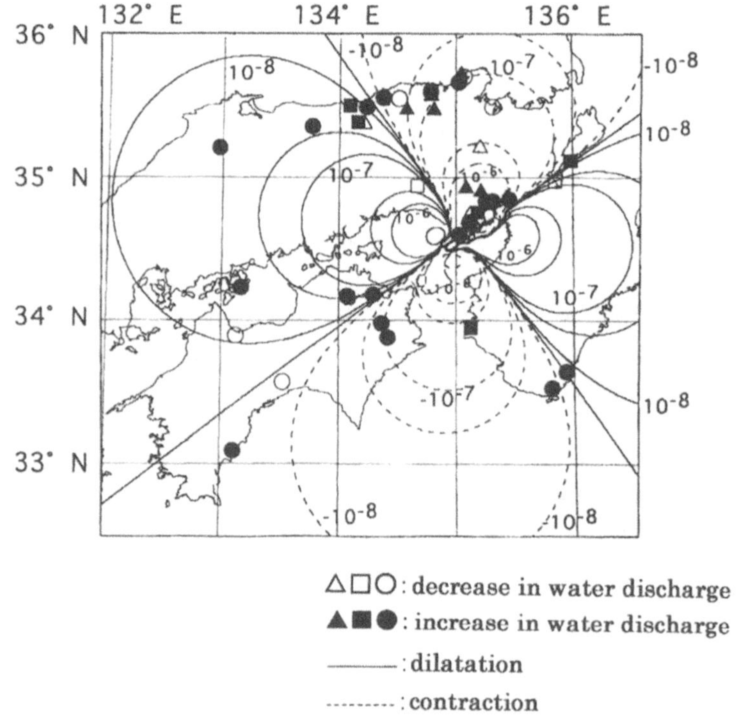

Figure 13
Change in water discharge in hot springs or mineral springs after the 1995 Hyogo-ken Nanbu earthquake (after KOIZUMI *et al.*, 1996). Coseismic volumetric strain change, (thin and thin dashed lines). Solid symbols and open symbols represent increases and decreases in water discharge, respectively. An increase in water discharge was seen at more than 80 percent of the investigated sites, although sites with increased discharge were not necessarily located in the contraction area.

studies have attempted to account for such seismicity under the negligible influence of water (e.g., GOMBERG, 1996; GOMBERG *et al.*, 1998) and some of these studies may possibly apply to same extent in the present case. However, these studies have not considered all possible phenomena, in particular *in situ* data from the immediate vicinity of the hypocentral faults. The typical distance from a surface monitoring station to the hypocenter of a natural earthquake is normally more than 10 km. We think that we will not fully understand the response of the fault to dynamic strain until we have complied *in situ* data from close to the hypocenter of a triggered event. The Research Group for Semi-controlled Earthquake Generation Experiments in South African Gold Mines initiated underground monitoring of normal strain and shear strain several meters from a fault. There is negligible water near this fault, and an event of $M > 2$ is exptected as a result of mining activities (OGASAWARA *et al.*, 2001). Strains reaching 10^{-4} are continuously monitored without clipping, and minute changes are monitored with 24-bit, 25-Hz sampling. This research group expects to obtain interesting *in situ* data, which may allow them to understand not only the response of a fault to dynamic strain, but also the initiation process of an earthquake.

Acknowledgments

We thank Mitsubishi Material Co. Ltd., Silver Ikuno Co. Ltd., and Techno Ohte Co. Ltd. for kindly giving us permission to carry out multidisciplinary observations in the Ikuno mine. We are grateful to Mr. S. Ogata and Mr. H. Matsumoto of Techno ohte Co. Ltd. for kindly supplying us water table data for the Koei shaft and for helping us to conduct our observations. We express our appreciation to Mr. S. Kotani of Mitsubishi Material Co. Ltd. and to colleagues at Kyoto University, who also assisted us to with observations. Comments by Drs. J. Gomberg, I. Beresnev and C. Trifu aided us to greatly improve our manuscript. We thank Prof. M. Yoshida for his help in statistical analysis and Mr. S. Matsuo for his help in preparation of electrodes.

REFERENCES

ANDERSON, J. G., BRUNE, J. N., LOUIE, J. N., ZENG, Y., SAVAGE, M., YU, G., CHEN, Q., and DEPOLO, D. (1994), *Seismicity in the Western Great Basin Apparently Triggered by the Landers, California, Earthquake, 28 June 1992*, Bull. Seismol. Soc. Am. *84*, 863–891.

DIETERICH, J. H. (1972), *Time-dependent Friction in Rocks*, J. Geophys. Res. *77*, 2690–2697.

DIETERICH, J. H. (1978), *Time-dependent Friction and the Mechanics of Stick Slip*, Pure appl. geophys. *116*, 790–806.

DISASTER PREVENTION RESEARCH INSTITUTE, KYOTO UNIVERSITY (1995), *Activation of Seismicity in the Northern Kinki District*, Rep. Coordinating Committee Earthq. Prediction *54*, 517–521 (in Japanese).

DISASTER PREVENTION RESEARCH INSTITUTE, KYOTO UNIVERSITY (1996), *Seismicity Aroung the Rupture Zone of the Hyogoken-Nanbu Earthquake*, Rep. Coordinating Commettee Earthq. Prediction *55* 508–515 (in Japanese).

EBERHART-PHILLIPS, D. and OPPENHEIMER, D. H. (1984), *Induced Seismicity in The Geysers Geothermal Area, California*, J. Geophys. Res. *89*, 1191–1207.

FURUSAWA, T., *The preceding crustal movement*. In The 1995 Hyogoken-Nanbu Earthquake (ed. Research Center for Earthquake Prediction, Disaster Prevention Research Institute, Kyoto University 1995), pp. 22–24 (in Japanese).

GEOGRAPHICAL SURVEY INSTITUTE, JAPAN (1995), *Crustal Movements in the Kinki District*, Rep. Coordinating Committee Earthq. Prediction *54*, 663–521 (in Japanese).

GOMBERG, J. (1996), *Stress/Strain Changes and Triggered Seismicity Following the M_w 7.3 Landers, California, Earthquake*, J. Geophys. Res. *101*, 751–764.

GOMBERG, J. and BODIN, P. (1994), *Triggering of the $M_s = 5.4$ Little Skull Mountain Nevada, earthquake with Dynamic Strains*, Bull. Seismol. Soc. Amer. *84*, 844–853.

GOMBERG, J. and DAVIS, S. (1996), *Stress/Strain Changes and Triggered Seismicity at The Geysers, California*, J. Geophys. Res. *101*, 733–749.

GOMBERG, J., BLAMPIED, M. L., and BEELER, N. M. (1997), *Transient Triggering of near and Distant Earthquakes*, Bull. Seismol. Soc. Am. *87*, 294–309.

GOMBERG, J., BEELER, N. M., BLANPIED, M. L., and BODIN, P. (1998), *Earthquake Triggering by Transient and Static Deformations*, J. Geophys. Res. *103*, 24,411–24,426.

HASHIMOTO, M. (1995), *Static Stress Changes associated with the Kobe Earthquake: Calculation of Changes in Coulomb Failure Function and Comparison with Seismicity Change*, Zisin *2*(48), 521–530 (in Japanese with English abstract).

HASHIMOTO, M. (1997), *Correction to Static Stress Changes Associated with the Kobe Earthquake: Calculation of Changes in Coulomb Failure Function and Comparison with Seismicity Change*, Zisin *2*(50), 21–27 (in Japanese with English abstract).

HASHIMOTO, M., SAGIYA, H., TSUJI, H., HATANAKA, Y., and TADA, T. (1996), *Co-seismic Displacements of the 1995 Hyogo-ken Nanbu Earthquake*, J. Phys. Earth *44*, 255–279.

HILL, D. P., REASENBERG, P. A., MICHAEL, A., ARABAZ, W. J., BEROZA, G., BRUMBAUGH, D., BRUNE, J. N., CASTRO, R., DAVIS, S., DEPOLO, D., ELLSWORTH, W. L., GOMBERG, J., HARMSEN, S., HOUSE, L., JACKSON, S. M., JOHNSTON, M. J. S., JONES, L., KELLER, R., MALONE, S., MUNGUIA, L., NAVA, S., PECHMANN, J. C., SANFORD, A., SIMPSON, R. W., SMITH, R. B., STARK, M., STICKNEY, M., VIDAL, A., WALTER, S., WONG, V., and ZOLLWEG, J. (1993), *Seismicity Remotely Triggered by the Magnitude 7.3 Landers, California Earthquake*, Science *260*, 1617–1623.

ISHIDO, T. and MIZUTANI, H. (1981), *Experimental and Theoretical Basis of Electrokinetic Phenomena in Rock-water Systems and its Applications to Geophysics*, J. Geophys. Res. *86*, 1763–1775.

KATAO, H. and ANDO, M. (1996), *Seismicity Before and After the Hyogoken-Nanbu Earthquake*, Kagaku *66*, 78–85 (in Japanese).

KOIZUMI, N., KANO, Y., KITAGAWA, Y., SATO, T., TAKAHASHI, M., NISHIMURA, S., and NISHIDAS, R. (1996), *Groundwater Anomalies Associated with the 1995 Hyogo-ken Nanbu Earthquake*, J. Phys. Earth *44*, 373–380.

KUWABARA, Y. (1998), *The postseismic change of the 1995 Hyogo-ken Nanbu Earthquake Detected at old Ikuno Mine (Δ = 67 km)* and its interpretation, Master's Thesis, Faculty of Science and Engineering, Ritsumeikan Univ. (in Japanese).

KUWABARA, Y., OGASAWARA, H., FUJIMORI, K., and KOIZUMI, M. (1997a), *The changes in Electric self-potential before and after 1995 Hyogo-ken Nanbu earthquake*, Abstracts, International Workshop on Seismo-Electromagnetic, March 3, 1997, Chofu, Tokyo, Japan, pp. 133–134.

KUWABARA, Y., OGASAWARA, H., KOYAMA, T., MIWA, T., FUJIMORI, K., and KOIZUMI, N. (1997b), *Changes before and after the Hyogo-ken Nanbu earthquake detected by multidisciplinary observations at Old Ikuno Mine (Δ = 67 km)*, 1997 *Proc. Cond. Anomal. Res. Meeting*, 29–36 (in Japanese with English abstract).

NISHIDA, R. and TANAKA, Y. (1972), *Ground Tremors Caused by Rock Bursts in the Ikuno Mine (continued)*, Disast. Prev. Res. Inst., Kyoto Univ. *15-B*, 43–51 (in Japanese with English abstract).

NISHIGAMI, K. and TSUKUDA, T. (1985), *Occurrence Process of the Earthquake – Symposium on the Yamasaki Fault*, The Earth Monthly. *67*, 43–48 (in Japanese).

NISHIGAMI, K., NAKAMURA, M., YABE, M., WATANABE, K., and MATSUMURA, K. (1995), *Change of Seismicity in the Eastern Chugoku Region after the 1995 Hyogoken-Nanbu Earthquake*, Abstract, Seismol. Soc. Jpn. *2*, A50 (in Japanese).

OGASAWARA, H. (2001), *A Simple Analogue Experiment to Account for a Power-law and an Exponential Decay of Earthquake Sequence*, Pure appl. geophys. *159*, 309–343.

OGASAWARA, H. and FUJIMORI, K. (1995), *Observations of Electric Self-potential at a Closed Old mine*, Tech. Note Natl. Res. Inst. Earth Sci. Disast. Prev. *166*, 177–197 (in Japanese with English abstract).

OGASAWARA, H., FUJIMORI, K., KOIZUMI, N., FUJIWARA, S., NAKAO, S., NISHIGAMI, K., TANIGUCHI, K., OTSUKA, S., HIRANO, N., NISHIDA R., and IIO, Y. (1992), *Comprehensive Observation System at Old Ikuno mine for Earthquake Prediction*, Ann. Disast. Prev. Res. Inst., Kyoto Univ. *35B-1*, 389–400 (in Japanese with English abstract).

OGASAWARA, H., FUJIMORI, K., KOIZUMI, N., HIRANO, N., FUJIWARA, S., OTSUKA, S., NAKAO, S., NISHIGAMI, K., TANIGUCHI, K., IIO, Y., NISHIDA, R., OIKE, K., and TANAKA, Y. (2001), *Microseismicity Induced by Heavy rainfall around Flooded Vertical Ore Veins of 1 km deep*, Pure appl. geophys. *159*, 91–109.

OGASAWARA, H., SATO, S., NISHII, S., ISHII, H., IIO, Y., NAKAO, S., ANDO, M., NAGAI, N., OHKURA, T., CICHOWICZ, A., KAWAKATA, H., KUSUNOSE, K., SATOH, T., CHO, A., SUMITOMO, N, and GREEN, R. W. K. (2000), *Semi-controlled Seismogenic Experiment in South African Deep Gold Mines*, Abst. AGU Fall meeting, EOS, 81, No. 48, F1172.

OIKE, K. and MATSUMURA, K. (1985), *Triggers of Earthquake Occurrence*, Chikyu monthly, *7*, 15–19 (in Japanese).

OKADA, Y. (1992), *Internal Deformation due to Shear and Tensile Faults in a Half Space*, Bull. Seismol. Soc. Am. *82*, 1018–1040.

OKADA, Y. and YAMAMOTO, E. (1991), *A Model for the 1989 Seismo-volcanic Activity Off Ito, Central Japan, Derived form Crustal Movement Data*, J. Phys. Earth *39*, 177–195.

RICE, J. R. and CLEARY, M. P. (1976), *Some basic Stress Diffusion Solutions for Fluid Saturated Elastic Porous Media with Compressible Constituents*, Rev. Geophys. Space Phys. *14*, 227–241.

RUINA, A. (1983), *Slip Instability and State Variable Friction Laws*, J. Geophys. Res. *88*, 10359–10370.

PUSCH, R., *Rock Mechanics on a Geological Base* (Elsevier 1995) p. 69.

SETIN, R. S. and LISOWSKI, M. (1983), *The 1987 Homestead Valley earthquake Sequence, California: Control of Aftershocks and Postseismic Deformation*, J. Geophys. Res. *88*, 6477–6490.

SASSA, K. and NISHIMURA, E. (1951), *On Phenomena Forerunning Earthquakes*, Trans. Am. Geophys. Union *32*, 1–6.

SLEEP, N. H. and BLANPIED, M. L. (1992), *Creep, Compaction and the Weak Rheology of Major Faults*, Nature *359*, 687–692.

SLEEP, N. H. and BLANPIED, M. L. (1994), *Ductile Creep and Compaction: A Mechanism for Transiently Increasing Fluid Pressure in Mostly Sealed Fault Zones*, Pure appl. geophys. *143*, 9–40.

STARK, M. A. and DAVIS, S. D. (1996), *Remotely Triggered Microearthquake at The Geysers Geothermal Field, California*, Geophys. Res. Lett. *23*, 945–948.

TABEI, T. (1987), *On the Postseismic Deformation at Ikuno Immediately after the 1984 Yamasaki Earthquake*, J. Phys. Earth *35*, 225–239.

TABEI, T., FUJIMORI, K., and TANAKA, Y. (1985), *Observation of Crustal Movements at Ikuno (1977–1983)*, J. Geod. Soc. Jpn. *31*, 189–201 (in Japanese with English abstract).

TADA, T. and HASHIMOTO, M. (1988), *On the Cause of Abnormal Crustal Deformation in the Northeastern Izu Peninsula (2) – Magma Chamber and Open Crack*, Abst. Seismol. Soc. Jpn. *1*, 139 (in Japanese).

TADA, T. and HASHIMOTO, M. (1989), *On the Cause of Abnormal Crustal Deformation in the Northeastern Izu Peninsula (3) – Open Crack Model and Activity in the Summer of 1988*, Abst. Seismol. Soc. Jpn. *1*, 41 (in Japanese).

TAKEUCHI, Y. (1981), *Analyses of Thermal Stresses*, Nisshin-Shuppan, Tokyo, Japan, pp. 346 (in Japanese).

TANAKA, Y. and NISHIDA, R. (1971), *Ground Tremors Caused by Rock Bursts in the Ikuno Mine*, Bull. Disast. Prev. Res. Inst., Kyoto Univ. *14-A*, 149–164 (in Japanese with English abstract).

TANAKA, Y. and NISHIDA, R. (1977), *Generation mechanism of rock bursts and microearthquakes induced by water-injection into a deep abandoned mine*, In Proc. *5th Jpn. Symp. Rock Mech.*, pp. 91–96 (in Japanese).
TANAKA, Y. and OKA, Y. (1979), *Generation Mechanism of Rock Bursts and Water-induced Earthquakes Under the Tectonic Stress Field*, Rock Mech. Jpn. *3*, 71–73.

(Received March 16, 2000, revised/accepted October 19, 2000)

Pure appl. geophys. 159 (2002) 091–109
0033–4553/02/030091–19 $ 1.50 + 0.20/0

Pure and Applied Geophysics

Microseismicity Induced by Heavy Rainfall Around Flooded Vertical Ore Veins

HIROSHI OGASAWARA,[1] KUNIO FUJIMORI,[2] NAOJI KOIZUMI,[3] NORIO HIRANO,[4] SATOSHI FUJIWARA,[5] SHIGEAKI OTSUKA,[6] SETSURO NAKAO,[4] KIN'YA NISHIGAMI,[7] KEISUKE TANIGUCHI,[8] YOSHIHISA IIO,[9] RYOHEI NISHIDA,[10] KAZUO OIKE,[2] and YUTAKA TANAKA[2]

Abstract—Microseismicity ($M < 0$) induced by heavy rainfall was investigated around the flooded, vertically dipping Tertiary ore veins with dimensions of about 1 km × 1 km in the Ikuno mine, Japan. The ore veins had rock bursts ($M < 3$) before the mine was closed in 1973, as well as seismic events ($M < 3$) during flooding after it was closed down. The stress state is therefore critical to failure, at least within one stress drop of a seismic event. Since 1987, when the veins had become mostly flooded, 56 mine tremors ($M < 0$) were observed over a 5-year period. Several times during this five-year period the mine sustained heavy rainfall of several tens of millimeters per day, and the water table flooding over the ore veins was elevated by several meters. Significant changes in strain larger than 10^{-6} were also monitored at a crustal movement monitoring station located several hundred meters from the veins. It was found that the opening of the vertical ore veins primary led to significant strain and tilt, but not to seismicity, because the delay and the longer duration of the seismicity were significant. Most seismic events involve thrusting mechanisms that are consistent with the present stress state of E-W-oriented tectonic compression, but are not consistent with the opening of the deepest ore vein. Interstingly, all the events within a few months of the heavy rainfall occurred near the faults that offset the deepest ore veins, wheareas all those events located away from the deepest ore veins occurred many months after the heavy rainfall. Consequently, the delayed diffusion of water appears to have played a dominant role in reducing rock strength, which led to seismicity in the Ikuno mine.

---

[1] Faculty of Science and Engineering, Ritsumeikan University, 1-1-1 Noji Higashi, Kusatsu, 525-8577, Japan. E-mail: ogasawar@se.ritsumei.ac.jp

[2] Graduate School of Science, Kyoto University, Sakyo-ku, Kyoto, 606-8502, Japan.

[3] Geological Survey of Japan, 1-1-3, Higashi, Tsukuba, Ibraki, 305-8567, Japan.

[4] Research Center of Earthquake Prediction, Disaster Prevention Research Institute, Kyoto University, Gokasho, Uji, 611-0011, Japan.

[5] Geographical Survey Institute, 1 Kitasato, Tsukuba, 305-0811, Japan.

[6] Faculty of Humanities and Sciences, Kobe Gakuin University, 518 Arise, Ikawadani-cho, Nishi-ku, Kobe, 651-2180, Japan.

[7] Research Center for Disaster Reduction Systems, Disaster Prevention Research Institute, Kyoto University, Gokasho, Uji, Kyoto 611-0011, Japan.

[8] Kyoto University of Education, 1 Fukakusa-Fujinomori-cho, Fushimi-ku, Kyoto, 612-8522, Japan.

[9] Earthquake Research Institute, the University of Tokyo, 1-1-1 Yayoi Bunkyo Tokyo 113-0032, Japan.

[10] Department of Civil Engineering, Faulty of Engineering, Tottori University, 4-101, Koyama-Minami, Tottori, 680-8552, Japan.

Key words: Seismicity, flooded mine, deep vein, faults, heavy rain, multidisciplinary monitoring.

1. Introduction

Induced or triggered earthquakes may be caused by changes in stress and/or pore pressure, as reviewed in McGarr and Simpson (1997). We discuss the contribution of each type of change, using available data.

One of the earliest examples of seismicity caused by fluid injection occurred in Colorado ($M < 5$) at a pressure on the order of 10 MPa (Evans, 1996; Healy *et al.*, 1968). The seismicity was both simultaneous and delayed with respect to the injection. It was later found that an increase in pore pressure of 1 to 5 MPa was enough to trigger the seismicity (Fletcher and Sykes, 1977).

The impoundment of water by huge reservoirs also contributes to significant seismicity (Raleigh *et al.*, 1976). However, once a critical state of stress has been established by impounding the water, much smaller fluctuations of water than those involved in injection experiments have sometimes simulated the seismicity. For example, Simpson and Negamatullaev (1981) found that a drop in water level of a few meters (on the order of 0.01 MPa) and a rapid decrease in the reservoir-filling rate (0.5 m/day) stimulated seismicity. Simpson *et al.* (1988) showed that reservoir-triggered seismicity can be divided into two types – a rapid reaction related to instantaneous elastic response, and a delayed reaction related to fluid diffusion.

The studies noted above have mainly been concerned with seismicity, and with descriptions of tectonic settings and the magnititude of water pressure. The stress state in seismic source areas has been inferred in some studies however mostly this has been based on numerical estimation (Simpson and Negamatullaev, 1981). Few studies have tried to infer stress states in source areas using *in situ* strain and tilt data.

A deep, flooded mine can be regarded as a huge water reservoir in which the enclosing rocks have generally been well investigated. The Ikuno mine is one of the deepest flooded mines in Japan with vertically dipping ore veins. It was closed down in 1973 because of severe rock bursts (maximum $M = 2.4$; Tanaka and Nishida, 1971, 1977; Nishida and Tanaka, 1972), and flooding subsequent to the closure induced seismicity (maximum $M = 2.9$; Tanaka and Nishida, 1977; Tanaka and Oka, 1979). The increase in water level was closely monitored, and it was shown that there was a good correlation between water level and seismicity. Unfortunately, only one seismometer with a rotating-drum recording system was installed in the mine. This seismometer counted events with small S-P times, and more detailed measurements were not taken. A similar case history in flood-induced seismicity in a hard rock mine in Canada (Wetmiller *et al.*, 1993) yielded seismograms like multiplets, suggesting similar mechanisms during flooding stages. However, reliable water-level data were few and there were no seismic stations in the mine.

By 1976 the Ikuno mine was mostly flooded. Ikuno sometimes experiences heavy rainfall amounting to several tens of millimeters per day, with the resulting water table elevated by several meters. The Ikuno mine has an underground crustal movement monitoring station on intact rock, so that *in situ* strain and tilt data are available. In addition, we deployed a small underground seismic array on intact rock within the mine to detect seismicity. In this paper we present a case study of seismicity involving a deep, flooded, vertical ore vein, and discuss the relationship between the seismicity and heavy rainfall.

2. *Ikuno Mine with Flooded Vertical Veins and a History of Induced Seismicity*

The Ikuno mine is adjacent to an area of naturally high seismicity in southwestern Japan, as shown in Fig. 1(a), and minor active faults (RESEARCH GROUP for ACTIVE FAULT, 1990) have been studied near the Ikuno mine (Fig. 1(b)). However, very few natural earthquakes ($M > 1$) have been measured in the years 1965–1984 by the microearthquake monitoring network of Kyoto University in an area of several km^2 immediately surrounding the Ikuno mine, (Fig. 1(b)).

The ore veins in the Ikuno mine originated from Tertiary hydrothermal activity. Most ore veins were intruded nearly vertically into Tertiary volcanics overlying Paleozoic sedimentary rocks (Fig. 2).

The fault system in Ikuno is characterized by NE-SW-striking, nearly vertical faults (the Nendo and Sanjugo faults), as well as by conjugate NW-SE-striking faults dipping norhteast (the Nii and Juichii faults). As noted in Table 1, the former faults are relatively significant and their associated fissure zones are well developed, resulting in a NE-SW-striking valley about 10 km long (Figs. 2 and 3). The latter faults are relatively minor, but also have significant lateral offset (the Senjuhon-hi vein is right-laterally offset by the Nii fault in Fig. 2 (upper)). The activity of these faults began after the intrusion of the ore veins and therefore the ore veins have clear offsets as shown in Table 1, consistent with the Tertiary N-S compressive tectonic stress. Although the present principal tectonic stress state is E-W compression (TANAKA *et al.*, 1989), these faults are still oriented to cause seismic events.

The Ikuno mine has a long history that began in 806 A.D., more than a thousand years ago. Numerous small tunnels, called '*raccoon dog's holes*,' were excavated by chisel and hammer in the early historic period. By the late 1960s, the excavation had reached a depth of about 1 km in the Kanagase area. The Kinsei-hi vein, Senju-hon-hi vein and Senju-zen-hi vein were the deepest and were excavated to a depth of about 1 km (−500 m altitude in Figs. 2 and 3). The largest rock bursts of about $M = 2$ occurred at the Kinsei-hi and Senjuhon-hi veins from 1970 to 1972 (Figs. 2 and 4); TANAKA and NISHIDA, 1971; NISHIDA and TANAKA, 1972), and were measured and located by the microearthquake monitoring network of the Disaster Prevention Research Institute of Kyoto University (Fig. 1(c)). These rock bursts forced closure of the Ikuno mine in 1973, as noted. The drain pumps at the deepest level of the mine were stopped, and the excavation began to flood. The total amount of water flooding the mine by the late 1970s was in excess of several tens of millions of tons. TANAKA and NISHIDA (1977) and TANAKA and OKA (1979) installed a

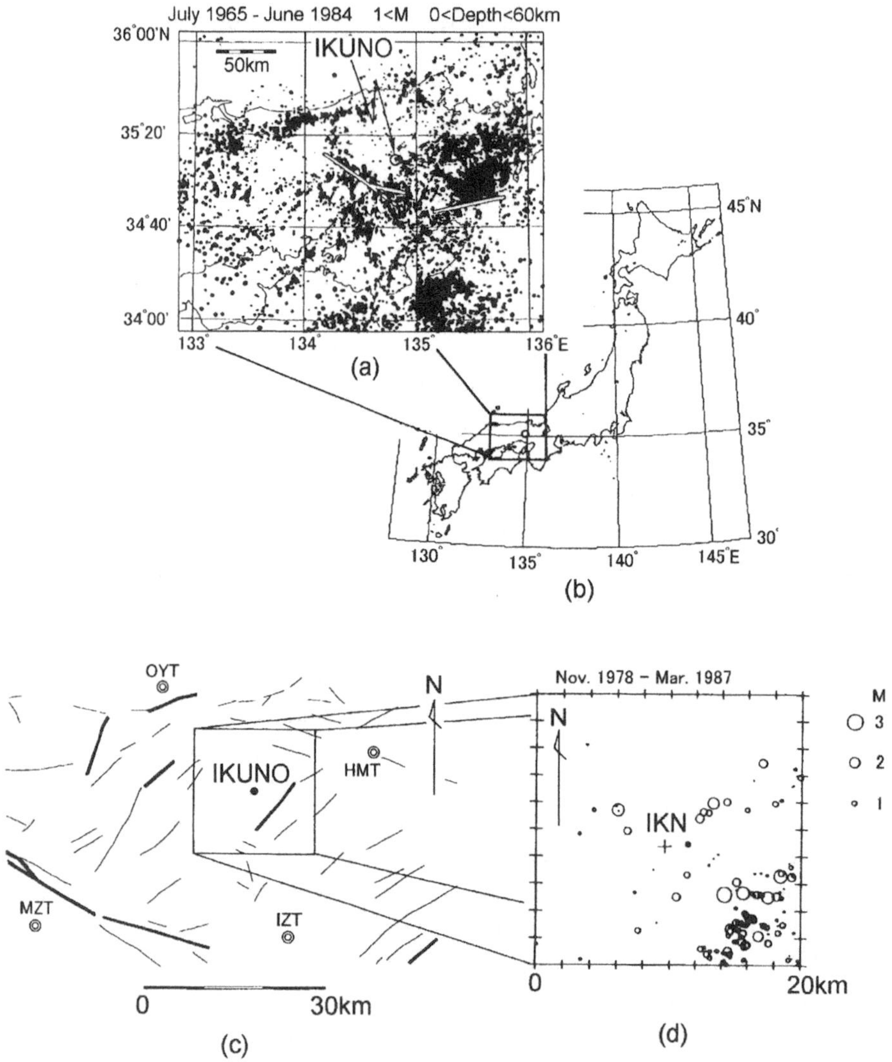

Figure 1
Location of the Ikuno mine, background seismicity and active faults. (a and b) Location of the Ikuno mine and regional seismicity around the mine. (c) Active faults (thick lines) and lineaments (thin lines). Microseismic monitoring stations of the Disaster Prevention Research Institute, Kyoto University are shown by double circles. (d) Enlarged view of the distribution of microearthquakes around the Ikuno mine. (The microseismicity data from the Disaster Prevention Research Institute, Kyoto University; fault data from the Research Group for Active Faults, 1990).

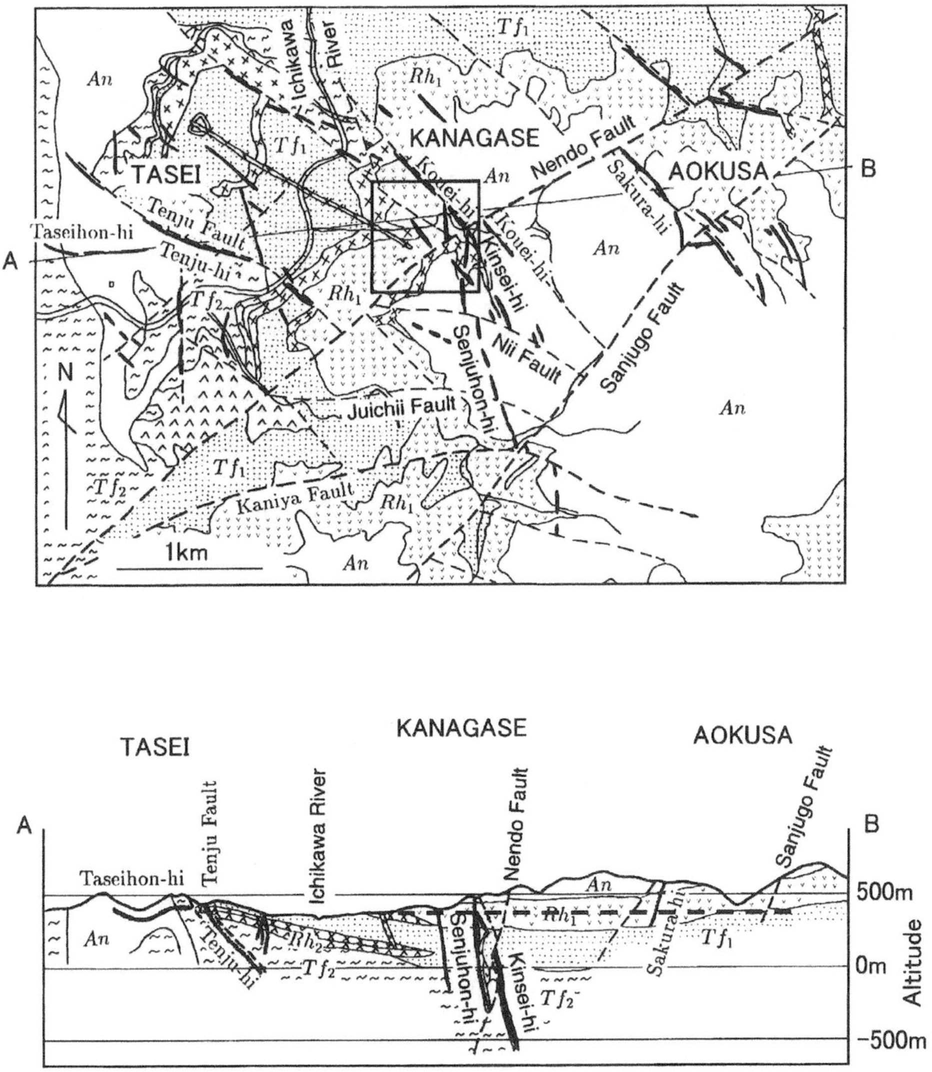

Figure 2

Geology of the Ikuno mine. Upper: plan view, lower: cross section. An, Rh and Tf: andesite, rhyolitic welded tuff and tuff, respectively. The Ikuno mine consists of three areas: Tasei, Kanagase, and Aokusa. The Kanagase area has the deepest ore veins. Note that ore veins are offset significantly by the Nendo, Nii and Sanjugo faults. The 750 m × 750 m square in the Kanagase area in the upper figure shows where we installed instruments (see Fig. 3). A horizontal dashed line in the lower figure represents an underground level where a geophone array is deployed and a crustal movement monitoring station is located.

seismometer with a rotating-drum recording system and counted the events with short S-P times. While the mine was being flooded, these instruments detected the water-induced earthquakes, the maximum magnitude being about three (Fig. 4). More-detailed seismicity was not investigated at the time, however. By 1976, most of

Table 1

Major faults in the Kanagase area, Ikuno mine

Fault	Strike	Dip	Lateral offset (m)	Thrust offset (m)
Nendo	NE-SW	85	100 (left)	40
Sanjugo	NE-SW	85	170 (left)	10
Nii	NW-SE	68	70 (right)	20
Juichii	NW-SE	67	40 (right)	10

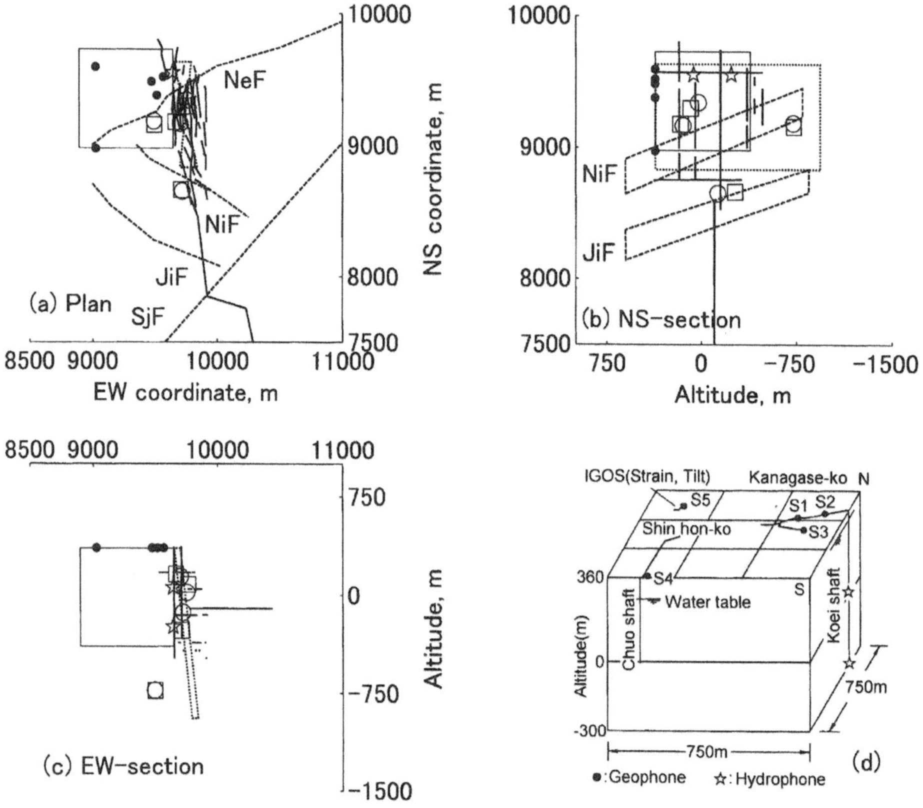

Figure 3

Configuration of faults, flooded tunnels and instruments around the ore veins. Faults are shown by dashed lines in (a) and (b). NeF: Nendo fault, NiF: Nii fault, JiF: Juichii fault, SjF: Sanjugo fault. Flooded tunnels at the deep, mined ore veins are shown by thick, solid lines in (a), (b) and (c). The dotted rectangle in (a), (b) and (c) is the location of the hypothesized tabular cavity discussed in the text as the source of the observed changes in strain and tilt. A 750 m × 750 m square in (a), (b) and (c) represents a cubic area where we installed the instruments shown in (d). Geophones and hydrophones are shown by dots and stars, respectively. Circles and small squares show the location of four events determined by both geophones and hydrophones, and by geophones only, respectively. Extensometers and tiltmeters are installed at the Ikuno Geophysical Observation Station (IGOS). The level of water flooding the mined ore veins was measured in the Koei shaft.

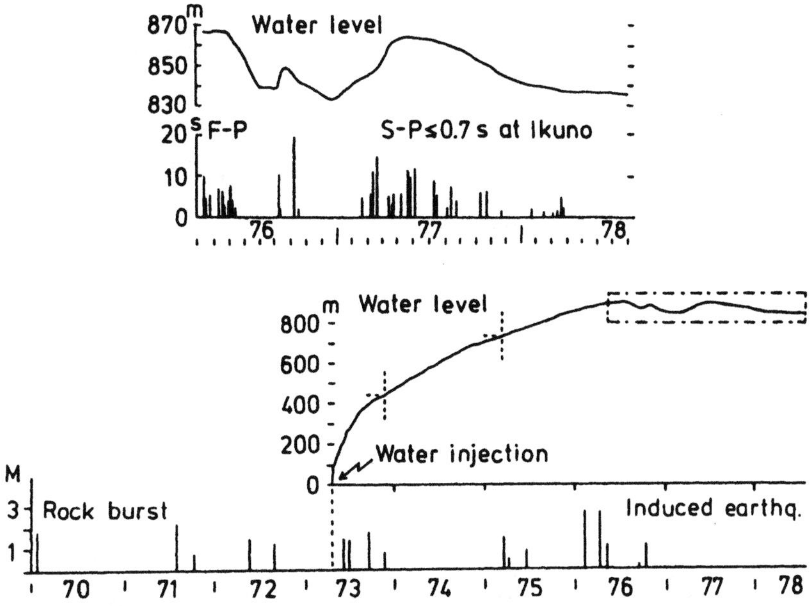

Figure 4
Water table elevations and induced earthquakes from 1970 to 1978 (after TANAKA and OKA, 1979). (Lower): The water level above the bottom of the excavation and the magnitude of rock bursts and earthquakes. (Upper): Enlarged. The Ikuno mine was closed down in 1973 and experienced large induced earthquakes after most of it flooded by 1976; from 1977 on, seismicity was associated with a rise in the water table from approximately 835 m to some 860 m.

the excavated area was flooded. To prevent the leakage of polluted water from a number of '*raccoon dog's holes*', drain pumps were installed and began to lower the water table to about 50 m below the level of the Ichikawa River (Fig. 2), the seismicity then decreased.

3. Rainfall and the Underground Water System

The level of the underground water that floods all the deeply excavated cavities in the Kanagase area (Figs. 2 and 3) is measured at the Koei shaft (Fig. 3(d)) by an engineer with a measuring device once a day. Water is then drained by pumps in this shaft, if the water level is relatively high. The drain pumps operate for several days, hence we can see variations in the water level with an amplitude of a few meters and making plots that look like saw teeth when it rains slightly (Fig. 5(f)). Since the Ikuno mine has open pits as well as numerous historic '*raccoon dog's holes*', rainfall can easily flow into the excavated cavities. Heavy rainfall of several tens of millimeters per day (Fig. 5(c)) has sometimes caused rapid elevation of the water level (Fig. 5(e)) that the pumps could not reduce. The resulting elevation of the water level has been as high as several meters (Fig. 5(f)).

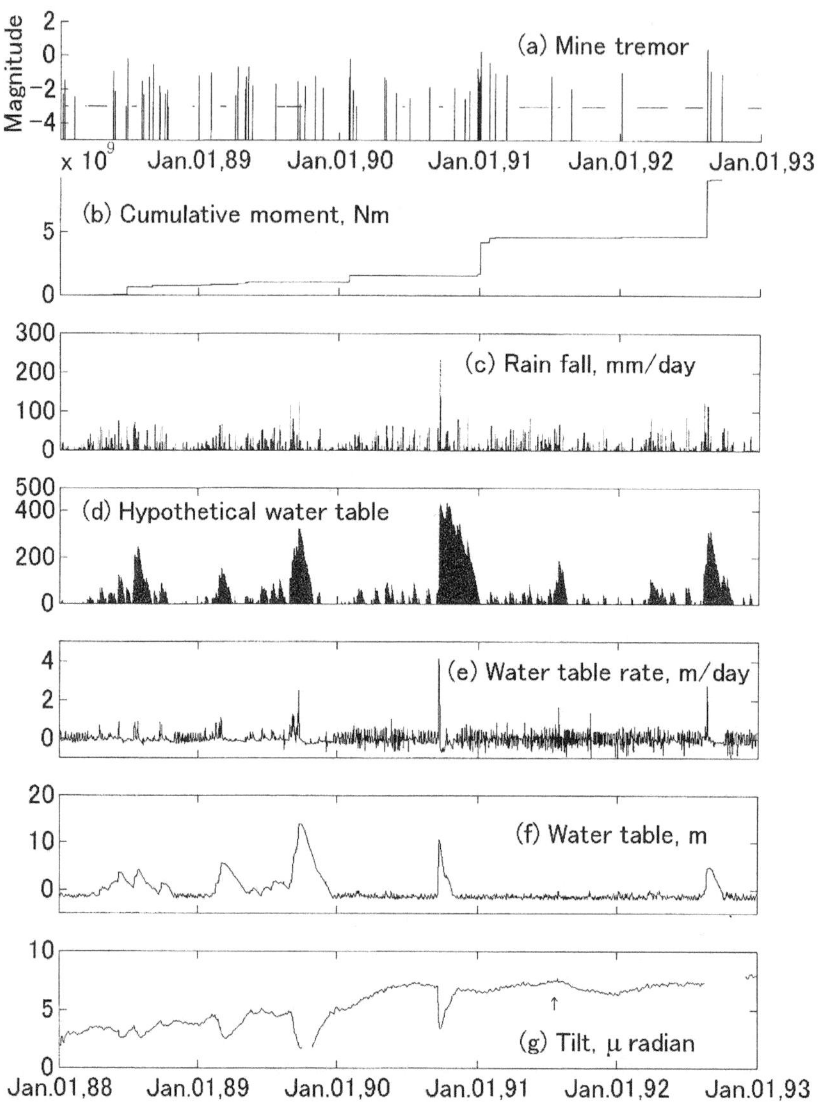

Figure 5

Relationship between Seismicity (a), cumulative seismic moment (b), rainfall (c), underground water table (f) and tilt (g) for five years starting in 1988. Synthetic water level is shown in (d) for comparison, obtained by assuming the accumulation of rainfall and a constant drainage rate best-fit to the observed water table fluctuation until 1990, just before the drain pumps were replaced by more efficient ones. Rate of water table rise is calculated from (f) and is also shown in (e). Note the immediate, significant response of the tilt to the water level, and the slow, small response of the tilt to the large cumulative rainfall (d) in the middle of 1991 (shown by an arrow), which continued for several months.

To understand how the water table responded to rainfall, we compared the observed water table (Fig. 5(f)) with a hypothetical water table (Fig. 5(d)). To obtain the hypothetical water table (Fig. 5(f)), we first subtract the pumps' effects from the cumulative rainfall, assuming that the pumps are operating only during positive cumulative rainfalls. A constant rate of pump drainage was assumed, in order to get a best-fit of the hypothetical water table up to 1990. As seen in Fig. 5(d), the hypothetical water table matches the observed water level very well until the end of 1989. Therefore, we can regard the water system that floods the ore veins basically as a huge reservoir with input from rainfall and output by pumping. The pumps were replaced by more efficient ones in 1990, so the subsequently observed water level (Fig. 5(f)) has been relatively low, compared with the hypothetical water table (Fig. 5(d)). There was no significant elevation of water level in the middle of 1991, despite rainfall as significant as in 1989, owing to the more efficient pumps.

4. Strain and Tilt, and its Relation to the Water System

a) Observations and Data

The Ikuno Geophysical Observation Station (hereinafter IGOS; Fig. 3(d)), which monitors crustal movement, is several hundred meters to the WNW of the flooded vertical ore veins, and in hard rock (rhyolitic welded tuff) with an overburden of hard rock about 40 m thick. In IGOS, four extensometers (EXT) with super-Invar pipes 5–10 m long, and water tube tiltmeters (WTT) with four pots at intervals of 10 m, are installed in a redundant configuration to reliably monitor the strain and tilt field most. Four horizontal pendulum-type tiltmeters (HPT) are also in operation however we have used their records only to check the sense of tilt observed by the WTTs, because the HPTs tend to be significantly affected by local effects. Strain and tilt have been monitored with resolution sufficient for detecting earth tides as small as 10^{-8} (TABEI *et al.*, 1985) and calibrated well with theoretical earth tides (KOYAMA, 1988).

b) Elastic Response to Water Level

Figure 5(g) shows tilting in the N28W–S28E direction. Comparing Figure 5(g) with Figure 5(f), it is clear that the tilt correlates well to fluctuations of the water table. Also note the tilt in the middle of 1991 (an arrow in Fig. 5(g)); a significant increase is seen in the hypothetical water table (Fig. 5(d)), while no *prima facie* changes are seen in either the water table or the tilt. Thus the tilt always responds to the water level, although not always to the hypothetical water level. Because the water table reflects water pressure, elastic response to the increase in water pressure can account well for such changes in tilt. Since a similar correlation with the water table can be seen in strain, the elastic response can also account well for the change in strain.

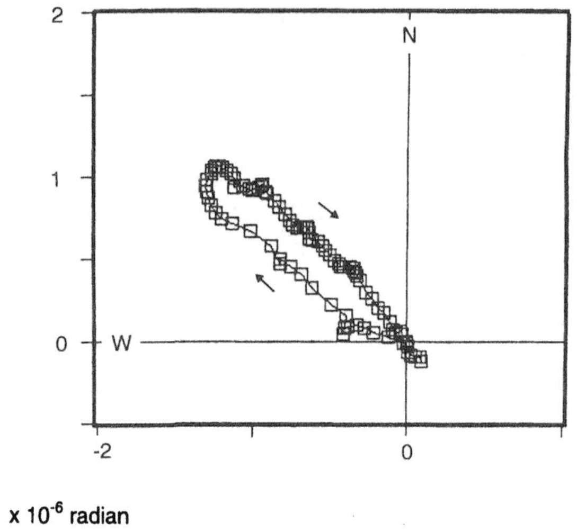

Figure 6
Typical example of tilt response to significant elevation of the water level (from February 16 to May 16, 1989). The direction of maximum tilt and the magnitude of the tilt are represented by a vector from origin to square plotted for each day. Immediately after the rainfall of May 16 a tilt began toward the northwest with an increase in the water level, and a reversal of the tilt with a decrease of the water level. Since the deep ore veins are located southeast of IGOS, uplift is required at the ore veins to account for the observed tilt.

Figure 6 shows a typical example of the tilt response to a significant elevation of water level for about four months in 1989. It rained 81 mm in three days starting February 16, and 238 mm in ten days starting February 24, consequently the water level was elevated about 6 m higher than usual. The ground tilted down to the northwest along with the elevation of the water table, and recovered as the water table lowered.

Strain changes associated with the rainfall in February 1989 are shown in Table 2. The sense of strain is different in different directions, suggesting that significant shear strain occurred in association with the rainfall.

As shown in Table 2, subsidence measures by water weight will lead to tilt or extension in a southeast direction, because the ore veins are located to the southeast of IGOS. Therefore, the weight of water cannot account for the tilt and strain.

Here we will consider if the opening of a vertical tabular cavity in the deepest ore veins can account for the observed strain and tilt, because rainfall will increase water pressure and the associated opening will produce vertical deformation on the surface, as shown in Figure 7. The strain and tilt were estimated by dislocation theory and computer code by OKADA (1992). The widest tabular cavity (1300 m deep, 800 m long and 1300 m wide) that covers all areas of the deepest veins and hypocenters was initially used for estimation, as denoted by a dotted square in Figures 3(a–c). By changing the depth of the upper edge of the tabular cavity, we tried to find the best-case solution. The best-case solution thus obtained is shown in the rightmost columns in Table 2. Because the correlation coefficient is 0.86, we conclude that a cavity with its upper edge at an altitude of 210 m and an opening of 1.6 cm is the best model accountable for the observed strain and tilt.

Table 2

The observed strain and tilt ($\times 10^{-6}$) *and comparison with the synthesized ones. One of the assumed tabular cavities (1300 m deep, 800 m long and 1300 m wide; upper edge at a depth of 360 m) is shown in Figure 3 by a dotted line*

	Observed	Subsidence model	Cavity opening model		
			Opening Upper edge	0.39 cm 360 m	1.4 cm 110 m
Tilt (W down)	1.4	−		0.22	0.26
Tilt (N down)	1.1	−		0.24	0.91
Strain (N28W)	−0.3	+		−0.8	−0.33
Strain (N73W)	−0.3	+		−0.09	−0.02
Strain (N17E)	0.42	−		−0.12	0.9
		Correlation coefficient =		0.52	0.86

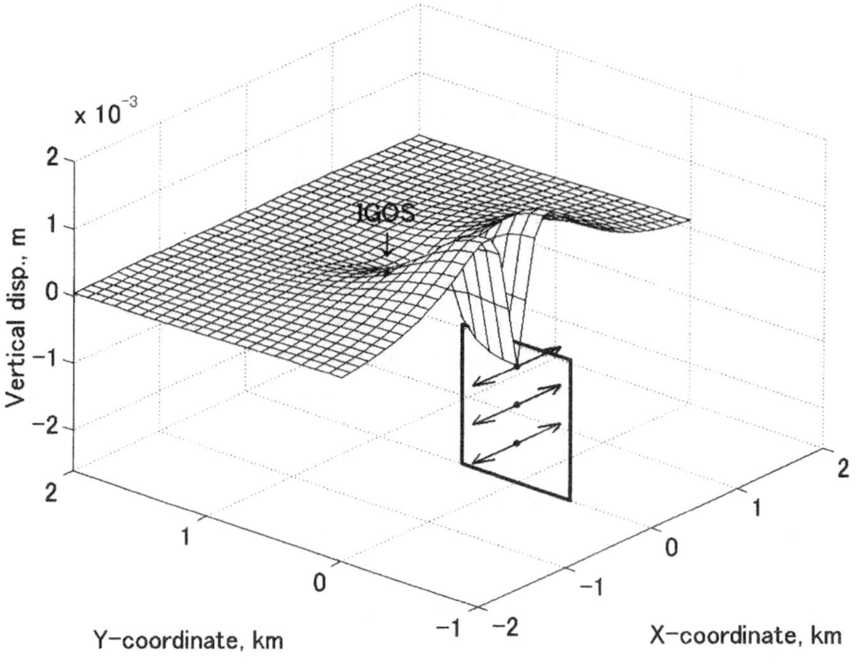

Figure 7
Schematics showing a model that accounts for uplift in the direction of the ore veins. If we assume that a vertical tabular cavity opens, and that IGOS is located as shown in this figure, then uplift in the direction of the ore veins is observed at IGOS.

c) Slow and Delayed Response to Rainfall

If we attend more to tilt in the middle of 1991 (an arrow in Fig. 5(g)) however, there was a gradual change in tilt that steadily decreased by about 2×10^{-6} until the end of 1991 and recovered in April 1992. We also noted such changes starting after heavy rainfalls and continuing for several months. Therefore, we think that the diffusion of underground water may be responsible for these changes. Most cavities made by mining had already been filled with sand or slime, and the mine has faults with high permeabilities. Consequently, poroelastic response associated with the diffusion of water may be responsible for the gradual change and recovery.

5. Seismic Events Associated with Heavy Rainfall

a) Monitoring and Location of Mine Tremors

Seismic events in the mine have been monitored since 1987 using geophones at five sites, and temporary hydrophones in the flooded Koei shaft (Figs. 3(d); OGASAWARA *et al.*, 1992) using a PC recording system with a 12-bit, 2-kHz sampling A/D board. All geophones (Mark Products L22-D) were installed on intact rock (rhyolitic welded tuff) in haulage tunnels at the 360-m altitude level (shown by a dashed line in Fig. 2 (lower)) with an overburden of intact rock of from a few tens to a hundred meters. Before installing geophones we struck the rock with a hammer to carefully ascertain that it was neither floated nor fissured. All geophones were glued to the rock with cement. A 3-component geophone was installed at site S3 and only vertical components at other sites. As shown in Figure 8, events with magnitudes as low as −2.5 were clearly detected by the system. Events that simultaneously triggered at least three geophone sites were recorded and used for analysis. Hydrophones were not used as a trigger source, so there was no difference in the trigger threshold of the recording system whether the hydrophones were in operation or not.

Errors in *P*-reads were within a few ms. *S*-reads at site S3 and other available sites were used for hypocenter locations. $V_p = 4.65$ km/s was determined by blasting at a quarry about 1 km to the west of our seismic array. A homogeneous isotropic elastic medium and $V_s = V_p/\sqrt{3}$ was asumed for the hypocenter location. A hydrophone at a deeper level in the Koei shaft (at an altitude of −300 m; Fig. 3) was in operation for only a month from the end of December 1987, and the shallower one for a considerably shorter period. However, one event was recorded by 2 hydrophones and 4 geophones, and three more events were recorded by a deeper hydrophone and 5 geophones. Using these records we determined station corrections to minimize residuals. The maximum station correction was about 10 ms for *P*-reads for geophone sites within ore veins. Corrections for hydrophones were plus or minus 5 ms, being independent of depth. The locations of the four events determined with

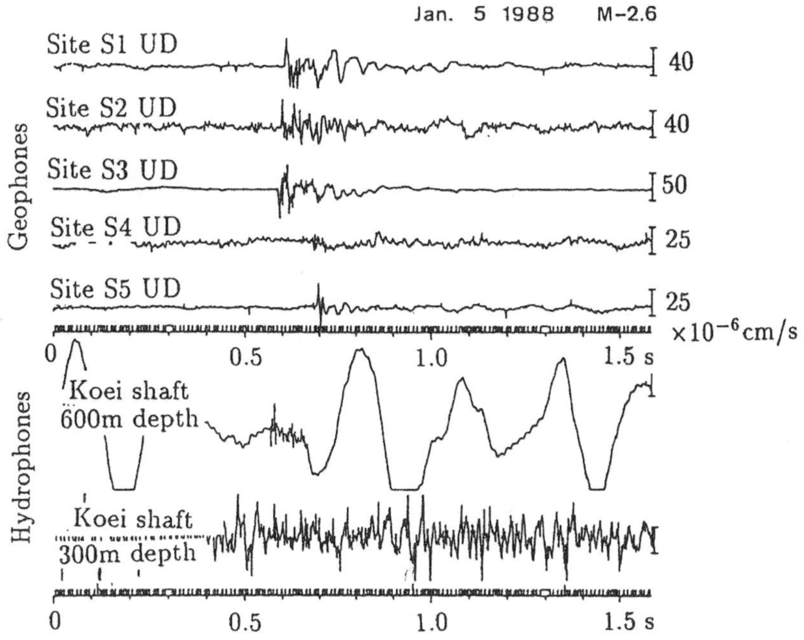

Figure 8
Examples of seismograms ($M = -2.6$). Refer to Fig. 3(d) for the instrument configuration.

station corrections and by the hydrophones are shown by circles in Figs. 3(a–c), with all events in locations relevant to the present study. With respect to these, the hypocenters determined only from station corrections and geophones (the squares in Figs. 3(a–c)) do not differ by several tens of meters. Consequently, a homogeneous structure is a reasonable velocity model with acceptable accuracy for locations in this study if we take into account the station corrections. Since only four events occurred during hydrophone operation, most events were located using only geophone data and station corrections. The magnitude was determined using maximum velocity amplitude according to the empirical relationship of WATANABE (1971).

b) Relationship Between Seismic Events and Geology

A total of 56 events were detected in the Ikuno mine over a five-year period (Fig. 5(a)), the largest one about $M = 0$. It is important to note that all events occurred in the Kanagase area within deep ore veins (Fig. 9), while no seismic events occurred within shallow ore veins in other areas, such as Tasei or Aokusa (Fig. 2).

Seismicity in the Kanagase area can be characterized as follows:

(1) The area with the deepest ore veins, between the Nendo and Nii faults (NeF and NiF in Fig. 9(a)), is the most active.

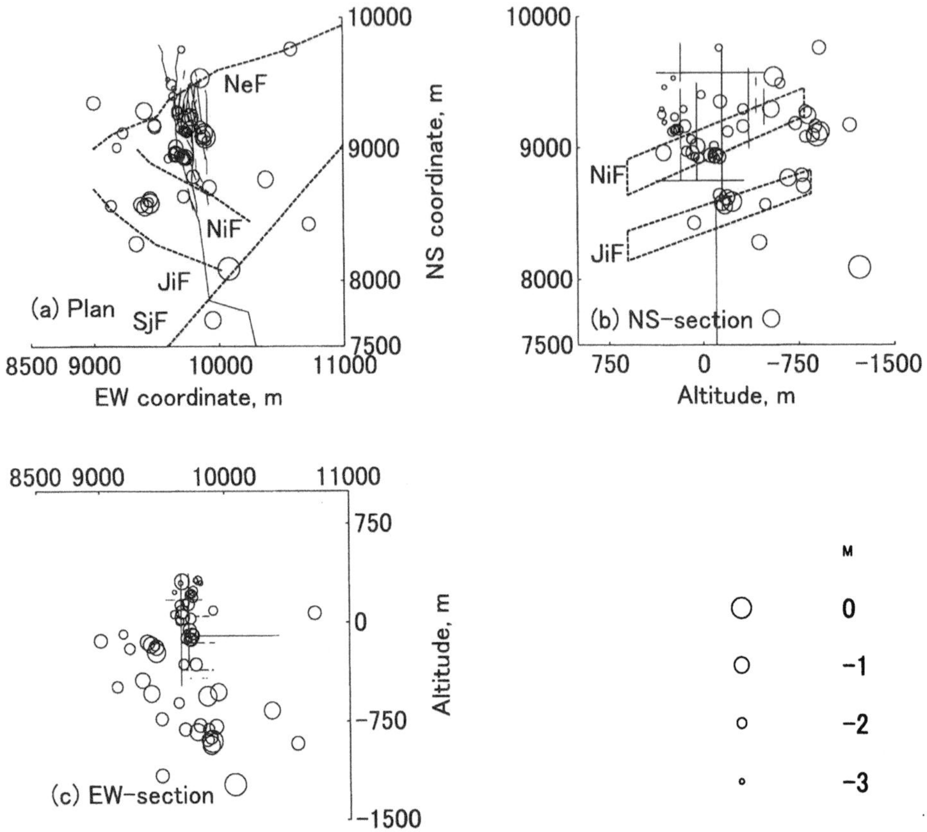

Figure 9
Relationship between hypocenters (o), faults (thick, dashed lines) and flooded tunnels through major ore veins (thin veins). NeF: Nendo fault, NiF: Nii fault, JiF: Juichii fault, SjF: Sanjugo fault. (a): Plan view of epicenters and surface traces of faults. (b) and (c): Hypocenters projected on N-S cross section and an E-W cross section, respectively.

(2) The highest activities occur at an altitude of about 0 m, where the Nii and Juichii faults offset the ore veins (see NiF and JiF in NS-section; Fig. 9(b)).
(3) The events occur at an altitude of about −750 m, substantially deeper than the excavated veins and near the faults (Fig. 9(b));
(4) Seismicity occurs around the upper edge of a vertical ore vein, although it consists of small events (Figs. 9(b and c));
(5) Activity can be observed along the Nendo fault (Fig. 9(a)).

c) Composite Focal Mechanism

Figure 10 displays the composite focal mechanism of all 56 seismic events shown in Figure 9. In the case of distant hypocenters in layered structures with significant differences in velocity, direct arrivals will generally depart downward from sources,

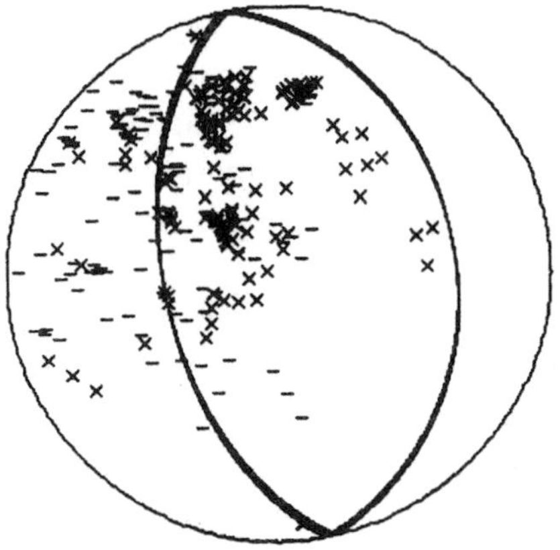

Figure 10
Composite focal mechanism for events in the Kanagase area. The composite pattern of initial push-pull of the *P* wave for all events detected over a 5-year period is represented by '+' and '–', respectively, on the upper hemisphere of the Wolf net. We assumed that direct arrivals departed from sources upward because the geophone arrays were deployed on underground intact rock and velocity seems not to dependent on depth. The mechanism is not for a single event, therefore the nodal planes are drawn so as to maximize the difference between the number of cases of push and pull.

and an inappropriately assumed source depth or take-off angles could lead to misunderstandings of the focal mechanism. However, the scope of our interest can be regarded as primarily a homogeneous velocity structure; our geophone array is deployed in underground intact rock, and hypocenters are near and below our array. We therefore assumed that all direct arrivals departed sources in an upward direction and came straight toward our array.

Figure 10 indicates thrust faulting consistent with the nearly horizontal E-W compressive tectonic stress state (TANAKA *et al.*, 1989) Different mechanisms may be present in the composite mechanism and we can only observe the average stress state that induces seismicity. Nonetheless, the E-W compression is inconsistent with many events which occur around the upper edge of the vertical cavity, if we assume the opening of a vertical cavity as a cause of seismicity. This is because an E-W tensile stress state is expected around the edge of the cavity, as inferred from Figure 7. Consequently, we suggest that elastic response cannot account for the composite mechanism.

d) Correlation of Seismicity with the Amount of Heavy Rainfall

Figure 5(a) shows the magnitude of each event plotted against time. The horizontal lines in Figure 5(a) represent periods with no available data. Although the periods without data are not short, the correspondence of seismicity to the rise in the water table Figure 5(f) can be seen *prima facie*. However, the seismicity corresponds imprecisely to the water level, since event frequency does not always peak when the water level peaks. The frequency increases with some delay from the onset of heavy rainfall, or the peak water level, and the event frequency will continue to be high even after any significant changes in water and strain have ceased. This clearly shows that the elastic response of the ground to the water load is not directly responsible for the seismicity.

Figure 5(b) shows the cumulative seismic moment versus time. The seismic moments were calculated from magnitudes according to the magnitude-moment relationship of HANKS and KANAMORI (1979). As can be seen in Figure 5(b), larger events appear to occur associated with heavy rainfall, although with significant delay.

Figure 11 represents the hypocenter distribution with diamonds scaled by time lapse since the onset of heavy rainfall. Interestingly, all the events within a few

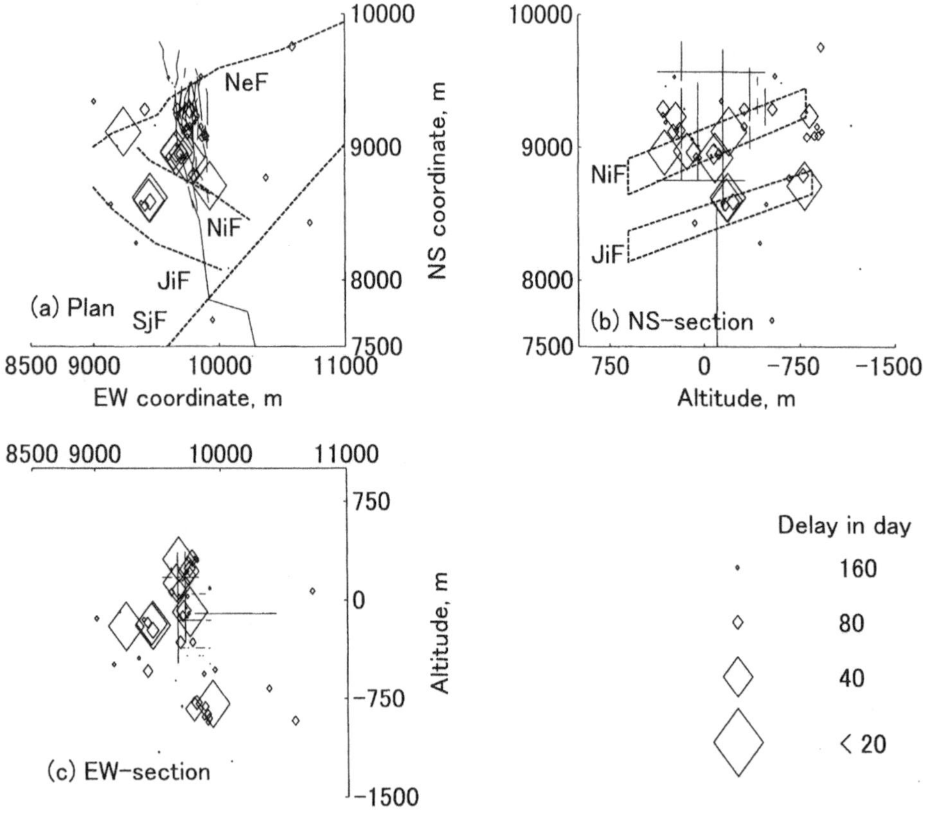

Figure 11
Relationship between the delay in time after heavy rainfalls and the location and depth of seismicity. Diamonds: hypocenters scaled by the delay of the seismic event compared to the time of heavy rainfall. Dashed line: faults, thin line: flooded tunnels near ore veins. NeF, NiF, JiF and SjF: Nendo, Nii, Juichii and Sanjugo faults, respectively.

months after the heavy rainfall occurred near the faults that offset the deepest ore veins, whereas all those located away from the deepest ore veins took place many months after the heavy rainfall. The shallower the hypocenters, the faster the response. Consequently, it is strongly suggested that diffusion controls the occurrence of the events.

6. Conclusions

Fifty-six microearthquakes recorded over a 5-year period were located around the mined and flooded, vertical ore veins that extend to 1 km deep (Figs. 2, 5(a), 9 and 11). During this 5-year period, heavy rainfalls of several tens of millimeters elevated the water table several times and flooded the vertical ore veins to a depth of several meters (Figs. 5 (c) and (f)). Significant changes in strain and tilt larger than 10^{-6} (Fig. 5(g)) were also monitored at the crustal movement monitoring station located several hundred meters from the veins. By simply multiplying strains larger than about 2×10^{-6} by an elastic modulus of about 50 GPa, the strain changes corresponded to stress changes of about 0.1 MPa. We infer that the opening of the deep tabular cavities led directly to the significant strain and tilt we observed (Fig. 3 and Table 2), but not directly to the seismicity, for the following reasons:

(1) The seismicity seemed to respond to water level elevation as well as to strain and tilt, although the delay in the seismicity was significant, and the seismicity tends to persist even after the change in strain and tilt had ceased.

(2) After more-efficient drain pumps were introduced, the seismicity seemed to respond to heavy rainfall, even though elevation of the water table did not take place.

(3) Associated with heavy rainfalls, the moderate and slow changes in strain and tilt corresponding to a stress change of 0.1 MPa ($= 2 \times 10^{-6} \times 50$ GPa) persisted for several months, suggesting the diffusion of water underground.

(4) Some seismic events had focal mechanisms inconsistent with the stress state expected from the opening of a vertical tabular cavity.

The faults in the Ikuno mine have preferred orientations in accord with seismic events in the present stress state; direct elastic response to the load or water pressure was significant and larger than 0.1 MPa. However, this direct rise in pressure did not stimulate seismicity around the faults in the Ikuno mine, probably because the faults were not optimally oriented with respect to the stress increase. Consequently, the localized and delayed diffusion of water played the dominant role in stimulating seismicity in the Ikuno mine, though the change in pore pressure was slow and decidedly less significant than the direct change in stress due to an increase in the water load.

Acknowledgments

We express our sincere thanks to Mitsubishi Material Co. Ltd., Silver Ikuno Co. Ltd. and Techno Ohte Co. Ltd., who kindly gave us permission to carry out our multidisciplinary observations in the closed mine and assisted in the continuation of the observations. We are grateful to Mrs. S. Ogata and H. Matsumoto of Techno Ohte Co. Ltd., who kindly supplied us with water table data for the Koei shaft and assisted us in carrying observations. We are grateful to Mr. S. Kotani of Mitsubishi Material Co. Ltd. and to colleagues at Kyoto University, who have helped us to conduct the multidisciplinary observations.

Lastly, we thank Dr. C. Trifu and two anonymous reviewers for their kind and valuable suggestions for improving our manuscript.

References

Evans, D. M. (1966), *Man-made Earthquakes in Denver*, Geotimes *10*, 11–17.

Fletcher, J. B. and Sykes, J. R. (1977), *Earthquakes Related to Hydraulic Mining and Natural Seismic Activity in Western New York State*, J. Geophys. Res. *82*, 3767–3780.

Hanks, T. C. and Kanamori, H. (1979), *A Moment Magnitude Scale*, J. Geophys. Res. *84*, 2348–2350.

Healy, J. H., Rubey, W. W., Griggs, D. T., and Raleigh, C. B. (1968), *The Denver Earthquakes*, Science *161*, 1301–1310.

Koyama, T. (1998), *An Interpretation of Tidal Change of Water Level at the Chuo Shaft at Ikuno Mine, Japan.* Master's Thesis, Graduate School of Science and Engineering, Ritsumeikan Univ. (in Japanese).

McGarr, A. and Simpson, D., *A broad look at induced and triggered seismicity*. In *Rockbursts and Seismicity in Mines* (eds Gibowicz, S. J. and Lasocki, S.) (Balkema, Rotterdam 1997) pp. 385–396.

Nishida, R. and Tanaka, Y. (1972), *Ground Tremors Caused by Rock Bursts in the Ikuno Mine* (continued), Disast. Prev. Res. Inst., Kyoto Univ. 15-B, 43–51 (in Japanese with English abstract).

Ogasawara, H., Fujimori, K., Koizumi, N., Fujiwara, S., Nakao, S., Nishigami, K., Taniguchi, K., Otsuka, S., Hirano, N., Nishida, R., and Iio, Y. (1992), *Comprehension Observatiion System at Old Ikuno Mine for Earthquake Prediction*, Ann. Dissast. Prev. Res. Inst., Kyoto Univ. 35B-1, 389–400 (in Japanese with English abstract).

Okada, Y. (1992), *Internal Deformation Due to Shear and Tensile Faults in a Half Space*, Bull. Seismol. Soc. Am. *82*, 1018–1040.

Raleigh, C. B., Healy, J. H., and Bredehoeft, J. D. (1976), *An Experiment in Earthquake Control at Rangeley, Colorado*, Science *191*, 1230–1237.

Simpson, D. W., Leith, W. S., and Scholz, C. H. (1988), *Two Types of Reservoir-induced Seismicity*, Bull. Seismol. Soc. Am. *78*, 2025–2040.

Simpson, D. W. and Negamatullaev, S. Kh. (1981), *Induced Seismicity at Nurek Reservoir*, Bull. Seismol. Soc. Am. *71*, 1561–1586.

Tabei, T., Fujimori, K., and Tanaka, Y. (1985), *Observation of Crustal Movements at Ikuno (1977–1983)*, J. Geod. Soc. Jpn. *31*, 189–201 (in Japanese with English abstract).

Tanaka, Y. and Nishida, R. (1971), *Ground Tremors Caused by Rock Bursts in the Ikuno Mine*, Bull. Disast. Prev. Res. Inst., Kyoto Univ. 14-A, 149–164 (in Japanese with English abstract).

Tanaka, Y. and Nishida, R. (1977), *Generation mechanism of rock bursts and microearthquakes induced by water-injection into a deep abandoned mine*, Proc. 5th Jpn. Symp. Rock Mech. 91–96 (in Japanese).

Tanaka, Y. and Oka, Y. (1979), *Generation Mechanism of Rock Bursts and Water-induced Earthquakes under the Tectonic Stress Field*, Rock Mech. Jpn. *3*, 71–73.

TANAKA, Y., FUJIMORI, K., and TAKEUCHI, T. (1989), *Comparison Between Measured Crustal Stress Changes and Observed Crustal Strain Changes*, Ann. Disas. Prev. Res. Inst., Kyoto Univ., 32B-1, 61–73 (in Japanese with English abstract).

WATANABE, H. (1971), *Determination of Earthquake Magnitude at Regional Distance in and near Japan*, Zisin (J. Seismol. Soc. Jpn.), Ser. 2, 24, 189–200 (in Japanese with English abstract).

Wetmiller, R. J., Galley, C. A. and Plouffe, M., *Post-closure seismicity at a hard rock mine*. In *Rockbursts and Seismicity in Mines 93* (ed. Young, R. P.) (Balkema, Rotterdam 1993) pp. 445–448.

(Received March 21, 2000, revised/accepted July 17, 2000)

Pure appl. geophys. 159 (2002) 111–130
0033–4553/02/030111–20 $ 1.50 + 0.20/0

Pure and Applied Geophysics

Moment Tensor Inversion of Regional Phases: Application to a Mine Collapse

V. Teyssoneyre,[1] B. Feignier,[1] J. Šílený,[2] and O. Coutant[3]

Abstract—The aim of our study consists of analyzing potentially non-double-couple seismic events recorded at regional distances. In order to define the nature of the seismic source, a moment tensor inversion is carried out as this method is general enough not to initially constrain the source mechanism. In this paper we present an application to a seismic event induced by a mine collapse which occurred near the town of Halle in Germany. Because of its induced nature, many parameters such as the location and geometry of this seismic source are known. This information allows us to test the influence of inadequate propagation modeling on the moment tensor obtained from the inversion. Green's functions have been computed with the reflectivity method in a flat layered medium, using the European model EurID (Du *et al.*, 1998; Dufumier *et al.*, 1997). From the inversion of *P*-wave seismograms recorded by the German Regional Seismic Network will, we obtained a source time function which can be decomposed into two subevents. The first one has a large isotropic part and a deviatoric mechanism with near vertical nodal planes. No volume change is observed for the second subevent, but a deviatoric component opposite of the first one. The addition of *S*-waves does not change the results of the inversion which are stable. Surface waves were not used because of their poor dispersion curves. Based on the moment tensor obtained from these inversions, the physical process at the source is compatible with a large cavity collapse.

Key words: Induced seismicity, regional phases, moment tensor inversion, isotropic sources.

Introduction

Moment tensor inversion is a widely used method to retrieve the nature of seismic sources. In most studies the solution is forced to solve for a double-couple mechanism which is adequate for most natural events. Nevertheless, many authors have also outlined non-double-couple behavior. Furthermore, in specific cases such as volcanic or induced events, several studies have pointed out a non-double-couple behavior of the seismic sources. For instance, Julian *et al.* (1997) have observed a significant implosive component in the Hengill-Grensdalur volcanic complex. According to these authors, this negative volume change was probably related to a

---

[1] Laboratoire de Détection et de Géophysique, CEA, BP 12, 91680 Bruyères le Chatel, France.
[2] Geophysical Institute, Academy of Sciences of the Czech Republic, Bocni II/1401, 1401,141 31 Praha 4, Czech Republic.
[3] Laboratoire de Géophysique Interne et Tectonophysique, BP 53, 38041 Grenoble Cedex 9, France.

cavity collapse. McGarr (1992) has observed significant isotropic components on some cases of mining-induced events in South Africa. Finally, in the field of nuclear explosion, Vasco and Johnson (1989) used moment tensor inversion to characterize the clear explosive part in the source mechanism. Thus moment tensor inversion has shown to be an efficient method to quantify the amount of deviatoric and isotropic components present in the seismic source.

Moment tensor inversion has been extensively used at the very local scale (Feignier and Young, 1992; Šilény *et al.*, 1996; Cespuglio *et al.*, 1996) and at the global scale (Julian *et al.*, 1997; Dziewonski and Woodhouse, 1983), however until recently, relatively few studies have applied this method to the regional scale (Nabelek and Xia, 1995; Dufumier *et al.*, 1997). It is essentially since the early 1990s that we see the development of numerous techniques for analyzing the regional signal. Many authors determined source parameters of regional earthquake using long-period waves (Ritsema and Lay, 1995; Nakanishi *et al.*, 1991; Giardini *et al.*, 1993), short-period surface waves (Thio and Kanamori, 1995), or even broadband body waves (Dreger and Helmberger, 1991, 1993), that are more difficult to model due to the complex crustal structure. For some regional networks, such moment tensor inversion methods are now routinely applied to near real-time data (Romanowicz *et al.*, 1993; Thio and Kanamori, 1995; Giardini *et al.*, 1993). The main difficulty of using regional waveforms is that we need a sufficient knowledge of the propagation medium to properly model the Green's functions. In other words, results derived from moment tensor inversion may be difficult to interpret as the potential isotropic nature of the source is, in some cases, due to the improper modeling of the Green's functions. In the present study we present our inversion method applied to a large mine-induced event. We determine the non-double-couple (NDC) component and determine if it corresponds to a real property of the source.

The Event of Halle

The studied event occurred on 11 September 1996 in the closed potash mine of Teutschenthal, near the town of Halle in Germany. An area of 2.7 km^2 collapsed, following the sudden breakage of carnallitite pillars at a depth of 700 m. This collapse induced a seismic event with a moment magnitude of 4.8 (Klinge, 1997). The location obtained from seismological data (BGR Seismological Bulletin) is 55.44 N and 11.84 E. This location corresponds to the observation of ground subsidence with a maximum amplitude of 45 cm (Klinge, 1997). All of this information provides us with the nature, geometry, and location of this event, which is consequently well defined. Therefore, this source is of particular interest in adjusting and testing our waveform inversion methodology.

Waveforms were retrieved from the three-component broadband seismic stations of the German Regional Seismic Network which have a sample frequency of 20 Hz

(Seismic Data Analysis Center of Hannover). The map in Figure 1 shows the location of stations and source. The distance between the source and stations lies between 80 km and 400 km. Using the data from the GRSN, the overall azimuthal coverage of the seismic source is approximately 270 degrees. Nevertheless, all the stations except RGN and BRNL are well distributed in the south-southeast of Halle, corresponding to a coverage of 180°. Waveforms display a very complex behavior, with multiple arrivals and important coda. Several seismograms exhibit large surface waves, as CLL in Figure 2. Generated by waves trapped in the crust, these surface waves are typical of a shallow source.

Wave propagation

Green's functions were computed by the AXITRA software (COUTANT, 1994; COTTON and COUTANT, 1997). This code uses the reflectivity method of MÜLLER (1985), combined with the discrete wave number method of BOUCHON (1979). We first tried to derive a seismic velocity model from the measurements of arrival times for regional waves. Indeed, graphs showing arrival times versus source-station

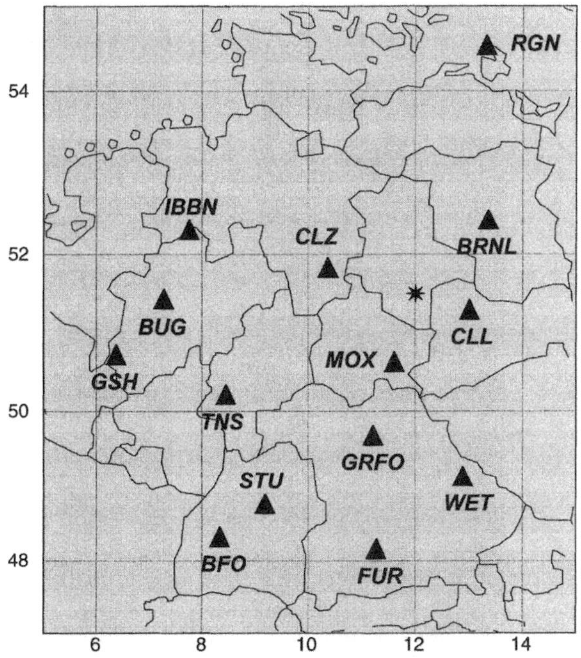

Figure 1
Station locations of the German Regional Seismic Network (triangles) and location of the mine collapse of Halle (star).

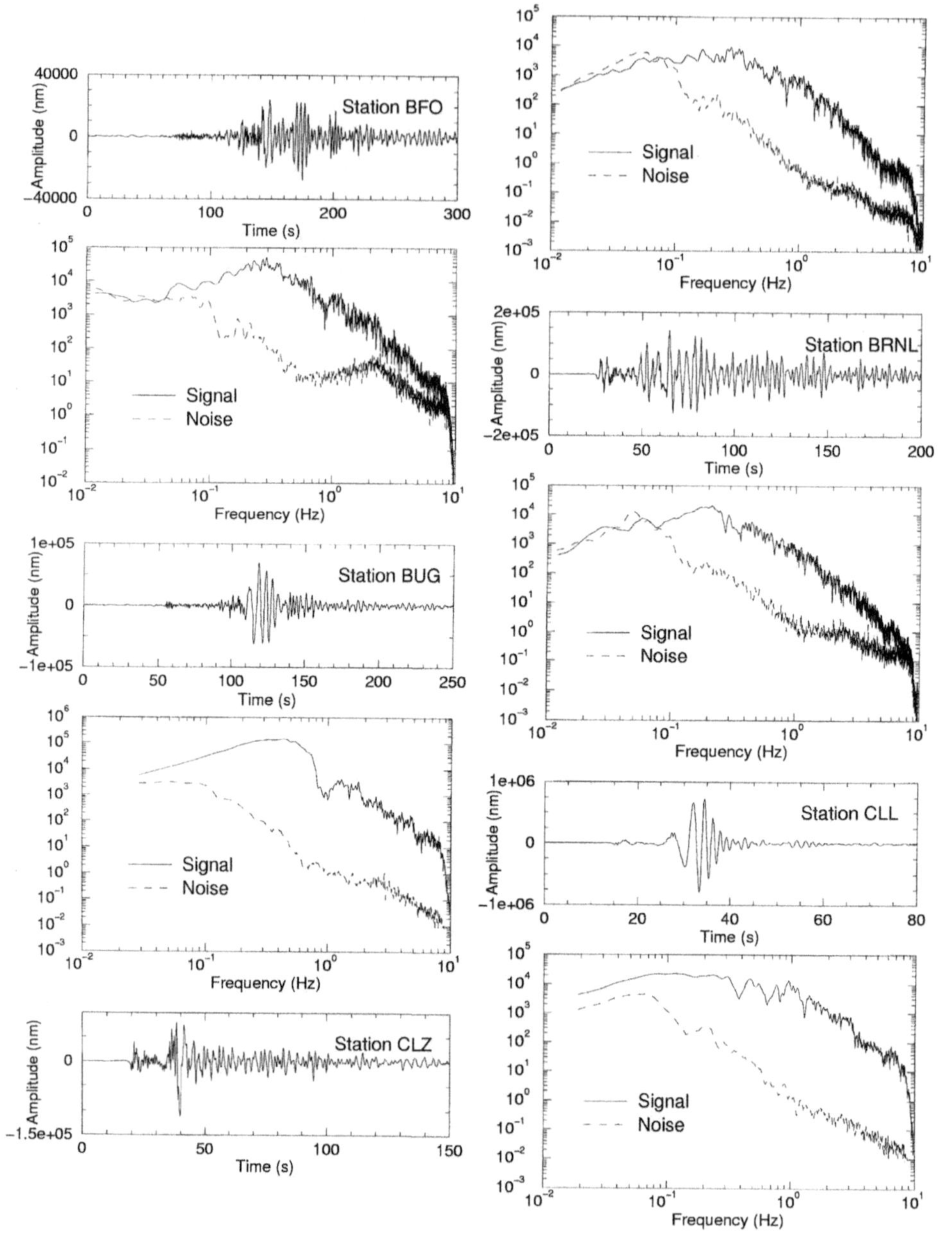

Figure 2

Vertical displacements and spectrums of signal and noise for all stations. Signals are very complex. Surface wave dominates for some stations such as CLL. In the inversion, data will be filtered between 0.1 Hz to 8 Hz where the noise is minor compared to the signal.

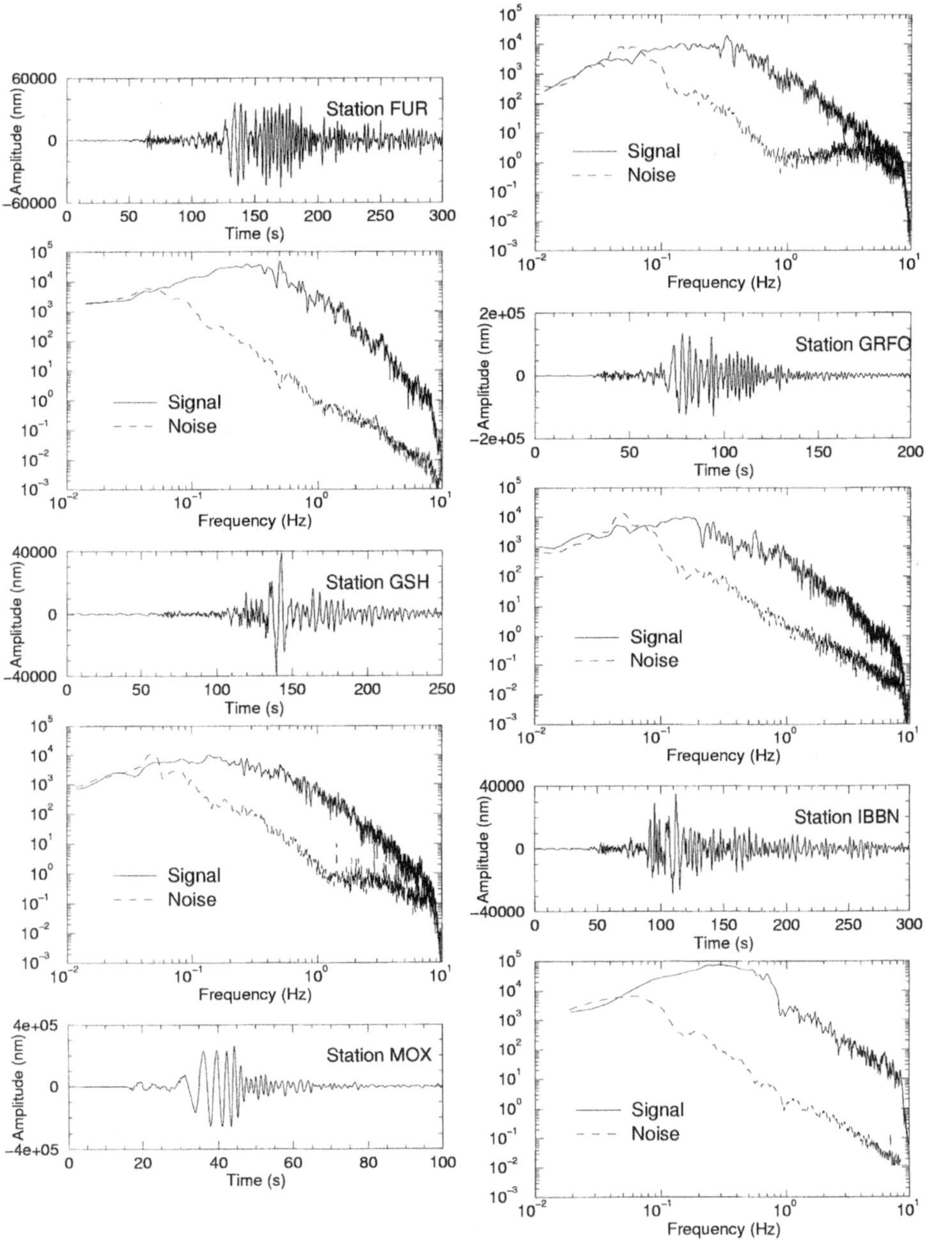

Figure 2
Continued.

distance for P_g, P_n, S_g, and Rayleigh waves, exhibit a linear behavior (Figure 3). Linear regressions give velocity values of 5918 m/s for P_g, 8354 m/s for P_n, 3606 m/s for S_g, and 2954 for Rayleigh waves. Therefore, our first idea was to assume a model

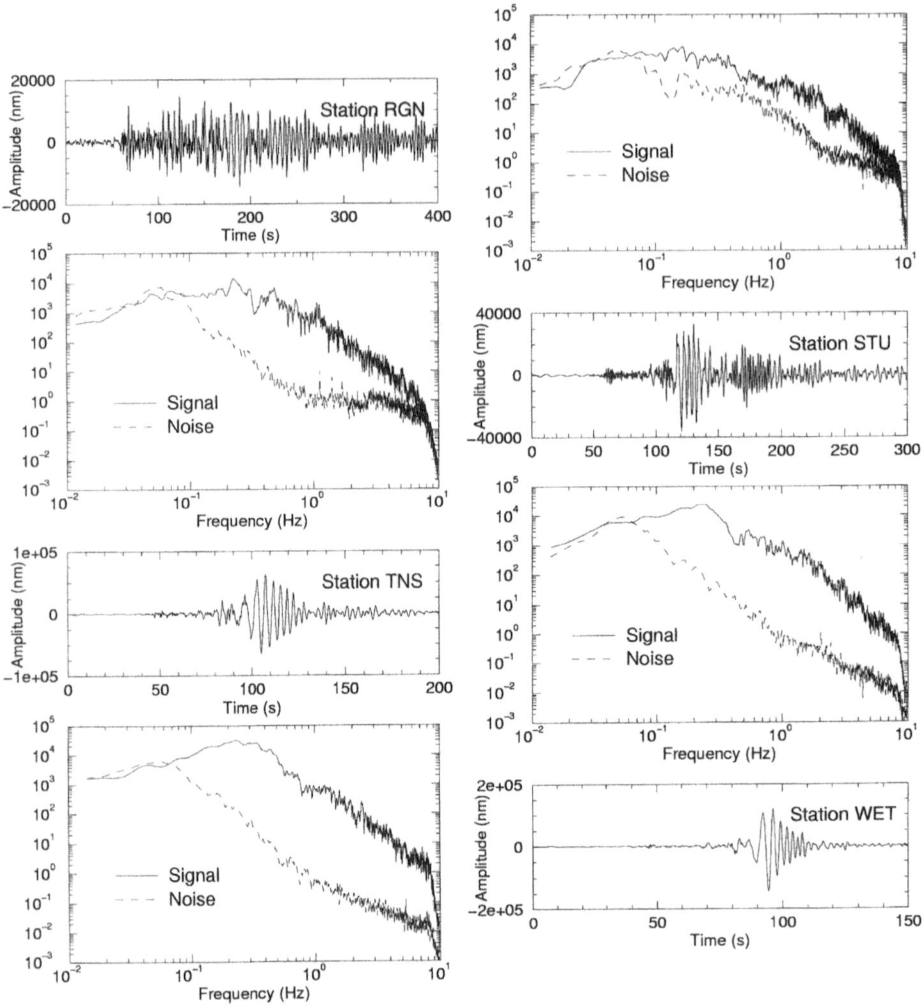

Figure 2
Continued.

made of a unique layer 30 km thick for the crust, above an infinite half-space. However, Green's functions computed with this model could not correctly estimate the complexity of real signals and simulate body and surface waves simultaneously. Another drawback was that this simple model did not give values for quality factors easily, without spectral analysis.

Therefore we looked for a more complex model and decided to compute Green's functions using a flat-layered medium derived from the European model EurID (Du *et al.*, 1998; Dufumier *et al.*, 1997). This model is a compilation of about 6000 profiles for all of Europe, from which 3-D models can be constructed by

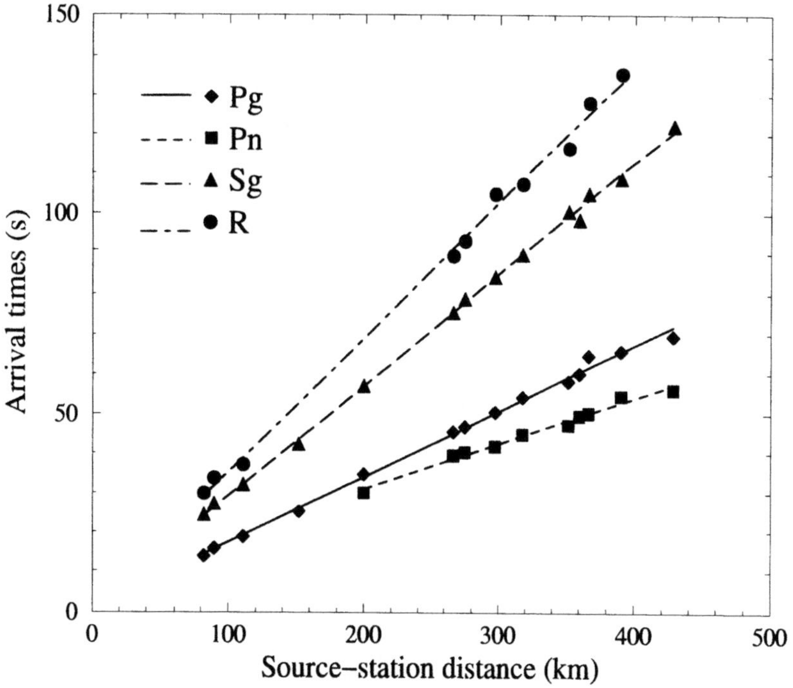

Figure 3
Pickings of P_g, P_n, S_g, and Rayleigh waves. The behavior is linear for various waves. A linear regressions generates velocity values of 5918 m/s for P_g, 8354 m/s for P_n, 3606 m/s for S_g, and 2954 for Rayleigh waves.

interpolation. For each given source-station path, a 1-D model can then be calculated by averaging the 3-D model. Due to the complexity of the waveforms, we decided to take a different model for each station. Table 1 gives an example of such a model for station CLL. For each layer P and S-wave velocities, density, and Q_p and Q_s quality factors are defined. The use of this precise propagation model allowed us to properly fit observed arrival times and waveforms.

Table 1

Velocity model for station CLL

Depth (km)	V_P (m/s)	V_S (m/s)	Density (kg/m^3)	Q_P	Q_S
0–1	2552	1474	2108	88	80
1–3	3257	1880	2124	70	64
3–5	5920	3418	2764	318	
5–7	5998	3463	2784	316	
7–9	6077	3509	2803		
9–13	6156	3554	2823		288
13–17	6413	3702	2891	317	
17–21	6418	3706	2893		
21–25	6424	3710	2895	316	
25–29	6881	3972	3019		

Inversion Method

The inversion code we used was developed by ŠILÉNY *et al.* (1992) following the scheme proposed by SIPKIN (1982), and works on waveforms in the time domain. It is based on an indirect parameterization of the point source: Seismic waveforms are inverted into moment tensor rate functions (MRTFs), which are subsequently reduced into the moment tensor and source time function. Each MRTF is expressed as a sum of N_t delayed overlapping triangle functions $T_{\Delta\tau}$ (NABELEK, 1984) with amplitudes determined by inversion:

$$F_m(t) = \sum_{n=1}^{N_t} F_{nm} T_{\Delta\tau}[t - (n-1)\Delta\tau] \tag{2}$$

where F_{nm} is the amplitude of the nth triangle of the mth moment rate tensor component. Finally, introducing a new index $p = m + 6(n-1)$, the displacement can be written:

$$d_k(t) = \sum_{p=1}^{6N_t} w_p A_{kp}(t) \tag{3}$$

where $w \leftrightarrow F_{nm}$ and $A_{kp}(t) \leftrightarrow T_{\Delta\tau}[t - (n-1)\Delta\tau] * \Phi_{km}(t)$.

Consequently, for all components of all stations, we must solve the equation system $Aw = d$. The inversion code solves the associated normal equation system $(A^T A)w = A^T d$, whose solution minimizes the L_2 norm of the difference between data and predicted displacement. After the inversion we obtain six independent MRTFs, from which synthetic seismograms are computed and compared with the data. MRTFs are then reduced into the moment tensor and the source time function. Thanks to considering the MRTFs as independent time functions in the first step of the procedure, the inversion of seismograms is linear and thus advantageously fast. The nonlinearity of the problem is moved to the second step, reduction of the independent MRTFs, where it is more easily tractable. Indeed, in traditional schemes of the direct parameterization, waveforms must be matched with synthetic seismograms at each iteration, which may be time-consuming because of their number and length. Whereas, with our two step algorithm only the six MRTFs, whose length is usually shorter, are modeled iteratively. Moreover, it was observed that this scheme is advantageous in case the medium is not known in full and thus the Green's functions we must use in the inversion are not completely consistent with the data. The indirect parameterization algorithm was found to perform well in the cases of computing the Green's functions in a medium with simplified inhomogeneities (KRAVANJA *et al.*, 1999), neglecting the anisotropy (ŠILÉNY and VAVRYCUK, 2000), and mismodeling an interface near the hypocenter (ŠILÉNY *et al.*, 2000).

Selection of Data

In the case of the Halle event, recordings displayed in Figure 2 clearly show the typical complexity of regional data. Therefore, we decided not to invert the whole seismograms but only portions of them where seismic phases could be precisely identified and modeled. Prior to any selection, velocity recordings were corrected for the sensor response and integrated into displacement recordings. Data were also bandpassed through a Butterworth filter, between 0.1 and 8 Hz, to ensure a signal-to-noise ratio greater than ten. Signals are very complex because various waves are intermixed. However, considering the simple layered model available, we are only able to model the direct waves. Thus we must take small time windows to ensure only one wave, predicted in the seismograms, is isolated at the same time. In our inversion we used P_g or P_n waves, depending on the source-station distance, except an S wave for the station BFO. Moreover, we must only keep stations with a clear P_g or P_n arrival, because a very important step in the preprocessing of data consists of adjusting Green's functions with the data. The code is very sensitive to this adjustment, and consequently a small delay between Green's function and the corresponding real signal can result in instability. Therefore, a precise picking of arrivals both of data and Green's functions is necessary to ensure a proper adjustment.

Inversion Results

We will now present the results for the inversion of body waves. We selected phases from the seven stations BFO, CLL, CLZ, FUR, MOX, TNS and WET, which have recorded the less complex signals as shown in Figure 2. Preliminary tests have shown that horizontal components were not well resolved. Indeed, for a given station, when we succeeded in adjusting the real signal to the Green's functions for the vertical component, horizontal seismograms and their corresponding Green's functions were so misaligned. In other words, there was a dephasing between the vertical component and the horizontal ones, which was not modeled by the Green's functions. This is probably due to the complexity of the medium itself, namely the presence of three-dimensional heterogeneities, anisotropy, or scattering. Consequently, further calculations were deemed performable using only vertical displacements for all stations. A trial-and-error process allowed us to define the proper parameters to use in the inversion, based on the minimization of the square difference between the data and the synthetics. We determined that 15 triangles were necessary for each component of the moment rate function. The width of time windows was fixed to 3 seconds with a sample step of 0.1 s, which leads to 30 points in time for each signal. As we wanted to take into account all 7 stations in the inversion process, we put weights on each station, in order to enhance the furthest ones, which have the lowest amplitudes due to attenuation, relative to the nearest ones. This is necessary to

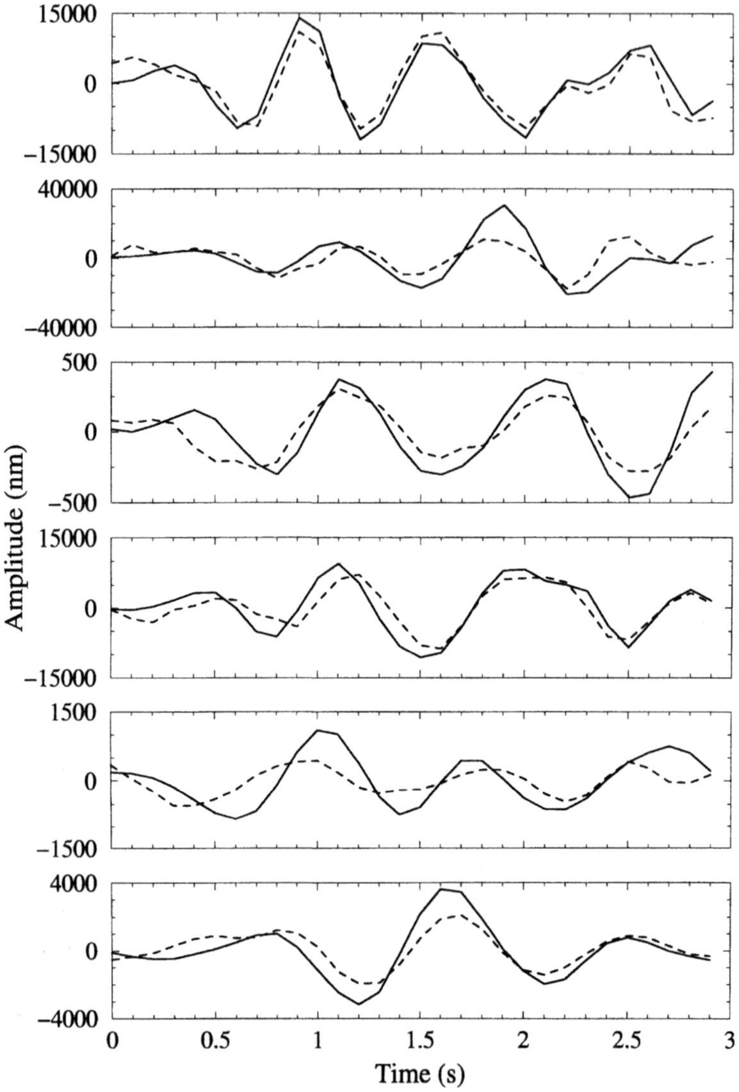

Figure 4
Synthetic vertical components compared to real data for stations CLL, CLZ, FUR, MOX, WET (*P* wave), and BFO (*S* wave). Each window endures 3 s and begins a few samples before the picking of the corresponding arrival.

stabilize the inversion. Otherwise, the covariance matrix becomes so singular that it is impossible to achieve a proper fit for the furthest stations. Accordingly, we gave a weight of 1 to the most distant station from the source. For any other station the weight was computed as the ratio of the maximum amplitude of the inverted phase for the furthest station to the maximum amplitude one for the current station.

The reconstructed synthetic seismograms we obtained from the inversion are in good agreement with the recorded signals (Fig. 4). The quality of this agreement is estimated through a fit value computed as the sum, for all components of all stations, of the difference in an absolute sense between a real signal and a synthetic one. The corresponding moment rate functions are stable regarding the number of stations. Indeed, we tested the stability of our results by taking six randomly chosen stations for inversion. Regardless of the choice of stations, results we obtained were quite similar, as presented in terms of fit in Table 2. This is presented on Figure 5 which

Table 2

Value of the fit between data and synthetics for inversion, using 6 among 7 stations. The fit value is computed as the sum, for all components of all stations, of the difference in an absolute sense between a real signal and synthetic one. A perfect fit corresponds to a value of 0, and a very bad fit to 1

Fit	Without station
0.6289	BFO
0.6690	CLL
0.5364	CLZ
0.6213	FUR
0.5651	MOX
0.5593	WET
0.5346	TNS

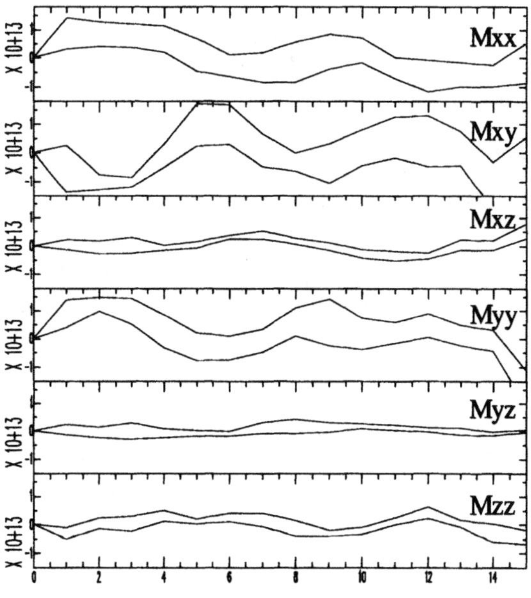

Figure 5
The six moment rate tensor components as a function of time. For each component the two curves represent the minimum and the maximum envelops for all the moment tensors obtained, changing some parameters of the inversion, such as the weights on stations. It provides us with an estimation of error.

delineates for each component of the moment rate tensor, two extreme temporal evolutions. For all the inversions we produced, varying stations and inversion parameters, we extract characteristics of the source which seems to be stable. The main feature is that the M_{xz}, M_{yz}, and M_{zz} components are small compared to the other components. This seems to indicate that equivalent forces at the source act in an horizontal plane. Moreover, we tried to factorize the source, in other words to express moment rate functions as a product of a moment rate tensor and a unique source time function. We clearly obtained a source time function which can be divided into two subevents: the first one from 0 to 0.5 s, and the second one from 0.5 to 1 s. Therefore, we factorized the source time function for each two subevents independently. We used the graphic display of HUDSON *et al.* (1989) to represent the results of this factorization (Fig. 6). For the first subevent (Fig. 7) we obtained a source time function with a large positive isotropic part of approximately 30%, and a deviatoric part nearer a tensile crack or a compensated linear vector than a double-couple. The second one (Fig. 8) corresponds to a nearly pure double-couple mechanism with the same nodal planes as the deviatoric part of the first subevent, but with opposite polarities. In our case, we expected a dominant implosive component coherent with a breakage of pillars, as has been observed for many shallow collapses (MCGARR, 1992; PATTON and ZANDT, 1991). Such an implosional source mechanism

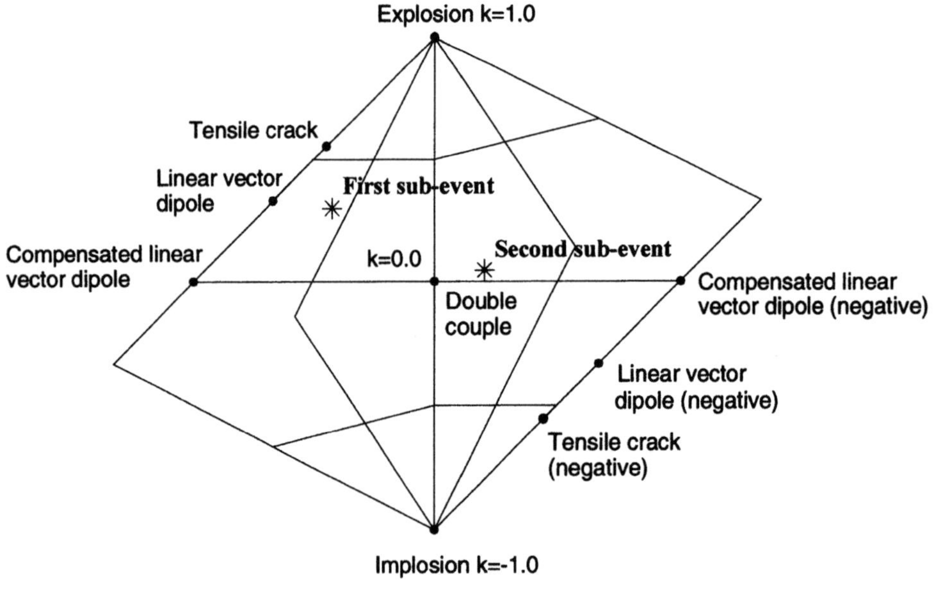

Figure 6
Source type plot for the two subevents. The k axe represents the proportion of isotropic part in the source, whereas the second T axe gives the nature of the deviatoric part. The first one exhibits a large explosive part (about 30%), whereas the second has no isotropic component and a deviatoric mechanism near a pure double-couple.

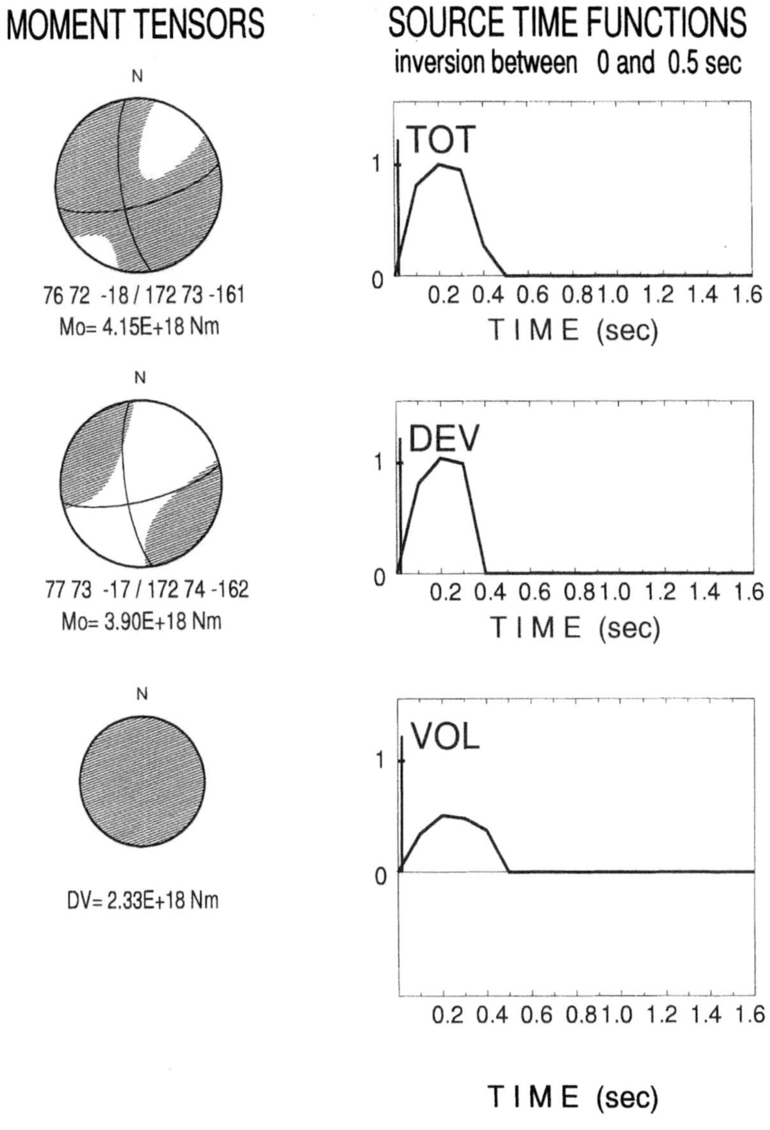

Figure 7
Factorization of the first subevent: Mechanism and source time function for the total (top), the deviatoric part (middle), and the positive isotropic significant part of the source.

was found to be consistent with a near horizontal closing crack source representing the closure of the mine cavity (PECHMANN *et al.*, 1995; WALTER *et al.*, 1997), and was modeled by TAYLOR (1994) as a tabular excavation collapse. Nevertheless, we found here an explosive component. Such a characteristic has been observed in the case of an explosively induced mine collapse studied by YANG *et al.* (1998). These authors proposed a source process beginning with an opening crack due to the sudden loading of the collapsed roof mass after pillar removal by explosion. In our case,

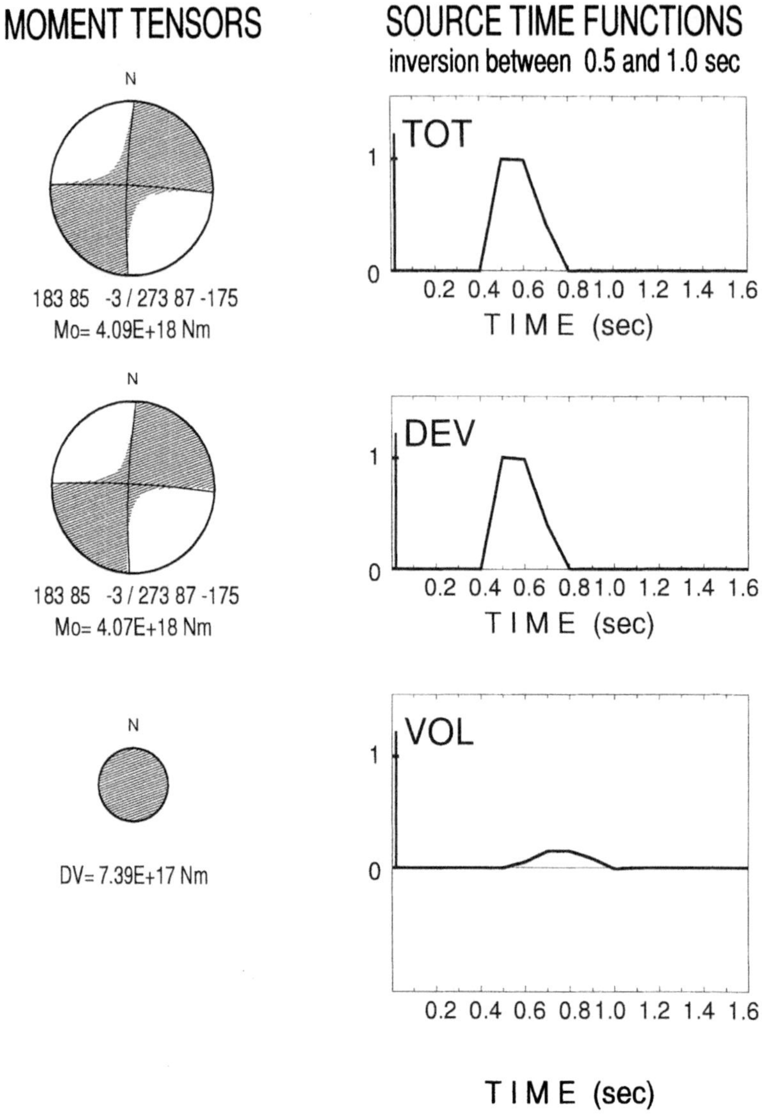

Figure 8
Factorization of the second subevent: mechanism and source time function for the total (top), the deviatoric part (middle), and the isotropic part (bottom) of the source. We obtain a nearly pure double-couple which has the same nodal planes as the deviatoric part of the first subevent, and an opposite polarity.

▶

Figure 9
Dispersion curves for stations CLL and BFO. Eight stations have a behavior similar to CLL, although only four can be useful as an inversion of the velocity model for surface waves. Other stations which have had a dispersion curve, such as BFO, cannot be used.

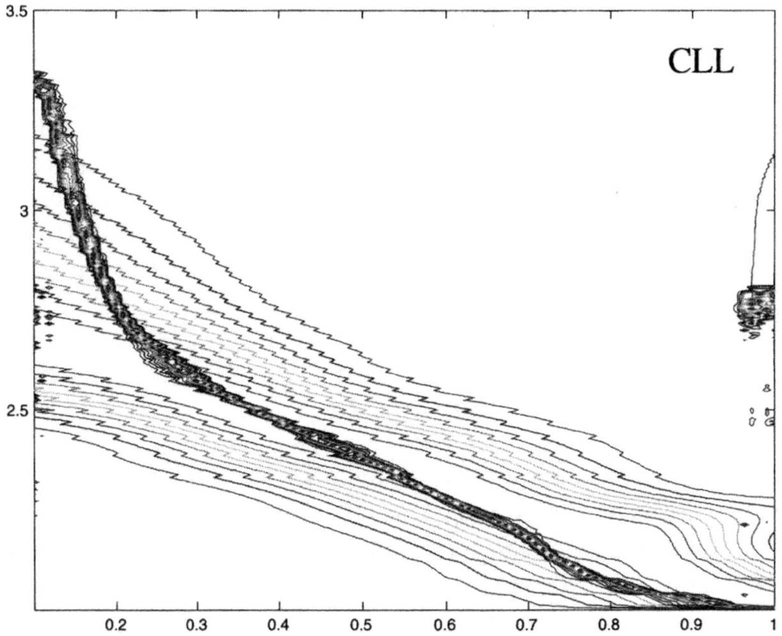

Frequency (Hz)

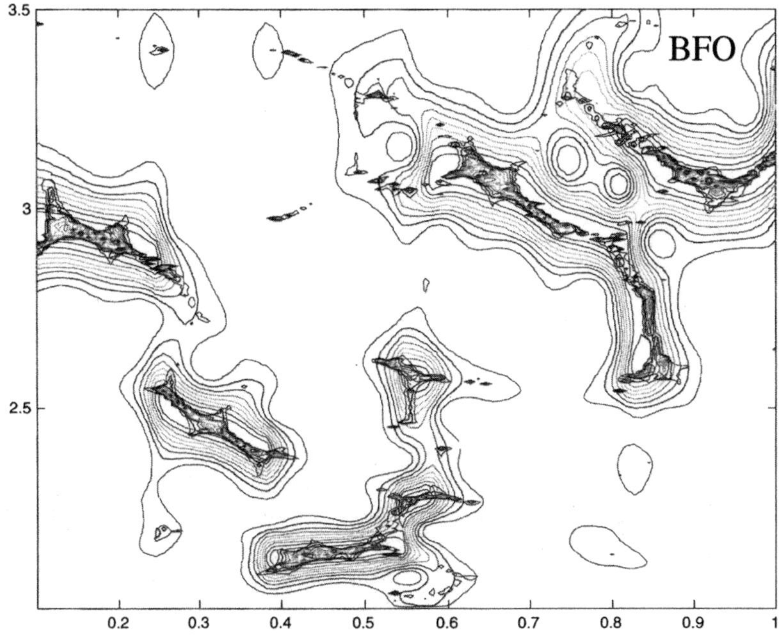

Frequency (Hz)

where collapse was not induced by an explosive breakage of pillars, we can imagine that a similar crack slowly grew under local stresses. When the weight of the slice of rock became to heavy, the crack suddenly propagated, generating a positive isotropic component. Such a mechanism of roof delamination has been proposed by PRUGGER and GENDZWILL (1993). The collapse of the cavity roof on the floor could have produced a downward vertical force, appearing in the source mechanism as an explosive component also, as proposed by HASEGAWA *et al.* (1989). However, it would have probably not produced enough energy to explain the 4.8 event magnitude.

Surface Waves

We also tried to add surface waves to the inversion. For that reason, we first computed the dispersion curves using a time-frequency analysis (BAUMONT, 1999; COTTE *et al.*, 1999). We obtained curves with a classical evolution of group velocity decreasing with frequency, for eight stations. Figure 9a supplies an example of such a curve for station CLL. For the other stations, as for BFO (Fig. 9b), there was no clearly identified dispersion in the frequency range 0.1–1 Hz of which we are interested. For the first eight stations we then tried to invert the dispersion curves to derive a model specific to the surface waves. Unfortunately, the inversion was

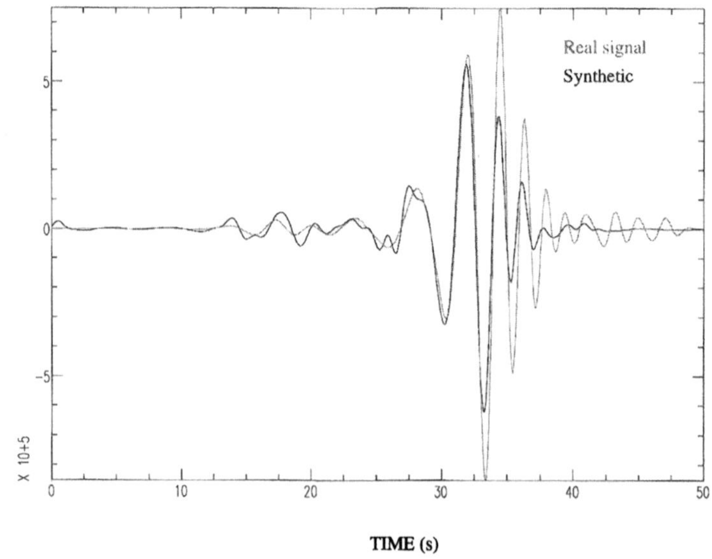

Figure 10
Synthetic seismogram compared to a real signal for surface waves on the vertical component of station CLL, filtered between 0.2 Hz and 0.7 Hz, the range used to compute the dispersion curves. There is good agreement between synthetics and real signals.

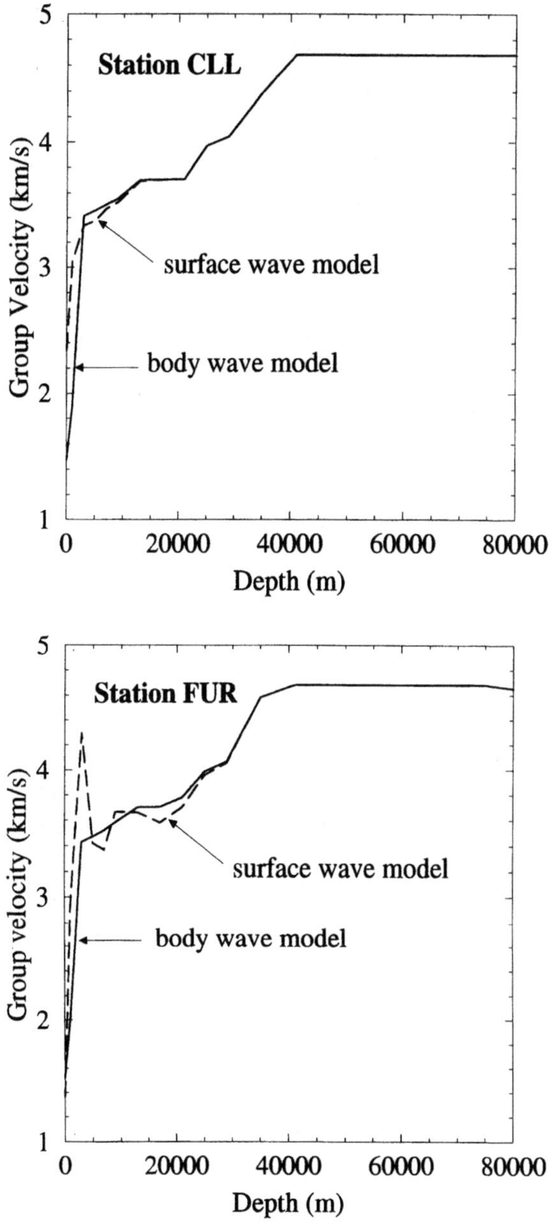

Figure 11
Surface wave models obtained for stations FUR and CLL. The model for FUR exhibits a superficial low-velocity layer inconsistent with a real velocity structure.

successful for only the CLL station. Models obtained for other stations all exhibit a superficial low-velocity layer inconsistent with a real velocity structure (Fig. 11a). Therefore we had insufficient stations for which surface waves could be inverted so as to retrieve a quantitative moment tensor. Nevertheless, we used the new layered model obtained for CLL surface waves (Fig. 11b) to compute synthetic seismograms. Indeed, we convolved the Green's functions for this model with the first subevent moment tensor obtained from the inversion of body waves. Figure 10 shows the synthetic and the data for the vertical component of CLL, filtered between 0.2 Hz to 0.7 Hz, where the dispersion curve is well defined. We observe that a large part of the synthetic seismogram is in good agreement with the recorded surface waves. So, the mechanism obtained from the body waves allows a good forward modeling of surface waves. Moreover, there are also features which can furnish information regarding the source radiation pattern. Particularly stations MOX, WET and CLL, which are located southeast of the event, exhibit large surface waves predominating body waves, whereas such is not the case for other stations. Consequently, the event seems to have a lobed Rayleigh wave pattern that is not expected for a volumetric source. Furthermore, because no strong Love waves are observed, the signals are not consistent with a pure double-couple source. Thus surface waves favor a non-double-couple mechanism not exclusively isotropic.

Conclusion

We succeeded in retrieving a mechanism and a source time function for the mine collapse of Halle, from regional body wave data. We showed that there were in fact two subevents: the first one containing a large positive isotropic component, and the second with a deviatoric mechanism close to a pure double-couple. Unfortunately, due to the low quality of the dispersion curves, we were not able to use surface waves in the inversion. Nevertheless, the synthetics seismograms using the moment tensor obtained from body waves are in good agreement with the recorded surface waves. In due course, it would be interesting to seek a mechanical model of the source which agrees with the seismic energy distribution of each kind of wave.

References

Baumont, D. (1999), *Caractérisation sismologique de la structure lithosphérique des Andes Centrales (17°–20°S)*, Ph.D. Thesis of the Joseph Fourier University, Grenoble I, France.

Bouchon, M. (1979), *Discrete Wave Number Representation of Elastic Wavefields in Three-space Dimensions*, J. Geophys. Res. *84*, 3609–3614.

Cespuglio, G., Campus, P., and Sileny, J. (1996), *Seismic Moment Tensor Resolution by Waveform Inversion of a Few Local Noisy Records – II Application to the Phlegraean Fields (Southern Italy) Volcanic Tremors*, Geophys. J. Int. *126*, 605–619.

Cotte, N., Pedersen, H. A., Campillo, M., Ni, J., Kind, R., Sandoval, E., and Zhao, W. (1999), *Determination of the Crustal Structure in South Tibet by Dispersion and Amplitude Analysis of Surface Waves*, Geophy. J. Int. *138*, 809–819.

Cotton, F. and Coutant, O. (1997), *Expression of the Green's Functions in Cylindrical Coordinates Used with a Reflectivity Method*, Geophys. J. Int. *128*, 676–688.

COURBOULEX, F., VIRIEUX, J., and GIEBERT, D. (1996), *On the Use of Simulated Annealing Method and Cross-validation Theory for Deconvolution of Seismograms*, Bull. Seismol. Soc. Am. *86*, 1187–1193.

COUTANT, O. (1994), *Expression of the Green's Functions in Cylindrical Coordinates Used with a Reflectivity Method*, Technical Report, LGIT, Grenoble.

DREGER, D. S., and HELMBERGER, D. V. (1991), *Source Parameters of the Sierra Madre Earthquake from Regional and Local Body Waves*, Geophys. Res. Lett. *18*, 2015–2018.

DREGER, D. S., and HELMBERGER, D. V. (1993), *Determination of Source Parameters at Regional Distances with Three-component Sparse Network Data*, J. Geophys. Res. *98*, 8107–8125.

DU, Z. J., MICHELINI, A., and PANZA, G. F. (1998), *EurID: A Regionalized 3-D Seismological Model of Europe*, Phys. Earth Planet. Int. *2*.

DUFUMIER, H., MICHELINI, A., DU, Z., BONDÁR, I., ŠILÉNY, J., MAO, W., KRAVANJA, S., and PANZA, G. F. (1997), *Regional Structure Modelling and Source Inversion for the 1992 Roermond Earthquake*, J. Seismol. *1*, 321–340.

DZIEWONSKY, A. M. and WOODHOUSE, J. H. (1983), *An experiment in systematic study of global seismicity: CMT solutions for 201 moderate and large earthquakes of 1891*, J. Geophys. Res. *88*, 3247–3271.

FEIGNIER, B., and YOUNG, R. P. (1992), *Moment Tensor Inversion of Induced Microseismic Events: Evidence of Non-shear Failures in the* $-4 < M < -2$ *Moment Magnitude Range*, Geophys. Res. Lett. *19*, 1503–1506.

GIARDINI. D., BOSCHI, E., and PALOMBO, B. (1993), *Moment Tensor Inversion from Mednet Data (2), Regional Earthquakes of the Mediterranean*, Geophys. Res. Lett. *20*, 273–276.

HARA, T., KUGE, K., and KAWAKATSI, H. (1995), *Determination of the Isotropic Component of the 1994 Bolivia Deep Earthquake*, Geophys. Res. Lett. *22*, 2265–2268.

HASEGAWA, H. S., WETMILLER, R. J., and GENDZWILL, D. J. (1989), *Induced seismicity in mines in Canada – An overview*. In *Seismicity in Mines* (ed. Gibowicz S. J.) Pure appl. geophys. *129*, 423–453.

HUDSON, J. A., PEARCE, R. G., and ROGERS, R. M. (1989), *Source Time Plot for Inversion of the Moment Tensor*, J. Geophys. Res. *94*, 765–774.

JULIAN, B. R., MILLER, A. D., and FOULGER, G. R. (1997), *Non-double-couple Earthquake Mechanisms at the Hengill-Greensdalur Volcanic Complex, Southwest Iceland*, Geophys. Res. Lett. *24*, 743–746.

KLINGE, K. (1997), *Broadband recordings of strong mining-induced seismic events in Central Germany*, The 29th General Assembly of the International Association of Seismology and Physics of the Earth's Interior, Thessaloniki, Greece.

KRAVANJA, S., PANZA, G. F., and ŠILÉNY, J. (1999), *Robust Retrieval of a Seismic Point Source Time Function*, Geophys. J. Int. *136*, 385–394.

KUGE, K. (1994), *Data-dependent Non-double-couple Components of Shallow Earthquake Source Mechanisms: Effects of Waveform Inversion Instability*, Geophys. Res. Lett. *21*, 9–12.

MCGARR, A. (1992), *An Implosive Component in the Seismic Moment Tensor of a Mining-induced Tremor*, Geophys. Res. Lett. *19*, 1579–1582.

MÜLLER, G. (1985), *The Reflectivity Method: A Tutorial*, J. Geophys. *58*, 153–174.

NABELEK, J. L., and XIA, G. (1995), *Moment Tensor Analysis Using Regional Data: Application to the 25 March 1993, Scotts MILLS, Oregon, Earthquake*, Geophys. Res. Lett. *22*, 13–16.

NABELEK, J. L. (1984), *Determination of Earthquake Source Parameters from Inversion of Body Waves*, Ph.D. Thesis, MIT.

NAKANISHI, I., HANAKAGO, Y., MORIYA, T., and KASAHARA, M. (1991), *Performance Test on Long-period Moment Tensor Determination for Near Earthquakes Recorded by a Sparse Local Network*, Geophys. Res. Lett. *18*, 223–226.

PASYANOS, M. E., DREGER, D. S., and ROMANOWICZ, B. (1996), *Toward Real-time Estimation of Regional Moment Tensors*, Bull. Seismol Soc. Am. *86*, 1255–1269.

PATTON, H. J., and ZANDT, G. (1991), *Seismic Moment Tensors of Western U.S. Earthquakes and Implications for the Tectonic Stress Field*, J. Geophys. Res. *96*, 18,245–18,259.

PECHMANN, J. C., WALTER, W. R., NAVA, S. J., and ARABASZ, W. J. (1995), *The February 3, 1995 ML 5.1 Seismic Event in the Trona Mining District of Southwestern Wyoming*, Seism. Res. Lett. *66*, 25–34.

PRUGGER, A. F. and GENDZWILL, D. J., *Fracture mechanism of microseisms in Saskatchewan potash mine*. In *Rockbursts and Seismicity in Mines* (ed. Young), (Balkema, Rotterdam 1993).

RITSEMA, J., and LAY, T. (1995), *Long-period Regional Wave Moment Tensor Inversion for Earthquakes in the Western United States*, J. Geophys. Res. *100*, 9853–9864.

ROMANOWICZ, B., DREGER, D., PASYANOS, M., and URHAMMER, R. (1993), *Monitoring of Strain Release in Central and Northern California Using Broadband Data*, Geophys. Res. Lett. *20*, 1643–1646.

ŠILÉNY, J., CAMPUS, P., and PANZA, G. F. (1996), *Seismic Moment Tensor Resolution by Waveform Inversion of a Few Local Noisy Records – I. Synthetic Tests*, Geophys. J. Int. *126*, 605–619.

ŠILÉNY, J., PANZA, G. F., and CAMPUS, P. (1992), *Waveform Inversion for Point Source Moment Tensor Retrieval with Variable Hypocentral Depth and Structural Model*, Geophys. J. Int. *109*, 259–274.

ŠILÉNY, J., and VAVRYCUK, V. (2000), *An Approximate Retrieval of Point Source in Anisotropic Media: Numerical Study by Indirect Parameterization of the Source*, Geophys. J. Int., submitted.

ŠILÉNY, J., PSENCIK, and YOUNG, R. P. (2000), *Point Source Inversion Neglecting Nearby Free Surface: Simulation of the Underground Research Laboratory, Canada*, Geophys. J. Int., submitted.

SIPKIN, S. A. (1982), *Estimation of Earthquake Source Parameters by the Inversion of Waveform Data: Synthetic Waveforms*, Phys. Earth Planet Inter. *30*, 242–259.

TAYLOR, S. R. (1994), *False Alarms and Mine Seismicity: An Example from the Gentry Mountain Mining Region, Utah*, Bull. Seismol. Soc. Am. *84*, 350–358.

THIO, H. K. and KANAMORI, H., (1995), *Moment-tensor inversions for local earthquakes using surface waves recorded at TERRAscope*, Bull. Seismol. Soc. Am. *85*, 1021–1038.

VASCO, D. W., and JOHNSON, L. R. (1989), *Inversion of Waveforms for Extreme Source Models with an Application to the Isotropic Moment Tensor Component*, Geophys. J. *97*, 1–18.

WALTER, W. R., HEUZE, F., and DODGE, D. (1997), *Seismic signals from underground cavity collapses and other mining-related failures*. In Proc. 19th Annual Seismic Research Symposium on Monitoring a Comprehensive Test Ban Treaty, pp. 678–687.

YANG, X., STUMP, B. W., and PHILLIPS, W. S. (1998), *Source Mechanism of an Explosively Induced Mine Collapse*, Bull. Seismol. Soc. Am. *88*, 843–854.

(Received May 18, 2000, revised January 2, 2001, accepted January 3, 2001)

Pure appl. geophys. 159 (2002) 131–144
0033–4553/02/030131–14 $ 1.50 + 0.20/0

Pure and Applied Geophysics

Source Time Function of Seismic Events at Rudna Copper Mine, Poland

B. DOMAŃSKI,[1] S. J. GIBOWICZ[1] and P. WIEJACZ[1]

Abstract—The empirical Green's function deconvolution technique is used to retrieve the source time functions from the records of *P* waves of seven seismic events that occurred at the Rudna copper mine in 1996 and were located in the middle of the underground network. Their moment magnitudes ranged from 2.1 to 2.9. The records of smaller events from the same area and with similar source mechanism, with moment magnitudes ranging from 1.5 to 2.0, were accepted as empirical Green's functions. The relative source time functions were successfully retrieved at a number of stations for six events. Directivity effects, implying unilateral rupture propagation, were observed in five cases. The azimuth of rupture propagation direction and the rupture velocity were estimated from the distribution of pulse widths and pulse maximum amplitudes as a function of the cosine of station azimuths. The rupture propagated approximately either from south to north or from north to south. The rupture velocity was low, ranging from 0.25 to 0.54 of the shear-wave velocity. The source dimensions, represented by the fault length, were also small in comparison with those estimated in the frequency domain and ranged from 80 to 250 m.

Key words: Induced seismicity, empirical Green's function, source time function, directivity effects, rupture direction, rupture velocity.

Introduction

The source processes of seismic events induced by mining have been extensively studied, by means of both spectral analysis of the records of seismic waves in order to estimate source parameters and by the moment tensor inversion technique to investigate source mechanisms (see, e.g., GIBOWICZ and KIJKO, 1994). New approaches to studying the sources of seismic events in mines, however, are needed for a better understanding of their properties and their generation mechanisms. Determining the source time function is just such a new approach in studies of seismicity in mines, although this procedure has long been used in studies of natural earthquakes.

[1] Institute of Geophysics, Polish Academy of Sciences, ul.Ks.Janusza 64, 01-452 Warsaw, Poland.
E-mail: gibowicz@igf.edu.pl

The source time function (STF) of a seismic event describes the seismic moment release and the rupture evolution in the source. It can provide information on rupture type, velocity and propagation direction when directivity effects are observed. Source parameters can be estimated in the time domain and compared with those estimated in the frequency domain.

The extraction of STF from seismograms, however, requires a separation of the source effects from those of the path, the site, and the instrumental response. For this a deconvolution technique based on empirical Green's functions (EGF) can be used. A knowledge of earth structure, attenuation effects, and instrumental response are then not required (e.g., MUELLER, 1985; LI and THURBER, 1988; MORI, 1993; LI *et al.*, 1995; BOUR and CARA, 1997; FLETCHER and SPUDICH, 1998; JOST *et al.*, 1998). In the case of two seismic events of different strength, located closely to each other and with similar focal mechanism, the record of the smaller event can be considered as the EGF (HARTZELL, 1978). It can be deconvolved by spectral division in the frequency domain from the record of the larger event to obtain a relative source time function (RSTF) at a given station for the larger event (MUELLER, 1985). Such deconvolution should result in a RSTF that is corrected for the path, the site, and the instrumental effects.

Recently, new methods for determination of the source time function by the empirical Green's function approach in the time domain have been proposed. They are based on the search by successive iterations for the solution of integral equations describing the observed seismograms as the convolution of the source time function with the empirical Green's function. An interesting and effective technique, in this respect, seems to be the projected Landweber method which is an iterative nonlinear technique allowing the introduction of physical constraints (positivity, causality, etc.) on the final STF (BERTERO *et al.*, 1997; PIANA and BERTERO, 1998). A nontrivial modification of an iterative blind-deconvolution method used for image identification has also been proposed as a novel technique to retrieve the STF (BERTERO *et al.*, 1998).

The EGF approach should be especially convenient for seismic source studies in mines, where underground seismic networks are situated in the source area and are often composed of a large number of sensors. Furthermore, the seismic events there are located to a high degree of accuracy, having standard errors not exceeding a few tens of meters. DOMAŃSKI and GIBOWICZ (1999) recently described the EGF classic deconvolution procedure in the frequency domain, based on numerous trials and tests performed on real data from Polish mines. The aim of this paper is to make the first attempt to recover the STF of seismic events induced by mining. For this, the records of selected seismic events which occurred at Rudna copper mine in Poland were processed, using the classic deconvolution procedure. The time domain approach will be used in subsequent studies.

Data

The Rudna copper mine, situated in the Lubin Copper Basin in southwestern Poland, has a local digital seismic network composed (by 1996) of 32 vertical sensors located underground at a depth from 550 to 1150 m. Willmore MK-II and MK-III and Russian SM3 seismometers are used; their horizontal distribution is shown in Figure 1. The seismic signals are transmitted to the central site at surface in analog form with FM modulation by standard cables used in mining. The signals are digitized at the central site; a 14-bit converter is used. The frequency band is from 0.5 to 150 Hz and the system dynamics is about 70 dB. The sampling frequency is 500 Hz. The system works in a triggering mode.

Seven larger events, with moment magnitudes (HANKS and KANAMORI, 1979) ranging from 2.1 to 2.9, which occurred in 1996 in two sectors G-13/3 and G-1/7 of the mine were selected for this study. Five smaller events from the same area, with moment magnitudes ranging from 1.5 to 2.0, were selected as EGF events. The selected events were located in the middle of the underground seismic network (Fig. 1). The distribution of the main and EGF events is shown in Figure 2, where the main events are marked by uppercase letters from **A** to **G** and the EGF events are

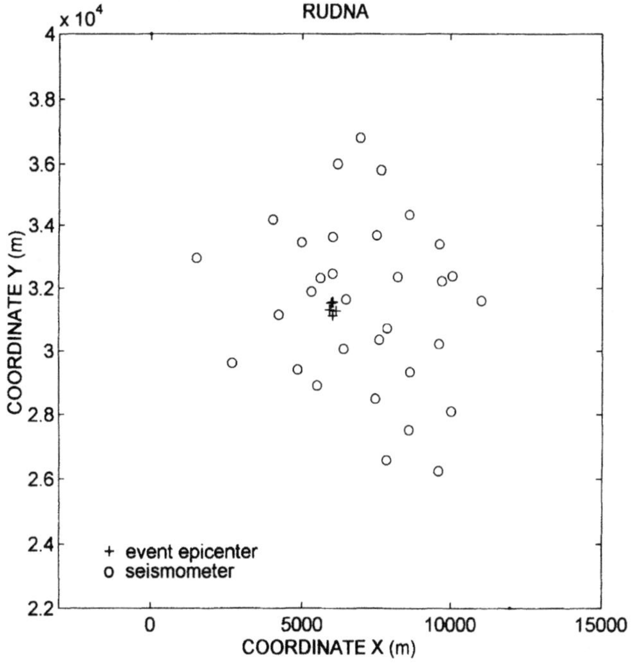

Figure 1
Horizontal distribution of seismic stations (open circles) and selected seismic events (crosses) at Rudna copper mine. Local coordinate system is used.

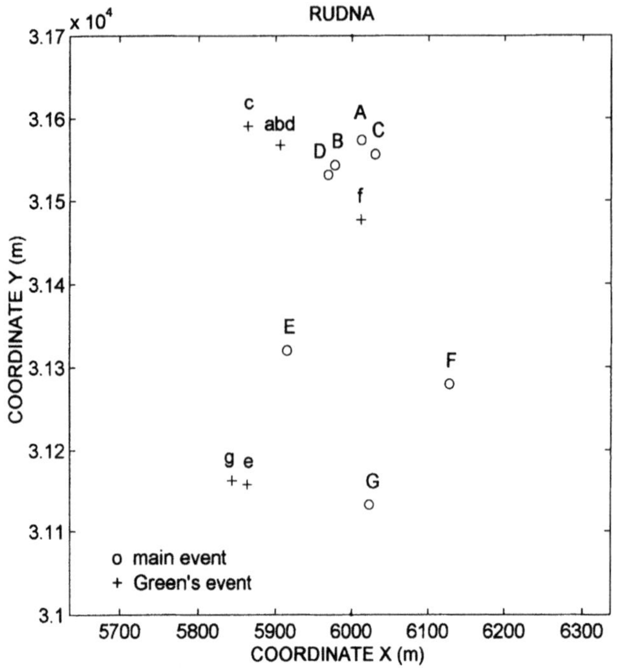

Figure 2
Distribution of the selected larger events, marked by open circles and uppercase letters, and the smaller events, marked by plus signs and lowercase letters, used as Green's functions.

marked by lowercase letters from **a** to **g**. The records of one EGF event were used to retrieve the STF of three main events and this event is marked as **abd**.

The list of selected events is given in Table 1, where the values of their seismic moment and seismic energy are also listed. The seismic moment was calculated from the spectral low-frequency level and the seismic energy from the energy flux of P and S waves recorded by several stations, using standard relations (e.g., GIBOWICZ and KIJKO, 1994). The distance between the main and EGF events ranged from 78 to 274 m (Table 2).

To investigate the focal mechanism of selected events, moment tensor inversion in the time domain was performed. The calculation procedure was developed and described by WIEJACZ (1992). The calculations were performed twice for each event using norms L_1 and L_2. For each norm, a full moment tensor, a constrained deviatoric solution and a double-couple solution were calculated. After careful inspection of all solutions, the L_2-norm solutions were accepted for further consideration. In general, the focal mechanisms of main and EGF events were found to be similar. More precisely, those events that have a focal mechanism similar to that of the main events were selected as EGF events. As an example, the graphical and numerical presentations of general, deviatoric and double-couple moment tensor solutions for main event **C** are shown in Figure 3, and those for EGF event **c** are presented in Figure 4.

Table 1

Selected seismic events, marked by uppercase letters, and corresponding smaller events, marked by lowercase letters, used as Green's functions from sectors G-13/3 and G-1/7 of the Rudna mine

Event symbol	Date	Local time (h m s)	Local coordinates X (m)	Y (m)	Z (m)	Seismic moment (N·m)	Seismic energy (J)
Sector G-13/3							
A	96.06.14	15:58:59	6013	31573	650	3.2E12	4.2E6
a, b, d	96.11.09	06:47:55	5906	31567	800	2.4E11	5.1E4
B	96.09.14	17:10:30	5978	31543	780	1.7E12	1.7E6
C	96.10.27	14:06:34	6031	31556	740	1.3E13	8.1E7
c	96.12.15	04:30:47	5864	31590	800	8.5E11	3.2E5
D	96.10.27	14:15:04	5969	31531	720	1.4E12	1.0E6
Sector G-1/7							
E	96.04.23	17:25:30	5915	31320	790	7.1E12	1.1E7
e	96.12.09	00:40:44	5863	31158	680	3.1E11	2.0E5
F	96.11.07	21:48:50	6128	31279	640	2.4E13	7.3E7
f	96.06.04	19:09:36	6012	31477	790	1.3E12	5.2E5
G	96.12.19	11:02:38	6023	31133	790	2.8E12	1.6E6
g	96.09.23	15:52:20	5843	31163	750	2.2E11	5.0E4

Table 2

Results of the analysis of source time functions derived for the selected seismic events

Event symbol	M_w Main	M_w EGF	Δ (m)	Number of stations	M	A_o (deg)	V_r/V_s	L (m)	R_c
A	2.3	1.5	184	16	2.2	207	0.28	81	−0.722
B	2.1	1.5	78	17	2.0			130*	
C	2.7	1.9	181	17	2.5	19	0.48	128	−0.810
D	2.1	1.5	108	15	2.0	176	0.25	76	−0.822
F	2.9	2.0	274	14	2.7	24	0.54	249*	−0.871
G	2.3	1.5	187	14	2.3	212	0.31	118*	−0.628

Δ is the distance between the main and EGF events, M is the main event magnitude calculated from the source pulse area, A_o is the azimuth of the rupture direction, V_r is the rupture velocity, V_s is the velocity of shear waves, L is the fault length (estimated from the rise time when marked by an asterisk), and R_c is the correlation coefficient.

Deconvolution Procedure

MUELLER (1985) developed the spectral division deconvolution technique to separate the STF from seismic records. The observed seismogram $U(t)$ of a given larger event can be expressed as the convolution of its source time function $S(t)$, the impulse response path $P(t)$, the recording site $R(t)$, and the recording instrument $I(t)$:

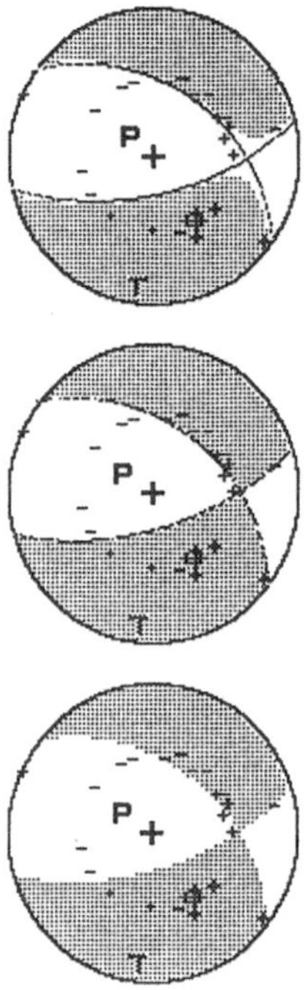

```
L2 tensor for : 14971 4.0E6 G-13/3 Oct27/14:06

 Full solution :
              .875E+13     .209E+13    -.389E+13
              .209E+13    -.182E+13     .336E+13
             -.389E+13     .336E+13    -.808E+13
 T0=    .0262  M0= .102E+14  MT= .102E+14  ERR= .695E+12
 Expl.=  -3.6%   CLVD.=    -.1%   DBCP.=  96.3%
             ΦA=  77.05  δA= 60.28  ΦB= 304.50  δB= 40.17
             P-axis: trend= 299.34  plunge= 62.99
             T-axis: trend= 187.07  plunge= 10.93
             B-axis: trend=  92.05  plunge= 24.37
             Quality index =  69.9  normal  fault

 Trace null solution :
              .855E+13     .199E+13    -.369E+13
              .199E+13    -.187E+13     .356E+13
             -.369E+13     .356E+13    -.668E+13
 T0=    .0262  M0= .952E+13  MT= .953E+13  ERR= .704E+12
                   CLVD.=    -.2%   DBCP.=  99.8%
             ΦA=  73.78  δA= 62.08  ΦB= 307.10  δB= 41.58
             P-axis: trend= 296.24  plunge= 59.32
             T-axis: trend= 186.36  plunge= 11.41
             B-axis: trend=  90.18  plunge= 28.06
             Quality index =  72.8  normal  fault

 Double couple solution :
              .855E+13     .199E+13    -.368E+13
              .199E+13    -.189E+13     .355E+13
             -.368E+13     .355E+13    -.666E+13
 T0=    .0262  M0= .952E+13  MT= .952E+13  ERR= .704E+12
                                    DBCP.= 100.0%
             ΦA=  73.73  δA= 62.10  ΦB= 307.15  δB= 41.62
             P-axis: trend= 296.18  plunge= 59.25
             T-axis: trend= 186.36  plunge= 11.41
             B-axis: trend=  90.16  plunge= 28.13
             Quality index =  76.1  normal  fault
```

Figure 3

Graphical and numerical presentation of full, deviatoric and double-couple moment tensor solutions, displayed on the lower hemisphere equal-area projections, for event **C** of moment magnitude 2.7 that occurred on October 27, 1996. The areas of compression are shaded, the P and T axes are marked, and the nodal lines for the best double-couple solutions are shown.

$$U(t) = S(t) * P(t) * R(t) * I(t) \ , \tag{1}$$

where symbol $*$ represents the convolution operation. In the frequency domain the convolution becomes the operation of multiplication:

$$U(\omega) = S(\omega)P(\omega)R(\omega)I(\omega) \ , \tag{2}$$

where ω is an angular frequency.

The seismogram of a smaller event $G(t)$, whose source time function is assumed to be a delta function $\delta(t)$, can be considered as an EGF

$$G(t) = \delta(t) * P(t) * R(t) * I(t) \ . \tag{3}$$

The deconvolution is performed by spectral division in the frequency domain

$$\frac{U(\omega)}{G(\omega)} = \frac{S(\omega)P(\omega)R(\omega)I(\omega)}{\delta(\omega)P(\omega)R(\omega)I(\omega)} = S(\omega) \ , \tag{4}$$

```
L2 tensor for : 17681 1.3E4 G-13/3 Dec15/04:30

Full solution :
                    .407E+13     .754E+12      .116E+13
                    .754E+12    -.126E+12      .286E+13
                    .116E+13     .286E+13     -.872E+13
T0=    .0273   M0= .751E+13   MT= .920E+13   ERR= .526E+1
Expl.= -11.6%    CLVD.= -14.8%    DBCP.=   73.6%
                 ΦA= 297.26   δA= 54.61   ΦB=   87.60   δB= 39.2
                 P-axis: trend= 257.00   plunge= 73.13
                 T-axis: trend=  14.34   plunge=  7.93
                 B-axis: trend= 106.45   plunge= 14.80
                 Quality index =   62.0  normal   fault

Trace null solution :
                    .168E+13     .582E+12      .150E+13
                    .582E+12     .193E+11      .850E+12
                    .150E+13     .850E+12     -.169E+13
T0=    .0273   M0= .241E+13   MT= .248E+13   ERR= .619E+12
                 CLVD.=    5.1%   DBCP.=   94.9%
                 ΦA= 296.21   δA= 67.22   ΦB=   97.78   δB= 23.88
                 P-axis: trend= 219.59   plunge= 66.98
                 T-axis: trend=  20.61   plunge= 21.89
                 B-axis: trend= 113.35   plunge=  6.78
                 Quality index =   53.1  normal   fault

Double couple solution :
                    .162E+13     .484E+12      .145E+13
                    .484E+12     .905E+11      .837E+12
                    .145E+13     .837E+12     -.171E+13
T0=    .0273   M0= .241E+13   MT= .241E+13   ERR= .619E+12
                                  DBCP.=  100.0%
                 ΦA= 295.68   δA= 67.16   ΦB=   96.62   δB= 24.02
                 P-axis: trend= 219.59   plunge= 66.98
                 T-axis: trend=  19.87   plunge= 21.80
                 B-axis: trend= 112.70   plunge=  7.04
                 Quality index =   56.6  normal   fault
```

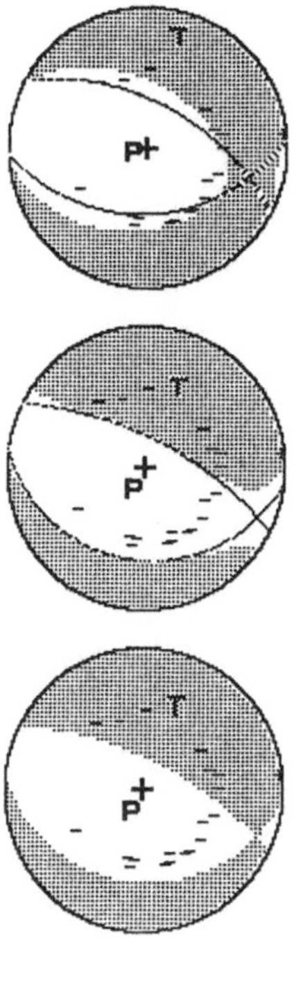

Figure 4

Graphical and numerical presentation of moment tensor solutions for event **c** of moment magnitude 1.9 that occurred on December 15, 1996.

where $\delta(\omega) = 1$. Then, after taking the inverse Fourier transform, the STF of the larger event is obtained. The STF of the smaller event is in practice never equal to a delta function, nonetheless its spectrum can be approximated by a constant up to its corner frequency.

Although the notion of the EGF deconvolution procedure in the frequency domain is rather simple, the application can be complicated. There are numerical instabilities resulting from spectral division and several other problems inherent in the method. The computational procedure used here is based on numerous tests performed on real data.

Only *P*-wave records were analyzed, since only vertical sensors are in operation at the Rudna mine. The vertical sensors are mainly triggered by *P* waves refracted in the high-velocity dolomite layer situated immediately above the working area, where the seismic network is located. The selected waveform intervals of the larger events were first filtered by a sixth-order Butterworth bandpass filter with a frequency band from 1 to 30 Hz in order to reduce the high-frequency noise. The *P* waveforms of both the larger event and the EGF event were windowed to a length of about 0.3 s, processed

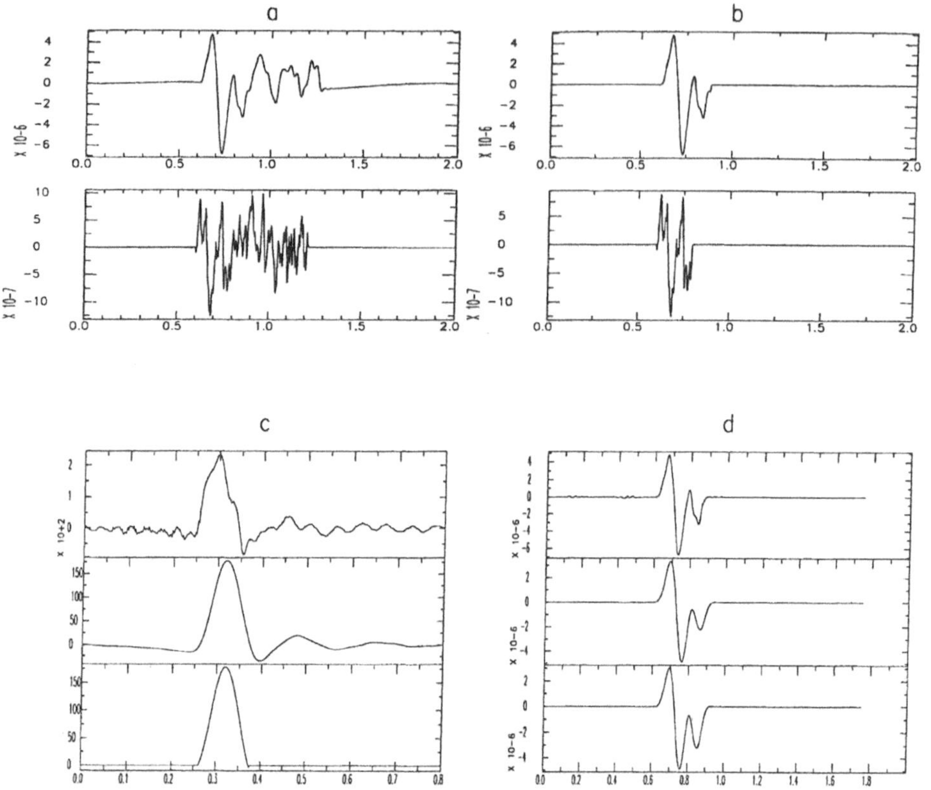

Figure 5
An example of the relative source time function of event **C**, derived from *P* waves by using the record of event **c** as a Green's function. (a) *P*-wave records of the main event (top) and the Green's event (bottom); time is in s and amplitudes in m/s. (b) Selected windows for processing. (c) The original (top), the smoothed (middle), and the windowed (bottom) relative source time functions. (d) Results of the convolution of the three source time functions with the record of Green's event.

to remove the DC component, and then Fourier-transformed to the frequency domain. To stabilize the deconvolution procedure, a "water level" equal to 0.01 of the maximum spectral amplitude was applied to the EGF spectrum. The amplitude spectrum of the larger event was divided by that of the EGF event and the phase spectrum of the EGF event was subtracted from that of the larger event. The resulting amplitude and phase spectra were then inverse-transformed to the time domain to obtain the Relative Source Time Function (RSTF). After deconvolution, a sixth-order Butterworth low-pass filter with a corner frequency of 20 Hz was used to reduce the high-frequency noise. The filtered RSTF impulse was additionally smoothed by an averaging window procedure; the window comprised 10 samples and moved forward after each single sample. A positive impulse was then cut out of the smoothed time series and accepted as the final RSTF for a given event at a given

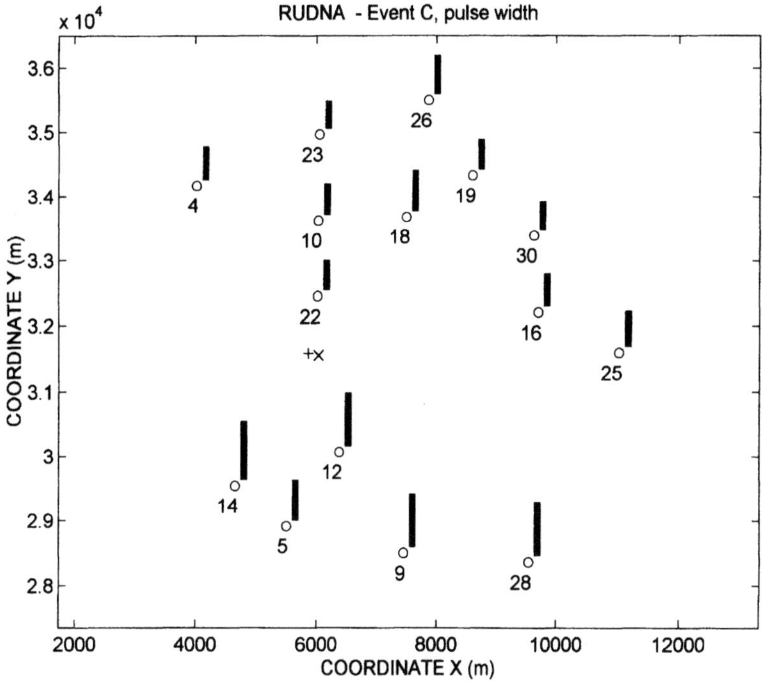

Figure 6

Spatial distribution of pulse widths of the relative source time functions for event **C**. The locations of the main and Green's events are shown by a cross and a plus sign, respectively, the stations are marked by open circles and numbered, and the pulse widths, normalized to the largest value, are shown by vertical thick bars.

station. In the final stage, the three successive RSTFs (original, smoothed, and windowed) were convolved with the corresponding EGF to generate the original waveforms of the larger event as a final check-up for the procedure.

The procedure was not successful for the main event **E** and its EGF event **e** (Table 1), despite the fact that their selection criteria (focal mechanism and distance) were well observed. For the other events the RSTFs were retrieved from the records of many stations, ranging from 14 to 17 (Table 2). An example of the RSTF retrieved for event **C** is shown in Figure 5.

Directivity Effects and Source Parameters

After retrieving RSTF pulses from the records of several stations, it is possible to estimate whether they are symmetric or asymmetric, and in the latter case to consider their dependence on the station azimuth. If the pulse widths of the RSTF depend on the station azimuth in a regular manner, then it could be an indication that the rupture propagated unilaterally. For a unilaterally propagating rupture, the source

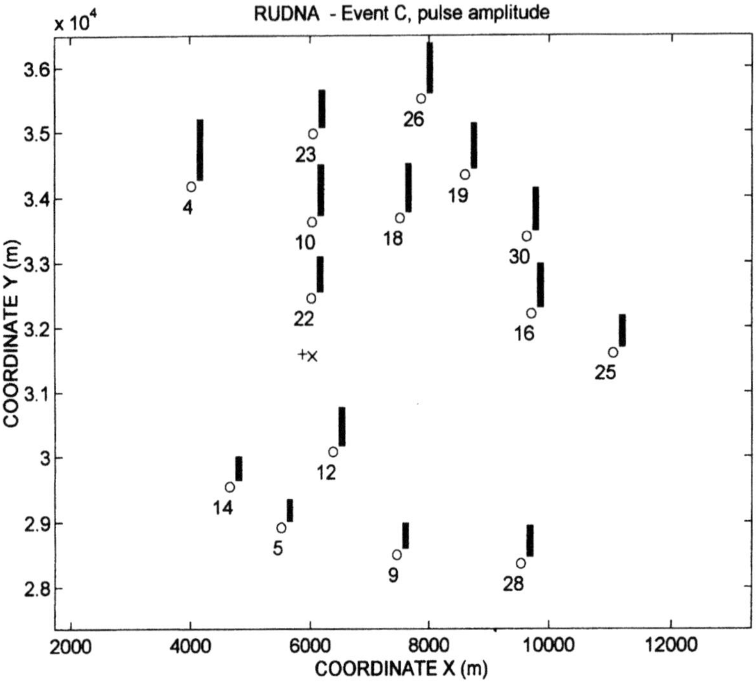

Figure 7
Spatial distribution of maximum amplitudes of the relative source time functions for event **C**.

duration is narrowest in the direction of rupture propagation and widest in the opposite direction. Similarly, the maximum amplitude of the source pulse would be largest in the direction of rupture propagation and smallest in the opposite direction. The surface of RSTF pulses should be approximately the same at any station. It is a measure of seismic moment in the time domain. More precisely, this surface represents a ratio of seismic moment of the main event over seismic moment of the Green's event.

The RSTF pulses of all the events studied, except event **B**, display directivity effects, implying unilaterally propagation ruptures, however the spatial distribution of their widths and amplitudes is not uniform in all cases. The widths and maximum amplitudes of the RSTF pulses of events **A**, **C**, and **D** show distinct dependence on the station azimuth and are in excellent agreement with each other. The spatial distribution of pulse widths of the RSTF for event **C** is shown in Figure 6, and that of pulse amplitudes is presented in Figure 7. The directivity effects from events **F** and **G** are displayed distinctly by the spatial distribution of pulse maximum amplitudes only and not by pulse widths. These events seem to be complex ones and their pulse widths are not regularly distributed in space.

The rupture velocity and the rupture propagation direction can be determined simultaneously by a least-squares technique either from the distribution of the RSTF

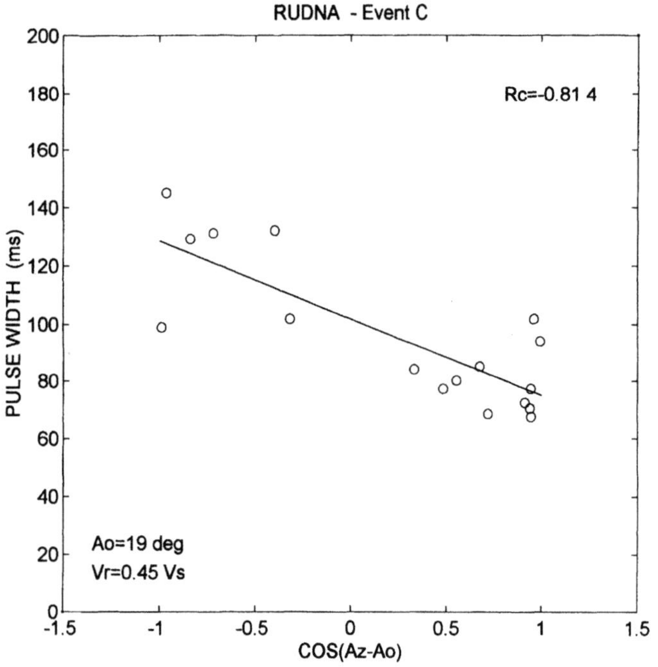

Figure 8
The best linear fit of the pulse widths of relative source time functions against the cosine of the difference between the station azimuth A_z and the rupture propagation direction A_o for event **C**. The equation of this straight line is used to derive the rupture velocity V_r as a fraction of the shear-wave velocity V_s. The correlation coefficient R_c is also shown.

pulse widths (e.g., LI and THURBER, 1988) or from the distribution of their maximum amplitudes (LI *et al.*, 1995). In the latter case, unilateral and bilateral ruptures can be distinguished. For a finite moving source, the relation between the pulse width τ and the azimuth angle A_o, measured with respect to the propagation direction, can be written as (e.g., BEN-MENAHEM, 1962)

$$\tau = (L/V_r) - (L/V_p)\cos(A_z - A_o) , \tag{5}$$

where L is the length of the fault, V_r and V_p are the rupture and P-wave velocity, respectively, and A_z is the azimuth of a given station. The rupture direction is estimated by fitting the pulse duration at stations with different azimuths to obtain a best straight line. Then the slope and intercept of this straight line are used to estimate the rupture velocity and the fault length. A similar procedure can be applied to the distribution of pulse maximum amplitudes to estimate the rupture direction and rupture velocity independently (for details, see LI *et al.*, 1995).

An example of such a procedure for the pulse widths from the event **C** is shown in Figure 8 and that for the pulse maximum amplitudes from the same event **C** is presented in Figure 9. Both procedures provide very similar results. Source

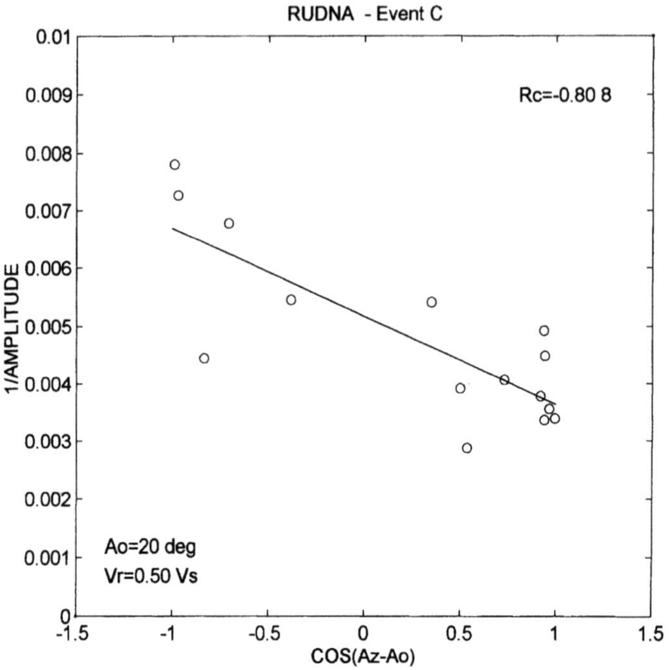

Figure 9
The best linear fit of the maximum amplitude reciprocals of relative source time functions against the cosine of the difference between the station azimuth A_z and rupture propagation direction A_o for event **C**.

parameters calculated from the STF of the investigated seismic events are listed in Table 2, where correlation coefficients related to the best straight line fitting are also given. A *P*-wave velocity in the source area equal to 4800 m/s and a *S*-wave velocity equal to 2800 m/s were accepted for the calculations.

The moment magnitude *M* estimated from the source pulse area is only slightly smaller than M_w estimated from the spectral level. The fault length *L* for events **B**, **F** and **G** (marked by asterisks in Table 2) was estimated from the mean value of the RSTF rise time, ranging from 0.054 to 0.065 s, and the corresponding rupture velocity using the suitable relation given by BOATWRIGHT (1980). In general, the fault length of seismic events from the Rudna mine estimated from the STF is several times smaller than that estimated from corner frequencies through the model-dependent corner frequency-source radius relations, commonly used in studies of source parameters (e.g. GIBOWICZ, 1997).

The rupture propagated approximately either from south to north or from north to south, within about 25 degrees. The rupture velocity was low in comparison with that from natural earthquakes, only between 0.25 to 0.54 of the shear-wave velocity. It seems that the shallow ruptures induced by mining, at least in the Lubin Copper Basin, propagate with distinctly less velocity than the ruptures generated by tectonic processes. GIBOWICZ *et al.* (1989) have drawn similar conclusions for the same area.

Furthermore, the rupture velocities from seismic events at Strathcona mine, Sudbury, Ontario, found by Urbancic et al. (1993) from the velocity pulse rise time measurements, were almost identical to those found here, ranging from 0.3 to 0.6 of the shear wave velocity.

Conclusions

The use of the empirical Green's function deconvolution technique is especially convenient for studying source time functions of seismic events induced by mining. Underground seismic networks in Polish copper mines are located close to the source area and constitute numerous sensors.

The shape of the relative source time functions for five events out of six retrieved at several stations at Rudna copper mine depends on the station azimuth, which indicates that the ruptures propagated unilaterally. In fact, they propagated approximately either from south to north or from north to south. The rupture velocity was low in comparison with that from natural earthquakes, ranging from 0.25 to 0.54 of the shear-wave velocity.

The moment magnitudes estimated from the source pulse area are only slightly smaller than those estimated from the spectral level. The fault length, however, estimated from the source time function is several times smaller than that estimated from corner frequencies through the model-dependent corner frequency-source radius relations.

The procedure applied for deriving the source time function provides information on source processes not available from other techniques currently employed in studies of seismic events induced by mining.

Acknowledgment

We are grateful to two anonymous reviewers whose comments aided considerably the text enhancement of this work. We are also grateful to the management and professional staff of the Rudna copper mine for providing digital records of seismic events, their location and instrumental information from the seismic network. This work was financially supported by the Polish Committee for Scientific Research, Grant No. 9 T12B 026 17.

References

BEN-MENAHEM, A. (1962), *Radiation of Seismic Body Waves from a Finite Moving Source in the Earth*, J. Geophys. Res. *67*, 345–350.

BERTERO, M., BINDI, D., BOCCACCI, P., CATTANEO, M., EVA, C., and LANZA, V. (1997), *Application of the Projected Landweber Method to the Estimation of the Source Time Function in Seismology*, Inverse Problems *13*, 465–486.

BERTERO, M., BINDI, D., BOCCACCI, P., CATTANEO, M., EVA, C., and LANZA, V. (1998), *A Novel Blind-deconvolution Method with an Application to Seismology*, Inverse Problems *14*, 815–833.

BOATWRIGHT, J. (1980), *A Spectral Theory for circular Seismic Sources: Simple Estimates of Source Dimension, Dynamic Stress Drop, and Radiated Energy*, Bull. Seismol. Soc. Am. *70*, 1–28.

BOUR, M. and CARA, M. (1997), *Test of a Simple Green's Function Method on Moderate-sized Earthquakes*, Bull. Seismol. Soc. Am. *87*, 668–683.

DOMAŃSKI, B. and GIBOWICZ, S. J. (1999), *Determination of Source Time Function of Mining-induced Seismic Events by the Empirical Green's Function Approach*, Publ. Inst. Geophys. Polish Acad. Sci. *M-22 (310)*, 5–14.

FLETCHER, J. B. and SPUDICH, P. (1998), *Rupture Characteristics of the Three M ~ 4.7 (1992–1994) Parkfield Earthquakes*, J. Geophys. Res. *103*, 535–554.

GIBOWICZ, S. J. (1997), *Scaling Relations for Seismic Events at Polish Copper Mines*, Acta Geophys. Pol. *45*, 169–181.

GIBOWICZ, S. J. and KIJKO, A., *An Introduction to Mining Seismology* (Academic Press, San Diego 1994).

GIBOWICZ, S. J., NIEWIADOMSKI, J., WIEJACZ, P., and DOMAŃSKI, B. (1989), *Source Study of the Lubin, Poland, Mine Tremor of 20 June 1987*, Acta Geophys. Pol. *37*, 111–132.

HANKS, T. C. and KANAMORI, H. (1979), *A Moment Magnitude Scale*, J. Geophys. Res. *84*, 2348–2350.

HARTZELL, S. H. (1978), *Earthquake Aftershocks as Green's Functions*, Geophys. Res. Lett. *5*, 1–4.

JOST, M. L., BUSSELBERG, T., JOST, O., and HARJES, H.-P. (1998), *Source Parameters of Injection-induced Microearthquakes at the KTB Deep Drilling Site, Germany*, Bull. Seismol. Sol. Am. *88*, 815–832.

LI, Y. and THURBER, C. H. (1988), *Source Properties of Two Microearthquakes at Kilauea Volcano, Hawaii*, Bull. Seismol. Soc. Am. *78*, 1123–1132.

LI, Y., DOLL, C., Jr., and TOKSÖZ, M. N. (1995), *Source Characterization and Fault Plane Determinations for $M_{bLg} = 1.2$ to 4.4 Earthquakes in the Charlevoix Seismic Zone, Quebec, Cananda*, Bull. Seismol. Soc. Am. *85*, 1604–1621.

MORI, J. (1993), *Fault Plane Determinations for Three Small Earthquakes along the San Jancino Fault, California: Search for Cross Faults*, J. Geophys. Res. *98*, 17,711–17,722.

MUELLER, C. S. (1985), *Source Pulse Enhancement by Deconvolution of an Empirical Green's Funciton*, Geophys. Res. Lett. *12*, 33–36.

PIANA, M. and BERTERO, M. (1997), *Projected Landweber Method and Preconditioning*, Inverse Problems *13*, 441–463.

URBANCIC, T. I., TRIFU, C-I. and YOUNG, R. P. (1993), *Stress release estimates, scaling behaviour, and source complexities of microseismic events* (R. P. Young, ed.), A. A. Balkema, Rotterdam, 255–260.

WIEJACZ, P. (1992), *Calculation of Seismic Moment Tensor for Mine Tremors from the Legnica-Głogów Copper Basin*, Acta Geophys. Pol. *40*, 103–122.

(Received February 3, 2000, revised/accepted April 20, 2000)

Pure appl. geophys. 159 (2002) 145–164
0033–4553/02/030145–20 $ 1.50 + 0.20/0

Pure and Applied Geophysics

Reliability of Seismic Moment Tensor Inversions for Induced Microseismicity at Kidd Mine, Ontario

C. I. TRIFU[1] and V. SHUMILA[1]

Abstract — A novel seismic moment tensor inversion approach is applied to microseismic events from the Kidd mine with moment magnitudes ranging between −1.2 and 0. Data consist of 35 events recorded on 8 triaxial accelerometers installed underground. Reliable solutions are obtained for 21 events, of which 14 represent pure shear mechanisms, whereas the remaining 7 exhibit a significant positive volumetric component (72–76%), along with some pure shear failure (15–20%). Further analysis indicates that 6 of the events characterized by high volumetric components are located within a sill pillar on the 4,700 level and have subvertical P and subhorizontal T axes. This is in agreement with the presence of tensile cracks close to openings for incipient pillar bursting. The pure shear events are located outside the sill pillar, between the 4,600 and 4,800 levels, and on the 5,600 level within a highly fractured rockmass. For the latter events, the subvertical nodal planes are found to match closely the orientation of subvertical NW-SE fractures aligned parallel to the major faults in the area.

Key words: Induced seismicity, seismic moment tensor, tensile mechanism.

Introduction

Numerous studies have outlined that most of the mining-induced seismic events exhibit pure-shear mechanisms, resulting from slip along pre-existing geological features such as faults, dykes and joint sets (e.g., SPOTTISWOODE and MCGARR, 1975; GIBOWICZ, 1990; TRIFU and URBANCIC, 1997). Nonetheless, the presence of underground excavations like stopes and drifts allows for the potential of volumetric components of failure to occur (JOUGHIN and JAGER, 1983). For instance, application of moment tensor inversion technique has enabled the identification of such components in the well monitored environment of the Underground Research Laboratory in Pinawa, Manitoba (FEIGNIER and YOUNG, 1992; BAKER and YOUNG, 1997; COLLINS and YOUNG, 2000). The methodologies employed in these studies, however, are time-consuming and therefore impractical for day to day seismic monitoring required by the mining industry.

[1] Engineering Seismology Group, Kingston, Ontario K7K 7G3, Canada.

Kidd Mining Division of Falconbridge Ltd. operates a deep copper-zinc mine near Timmins, Ontario. Currently, mining extends to a depth of 2,100 m (6,800 ft), but approval has been given to further the operations to 3,200 m (10,400 ft) below surface. Monitoring of the deeper section of the mine is presently carried out with an underground microseismic array of 20 uniaxial and 8 triaxial accelerometers. At such depths the evaluation of microseismic failure mechanisms is considered critical at Kidd for a better understanding of the rockmass response to mining and assessment of the ground conditions.

In a recent study TRIFU *et al.* (2000) proposed a rapid evaluation of both shear and non-shear failure components of induced seismic sources by employing *P*- and *S*-wave low frequency spectral amplitude estimates determined directly in the time domain, with first-wave polarities attached. Based on extensive testing on synthetic data, they concluded that the methodology was robust and proposed it for applications on real data. The purpose of this study is to apply the above methodology to a mining-induced microseismicity data set recorded at Kidd, evaluate the reliability of the results derived, and conclude upon the possibility of employing this technique on a routine basis at the mine.

Methodology

At great distances from the source and for wavelengths much greater than its linear dimensions, a fault can be represented as a point source. A second rank moment tensor can describe, to first-order approximation, the equivalent forces of a general seismic point source and a variety of seismic source models. Assuming that all components of the time-dependent seismic moment tensor have the same time dependence $s(t)$ and neglecting higher-order terms, the displacement field is approximated by

$$u_i(\mathbf{x}, t) = M_{jk}[G_{ij,k} * s(t)] \quad (1)$$

where $M_{jk}(\equiv\mathbf{M})$ is the seismic moment tensor, $G_{ij,k}(\equiv\mathbf{G}')$ is the first derivative of the Green's function, and indicial summation is implied. The seismic moment tensor describes both the scalar moment strength (M_0) and the radiation pattern of the source.

Due to the conservation of kinetic moment, the tensor $\mathbf{M}$ is symmetric, and so contains only six independent components. For a homogeneous and isotropic medium each component is equivalent to a force couple acting in a particular direction. When the force and the moment arm are similarly oriented, the component expresses a volume change. All other possible couples represent true deviatoric components. There are numerous ways of decomposing the general moment tensor $\mathbf{M}$. An interesting decomposition appears to be one that describes both volumetric and pure shear or double-couple components (DC). This decomposition will additionally require the presence of a components linear vector dipole (CLVD; KNOPOFF and RANDALL, 1970), which according to FROHLICH *et al.* (1989) may be indicative of source complexity (second term below)

$$\mathbf{M} = 1/3(\lambda_1 + \lambda_2 + \lambda_3)\mathbf{I} + \lambda_3^D F(2\mathbf{a}_3^2 - \mathbf{a}_2^2 - \mathbf{a}_1^2) + \lambda_3^D(1 - 2F)(\mathbf{a}_3^2 - \mathbf{a}_2^2) \quad (2)$$

where $\mathbf{I}$ is the identity matrix, λ_1 to λ_3 the eigenvalues, $\mathbf{a}_1$ to $\mathbf{a}_3$ the eigenvectors, $F = -\lambda_1^D/\lambda_3^D, \lambda_i^D = \lambda_i - (\lambda_1 + \lambda_2 + \lambda_3)/3$, and $|\lambda_3^D| \geq |\lambda_2^D| \geq |\lambda_1^D|$ (JOST and HERRMANN, 1989). The relative contribution of each failure component to the above decomposition can be evaluated by normalizing each coefficient in equation (2) – in absolute value for the deviatoric terms – to the sum of the absolute values of all coefficients.

Assuming a step time function for the moment tensor components, the theoretical or expected spectral displacement field can be written as

$$\bar{\mathbf{u}} = \mathbf{G}'\mathbf{M} = c\mathbf{F}\mathbf{M} \quad \text{or} \quad \bar{u}_i = \sum_{k=1}^{6}(c^i f_k^i m_k) \quad (3)$$

where $c = 1/(4\pi\rho v^3 R)$, with ρ the density of the homogeneous medium, v the wave velocity, R the source-sensor distance, and m_k are the components of the moment tensor, in a vectorial notation, $m_1 = M_{11}$, $m_2 = M_{12} = M_{21}$, $m_3 = M_{22}$, $m_4 = M_{13} = M_{31}$, $m_5 = M_{23} = M_{32}$, $m_6 = M_{33}$. For each wave type, the components of the excitation matrix f_k, $f_1 = F_{11}$, $f_2 = F_{12} + F_{21}$, $f_3 = F_{22}$, $f_4 = F_{13} + F_{31}$, $f_5 = F_{23} + F_{32}$, $f_6 = M_{33}$ (STRELITZ, 1978), depend on the take-off and azimuth angles (TRIFU *et al.*, 2000). The notation 'i' is used to indicate both a specific location and a wave type (P, SV, SH).

Considering that the differences between observational (u_i) and theoretical ($\bar{u}_i$) displacements are normally distributed (with dispersion σ_i^2), the maximum likelihood function can be defined based on the probability density distribution function (ψ) as

$$L(u_i|\mathbf{m}) = \sum_i \log(\psi_i) = -1/2\sum_i[(u_i - \bar{u}_i)^2/\sigma_i^2] - \sum_i \log\sigma_i^2 + \text{constant} \ . \quad (4)$$

At the optimal point, where L is maximum, the first derivatives with respect to model parameters must be zero

$$\partial L(\mathbf{m})/\partial m_j = \sum_i[(u_i - \bar{u}_i)c^i f_j^i/\sigma_i^2] = 0 \ . \quad (5)$$

After substituting the theoretical displacements based on equation (3), this generates a system of linear equations with respect to the unknown model parameters (m_k, $k = 1$ to 6). For simplicity, the dispersion can be assumed constant for all data, $\sigma_i^2 = \sigma_0^2$, and evaluated as a parameter in equation (4). The maximum likelihood estimates from equation (5) will have an asymptotically normal distribution with the

Table 1

Possible source models

Model	Eigenvalues	Independent parameters	Comments
Isotropic	$(\lambda_1, \lambda_1, \lambda_1)$	1	volume change, i.e., implosion or explosion
Double-couple	$(\lambda_1, -\lambda_1, 0)$	4	pure shear
Deviatoric	$(\lambda_1, \lambda_2, -\lambda_1-\lambda_2)$	5	complex shear, i.e., pure shear and compensated linear vector dipole
General	$(\lambda_1, \lambda_2, \lambda_3)$	6	complex shear and volume change

mean value **m** and a covariance matrix defined as the inverse of the Fisher matrix (COX and HINKLEY, 1986)

$$C_{ij} = [B(\mathbf{m})]^{-1} = [\overline{-\partial^2 L(\mathbf{m})/\partial m_i \partial m_j}]^{-1} \ . \tag{6}$$

Seismic source models, best characterized by the eigenvalues of the associated diagonalized moment tensor (Table 1), impose restrictions on the seismic moment tensor components. The inversion for any particular source model other than the general case will imply constraints on the inversion process. For example, in the deviatoric model the volume change is nil and thus the trace of the tensor must be zero

$$\mathrm{Tr}(\mathbf{M}) = m_1 + m_3 + m_6 = 0 \ . \tag{7}$$

The imposed restriction is that $m_6 = -(m_1 + m_3)$, with which the system of equations (5) remains linear with respect to the unknowns ($m_k, k = 1$ to 5).

For linear systems of equations, as in the case of general and deviatoric models, the preferred method of inversion is that of single value decomposition (SVD). This technique is typically used in cases when the amount of data is small compared to the number of model parameters, and when the associated matrices are ill-defined. In SVD the excitation matrix F in equation (3) is decomposed as

$$\mathbf{F} = \mathbf{W}\,\Lambda\,\mathbf{V}^T \tag{8}$$

where **W** contains the eigenvectors corresponding to the non-zero eigenvalues of $\mathbf{F}\mathbf{F}^T$, Λ is a diagonal matrix of the positive square-roots of the non-zero eigenvalues of $\mathbf{F}^T\mathbf{F}$ and **V** contains the eigenvectors corresponding to the non-zero eigenvalues of $\mathbf{F}^T\mathbf{F}$ (STUMP and JOHNSON, 1997). The generalized inverse of **F** is then defined by

$$\mathbf{F}^{-1} = \mathbf{V}\,\Lambda^{-1}\,\mathbf{W}^T \tag{9}$$

and the solution to the seismic moment tensor is

$$\mathbf{M} = \mathbf{F}^{-1} c_1 \mathbf{u} = (\mathbf{V}\,\Lambda^{-1}\,\mathbf{W}^T) c_1 \mathbf{u} \tag{10}$$

where $c_1 = 1/c$, and **u** is the vector of observations. When the solution is unique and the model fits the data, this reduces to the least-squares solution of equation (5).

For the double-couple model the restriction is nonlinear, requiring not only the trace to be zero, but one of the three eigenvalues to be zero too (Table 1). This nonlinearity is more easily seen by stating that the determinant of the moment tensor be zero

$$\det|\mathbf{M}| = \Phi(\mathbf{m}) = 0 \tag{11}$$

or that the product of its eigenvalues is zero (STRELITZ, 1978). In order to find the model parameters, the maximum of the likelihood function must be searched under nonlinear restrictions of its arguments. The likelihood function to be maximized is

$$L(\mathbf{m}) = L_0(\mathbf{m}) + \delta\Phi(\mathbf{m}) + \gamma\Phi^2(\mathbf{m}) \tag{12}$$

where L_0 is the unrestricted likelihood function given in equation (4), and δ and γ are undetermined multipliers. The last two terms in equation (12) act as a penalty function which restricts the search area to the values **m** from the equation (11). At the optimal point the first derivatives with respect to model parameters must be zero, similar to the condition in equation (5), but the resulting system of equations is now nonlinear in model parameters. The numerical algorithm iteratively changes the undetermined multipliers to ensure the condition (11) is satisfied.

For a nonlinear system of equations the maximum likelihood is linearized through a parabolic approximation. The problem is then solved using the Newton-Raphson method which requires the evaluation of both the function and its derivative at arbitrary points x (PRESS *et al.*, 1989). In terms of the likelihood function, the solution search is represented by the equation

$$\mathbf{m}_n = \mathbf{m}_{n-1} + [B(\mathbf{m}_{n-1})]^{-1}\nabla L(\mathbf{m}_{n-1}) \ . \tag{13}$$

Model Evaluation

There are several statistical parameters that must be investigated for a correct assessment of the derived model parameters. The covariance matrix (equation 6) evaluates the linear independence of the obtained estimates of the moment tensor components. Values close to 0 imply little or no linear dependence between tensor components, whereas values close to $|1|$ imply a strong linear dependence.

The overall dispersion is determined from

$$\sigma_0^2 = 1/(K-p)\sum_i (u_i - \bar{u}_i)^2 \tag{14}$$

where K is the total number of observations, and p corresponds to the number of independent components of the moment tensor ($p = 6$ for the general model, $p = 5$ for the deviatoric model, and $p = 4$ for the double-couple model). The standard dispersion estimates the overall size of deviation between the observed and modeled values.

Another estimate of the model fit is the multiple correlation coefficient

$$R^2 = \left[\sum_i (\bar{u}_i - u^*)^2\right] \Big/ \left[\sum_i (u_i - u^*)^2\right] \tag{15}$$

where u^* is the average observed displacement. R^2 values close to 1 indicate a good fit and close to 0 indicate a poor fit between the model and observed values.

Based on the seismic moment tensor decomposition proposed in equation (2), for each event there are three possible source models from which to choose: general, deviatoric, and double-couple. In the presence of errors the solution favors the model with the largest number of independent parameters (p). In other words, although the maximum likelihood estimate may favor the general model, a hypothesis testing approach is needed in order to evaluate the relevance of the model solution. This is carried out using the F-test. To examine the model solution the ratio of the model fit dispersion σ_0^2 (equation 11) to the data dispersion σ^2,

$$\sigma^2 = 1/(K-1) \sum_i (u_i - \bar{u}_i)^2 \tag{16}$$

is calculated. This ratio leads to the significance level at which the hypothesis that the model fit has smaller variance than that of the data can be rejected (PRESS *et al.*, 1989). A large value implies a high confidence that the hypothesis can be rejected. This indicates that it is less probable one will obtain the same result by repeating experiments.

To compare the significance of different models, the dispersion of one model fit to another is calculated. The smaller of the two dispersions is placed in the numerator to avoid possible singularities. A low value for the ratio indicates that the prevalence in the preferred model being caused by purely random factors is small, whereas high values indicate that the prevalence is most likely due to random factors.

Kidd Mining Division

The Kidd Mining Division of Falconbridge Ltd. in Timmins, Ontario (Fig. 1) produces approximately 2.5 million tons of copper-zinc ore per year, using open stope blast hole mining with delayed consolidated rock fill. Mining operations began some 35 years ago with an open-pit and proceeded underground in stages. No. 1 Mine extends to the 850 m depth (2,800 ft level), No. 2 Mine to 1,400 m (4,600 level),

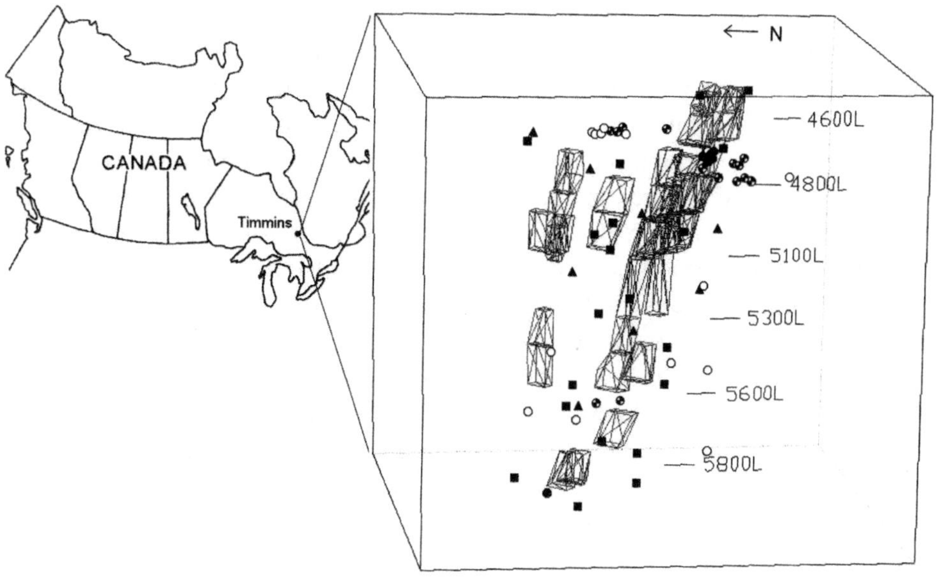

Figure 1
Screen shot depicting a 3-D view of the study volume at the Kidd mine: Locations of the uniaxial sensors are marked by squares, and those of triaxial sensors by triangles; solid circles denote events for which a failure mechanism with a strong volumetric component was derived, beach-ball circles those for which pure-shear failure was found, and open circles the events for which no reliable mechanism solution was obtained. The mined out and refilled stopes indicate the geometry of the ore body.

and No. 3 Mine to 1,830 m (6,000 level). Phase 2 of No. 3 Mine, from 1,830 to 2,070 m began production in early 2001. An extensive deep drilling program has identified a large mineral resource below 2,070 m to 3,050 m, and approval has been given to develop "D" Mine. Shaft sinking will start in 2001 from the 1,400 to 3,170 m depth, targeting a 2 million ton per year extraction rate using the same open stope blast hole mining technique. This will make the Kidd Mining Division the deepest base metal mine in the world.

The complex geology includes low strength ultramafic rocks in the hangingwall, felsic volcanics which are the host to the orebody and comprised of folded rhyolite, and very brittle, competent massive sulphide ore. The footwall consists of two geological units: an andesite-diorite that has a compressive strength similar to the sulphides, with many subhorizontal and subvertical quartz-filled joint sets; and a pillow mafic flow (dacite) that has properties similar to the andesite-diorite. The stratigraphy trends north-south, is overturned and dips steeply to the east. Three phases of folding have shaped the orebody, resulting in a tight, steeply northeast plunging rod. There are two systems of faults that affect the mine-wide stability: the Gauge and Splay Faults which have an impact on hangingwall behavior, and a family of steeply south dipping echelon faults, associated with some of the larger seismic events at Kidd.

Intuitively, rockbursts and mining-induced seismicity occur more often in deeper mines due to the presence of higher stresses. In general, rockbursts also depend on the shape of the orebody and mining configuration. At Kidd, the majority of the induced seismicity is not randomly distributed nor does it dramatically increase with depth. The activity is mainly concentrated in the sill pillars that form the boundaries between Nos. 1 and 2 Mines, an internal sill at 1,220 m depth in No. 2 Mine and, to a lesser extent, between Nos. 2 and 3 Mines. Since it has been observed that seismicity decreases steadily after destressing of these pillars, the strategy has been to mine through the sills early in the mining cycle before they become subject to huge induced stresses (BRUMMER *et al.*, 1996).

The plug-shaped orebody at Kidd has a significant influence on the size and rate of seismicity. This shape, almost round in cross section, ensures a stiff loading where the load from the host rock is easily transferred to the abutments as the ore yields. The pillars are relatively soft except for some in the sulphide zone and tend to fail before storing high stresses. This is in strong contrast with the situation encountered in the case of tabular ore bodies (horizontally long and very thin in cross section) where loads are not so easily transferred since the abutments are remotely located from the center. The latter geometry leads to a soft loading that acts as a continuous transfer of the stress from the host rock to the pillars as the ore is extracted. For most seismically active mines, pillars are very stiff and tend to fail violently resulting in very unstable conditions.

Kidd has taken advantage of the information provided by the microseismic monitoring over the past 20 years. For daily operation as well as a guide for mine design purposes, the identification of microseismic sources in real time is a routine procedure at the mine. As mining activity is scheduled to deepen, microseismic monitoring is expected to provide additional information on the integrity of sill pillars and the activation of known structures such as faults, dykes and joint sets. The evaluation of the failure mechanism of small events can be instrumental for a quantitative assessment of the above conditions. Conceptually, the seismic moment tensor inversion shows great promise in this regard. Due to the complexity of this technique however, a study of the reliability of inversion results on real data recorded at Kidd is warranted prior to applying this technique as a routine procedure at the mine.

Instrumentation and Data Analysis

The study region consists of the upper part of the No. 3 Mine, a volume of approximately 300 m × 300 m × 400 m extending down between the 4,600 and 6,000 levels. This rockmass is monitored by an ESG's Hyperion underground microseismic array of 20 uniaxial and 8 triaxial sensors homogeneously distributed throughout (Fig. 1). The sensor types are the A1030 uniaxial accelerometer with a flat response (±3 dB) between 50 Hz and 5 kHz and a sensitivity of 30 V/g, and the A3005 triaxial accelerometer with a flat response (±3 dB) from 1 Hz to 5 kHz at 0.3 V/g. The uniaxial sensors are cemented uphole, whereas the triaxial sensors are cemented downhole, away from the openings. The sensors are connected to an underground central multiplexor via copper wires where they are sampled at 20 kHz with 16-bit resolution and channeled to surface via fiber optic cable. Signals are there demultiplexed and stored to disk.

The use of low frequency displacements in the evaluation of failure components offers the possible estimation of the general failure geometry without interference from rupture details. For a Brune source the low frequency spectral amplitude plateau can be evaluated as

$$|u_{0i}| = 2S_{D2_i}^{3/4}/S_{V2_i}^{1/4} \tag{17}$$

where S_{D2} and S_{V2} are the time integrals of the squared displacement and velocity, respectively,

$$\begin{aligned} S_{D2} &= \int_{t_1}^{t_2} D^2(t)\mathrm{d}t \\ S_{V2} &= \int_{t_1}^{t_2} V^2(t)\mathrm{d}t \end{aligned} \tag{18}$$

and i denotes, for simplicity, both the data location and the particular P, SV, and SH wavetrain. The limits of the above time integrals correspond to the start and end of each particular wavetrain. The P wavetrain is considered to last until the arrival of the $S-$ wave, whereas the S wavetrain is assumed to be twice as long. Note that a complete description of the radiation pattern requires however, both amplitude and phase information. Synthetic tests have clearly shown that for a small data set the lack of phase information will lead to a loss of linearity in equation (5) that can negatively affect the inversion process (TRIFU *et al.*, 2000). As such, low-frequency displacement amplitudes must be supplemented with first motion polarities for each wave type (Pol_i)

$$u_i = Pol_i|u_{0i}| \ . \tag{19}$$

The evaluation of low-frequency displacements and first motion polarities requires the identification of each wave type after rotating the three-component waveforms into their local ray coordinate system. The radial, transverse vertical and transverse horizontal components correspond to the $P-$, $SV-$, and $SH-$ wave, respectively (Fig. 2). Rotations have been carried out based on known event locations, and assuming linear raypaths. An indirect measure of the deviations from linear paths

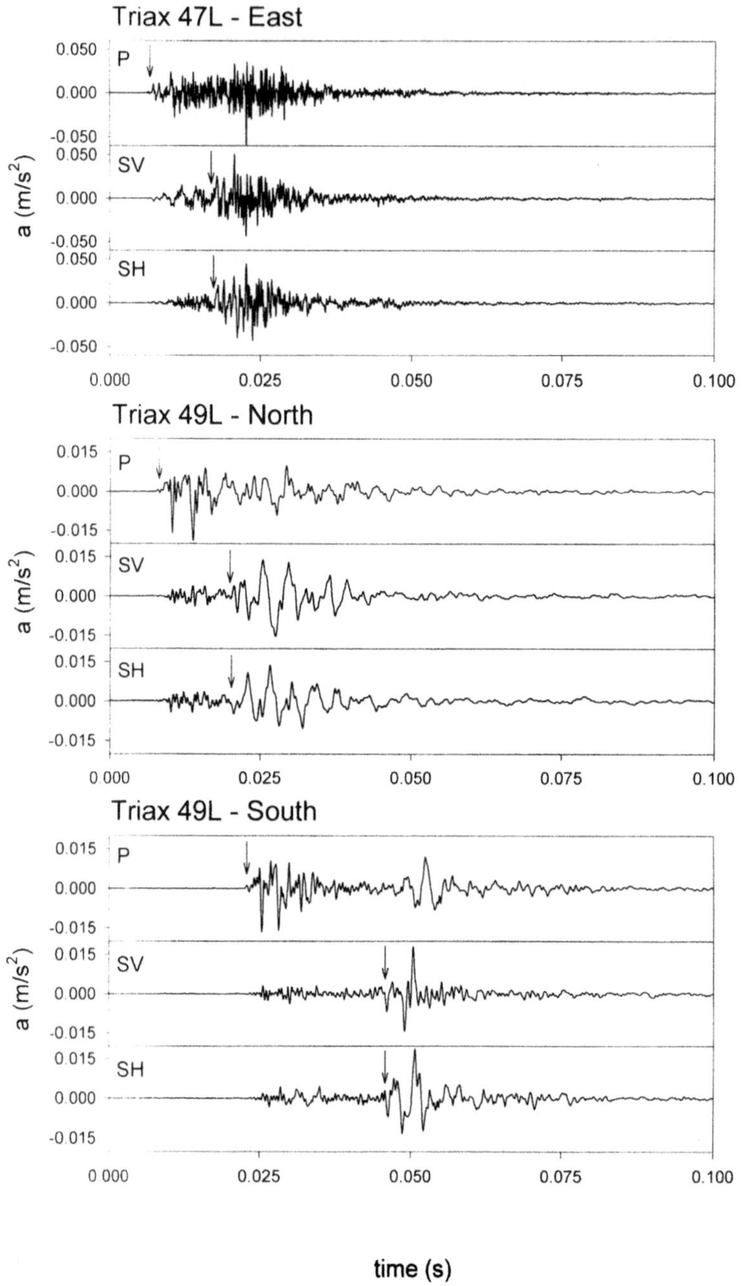

Figure 2
Examples of rotated waveforms: event #27 (see Table 2, in mine coordinates: northing 65,586 m; easting 65,774 m; depth 1,447 m), recorded by 3 triaxial accelerometers located on the 4,700 level-east (65,780 m; 65,814 m; 1,433 m), 4,900 level-north (65,664 m; 65,774 m; 1,514 m) and south (65,583 m; 65,729 m; 1,523 m), respectively.

can be obtained by employing a polarization filter. Note that wavetrains containing minimal wave energy or low signal-to-noise ratio (below 4) were disregarded prior to this analysis.

For rotation, a covariance matrix of all the components is defined for a selected time window around the *P*-wave arrival, usually at least one cycle in length (MONTALBETTI and KANASEWICH, 1970)

$$\mathbf{C} = \begin{pmatrix} \mathrm{Var}[x_1] & \mathrm{Cov}[x_1,x_2] & \mathrm{Cov}[x_1,x_3] \\ \mathrm{Cov}[x_1,x_2] & \mathrm{Var}[x_2] & \mathrm{Cov}[x_2,x_3] \\ \mathrm{Cov}[x_1,x_3] & \mathrm{Cov}[x_2,x_3] & \mathrm{Var}[x_3] \end{pmatrix} \tag{20}$$

where (x_1,x_2,x_3) are the components of motion in the (east, north, up) coordinate system, and

$$\mathrm{Var}[x] = \frac{1}{(n-1)}\sum_{i=1}^{n}(x_i - x^*)^2, \quad \mathrm{Cov}[x,y] = \frac{1}{(n-1)}\sum_{i=1}^{n}[(x_i - x^*)(y_i - y^*)] \tag{21}$$

where * denotes the average amplitude within the time window, and n is the number of sample points. The above matrix can be diagonalized by solving an eigenvalue and eigenvector problem, and an estimate of the linearity of the *P*-wave polarization calculated as

$$p = \frac{(p_1 - p_2)^2 + (p_1 - p_3)^2 + (p_2 - p_3)^2}{2(p_1 + p_2 + p_3)^2} \tag{22}$$

where p_1, p_2, and p_3 are the eigenvalues. This coefficient will vary between 1 for complete linear motion ($p_1 = 1$, $p_2 = p_3 = 0$) and 0 for unpolarized motion ($p_1 = p_2 = p_3$). Only coefficient values above 0.9 were retained in the analysis. The axis of polarization, corresponding to the *P*-wave motion, is determined from the eigenvector associated with the largest eigenvalue, and is characterized in terms of the take-off (ϕ) and azimuth (φ) angles. The components of motion in the ray coordinate system are obtained as

$$\begin{pmatrix} u_P \\ u_{SV} \\ u_{SH} \end{pmatrix} = \begin{pmatrix} -\sin\phi\sin\varphi & -\sin\phi\cos\varphi & \cos\phi\sin\varphi \\ \cos\phi\sin\varphi & \cos\phi\cos\varphi & \sin\phi \\ -\cos\varphi & \sin\varphi & 0 \end{pmatrix}\begin{pmatrix} x_1 \\ x_2 \\ x_3 \end{pmatrix} \tag{23}$$

Results

The induced microseismicity at Kidd comprises an average of 30–40 events per day, ranging in moment magnitude form about –2.5 to 0.5. However, the moment tensor technique requires a correct identification of various wavetrains, thus only triaxial recordings will be used. With a limited number of 8 triaxial sensors installed, only events capable of triggering virtually all these sensors were retained. As such, a data set of 35 events with magnitudes between −1.2 and 0 was selected for this analysis. These events occurred between June 1998 and May 1999, at a rate of 1–4 events per month (Fig. 1 and Table 2). The event locations were obtained using a Simplex–Geiger algorithm, using manually picked first *P*-wave arrivals at 16 to 26 sensors and assuming a uniform *P*-wave velocity of 5,980 m/s. Location errors range between 6 and 18 m (in a vectorial sense) with an average of 12 m. Synthetic tests indicate that a location error of about 2% of the array diameter will have virtually no effect on the mechanism results (TRIFU *et al.*, 2000).

For each event, moment tensor inversions were carried out based on three source models (i.e., general, deviatoric, and double-couple) along with significant tests to discriminate between alternative failure mechanisms. Figure 3 shows the inversion results obtained for event #27. The significance tests indicate that the general model is prefered to within 95% confidence level. A multiple correlation coefficient of 0.9 is found for the general model inversion.

Table 2 shows that reliable solutions ($R^2 \geq 0.6$) were derived for 21 events or 60% of the analyzed data set. These solutions are based on inversions from 13 to 24 *P* and *S* wavetrains. The standard deviations in the azimuth angles of the resulting *P*, *B* and *T* axes are between 0 and 60°, whereas those in the plunge angles are between 0 and 40°, with an average value of 10° for each of the two angles. Only 7 events or one third of the solutions favor a general model to within a significance level higher than 95%, and a correlation coefficient usually greater than 0.9, whereas the remaining 14 events or two thirds of the solutions indicate a double-couple model. All the events described as having a general failure mechanism exhibit a major proportion (72–76%) of positive tensile or isotropic components supporting a mechanism similar to fresh rock opening. Note that the largest of the remaining failure components for these events is of the double-couple type (15–20%).

An analysis of the spatial distribution of the events for which statistically reliable solutions were derived provides additional support for the moment tensor inversion results. For example, 6 of the 7 events characterized by a high isotropic component are located within a few meters from each other, at the north-west edge of a sill pillar on the 4,700 level (Fig. 4). Their position is consistent with a fresh rock tensile mechanism for incipient pillar bursting. Worth noting, the event mechanisms exhibit similar subvertical and subhorizontal orientations of the *P* and *T* axes, respectively (Fig. 5). Although the far-field principal stresses T_1 and T_3 are subhorizontal and subvertical, respectively, since the pillar is mined out all around the stress flow will make them flip by 90° within the pillar, to orientations in agreement with the mechanism results.

The events associated with a double-couple source tend to be far more spread out (Fig. 1). Most of these events are located between the 4,600 and 4,800 levels outside the sill pillar on the 4,700 level. Two events however are located on the 5,600 level about 300 m deeper and oriented SW-NE inside a remnant (Fig. 6). The mechanism

Table 2

Location of the events used in the study (in mine coordinates), and the derived failure mechanism results; Err is the location error, NP the number of P-wave arrivals used in location, M_w the moment magnitude, NM the number of phases used in moment tensor inversion, MOD the derived source model (with DC and G denoting the double-couple and general models, respectively), R^2 the correlation coefficient for the resulting model fit, and VOL, |CLVD|, and |CD| the relative percentage of each failure component

Ev #	Date (ddmmyy)	Time (hh:mm:ss)	North (m)	East (m)	Depth (m)	Err (m)	NP	M_W	NM	MOD	R^2 (%)	VOL (%)	|CLVD| (%)	|DC| (%)
1	270698	12:54:45	65683	65771	1719	6	18	−0.6	19	DC	82			
2	270698	16:22:44	65593	65763	1443	13	18	−0.7	14	G	93	73	7	20
3	260798	21:16:00	65710	65740	1411	10	25	−0.4	22					
4	280798	14:18:52	65724	65736	1416	9	21	−0.6	19					
5	280798	14:42:42	65719	65716	1412	7	24	−0.6	24					
6	310798	16:01:41	65706	65742	1414	11	19	−0.5	24	DC	60			
7	040898	02:24:45	65592	65776	1596	11	22	−0.5	18					
8	060898	12:10:57	65615	65831	1694	15	18	−0.5	15					
9	250898	05:02:33	65698	65723	1413	8	22	−0.6	19	DC	64			
10	260898	22:45:04	65703	65717	1410	7	22	−0.6	22	DC	74			
11	130998	18:51:03	65595	65711	1673	13	18	−0.5	14					
12	170998	18:42:04	65551	65751	1471	15	17	−0.4	21	DC	77			
13	021098	14:13:52	65591	65768	1450	9	18	−0.7	16	G	93	76	8	16
14	131198	03:51:05	65766	65750	1817	16	23	−0.3	19	G	74	74	11	15
15	181198	20:54:31	65561	65763	1475	11	20	−0.6	20	DC	70			
16	201198	22:24:24	65697	65725	1414	6	21	−0.6	20					
17	071298	22:21:55	65594	65765	1449	15	18	−0.7	15	G	93	72	9	19
18	121298	23:06:37	65565	65750	1453	12	17	−0.6	17	DC	63			
19	171298	22:04:35	65587	65767	1444	12	19	−0.5	21					

Table 2

Continued

Ev #	Date (ddmmyy)	Time (hh:mm:ss)	North (m)	East (m)	Depth (m)	Err (m)	NP	M_W	NM	MOD	R^2 (%)	VOL (%)	|CLVD| (%)	|DC| (%)
20	291298	13:44:49	65668	65615	1392	16	20	−0.6	21	DC	70			
21	300199	04:40:47	65550	65723	1469	18	16	−1.2	17	DC	82			
22	020299	04:27:54	65586	65750	1776	15	25	−0.5	20					
23	240299	13:15:20	65521	65650	1451	18	26	−1.1	18					
24	260299	06:21:03	65581	65752	1470	16	17	−1.0	17	DC	83			
25	260299	07:39:15	65590	65767	1452	12	19	−0.6	19	G	83	73	12	15
26	130399	17:49:59	65558	65752	1455	12	22	−0.8	15	DC	79			
27	260399	04:38:42	65586	65774	1447	14	19	−0.4	17	G	90	72	11	16
28	020499	03:33:58	65787	65757	1729	6	18	−0.6	18					
29	130499	03:23:25	65778	65676	1645	8	23	−0.9	20					
30	220499	15:32:40	65736	65749	1737	13	23	−0.5	16					
31	230499	02:36:22	65707	65782	1726	11	24	−0.2	16	DC	89			
32	120599	11:25:01	65723	65737	1416	8	24	0.0	20					
33	130599	10:01:17	65553	65765	1451	14	19	−0.7	13	DC	72			
34	130599	20:15:05	65587	65761	1443	12	22	−0.2	15	G	95	74	15	11
35	140599	08:35:36	65596	65772	1459	9	18	−0.7	15	DC	62			

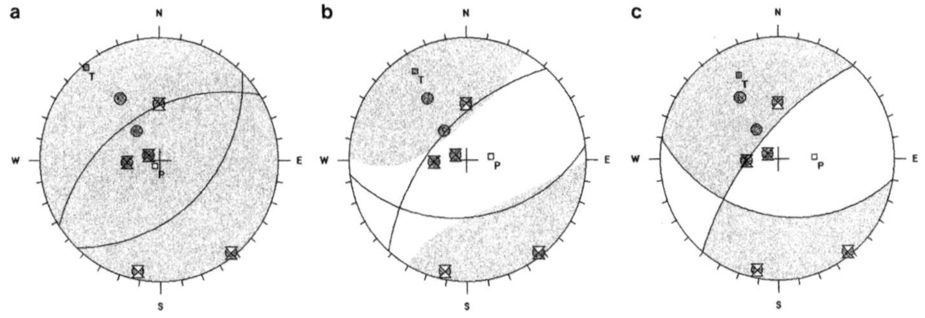

Figure 3
Moment tensor inversion results for event #27: (a) general model; (b) deviatoric model; (c) double-couple model. The general model solution is found to best describe the failure mechanism to within 95% confidence level. *P*-wave data are denoted by circles, *SV* data by inverted triangles, and *SH* data by upright triangles: Filled symbols correspond to positive first-motions, open symbols to negative ones, and a lower hemisphere projection was used throughout the study.

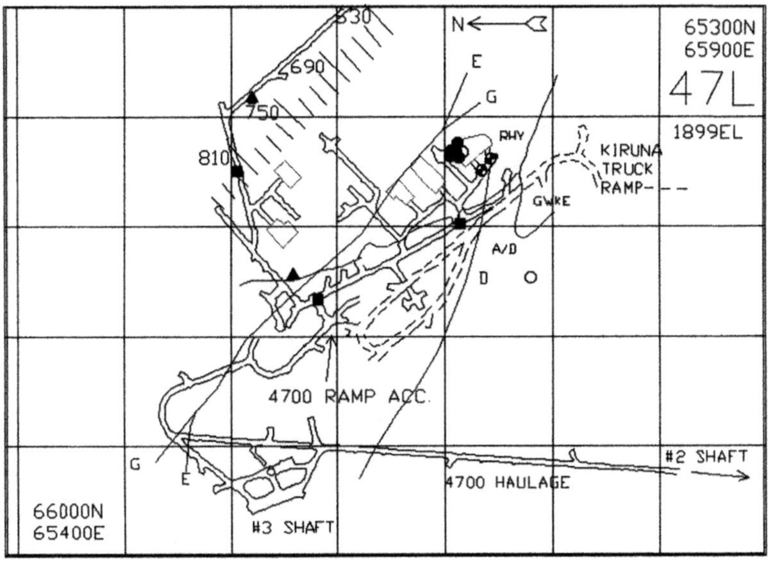

Figure 4
Plane view of the event locations within close proximity to the 4,700 level (see Fig. 1 for a description of symbols). There are six events, all located at the northern edge of the pillar between No. 2 and 3 Mines that are characterized by a large positive volumetric component, in agreement with incipient pillar bursting.

solutions outline the presence of subvertical nodal planes with subhorizontal *P* and *T* axes, approximately oriented E-W and N-S (Fig. 7), in agreement with the orientation of the principal stresses in this area. The complex local conditions however easily allow σ_1 and σ_3 to change places, and therefore both normal and

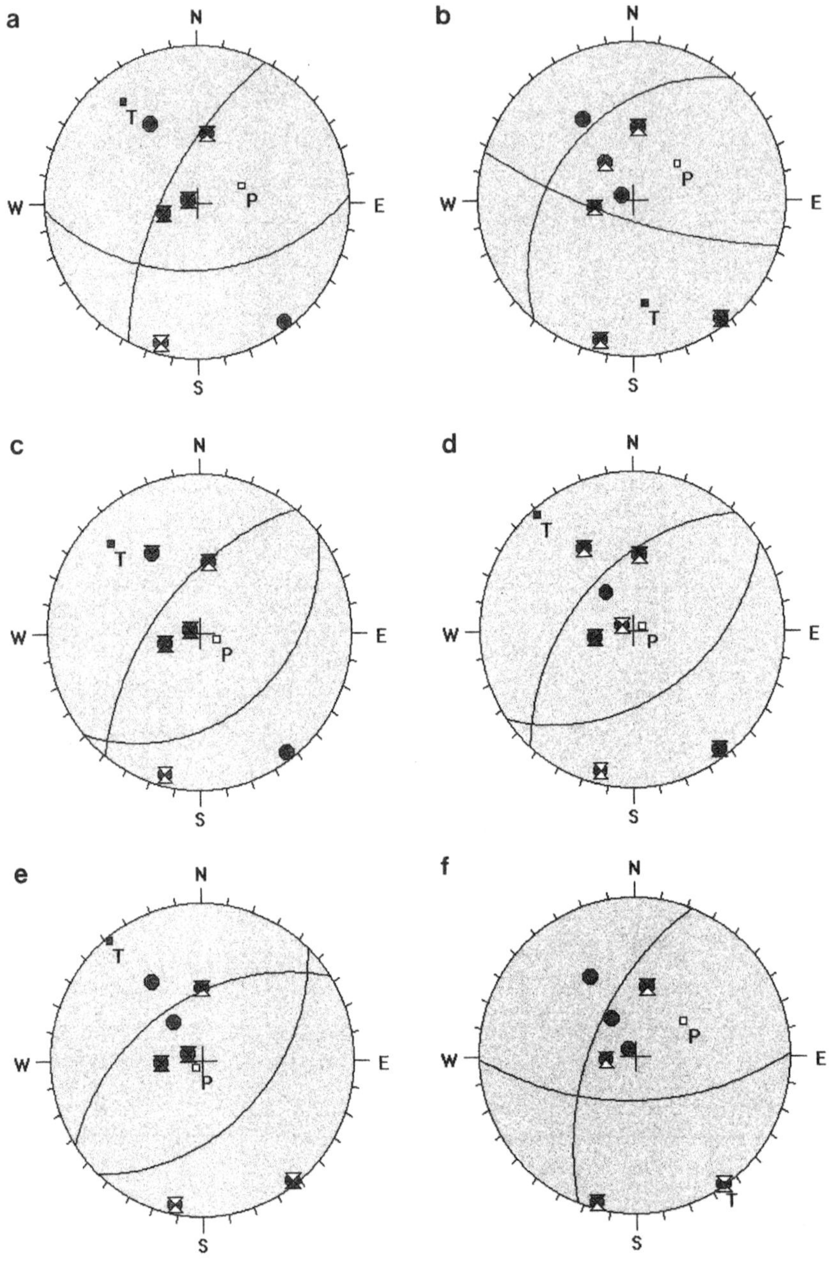

Figure 5

The failure mechanisms of events located within the sill pillar on the 4,700 level. All these events exhibit large positive volumetric components and small pure shear components. The distributions of the *P* and *T* axes match expected trends for the maximum and minimum principal stresses within this pillar: Solutions correspond to events #2, 13, 17, 25, 27, and 34 in Table 2, from (a) to (f), respectively.

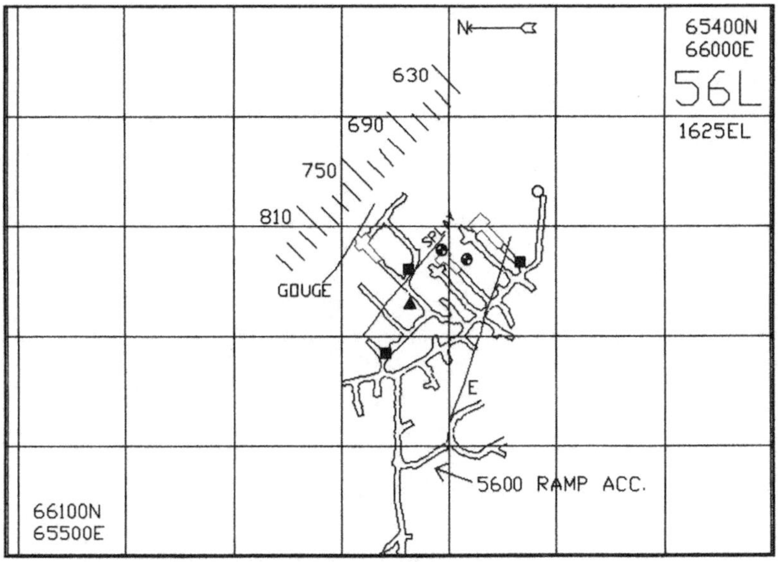

Figure 6
Plane view of the event locations within close proximity to the 5,600 level (see Fig. 1 for a description of symbols). The two events aligned along one of the drifts, number #31 and #1 from north to south (Table 2), exhibit pure shear failure mechanisms.

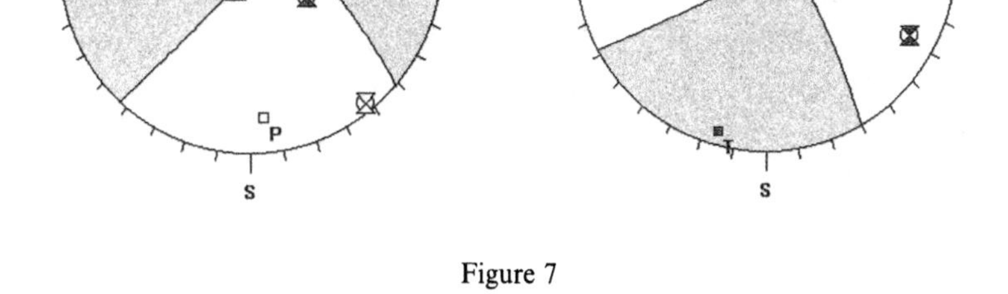

Figure 7
Failure mechanisms of (a) event #31 and (b) event #1 are of the pure shear type. The nodal planes are oriented approximately parallel and perpendicular to the nearby major faults.

reverse faulting are expected. Worth noting, the NW-SE orientation of one of the nodal planes for each solution matches the orientation of a series of subvertical fractures that are parallel to the major faults in the area (Fig. 8). These fractures have generated considerable microseismicity within this remnant, and are likely respon-

Figure 8
A section of the fractured remnant that hosted the pure-shear events which occurred on the 5,600 level. The subvertical fractures are viewed from the entrance into the drift located to the NW; these fractures are oriented NW-SE, parallel to the major faults in the area.

sible for these events as well. As such, *in situ* data support the mechanism results, namely that shear faulting along subvertical NW-SE fractures is responsible for the mechanisms of the events which occurred on the 5,600 level.

Conclusions

The evaluation of failure mechanisms for a data set of 35 microseismic events which occurred at Kidd mine in Timmins, Ontario was carried out using the inversion of the seismic moment tensor. The methodology consists of the time-domain calculation of low frequency spectral amplitudes of direct *P*, *SV* and *SH* waves using first-wave polarities and assumes a straight ray approximation in a homogeneous and infinite medium. These events, with magnitudes ranging between -1.2 and 0, were recorded by an underground microseismic array of

20 uniaxial and 8 triaxial accelerometers. Based on a maximum likelihood approach, inversions were performed employing three possible models: pure shear, complex shear, and a mechanism that includes both complex shear and tensile or volumetric components of failure. The F-test and multiple correlation coefficient were used to discriminate among alternative models and measure the data fit to the model.

Failure mechanism solutions for 21 events (60% of the data set) are found to have a multiple correlation coefficient greater than 0.6. Of these, 14 events are characterized by pure shear mechanisms, whereas 7 events exhibit a major positive volumetric component (72–76%) along with pure shear (15–20%). The analysis indicates that six of the latter seismic sources are located within a sill pillar on the 4,700 level and the distribution of their subvertical P and subhorizontal T axes is in good agreement with the expected orientation of the principal stresses at the edge of the pillar. Most pure shear events are located between 4,600 and 4,800 levels outside the mentioned pillar. Worth noting, however, are two pure shear events located on the 5,600 level inside a fractured remnant. Event mechanism results correlate well with *in situ* mapping and indicate that these events occurred along subvertical NW-SE fractures parallel to the major faults in the area.

The presence of large volumetric failure components for the microseismicity located within the sill pillar on the 4,700 level is in agreement with fresh rock tensile cracks, a mechanism typically encountered close to openings, in incipient pillar bursting. Also, the pure shear mechanism derived for the events occurring on the 5,600 level is in accordance with the typically expected failure mechanism within a highly fractured rock mass. These results indicate that the application of the above moment tensor inversion approach allows for a reliable evaluation of the failure mechanisms of microseismic events at Kidd. It is therefore reasonable that the methodology be taken one step further towards routine on-site application. Thus, as the mine operation develops to greater depths, the analysis of failure components of induced microseismic events will likely assist mine design by providing additional information on the effects of stress change in the rockmass.

Acknowledgments

The authors would like to thank Kidd Mining Division, Falconbridge Ltd., for both the continuous assistance received from their personnel during this study and the permission to publish the results. We are particularly indebted to Norm Disley for providing us with the selected data set, information concerning the on-going mining operations, and his feedback on the results obtained. Shawn Seldon and Graham Swan are also acknowledged for their enthusiastic support of this study and their comments on the manuscript. Finally, we thank Doug Angus for discussions and suggestions on the manuscript.

References

BAKER, C. and YOUNG, R. P. (1997), *Evidence for Extensile Crack Initiation in Point Source Time-dependent Moment Tensor Solutions*, Bull. Seismol. Soc. Am. *87*, 1442–1453.

BRUMMER, R., BOARD, M., and SELDON, S. (1997), *Risk Assessment of Deep Mining at Kidd Mine*, Report to Kidd Mining Division, Falconbridge Ltd., Timmins, 16 pp.

COLLINS, D. S. and YOUNG, R. P. (2000), *Lithological Controls on Seismicity in Granite Rocks*, Bull. Seismol. Soc. Am. *90*, 709–723.

COX, D. R. and HINKLEY, D. V., *Theoretical Statistics* (Chapman and Hall, New York 1986).

FEIGNIER, B. and YOUNG, R. P. (1992), *Moment Tensor Inversion of Induced Microseismic Events: Evidence of Non-shear Failures in the* $-4 \leq M \leq -2$ *Moment Magnitude Range*, Geophys. Res. Lett. *19*, 1503–1506.

FROHLICH, C., RIEDESEL, M. A., and APPERSON, K. D. (1989), *Note Concerning Possible Mechanisms for Non-double-couple Earthquake Sources*, Geophys. Res. Lett. *16*, 523–526.

GIBOWICZ, S. J., *The mechanism of seismic events induced by mining – A review*. In *Rockburst and Seismicity in Mines* (ed. Fairhurst, C.) (A. A. Balkema, Rotterdam 1990) pp. 3–27.

JOST, M. L. and HERRMANN, R. B. (1989), *A Student's Guide to and Review of Moment Tensors*, Seismol. Res. Lett. *60*, 37–57.

JOUGHIN, N. C. and JAGER, A. J. (1983), *Fracture of rock at stope faces in South african gold mines*. In *Proc. Symp. Rockbursts: Prediction and Control*, Inst. Min. Metal., London, pp. 53–66.

KNOPOFF, L. and RANDALL, M. J. (1970), *The Compensated Linear-vector Dipole: A Possible Mechanism for Deep-earthquakes*, J. Geophys. Res. *75*, 4957–4963.

MONTALBETTI, J. M. and KANASEWICH, E. R. (1970), *Enhancement of Teleseismic Body Waves with a Polarization Filter*, Geophys. J. R. astr. Soc. *21*, 119–129.

PRESS, W. H., FLANNERY, B. P., TEUKOLSKY, S. A., and VETTERLING, W. T., *Numerical Recipes: The Art of Scientific Computing (Fortran Version)* (Cambridge University Press, Cambridge, UK 1989).

SPOTTISWOODE, S. M. and MCGARR, A. (1975), *Source Parameters of Tremors in a Deep Level Gold Mine*, Bull. Seismol. Soc. Am. *65*, 93–112.

STRELITZ, R. A. (1978), *Moment Tensor Inversions and Source Models*, Geophys. J. R. astr. Soc. *52*, 359–364.

STUMP, B. W. and JOHNSON, L. R. (1977), *The Determination of Source Parameters by the Linear Inversion of Seismograms*, Bull Seismol. Soc. Am. *67*, 1489–1502.

TRIFU, C-I. and URBANCIC, T. I. (1997), *Characterization of Rockmass Behavior Using Mining induced Microseismicity*, CIM Bull. *90*, 62–68.

TRIFU, C-I., ANGUS, D., and SHUMILA, V. (2000), *A Fast Evaluation of the Seismic Moment Tensor for Induced Seismicity*, Bull. Seismol. Soc. Am. *90*, 1521–1527.

(Received November 11, 2000, revised/accepted January 15, 2001)

Pure appl. geophys. 159 (2002) 165–177
0033–4553/02/030165–13 $ 1.50 + 0.20/0

Pure and Applied Geophysics

Effect of the Rock Properties on Mining-induced Seismicity Around the Ventersdorp Contact Reef, Witwatersrand Basin, South Africa

A. M. Milev[1] and S. M. Spottiswoode[1]

Abstract—The Ventersdorp Contact Reef is characterised by a variety of different rock types that are found above and below the reef in different mining areas. These variations in the host rock type have been classified into six main geotechnical areas. Knowledge of the difference in seismicity related to these geotechnical areas could influence the design of deep level stopes and optimise the support systems.

In this study, the total seismicity was evaluated using the relationship between the cumulative seismic moment and volume of convergence. In total 300,000 seismic events, occurred at Deelkraal, Elandsrand, East Driefontein, Kloof and Leeudoorn mines. Kloof and Leeudoorn gold mines were divisions of Kloof Gold Mine Co. Ltd. Mponeng was previously called Western Deep Levels Ltd.-South Mine.

The seismicity generated in areas with different mining conditions and geology has been compared. Special attention was paid to evaluate the difference in seismicity associated with different geotechnical classifications. The following variations were found: (i) geotechnical areas with *soft (Westonaria) lava* hangingwall and *quartzite-conglomerate* footwall are less active than the geotechnical areas with *hard (Alberton) lava* hangingwall and *quartzite-conglomerate* footwall at East Driefontein, Kloof and Leeudoorn Mines, (ii) geotechnical areas with *soft lava* hangingwall and *Jeppestown Shale* footwall are less active than geotechnical areas with *soft lava* hangingwall and *quartzite-conglomerate* footwall at East Driefontein Mine, (iii) geotechnical areas with *hard lava* hangingwall and *quartzite-conglomerate* footwall and *hard lava* hangingwall and *Booysen's Shale* footwall at Western Deep Levels-South Mine do not indicate differences in the associated seismicity, (iv) the variations in the properties of the different types of quartzite forming the footwall at Deelkraal, Elandsrand, Mponeng, East Driefontein and Kloof Gold Mines do not appear to influence the level of seismicity.

Key words: Mining-induced seismicity, rock properties, convergence, deep-level mining.

Introduction

The ventersdorp Contact Reef (VCR) is one of the most important orebodies in the Witwatersrand Basin and is located in its northern and northwestern parts. The Reef separates the predominantly sedimentary Witwatersrand lithologies from the

[1] CSIR Mining Technology, P.O. Box 91230, Auckland Park 2006, Johannesburg, South Africa. E-mail: amilev@csir.co.za or sspottis@csir.co.za

volcanic Ventersdorp Supergroup. The combination of three footwall and two hangingwall types of rocks results in delineation of six geotechnical areas across the orebody (ROBERTS and SCHWEITZER, 1999). Quartzite, conglomerate and shale are the major rock types forming the footwall. Jeppestown shale and Booysens shale, however, have different geological and rock engineering properties, and therefore have been considered independently. Volcanic rocks of the Westonaria ("soft lava") and Alberton ("hard lava") formations are the major hangingwall types. The following abbreviations were used in the classification of the geotechnical areas: *SL* – soft (Westonaria) lava; *HL* – hard (Alberton) lava; *QC* – quartzite-conglomerate; *JSH* – Jeppestown Shale; *BSH* – Booysens Shale.

Mining-induced seismicity around deep level stopes is known to depend on the following factors: (i) the amount of mining; (ii) method of rock excavation – mining method; (iii) the geological and rock features; (iv) virgin stresses acting on the rock mass.

Various techniques have been suggested for quantifying the amount of seismicity and identifying the influence of these factors on total seismicity. MCGARR (1976) showed a close relationship between seismic moment and volume of convergence of the mined-out area. The volume of convergence can be calculated from the sequence of mining and the principal stresses, and compared to the amount of seismicity. This technique was applied by MCGARR and WIEBOLS (1977). The cumulative volume of convergence is also closely related to Energy Release Rate (ERR), the most widely used criterion for design of mining layouts in the South African gold mines.

Following this lead, the mining-induced seismicity was analysed based on the ratio of cumulative seismic moment ($\sum M_0$) to the volume of convergence (ΔV_E). It was also anticipated that there might be a variation in mining-induced seismicity associated with the geotechnical classifications. In summary, the aim of this study was to investigate this concept.

Method of Analysis

Mining of deep-level hard-rock deposits results in seismicity and occasionally rockbursts. The seismic energy released was empirically related to the volume of convergence by MCGARR (1976). This relationship can be expressed as:

$$\gamma_E = \sum M_0 / G\Delta V_E \ , \tag{1}$$

where $\sum M_0$ is the cumulated seismic moment, G is an average modulus of rigidity of the rock types involved (for the example given below G was taken as 3.01×10^{10} N/m^2), and ΔV_E is the volume of elastic convergence due to mining. In this paper we calculated M_0 from the local magnitude M_L using the HANKS-KANAMORI relationship (1979):

$$\text{Log}\, M_0 = 9.1 + 1.5\, M_L \ . \tag{2}$$

Generally the γ_E parameter varies between 0 and 1 and represents the level of seismic activity due to mining. This factor is, however, meaningful only for seismogenic areas for which the measured seismicity can be confidently associated with a certain volume of convergence.

Large-scale elastic stresses around deep tabular mining excavations can be efficiently estimated using the Boundary Element Method (BEM) (CROUCH and STARFIELD, 1983; RYDER and NAPIER, 1985). In this analysis, a two-dimensional mesh represents areas that have been mined at any stage. Displacement Discontinuities (DDs) representing convergence and ride between the hanging- and footwalls are calculated by iterating with Successive Over Relaxation (SOR) until the open mined-out areas carry no normal or shear stress.

The total volume of convergence across a mine can then be easily obtained from summation of all the DDs. To obtain localised contributions to the changes in the volume of convergence (ΔV_E), it is convenient to use the Energy Release Rate (ERR), defined in the mining context as:

$$ERR = \sigma D/2 \ , \tag{3}$$

where σ is the stress before an element is mined and D is the convergence between the hanging- and footwall after the element is mined. As the *ERR* represents the energy released (or stored) per area mined (SALAMON, 1984) and we are dealing with linear elasticity, its calculation should not be a function of the path taken between any two mining layouts. Indeed, we have found that our code satisfies this condition and the equivalent value of ΔV_E is then determined using:

$$\Delta V_E = \sum \sigma_i D_i / \sigma_v \ , \tag{4}$$

where summation occurs over all elements mined in any area and time, and σ_V is the virgin stress before mining.

The error in the calculation of ΔV_E is principally controlled by the accuracy of representing the area mined. For N elements mined in any step the error in estimating ΔV_E therefore varies as $\sqrt{1/N}$. In our studies, $N > 100$ and therefore the error in ΔV_E was less than 10%.

The time interval used when calculating the volume of convergence due to mining corresponds to that used for the selection of seismic data in each seismogenic area, from which the cumulative seismic moment is also generated. In this approach, seismogenic areas associated with consecutive mining steps may clearly overlap (the term "mining step" has been defined here as a certain amount of mining over a certain period of time). The calculation of the volume of convergence was performed using a computer program which calculates ΔV_E from step-wise mining as described by SPOTTISWOODE (1988, 1999).

The digitised mine outlines were used to prepare a pattern to describe the history of mining of the reef on a coarse scale of 64 by 64 elements. Digitising was done by hand from mine plans, as suitable digital data were not available. The solver then used the percentage-mined description on each element to obtain better resolution when each element was enlarged to form 16 (4 by 4) elements, bringing the "problem" size to 256 by 256 elements. Each element was then approximately 10 m on a side, allowing the simulation of area with reasonable accuracy across areas 2.5 km by 2.5 km. Element sizes of 10 m or less were required to provide sufficient accuracy for stresses on typical pillar sizes of 40 m and more. Typical values measured for the Witwatersrand quartzite were used: Young's modulus $E = 70$ GPa and Poisson's ratio $\nu = 0.2$.

Each mine was divided into seismogenic areas. These were chosen based on the pattern of seismic locations, mining steps and local geology so that we could search for differences in behaviour among areas.

The seismic deformations in a particular mining region for a given time period could also be statistically described by a Gutenberg-Richter frequency-magnitude relationship express as:

$$\text{Log } N = a - bM \quad , \tag{5}$$

Where N is the total number of events with magnitudes M or greater; M is the local Richter magnitude and a and b are constants. The constant b, widely known as a b value, is of interest as it is controlled by the ratio of the number of smaller events to the number of larger events in a given area.

A minimum magnitude of $M_{\text{min}} \geq 0.5$ was used in this study to calculate the b value for each seismogenic area using the maximum likelihood method of AKI (1965). Therefore, both parameters γ_E and b value could be related to rock properties which tended to seismic deformations during the process of relaxation.

Data

The seismic data from six VCR mines have been accumulated and processed. In total, the data set comprises 300,000 seismic events recorded underground by either the Integrated Seismic Systems (ISS) – AngloGold and Gold Fields of South Africa (GFSA) from 1983 to 1996. Mining plans with a scale of 1:1000 and 3-monthly face positions for Deelkraal, Elandsrand, Mponeng, East Driefontein, Kloof and Leeudoorn mines were processed. The average depth of mining at each of these mines was more than 2000 m below the surface.

The local magnitudes for the larger events $M_L \geq 2.0$ supplied by each mine were compared to those reported by the South African Geological Survey (SAGS). The average and the standard deviation of the magnitude residual for each of the mines is given in Table 1.

Table 1

Residual of the local magnitudes and magnitudes calculated by South African Geological Survey

Mine	Seismic network	Number of events	Magnitude range	Magnitude residual ($M_{MINE} - M_{SAGS}$)	
				Average	Standard deviation
Deelkraal	GFSA	10 000	0.0 to 3.9	−0.35	0.44
Elandsrand	ISS	70 000	−2.0 to 3.8	0.03	0.21
Mponeng	ISS	80 600	−2.0 to 3.7	−0.06	0.29
East Driefontein	GFSA	3 200	−2.1 to 3.9	−0.11	0.20
Kloof	GFSA	32 400	−2.0 to 4.0	−0.14	0.31
Leeudoom	ISS	90 000	−2.5 to 3.7	0.05	0.25

The magnitude residuals were then used to adjust the magnitudes recorded by the mine network before applying equation (2). The magnitudes reported by GFSA networks were also corrected for the bandwidth limitation (4.5 Hz to 200 Hz) of their recording system. This correction improved the magnitude values in the lower part of the magnitude scale and removed considerable curvature in the frequency-magnitude plots for $M_L < 1.0$.

After a preliminary analysis including quality evaluation, spatial and temporal distribution in terms of the face advance and digitising of the mining plans, the seismic locations were referred to the corresponding mining steps defined by quarterly face positions. The entire mining area of each mine was divided into seismogenic sub-areas using location pattern, depth profile, time intervals and local geology. Parameters of seismicity such as γ_E and b value were calculated for each of these subdivisions. Comparative analysis was done on the seismicity over entire mine areas and on the seismicity associated with the different geotechnical classifications.

Results and Discussion

Two of the mines, East Driefontein and Kloof, contained the most complex combination of diverse lava in their hangingwalls and were therefore chosen for detailed study. The seismicity of these mines is discussed below in detail.

East Driefontein Mine

The distribution of the seismic events with local magnitude $M_L \geq 1.0$ is shown in Figure 1a. It can be seen that there are several clusters of seismic events around areas with active mining. Increased seismic activity is observed in the southwestern part of the mine where the seismicity was formed by mining of three reefs: VCR, the Main Reef and the Carbon Leader Reef.

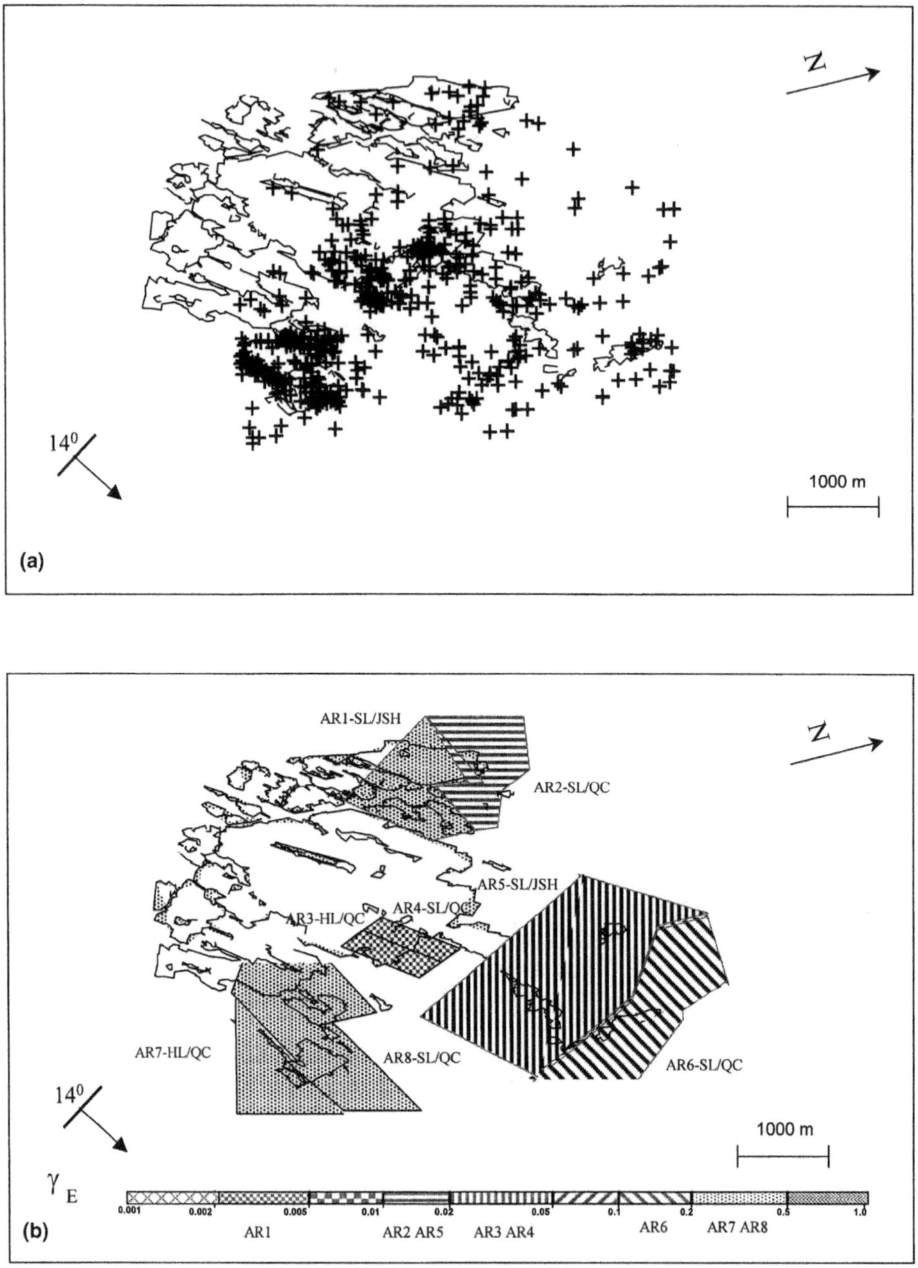

Figure 1

East Driefontein Mine: (a) hypocentral map for seismic events with local magnitude ($M_L \geq 1.0$); (b) mining induced seismicity γ_E for areas corresponding to different geotechnical classifications.

East Driefontein Mine contains the largest variety of geotechnical classification of the local VCR mines. Eight seismic areas were identified at East Driefontein and

Table 2

Seismic parameters for selected seismogenic areas at East Driefontein Mine

Area name	Geotechnical class	Time period	Number of events	M_{max}	γ_E	b value
AR1	SL/JSH	JUL 88–AUG 94	62	2.1	0.003	0.77 ± 0.10
AR2	SL/QC	MAY 90–OCT 94	65	2.2	0.01	0.82 ± 0.10
AR3	HL/QC	MAY 89–DEC 93	93	3.0	0.05	0.64 ± 0.07
AR4	SL/QC	JAN 90–OCT 94	153	2.5	0.04	0.80 ± 0.06
AR5	SL/JSH	JAN 93–OCT 94	223	2.7	0.03	0.91 ± 0.06
AR6	SL/QC	MAR 93–OCT 94	59	2.9	0.13	0.79 ± 0.10
AR7	HL/QC	JAN 88–MAR 93	50	2.7	0.49	0.53 ± 0.07
AR8	SL/QC	JAN 88–OCT 94	1479	3.2	0.55	0.78 ± 0.02
	Standard deviation				0.22	0.12

associated with the following geotechnical areas: *HL* hangingwall and *QC* footwall; *SL* hangingwall and *QC* footwall; and *SL* hangingwall and *JSH* footwall (Fig. 1b).

Seismic activity for several adjacent areas with different geotechnical classifications, *SL*/*JSH* and *SL*/*QC*, manifests differences in γ_E with the *SL*/*JSH* area being less active. The areas with *SL*/*QC* display lower seismic activity when compared to adjacent areas with *HL*/*QC*. However, the differences in γ_E values in this portion of the mine are small. They seem to be influenced by mining of Main Reef and Carbon Leader Reef in proximity to the VCR. The summary of the results is given in Table 2.

A brief discussion of b values will follow below.

Kloof Mine

The location of the seismic events with local magnitude $M_L \geq 2.0$ is shown in Figure 2a. Most of the events were related to the areas with active mining while others were associated with areas highly stressed by previous mining. On average, the southern part of the mine shows higher seismic activity, by factor of two, when compared to its northern part. The smaller mining spans in the northern part apparently played an important role in reducing the level of seismicity. Smaller spans have lower *ERR* and therefore less convergence. The resulting lower values of ΔV_E per area mined therefore do not adequately account for the benefit of smaller spans.

Kloof Mine is a complex of several different geotechnical classifications; *HL* hangingwall with *QC* footwall, and *SL* hangingwall with *QC* footwall occur most widely. A total of 11 seismogenic areas were delineated in such a way that each area contained a single geotechnical classification. The position of the areas and distribution of the γ_E are illustrated in Figure 2b.

Two pairs of adjacent seismogenic areas for geotechnical classes of *SL*/*QC* and *HL*/*QC* located in the southern and northern portions of the mine indicate differences in the level of seismicity. The areas with *SL*/*QC* are less active than the areas with *HL*/*QC*. The summary of the results is given in Table 3.

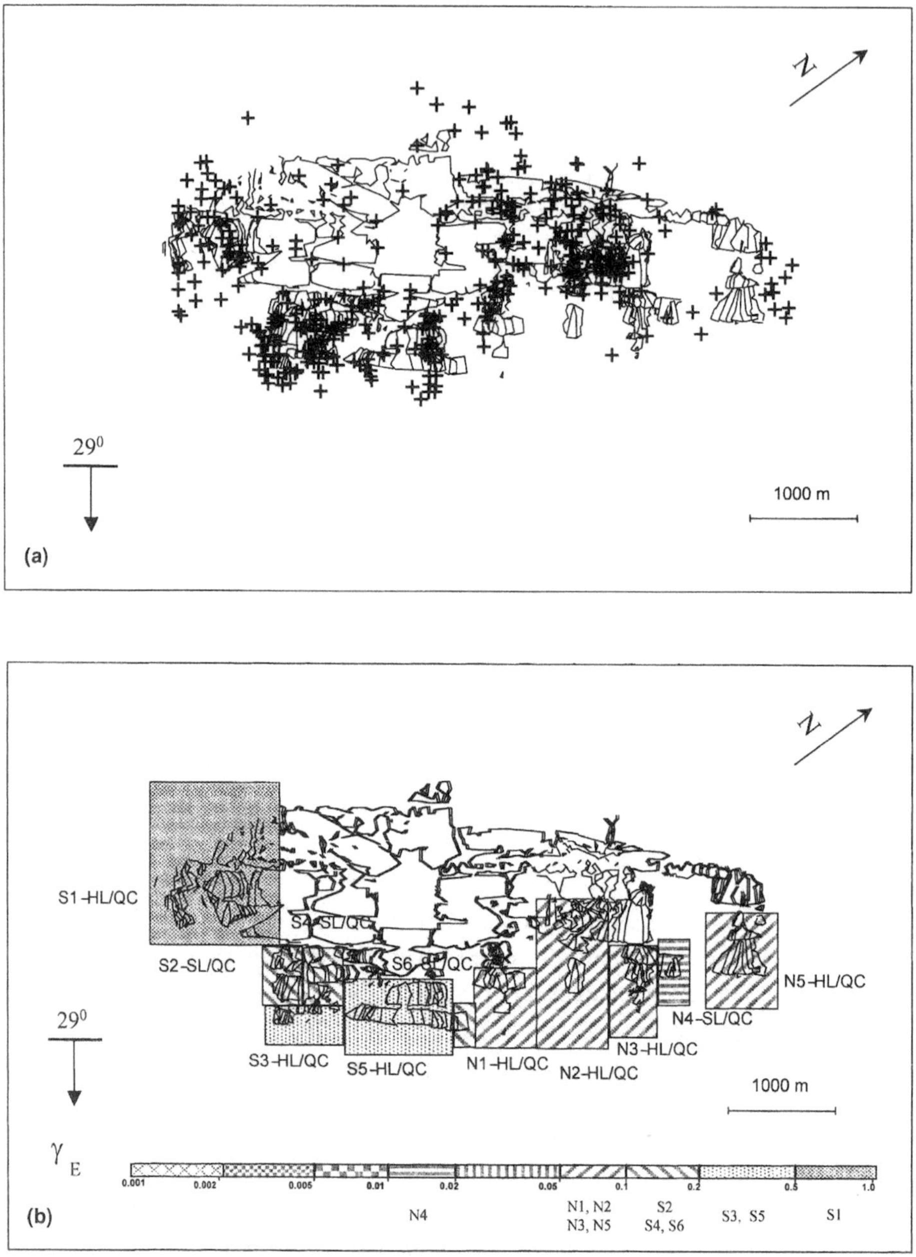

Figure 2
Kloof Mine: (a) hypocentral map for seismic events with local magnitude $M_L \geq 2.0$; (b) mining induced seismicity γ_E for selected seismogenic areas.

Figure 3 is a plot of the cumulative distributions of γ_E for the two hangingwall types, hard lava and soft lava overlying a quartzite-conglomerate footwall, analysed at East Driefontein and Kloof mines.

Table 3

Seismic parameters for selected seismogenic areas at Kloof Mine

Area name	Geotechnical class	Time period	Number of events	M_{max}	γ_E	b value
S1	HL/QC	JAN 87–AUG 95	4100	4.0	0.57	0.87 ± 0.01
S2	SL/QC	JAN 87–AUG 91	720	3.3	0.12	0.85 ± 0.03
S3	HL/QC	JAN 90–AUG 95	550	3.0	0.23	0.69 ± 0.03
S4	SL/QC	JAN 87–AUG 91	1100	3.2	0.13	0.82 ± 0.02
S5	HL/QC	JAN 87–AUG 95	3520	3.8	0.21	0.76 ± 0.01
S6	SL/QC	JAN 93–AUG 95	85	1.9	0.13	0.99 ± 0.11
N1	HL/QC	JAN 90–AUG 95	1300	3.1	0.05	0.82 ± 0.02
N2	HL/QC	JAN 87–AUG 95	2850	3.4	0.07	0.86 ± 0.02
N3	HL/QC	JAN 91–AUG 95	390	3.2	0.07	0.78 ± 0.04
N4	SL/QC	JAN 87–AUG 95	190	2.1	0.01	0.86 ± 0.06
N5	HL/QC	JAN 89–AUG 95	240	3.5	0.09	0.76 ± 0.05
Standard deviation					0.15	0.08

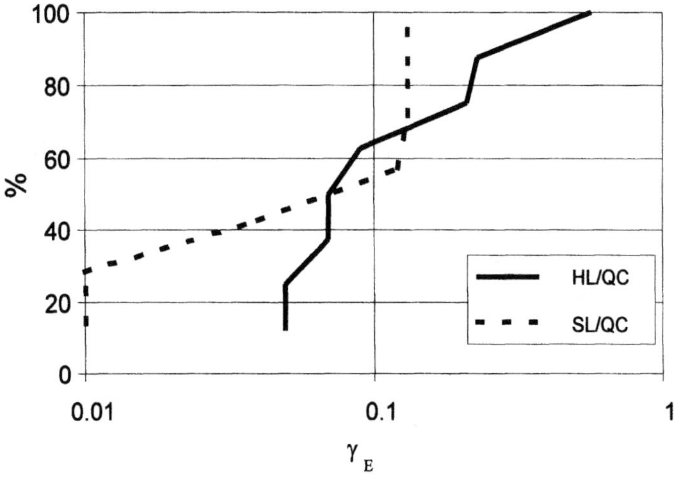

Figure 3
The cumulative frequency curves of the γ_E values calculated for areas with soft and hard lava hangingwall at East Driefontein and Kloof Mines.

It can be seen that the areas covered in hard lava were several times more seismically active, when normalised by ΔV_E, than the areas covered by soft lava. This result is in agreement with the conclusion of McGarr *et al.* (1975) who attributed the concentration of seismic events in the hangingwall to higher strength of hangingwall rocks in their study area.

In summary, the γ_E on two mines was found to be slightly higher for stronger hangingwall rocks and stronger footwall rocks, although the contrast in the values of γ_E are not statistically significant, with each population showing large deviations of their average, both in a linear and logarthmic sense (Table 4).

Table 4

Statistical evaluation of γ_E for seismogenic areas with hard lava and soft lava hangingwall types overlying a quartzite-conglomerate footwall

Geotechnical class	Number of areas	Linear		Geometric	
		Average	Standard deviation	Average	Standard deviation
HL/QC	8	0.168	0.177	0.115	2.40
SL/QC	7	0.081	0.058	0.052	3.34
SL/JSH	2	0.017	0.019	0.009	5.09
All areas studied	65	0.241	0.294	0.084	5.79

The deviations in the values of γ_E of factors of two and more are of great importance, as control of ΔV_E is achieved through leaving 15% or more of the ground behind as regional stability pillars. These pillars are typically designed with the objective of reducing seismicity by a factor of about three.

The conclusion of McGarr and Wiebols (1977) that seismicity at ERPM mine was reduced in accordance with the reduction in ΔV_E is not supported in this paper for the two mines studied. The large pillars (60 m wide) and the regular mining geometry at ERPM mine might have led to behaviour closer to ideal elastic. The mine layouts studied in this paper (Figs. 1a and 2a) were considerably more complex than the ERPM practice. Further work is underway to investigate whether these discrepancies can be attributed to non-linear effects.

An inverse relationship between γ_E and b values was pointed out by McGarr (1986). He commented "...it is of considerable interest to find out if there are any geological factors that may cause low values of γ_E and high values of b". Figure 4 shows the cumulative frequency curves of the b values calculated for areas with soft and hard lava hangingwall at East Driefontein and Kloof Mines.

In contrast with Figure 3, the graphs plotted on Figure 4 clearly indicated that the b values calculated for areas with soft lava hangingwall are higher then the b values calculated for areas with hard lava. Higher strength rocks can build up more strain energy before failure and then fail more violently. As low b values implies a high ratio of large to small seismic events and since the large seismic events cause most of the seismic deformations, the high γ_E values in the areas with hard lava are physically well supported.

The other mines studied: Deelkraal, Elandsrand, Mponeng and Leeudoorn have insignificant variations in terms of geotechnical classes. The variations of γ_E for these mines were summarized in Figure 5 and Table 4.

Considerable effort and expense is applied to reduce the volume of convergence by leaving pillars. Therefore, γ_E should have the same values for all seismogenic areas. However, Figure 5 shows significant scatter in γ_E values. This indicates factors other than the country rock type might significantly contribute to the formation of

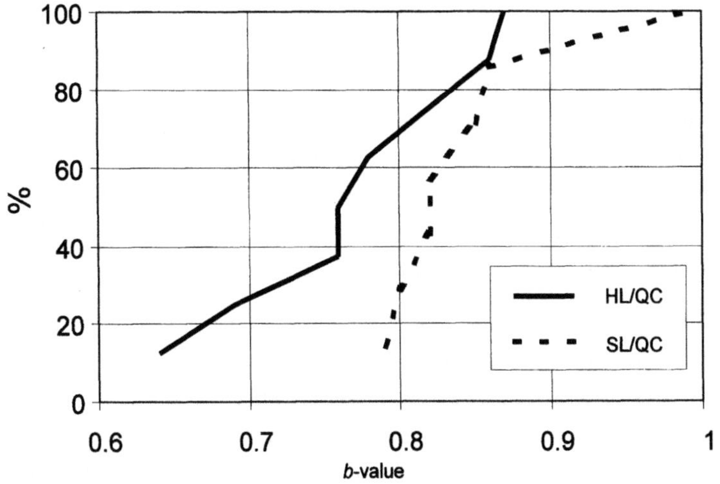

Figure 4
The cumulative frequency curves of the b values calculated for areas with soft and hard lava hangingwall at East Driefontein and Kloof Mines.

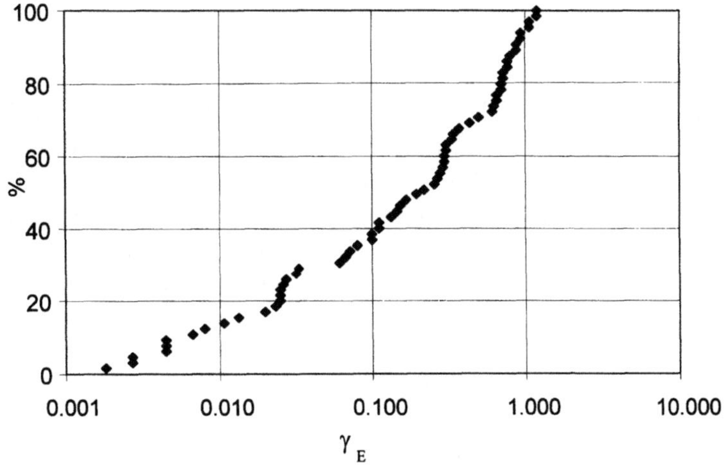

Figure 5
The cumulative frequency curves of the γ_E values calculated for all areas at Deelkraal, Elandsrand, Mponeng and Leeudoorn mines with hard lava hangingwall and quartzite-conglomerate footwall.

γ_E. For example, the stope spans at Kloof seemed to have some control on γ_E. In addition, as suggested by MILEV *et al.* (1995), "inherited" seismicity of previously mined-out areas might also provide an important control on γ_E values. Although the large scatter observed in γ_E places considerable doubt on the value of convergence control as currently practiced. Further research on this topic is ongoing. The b values also show a significant spread, from 0.53 to 1.72, throughout all areas and this indicates that changes in the mine production strongly effect seismicity.

Conclusions

There appeared to be some control of rock types on the ratio γ_E between the cumulative seismic moment and the volume of convergence across two mines that reflected the most variation in country rock types.

- The γ_E in areas with soft lava in the hangingwall was about 50% of γ_E where only hard lava was present.
- The γ_E was less for areas with shale footwall.

Conclusions drawn from a larger set of six mines were:

- The γ_E was lower in areas of lower span.
- Different types of quartzite footwall did not appear to influence γ_E.
- The standard deviation of a factor of 5.8 in all the 65 selected areas, for which values of γ_E were measured, is higher than the factor-of-three benefit expected from leaving pillars.

More accurate models are needed for estimating the anticipated level of seismicity around deep tabular mining.

Acknowledgments

This work was the part of the research program of the CSIR Mining Technology supported by the Safety in Mines Research Advisory Committee (SIMRAC) under the projects GAP 102, GAP 612 and GAP 722. The authors thank the management of: Deelkraal, Elandsrand, Mponeng, East Driefontein, Kloof and Leeudoorn Gold Mines for providing them with mining and seismic data. The authors also appreciate the supportive review and suggestions by Dr. J.A.L. Napier.

References

AKI, K. (1965), *Maximum Likelihood Estimate of b in the Formula* Log $N = a - bm$ *and its Confidence Limits*, Bull. Earthquake Res. Inst., Tokyo Univ. *34*, 237–239.

CROUCH, S. L. and STARFIELD, A. M., *Boundary Element Methods in Solid Mechanics* (G. Allen and Unwin., London 1983).

HANKS, T. C. and KANAMORI, H. (1979), *A Moment Magnitude Scale*, J. Geophys. Res. *84*, 2348–2350.

MCGARR, A., SPOTTISWOODE, S.M., and GAY, N. C. (1975), *Relationship of Mines Tremors to Induced Stress and to Rock Properties in the Focal Region*, Bull. Seismol. Soc. Am. *65*, 981–993.

MCGARR, A. (1976), *Seismic Moment and Volume Changes*, J. Geophys. Res. *81*, 1487–1494.

MCGARR, A. and WIEBOLS, G. A. (1977), *Influence on Mine Geometry and Closure Volume on Seismicity in Deep Level Mine*, Int. J. Rock Mech. Min. Sci. & Geomech. Abstr. *14*, 139–145.

MCGARR, A. (1986), *Some Comments on the Nature of Witwatersrand Mine Tremors*, Unpublished report.

MILEV, A. M., SPOTTISWOODE, S. M., and NOBLE, K. R. (1995), *Mine-induced Seismicity at East Rand Proprietary Mines*, Int. J. Mech. Min. Sci & Geomech. Abstr. *32*, 629–632.

ROBERTS, M. K. and SCHWEITZER, J. K. (1999), *Geotechnical Areas Associated with the Ventersdorp Contact Reef, Witwatersrand Basin, South Africa*, J. of S. Afr. Inst. Min. and Metall. May/June, 157–166.

RYDER, J. A. and NAPIER, J. A. L. (1985), Error analysis and design of large-scale tabular mining stress analyser. In P*roc. of the 5th Int. Conf. on Numerical Methods in Geomechanics* Nagoia, Japan, 1985, pp. 1549–1555.

SALAMON, M. D. G. (1984), *Energy Considerations in Rock Mechanics: Fundamental Results*, J. S. Afr. Inst. Min. Metall. *84*, 233–246.

SOUTH AFRICA GEOLOGICAL SURVEY (1986–1994), *Seismological Bulletins*, Pretoria, South Africa.

SPOTTISWOODE, S. M. (1988), *Total Seismicity, and the Application of ESS Analysis to Mine Layouts*, J. S. Afr. Inst. Min. and Metall. *88*, 109–116.

SPOTTISWOODE, S. M. (1999), *Modelling seismicity and creep on multiple stope-parallel layers* In *Proc. 2nd Southern African Rock Engineering Symposium*, (ed. Hagan, T.O) pp. 252–258.

(Received June 21, 2000, revised/accepted January 18, 2001)

Pure appl. geophys. 159 (2002) 179–195
0033–4553/02/030179–17 $ 1.50 + 0.20/0

Pure and Applied Geophysics

The Assessment of Damage Around Critical Engineering Structures Using Induced Seismicity and Ultrasonic Techniques

Will S. Pettitt,[1] Calum Baker,[1] R. Paul Young,[2]
Lars-Olof Dahlström,[3] and Gunnar Ramqvist[4]

Abstract—Two large-diameter boreholes have been excavated vertically from the floor of a tunnel at the Äspö Hard Rock Laboratory, Sweden. The two deposition holes will have simulated high-level radioactive waste canisters installed in them in an experiment undertaken to test the retrievability of waste from a proposed repository. Induced seismicity and other acoustic monitoring techniques have been used to investigate the Excavation Damaged Zone (EDZ) around the two holes. High-frequency acoustic emission (AE) monitoring has been used to delineate regions of stress-induced microfracturing on the millimetre scale. This has been shown to locate in clusters around the perimeter of the deposition hole at azimuths orthogonal to the far-field maximum principal stress. Three-dimensional velocity surveys have been conducted along ray paths that pass through the damaged region and through a stress-disturbed zone around the excavation. Induced microfracturing and stress disturbance have been observed as sharp decreases in velocity as the excavation proceeds through the rock mass. The combination of the high-resolution velocity measurements and the AE source locations has allowed the linking of the velocity measurements to a volume of excavation damaged rock. This has provided a quantitative estimate of the effect of the EDZ on the rock mass.

Key words: Excavation damage, microcracking, borehole breakout, acoustic emission, micro-velocity.

Introduction

Induced seismicity, among other acoustic techniques, is used as a tool for examining the localisation and mechanics of damage in the Excavation Damaged Zone (EDZ) around an excavation. This damage is induced by the excavation method itself (explosive blasting or action of a cutter head) and by the redistribution of stresses (loading or unloading) resulting from the void. The spatial extent, temporal dependence and scale of fracturing and the effect of this on the overall stability of the rock mass, is of high interest to engineers, particularly when the

[1] Applied Seismology Consultants, 10 Belmont, Shrewsbury, UK. E-mail: will@seismology.org
[2] Applied Seismology Laboratory, Department of Earth Sciences, Liverpool University, Liverpool, UK. E-mail: r.p.young@liverpool.ac.uk
[3] NCC Tecnik, Gothenburg, 40514 Sweden. E-mail: lars-olof.dahlstrom@ncc.se
[4] SKB Äspölaboratoriet, Figeholm, 57295 Sweden. E-mail: gunnar.ramqvist@skb.se

integrity of a structure is critical to the performance of a project. Acoustic techniques are particularly adept at assessing the EDZ as they allow it to be mapped spatially and temporally with high resolution, and they allow the effect on the rock mass to be quantifiably measured. This provides an interpretation of the significant mechanisms that are driving the damage that can then be incorporated into the design and excavation methodology used in future projects. Furthermore, acoustic techniques allow investigations to be conducted remotely, without the need for potentially damaging coring, and they allow measurements to be conducted in inaccessible environments before, during and after the excavation procedure.

Acoustic monitoring using microseismic techniques (0.5–10 kHz) has been researched extensively, and is now used routinely, for example at AECL's Underground Research Laboratory, Canada (URL) (YOUNG and MARTIN, 1993). These techniques investigate fracturing, with source dimensions of 10 s of centimetres, within volumes of 100 × 100 × 100 m scale. Ongoing developments in processing procedures have allowed complexities in the micromechanics of fracturing in the EDZ to be resolved (e.g., BAKER and YOUNG, 1997).

AE and ultrasonic techniques, using frequencies ranging from 35–350 kHz, examine smaller-scale damage, with source dimensions of millimetres, within volumes of 10 × 10 × 10 m scale. FALLS and YOUNG (1998) give a review of AE and ultrasonic results from numerous excavation experiments conducted in different underground environments. Recent research at the URL has shown locations of AEs recorded in the EDZ around a tunnel bulkhead (YOUNG and COLLINS, 1999). Changes in ultrasonic propagation (velocities and signal amplitudes) between an array of ultrasonic transmitters and receivers have allowed a quantification of the accumulated damage and its progression through time. This has been extremely useful in understanding the behaviour of the rock mass under changing experimental conditions. The source mechanics of the fracturing associated with the AEs has also been investigated employing a moment tensor technique (PETTITT, 1998). These have then been related to the source mechanics of microseismic events that are located closely in both space and time, but which are approximately four orders of magnitude larger (COLLINS *et al.*, 2001). Through this approach the micromechanics of fracturing associated with both the AEs and microseismic events, which ultimately causes the rock strength and mechanical properties to degrade, may be further understood.

We demonstrate the use of AE and ultrasonic techniques with data from Sweden's Äspö Hard Rock Laboratory (HRL) operated by SKB. Two large-diameter boreholes (deposition holes) have been excavated as part of the *Canister Retrieval Test* (Fig. 1). Each deposition hole is 8.8 m in length and 1.75 m in diameter. This experiment will continue over several years in order to test the retrievability of high-level waste canisters from an underground repository.

An ultrasonic array was installed around the volume through which each deposition hole was excavated. AE monitoring has been used with the aim of

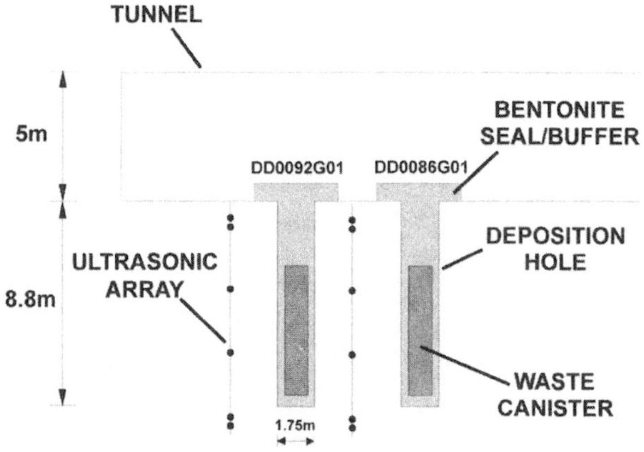

Figure 1
A schematic cross section of the Canister Retrieval Tunnel with excavated deposition holes (denoted DD0092G01 and DD0086G01). Mock waste canisters are shown for illustration. An ultrasonic array is shown installed around DD0092G01 (filled circles are approximate locations of transducers).

delineating zones of stress-related fracturing around the deposition hole perimeter, and to observe the effect of excavation on pre-existing macroscopic fractures. Changes in ultrasonic velocities, measured every hour, have been used to investigate stress disturbance of the rock mass over a broader time and volume than the AE scale, and to quantitatively measure the accumulation of fracturing in the damaged zone. We emphasise the high resolution that is available in the velocity measurements and how these can be linked to stress-induced damage observed in the AE measurements. This provides a quantitative interpretation of the mechanics of induced damage in the rock mass around the excavations.

Experiment Configuration

The Canister Retrieval Tunnel is approximately 6 m in diameter, horizontal, and situated at 420 m depth. It was excavated using a drill and blast technique. The rock mass at this level is predominantly massive Äspö diorite with sparse *en echelon* fractures orientated in two main discontinuous sets; a Northwest, steeply dipping, set is regarded as the main water-bearing set. An average stress tensor for this test location, based on more than 20 triaxial overcore measurements, is given in Table 1. An *in situ* stress ratio ($\sigma_1{:}\sigma_3$) of 3:1 exists.

A 24-channel ultrasonic array was installed around the volume of each deposition hole (approximately a $5 \times 5 \times 10$ m volume) using four vertical 76 mm-diameter instrumentation boreholes. Monitoring was conducted from one week prior to excavation in order to assess the unexcavated rock mass. The array was mounted within four borehole sondes (Fig. 2) with each sonde containing four receivers spaced

Table 1

Principal stress values for the 420 m level used by Young *et al. (1996). The values represent an average tensor calculated from data reported by* Leijon *(1995)*

Stress component	Magnitude (MPa)	Trend (°)	Plunge (°)
σ_1	32	131	0
σ_2	17	41	25
σ_3	10	229	65

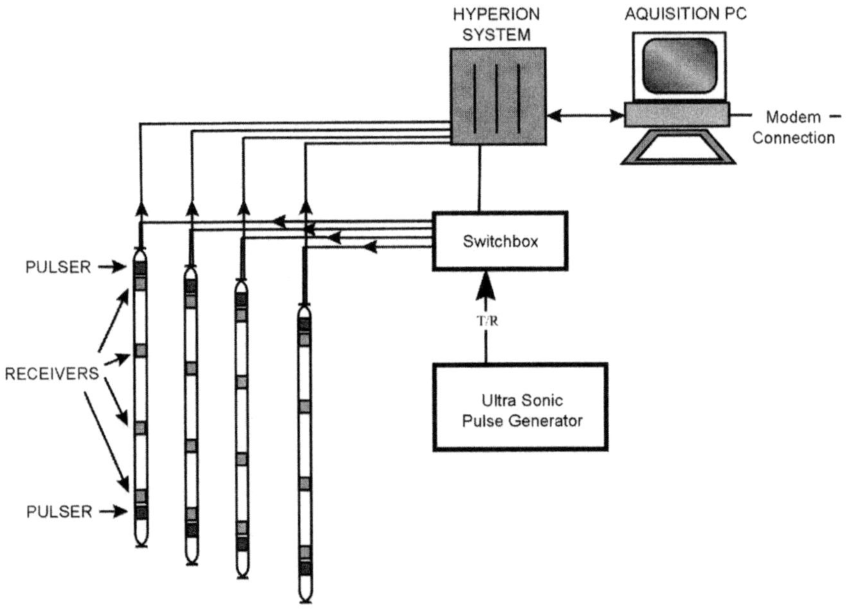

Figure 2
A schematic illustration of the instrumentation used for acoustic monitoring.

at 2.5 m intervals and a transmitter at either end, totaling 16 receivers and 8 transmitters. The array geometry for deposition hole DD0092G01 is shown in Figure 3. The first receiver on each sonde was located at a depth of about 1.65 m and the last receiver at about 9.15 m depth below the floor of the tunnel and 0.35 m below the deepest extent of each deposition hole.

A modified tunnel-boring machine (TBM) was used to excavate each deposition hole from the floor of the tunnel. Excavation was performed in eleven 0.8 m drilling steps with each step taking between 2 to 4 hours to complete. No monitoring could be conducted during excavation itself due to drill-bit noise. AE monitoring was switched on immediately before the completion of excavation of each step and switched off immediately after the start of the subsequent one. The drilling schedule was modified so that a 2-hour quiet period was observed after each step. No drilling

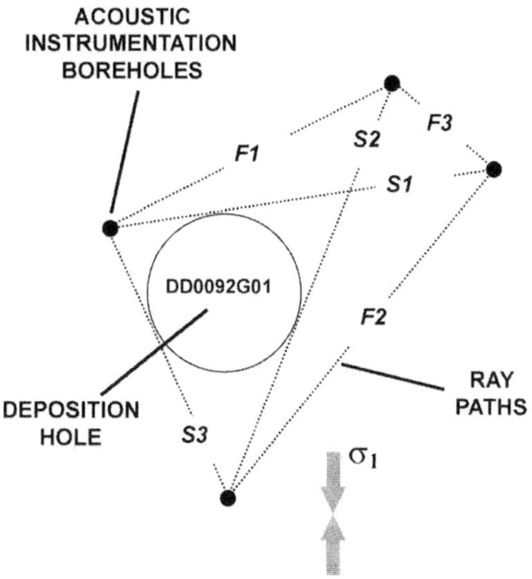

Figure 3
Plan view of the array geometry for deposition hole DD0092G01. Ray paths for ultrasonic surveys are split into six categories; *S1*, *S2*, *S3* skim the deposition hole wall.

or maintenance could be performed in this period. For all other times, any drill and maintenance noise was manually removed from the AE data during post-processing.

Ultrasonic surveys were conducted hourly in order to measure changes in *P*- and *S*-wave signals as the deposition hole excavation progressed. Figure 3 shows a plan view of the possible 128 ray paths between transmitters and receivers. Six ray-path categories are defined. The array was designed so that three of the ray-path categories skimmed the deposition hole wall at approximately 2–3 cm.

Array Instrumentation and Processing

The ultrasonic transducers used in the array registered a frequency response of approximately 35–350 kHz. Each had a hemispherical brass cap fixed over its active face and was then spring-loaded against the borehole surface so as to obtain good coupling to the rock mass. Signals were amplified at 40 dB close to the sensor after which full waveforms were captured using an ESG Hyperion Ultrasonic System placed at distance from the array. Acquisition of AEs used preset trigger thresholds on a specified number of receivers in the array. Waveforms were stored with lengths of 4096 data points (allowing *P*- and *S*-wave arrivals) at 14-bit, 1 MHz sampling.

Calibration surveys conducted post excavation showed AE locations to have an estimated uncertainty of 10 cm. The sensitivity of the array was such that the array was able to record, and then locate, a pencil lead break on the wall of the deposition hole. This corresponds to an AE source dimension on the millimetre scale (grain size). Subsequent analysis showed this to be equivalent to the smallest AEs recorded during the experiment. Waveform processing was conducted post-excavation using an automatic picking routine. *P*- and *S*-wave arrival times were then inverted to calculate AE locations. A downhill-simplex routine was used with an input velocity model derived from the ultrasonic surveys. Only AEs with >6 *P*-wave picks from transducers on at least three sondes were located.

Ultrasonic surveys utilised a Panametrics signal generator to produce a high-frequency, high-voltage spike that was sent to each of the eight transmitters in turn. The signal emitted from each transmitter was then recorded over the array of receivers, allowing a fully three-dimensional analysis to be obtained through the rock volume. As surveys were conducted hourly, 300–400 separate surveys were obtained in total for monitoring of each deposition hole. These were processed using a cross-correlation technique.

P- and *S*-wave arrivals were manually picked for a base survey recorded before excavation. Ultrasonic velocities were calculated with an estimated maximum uncertainty of ± 30 m s^{-1} in any one measurement. This is approximately ± 3 data points along the shortest ray paths of 4 m. The main reason for this uncertainty is the dependency of manual picking on the user's judgement of the point of arrival. This can usually be quite haphazard because of random noise superimposed on the first few data points of the first break.

The cross-correlation procedure has been developed during research conducted at the Äspö HRL and at the URL. In these studies it has been extensively tested and has been shown to provide robust measures of very small changes in *P*- or *S*-wave arrival times depending on the experimental conditions. The procedure takes the waveform recorded from one transmitter-receiver pair and cross-correlates it with the waveform from the base survey. In this study the technique yields an estimated maximum uncertainty of ± 2 m s^{-1} between measurements along the same ray path, because the effect of manual picking is removed. This allows for a high-resolution analysis to be performed and hence slight changes in velocity to be observed. This is extremely important when changes in rock properties occur over only a small section (5%) of the ray path.

The cross-correlation procedure is designed to be very sensitive to a shift in the first half-cycle of the arrival (either *P* or *S* wave). Figure 4 shows *P*-wave arrivals along one example ray path. The waveform is first smoothed using a natural-spline function (e.g., Press *et al.*, 1994) to give a waveform with 5× oversampling compared to that recorded in the field. At 1 MHz sampling there is approximately 10 data points in one wavelength. By increasing the sampling to 5 MHz in this manner the first cycle has a much clearer definition. The waveform is then automatically picked to obtain an estimate of the *P*-wave or *S*-wave arrival. A window of approximately 4 wavelengths is then automatically defined around the arrival and a bell function is

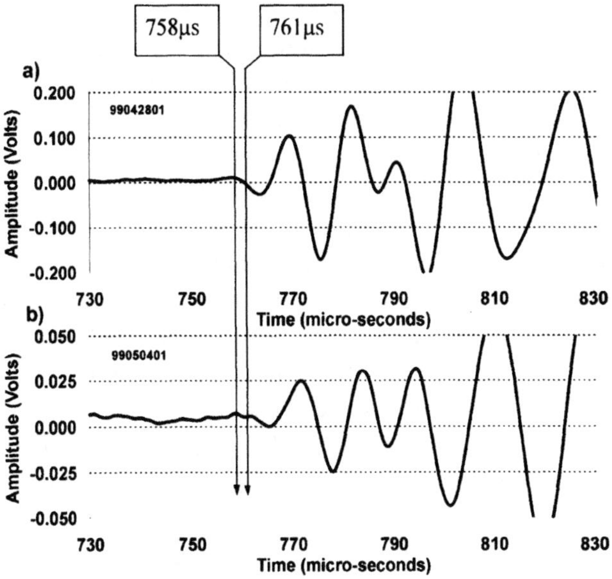

Figure 4
P-wave arrivals recorded along the ray path shown in the uppermost plot of Figure 9. (a) Before excavation of the deposition hole (at 0100 hrs on 28th May 1999). (b) After excavation. Manually picked first arrivals are shown. The time shift in the arrivals, measured using automatic cross-correlation, is 2.20 μs. The ray path length is 4.409 m.

applied, centred on the automatic pick. The data at the ends of the window then have a significantly smaller effect on the cross-correlation. The windowed data are then cross-correlated (e.g., TELFORD *et al.*, 1990) with a similar window constructed around the arrival on the base survey. The change in arrival time is then converted to a change in velocity knowing the manually-picked arrival time for the base survey. The precision in the procedure is a change in the two arrival times of one oversampled data point. This is shown by the scatter in the results obtained over one day of monitoring before excavation. Waveforms that do not provide automatic picks are not cross-correlated. This results in an automatic discrimination of signals that have very poor signal-to-noise ratios and could give spurious cross-correlation results from poor discrimination of the first arrival.

Stress Disturbance

Studies of the EDZ (including other acoustic experiments), around a similar drill and blast tunnel situated at this level of the Äspö HRL, have shown the damage to extend radially a maximum of 1 m from the tunnel perimeter (FALLS and YOUNG, 1996; EMSLEY *et al.*, 1997). This damage was principally attributed to the excavation method itself rather than as a result of the *in situ* stresses. Another tunnel excavated at this level, using a 5-m-diameter TBM, showed that damage was restricted to only a few centimetres radial distance. The laboratory uniaxial compressive strength of the Äspö diorite is approximately 170 MPa and a laboratory uniaxial crack-initiation stress is reported as 64 MPa (YOUNG *et al.*, 1996).

Although stress magnitudes at the Äspö HRL are relatively small compared to the strength of the rock, the effect of the tunnel as shown by a simple Kirsch solution (e.g., YOUNG *et al.*, 1996), is likely to increase the deviatoric stress, $\sigma_1 - \sigma_3$, observed in the tunnel floor to slightly over the crack-initiation stress. In this experiment there is then an additional effect of stress distribution around the large-diameter deposition holes. This multiplication of stresses may result in stress-induced fracturing around the perimeter of the holes. Similar observations have been made at the URL (e.g., READ and MARTIN, 1996) but in this case the *in situ* stress ratio is approximately 6:1.

EMSLEY *et al.* (1997) also discuss stress disturbance of pre-existing fractures as a possible mechanism for AE occurrence at the Äspö HRL. Microcrack-extension models using σ_3 unloading have been proposed to explain AEs occurring in lower stressed regions of the rock mass. Stress disturbance on pre-existing macroscopic fractures (common in the Äspö HRL) may also induce slip on these if the orientation of the fracture supports a decrease in normal stress and/or an increase in shear stress acting on the fracture.

Obvious parallels between excavation of the deposition holes in this study and that of the TBM tunnel should indicate a similar EDZ resulting from the excavation

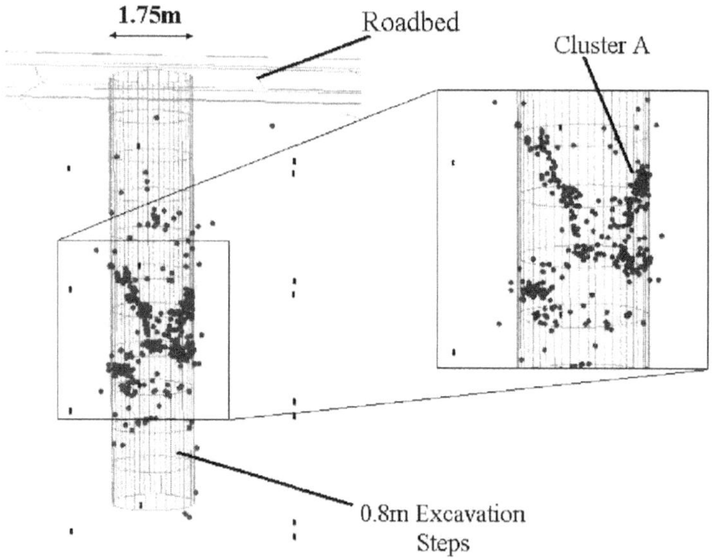

Figure 5
AE locations (round markers) from monitoring of deposition hole DD0092G01. Square markers are transducer locations. View is 260° east of north, 10° plunge.

method itself. Any recorded AEs are then principally a response of the rock to stress disturbance rather than the excavation method, although this may have pre-weakened the rock mass during drilling. The skimming ray paths, sampled during the ultrasonic surveys, passed through any regions of damage due to the excavation method, as well as damage caused by excavation-induced stresses. All the ray paths were also sensitive to closing or opening micro-fractures, whether pre-existing or induced, caused by stress loading or unloading.

Results

Acoustic Emission Data

Results from both deposition holes were similar. A total of 1531 system triggers occurred during excavation of the first deposition hole (denoted DD0092G01) and 1215 during the second (DD0086G01). Figure 5 shows 490 automatic AE locations around deposition hole DD0092G01. Note that very little activity occurred in the upper few steps of the excavation. Activity increased considerably in step six (4.8 m depth) onwards. The AE locations have strong spatial clusters that appear to fit with one or two linear features. These features are believed to be associated with excavation of the hole through pre-existing macroscopic fractures that are common

(a)

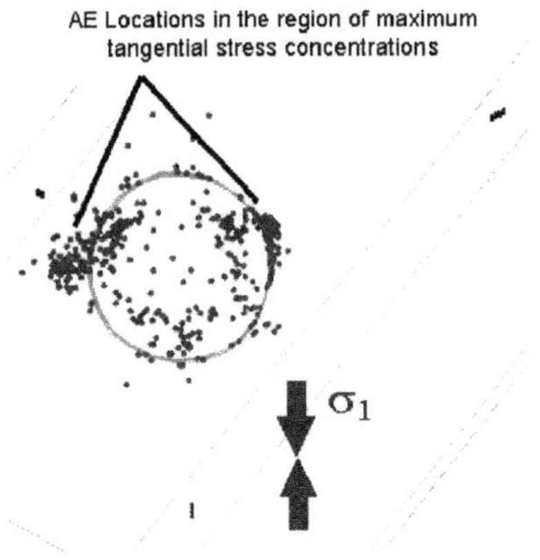

(b)

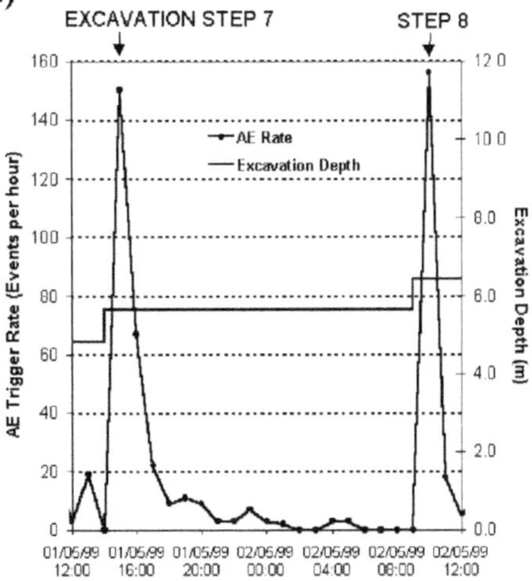

Figure 6
(a) Plan view of AE locations. AEs appearing to locate inside the deposition hole occur ahead of the advancing face. (b) Decay of AE triggers after step seven in the excavation.

in this rock mass. Future comparison of these results with geological mapping of the deposition hole will analyse this relationship further. The immediate region around the fracture probably acted as a weaker region of the rock mass around which stresses were sufficiently disturbed to create additional damage or generate differential movement. Action from the TBM cutter head may have exacerbated this.

In plan (Fig. 6a) the AEs do not cluster around the complete circumference but are clustered into two diametrically opposite regions. These are orthogonal to the maximum principal stress described in Table 1 and hence represent classic "borehole-breakout" regions where induced compressive stresses are high. It should be noted though that this damage is estimated to be of the order of the grain size in scale and hence is likely to be only visible under magnification. The term "breakout" can only be loosely applied here as solely where the AEs are intensely clustered on the deposition hole wall is there likely to be obvious physical damage. The observation of the damage at all is due to the sensitivity of the instrumentation. The vast majority of the AEs cluster within the first 20 cm of the deposition hole wall. This is taken as a reliable estimate of the maximum extent of the EDZ. The time history of AE triggering observed after step seven (Fig. 6b) is shown as an example. It describes a classic exponential type decay with triggering reducing to < 5 per hour within six to eight hours after excavation. The spatial-temporal relationship of the AEs is described in Figure 7. AEs locating in proximity of the macroscopic fractures continue to occur for many days after excavation through this region.

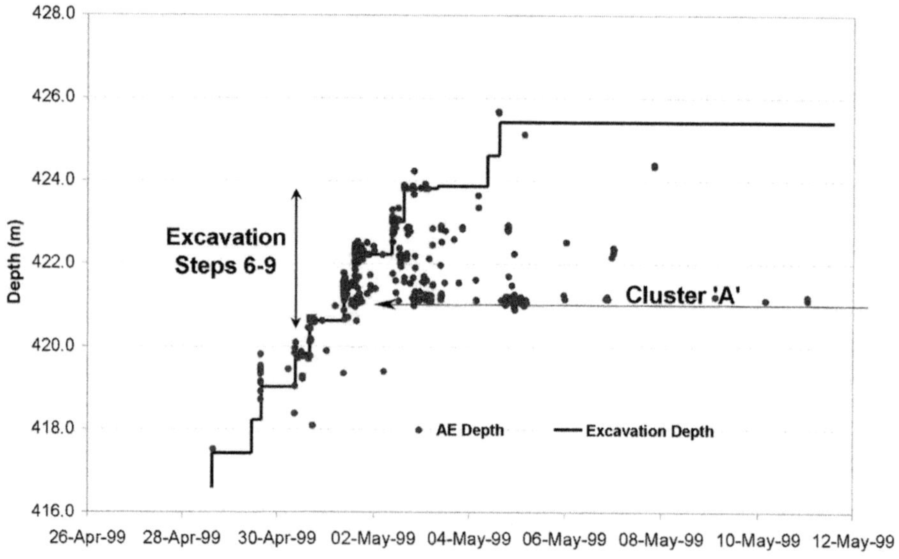

Figure 7
Spatial-temporal dependency of AE activity. Note the time dependency of the fracturing associated with Cluster A marked on Figure 5.

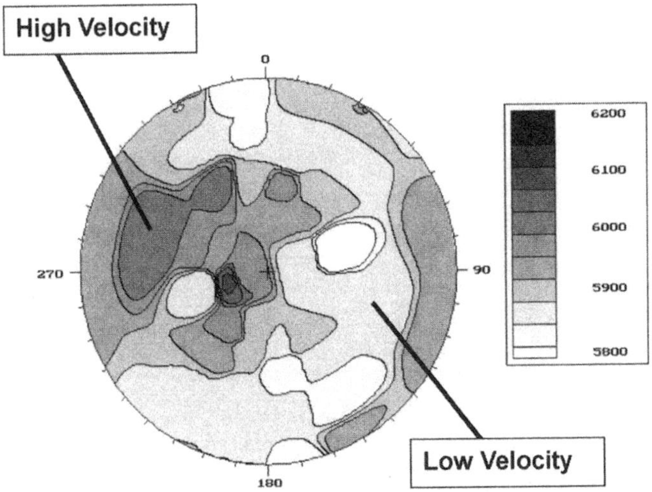

Figure 8
Lower hemisphere stereonet showing measured absolute velocities from the base survey prior to excavating deposition hole DD0092G01.

Ultrasonic Velocity Data

Shown in Figure 8 is a lower-hemisphere stereonet of the absolute velocity anisotropy measured prior to excavation. A mean velocity of 5903 m s^{-1} was measured, although the rock mass exhibits a slight anisotropy (approximately 3%) with the fast direction orientated at approximately 45° down to the Northwest. This may be attributed to the interaction of the principal stress with a pre-existing microstructure in the rock mass. A consistent result was obtained for the second deposition hole and both were consistent with previous velocity measurements conducted at this level of the Äspö HRL (FALLS and YOUNG, 1996).

Figure 9 describes the change in *P*-wave velocity measured along four ray paths in category *S1*. The fact that surveys were performed every hour allows for a high temporal resolution. The processing technique employed bears an estimated maximum uncertainty of ± 2 m s^{-1} between measurements. As can be seen, all the ray paths exhibit a sharp drop in velocity as the excavation passes by. The velocity changes are restricted to the few hours during and immediately after excavation. Further drops in velocity often occur during subsequent excavation steps. *S*-wave velocities could only be calculated for a few ray paths due to the ray path orientations relative to the radiation pattern of the transducers. The *S*-wave velocities show similar trends to the *P* waves. The velocity decreases are also accompanied by a reduction in signal amplitude of generally 50 to 75%. For ray paths that passed beneath the deposition hole, no significant change in velocity or

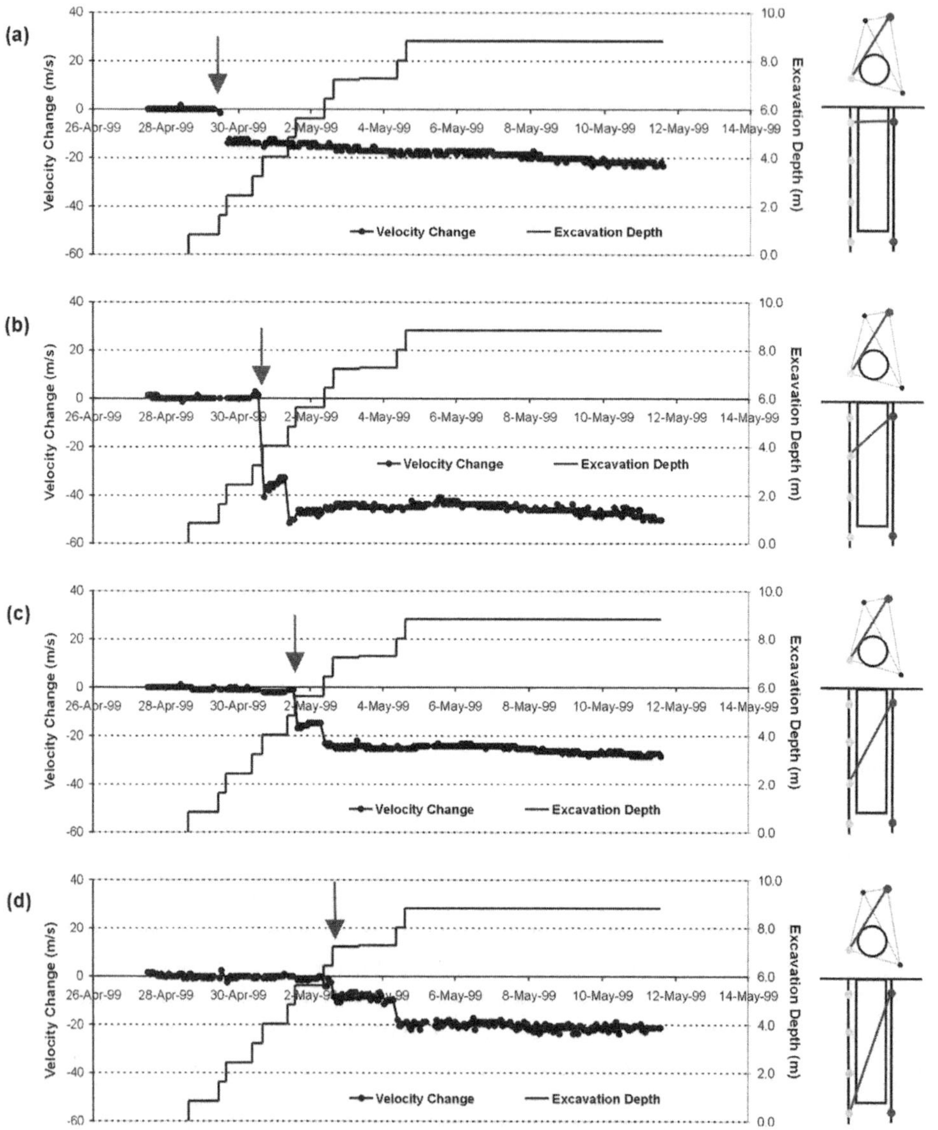

Figure 9
Example changes in *P*-wave velocity through excavation of deposition hole DD0092G01. The ray path orientations are illustrated in the right-hand margin. The arrow indicates when the excavation passes the ray path.

amplitude is observed. The magnitude of the velocity decrease is usually no larger than 30 m s^{-1}, and varies systematically with ray path orientation around the excavated deposition hole.

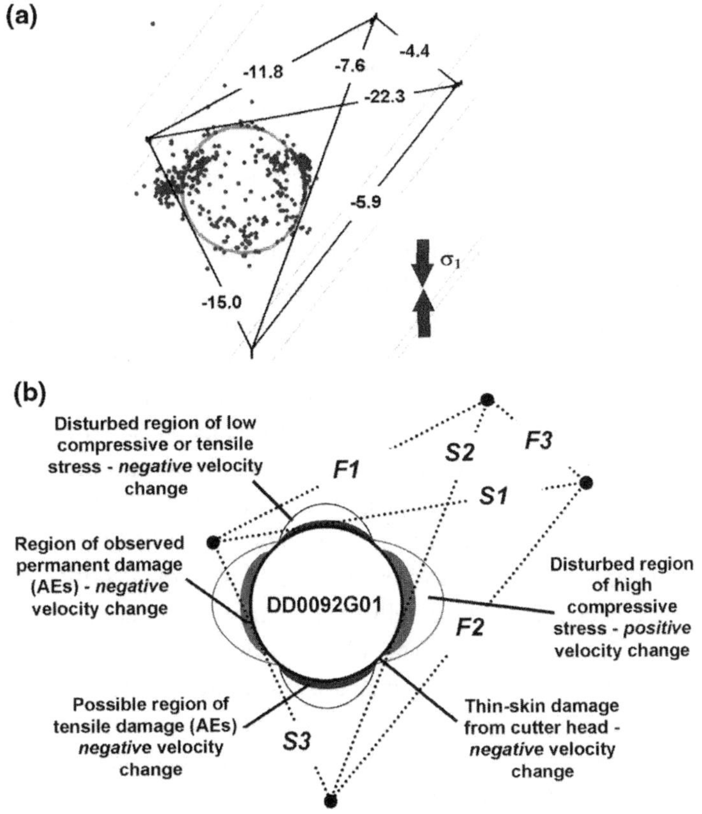

Figure 10
(a) Plan view of the deposition hole with AE source locations. Superimposed onto this are the six ray path categories (Fig. 3) with mean observed changes in velocity. (b) A simple stress-disturbance/damage model for observed AEs and velocity changes.

Rock Mechanical Interpretation

Figure 10a shows a plan view of the AE locations and mean velocity changes along the six ray path categories. Again consistent results have been obtained for the second deposition hole. In Figure 10b a simplistic model of the induced damage is shown. The relative stress changes are those expected from a Kirsch solution and the orientation of the maximum principal stress (σ_1). A small tensile stress would be anticipated parallel to σ_1 if the local principal stress ratio, $K = \sigma_1/\sigma_3 > 3$. Otherwise this region will experience a reduction in the compressive stress. Regions orthogonal to σ_1 will experience increased compressive stress. This culminates in the AEs discussed above. The entire deposition hole circumference is likely to have a thin skin of a few centimetres of damage inflicted by the action of the TBM cutter head and not imaged in the AEs.

The largest mean change in velocity (22 m s^{-1}) is observed on the ray path category *S1* that passes through the tensile/low compression region. This change must be attributed to opening microfractures in this region; a phenomenon that is well documented in laboratory results (e.g., KING *et al.*, 1997). The ray path will hence be effected by: (a) accumulated opening (relaxation) of preferentially orientated pre-existing microfractures in the stress-disturbed zone; (b) new microfractures in the damaged zone generated by the cutter head. Ray path *F1* also exhibits a drop in velocity although of smaller magnitude (12 m s^{-1}). As this does not skim the deposition hole, it is only effected by relaxation in the stress-disturbed zone. It should be noted that this relaxation may be quite small but a significant portion of the ray path passes through this relaxed zone. The result is an accumulated effect. In contrast, ray path *S2* displays only a modest decrease in average velocity, even though it passes through the damaged region induced by the cutter head and through a region of stress-induced damage imaged in the AEs. These effects would both act to reduce the velocity. However, a high proportion of the ray path also passes through a stress-disturbed zone that is under increased compression. This is likely to close preferentially orientated microfractures, causing an accumulated increase in velocity that offsets any decrease. Similar velocity structures have been imaged elsewhere (MAXWELL and YOUNG, 1995). The observed time-dependency in the velocity changes is related to a stress-induced increase in the microfracture density, or change in microfracture apertures, as the excavation progresses to the next step.

The velocity measurements allow an estimation of the change in rock properties in the damaged zone alone. Figure 11 shows a simple model of the damaged zone around the deposition hole perimeter. A section of the ray path (of total length *d*) between the transmitter and receiver samples the damaged zone (length d_D) and a

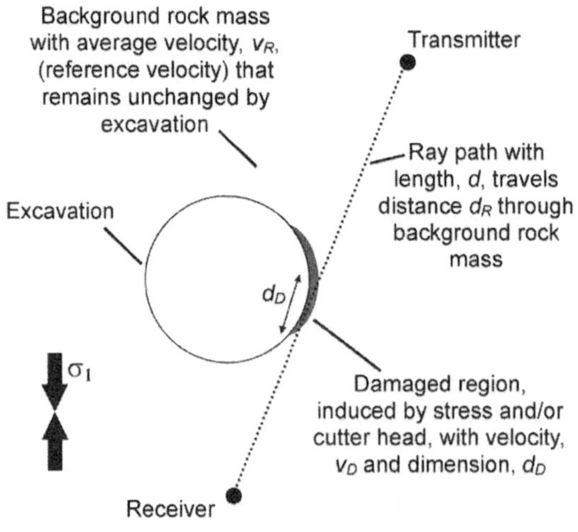

Figure 11
A simple model used for estimating the change in rock properties in the damaged zone.

section samples the remainder of the rock mass (length d_R). From the velocity measurements, we estimate the damaged zone causes a *decrease* in observed velocity (Δv_m) along the entire ray path of 15 m s^{-1}, and is similar for both *P* and *S* waves. The velocity of the rock in the damaged zone alone, v_D, is then given by Equation (1). This assumes that the remainder of the rock mass beyond the damaged zone has an average velocity that is unchanged by the excavation, and that the ray path remains a direct ray. The average rock mass velocity is termed the "reference velocity", v_R, and is measured during surveys performed prior to excavation ($v_p = 5900$ m s^{-1} and $v_s = 3350$ m s^{-1} for the *P* and *S* waves, respectively).

$$(v_R - v_D) = \frac{\Delta v_m \cdot v_R \cdot d}{d_D v_R + \Delta v_m d_R} \quad . \tag{1}$$

Assuming that the measured velocity change is small compared to the reference velocity and that the ratio d_D/d_R is not $\ll 1$, then Equation (1) approximates to Equation (2). That is, the difference between the velocity in the damaged zone and the velocity outside it is a simple function of the relative ray path lengths, d and d_D (how much of the ray path is damaged).

$$(v_R - v_D) \approx \Delta v_m \cdot \frac{d}{d_D} \quad . \tag{2}$$

We estimate that 5% of a skimming ray path typically passes through the damaged region. Typical ray path lengths are approximately 4 m, and we assume that the damage zone is approximately 20 cm of the ray path as defined by the AE observations above. Then, using Equation (2), the *decrease* in *P*- and *S*-wave velocities in the damaged zone is $(v_R - v_D) = 300$ m s^{-1} i.e., a drop of 20× the measured velocity change. Using a rock mass density of $\rho = 2650$ kg m^{-3}, then dynamic elastic moduli can be calculated for the intact and damaged rock. This results in Dynamic Young's moduli, $E_R = 75$ GPa and $E_D = 64$ GPa. The damaged region hence experiences a reduction in Young's modulus of approximately 15%.

Discussion

Induced seismicity has been recorded around two deposition holes excavated as part of an ongoing experiment investigating radioactive waste storage in deep repositories. Acoustic monitoring techniques have utilised AEs to investigate microfracture development in the EDZ. Damage is observed in the regions of tangential stress concentrations, i.e., potential "breakout" regions, orthogonal to the measured far-field maximum *in situ* stress at this level of the Äspö HRL. The AEs relate to small-scale adjustments on microfractures of millimetres in dimension and are interpreted as stress-induced degradation of the rock mass properties around pre-existing fractures. This is observed to occur for some days after the excavation has passed through the fracture, although activity dies away very quickly after excavation ceases. Damage induced by the action of the TBM cutter head is likely to be a thin skin of a few centimetres dimension around the whole circumference, as observed around similar excavations at this level. This damage may exacerbate the subsequent stress-induced micro-fractures by weakening the rock mass.

Ultrasonic velocity measurements have been conducted along ray paths that pass within a few centimetres of the deposition hole wall. Depending upon their orientation, these pass through damage inflicted by the action of the TBM cutter head and through stress-induced damage. They are also effected by induced stresses acting to open or close microfractures in relaxed or compressed regions of stress disturbance. Velocity measurements show a sharp decrease in velocity during and immediately after excavation. This is primarily associated with a relaxation of the rock mass and accumulated damage close to the hole. Measurements describe smaller velocity changes along ray paths that also pass through stress-disturbed zones with induced compressional stresses. These act to close microfractures.

The combined interpretation of the AE and ultrasonic data allows for a simplistic rock mechanical model of damage and stress disturbance to be constructed. The high-resolution velocity measurements can then be linked to a volume of rock close to the deposition hole wall in which excavation damage is observed. This results in a quantitative estimate of the change in rock properties in the excavation damaged zone.

This paper has presented an example of the use of induced seismicity and other acoustic methods as a tool for examining changing rock properties in engineering scenarios. Advances in acquisition instrumentation and processing techniques, for AE and ultrasonic monitoring in particular, are allowing higher-precision real-time measurements that are of great benefit in the design and performance evaluation of engineered structures. It is clear from the results presented in this paper that the use of these techniques, in combination with other methods, at experimental sights such as the Äspö HRL, can further understanding and then quantify the effects of excavation disturbance. The techniques also provide a method of remotely monitoring changing rock mass conditions during further loading or unloading, saturation or heating. For the storage of high-level radioactive waste, monitoring techniques such as these will ultimately lead to the use of optimised design and excavation methods, providing a higher confidence in the engineered solution, and may also provide a real-time on-line method for safety assessment during operation.

Acknowledgements

We extend appreciation to Olle Olsson for his continued support of acoustic monitoring techniques and to the staff at the Äspö Hard Rock Laboratory for their valued assistance with this work. Thanks to Dave Collins for fieldwork assistance and lengthy discussions. Also, we are grateful to Steve Falls for the use of his plotting program, and members of the Liverpool Research Group for their continuing support.

References

Baker, C. and Young, R. P. (1997), *Evidence for Extensile Crack Initiation in Point Source Time-dependent Moment Tensor Solutions*, Bull. Seismol. Soc. Am. *87*(6), 1442–1453.

Collins, D. S., Pettitt, W. S., and Young, R. P. (2001), *High-resolution Mechanics of a Micro-earthquake sequence*, this issue.

Emsley, S., Olsson, O., Stenberg, L., Alheid, H.-J., and Falls, S. (1997), *ZEDEX – A Study of Damage and Disturbance from Tunnel Excavation by Blasting and Tunnel Boring*, SKB Technical Report #97-30, SKB, Stockholm, Sweden.

Falls, S. D. and Young, R. P., *Examination of the excavation disturbed zone in the Swedish ZEDEX tunnel using acoustic emission and ultrasonic velocity measurements*, Proc. Eurock '96, Turin, Italy, 1996 (Balkema, Rotterdam 1996) pp. 1337–1344.

Falls, S. D. and Young, R. P. (1998), *Acoustic Emission and Ultrasonic-velocity Methods Used to Characterise the Excavation Disturbance Associated with Deep Tunnels in Hard Rock*, Tectonophysics *289*, 1–15.

King, M. S., Shakeel, A., and Chaudhrey, N. A. (editors), *Acoustic Wave Propagation and Permiability in Sandstones with Systems of Aligned Cracks, Developments in Petrophysics*, 69–85 pp. (Geological Society Special Publication No. 122, 1997).

Leijon, B. (1995), *Summary of Rock Stress Data from Äspö*, Äspö Hard Rock Laboratory Progress Report 25-95-15, SKB, Sweden.

Maxwell, S. C. and Young, R. P. (1995), *A Controlled in situ Investigation of the Relationship Between Stress, Velocity and Induced Seismicity*, Geophys. Res. Lett. *22*, 1049–1052.

Pettitt, W. S. (1998), *Acoustic Emission Source Studies of Microcracking in rock*, Ph.D. Thesis, Keele University, Keele, Staffordshire, UK.

Press, W. H., Flannery, B. P., Teukolsky, S. A., and Vetterling, W. T. *Numerical Recipes in C: The Art of Scientific Computing, 2nd edition* (Cambridge University Press, Cambridge (1994)).

Read, R. S. and Martin, C. D. (February 1996), *Technical Summary of AECL's Mine-by Experiment Phase 1: Excavation Response*, AECL, Whiteshell Laboratories, Pinawa, Manitoba, Canada.

Telford, W. M., Geldart, L. P., and Sheriff, R. E., *Applied Geophysics* (Cambridge University Press, Cambridge, UK, 1990).

Young, R. P. and Collins, D. S., *Monitoring an experimental tunnel seal in granite using acoustic emission and ultrasonic velocity*. In *Rock Mechanics for Industry*, Proc. 37th U.S. Rock Mechanics Symposium (Amadei, Kranz, Scott and Smeallie, eds.) Vail, Colorado (Balkema, Rotterdam 1999).

Young, R. P. and Martin, C. D. (1993), *Potential Role of Acoustic Emission/Microseismicity Investigations in the Site Characterisation and Performance Monitoring of Nuclear Waste Repositories*, Int. J. Rock Mech. Min. Sci. and Geomech. Abstr. *30*(7), 797–803.

Young, R. P., Martin, C. D., Murdie, R., Alcott, J., Falls, S., Stimpson, I., and Yazici, S. (1996), *Numerical Modelling, Acoustic Emission and Velocity Studies of the Excavation Disturbed Zone at the Hard Rock Laboratory*, Äspö Hard Rock Laboratory Technical Note, SKB, Sweden.

(Received March 22, 2000, revised December 29, 2000, accepted December 30, 2000)

Pure appl. geophys. 159 (2002) 197–219
0033–4553/02/030197–23 $ 1.50 + 0.20/0

Pure and Applied Geophysics

High-resolution Mechanics of a Microearthquake Sequence

David S. Collins,[1] Will S. Pettitt,[1] and R. Paul Young[2]

Abstract—Unique data from microseismic (MS) and acoustic emission (AE) systems monitoring a common rock volume have been analyzed from the Underground Research Laboratory in Canada. The two 16 sensor systems were installed to investigate stress-induced microcracking around clay and concrete bulkheads as part of the Tunnel Sealing Experiment. Single and double MS events are found to be spatially associated with clusters of between 19–86 higher frequency AE events. Each AE cluster is elongated with the longest axis between 15 and 50 cm. The AE events occurring before the associated MS event are termed foreshocks, and for two of the three analyzed AE clusters an acceleration in the AE frequency and cumulative magnitude occurs upto the time of the MS event, however with one AE cluster very few foreshocks are recorded, possibly indicating a more homogeneous failure plane. Time independent moment tensors were determined with the MS events showing significant deviatoric sources. The majority of the AE events have deviatoric mechanisms with a few crack opening and crack closure type events also determined. This study highlights the benefits of using small-scale seismic systems to investigate the temporal fracture mechanics of microcrack formation, and allows comparisons with larger more damaging seismicity.

Key words: Source mechanics, foreshocks, aftershocks, acoustic emissions, moment tensors, microcracks.

1. Introduction

Dodge *et al.* (1996) note that foreshocks are the most obvious manifestation of earthquake nucleation. Foreshocks generally occur due to the breaking of asperities and the sequences often cluster in the region of an eventual large magnitude earthquake. Earthquake prediction, however, is not easy since foreshock sequences may increase in frequency (low b value) before a large earthquake or instead may decrease with a period of quiescence before a large stress drop. Regardless of whether the foreshocks are incidental to the earthquake rupture or actually trigger the rupture, it is essential to study the locations and mechanics of foreshocks to aid in understanding the failure evolution leading up to the main shock.

[1] Applied Seismology Consultants, 10 Belmont, Shrewsbury, UK. E-mail: dave@seismology.org
[2] Applied Seismology Laboratory, Department of Earth Sciences, Liverpool University, Liverpool, UK. E-mail: r.p.young@liverpool.ac.uk

The number of foreshocks recorded partially depends upon the sensitivity of the seismic monitoring system. At the Underground Research Laboratory (URL) in Canada a unique situation exists in which two seismic systems monitor over a common rock-mass volume. The two seismic systems provide a large dynamic range of possible recorded seismicity (between ~−7 and −2 moment magnitude M_w), and spatial and temporal results by YOUNG and COLLINS (1999) indicate single low frequency events to often be associated with clusters of higher frequency events. The rock-mass at depth at the URL has been highly characterized for geomechanical, stress, and geophysical properties, and is a relatively unfractured homogeneous volume. Additional to this, the three-dimensional coverage of the two seismic systems and underground access to the experimental volume produces an excellent condition for advanced processing of the recorded seismic data, using methods such as moment tensor (MT) inversion. The MT method is routinely used for the study of global earthquakes (e.g., DZIEWONSKI *et al.*, 1997) and has also become popular for the study of smaller magnitude events where a local seismic network is present (e.g., SPOTTISWOODE, 1984; FEIGNIER and YOUNG, 1992; MCGARR, 1993). More recently, laboratory studies (OHTSU, 1991; DAHM *et al.*, 1998; PETTITT, 1998) have calculated MTs of high frequency events recorded in laboratory rock experiments. The MT method is particularly suited to seismic events occurring near voids where non-double-couple (DC) sources may be expected.

This paper provides insight into the temporal fracture mechanics of small-scale microcrack formation, and allows comparisons with the development of larger more damaging mining and tectonic induced seismicity.

2. Experimental Setup

The two seismic systems were installed to record stress induced microcracking as part of the Tunnel Sealing eXperiment (TSX). The TSX is a major international experiment taking place at 420 m depth at the URL, demonstrating full-scale tunnel sealing technologies (MARTINO and CHANDLER, 1999). The main objective of the TSX is to engineer barriers that will prevent any fluid flow from passing through the excavation damage zone (EDZ) in the host Lac du Bonnet granite. The granite (grain size ~0.5–10 mm) below 240 m at the URL is essentially unfractured so the only high permeable regions are the excavations themselves and the EDZ that forms around the excavations. Figure 1 is a schematic of the TSX showing the 13.5-meter long test chamber and the two bulkhead seals keyed into the rock mass. The URL is characterized by having high horizontal principal stresses (σ_1 and σ_2) and a lower magnitude subvertical principal stress (σ_3) as listed in Table 1. Due to this, high compressive stress concentrations exist in the roof and floor of subhorizontal tunnels which coincide with the location of the majority of recorded AE and MS activity.

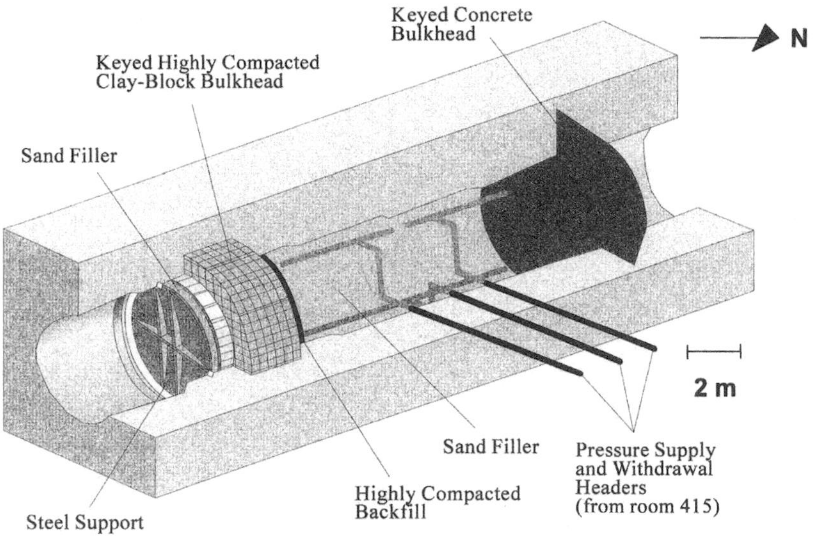

Figure 1
Schematic drawing of the Tunnel Sealing Experiment (MARTINO AND CHANDLER, 1999).

Table 1

Principal stress magnitudes and orientations

Principal stress	Magnitude (MPa)	Trend (°)	Plunge (°)
σ_1	60 ± 3	145	11
σ_2	45 ± 4	54	8
σ_3	11 ± 2	209	77

Three-dimensional stress modelling by MARTIN (1993) has shown the location of the seismicity to correspond to an average deviatoric stress ($\sigma_1 - \sigma_3$) equal to 70 MPa, which is termed the crack initiation stress. The two keyed regions of the TSX are shown by DZIK (1997) to completely interrupt the regions of high deviatoric stress to a value below the crack initiation stress.

The first system is termed the MS (microseismic) system and operates in the frequency range 0.5–10 kHz monitoring a region of approximately 10^6 m^3 surrounding the TSX. Within this volume an AE (acoustic emission) system monitors a 10^3 m^3 region in the frequency range 40–400 kHz. The locations of the sensors for the two seismic systems are shown in Figure 2 along with the excavated tunnels associated with the TSX. The two systems are fully described by YOUNG and COLLINS (1999). The 16 MS sensors are triaxial accelerometers grouted at the ends of boreholes in the rock mass. The 16 AE sensors are mounted in AE sondes and are spring-loaded against the sides of short boreholes into the rock. The AE system monitors a region around the lower half of the clay bulkhead seal. Mercury

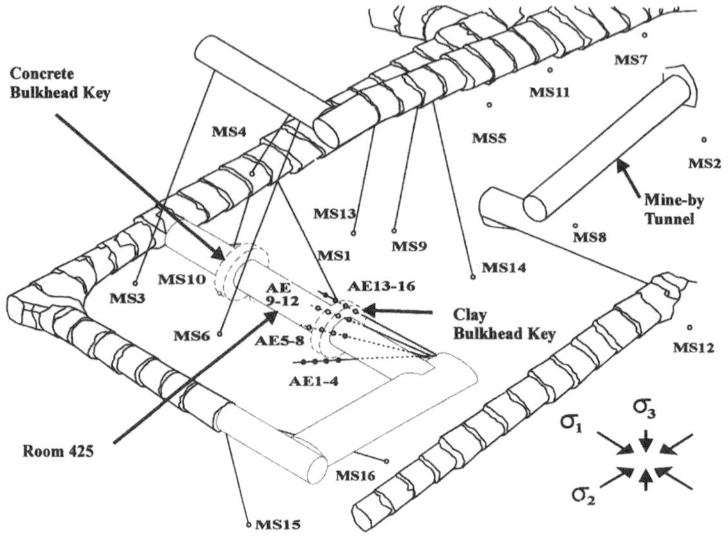

Figure 2
The tunnels associated with the TSX, relative to the principal stress directions. The 16 AE and 16 MS sensor locations are shown. Modified from AECL source.

tilt switches contained in the MS sensors and AE sondes were monitored during the installation process, resulting in the sensors being oriented to an accuracy of 2°.

The majority of the acquisition hardware and software is manufactured by the Engineering Seismology Group (ESG) in Canada (ESG, 2000), although the borehole sondes for the AE sensors were designed and built by the authors. Full waveforms of events are digitized when user-set trigger logic criteria are met at 50 kHz sampling frequency and 12 bit resolution for the MS system, and 1 MHz and 14 bit resolution for the AE system.

The MS system has been monitoring fairly continuously since January 1997 when the excavation of the main tunnel, room 425, started. The clay key was excavated between July and September 1997, and the AE system started recording seismicity a few weeks before this. The AE system has been monitoring fairly continuously to the present, though due to its high sensitivity is limited to 100 events/hour. The computer clocks of the two seismic systems are GPS time synchronized.

3. Seismic Processing Methodology

The AE sensors, model R6 and R6i manufactured by Physical Acoustics Corporation, are proportional to velocity (PAC, 1997) and Figure 3 shows an example recorded waveform with clearly identifiable *P* and *S* waves. Example

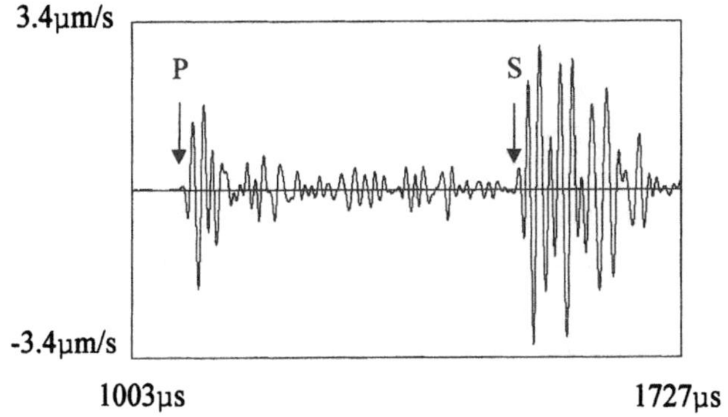

Figure 3
A typical waveform recorded by the AE system with the *P*- and *S*-wave arrival times identified.

waveforms recorded from the MS system are displayed by COLLINS and YOUNG (2000).

AE and MS source locations were calculated by a Simplex/Geiger method (ESG, 2000) using *P*-wave arrival times, a homogeneous velocity model, and assuming straight ray paths. A Schmidt hammer calibration survey was performed in July 1998 at different locations in the TSX with energy recorded on both seismic arrays. The difference between the source and surveyed locations was found to be ± 0.6 m for the MS system and ± 0.05 m for the AE system.

The mechanisms of the AE and MS events were determined using a time domain moment tensor (MT) inversion technique (PETTITT, 1998). The procedure is based upon the method used by STUMP and JOHNSON (1977), DE NATALE and ZOLLO (1989), FEIGNIER and YOUNG (1992), and OHTSU (1991). The method uses amplitude data taken from the recorded waveforms and an estimate of the source-receiver propagation effects (Green's functions) to invert for the nine equivalent forces at the source (JOST and HERRMAN, 1989). These are represented as a symmetric 3×3 matrix (M) which can be diagonalised into three principal forces (eigenvalues and eigenvectors). The Green's functions assume a far-field radiation pattern (the receivers are checked to be >10 wavelengths distance from the source), and a homogeneous and isotropic medium. *P*-wave amplitudes from the time domain are only considered due to the uniaxial nature of the AE sensors. A triaxial rotation of the MS waveforms was performed to determine the full *P*-wave trace. *S* waves are currently not incorporated into the MT inversion software. Each amplitude is corrected for geometrical spreading, anelastic attenuation, free-surface amplification at the transducer face (AKI and RICHARDS, 1980) and receiver coupling. A quality factor $Q = 223 \pm 37$ was found for the undisturbed URL granite at MS frequencies (FEUSTEL, 1995). Due to the increased fracturing in the localised region around the TSX tunnel and key, a lower $Q = 180$ is assumed in the inversion for the AE data. PETTITT (1998) accurately measured the azimuthal response of the PAC R6 sensor using a semicylindrical aluminum block, and determined that following the free-surface correction, the amplitude is flat within ±5% in the range of −80° to 80° angle from the axis of the sensor. A similar study with numerous sensors of varying piezoelectric crystal shapes and sizes by PETTITT (1998) has shown that azimuthal corrections are necessary for certain piezoelectric sensors dependent upon the crystal size and frequency response. Coupling factors of the AE sensors to the borehole wall were determined from the daily velocity surveys performed, and found to vary between 0.7 and 1.3. Due to the MS sensors being grouted into the rock mass, a unit coupling factor was given, and no free-surface correction was applied.

The MT assumes a point source and a delta type source-time function, and the inversion is determined using Singular Value Decomposition. Decomposition of the MT was chosen using the source type plot of HUDSON *et al.* (1989) which allows the resultant source mechanisms to be displayed as data points and their distribution to be correlated to known source types. This method uses the eigenvalues of M to calculate two parameters, T, relating to the deviatoric component, and k relating to the isotropic (volumetric) component. Organising the eigenvalues such that $m_1 > m_2 > m_3$, the equations for the T and k values are dependent upon the sign of the deviatoric eigenvalue $m_i^* = m_i - \frac{1}{3}\mathrm{tr}(M)$, as shown below (HUDSON *et al.*, 1989):

$$k = \frac{\frac{1}{3}\mathrm{tr}(M)}{\left(|\frac{1}{3}\mathrm{tr}(M)| - m_3^*\right)}, \quad \text{for } m_2^* \geq 0 \ , \tag{1}$$

$$k = \frac{\frac{1}{3}\mathrm{tr}(M)}{\left(|\frac{1}{3}\mathrm{tr}(M)| + m_1^*\right)}, \quad \text{for } m_2^* < 0 \ , \tag{2}$$

$$T = \frac{-2m_2^*}{m_3^*}, \quad \text{for } m_2^* > 0 \ , \tag{3}$$

$$T = 0, \quad \text{for } m_2^* = 0 \ , \tag{4}$$

$$T = \frac{2m_2^*}{m_1^*}, \quad \text{for } m_2^* < 0 \ . \tag{5}$$

Both T and k vary between −1 and 1 (see Figure 7a for an example of the $T - k$ source type plot with common source types marked).

To allow a comparison with previous studies, the isotropic percentage, R, is also calculated as (FEIGNIER and YOUNG, 1992):

$$R = \frac{\mathrm{tr}(M) \times 100}{\left(|\mathrm{tr}(M)| + \sum |m_i^*|\right)} \ , \tag{6}$$

where R varies from 100 (pure explosion) to −100 (pure implosion).

The quality of the inversion is determined in three ways. Individual amplitude residuals (difference between corrected and calculated amplitude which can vary between 0.0 and 2.0) and a mean residual are outputted for each event, allowing amplitudes with high residuals to be rechecked. Event mechanisms are only kept with a mean residual <0.2, a minimum of eight high confidence amplitudes, and a good focal-sphere coverage (reasonably even distribution of data points over lower hemisphere). Secondly, standard deviations are calculated for the measured amplitudes proportional to the measured signal-to-noise on each channel and an estimated percentage error for each of the correction factors applied. These values are used to calculate a standard deviation ΔM_j for the 6 MT elements M_j following STUMP and JOHNSON (1977). 729 MTs can be formed using all combinations of $M_j \pm \Delta M_j$, and these combinations are used to sample the error space around each $T - k$ value and thus calculate a standard deviation. Thirdly, the robustness of the inversion is determined by the condition number.

Source parameters were calculated for the MS events using P and S waveforms and the automated time-domain method of URBANCIC *et al.* (1996). A similar method was used for the AE events to calculate seismic moment from the corrected P-wave amplitudes.

A significant amount of stress-induced seismicity was recorded in 1997, mainly due to the blast excavation of room 425 and the mechanical excavation of the two keyed regions. In total 12000 MS events and 21900 AE events were processed and tabulated (YOUNG and COLLINS, 1999). The majority of events are induced within a 1 meter region of room 425 and the keyed regions, although, significant activity also takes place in a 1 meter region around older existing tunnels of the 420 Level, and this is believed to be a triggered stress transfer process. This is discussed later in the paper.

An initial analysis of the AE/MS event database searched for "clusters" of AE events associated with MS events. Groups of four or more AE events occurring within a one minute interval were investigated to see whether they spatially corresponded with an MS event in the same minute. From this, three clusters were chosen to be analysed for source mechanisms in this study. Centered about each cluster time, additional events occurring in the cluster were found by searching over a minimum 2-hour interval. All AE and MS events occurring in this time frame were automatically processed and the P-wave arrival times manually checked. The new source locations were then spatially analyzed to see whether they occurred in the cluster region. Due to the large number of AE events, approximately half were chosen for MT analysis and a subset of these had successful MT solutions. Three of the four MS events had successful MT solutions. The main reason for unsuccessful MT solutions was due to the polarity of P waves being difficult to measure because of high background noise due to effects such as multiple events being present on the recorded traces.

4. Cluster Activity

The AE locations of the three analysed clusters are shown in Figure 4. Clusters 1 and 2 occur before the excavation of the clay key, and locate just below the floor of room 425. Cluster 3 occurs when the clay key has been approximately 3/4 excavated and locates at the 90° angle of the lower part of the key. One MS event is associated with each of Clusters 1 and 3, two MS events with Cluster 2, and in total 155 AE events. Table 2 documents the date and time of the AE/MS clusters. During 1997, a problem occurred with the GPS connection to the AE system, resulting in a linear drift away from the correct GPS time. This has been corrected for in Table 2 and therefore it is believed that the times of the AE and MS events can be correlated to within 1 second accuracy.

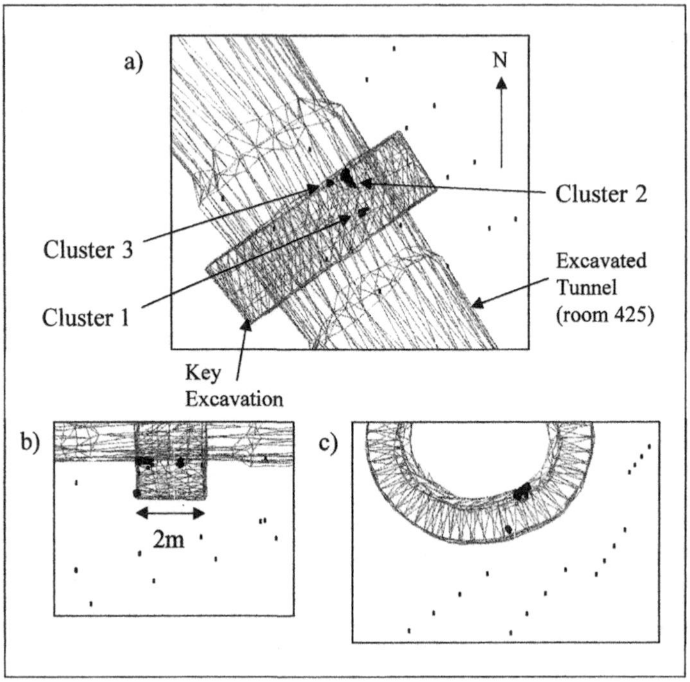

Figure 4
The locations of the three AE clusters relative to the excavation, on (a) plan view, (b) side view (azimuth 55°, dip −20°), and (c) section view (azimuth 325°, dip 0°). Note Clusters 1 and 2 occurred before the key excavation. The 16 AE sensors are shown as black rectangles.

Table 2

Details of the 3 AE/MS clusters including date, time, and number of events

	Date/time of MS events – in brackets the time adjustment to AE computer clock	Date/time of AE events	Number of AE events
Cluster 1	06-21-1997 04:32:30 (−3 seconds)	06-21-1997 04:26:45–05:27:08	19
Cluster 2	06-21-1997 05:29:01 and 05:29:03 (−3 seconds)	06-21-1997 00:02:49–05:29:36	86
Cluster 3	08-19-1997 04:16:10 (−57 seconds)	08-19-1997 03:16:08–05:11:05	50

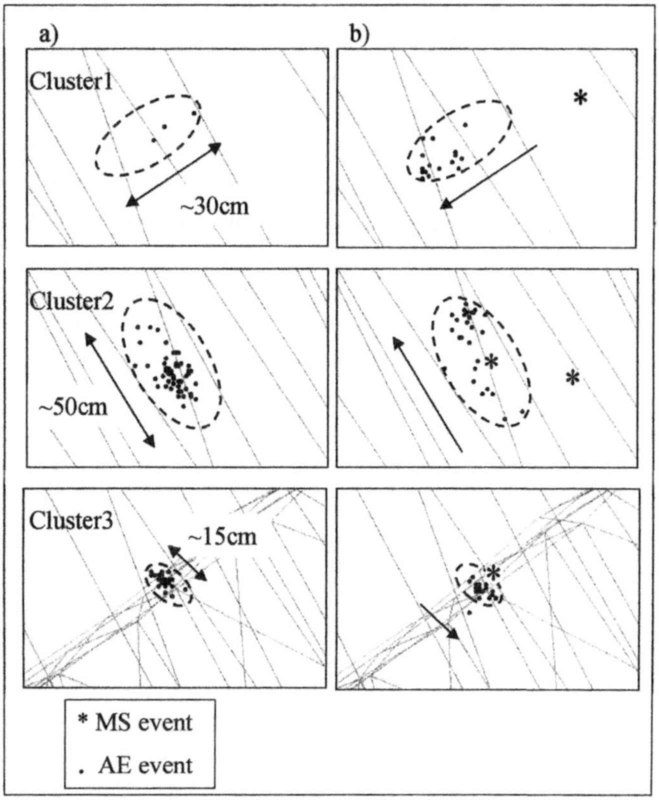

Figure 5
Plan views of the three clusters showing (a) the AE events which occur before the time of the MS event(s), and (b) the AE events which occur during and after the MS event(s). The dashed ellipse marks the extent of each AE cluster and the length of the long axis is marked. The arrow marks the direction of the temporal movement of the AE event locations. The accuracy in the AE locations is ± 5 cm and in the MS locations is ± 60 cm.

4.1 Spatial and Temporal Trends

Spatially it is observed that all three AE clusters are significantly elongated in one sub-horizontal direction. This is shown on a plan view (Fig. 5), where an ellipse is found to be the best shape to approximately enclose each AE cluster, with its longest axis between 15 and 50 cm. Due to the fact that the clusters are elongated subhorizontally, depth is not considered in this analysis, and the spatial trends are analyzed in two dimensions only. Within error bars, the MS locations fit inside the AE cluster ellipses, and it is likely that the movement associated with the MS event occurred inside this region. We therefore interpret the AE cluster to spatially delineate the failure region of the MS event. The temporal movement of the AE locations may then indicate the MS slip direction. Figure 5a presents the AE locations up to the time of the MS event(s), and Figure 5b the AE locations during and after the MS event(s). In all three cases a clear difference is seen in the AE cluster locations, and arrows have been marked in Figure 5b indicating the interpreted temporal movement trends.

Figure 6 presents the cumulative seismic moment of the AE events versus time for the three clusters, with the times of the MS events marked. The MS event in Cluster 1 coincides in time with the fourth AE event, and it can be seen that only three small

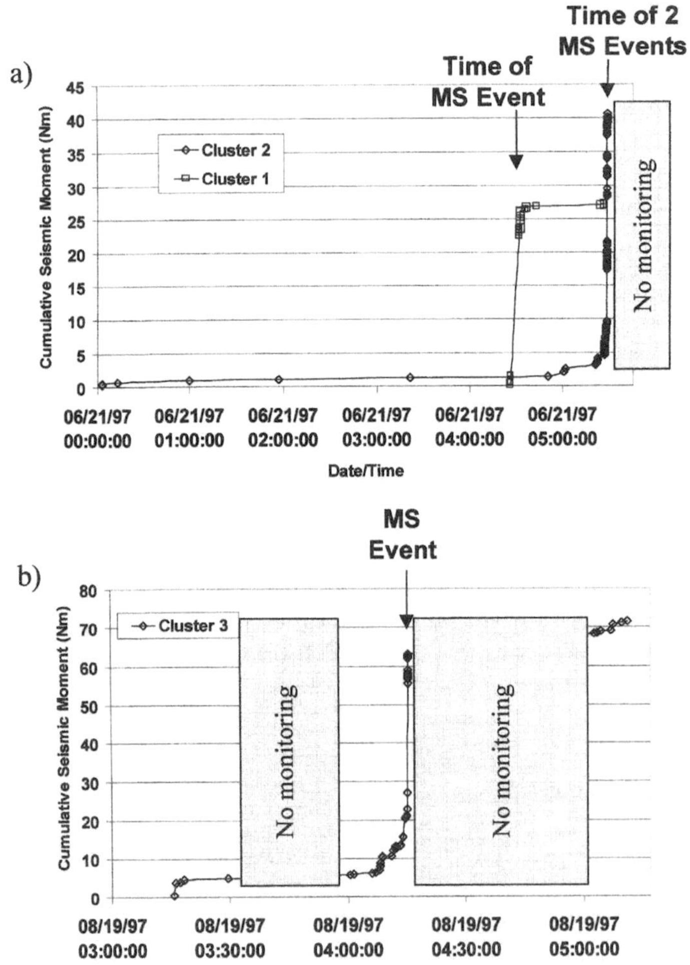

Figure 6
The cumulative seismic moment versus time for the AE events in (a) clusters 1 and 2, and (b) cluster 3. The times when the AE system was not monitoring are marked "No Monitoring" (due to a maximum limit of 100 events/hour). The arrows indicate the times of the MS events.

magnitude AE events occur about 6 minutes before the MS event. A total of 11 AE events occur within a 20 second interval of the MS event, and a further 5 over the next hour. In contrast, the AE events in Cluster 2 provide considerably more warning of the occurrence of the two MS events. Seven events occurred between midnight and 05:00, followed by nine events between 05:00–05:26, then 13 events at 05:27, and 30 events at 05:28, before the occurrence of the two MS events separated by two seconds. A further 27 events occur during and after the time of the MS events over a 1 minute interval until the AE system reaches its maximum limit. In Cluster 3, five AE events occur between 03:15–03:30, followed a period of no monitoring, then fourteen between 04:00–04:12, followed by four AE events at 04:13, six at 04:14, then one at 04:15 before the MS event. Thirteen events occur at 04:15 during and following the MS event before the AE system reaches its maximum limit. Seven small magnitude AE events occur between 05:00–05:12.

4.2 Mechanisms

Table 3 documents the date, time and moment magnitude of the AE/MS events with successful MT solutions, as well as details of the calculated moment tensors, error estimates and decomposition values. The (T, k) values are given along with their error estimates (standard deviations). It is noticed that the T-values are more highly variable ($\sim\pm 0.1$–0.3) than the better constrained k values ($\sim < \pm 0.05$). It is likely that the variation in the T and k values in Table 3 underestimates the true error in the solutions. O'Connell and Johnson (1988) note that significant isotropic and compensated linear vector dipole (CLVD) elements could be introduced by event mislocation and velocity model errors. In extensive error analysis studies of MS event mechanisms at the URL, Feignier and Young (1992) and Baker and Young (1997) use a value of 30% to indicate the division between a significant and insignificant isotropic (volumetric) component:

$$-30\% > R > 30\% \ , \tag{7}$$

which is equivalent to:

$$0.22 > k > 0.22 \ . \tag{8}$$

In this study this relation is used for both the AE and MS data to suggest whether an event is considered significantly isotropic. Figure 7 presents the $T - k$ source type plots for the three clusters. Common source types including double-couple (DC), CLVD, explosion, and implosion are marked on Figure 7a. Additionally the location of a pure tensile event is displayed assuming a Poisson's Ratio of 0.25. In Figure 7, the region of the $T - k$ plot where events are considered to be deviatoric in nature with no significant volumetric component is marked in grey. The $T - k$ plots show that although the majority of analysed events are deviatoric in nature, 26% have significant volumetric components with at least one occurring in each cluster.

Table 3

Mechanism results for the AE and MS events in the three clusters. The columns are: Event name; date; time; moment magnitude, number of data values, three principal eigenvalues, percentage decomposition into isotropic component R, T and k decomposition with calculated standard deviations (σ); mean residual, and SVD inversion condition number

Ev. #	Date	Time	M_w	N	m_1 (T)	m_2 (B)	m_3 (P)	R (%)	T	$\pm\sigma$	k	$\pm\sigma$	Mean resid.	C#
MS1	06/20/1997	4:32:27	−3.40	9	0.87	−0.15	−0.47	13	−0.60	0.13	0.09	0.15	0.02	7.8
AE1a	06/21/1997	4:32:29	−6.29	13	0.73	−0.13	−0.67	−4	−0.29	0.08	−0.03	0.02	0.15	3.1
AE1b	06/21/1997	4:32:31	−6.50	8	0.80	0.19	−0.56	23	0.12	0.16	0.17	0.04	0.04	3.6
AE1c	06/21/1997	4:32:32	−6.41	9	0.81	−0.03	−0.59	11	−0.26	0.12	0.08	0.02	0.04	3.5
AE1d	06/21/1997	4:32:35	−6.41	10	0.82	0.57	0.02	61	0.40	0.22	0.51	0.02	0.09	2.7
AE1e	06/21/1997	4:32:47	−6.48	10	0.79	0.11	−0.60	17	0.02	0.16	0.12	0.03	0.06	3.2
AE1f	06/21/1997	4:43:43	−6.50	11	0.75	0.11	−0.65	13	0.11	0.14	0.09	0.03	0.07	3.2
AE2a	06/21/1997	5:27:45	−6.56	9	0.81	0.59	0.07	64	0.39	0.29	0.53	0.06	0.05	3.8
AE2b	06/21/1997	5:28:26	−6.19	13	0.80	−0.04	−0.60	9	−0.24	0.14	0.06	0.03	0.09	2.9
AE2c	06/21/1997	5:28:30	−6.61	9	0.81	0.57	0.10	65	0.34	0.32	0.55	0.05	0.05	3.7
AE2d	06/21/1997	5:28:46	−6.35	10	0.56	−0.02	−0.83	−16	0.20	0.09	−0.11	0.03	0.10	3.2
AE2e	06/21/1997	5:28:55	−6.50	10	0.72	0.09	−0.69	7	0.14	0.11	0.05	0.03	0.13	3.2
MS2a	06/21/1997	5:28:58	−3.92	8	0.48	−0.22	−0.85	−30	−0.08	0.35	−0.22	0.11	0.05	26.3
AE2f	06/21/1997	5:29:00	−6.19	13	0.77	−0.01	−0.63	8	−0.15	0.11	0.05	0.03	0.09	3.0
AE2g	06/21/1997	5:29:01	−5.88	10	0.24	−0.45	−0.86	−47	−0.30	0.13	−0.37	0.04	0.07	5.4
AE2h	06/21/1997	5:29:03	−6.37	10	0.55	0.09	−0.83	−11	0.40	0.05	−0.08	0.02	0.15	2.7
AE2i	06/21/1997	5:29:05	−6.32	11	0.48	−0.15	−0.86	−28	0.08	0.12	−0.20	0.03	0.14	2.8

AE2j	06/21/1997	5:29:18	−6.51	10	0.57	0.15	−0.81	−5	0.45	0.11	−0.03	0.04	0.07	3.8
AE2k	06/21/1997	5:29:32	−6.35	12	0.22	−0.55	−0.80	−49	−0.57	0.12	−0.39	0.03	0.05	2.9
AE3a	08/19/1997	4:00:23	−5.99	12	0.51	−0.01	−0.86	−20	0.29	0.07	−0.14	0.02	0.15	2.6
AE3b	08/19/1997	4:08:01	−5.96	11	0.58	−0.14	−0.80	−20	−0.07	0.08	−0.14	0.02	0.18	2.4
AE3c	08/19/1997	4:08:36	−5.88	12	0.38	−0.20	−0.90	−35	0.12	0.07	−0.26	0.01	0.15	2.4
AE3d	08/19/1997	4:11:25	−5.82	10	0.72	−0.01	−0.70	1	−0.03	0.09	0.00	0.03	0.11	2.5
AE3e	08/19/1997	4:11:53	−6.31	9	0.80	−0.12	−0.59	5	−0.39	0.07	0.03	0.03	0.16	4.6
AE3f	08/19/1997	4:12:04	−6.11	9	0.81	−0.04	−0.58	11	−0.27	0.09	0.08	0.03	0.17	4.4
AE3g	08/19/1997	4:13:11	−6.35	9	0.80	0.27	−0.53	27	0.25	0.12	0.20	0.02	0.06	3.7
AE3h	08/19/1997	4:13:57	−6.35	8	0.84	0.16	−0.52	26	0.00	0.11	0.19	0.03	0.19	3.6
AE3i	08/19/1997	4:14:31	−5.54	11	0.43	−0.16	−0.89	−31	0.13	0.09	−0.23	0.02	0.04	3.1
AE3j	08/19/1997	4:14:59	−5.83	10	0.99	0.12	−0.05	45	−0.72	0.08	0.35	0.02	0.05	2.3
AE3k	08/19/1997	4:15:08	−5.57	11	0.72	0.08	−0.69	8	0.12	0.10	0.05	0.03	0.10	2.8
MS3	08/19/1997	4:15:13	−2.08	9	0.86	−0.25	−0.44	10	−0.77	0.08	0.07	0.03	0.08	9.1
AE3l	08/19/1997	4:15:13	−5.03	10	0.65	−0.28	−0.70	−18	−0.45	0.08	−0.13	0.02	0.13	4.1
AE3m	08/19/1997	4:15:15	−6.29	9	0.77	0.26	−0.59	23	0.30	0.09	0.17	0.02	0.08	3.6
AE3n	08/19/1997	4:15:16	−6.25	9	0.86	0.15	−0.49	28	−0.08	0.12	0.20	0.03	0.07	3.6

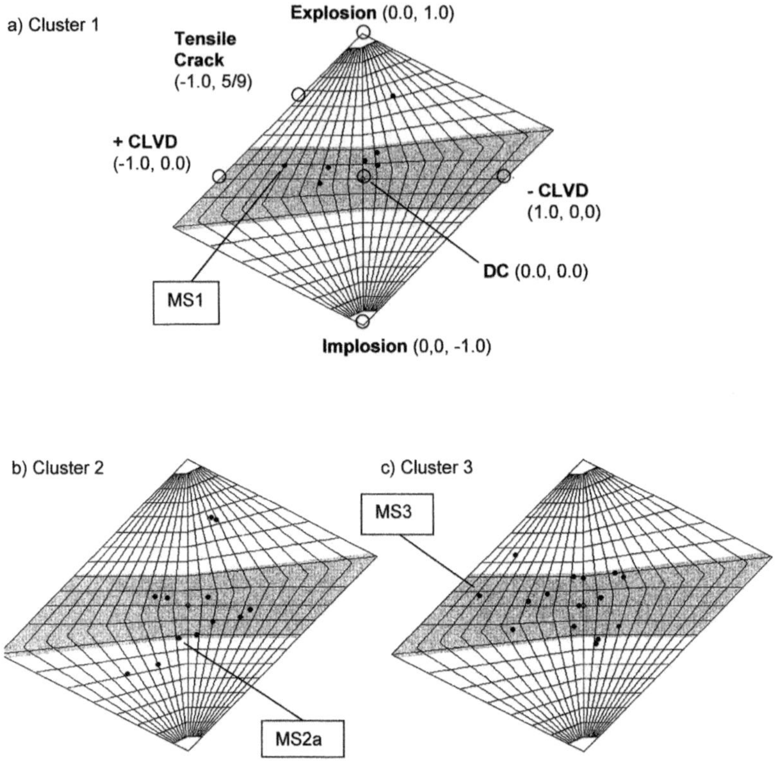

Figure 7
The '$T - k$ source type plot' after Hudson *et al.* (1989). The filled circles present the source mechanism types for the AE and MS events from the three clusters. In (a) commonly identified source types are marked as unfilled circles. The shaded zone identifies the region where events are considered to be significantly deviatoric in nature (see text for details).

The MT orientations are shown for all events in Figures 8–10, highlighting the positive (filled) and negative (empty) first motion data points as well as the calculated P-wave radiation pattern. The P-, B-, and T-axis orientations are inferred from the MT eigenvalues and are interpreted as the local force directions acting at the source (Jost and Herrmann, 1989). Two nodal planes are marked if $\%DC \geq 50\%$, while a single plane perpendicular to the largest force direction is marked where $\%DC < 50\%$, using the equation (Dziewonski *et al.*, 1981):

$$\%DC = 100 - \frac{|m^*|_{\min}}{|m^*|_{\max}} * 200 \ . \tag{9}$$

Many studies consider the P and T axes to be reasonably similar to the maximum and minimum compressive stress directions at the source, and one justification for this assumption is that the associated nodal planes represent orientations of

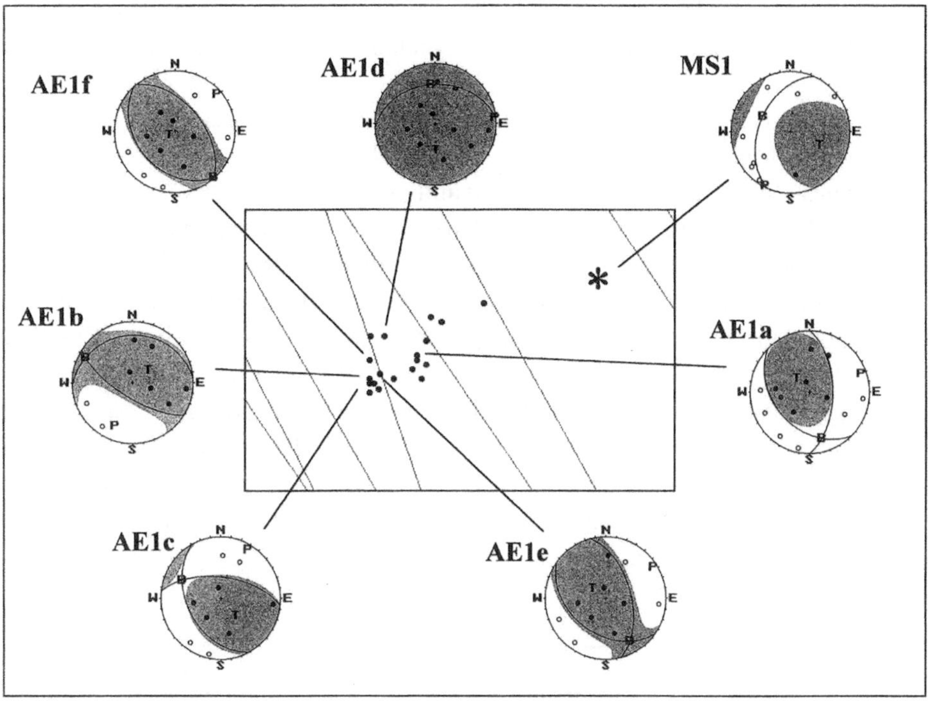

Figure 8
Stereonet representations of moment tensor solutions of AE and MS events from Cluster 1 (see text for details). The AE events are labelled in order of occurrence (see Table 2). The location of the events are shown in plan view with a star symbol marking the MS event.

maximum shear stress which might be reasonable locations for new fractures (GEPHART and FORSYTH, 1984). However it should be noted that the *P* and *T* axes for single mechanism solutions may vary significantly from the principal stress directions (MCKENZIE, 1969), especially in situations where slip occurs on a pre-existing plane of weakness.

In Cluster 1 (Fig. 8) the AE and MS MT solutions all have very consistent subhorizontal *P* axes indicating a local maximum stress striking ~SW-NE. One of the AE events (AE1d) has a high positive volumetric component and the orientation suggests subhorizontal crack opening. The other 5 (AE1a, b, c, e, f) show very consistent MT solutions with nodal planes striking approximately in the direction of the tunnel. Slots cut into tunnels at the URL have identified cracking parallel to the free surface to be a dominant mechanism of the excavation damage zone. The tunnel perimeter above Cluster 1 dips to the SW at 30°, and it is therefore suggested that the shallow SW dipping nodal planes of 4 of these 5 events represent the main failure plane. The MS event occurs before the 6 AE events and is significantly deviatoric in nature with similar principal axis directions to many of the deviatoric AE events. On the $T - k$ plot, the MS event locates to the left of the deviatoric AE events, nearer the +CLVD source, with a T-axis direction similar to that of the volumetric event AE1d.

The AE mechanisms in Cluster 2 (Fig. 9) can be divided into Pre- and Post-MS (see times in Table 3). During Pre-MS there are three AE events (AE2b, d, e) which are significantly deviatoric and two dominantly volumetric (positive) ones (AE2a, c). All five AE events have subhorizontal B axes, and it can be suggested that opening and slip on subhorizontal cracks has occurred. For Post-MS, four AE events (AE2f, h, i, j) have significant deviatoric components and two AE events (AE2g, k) indicate crack closure mechanisms, suggesting possible mineral grain crushing. From the nodal planes, as with Cluster 1, it can be proposed that slip is occurring on a shallow SW dipping failure plane, parallel to the tunnel perimeter. The MS mechanism is dominantly deviatoric near the DC source on the $T - k$ plot, and again the subhorizontal nodal plane is proposed to be the failure plane. The slip vector associated with this nodal plane is in the NW direction, consistent with the temporal AE cluster movement (Fig. 5).

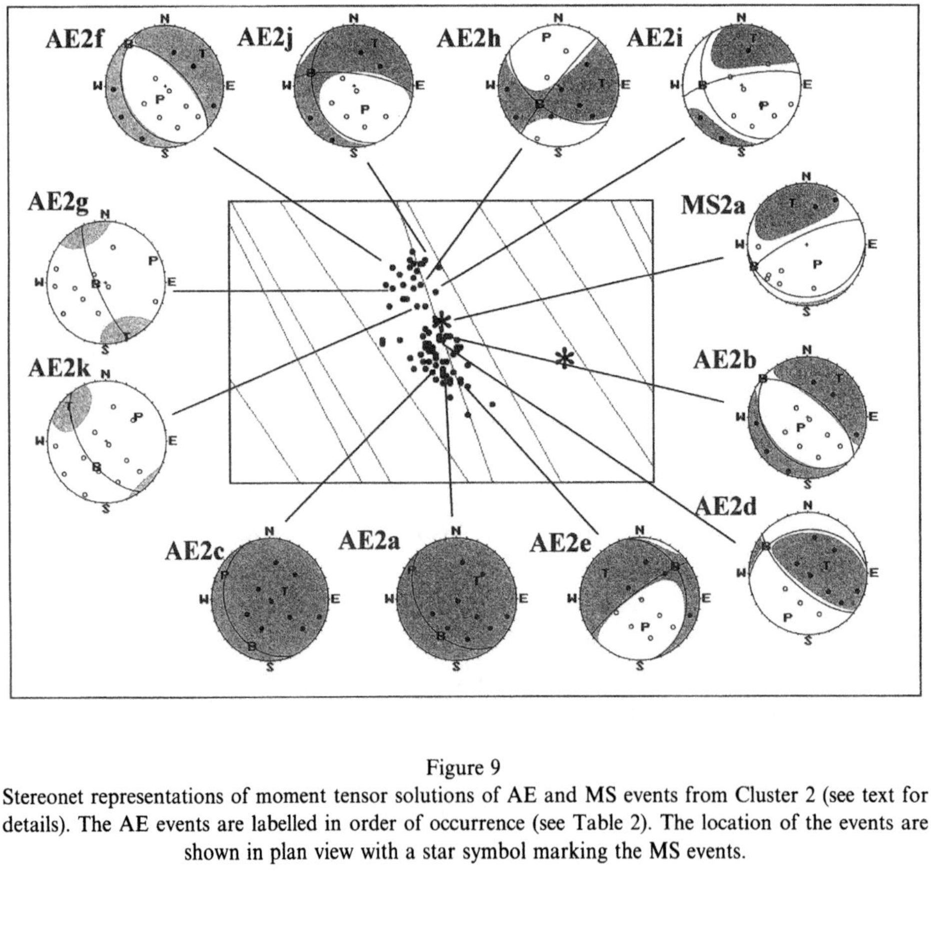

Figure 9

Stereonet representations of moment tensor solutions of AE and MS events from Cluster 2 (see text for details). The AE events are labelled in order of occurrence (see Table 2). The location of the events are shown in plan view with a star symbol marking the MS events.

Cluster 3 occurs on the corner of the clay key and therefore it is difficult to predict the local stresses acting or the orientation of any pre-existing cracks. Interestingly, the majority of the AE mechanisms (Fig. 10) indicate T axes that are consistently dipping to the SE. During Pre-MS, 8 of the AE events (AE3a, b, d, e, f, g, h, k) are dominantly deviatoric, while one (AE3j) has a significant positive volumetric component and two (AE3c, i) have significant negative volumetric components. The two Post-MS events are dominantly deviatoric. One of the AE events is believed to occur during the time of the MS event, and the mechanics of these two events (AE3l and MS3) display very similar mechanisms although they are different by three orders of magnitude. The MS event indicates dominant +CLVD movement in a steeply dipping direction to the SE and this movement direction is consistent with the temporal movement of the AE cluster (Fig. 5).

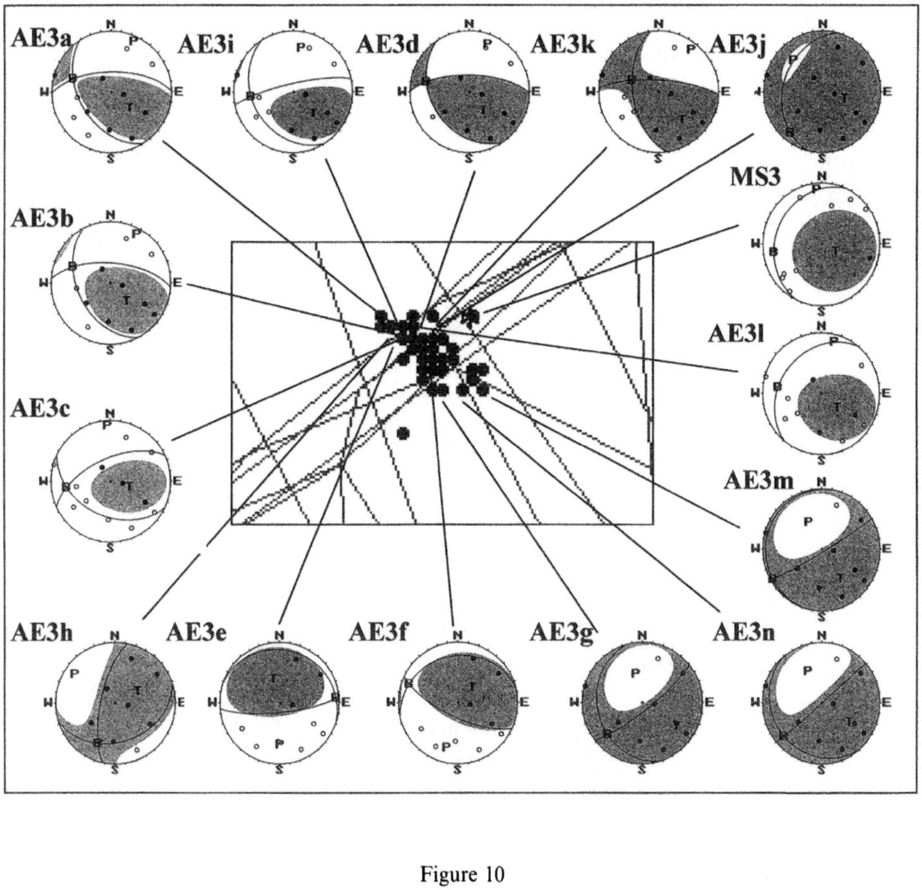

Figure 10
Stereonet representations of moment tensor solutions of AE and MS events from Cluster 3 (see text for details). The AE events are labelled in order of occurrence (see Table 2). The location of the events are shown in plan view with a star symbol marking the MS event.

5. *Discussion*

A schematic interpretation of the AE/MS relationship is given in Figure 11, showing the initial AE events as foreshocks which occur at highly stressed regions along the eventual MS (main shock) fracture plane. Subsequent AE events are considered aftershocks generated over the restabilising fracture and in the surrounding region.

BREHM and BRAILE (1998) use foreshocks recorded over a 20-year period by a microseismic system in the New Madrid area to test the time-to-failure method (VARNES, 1989) which fits a model acceleration curve to a graph of cumulative foreshock energy versus time. In most cases the times of failure and magnitudes of the actual main shocks were predicted reasonably well. On a much smaller time scale, in this AE/MS study, two of the three clusters have significant precursory seismic activity occurring for several hours before the time of the main shock, and additionally an acceleration in the frequency of activity occurs near the main shock agreeing with the time-to-failure model. Figure 12 is an expansion of Figure 6a which shows the detailed behaviour of Cluster 2 to be an irregular increase in cumulative seismic moment up to the time of the 2 MS events, followed by a sudden change in slope signifying the start of the AE aftershocks. A similar increase is found for Cluster 3, however with Cluster 1 very little warning of the main shock is given. One reason that has been suggested to explain why some earthquakes are preceded by foreshocks and others are not, is fault zone heterogeneity. DAS and SCHOLZ (1981)

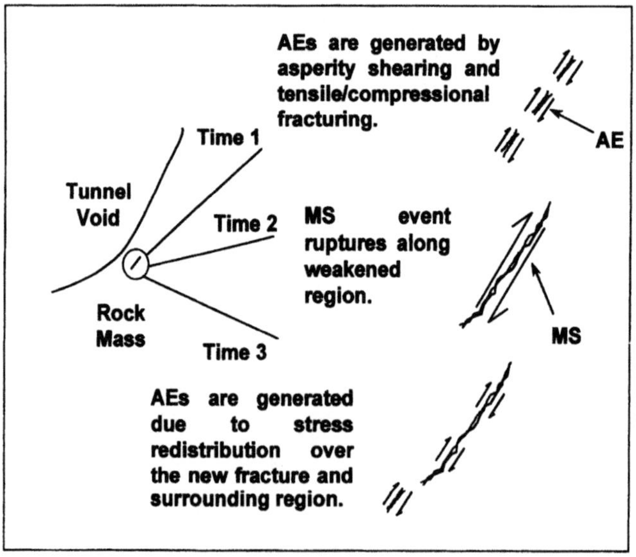

Figure 11
A schematic interpretation of the AE/MS relationship (PETTITT, 1998).

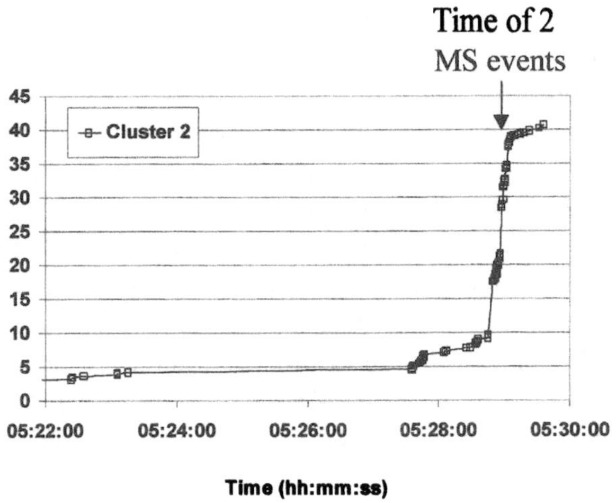

Figure 12
The cumulative seismic moment of AE events in Cluster 2 in an 8-minute interval surrounding the time of the 2 MS events. AE events before and after the MS events are considered foreshocks and aftershocks, respectively.

note that if a fault zone is homogeneous in strength and stress, there are no foreshocks.

It is interesting to examine the size of the AE cluster regions compared to the source parameters of the MS events (Table 4). The results are opposite to what might be intuitively thought. Cluster 3, with the smallest diameter of 15 cm, has an associated MS event with the largest magnitude and stress release, whereas Cluster 2, with the largest AE cluster diameter of 50 cm, has 2 MS events with the lowest magnitudes and stress releases. An explanation for this may be found by looking at the processes that control these seismic events using the "induced" versus "triggered" notation of McGarr and Simpson (1997). McGarr and Simpson define "induced" as seismicity resulting from a substantial change in stress from its ambient state, and "triggered" when the rock mass is sufficiently close to failure that a small change in stress or pore pressure stimulates seismicity. Temporal MS behaviour has been recorded around existing tunnels at the URL for many years and its cause is not fully understood. Developments in the dynamic modelling

Table 4

Source parameters of the MS events in the 3 clusters

MS event	Apparent stress release (Pa)	Moment magnitude M_w
1	5.18e+3	−3.4
2a	6.74e+2	−3.9
2b	2.12e+3	−3.7
3	9.88e+4	−2.1

package PFC (POTYONDY and CUNDALL, 1998) have modelled the temporal activity at the URL using a stress corrosion function applied to the mineral grain bond. Time-dependent processes have been noted by others including KRANZ and ESTEY (1996) who recorded seismic activity in a hard rock mine for more than 400 days after mining activity had ceased. The frequency and upper limit of source parameters of MS events are known to decrease over the months following a new excavation at the URL (COLLINS and YOUNG, 2000). The MS events in Clusters 1 and 2 occur approximately four months after the room 425 tunnel excavation and have relatively small magnitudes compared to events recorded directly following the excavation. In contrast, Cluster 3 occurs within a few hours of an excavation stage of the clay key. The associated MS event is of a similar magnitude to that of MS events recorded directly following excavations at the URL. Due to the predicted large stress change from the key excavation, it is suggested that Cluster 3 is induced, as opposed to Clusters 1 and 2 which are considered triggered. Interestingly, this trend is in contradiction with microseismic studies from a number of mines, where triggered events (such as rockbursts) occurring on pre-existing fault planes exhibit larger magnitudes than induced ones.

MCGARR (1994) performed a comparison of laboratory, mining-induced, and crustal earthquakes ($-4 < M_w < +8.7$) and found that in most cases events had values of stress release between 0 and 10 MPa, indicating that major crustal earthquakes appear to be the same "phenomena" as small induced tremors. The MS events are within this range, however the AE events are of smaller magnitude than this and have not currently been analyzed for stress release. DODGE *et al.* (1996) used foreshock sequences to analyse the evolution of six California earthquakes. Due to the similarity of the foreshock locations and focal mechanisms, very accurate relative locations were determined using a cross-correlation method. It was found that the foreshock sequences did not all occur on a single fault plane, but seemed to concentrate near structural discontinuities in the fault, and the foreshocks may themselves be a product of an aseismic process. It was also found that in some cases the foreshocks acted to de-stress the main-shock region. In our study it is not possible to say whether the AE cluster events occur on a single failure plane, though further investigations will be made using a relative location method in an attempt to produce a 3-D topography of the macrofracture, as well as determining and modelling stress release values. Further analysis of the MS mechanisms using a time-dependent moment tensor method (e.g., BAKER and YOUNG, 1997) would also aid the interpretation.

Without hindsight it is impossible to distinguish a foreshock from any other seismic event, thus making prediction very difficult. Only when a "large" magnitude event (main shock) occurs can earlier events be identified as foreshocks and subsequent events as aftershocks. This study however suggests that prediction may be possible in some cases if spatial regions can be identified quickly where small magnitude events begin to cluster. Also, aftershocks generally occur in regions that are stressed up by a main shock (HARRIS, 1998), and this study has shown spatial movement of the small magnitude AE cluster in the slip direction of the large magnitude MS event, thus suggesting it may be possible to predict the region of aftershocks from the MS event mechanism. The problem in many earthquake areas may be that seismic systems are not sensitive enough to foreshocks, i.e., a region appears to be relatively aseismic before a mainshock. It is hoped that further studies at this small magnitude high resolution scale can advance the understanding of the dynamics and temporal evolution of larger more damaging earthquakes.

6. Conclusion

This paper has analysed data from a unique experiment in which two different frequency seismicity systems (low frequency MS system and higher frequency AE system) monitor over a common rock volume, thus failure is observed over two distinct but interacting scales. Single or double MS events are found to be spatially associated with clusters of between 19–86 AE events. The analysed AE events occur up to 6 hours before the associated MS event and up to 1 hour after, and are termed foreshocks and aftershocks, respectively. Each AE cluster is elongated in shape with the long axis between 15 and 50 cm. Temporal movement of the AE foreshocks and aftershocks is used to infer the MS slip direction which in all cases is in this long axis direction. In two of the three AE clusters considerable foreshocks are recorded and an acceleration in the AE frequency and cumulative magnitude occurs up to the time of the MS event. With the other AE cluster very few foreshocks are recorded possibly indicating a more homogeneous failure plane. In all cases a significant number of aftershocks are recorded.

The time-independent moment tensors of the AE and MS events were determined with all three of the clusters having MS events showing dominantly deviatoric mechanisms. On the $T - k$ source type plot, two of the MS events locate relatively near to the +CLVD source while the other MS event is considered to be a dominantly DC source. The slip vector for the DC event agrees with the temporal movement of the associated AE event cluster locations. The majority of the AE events have deviatoric mechanisms and are interpreted to occur on cracks sub-parallel to the tunnel perimeter. Additionally each cluster had at least one AE event with significant positive volumetric components which is interpreted as tensile opening, and four AE events were recorded with significant crack closure mechanisms possibly relating to the crushing of asperities on the fracture. Remarkable similarities in the AE mechanics are observed in specific regions of the clusters, and it is hoped that future relative relocation of the AE events will provide a more accurate description of the asperities and grain boundary cracks along the macrofailure plane.

Acknowledgements

This research was made possible by funding from the European Union under the INCO-Copernicus programme. The authors would also like to thank members of the URL for their continued assistance with the MS and AE monitoring, backup, and maintenance, as well as frequent communication during this study. Additional thanks to members of the Engineering Seismology Group for software and hardware support and Calum Baker and Jim Hazzard for discussions regarding the manuscript. We are grateful to members of the Applied Seismology Laboratory at Keele University (now at Liverpool University), including Steve Falls and Malcolm Wright for their participation in the development of the borehole sondes.

References

Aki, K., and Richards, P. G., *Quantitative Seismology: Theory and Methods* (W. H. Freeman and Co., New York 1980).

Baker, C., and Young, R. P. (1997), *Evidence for Extensile Crack Initiation in Point Source Time-dependent Moment Tensor Solutions*, Bull. Seismol. Soc. Am. *87*, 1442–1453.

Brehm, D. J., and Braile, L. W. (1998), *Intermediate-term Earthquake Prediction Using Precursory Events in the New Madrid Seismic Zone*, Bull. Seismol. Soc. Am. *88*, 564–580.

Collins, D. S., and Young, R. P. (2000), *Lithological Controls on Seismicity in Granitic Rocks*, Bull. Seismol. Soc. Am. *90*(3), 709–723.

Dahm, T., Manthei, G., and Eisenblatter, J. (1998), *Relative Moment Tensors of Thermally Induced Microcracks in Salt Rock*, Tectonophysics *289*, 61–74.

Das, S., and Scholz, C. H. (1981), *Theory of Time-dependent Rupture in the Earth*, J. Geophys. Res. *86*, 6039–6051.

De Natale, G., and Zollo, A., *Earthquake focal mechanisms from inversion of P and S wave motions.* In *Digital Seismology and Fine Modelling of the Lithosphere* (eds. Cassinis, R., Nolet, G., and Panza, G. F.) (Plenum Press, New York and London 1989) pp. 399–419.

Dodge, D. A., Beroza, G. C., and Ellsworth, W. L. (1996), *Detailed Observations of California Foreshock Sequences: Implications for the Earthquake Initiation Process*, J. Geophys. Res. *101*, 22,371–22,392.

Dziewonski, A. M., Ekström, G., and Nettles, M., *Harvard Centroid-Moment tensor solutions 1976–1996: Significance of the non-double couple component.* In *Rockbursts and Seismicity in Mines* (eds. Gibowicz and Lasocki) (Balkema, Rotterdam 1997) pp. 3–16.

Dzik, E. J. (1997), *The Potential for Crack Formation Around the URL's Tunnel Sealing Experiment Test Tunnel*, Internal Report to Atomic Energy of Canada Limited, AECL-11582.

ESG, (2000), *HYPERION Hardware and Software User's Guide Version 6.0*, Engineering Seismology Group Inc., Kingston, Canada.

Feignier, B., and Young, R. P. (1992), *Moment Tensor Inversion of Induced Microseismic Events: Evidence of Non-Shear Failures in the $-4 < M < -2$ Moment Magnitude Range*, Geophys. Res. Lett. *19*, 1503–1506.

Feustel, A. J. (1995), *Seismic Attenuation in Underground Mines: Measurement Techniques and Applications to Site Characterization*, Ph.D. Thesis, Queen's University, Kingston, Canada.

Gephart, J. W., and Forsyth, D. W. (1984), *An Improved Method for Determining the Regional Stress Tensor Using Earthquake Focal Mechanism Data: Application to the San Fernando Earthquake Sequence*, J. Geophys. Res. *89*(B11), 9305–9320.

Harris, R. A. (1998), *Stress Triggers, Stress Shadows, and Implications for Seismic Hazard*, J. Geophys, Res. *103*, 24,347–24,358.

HUDSON, J. A., PEARCE, R. G., and ROGERS, R. M. (1989), *Source Type Plot for Inversion of the Moment Tensor*, J. Geophys. Res. *94*, 765–774.

JOST, M. L., and HERRMANN, R. B. (1989), *A Student's Guide to and Review of Moment Tensors*, Seis. Res. Lett. *60*, 37–57.

KRANZ, R. L., and ESTEY, L. H., *Listening to a mine relax for over a year at 10 to 1000 meter scale*. In *Proc 2nd North American Rock Mechanics Symposium* (Balkema 1996) pp. 491–498.

MARTIN, C. D. (1993), *The Strength of Massive Lac du Bonnet Granite Around Underground Openings*, Ph.D. Thesis, University of Manitoba, Winnipeg, Canada.

MARTINO, J. B., and CHANDLER, N. A., *Rock mechanics and the construction of the tunnel sealing experiment*. In *Rock Mechanics for Industry* (eds. Amadei, B., Kranz, R. L., Scott, G. A., and Smeallie, P. H.) (Balkema 1999) pp. 885–892.

MCGARR, A., and SIMPSON, D., *A broad look at induced and triggered seismicity*. In *Rockbursts and Seismicity in Mines* (ed. Gibowicz, and Lasocki) (Balkema 1997) pp. 385–396.

MCGARR, A. (1994), *Some Comparisons Between Mining-induced and Laboratory Earthquakes*, Pure appl. geophys. *142*, 467–489.

MCGARR, A. (1993), *Moment Tensors of 10 Witwatersrand Mine Tremors*, Pure appl. geophys. *139*, 781–800.

MCKENZIE, D. P. (1969), *The Relation Between Fault Plane Solutions for Earthquakes and the Directions of the Principal Stresses*, Bull. Seismol. Soc. Am. *59*(2), 591–601.

OHTSU, M. (1991), *Simplified Moment Tensor Analysis and Unified Decomposition of Acoustic Emission Source: Application to In Situ Hydrofracturing Test*, J. Geophys. Res. *96*, 6211–6221.

O'CONNELL, D. R. H., and JOHNSON, L. R. (1988), *Second-order Moment Tensors of Microearthquakes at the Geysers Geothermal Field, California*, Bull. Seismol. Soc. Am. *78*, 1674–1692.

PAC (1997), *Physical Acoustics Corporation Acoustic Emission Sensors Product Bulletin*.

PETTITT, W. S. (1998), *Acoustic Emission Source Studies of Microcracking in Rock*, Ph.D. Thesis, Keele University, Keele, UK.

POTYONDY, D. O., and CUNDALL, P. A., *Modelling notch-formation mechanisms in the URL mine-by test tunnel using bonded assemblies of circular particles*. In *Proc. 3rd North American Rock Mechanics Symposium* (Balkema 1998).

SPOTTISWOODE, S. M., *Source mechanism of mine tremors at Blyvooruitzicht Gold Mine*. In *Rockbursts and Seismicity in Mines* (eds. Gay, N. C., and Wainwright, E. H.) (Johannesburg 1984), pp. 29–37.

STUMP, B. W., and JOHNSON, L. R. (1977), *The Determination of Source Properties by the Linear Inversion of Seismograms*, Bull. Seismol. Soc. Am. *67*, 1489–1502.

URBANCIC, T. I., TRIFU, C.-I., MERCER, R. A., FEUSTEL, A. J., and ALEXANDER, J. A. G. (1996), *Automatic Time-domain Calculation of Source Parameters for the Analysis of Induced Seismicity*, Bull. Seismol. Soc. Am. *86*(5), 1627–1633.

VARNES, D. J. (1989), *Predicting Earthquakes by Analyzing Acceleration Precursory Seismic Activity*, Pure appl. geophys. *130*, 661–686.

YOUNG, R. P., and COLLINS, D. S., *Monitoring an experimental tunnel seal in granite using acoustic emission and ultrasonic velocity*. In *Rock Mechanics for Industry* (eds. Amadei, B., Kranz, R. L., Scott, G. A., and Smeallie, P. H.) (Balkema 1999) pp. 869–876.

(Received March 22, 2000, revised/accepted January 12, 2001)

Pure appl. geophys. 159 (2002) 221–245
0033–4553/02/030221–25 $ 1.50 + 0.20/0

Pure and Applied Geophysics

Simulation of Unstable Fault Slip in Granite Using a Bonded-particle Model

James F. Hazzard,[1] David S. Collins,[2] William S. Pettitt,[2] and R. Paul Young[1]

Abstract—A bonded-particle model is used to simulate shear-type microseismic events induced by tunnel excavation in granite. The model represents a volume of granite by an assembly of 50,000 individual particles bonded together at points of contact. A plane of weakness is included in the model and this plane is subjected to increasing shear load while the normal load across the plane is held constant. As shear stress in the model increases, bonds begin to break and small acoustic emissions (AE) result. After enough bonds have broken, macro-slip occurs across the large portions of the fault in an unstable manner. Since the model is run dynamically, seismic source information can be calculated for the simulated AE and macro-slip events. This information is compared with actual results obtained from seismic monitoring around an underground excavation. Although the modelled events exhibit larger magnitudes than the actual recorded events, there are many similarities between the model and the actual results, namely the presence of foreshocks before the macro-slip events and the patterns of energy release during loading. In particular, the model provides the ability to examine the complexity of the slip events in detail.

Key words: Numerical modelling, PFC, source complexity, acoustic emissions, microcrack, dynamic triggering.

Introduction

Unstable stick-slip motion commonly observed in the frictional sliding of rock was first proposed as a mechanism for earthquakes by Brace and Byerlee (1966). Since then, many experiments have been conducted to map out the conditions that cause unstable sliding and therefore earthquakes. It has been found that the factors controlling the stability of slip are quite complex and involve a variety of possible micromechanisms such as interlocking, riding up and shearing of asperities (see Scholz, 1990).

Recently the increase in computer power has enabled the creation of full micromechanical/discontinuum numerical models to examine the mechanics involved

[1] Department of Earth Sciences, University of Liverpool, 4 Brownlow St., Liverpool, L69 7GP, U.K. E-mail: hazzard@liv.ac.uk
[2] Applied Seismology Consultants, 10 Belmont, Shrewsbury, Shropshire, SY1 1TE, U.K. E-mail: dave@seismology.org

with earthquakes and faulting. ANTONELLINI and POLLARD (1995) use the distinct element method to examine the formation of deformation bands in sandstone. In their work, the sand grains are represented by cylindrical elements that move independently of one another and they show that shear localisation occurs along discrete faults when external stresses are applied. MORGAN and BOETTCHER (1999) conduct similar experiments and examine the effects of grain size distribution and interparticle friction on the mechanics of localisation. The particle-based lattice solid model is used by MORA and PLACE (1998) to simulate earthquakes with gouge and to endeavor to explain the heat flow paradox, and distinct element models are used by SCOTT (1996) to explore the same issues. Finally CUNDALL (1999) uses a bonded particle model to explore the behaviour of rough joints in shear. All of these studies have contributed greatly to understanding the mechanisms of shear slip, however none have considered the dynamic effects, i.e., waves emitted when unstable slip occurs and how this released energy could be used to calculate seismic source information, and how these waves might affect the model behaviour.

The work presented here uses discontinuum models to simulate a microseismic event recorded near an underground excavation to provide insight into the complexities of unstable shear slip, and endeavor to explain the observed seismic results. A bonded particle model is used (ITASCA CONSULTING GROUP, 1999) so that cracking and faulting can occur spontaneously throughout the model. Individual cracks can form between model elements and these cracks are then able to coalesce into larger fractures or faults. In addition, the model is run dynamically so that each crack releases seismic energy and actual source magnitudes and mechanisms can be obtained.

Seismicity on Different Scales

Relationships between earthquakes of different sizes and magnitudes have been intensely studied with the intention of formulating scaling relationships (see SCHOLZ, 1990, pp. 180–189 for a summary). Due to the large range in earthquake magnitudes (> 10 orders of magnitude) it is difficult to compare different sized earthquakes because of the need for different measurement techniques and instrumentation. One avenue that shows considerable promise is the investigation of mining-scale microseismic events. Examining these small earthquakes is useful because of the well-constrained environment, the ability to directly observe the earthquake sources (e.g., slabbing around a tunnel) and the possibility of direct comparison with large well-controlled laboratory tests (e.g., MCGARR, 1994). An experiment in which microseismic events are monitored on two different scales is described in COLLINS *et al.* (this issue – hereafter referred to as COLLINS *et al.*, 2002) and that experiment is the basis for the numerical modelling presented in this paper (see below for a summary).

In the microseismic monitoring experiment (COLLINS *et al.*, 2002), many tiny foreshocks and aftershocks were recorded prior to, and following the microseismic events. It is proposed that by studying the locations and mechanics of the foreshocks, a better understanding of stress evolution leading up to the main shock could be obtained, and this may contribute to the long-term goal of large-scale earthquake prediction. By examining the locations and mechanisms of all of the recorded events, it was postulated that the foreshocks represent crack opening or the shearing and breaking of asperities along the eventual main-shock fracture plane, and that the aftershocks are small slip episodes that occur as the fracture restabilises. By reproducing these experiments with numerical models, it is hoped that further insight can be gained into the mechanics of unstable shear faulting, and that the models can be partially validated by comparison with the actual information obtained from the underground experiment.

Acoustic Emissions and Microseismic Events at the URL

Two scales of seismic monitoring are presently being performed simultaneously around a tunnel excavation in the Tunnel Sealing eXperiment (TSX) at Atomic Energy of Canada Ltd.'s Underground Research Laboratory (URL), Canada. Small scale, high frequency AEs (frequency range 30–300 kHz) are being recorded in a $10 \times 10 \times 10$ m volume using a 16-channel ultrasonic array (COLLINS and YOUNG, 1998) Three-dimensional source locations and full moment tensor analyses have been performed for many of these events to observe the AE source mechanisms (PETTITT, 1998).

The TSX tunnel is also one of a network of tunnels at the URL that is being monitored by a lower frequency (0.1–10 kHz) microseismic (MS) array covering a $100 \times 100 \times 100$ m volume (COLLINS and YOUNG, 2000). This combination of monitoring the same rock mass using both microseismic and ultrasonic techniques affords the unique ability to investigate fracturing over different magnitude scales. The combined AE and MS systems record seismicity with moment magnitudes between -7 and -2, with only a small magnitude gap where the systems have non-overlapping monitoring frequencies. Of particular interest here is the observation by COLLINS and YOUNG (1998) that AEs are locating in very tight spatial and temporal clusters around the TSX tunnel consisting of often ten or more events. Each AE cluster can be associated in both time and space with one or more microseismic hypocentres. Detailed analyses of these AE/MS clusters are described in COLLINS *et al.* (2002).

One cluster of seismicity recorded during the TSX experiment was chosen for numerical simulation. The cluster is made up of 86 AEs spanning a volume of approximately $50 \times 25 \times 25$ cm close to the floor of the TSX tunnel (Cluster 2 in COLLINS *et al.*, 2002). Associated with the AEs are two larger microseismic (MS) events that occurred within the time and space of the AE cluster (see Fig. 1). The MS mechanisms suggest that these events represent shear slip on a failure plane oriented parallel to the tunnel perimeter.

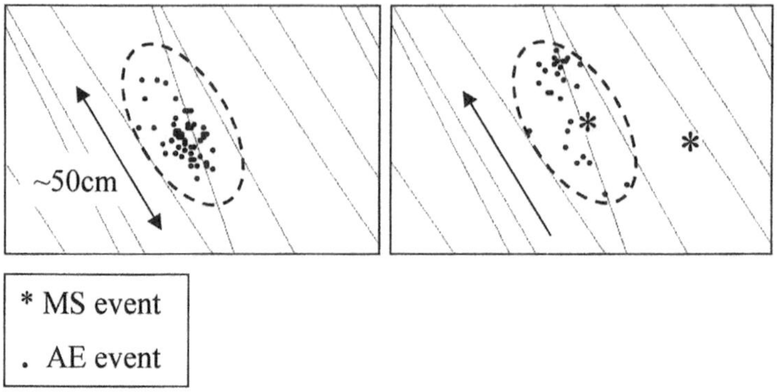

Figure 1
Plan views showing (left) the AE events which occur before the time of the MS events, and (right) the AE events which occur during and after the MS events. The arrow in the right plot denotes the direction of temporal AE migration. The thin grey lines have no physical meaning – they represent the construction lines of the 3-D CAD drawing of the tunnel on which the events are superimposed. (from COLLINS *et al.*, 2002).

Bonded Particle Models

Itasca Consulting Group's Particle Flow Code in Two Dimensions (PFC^{2D}) is used to numerically simulate a small volume of rock near the TSX tunnel. This code represents rock as a dense packing of circular particles (disks) with some random size distribution. The particles are assumed to be rigid (non-deformable), however overlap can occur at particle contacts (see Fig. 2). Contacts are assumed to exist only at a point and not over some finite surface area as would be the case with fully deformable particles. Particles are bonded together at points of contact to simulate a competent rock (rather than an assembly of grains).

Each contact can be envisaged as a pair of elastic springs with constant normal and shear stiffness. Each 'spring' also has a normal and shear strength that can be exceeded, causing the connecting bond to break. The normal force at a contact is given by:

$$\begin{aligned} &\text{Bond intact:} && F_i^n = K^n U^n n_i && \\ &\text{Bond broken:} && F_i^n = K^n U^n n_i && U^n > 0 \\ & && F_i^n = 0 && U^n < 0 \end{aligned} \tag{1}$$

where K^n is the normal stiffness of the contact, U^n is the amount of overlap (positive is compression and negative is tension), and n_i is the unit normal vector to the contact (see Fig. 2). The contact bond has infinite compressional strength although the tensile strength is finite. If the tensile force at the contact exceeds the tensile strength, then the bond breaks and the contact normal force becomes zero.

The shear force can be calculated in a similar way however when the shear strength of the bond is exceeded and the bond breaks, the shear force does not drop

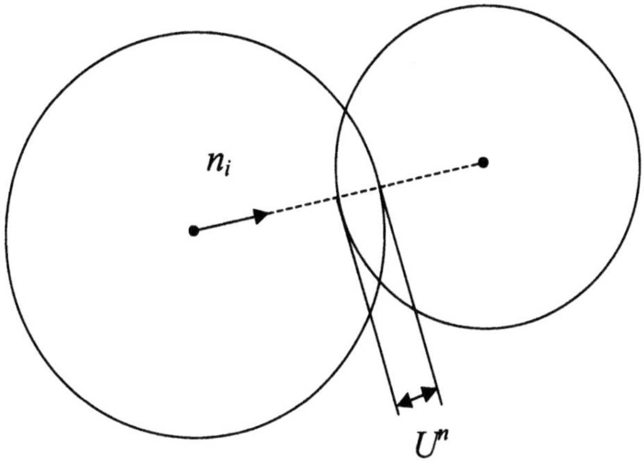

Figure 2
Ball–ball contact in PFC^{2D}.

to zero but instead frictional sliding takes over. The maximum shear force that can be supported at a contact is therefore given by:

$$\begin{aligned} &\text{Bond intact:} \quad F^s_{\max} = \text{shear bond strength} \\ &\text{Bond broken:} \quad F^s_{\max} = \mu |F^n_i| \end{aligned} \tag{2}$$

where μ is the coefficient of friction between the two particles. The shear contact behaviour can be thought of as possessing cohesion while the bond is intact and then becomes frictional after the bond breaks. The values of contact normal and shear stiffness and bond normal and shear strength are specified by the user and these values influence the overall stiffness (modulus) and strength of the modelled rock sample, respectively.

From the contact forces, particle accelerations can be calculated according to Newton's law of motion and from the accelerations, new particle positions can be obtained after each time step. The distinct element method (CUNDALL and STRACK, 1979) is employed to model the forces and motions of the particles within the assembly. This technique assumes that information cannot propagate further than a nearest neighbour in one calculation step. Therefore time steps are very small and many of them are required to reach equilibrium. This method is identical to that used in explicit finite-difference analyses and allows information to propagate dynamically through the system. For this reason, PFC^{2D} is a logical choice for modelling acoustic emissions and the resulting dynamic output.

In PFC, when the tensile or shear bond strength at a contact is exceeded and the bond breaks, contact forces drop instantaneously. A more realistic contact behaviour could be incorporated in which failure occurs over finite time, however this was not done for the models described here. Similarly, stress-corrosion or subcritical crack growth could be accounted for by making the strength of the bonds time and stress-

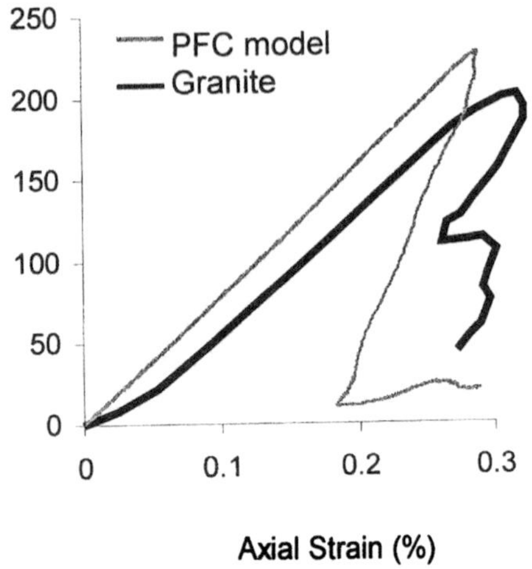

Figure 3
Comparison between the mechanical behaviour of a PFC model and an actual sample of Lac du Bonnet granite subjected to a uniaxial compression test.

dependent (see POTYONDY and CUNDALL, 1998), however for simplicity this was also not considered here.

Since particle micro-parameters cannot be determined directly, estimated values for the particle stiffnesses and bond strengths were assigned and uniaxial compression tests were simulated to determine the macro-strength and stiffness of the material. The micro-properties were then adjusted until the model reproduced the approximate stiffness and strength of Lac du Bonnet granite (see HAZZARD *et al.*, 2000). Note that the particle micro-stiffnesses are constant throughout the model but the bond strengths were assigned a Poisson's distribution about a mean value with a standard deviation of 25%. This ensured a more gradual sample failure with cracking occurring over a longer period of time than would be the case if all bonds were assigned the same strength. Figure 3 shows the mechanical behaviour of PFC model of a core sample (~8000 particles) during a uniaxial compression test compared with a laboratory test on a sample of Lac du Bonnet granite. The stiffness and strength of the material are well reproduced by the model. Note however that the initial curvature of the stress/strain curve is missing in the model. This is because there are no pre-existing cracks in the model that can close under stress. The effects of the different microparameters and particle packings on the macro-behaviour of PFC models have been thoroughly examined elsewhere (POTYONDY and CUNDALL, 1999) and will not be explored here. Further descriptions of bonded particle models and justification for their use in modelling competent rocks is given in HAZZARD *et al.* (2000).

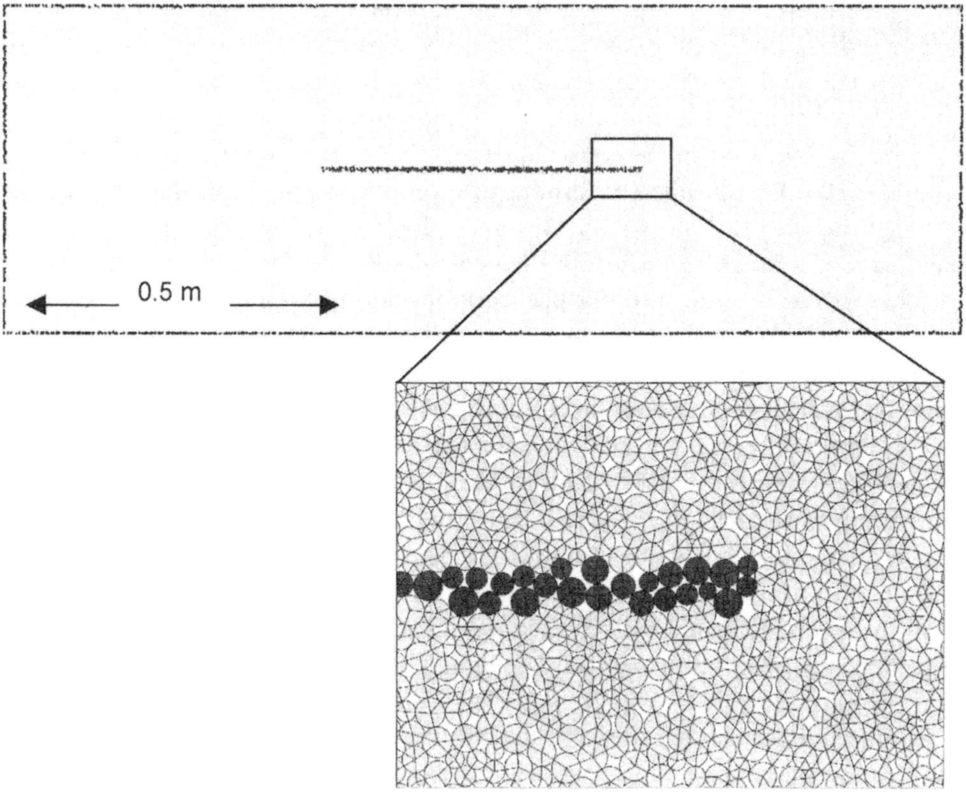

Figure 4
PFC^{2D} model of "shear box". Contact bonds are shown as black lines. Particles with three or more weakened bonds are darkened.

Simulation of Shear Fault

The shear-slip event was numerically simulated by creating a "shear box" model in PFC^{2D}. The box represents a $1.5 \times 0.5 \times 0.25$ m block of Lac du Bonnet granite with a 50 cm long plane of weakness in the centre corresponding to the extent of the recorded seismicity (see Fig. 4). The long axis of the model (and therefore the plane of weakness) is assumed to be parallel to the tunnel perimeter. The model comprises approximately 50,000 particles randomly placed within the rectangle (see Fig. 4). Note that because this is a 2-D simulation, the particles are actually disks or cylinders. The particle diameters range from 3 to 5 mm which is approximately the size of the grains in the granite. The particles (disks) are assigned thicknesses of 0.25 m so that the width of the fault zone is approximately the same as that measured for the actual microseismic event as described above.

The plane of weakness is simulated by weakening all bonds within 1 maximum particle diameter (5 mm) of the centre line. The strengths of these bonds were set to 10% of the bond strengths in the surrounding rock. The assumption of a weak fault is thought to be valid since many studies on tectonic-scale faults have shown faults to be significantly weaker than the surrounding rock. The specific value of 10% chosen for this modelled fault is somewhat arbitrary, however it can be shown that altering the strength between 10% and 50% of the surrounding rock strength has no major impact on the results. If the fault strength is set considerably weaker than 10% then stable sliding results. It is assumed that the weak bonds connecting particles on opposite sides of the fault represent interlocking asperities or weakly bonded fault segments. The bonds within 0.5 m of the left and right edges were not weakened to the end that the model simulates a fault of finite length embedded in a homogeneous rock mass.

Note that if bonds on the fault were assigned zero strength then the fault still exhibits inherent roughness. Shear resistance develops as interlocking particles ride up to move past each other. If zero strength and zero friction are assigned to the fault particles and the fault is extended to the sample edges, then a fault friction angle of 30° results. This suggests that fault roughness can be simulated without bonding (cohesion) on the fault, however when this was done the shear sliding was found to be stable (no stress drop) and therefore of little seismic interest. The minor episodes of energy release that result from particles jumping over one another could be considered tiny acoustic emissions, however this was not considered here.

The normal force across the plane of weakness was assumed to be 5 MPa. This value was determined by observing the minimum stress at the AE locations from a boundary element model of the TSX tunnel. For all of the AE, the minimum stress was less than 5 MPa and oriented approximately perpendicular to the tunnel perimeter. In the PFC model, this normal force is maintained by a servo-mechanism acting on the top layer of particles.

The shear force is increased from zero by applying a constant velocity to the boundary particles in the top half of the model. A velocity of 0.05 m/s was applied. This is about 5 orders of magnitude slower than the *P*-wave velocity of the material being modelled consequently it is assumed that this loading velocity is slow enough to prevent any dynamic effects (i.e., a wave propagating from the moving boundaries). The boundary particles in the lower half of the model are held fixed.

These boundary conditions are probably more representative of a laboratory shear box test than an actual *in situ* simulation. *In situ*, it is likely that the normal force will increase as the fault dilates (low compliance for normal motion) and the shear stress will be approximately constant instead of increasing monotonically. The "shear-box" boundary conditions are used here because slight cracking or slip would occur on the fault if the shear stress was held constant, unless some time-dependent stress-corrosion mechanism was included. As described above, this mechanism has not yet been implemented.

The microparameters (contact stiffness and strength) are chosen, based on previous PFC^{2D} models of Lac du Bonnet granite (POTYONDY *et al.*, 1996). The chosen micro-parameters cause the bonded particle assembly to exhibit approximately the same macro-strength and stiffness as actual Lac du Bonnet granite (see Fig. 3).

Results

Microseismic Events

The average shear stress on the fault is plotted against the average shear displacement across the fault in Figure 5. These values were obtained by taking the average of five stress and strain measurements at evenly spaced positions along the fault. The inset of Figure 5 shows the average shear slip plotted against time. The time has been normalised by dividing by T, the time taken for a compressional wave to propagate horizontally across the entire model. This normalisation technique was proposed by MORA and PLACE (1998) because the rate at which the stress is applied in the model must be unrealistically fast due to the small time-steps required. By

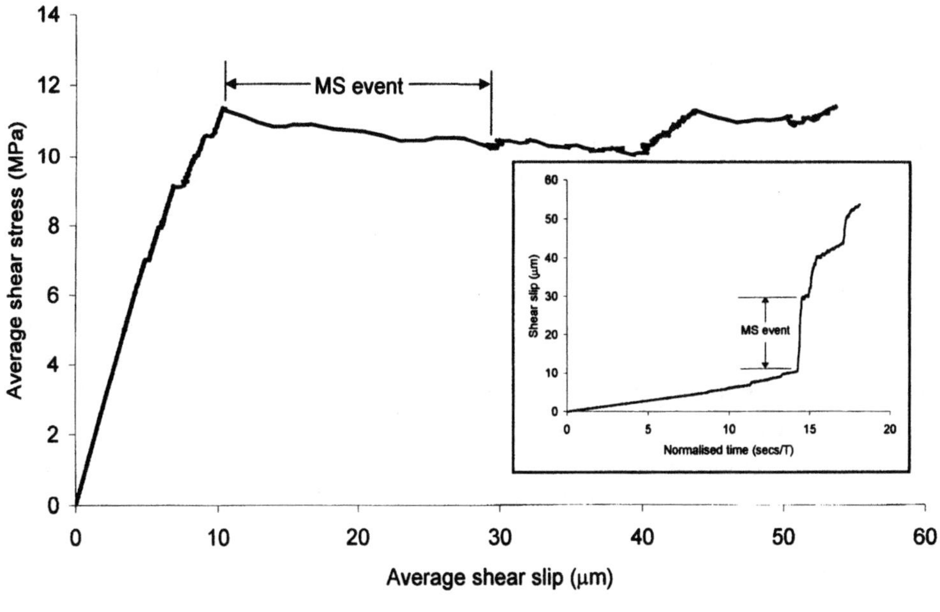

Figure 5
Stress/strain behaviour of the modelled fault. The inset shows the average shear slip plotted against time. The time has been normalised by dividing by T, the time taken for a compressional wave to propagate horizontally across the entire model.

normalising the time a more reasonable time scale results, similar to loading rates *in situ*.

Figure 5 shows the shear stress initially increasing in a linear fashion with only a slight increase in shear displacement. Since the fault is assumed to be "locked" initially, this displacement is attributed to elastic deformation and a few small cracks that occur before the major slip event. When the stress reaches about 11 MPa, rapid slip then occurs, accompanied by a drop in stress. This region represents unstable sliding and is accompanied by a jump in kinetic energy in the model – just as for an actual earthquake. This earthquake is assumed to correspond to the first MS event recorded at the actual TSX tunnel as described above.

The seismic moment of the modelled MS event can be calculated by:

$$M_0 = \mu DA \ , \tag{3}$$

where μ is the shear modulus of the rock, D is the shear slip and A is the fault area. The shear modulus was determined experimentally to be 25.5 GPa (near the actual value for Lac du Bonnet granite of 27.4 GPa), the fault slip was determined from Figure 5 to be 19.2 μm and the fault area is 0.125 m^2 (i.e., 0.5 m long $\times$ 0.25 m thick). This yields a value of $M_0 = 6.12 \times 10^4$ Nm. The moment magnitude can then be calculated by:

$$M_w = \frac{2}{3}\log M_0 - 6.0 \ . \tag{4}$$

This yields a magnitude of $M_w = -2.8$ for this event. This value is about an order of magnitude greater than that observed for the first MS recorded in the actual TSX experiment ($M_w = -3.9$). There are several possible reasons for this and these will be considered in the discussion section.

Figure 5 delineates that the first episode of slip is rapidly followed by another slightly smaller slip event. The magnitude of this event can be calculated as above, yielding a moment magnitude for this second event of $M_w = -3.0$. Comparing this to the magnitude of the second MS event recorded for this cluster at the TSX ($M_w = -3.7$), it can be seen again that the model produces a magnitude almost an order of magnitude larger than the actual event. A comparison of the seismicity obtained from the model and the seismicity recorded at the TSX is presented in Table 1.

If the modelled particle assembly continues to be loaded in shear, more slip events occur with decreasing magnitudes as the fault plane weakens (through more bond breakages) and approaches stable sliding conditions (see Fig. 5). For the events recorded at the TSX it is unlikely that a continual application of shear stress was present, therefore the model was stopped at some arbitrary time after the occurrence of a few events. Since only 2 MS events were recorded at the TSX, only the first two slip events produced by the model will be considered here.

Table 1

Comparison of modelled and actual seismicity

Event	Model magnitude	Actual magnitude
MS 1	−2.8	−3.9
MS 2	−3.0	−3.7
AE	−5.3 to −3.1 (78 events)	−6.9 to −5.4 (86 events)

Microcracks

During loading, many bond breakages occur in the PFC model. Each of these bond breakages may be considered an individual microcrack (HAZZARD *et al.*, 2000). The location, mechanism and orientation of the microcracks that form before, during and after the MS event are shown in Figure 6. Crack orientations are normal to the line connecting the centres of the two parent particles. The mode of failure for each

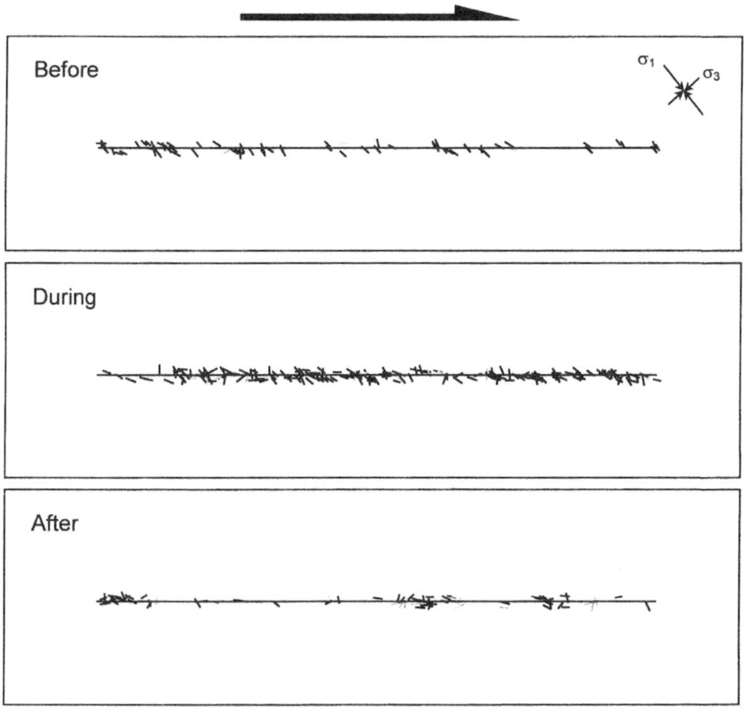

Figure 6
A close-up of the fault zone (the extent of which is indicated by a thin horizontal line) showing microcracks forming before, during and after the microseismic event. Tensile cracks are black and shear cracks are grey. The direction of shear is shown at the top. The principal stresses at the centre of the model just prior to the MS event are shown in the top right corner.

crack corresponds to whether the tensile or shear strength of the contact bond was exceeded.

It can be seen that the majority of cracks in all plots are tensile and are oriented approximately 45° to 50° clockwise from the orientation of the fault. These microcracks therefore correspond to tensile cracks opening in the direction of least principal stress, σ_3 and oriented parallel to the maximum stress, σ_1 (see Fig. 6). The mechanism of cracking agrees with what is generally observed in compression tests on intact crystalline rocks (TAPPONIER and BRACE, 1976). It is of interest that these tensile cracks contribute to unstable shear slip on the fault (the MS event). This will be addressed in the discussion section.

Figure 6 shows that before the MS event the cracking is comparatively sparse and the cracks locate rather evenly along the length of the fault. It is proposed that these cracks represent the foreshocks to the MS event(s) and help weaken parts of the fault to facilitate shear slip. It is also possible that these bond breakages shed stress to intact parts of the fault (concentrating stress on stronger asperities), bringing the system closer to macro-failure conditions.

During the first MS event there is considerably more cracking along the entire fault length. These cracks are obviously associated with the MS event, and it will be shown later that there is a somewhat complicated relationship between the microcracks and the shear slip on the fault. After the first MS event the cracks form in small, distinct clusters indicating where further small slip events are occurring (aftershocks). It is interesting to note that the orientations of the cracks which occur after the MS event are generally more parallel to the fault than the orientations of the foreshock cracks.

Acoustic Emissions

If a PFC model is run dynamically with low numerical damping, each time a bond breaks, the stored strain energy at the contact is released as kinetic energy in the form of seismic waves. In this model the numerical damping was set such that the rock has a seismic quality factor, Q, of approximately 100. Q is a common measure of attenuation or energy loss in real rocks and is defined as 2π times the ratio of stored energy to dissipated energy in one wavelength. FEUSTEL (1995) calculates the quality factor of Lac du Bonnet granite to be approximately 220. However, since the simulation presented here is close to a tunnel where much stress-relief cracking is thought to have occurred, then a lower value of $Q = 100$ is assumed.

Because this is a simulation of a fairly competent rock, and Q is quite high, there is very little loss in wave amplitude across the length of the model due to attenuation (~5%). If geometrical spreading is considered then a wave's amplitude will decrease by about 90% across the model. This presents the problem that waves may be reflected off the boundaries. In this model no attempt is made to simulate absorbing boundaries. A non-absorbing boundary may be realistic for one side of the model that is thought to represent the surface of the tunnel, however, this condition is unrealistic for the rest of the model in which the rock extends indefinitely away from the fault. The effect of these reflecting boundaries is that excessive energy is present in the system and more events with larger magnitudes may be occurring than would be the case for absorbing boundaries. Further study is required to formulate more realistic boundary conditions for dynamic models in PFC.

Previous work has shown that realistic seismic information can only be obtained from the PFC-generated cracks if they are clustered together into larger "events". If each crack is considered individually then all events will exhibit approximately the same magnitude. Therefore it is assumed that cracks occurring very close together in space and time make up one single acoustic emission. The criteria used to cluster the cracks into AEs and the way in which seismic information is obtained from the modelled AEs are described by HAZZARD and YOUNG (2000). This algorithm basically assumes that an AE is a small rupture propagating at 0.5 times the shear-wave velocity of the material. When a microcrack forms, the kinetic energy of all particles within a distance of three-particle diameters from the crack location is monitored. The kinetic energy of the source particles is monitored for the length of time it would take for a shear rupture to propagate from the microcrack location (the centre of the source) to the edge of the source (*time = (3 particle diameters)/(0.5×* shear-wave velocity)). If another crack forms within the source area during this time then it is considered part of the same seismic event. The source area is then expanded to include all particles within three diameters of the new crack. When no more cracks form within the source area, the AE magnitude is calculated from the peak kinetic energy attained by the particles in the source region over the duration of the event. In this way, microcracks occurring close together in space and time are considered part of the same rupturing event and a larger range of AE magnitudes is therefore possible.

When this AE recording technique is applied to the PFC model described here, a total of 78 AE are observed with magnitudes ranging from −5.3 to −3.1. This is roughly the same number of AE recorded in the actual TSX experiment (86), however the number of AE recorded in the PFC model will of course depend on how long the model is run and the clustering criteria used. To test the effect of the clustering criteria, the model was rerun with different clustering parameters. Two additional tests were run in which the initial source radius was set to one and two particle diameters. Figure 7 shows the magnitude distribution for the events when the initial source radius is set to one particle diameter and for the case when the initial source radius is set to three diameters (the case for two diameters is not shown for clarity but it falls in between the two plots shown). It can be seen that for smaller initial source areas less cracks are clustered and therefore there are more small magnitude events. The slopes of the straight part of the curves correspond to the Gutenberg-Richter *b* value, which is a quantitative indication of the magnitude

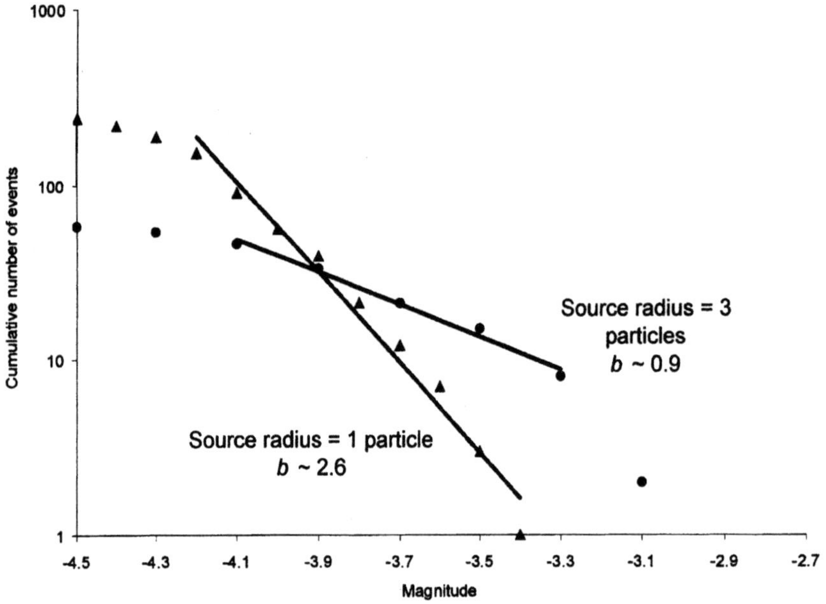

Figure 7
Distribution of event magnitudes for AE recorded in PFC model. Plots are shown for two different cases of crack clustering: 1. Initial source radius equals one particle diameter. 2. The initial source radius is three-particle diameters. The slope of the straight part of the curves is the Gutenberg-Richter b value.

distribution of events. In studies on actual earthquake series the b value is generally close to unity (LOCKNER, 1993). It can be seen from Figure 7 that using an initial source radius of 3 particle diameters provides the most realistic magnitude distribution, therefore only these simulations will be discussed further.

The magnitudes of the AE obtained from PFC are about an order of magnitude larger than those obtained from the actual experiment (see Table 1). One reason for this may be that there is a gap in the range of magnitudes that can be recorded at the TSX (due to the non-overlapping monitoring frequencies of the two systems) therefore events with magnitudes between -5 and -4 are not detected at the TSX. Other possible reasons for the magnitude discrepancy are outlined in the discussion.

The AEs are plotted in Figure 8. In this plot, AEs consisting of more than 2 microcracks are plotted as ellipses and the orientation and size of the ellipses are calculated from the positions of the cracks involved in the AEs. As expected, Figure 8 shows many small AEs occurring prior to the MS and then a few very large AEs occurring during the MS. After the MS the AEs are again quite small. It is interesting that the middle plot of Figure 8 (during the MS) shows that the MS event is actually made up of a few separate AEs instead of just one large event, indicating several separate episodes of cracking and/or slip constitute the actual MS event. This observation is not unrealistic, as it is known that many tectonic-scale earthquakes are

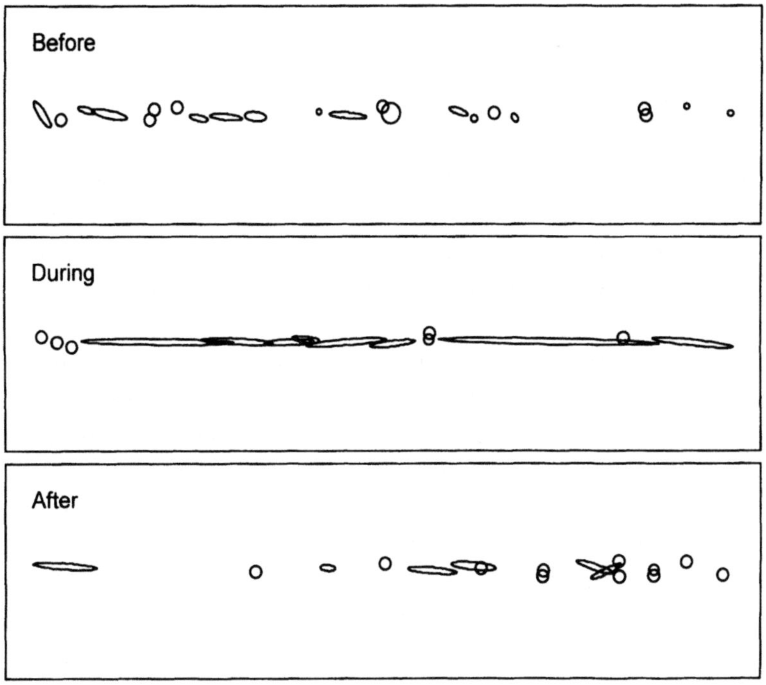

Figure 8
AEs measured in the PFC model before, during and after the first MS event. AEs consisting of more than two cracks are plotted as best-fit ellipses where the ellipse extent and orientation is determined from the locations of the microcracks making up the AE. The extent of the fault is not shown but it extends almost the width of the plots.

often made up of several pulses of slip or rupture (R. Abercrombie, personal communication). It is possible that these separate AEs could be attributed to the incorrect application of the crack clustering algorithm (i.e., if a larger initial source area was used, then more cracks would be clustered and only one large AE would result), however a close examination of the cracking on the fault during the MS reveals several distinct clusters of cracking spaced apart in time and space (see the figures in the discussion section), which indicates the interpretation of several small ruptures is correct.

In all three plots in Figure 8 it appears that the ellipses are generally oriented sub-parallel to the fault zone (and approximately parallel to the direction of maximum shear stress) contrasting with the individual cracks that were oriented approximately 50° to the fault (parallel to σ_1). It appears likely therefore that even though these events are generally composed of several small tensile cracks, the overall mechanism of the events, when motion is averaged over many particles, is probably shear (or some more complex source type). This agrees with the actual TSX results in which the majority of AE showed double-couple mechanisms with the failure plane oriented parallel to the tunnel perimeter.

Triggering

Previous work has shown that when PFC models are run dynamically with low numerical damping, waves emitted from one crack are capable of dynamically triggering more cracks (HAZZARD *et al.*, 2000). This mechanism is observed when a contact bond is stressed such that it is close to its maximum strength. If a transient (dynamic) stress pulse then passes by (caused by a nearby bond breakage) it may be sufficient to cause the contact bond to break. It can be shown that the dynamic stress pulse is often larger and more extensive than the long-term static change in stress caused by a bond breakage.

Figure 9 shows particle velocities and cracking after the start of localisation (after the peak stress in Fig. 5). This figure shows how cracking starts near the right edge of the fault and these cracks cause waves to propagate outwards. Further cracking along the fault then appears to occur as the wavefront passes by, suggesting that the wave may be dynamically triggering cracking on the fault. To investigate this further, the normal forces at some contacts were recorded before and after crack formation.

Figure 10 shows the normal force at a contact where a tensile crack forms. The bond is fixed to prevent breaking so that the total history of normal force at the contact can be observed. It is clear that the passing of a wave would cause the bond to break in this case and that the long-term (static) change in stress would be unlikely to cause bond breakage. Note however that in reality, the strength at a given contact may be frequency dependent. It has been shown that low stresses maintained for a sustained period can cause failure in granite samples and that if stresses are only applied for a short time then substantially higher magnitude stress changes are required to cause failure (see LOCKNER, 1998; SCHMIDTKE and LAJTAI, 1985). No consideration of this effect is made here and it is likely that if the bond strengths were assigned frequency dependence then the dynamic triggering effect would be diminished.

Discussion

It is clear that the shear failure of this simulated fault is very complex. Shear slip is preceded by the formation of many tensile microcracks that weaken the fault such that shearing can occur between unbonded grains. There appear to be three scales of seismicity occurring in this simulation: 1) tiny tensile microcracks oriented parallel to σ_1, 2) small AE oriented sub-parallel to the fault plane with magnitudes ranging from -5.3 to -3.1 and 3) the MS events of magnitudes > -3 representing slip on large portions of the fault. It is proposed that the AEs modelled here correspond to the AEs recorded in the actual experiment. This assumption is made because the AEs recorded in the experiment generally exhibit double-couple source mechanisms with failure planes oriented parallel to the failure plane of the MS event. Few tensile AEs were recorded in the *in situ* experiment, suggesting that the small opening events

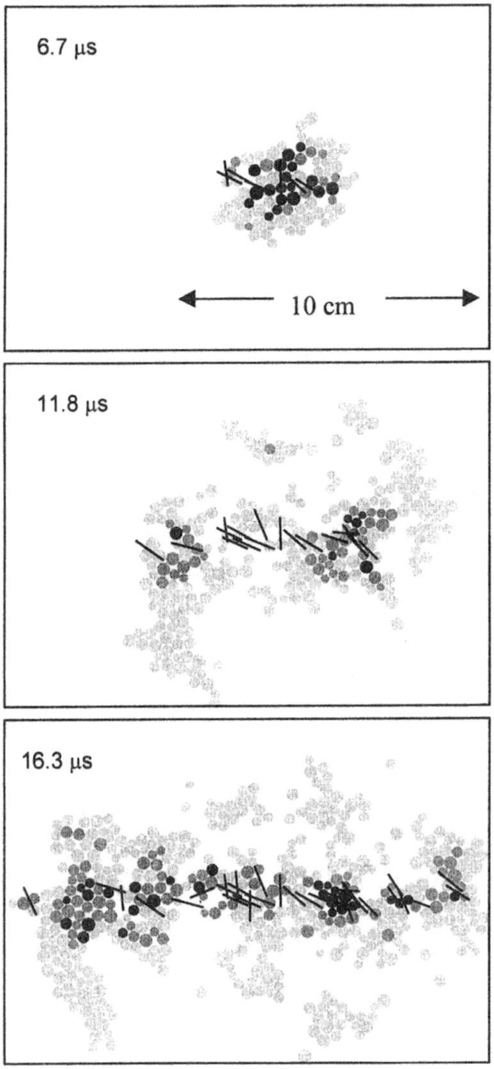

Figure 9
Three frames from a movie showing the velocity of particles and formation of cracks after the peak stress on a small portion of the fault. The shade of the particles indicates the magnitude of the velocity up to a maximum of 4 m/s (black). The time from the peak stress is shown in each frame. The crack sizes are exaggerated 2.5 times.

produced by the PFC model (microcracks) are too small to be picked up by the monitoring equipment. This interpretation is supported by observing that in laboratory AE experiments less than 1% of new microcracks result in recorded AE events (LOCKNER, 1993). Therefore it is possible that the small tensile cracks

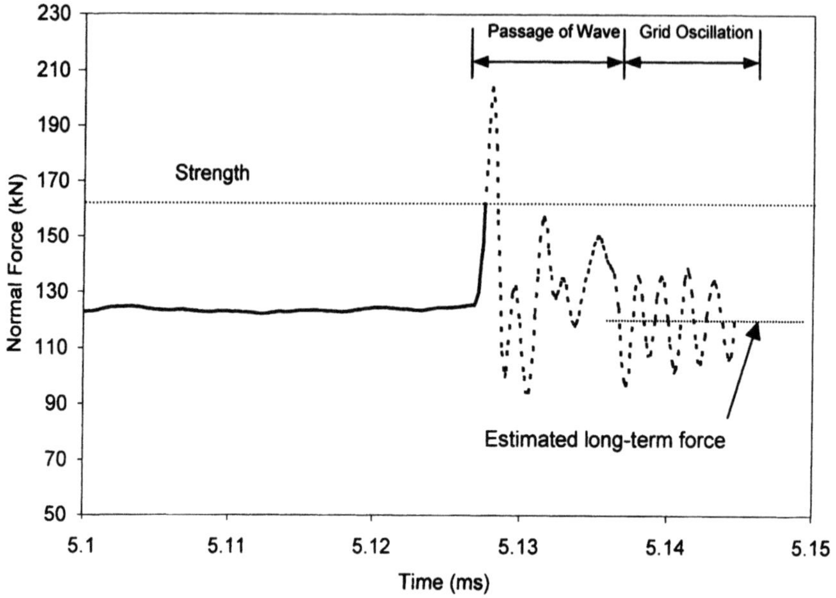

Figure 10
The normal force at a contact close to another bond breakage (tensile is positive). Since the force would normally drop to zero when the bond strength is exceeded, the bond was prevented from breaking so that the contact behaviour could be observed after the strength is exceeded (heavy dashed line).

generated by the PFC model equate to the large number of microcracks that form in rock without emitting detectable acoustic waves.

Another type of 'event' that occurs in the PFC models that has not been discussed is the small bumping and rolling of the round grains required to accommodate shear motion. Since the fault has inherent roughness (friction $\sim 30°$) then energy is required for shear motion to occur. As grains ride over other grains and then drop down again, small amounts of kinetic energy are released. These tiny energy bursts have not been quantified in this study, mostly because they will be much smaller than the events caused by bond breakages, however the observation of these small events would be a very interesting topic for future study.

This simulation shows other similarities with the actual recorded data. AEs are observed in the model before the MS event (foreshocks) and after the MS event (aftershocks) just as in the TSX experiment. The cumulative kinetic energy of the AE events is plotted against normalised time in Figure 11a. The cumulative moments for the actual recorded AEs are shown for comparison (Figure 11b). There are obvious similarities between the two graphs. There are initially a few low energy AEs that occur over a relatively prolonged period prior to the MS events. The number of AEs and the energy released by them then accelerates rapidly just prior to the MS events. The MS in both plots are accompanied by a few high energy (high moment) AEs occurring in rapid succession. The number of AEs and the

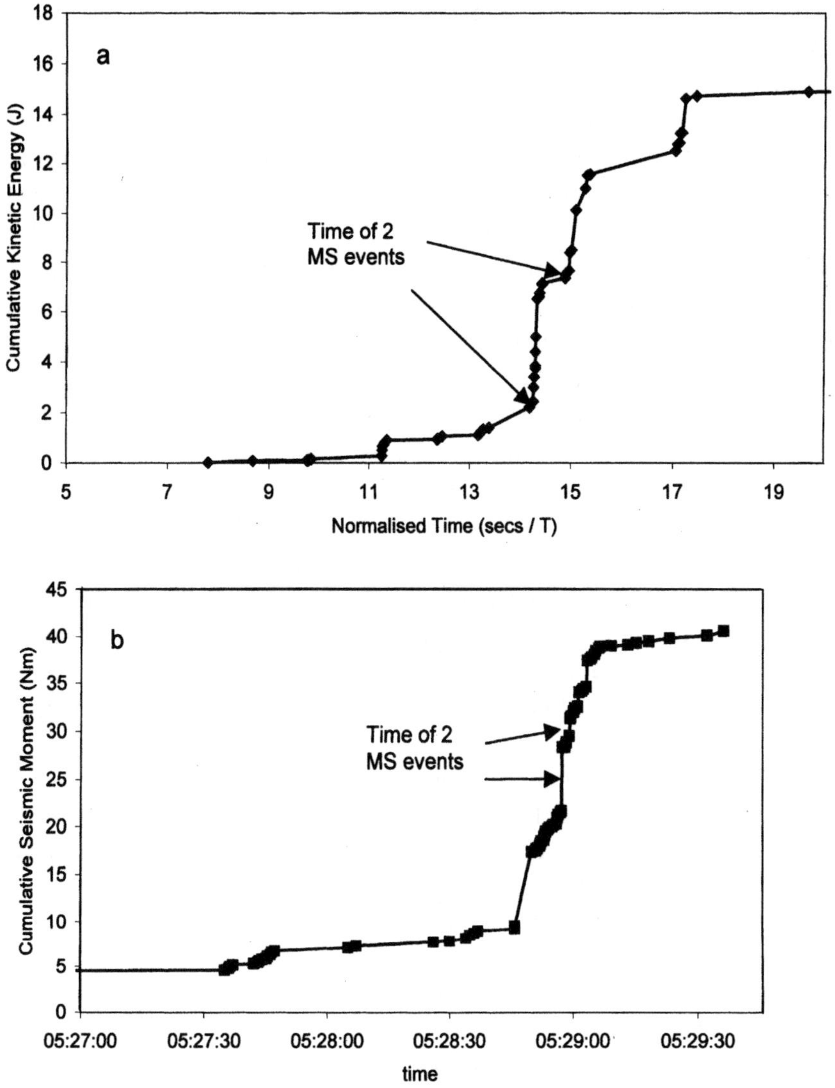

Figure 11
(a) Cumulative energy for all of the AE from the PFC model plotted against normalised time (as in Fig. 5). (b) Cumulative seismic moment for most of the AE recorded for cluster 2 from the TSX experiment (from COLLINS *et al.*, 2002).

energy released by them then decreases again following the two MS events. Note that for the modelled AEs, further AE clusters (associated with episodes of slip – see Fig. 5) continue to occur after the first two MS events because of the continued shear loading of the particle assembly. This continued loading is probably not present in the actual TSX experiment so these subsequent slip events are not observed. Note also that the time scales in Figures 11a and 11b do not correspond directly. This is because the small calculation time steps required by the model preclude the possibility of extending the seismic activity over minutes or hours as is observed in reality.

The relationships between the foreshocks and main shocks observed in Figure 11 agree qualitatively with what is observed in actual earthquake zones (see BREHM and BRAILE, 1998). The presence of the foreshocks before the main shocks and the accelerating nature of energy release may provide a clue in the pursuit of earthquake prediction. It is therefore encouraging that the numerical model presented here is capable of reproducing the foreshock activity observed in reality. DAS and SCHOLZ (1981) suggest that foreshock activity does not occur for homogeneous fault zones, therefore it is only through the use of numerical models with some type of random fault zone heterogeneity that the observed foreshock patterns can be simulated (e.g., random distribution of bond strengths and random particle packings in PFC).

One difference between the modelled seismicity and the actual recorded seismicity is that the modelled events exhibit magnitudes about an order of magnitude larger than the corresponding events recorded at the TSX. There are several possible reasons for this: (1) the standard seismological equations for calculating magnitude assuming a shear source are not applicable in this case where the source is quite complex, (2) slip is not occurring over the entire fault surface, (3) not enough energy is dissipated by the model, (4) the model is not a sufficiently accurate representation of the processes that are actually occurring. These issues will now be addressed in turn.

1 and 2. Reasons 1 and 2 are related because they both refer to the possible complexity of slip motion on the fault. The complexity of the modelled event can be seen in Figure 9 which shows that cracking during the MS event starts near the edge of the fault and propagates along the fault, perhaps with the help of some dynamic triggering mechanism. To examine how this cracking influences the slip history of the fault, four frames of a movie are shown in Figure 12. These frames show how shear slip starts at a single point on the fault and propagates outwards, following in the wake of the tensile cracking. Therefore it appears that cracking must occur before slip can take place between the unbonded grains. This figure also depicts how slip does not occur along the entire fault simultaneously and that the shearing actually propagates along the fault over some finite amount of time. The propagating nature of the modelled event agrees with the observation made in the actual experiment that the AE locations tend to migrate with time along the fault plane (the direction of the arrow in Fig. 1 indicates the temporal movement of the recorded AE). These observations suggest that our use of equations (3) and (4) to calculate the event magnitude may not be totally appropriate since they assume that slip occurs on the entire fault surface simultaneously. A more rigorous time-dependent method may be required to achieve truly accurate results (see BAKER and YOUNG, 1997).

3. Previous work by the authors (unpublished) has shown that seismic events in PFC are generally too efficient. MCGARR (1994) suggests that the efficiency of seismic events should be less than about 5% (where efficiency is the ratio of radiated energy

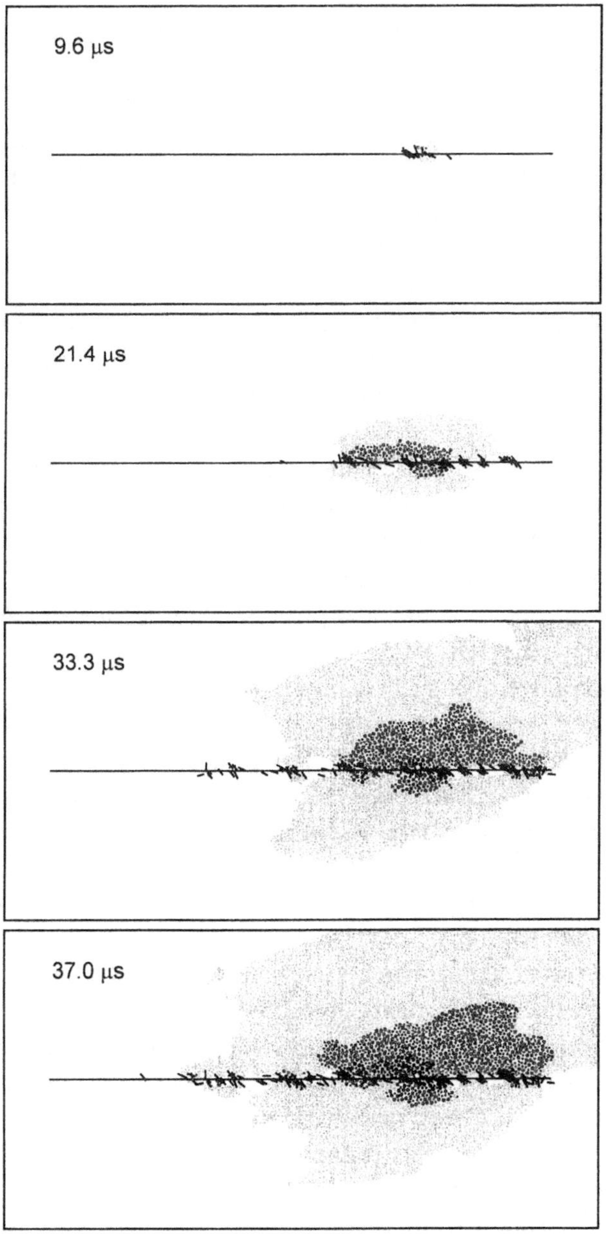

Figure 12
Horizontal displacements of particles from the time of peak stress (absolute value). The shade of the particles indicates the magnitude of displacement up to a maximum of 15 μm (black). Time since the peak stress is shown in each frame.

to total released energy). In the granite model presented here with a realistic amount of numerical damping (to account for wave attenuation) efficiencies approach 50%. This is likely because no energy loss occurs directly due to crack formation (i.e., fracture energy). When a bond breaks, all of the strain energy stored at a contact is converted into kinetic energy or dissipated through interparticle friction and numerical damping. Also, no gouge material exists and therefore insufficient energy dissipates in the breaking of asperities to form gouge and through frictional heating of gouge particles. Another possibility is that some of the kinetic energy used to calculate the AE magnitudes may actually be "incoherent" energy representing jostling and rolling of grains, and that not all of this energy would be radiated in a coherent way such that it could be recorded on a distant receiver (P. Cundall, personal communication). Means to account for these increased efficiencies in PFC are being investigated.

4. COLLINS *et al.* (2002) suggest that the event they recorded may actually be "triggered" instead of induced. This conclusion is drawn by observing that the event occurred approximately four months after the tunnel excavation (the major stress change). This suggests that the event we are considering was caused by a very small change in stress or pore pressure, and processes such as time-dependent creep and stress corrosion may be contributing to the rock failure. In contrast, the event in the PFC model was caused by a direct application of stress. COLLINS *et al.* (2002) show that an "induced" event occurring a few hours after excavation exhibits a considerably larger magnitude (−2.1). The processes causing this event may be closer to the processes simulated with PFC, which would explain why the PFC event magnitude is so large. Future models will endeavor to more accurately simulate the actual conditions by maintaining a constant shear stress across the fault and then subjecting the modelled rock to stress corrosion until failure occurs (see POTYONDY and CUNDALL, 1998).

The model will of course never exactly match the observations made underground owing to the many simplifications and assumptions inherent in the model (including the assumption that a 2-D model can replicate a 3-D process). Also, due to the random nature of the models (and indeed the actual rock), the exact nature of the earthquake dynamics will be different each time a model is run, therefore direct comparisons with certain aspects of the recorded data is not always enlightening. However, it is hoped that the model can highlight some of the possible mechanisms and complexities associated with induced seismicity and unstable shear-slip events. More accurate models could be created in three dimensions such that the entire excavation is considered, however this would involve tremendous computer power and is beyond the scope of the present work.

Although this model represents small-scale microseismicity on a relatively small fault, it is thought that these types of discontinuum models could be scaled up to represent tectonic-scale earthquakes so that large-scale earthquake processes could be examined in detail. Simple particle models of the San Andreas fault have already been created to try and explain the heat flow paradox (MORA and PLACE, 1998; SCOTT, 1996). If these models were run in a fully dynamic way with realistic levels of numerical damping (attenuation) it is possible that a variety of earthquake source information could be obtained from the models. In particular, a more detailed study of how dynamic triggering can contribute to event complexity, or even to remote triggering of other earthquakes would be of great interest (see HARRIS, 1998).

Conclusions

The excavation of a tunnel at Atomic Energy of Canada Ltd.'s Underground Research Laboratory has yielded many induced microseismic events. Because the microseismicity was recorded on two different scales (ultrasonic AE and lower frequency MS) it was thought that these events would provide a unique opportunity to examine the complexities of unstable shear slip in a brittle rock (granite). Several MS events associated with AE clusters have been thoroughly examined in COLLINS *et al.* (2002). To complement the seismic studies, a bonded-particle discontinuum model of one of the recorded AE clusters was created.

The model showed how the MS event was preceded by many small "foreshocks" and then followed by several discrete "aftershocks", exactly as observed underground. In addition, it was shown that the MS events themselves are quite complex. The model showed that slip on the fault was preceded by the formation of many small tensile cracks that facilitated slip between unbonded grains. The cracking/slip started at one point on the fault and propagated outwards at a finite speed until slip was occurring along the entire fault. In addition it was shown that the dynamic waves emitted by the tiny tensile cracks may have dynamically triggered other cracks on the fault, thereby adding to the complexity of the macro-fault slip.

AE and MS events produced by the model exhibited moment magnitudes about an order of magnitude larger than those actually recorded at the TSX. Possible reasons for this are that the complexity of the modelled events meant that the moment magnitude equations were inappropriately applied, and that the PFC events may not be dissipating enough energy during fracture formation and sliding. However, the dominant reason is probably that the shear strain was directly applied in the model causing fairly large magnitude events, whereas in the actual underground experiment it is likely that very small stress perturbations or a temporal decrease in rock strength triggered the recorded AE and MS events.

Future work will involve a more realistic simulation of the geometry and stress conditions around the TSX tunnel to endeavor to more accurately reproduce and explain the observed seismicity. In addition, future models could be scaled up to represent tectonic scale earthquakes and these models could be used as an attempt to answer some of the fundamental questions associated with earthquake dynamics and prediction.

Acknowledgements

We wish to thank David Lockner and another anonymous reviewer whose suggestions significantly improved the manuscript. This research was made possible by funding from the European Union under the INCO-Copernicus programme.

References

Antonellini, M. A. and Pollard, D. D. (1995), *Distinct Element Modeling of Deformation Bands in Sandstone*, J. Struct. Geology *17*, 1165–1182.

Baker, C. and Young, R. P. (1997), *Evidence for Extensile Crack Initiation in Point Source Time-dependent Moment Tensor Solutions*, Bull. Seismol. Soc. Am. *87*, 1442–1453.

Brace, W. F. and Byerlee, J. D. (1966), *Stick Slip as a Mechanism for Earthquakes*, Science *153*, 990–992.

Brehm, D. J. and Braile, L. W. (1998), *Intermediate-term Earthquake Prediction Using Precursory Events in the New Madrid Seismic Zone*, Bull. Seismol. Soc. Am. *88*, 564–580.

Collins, D. S. and Young, R. P. (2000), *Lithological Controls on Seismicity in Granitic Rocks*, Bull. Seismol. Soc. Am. *90*, 709–723.

Collins, D. S. and Young, R. P., *Acoustic emission and microseismic data from the excavation phase of the tunnel sealing experiment*. Report to Atomic Energy of Canada Limited, TSX-RP040 (Keele University, 1998).

Collins, D. S., Pettitt, W. S., and Young, R. P. (2002), *High Resolution Mechanics of a Microearthquake Sequence*, Pure appl. geophys., *this issue*.

Cundall, P.A., *Numerical experiments on rough joints in shear using a bonded particle model*. In *Aspects of Tectonic Faulting* (eds. Lehner, F. K. and Urai, J. L.) (Springer-Verlag, Berlin 1999) pp. 1–10.

Cundall, P. A. and Strack, O. (1979), *A Discrete Element Model for Granular Assemblies*, Geotechnique *29*, 47–65.

Das, S. and Scholz, C. H. (1981), *Theory of Time-dependent Rupture in the Earth*, J. Geophys. Res. *86*, 6039–6051.

Feustel, A. J. (1995), *Seismic Attenuation in Underground Mines: Measurement Techniques and Applications to Site Characterization*, Ph.D. Thesis, Queen's University, Kingston, Ontario, Canada.

Harris, R. A. (1998), *Introduction to Special Session: Stress Triggers, Stress Shadows, and Implications for Seismic Hazard*, J. Geophys. Res. *103*, 24,347–24,358.

Hazzard, J. F. and Young, R. P. (2000), Simulating Acoustic Emissions in Bonded-Particle Models of Rock, Int. J. Rock Mech. Min. Sci. *37*, 867–872.

Hazzard, J. F., Young, R. P., and Maxwell, S. C. (2000), *Micromechanical modelling of cracking and failure in brittle rocks*, J. Geophys. Res. *105*, 16,683–16,697.

Itasca Consulting Group, *Particle Flow Code in 2 Dimensions* (Itasca Consulting Group, Minneapolis, 1999).

Lockner, D. A. (1998), *A Generalized Law for Brittle Deformation of Westerly Granite*, J. Geophys. Res. *103*, 5107–5123.

Lockner, D. A. (1993), *The Role of Acoustic Emission in the Study of Rock Fracture*, Int. J. Rock Mech. Min. Sci. and Geomech. Abstr. *30*, 883–899.

McGarr, A. (1994), *Some Comparisons between Mining-Induced and Laboratory Earthquakes*, Pure Appl. Geophys. *142*, 467–489.

Mora, P. and Place, D. (1998), *Numerical Simulation of Earthquake Faults with Gouge: Toward a Comprehensive Explanation for the Heat Flow Paradox*, J. Geophys. Res. *103*, 21,067–21,089.

Morgan, J. and Boettcher, M. S. (1999), *Numerical Simulations of Granular Shear Zones Using the Distinct Element Method 1. Shear Zone Kinematics and the micromechanics of localization*, J. Geophys. Res. *104*, 2703–2719.

Pettitt, W. S. (1998), *Acoustic Emission Source Studies of Microcracking in Rock*, Ph.D. Thesis, Keele University, Staffordshire, UK.

POTYONDY, D. O. and CUNDALL, P. A. (1998), *Modeling notch-formation mechanisms in the URL mine-by test tunnel using bonded assemblies of circular particles*, Int. J. Rock Mech. Min. Sci., Special Issue (Proceedings of NARMS '98, 3rd North American Rock Mechanics Symposium, Cancun, Mexico, June/July 1998), *35*, Paper no. 067, 1998.

POTYONDY, D. O. and CUNDALL, P. A. (1999), *Modeling of notch formation in the URL mine-by tunnel: Phase IV – Enhancements to the PFC model of rock*, Report to Atomic Energy of Canada Ltd. by Itasca Consulting Group, Minneapolis, issued as Ontario Hydro report No. 06819-REP-01200-10002-R00.

POTYONDY, D. O., CUNDALL, P. A., and LEE, C., *Modeling of rock using bonded assemblies of circular particles*. In *Second North American Rock Mechanics Symposium – NARMS '96* (ed. Aubertin, M.) (Balkema, Rotterdam, 1996) pp. 1934–1944.

SCHMIDTKE, R. H. and LAJTAI, E. Z. (1985), The Long-term Strength of Lac du Bonnet Granite, Int. J. Rock Mech. Min. Sci. & Geomech. Abstr. *22*(6), 461–465.

SCHOLZ, C. H., *The Mechanics of Earthquakes and Faulting* (Cambridge University Press, Cambridge, 1990).

SCOTT, D. R. (1996), *Seismicity and Stress Rotation in a Granular Model of the Brittle Crust*, Nature *381*, 592–595.

TAPPONNIER, P. and BRACE, W. F. (1976), *Development of Stress-induced Microcracks in Westerly Granite*, Int. J. Rock Mech. Min. Sci. Abstr. *13*, 103–112.

(Received April 4, 2000, revised/accepted August 10, 2000)

Pure appl. geophys. 159 (2002) 247–276
0033–4553/02/030247–30 $ 1.50 + 0.20/0

Pure and Applied Geophysics

Modelling Seismic Waves Around Underground Openings in Fractured Rock

MARK WILLIAM HILDYARD[1] and R. PAUL YOUNG[2]

Abstract—The potential for large excavation-induced seismic events may be recognised, even if the timing of an event may be inherently unpredictable. In this case, modelling the wave propagation from a potential event could allow the dynamic motions around an excavation to be projected, and for areas of danger to be anticipated. However, the above and other potential applications require accurate models of wave interaction with the openings, as well as with the fractured rock which surrounds such excavations. This paper considers real recorded waveforms and how well these waveforms are modelled by explicit mechanical models of the source, the medium and the excavation. Models of experiments at three different scales of the problem are presented: small and large amplitude waveforms recorded around a deep-level mining tunnel in a synthetic rockburst experiment; waveforms from laboratory experiments of waves through plates of steel representing fractures; waveforms from active pulses in an acoustic emission experiment in a small volume of fractured rock at the surface of an underground excavation. The results show that elastic wave propagation around an excavation was a first approximation for small amplitude waves, but was less successful for modelling large amplitude waves and more fractured rock. Fractures in the models were represented explicitly with displacement discontinuities. Waveforms through known fracture geometries were particularly well-reproduced, and indicate the importance of fracture stiffness, the *in situ* stress state, and stress-dependence of the fractures in such models. Overall, the models are sufficiently successful at representing recorded behaviour, to be encouraging for the goal of representing accurate wave motions around excavations.

Key words: Wave, model, seismic, rock, rockburst, fracture.

1. Introduction

Rockbursts and seismicity in mines is a serious hazard and is widespread in the deep hard-rock mining districts of the world. Large numbers of induced seismic events occur in deep mines, many of which are potentially damaging. Although attempts are made to forecast such damaging events (MENDECKI, 1993), it is not in general possible to predict when damaging events will occur. Even knowing the location of a potential

[1] Department of Earth Sciences, Jane Herdman Laboratories, 4 Brownlow Street, University of Liverpool, Liverpool L697GP, United Kingdom. E-mail: mwh@liv.ac.uk and CSIR Division of Mining Technology, P.O. Box 91230, Auckland Park, 2006, Johannesburg, South Africa.
E-mail: mhildyar@csir.co.za

[2] Department of Earth Sciences, Jane Herdman Laboratories, 4 Brownlow Street, University of Liverpool, Liverpool L697GP, United Kingdom.

event, it is not obvious whether the event will cause damage and which regions of the excavation will be damaged. It is often reported that damage can occur quite distant from the source and in unexpected areas. This may be due to the influence of the excavation on wave propagation, as is illustrated by the example in Figure 1. If wave propagation could be modelled around excavations, then it may be possible to anticipate regions of likely damage for potential events. This could be used as important feedback into the safety design process.

The ability to model wave propagation around deep underground openings therefore has important potential applications affecting the design and stability of such openings. This is true both for mining where short-term stability is required, as well as for other fields such as nuclear waste disposal requiring extremely long-term

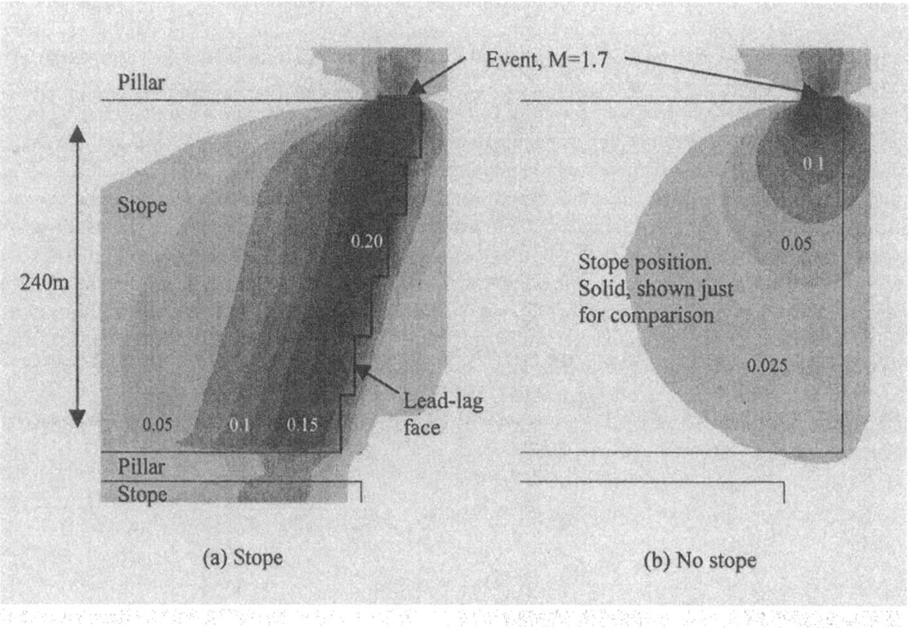

Figure 1
The projected influence of a mining layout on the wave propagation from a seismic event. Plan sections are shown through two three-dimensional elastic models, one containing a tabular mining excavation, the other a purely solid material. The contours indicate the maximum vertical velocity (in m/s), induced by a particular seismic event of magnitude 1.7, for a plane just below the excavation. The mining layout consists of a lead-lag stope, where the stope is a narrow excavations of 1.5 meters height but extending for hundreds of metres. Pillars are regions which have not been excavated, for stability reasons. The 'stepped' outline shows the mining face position or lead-leg face. The event was assumed to occur in a pillar parallel to the direction of mining advance, and just behind the face position of the mining. Further details regarding, the source are given in Appendix A, Section 1. The influence of the stope on the wave propagation causes velocities at a far pillar to be up to six times that of the solid model without the excavaion. This is due to wave propagation along the free surface, and the influence of the mining face. The above layout and spans of open excavation are quite typical in certain tabular mines (e.g., HANDLEY *et al.*, 1997; MALAN (1999); JAGER and RYDER, 1999).

stability. Such models allow studies to be made of the influence of a particular excavation on the wave propagation and distribution of peak motions, and on mechanisms that can lead to larger motions than expected. Furthermore if models can reproduce the wave propagation around the openings, then the likely motions and amplitudes could be projected for potential events. This would allow regions of likely damage to be identified, and for appropriate design action to be taken. Yet another possibility is that since the waves pass through fractured rock, matching recordings with models may provide information on the state of the rock at the skin of the excavation.

One feature of deep underground excavations is a highly fractured rock mass. This excavation-induced damage may extend many metres into the rockmass (e.g., ADAMS and JAGER, 1980; NAPIER *et al.*, 1997), complicating wave behaviour, and the task of modelling wave propagation. A wide body of literature deals with studies of waves through fractured rock. A displacement discontinuity has been investigated as a discrete representation of a fracture (SCHOENBERG, 1980; MYER, 1985), while experimental studies have indicated that this representation captures some of the frequency effects on waves due to fracturing (PYRAK-NOLTE *et al.*, 1990a, b). Numerical studies using this representation have tended to be made with assumptions of plane waves, two-dimensional modelling, or contain very limited numbers of fractures (e.g., GU *et al.*, 1996; CAI and ZHAO, 2000). Another approach attempts to encapsulate the effects of numerous fractures into the behaviour of the medium, in particular in terms of the effective wave-speed, but also attenuation effects (e.g., O'CONNELL and BUDIANSKY, 1974; SAYERS and KACHANOV, 1991; LIU *et al.*, 2000). These approaches assume low frequency. Most numerical studies of waves through fractures therefore make some of the following assumptions: two-dimensional wave propagation, plane wave propagation, small amplitude motions, low frequency relative to crack size, dilute crack concentration, uniform stress state. Many of these assumptions are poor in the context of real problems involved with wave propagation around underground openings. Excavations are firstly in a highly non-uniform stress field. In the case of deep-level mining, the excavation surface is highly fractured, wave amplitudes may be large, and sources are in the near-field and non-planar, while the full length of dynamic motion is important, and not just the initial arrivals. Microseismic and ultrasonic recordings used to monitor excavations and infer rock-mass state, include frequencies which do not satisfy low frequency or plane wave assumptions.

This paper begins to redress some of these deficiencies, by applying fully three-dimensional, full-wave modelling to waves recorded in excavations and through fractures. The modelling is unique in terms of the type and scale of problem to which it is applied.

Three case studies are reported covering different scales and aspects of the problem. The first study considers a real excavation problem, modelling both small and large amplitude waves originating in the near-field of a mine tunnel. The data modelled is from a rockburst experiment, where an artificial source was used to generate large amplitude, damaging seismic waves. Results cover attempts to model the waveforms with a purely elastic rock mass, although models with fractures were also examined. The second case study compares models with a laboratory experiment, where waves were passed through stacked steel plates, with the plate-to-plate boundaries representing fractures. The measured waveforms are reproduced in detail by the model, and the importance of the loading on the behaviour of the fractures is studied. The results demonstrate that this model of a fracture captures the essence of the wave behaviour. The third case study extends the successes of these fracture models to an unknown distribution of fractures in rock surrounding an excavation, nonetheless still with the advantages of a controlled and small-scale experiment. The data were recorded in velocity surveys from an acoustic emission experiment, performed in an underground tunnel as part of research into nuclear waste disposal. The modelling attempts to account for the recorded amplitudes and velocities of waveforms, relative to that expected in unfractured rock. The work touches on a second application requiring accurate models, which aids in interpreting the fracture state of the rock through which waves have passed.

2. *Methodology*

The cases studied in this paper intend to provide a measure of the ability of elastodynamic models to predict full wave propagation around openings and in fractured rock. Results are obtained by forward modelling using a known geometry and source. In all cases there are aspects of the source and geometry which are unknown. An effort is made to keep these assumptions physically reasonable and mechanistic. For example, it is possible to manipulate the spatial and time distribution of source in an attempt to better match recorded data. Instead, an important assumption is made that the source is a simple function, and that complexity is introduced by the geometry and fractures in the medium.

A number of different numerical methods exist for modelling elastodynamic behaviour, including variations of boundary element, finite element, finite difference and pseudo-spectral techniques. This work is performed with a finite difference model using the program WAVE (CUNDALL, 1992; HILDYARD *et al.*, 1995). One of the advantages is the size of three-dimensional problems which can be studied, including multitudes of fractures. The method used solves the coupled system of first-order equations obtained from the equation of motion, and the material constitutive equations relating stress and strain. The volume of interest is discretised on a staggered grid (c.f. Fig. 2), and the equations are solved using an explicit time-marching scheme. This approach is widely used for modelling seismic wave propagation (e.g., MADARIAGA, 1976; VIRIEUX, 1986; GRAVES, 1996), in part due to its computational efficiency. This efficiency is primarily due to it been readily

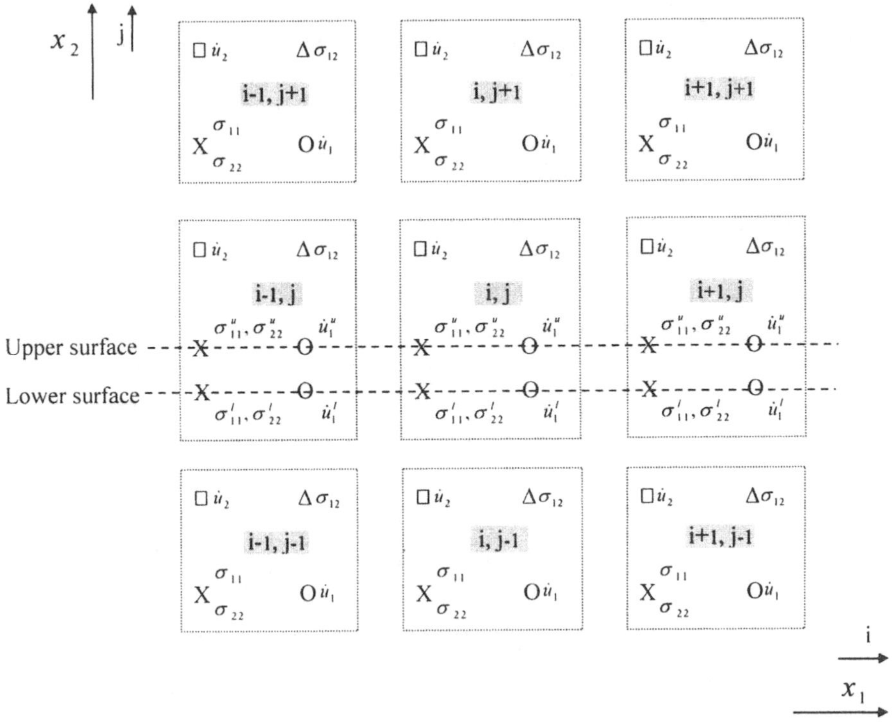

Figure 2
Portion of a two-dimensional staggered mesh showing the spatial distribution of components of velocity and stress. Representation of a two-dimensional horizontal crack shows surrounding grid variables and the appropriate dual-valued grid variables on the surfaces.

extendible to give higher order spatial accuracy (LEVANDER, 1988). The models are solved to fourth-order spatial accuracy, and second-order time accuracy.

Dynamic sources are implemented by prescribing the velocity or stress on grid-points. All boundaries impose absorbing conditions based on LYSMER and KUHLEMEYER (1969). Certain models solve the coupled problem of a static stress load with a transitory dynamic load. In these cases the static solution is first obtained through asymptotic solution of the wave equations to applied stresses at the boundaries, and by applying viscous damping in the mesh.

The main features used in the models in this paper are openings and cracks. Cracks are represented by displacement discontinuities, which are implemented by splitting a surface in the mesh into two coincident surfaces, each with its own set of grid-points (c.f. Fig. 2). Boundary conditions are enforced on these surfaces. The crack can be open, behaving like separate free surfaces which can interpenetrate, or coupled by a normal and shear fracture stiffness, which relate the normal and shear stress at the crack surfaces to the relative normal and tangential displacement of the crack surfaces. Other conditions such as tensile failure or frictional sliding were not used in the reported models. The surfaces are welded at the crack edges and behave as solid material. Openings are represented using open cracks. Narrow openings are modelled as a single open crack, while wide openings such as tunnels are modelled by enclosing a volume with open cracks, and isolating the enclosed grid-points.

HILDYARD *et al.* (1995) and NAPIER *et al.* (1997) contain previous applications of the two-dimensional code, including verification through comparison with photo-elastic experiments of wave interaction with openings and interfaces. We now derive selected grid and crack equations for three-dimensions (based on CUNDALL, 1992).

The method solves for velocity and stress on a staggered mesh, with the different components held at different positions in space (Fig. 2). Consider first a continuous solid mesh without any cracks. The time derivatives of the constitutive equations for a linear elastic isotropic material are:

$$\dot{\sigma}_{ij} = \delta_{ij}\left(K - \frac{2}{3}G\right)\dot{e}_{kk} + 2G\dot{e}_{ij}, \quad \text{where } \dot{e}_{ij} = \frac{1}{2}\left[\frac{\partial \dot{u}_i}{\partial x_j} + \frac{\partial \dot{u}_j}{\partial x_i}\right], \tag{1}$$

where K is the bulk modulus, G is the shear modulus, δ_{ij} is the Kronecker delta, σ_{ij} are components of the stress tensor, e_{ij} are components of the strain tensor, $\dot{u}$ is velocity, and a dot indicates a time derivative. Applying equation (1) at the σ_{22} position in cell (i,j) (assumed here to be in solid material), and approximating using second-order central finite differences, gives

$$\sigma_{22}^{t} = \sigma_{22}^{t-1} + E_1\frac{\Delta t}{\Delta x_2}\left(\dot{u}_2^{j} - \dot{u}_2^{j-1}\right)^{t-1/2} + E_2\frac{\Delta t}{\Delta x_1}\left(\dot{u}_1^{i} - \dot{u}_1^{i-1}\right)^{t-1/2} + E_2\frac{\Delta t}{\Delta x_3}\left(\dot{u}_3^{k} - \dot{u}_3^{k-1}\right)^{t-1/2} \tag{2}$$

where $E_1 = K + 4G/3, E_2 = K - 2G/3$. Superscripts i, j or k represent the spatial direction for the differencing, while superscript t indicates discrete time. A new value of σ_{22} at time t is calculated from σ_{22} at time $t-1$, and velocities at time $t-1/2$, so that velocities are staggered in time by $\Delta t/2$ with respect to stresses. Similar expressions are obtained for the five remaining stress components.

The equations of motion excluding body force are:

$$\rho\frac{\partial \dot{u}_i}{\partial t} = \frac{\partial \sigma_{ij}}{\partial x_j} \tag{3}$$

where ρ is density and $\dot{u}$ is velocity. Applying equation (3) at the $\dot{u}_2$ position in cell (i, j) in the staggered mesh, and approximating using second-order central finite differences, gives

$$\dot{u}_2^{t+\frac{1}{2}} = \dot{u}_2^{t-\frac{1}{2}} + \frac{1}{\rho}\frac{\Delta t}{\Delta x_2}\left(\sigma_{22}^{j+1} - \sigma_{22}^{j}\right)^{t} + \frac{1}{\rho}\frac{\Delta t}{\Delta x_1}\left(\sigma_{12}^{i} - \sigma_{12}^{i-1}\right)^{t} + \frac{1}{\rho}\frac{\Delta t}{\Delta x_3}\left(\sigma_{23}^{k} - \sigma_{23}^{k-1}\right)^{t}. \tag{4}$$

Similar expressions are obtained for the two remaining velocity components. Equations (2) and (4) and the related equations for other components provide the nine basic second-order grid equations for the solid material. Higher order spatial differencing is introduced by including more distant terms.

We now introduce a horizontal discontinuity into the mesh as illustrated for two dimensions in Figure 2. This consists of two coincident surfaces on which certain grid variables are controlled, and others allowed to be dual-valued. The grid variables locating on the crack surfaces have an upper and lower value i.e., σ_{22}^u, σ_{22}^l, σ_{11}^u, σ_{11}^l, σ_{33}^u, σ_{33}^l, σ_{13}^u, σ_{13}^l, $\dot{u}_1^u$, $\dot{u}_1^l$, $\dot{u}_3^u$ and $\dot{u}_3^l$. From continuity $\sigma_{22}^u = \sigma_{22}^l = \sigma_{22}$. However σ_{11}, σ_{33}, σ_{13}, $\dot{u}_1$ and $\dot{u}_3$ are dual-valued, and values for the upper and lower surfaces must be independently calculated. From equation (2) we can write two separate mesh equations for $\Delta\sigma_{22}$, which is the increment required to update σ_{22} from time-step t to $t + 1$, (for convenience, we now drop the t superscript notation), giving

$$\Delta\sigma_{22}^u = E_1 \frac{\Delta t}{\Delta x_2}(\dot{u}_2^j - \dot{u}_2^{(uf)}) + E_2 \frac{\Delta t}{\Delta x_1}(\dot{u}_1^{i(u)} - \dot{u}_1^{i-1(u)}) + E_2 \frac{\Delta t}{\Delta x_3}(\dot{u}_3^{k(u)} - \dot{u}_3^{k-1(u)}) \tag{5}$$

$$\Delta\sigma_{22}^l = E_1 \frac{\Delta t}{\Delta x_2}(\dot{u}_2^{(lf)} - \dot{u}_2^{j-1}) + E_2 \frac{\Delta t}{\Delta x_1}(\dot{u}_1^{i(l)} - \dot{u}_1^{i-1(l)}) + E_2 \frac{\Delta t}{\Delta x_3}(\dot{u}_3^{k(l)} - \dot{u}_3^{k-1(l)}) \tag{6}$$

where $\dot{u}_2^{(uf)}$ and $\dot{u}_2^{(lf)}$ are "fictitious" quantities, since $\dot{u}_2^{(uf)}$ falls below the upper surface and $\dot{u}_2^{(lf)}$ is above the lower surface of the crack. We now assume that the normal stress on the crack surface is coupled to the relative normal displacement by a linear stiffness k_n, such that

$$\Delta\sigma_{22} = \dot{u}_2^{\mathrm{rel}} k_n \Delta t = k_n \Delta t \left[\frac{\dot{u}_2^j + \dot{u}_2^{(uf)}}{2} - \frac{\dot{u}_2^{j-1} + \dot{u}_2^{(lf)}}{2} \right]. \tag{7}$$

Combining equations (5), (6) and (7), the fictitious stresses $\dot{u}_2^{(uf)}$ and $\dot{u}_2^{(lf)}$ are eliminated, giving

$$\Delta\sigma_{22} = E_1 \frac{E_1 k_n \Delta t}{E_1 + k_n \Delta x_2} \left[\left(\dot{u}_2^j - \dot{u}_2^{j-1} \right) + \frac{1}{2}\frac{E_2}{E_1}\frac{\Delta x_2}{\Delta x_1} \dot{u}_1^{\mathrm{diff}} + \frac{1}{2}\frac{E_2}{E_1}\frac{\Delta x_2}{\Delta x_3} \dot{u}_3^{\mathrm{diff}} \right] \tag{8}$$

where

$$\dot{u}_1^{\mathrm{diff}} = \dot{u}_1^{i(u)} - \dot{u}_1^{i-1(u)} + \dot{u}_1^{i(l)} - \dot{u}_1^{i-1(l)} \tag{9}$$

where

$$\dot{u}_3^{\mathrm{diff}} = \dot{u}_3^{k(u)} - \dot{u}_3^{k-1(u)} + \dot{u}_3^{k(l)} - \dot{u}_3^{k-1(l)} \tag{10}$$

σ_{11}^u, σ_{11}^l, σ_{33}^u and σ_{33}^l can then be calculated using

$$\Delta\sigma_{11}^u = \frac{2G(6K + 2G)\Delta t}{(3K + 4G)\Delta x_1}(\dot{u}_1^{i(u)} - \dot{u}_1^{i-1(u)}) + \frac{2G(3K - 2G)\Delta t}{(3K + 4G)\Delta x_3}(\dot{u}_3^{k(u)} - \dot{u}_3^{k-1(u)}) + \frac{E_2}{E_1}\Delta\sigma_{22} \tag{11}$$

$$\Delta\sigma_{11}^{l} = \frac{2G(6K+2G)\Delta t}{(3K+4G)\Delta x_1}(\dot{u}_1^{i(l)} - \dot{u}_1^{i-1(l)}) + \frac{2G(3K-2G)\Delta t}{(3K+4G)\Delta x_3}(\dot{u}_3^{k(l)} - \dot{u}_3^{k-1(l)}) + \frac{E_2}{E_1}\Delta\sigma_{22} \tag{12}$$

$$\Delta\sigma_{33}^{u} = \frac{2G(3K-2G)\Delta t}{(3K+4G)\Delta x_1}(\dot{u}_1^{i(u)} - \dot{u}_1^{i-1(u)}) + \frac{2G(6K+2G)\Delta t}{(3K+4G)\Delta x_3}(\dot{u}_3^{k(u)} - \dot{u}_3^{k-1(u)}) + \frac{E_2}{E_1}\Delta\sigma_{22} \tag{13}$$

$$\Delta\sigma_{33}^{l} = \frac{2G(3K-2G)\Delta t}{(3K+4G)\Delta x_1}(\dot{u}_1^{i(l)} - \dot{u}_1^{i-1(l)}) + \frac{2G(6K+2G)\Delta t}{(3K+4G)\Delta x_3}(\dot{u}_3^{k(l)} - \dot{u}_3^{k-1(l)}) + \frac{E_2}{E_1}\Delta\sigma_{22} . \tag{14}$$

Expressions for updating velocities $\dot{u}_1^u$, $\dot{u}_1^l$, $\dot{u}_3^u$ and $\dot{u}_3^l$ are calculated in a similar manner, by assuming that the surface shear stress and relative shear displacement are coupled by a linear stiffness k_s. i.e., by writing expressions for $\dot{u}_1^u$ and $\dot{u}_1^l$ using fictitious grid-points, and using the condition

$$\Delta\sigma_{12}^{\text{surf}} = \dot{u}_1^{\text{rel}} k_s \Delta t = k_s \Delta t(\dot{u}_1^u - \dot{u}_1^l) \tag{15}$$

where $\Delta\sigma_{12}^{\text{surf}}$ locates on the crack surface which is not a normal grid position for σ_{12}. Noting that σ_{12} is continuous across the crack, all fictitious points can be eliminated. The remaining crack values σ_{13}^u and σ_{13}^l can be calculated simply by ensuring that the velocities from the correct sides of the crack are used in the difference calculations. The special case of an open crack, with free surface conditions on both surfaces, is obtained by setting $k_n = 0$ and $k_s = 0$. Edge conditions are required at the ends of the crack to match the dual-valued crack calculations to the standard mesh calculations.

3. A Case Study: Modelling Seismic Waves from a Rockburst

This first section concerns the accuracy of wave modelling at the excavation-size scale for both small and large amplitude waves. It covers a case study from a real excavation. The data were obtained in an experiment to simulate a rockburst in a deep-level underground tunnel.

Although back-analysis of seismic events is widespread in studies of large earthquakes, sparse published work has attempted to apply numerical models to back-analyse the wave motions around underground openings due to rockbursts (e.g., HANDLEY *et al.*, 1996). Certain factors should be highlighted which contribute to the complexity of this task. The proximity of the excavation to the source means that waves begin to interact with the excavation before far-field conditions are reached. Fractured or damaged zones associated with deep excavations lead to more complex wave interactions. Finally, the primary interest is motion at the excavation surface so that meaningful back-analysis requires recordings within the excavation. In this case a simulated event is studied. The advantage of examining a simulated event rather than a natural event is *a priori* knowledge of the source and the excavation layout and condition, and the chance to prepare and receive good coverage of recordings, particularly at the surfaces of excavations.

The artificial rockburst experiment was performed in an underground tunnel at a deep-level gold mine in South Africa. The purpose of the experiment was to create and extensively monitor a controlled seismic event, and to induce and observe seismic damage in a nearby tunnel (MILEV *et al.*, 2000). Extensive numerical modelling of seismic wave propagation was used in the experiment (HILDYARD and MILEV, 1999). The forward analysis included the development of a source model for a propagating blast and the modelling of data from a small calibration blast. The back analysis applied the source model to data from the main experiment. Here however, we concentrate exclusively on the comparisons between the modelled waveforms and recordings for both large amplitude and small amplitude waves recorded in the experiment.

The experiment took place in a tunnel at a depth of 1600 m. The tunnel had an approximately square shape and relatively unfractured walls. A small (0.67 kg) calibration blast was made at one end of the tunnel, close to the tunnel surface. The main experiment was a larger (260 kg) explosion at the opposite end of the tunnel, comprised of five separate blastholes of 4 m to 7 m in length, deep in solid rock 6 metres away from the tunnel. Thirty-two geophones were in place with a sampling rate of 10 kHz and low-pass filtering from 750 Hz. For the main blast, three accelerometers were positioned on the tunnel surface where the largest motions were expected. These had a sampling rate of 500 kHz. Prior to the blast, fractures on the tunnel wall were mapped to determine the rock-mass state. The beddings were spaced between 150 mm and 600 mm and were generally closed. Joints were spaced between 0.5 m and 3.0 m, and were closed. Stress fractures associated with the development of the tunnel were generally open by up to 2 mm with a spacing of approximately 10 cm, and intersections defined wedge-shaped blocks (REDDY and SPOTTISWOODE, 2000).

The tunnel was modelled as a square cavity with a side of 3.6 metres, and 37 metres long. Figure 3 shows the model geometry and positions of the geophones and accelerometers used in comparisons. These were uniaxial and mounted along the surface of the near tunnel wall. A typical model used 800,000 grid-points with a 0.2 metre spacing, and fourth-order spatial differencing. Model details are given in Appendix A for Section 3.

The calibration blast had a 0.037 m diameter, a 0.65 m charge length, a 0.67 kg charge mass, and a detonation velocity of 4500 m/s. It was one metre from the tunnel and inclined away from the geophones at 75 degrees (nearly normal) to the tunnel surface. Figure 4 compares the modelled velocity seismograms with those recorded at varying distances along the tunnel near wall. The model is elastic and represents only the source, and the effect of the opening. Scales are the same for corresponding waveforms. The motions match well in a qualitative sense. In both cases the P wave is lower amplitude than the S wave, particularly with greater distance. Amplitudes are similar, and the decay with distance of both the P and the S waves is well modelled.

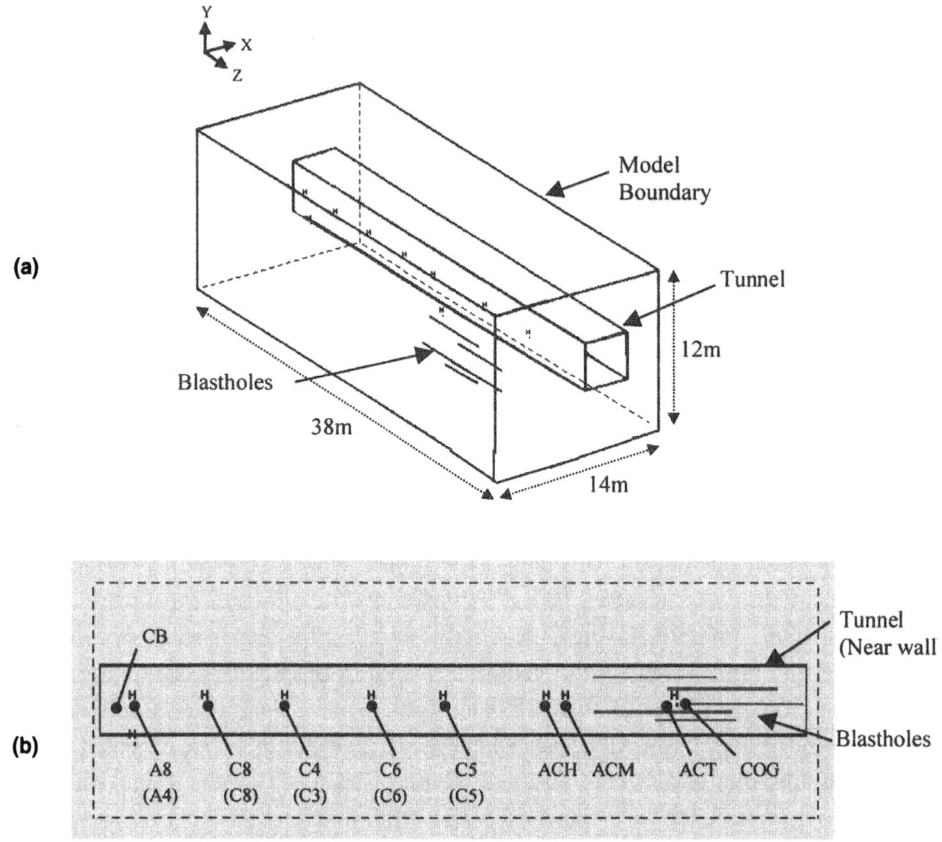

Figure 3
Model geometry for the rockburst experiment. (a) 3-D sketch, (b) front view showing positions of some of the geophones and accelerometers. 'COG' is the approximate centre of the main blast. 'CB' is the calibration blast. 'ACT', ACM' and 'ACH' are accelerometer positions, while other marked positions are geophones. Labels given in brackets are geophone positions for the calibration blast.

Arrivals are slightly quicker in the data, indicating faster wave speeds. At positions where triaxial data were available, not all components were well-matched.

The main experiment was designed to generate peak particle velocities in the tunnel of around 3 m/s, without any direct damage at the tunnel due to gas expansion. Five synchronous blasts were used to generate a large enough source. The five blastholes varied widely, but were approximately 6 metres from the tunnel, parallel to the tunnel, and vertically spaced with a spacing of 0.5 m. The diameter was 0.1 m, the average charge length 6 m, and the detonation velocity 3600 m/s, with detonation staggered by 70 μs for each blast-hole. The total charge mass was 260 kg of Anfo explosive. A number of the geophones were overdamped for the main experiment to allow recording of strong ground motion (MILEV *et al.*, 2000). In spite of the advantages of a known source position, considerable uncertainty exists for

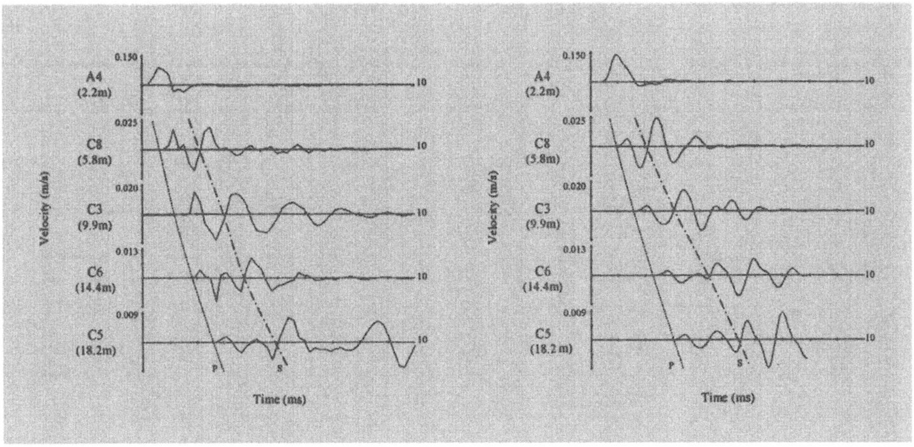

(a) Experiment **(b) Model**

Figure 4
Comparison of velocity seismograms for the calibration blast at varying distances along the tunnel near wall. Positions for geophones A4, C8, C3, C6, C5 are shown in brackets in Figure 3. Motion is normal to the tunnel surface. Approximate *P*-wave arrivals, and the position of secondary waves (not the *S*-wave arrival), are indentified.

wave propagation from a detonating rather than an instantaneous blast, and much of the modelling involved developing a suitable source model. A mechanistic representation of the source was used to account for the slow detonation, and because the proximity of the tunnel makes near-field wave propagation important. The source model involved a pressure propagating along the line of the blast hole at the velocity of detonation of the blast. Initial models assumed an elastic rock mass. Figure 5 compares the waveforms recorded at various distances along the surface of the tunnel near-wall for the experiment and for the model, respectively. The positions of these recordings are shown in Figure 3.

Considering firstly just the *P*-wave portion of the waveforms, Figure 5 shows that the model at least matches the data in the direction of the first motions. However, in the data the width of the first pulse varies between successive positions, and in fact there is little similarity in waveforms between successive recordings. In contrast, the modelled waveforms vary smoothly with distance. Moreover, Figure 6a shows that the *P*-wave amplitude decays smoothly in the model, while the decay is not monotonic in the data. Comparing Figure 5a with Figure 4a, the change and decay of waveforms with distance is also less coherent than that for small amplitude waves in the calibration blast. The waves therefore do not decay or change in the manner predicted by elastic wave propagation. The data set is small so this may simply reflect errors in the data due to the difficulty of recording large amplitude ground motion accurately. Two problems could lead to such errors in the data; either poor coupling, or the unusual damping where larger than normal resistors were used in the

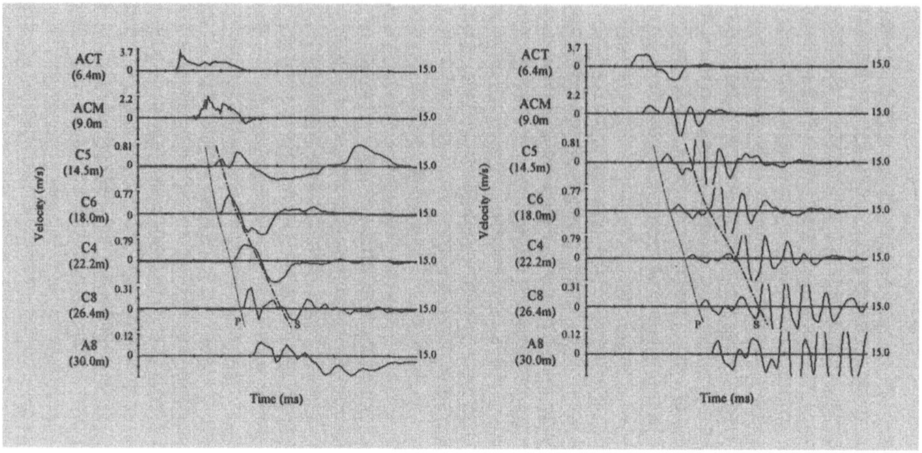

(a) Experiment **(b) Model**

Figure 5
Comparison of velocity seismograms for the main blast at varying distances along the tunnel near wall. Positions for accelerometers ACT and ACM, and geophones A8, C5, C6, C4 and C8 shown in Figure 3. Motion is normal to the tunnel surface. Approximate *P*- and *S*-wave arrivals are shown. ACT and ACM were integrated from accelerometer data. Recordings for ACT, ACM and A8 were not on the same time-base, and have been time-shifted to their probable arrival time relative to C5.

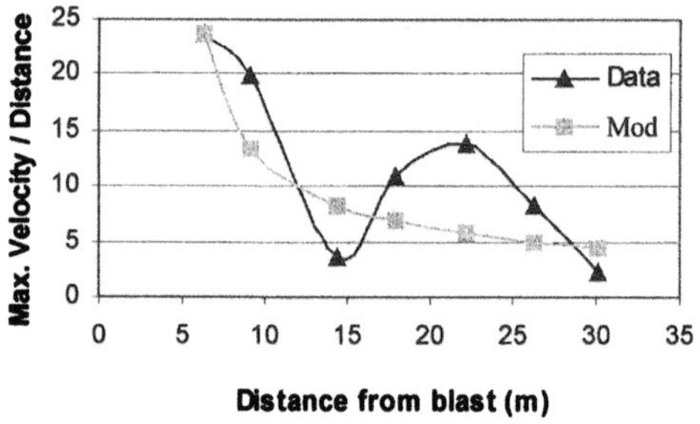

Figure 6
Decay in particle velocity with distance from the source for the recorded and modelled waveforms. The graph shows the amplitude of the first arrival scaled by distance from the source.

geophones to cope with velocities of up to 1 m/s (MILEV *et al.*, 2000). No physical evidence of poor coupling was reported for these geophones. Instead, the lack of coherence in the measured waveforms may indicate differences in the fracture state near geophones, and the differences in wave propagation for large amplitude waves. Such differences would occur due to the waves generating damage, or interacting with existing damage, or changing the loading and hence the rock properties. Although the recordings after accelerometer ACM were beyond the region of observed damage, the particle velocity in this region was still between 0.3 m/s and 0.81 m/s, inducing significant transitory stress changes.

The most fundamental difference in waveforms in Figure 5 is that the secondary waves in the model are much larger than the initial *P* wave, in contrast to the measured waveforms in which secondary waves were mostly of the same order as the initial arrivals. These differences could be poor source modelling. The large secondary waves in the model are in fact Rayleigh waves which develop due to large incident shear waves. This is a direct result of the source model for a propagating blast, which has been shown to generate large shear waves for a low detonation velocity (DAEHNKE, 1997; KOUZNIAK and ROSSMANITH, 1998). The source model is primarily based on theoretical rather than physical evidence, and physical evidence is required to confirm whether a shear wave can be expected at a distance from the blast. If the source model is valid, then an alternative explanation is that the differences again relate to the generation of or the opening of fractures. A model which allowed tensile failure on fractures near the tunnel surface was shown to reduce the relative content of the secondary waves by impeding development of a Rayleigh wave (HILDYARD and MILEV, 1999).

Two indicators from the model: maximum velocity and maximum induced tensile stress, were compared with regions of recorded damage. Unfortunately, these include the influence of the anomalously large secondary waves which make comparison somewhat misleading. Figure 7 compares the modelled maximum induced tensile stress for the near wall of the tunnel with recorded damage. Regions marked 'H' and 'L' indicate areas where high and low intensity damage was reported. Induced stress in σ_{yy} is concentrated opposite the blast holes (maximum 25 MPa) with a maximum of 10 MPa ahead of the blast. The induced tensile stress in σ_{zz} is also high opposite the blast holes (maximum 25 MPa), but there is a second highly tensile region ahead of the blast on the edge of the low intensity damage zone. The above stresses could lead to tunnel normal fracturing, however they should be seen in the context of the total stress state.

The above illustrates the value of accurate wave models. If real waveforms are known to be well-matched, then the model provides detail over a region, including induced stress state, which is not directly known from measurements. The waveforms in this case study however, are not sufficiently well-matched to infer such information. Notably, differences were observed between small and large amplitude waves. Modelling the small amplitude waves proved fairly successful with an elastic material, while modelling the large amplitude waves was not. Interaction with fracturing may account for these differences. The modelling of fractures is examined in the next sections.

4. *Wave Propagation through Fractures – Laboratory Experiments*

As has been discussed, one of the complications in modelling waves around deep excavations in a rock mass, is that the rock surrounding such openings is fractured.

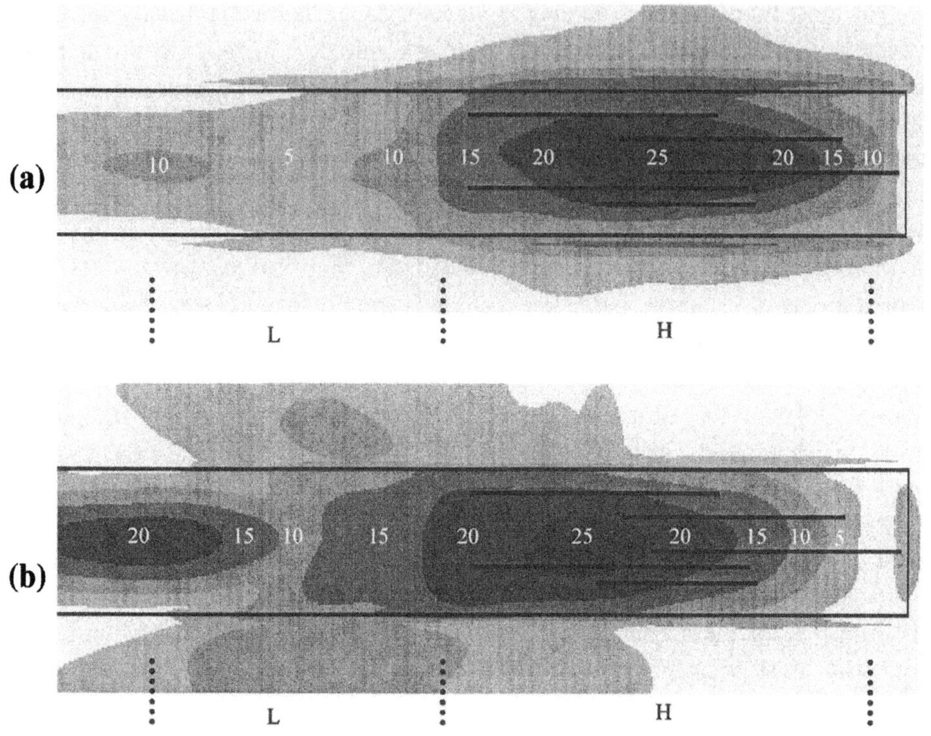

Figure 7
Distribution of maximum induced tensile stress (in MPa) at the tunnel near wall, for the 8 ms source, and for stress components (a) σ_{yy}, vertical, and (b) σ_{zz}, parallel to the tunnel. (i.e., the maximum recorded at a position over all time). Positions 'H' and 'L' show the regions where high and low intensity damage was reported in the experiment.

One approach to account for fracturing is to consider the fractured rock as an effective elastic medium, in which the elastic constants are related to the density of fracturing, typically leading to anisotropy in the seismic velocities (e.g., O'CONNELL and BUDIANSKY, 1974; CRAMPIN, 1981; SAYERS and KACHNOV, 1991). Expressions for the effective attenuation can also be calculated (HUDSON, 1981; LIU *et al.*, 2000). The main restriction is that this is valid only for wave propagation of wavelength considerably greater than the fracture size, although there are also restrictions on the density of fracturing.

An alternative approach is to model each fracture or group of fractures explicitly. A displacement discontinuity is such a model, where the displacements of the two surfaces of a zero-thickness interface are discontinuous, and the difference in displacements of the two surfaces is related to the stress across the interface. The stress and the discontinuity in displacement across the two surfaces are coupled by a fracture stiffness. This has been studied as a representation of a fracture (SCHOENBERG, 1980; MYER *et al.*, 1985), and shown to be consistent with experiments on natural dry fractures in rock, in terms of the frequency dependence of both wave speed and attenuation (PYRAK-NOLTE *et al.*, 1990a).

For modelling potentially damaging waves around underground openings, models of explicit fracturing seem attractive, as the effect of waves on the fractures can also be studied. This section evaluates the accuracy of fracture models based on the displacement discontinuity, by modelling experiments from PYRAK-NOLTE *et al.* (1990b). The experiments recorded waveforms for both *P* and *S* waves passed through a stack of parallel steel plates, representing a parallel set of fractures. Transmitting and receiving transducers contained both *P*- and *S*-wave piezoelectric elements of 22 mm diameter. Thirty-one mild steel plates were stacked to form a cube with a side of 90 mm. The plates were sandblasted before being stacked, to simulate fracture surfaces. The block was biaxially loaded with a force of 30 kN – one load clamping the plates and a second equal load parallel to the plates. A solid cylinder with an axis of 99 mm provided an unfractured control case. Figure 8a shows a schematic of the experiment.

This experiment was modelled in three dimensions, details of which are given in Appendix A, Section 4. *P*- and *S*-wave sources were inverted from the waveform through the unfractured cylinder, and these sources were then used for the fracture cases. Three *P*-wave cases and four *S*-wave cases were modelled, initially with fractures with uniform fracture stiffness of 6el3 Pa/m and 2el3 Pa/m for the normal and shear directions, respectively.

Figure 9 presents the comparison for *P*-wave transmission. Wave propagation parallel to the fractures (9c) results in a completely different waveform from that of

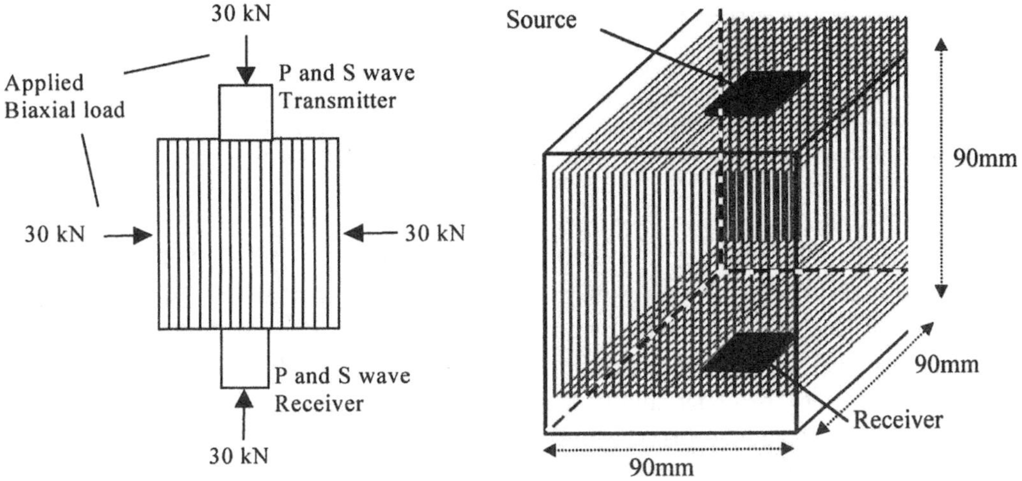

Figure 8
Sketch of the multiple fracture experiment. (a) The experimental system of packed steel plates, transmitter and receiver transducers, and bi-axial loading. (reproduced for PYRAK-NOLTE *et al.*, 1990b). (b) The 3-D WAVE model with displacement discontinuities representing interfaces between the steel plates.

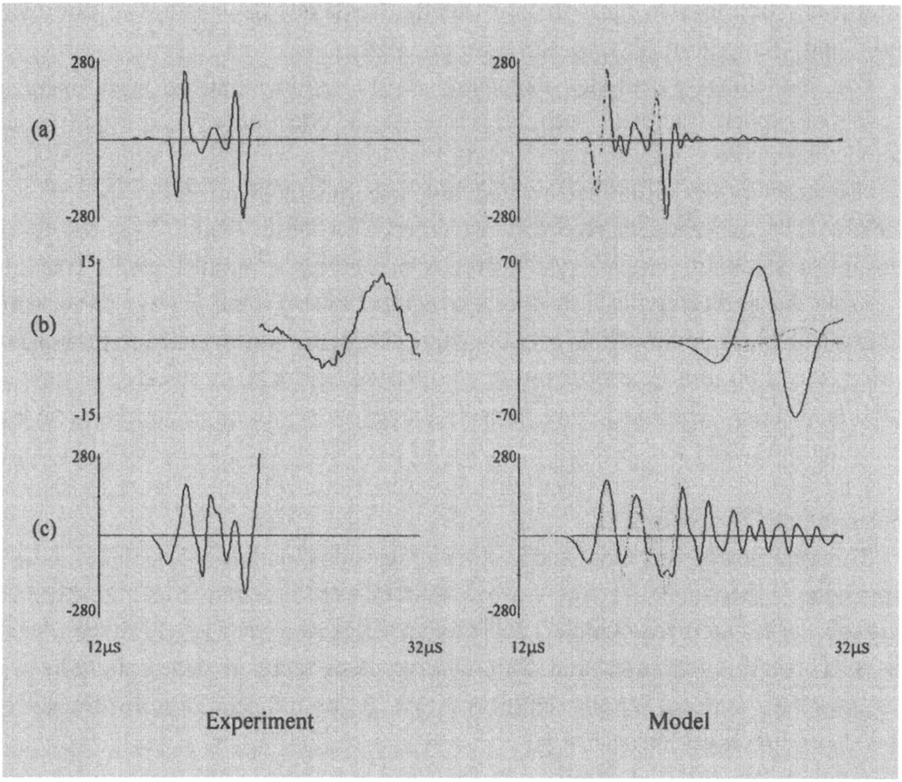

Figure 9
Comparison between experimental waveforms (left) and modelled waveforms (right) for *P*-wave transmission. (a) Solid cylinder. (b) Horizontal fractures with propagation transverse to the fractures. (c) Vertical fractures with propagation parallel to the fractures. The horizontal axis shows time (μs). Experimental waveforms are for voltage (mV), while modelled waveforms are for stress (kPa). Dotted lines are the experimental result superimposed over the model result. The cases for horizontal fractures are shown at different vertical scales due to significant attenuation. Experimental data from PYRAK-NOTLE *et al.* (1990b).

the solid case (9a), although this cannot be quantified simply by the effect on arrival or amplitude. The arrival, amplitude and the total waveform is well reproduced in the model (9c). Wave propagation across the fractures (9b) is significantly delayed and attenuated, with a lower dominant frequency than the solid case (9a). The arrival is approximately 50% later, and the amplitude is attenuated to one twentieth that of the solid case. Similar effects on arrival, amplitude and frequency are observed in the model however to a far lesser degree, where the wave is attenuated to just one quarter that of the solid case. Table 1 compares the experimental and the model times and amplitudes for the first peaks in each *P*-wave experiment, indicating that the most significant difference in the model is that it underestimates the attenuation for wave propagation across the fractures. Models of the shear-wave experiments yielded

Table 1

Amplitude and time for the first peak compared for the different P-wave experiments and for the uniform stiffness model and the stress-dependent stiffness model

		Unfractured		Horizontal fractures		Vertical fractures	
		Ampl. (mV or mm/s)	Time (μs)	Ampl. (mV or mm/s)	Time (μs)	Ampl. (mV or mm/s)	Time (μs)
Experiment		−204	16.9	−4.7	26.3	−102	16.3
Uniform crack stiffness		−191	16.8	−19.8	24.1	−95.5	16.25
	(% err)	6%	0.6%	320%	8%	6%	0.3%
Stress-dependent crack stiffness				−8.5	28.8	−97	16.7
	(% err)			80%	10%	5%	2%

similar results, where the wave propagation for two different polarizations parallel to the fractures were very well-reproduced, while for wave propagation across the fractures the effects on amplitude, arrival and frequency were consistent with the experiment, although attenuation in particular was too small.

The differences in wave amplitude could indicate the need for an additional dissipative mechanism in the displacement discontinuity model of a fracture to remove mechanical energy from the system. However, it is proposed that the differences result from a non-uniform loading in the experiment which leads to a non-uniform stiffness in the fractures. The biaxial load was applied over a portion of the block. Applying the load of 30 kN to a portion of the block (22 mm by 22 mm), leads to a highly non-uniform stress distribution. The distribution of stress normal to the fractures is shown in Figure 10. Fracture stiffness is related to the compression of a crack and can be expected to vary across the cracks if the normal stress distribution is non-uniform.

A model of stress-dependent fracture stiffness was then developed, based on the hyperbolic joint stiffness relation of BANDIS *et al.* (1983). Introducing this model and applying the loading conditions (details in Appendix A, Section 4), leads to a wide variation of the fracture stiffness in different fractures and within a single fracture. This model leads to considerably lower velocities towards the edges of the block, and greater scattering, resulting in much greater attenuation for wave propagation across the fractures. Figure 11 compares the results of the model of uniform fracture stiffness to those of stress-dependent fracture stiffness, showing that the stress-dependent fracture stiffness leads to substantially greater attenuation of the *P* waves for propagation across the fractures, while it has little effect on the waves parallel to the fractures. This implies that the displacement discontinuity model can likely account for the effects of fractures on waves without requiring further dissipative mechanisms.

This work highlights the importance of the fracture stiffness in modelling waves through fractures, and hence for modelling waves around deep underground

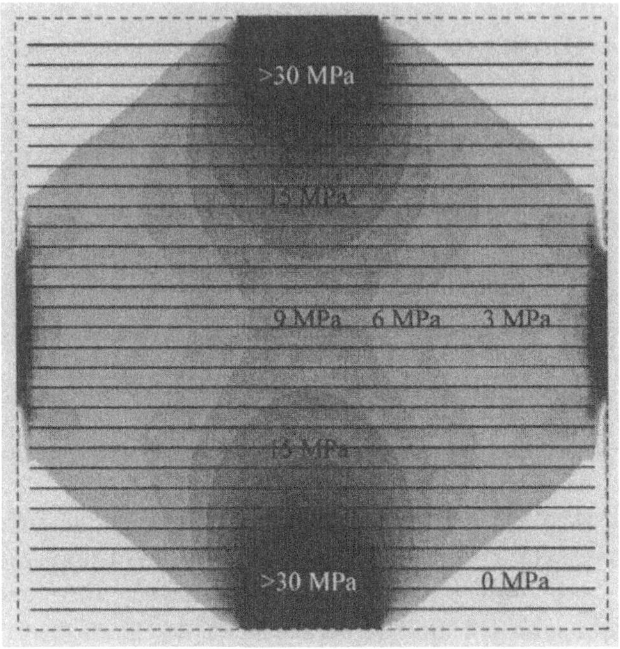

Figure 10
Cross section through the horizontal crack model showing the variation in crack normal stress (σ_{22}). The stress distribution is based on a biaxial load of a 30 kN, applied over the source area, a square region of 22 mm by 22 mm. The block faces are 90 mm by 90 mm.

openings. The fracture stiffness significantly effects the waveforms, delay, amplitude and frequency, causing entirely different behaviour from simply considering fractures as open or closed. In particular the stress dependence of the fracture can be accounted for with a stress-dependent fracture stiffness, and in a non-uniform stress field this stiffness may vary along a single continuous fracture. This has important consequences for modelling fractures around underground openings where stresses are highly non-uniform.

5. *Wave Propagation Though in situ Fractures*

This section extends the wave-fracture modelling to fractures in *in situ* rock at the surface of a deep tunnel. The data were collected in an acoustic emission experiment encompassing a small volume of rock (approximately 1 m^3) at the surface of the URL Mineby tunnel (CARLSON and YOUNG, 1992, 1993). An acoustic array was installed close to the face of the tunnel. Acoustic emissions were collected over a period of weeks during which there were three 0.5 metre face advances. A number of active velocity scans were also made.

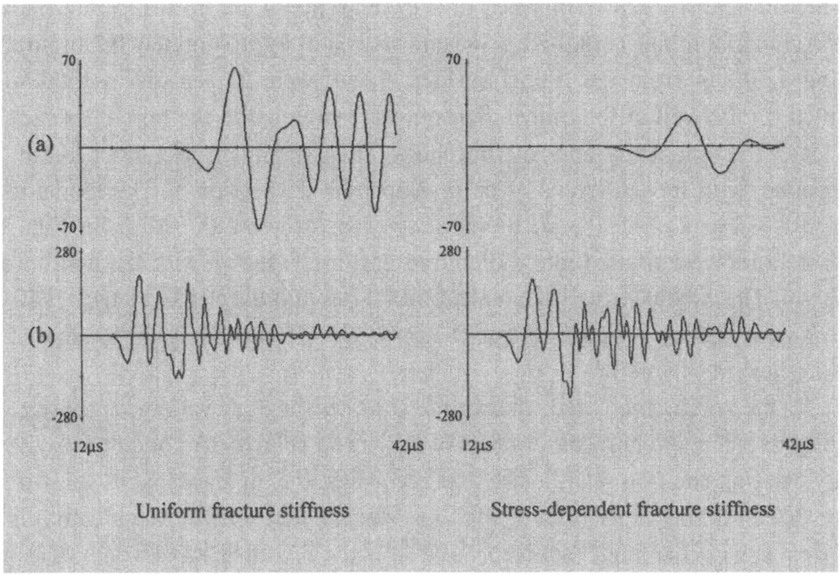

Figure 11
Comparison of waveforms from the uniform fracture stiffness and the stress-dependent fracture stiffness models. (a) *P*-wave response for horizontal fractures (propagation transverse to fractures). (b) *P*-wave response for vertical fractures (propagation parallel to fractures).

Analysis of the experiment (CARLSON and YOUNG, 1992) showed a 12% anisotropy in seismic velocities. The slow direction is orthogonal to the tunnel, while intermediate and maximum velocities are parallel to the tunnel. The anisotropy for compressional wave velocity prior to tunnel advance was calculated as 5810 m/s and 5080 m/s for the maximum and minimum directions, respectively. Crack densities were inferred from the measured seismic velocities and resulting Poisson's ratios, assuming a Poisson's Ratio of 0.2 for the uncracked rock. Crack densities varied with distance from the tunnel wall from 0.12 at the surface to 0.09 at one metre. Most of the recorded emissions were located outside of the array, however changes in velocity and in particular in amplitude were detectable in velocity scans before and after the tunnel advance.

Waveforms from the velocity scans in this experiment are being modelled with explicit representations of the cracking. Besides providing a further measure on how realistically such models capture the real wave behaviour, the modelling seeks to illuminate aspects about the fracturing not directly available from the seismic analysis. In particular the seismic velocities and crack density estimates do not provide a clear link to the actual fracture distribution and size of cracks. Also, since few new emissions were recorded inside the array volume, it is unclear whether changes in the waveforms with tunnel advance are due to new cracking or to a change in the stress field on the existing cracking. This section presents early results in modelling certain velocity scans.

The full acoustic array consisted of 4 parallel boreholes arranged in a diamond pattern. Each borehole contained 5 sensors spaced at approximately 0.2 m intervals, starting at 0.2 m from the tunnel surface. There were 23 sensors in total with 3 mounted on the wall of the tunnel. Sensors were oriented in the boreholes such that they faced the diagonally opposite borehole. The experiment was modelled in three dimensions, with model details given in Appendix A, Section 5. The model used a reduced 8×8 array and the positions of the sensors used are shown in Figure 12. The boreholes were rotated into a diamond pattern as shown with the first borehole at the top. The cube faces are the model boundaries and the element length is 10 mm. The tunnel is circular with a diameter of 3.5 m. The free surface is approximated with a flat surface in the model.

In all the results presented, the source is at sensor 3. A variety of sources were tested by exciting different components of stress and velocity. A good fit was given by simply exciting the normal velocity and recording this at the sensors, allowing for sensor orientation. Rather than trying to model the real source, which contains very high frequencies, the model source is based on the recorded waveforms where most high frequencies are completely attenuated. The amplitude of frequencies above 200 kHz is less than one tenth of the maximum amplitude, which occurs between 25 and 30 kHz. The model source peaks at approximately 60 kHz, and little content above 170 kHz. A typical limit for a fourth-order accurate finite difference scheme and an element size of 10 mm, is a maximum wavelength of 50 mm, which equates to

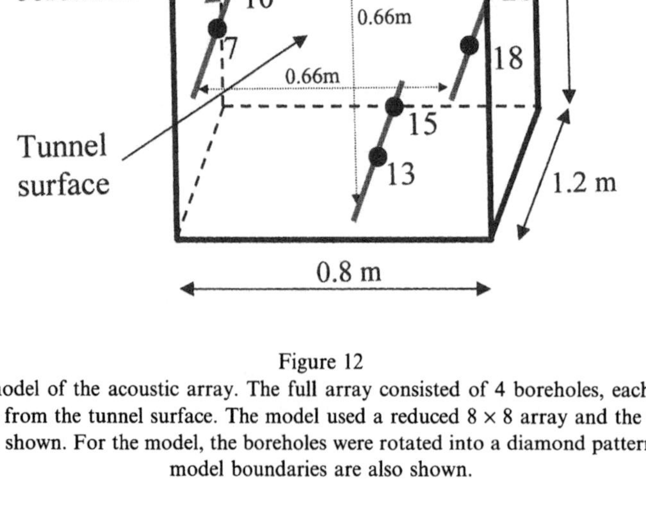

Figure 12
Sketch of the model of the acoustic array. The full array consisted of 4 boreholes, each with 5 sensors spaced at 0.2 m from the tunnel surface. The model used a reduced 8×8 array and the positions of the sensors used are shown. For the model, the boreholes were rotated into a diamond pattern as shown. The model boundaries are also shown.

maximum P- and S-wave frequencies of 120 kHz and 70 kHz. As a result some numerical dispersion can be observed in these models. The actual source wave shape is shown in Figure 13b, labelled "3:3".

A purely elastic model was constructed using P and S wavespeeds for the unfractured rock of 5890 m/s and 3425 m/s (YOUNG and COLLINS, 1997). Waveforms are compared with the recorded waveforms in Figure 13 for sensors 3, 7, 13 and 18. These all have wavepaths which are approximately parallel to the tunnel. P and S waves were found to match well, both in the arrival times and in relative amplitudes, although differences indicate a more complicated source waveshape. Figure 14 compares waveforms for wavepaths which are oblique to the tunnel (sensors 5, 10, 15 and 20). The modelled P and S arrivals are significantly earlier than the measured waveforms, and have greater amplitude errors than those parallel to the tunnel. This is consistent with the anisotropy shown in the seismic analysis (CARLSON and YOUNG, 1992). The arrival time and amplitude for the different sensors are compared in Table 2, and indicate larger errors in all the oblique paths.

It is proposed that the anisotropy is caused by micro-fracturing predominately parallel to the tunnel. Many models were investigated to account for the anisotropy with a fracture state. The smallest crack which can be represented explicitly in the model is 20 mm. The individual micro-cracks are expected to be of the order of a few

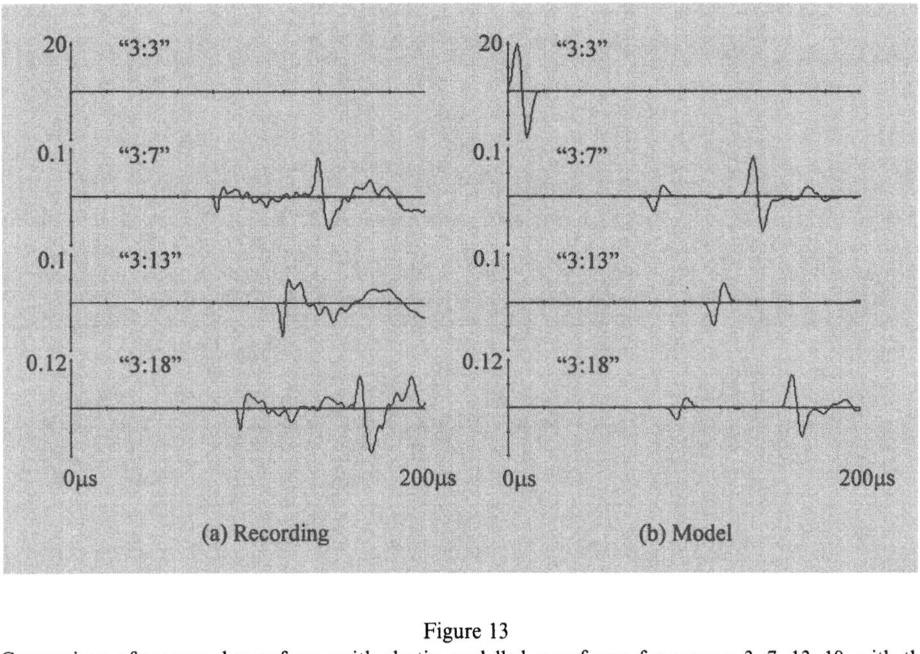

Figure 13
Comparison of measured waveforms with elastic modelled waveforms for sensors 3, 7, 13, 18, with the source at sensor 3. All waveforms are for a 200 μs window. Measured waveforms are in volts. Modelled waveforms are velocity but have been scaled so that amplitudes can be compared. The scaling factor was chosen so that the waveform for sensor 3 matches the amplitude of the experiment.

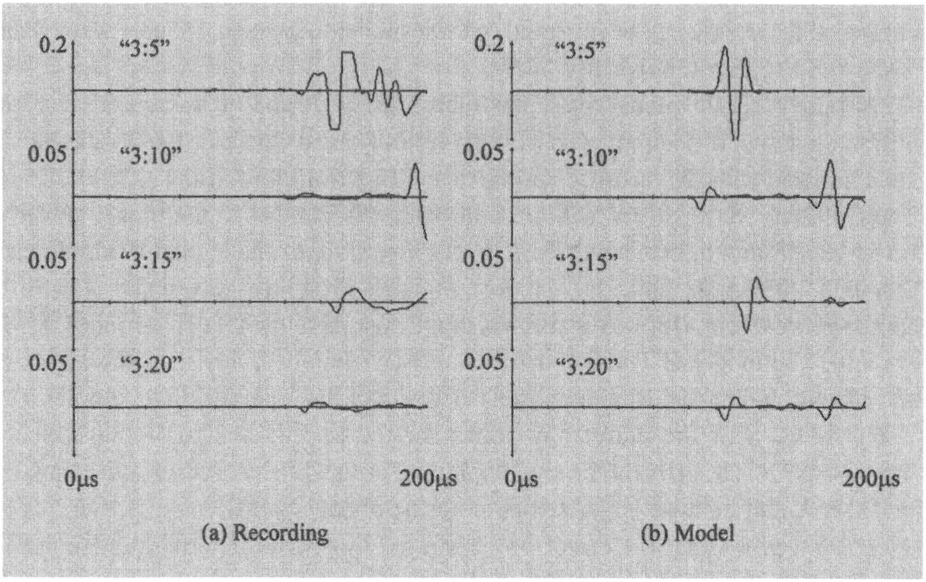

Figure 14
Comparison of measured waveforms with elastic modelled waveforms for sensors 5, 10, 15, 20, with the source at sensor 3. All waveforms are for a 200 μs window. Measured waveforms are in volts. Modelled waveforms are velocity but have been scaled so that amplitudes can be compared. The scaling factor was chosen so that the waveform for sensor 3 matches the amplitude of the experiment. Waveforms are labelled according to the source:receiver pair.

Table 2

Amplitude and time of first arrivals for both P and S waves, compared with the data from the acoustic emission experiment and the elastic model. The paths for sensors 7, 13 and 18 are parallel to the tunnel, while the paths for sensors 5, 10, 15 and 20 are oblique to the tunnel. (The error is expressed as the difference between the modelled and measured values as a percentage of the measured value.)

Sensor	Experiment				Model (%error)			
	P wave		*S* wave		*P* wave		*S* wave	
	Time (μs)	Ampl. (mV)	Time (μs)	Ampl. (mV)	Time (%err)	Ampl. (%err)	Time (%err)	Ampl. (%err)
3								
7	78.6	−31.3	133.5	83.4	−4	1	−1	−2
13	114.9	−72.0	–	–	−4	−31		
18	92.1	−54.6	158.5	83.0	−4	−43	−3	−0
5	73.6	−4.8	131.5	75.9	−11	−31	−13	150
10	117.9	−2.8	188.3	35.4	−15	370	−7	−12
15	142.5	−11.7	–	–	−11	180		
20	125.9	−5.4	–	–	−10	160		

millimetres, therefore it is not possible to represent micro-cracks in the full model. Two approaches were investigated. The first (Fig. 15a) is to use large extensive fractures with surfaces coupled by a fracture stiffness as studied in Section 3. This is based on the hypothesis that the wavelengths are substantially larger than the micro-cracks which can then be seen as the spacing of contacts within a larger crack. The peak frequency in the recorded waveforms corresponds to wavelengths of 200 mm for P waves and 120 mm for S waves.

The second approach is to suggest that the micro-fractures have coalesced into larger micro-fractures, and that these will have the greatest effect on the wave propagation, due to the long wavelengths involved. Crack distributions were generated to give a particular crack density, defined in O'CONNELL and BUDIANSKY (1974) as

$$\varepsilon = (2N/\pi)\langle A^2/P\rangle \ , \tag{16}$$

where the angular brackets denote an average, A is area, P is the perimeter and N is the number of cracks per unit volume, and ε is the crack density. In the models the cracks are rectangular, giving

$$\varepsilon = \frac{1}{\pi V}\sum \frac{A^2B^2}{A+B} \tag{17}$$

where A and B are the side lengths. It was verified that for square cracks this expression gives values in between the values for an inscribed circle and a

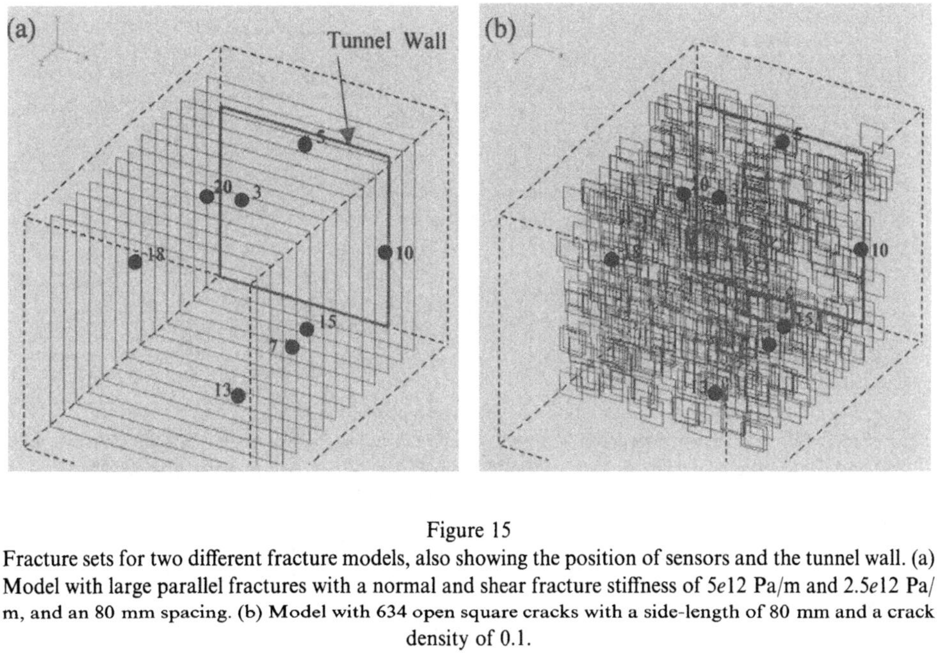

Figure 15
Fracture sets for two different fracture models, also showing the position of sensors and the tunnel wall. (a) Model with large parallel fractures with a normal and shear fracture stiffness of $5e12$ Pa/m and $2.5e12$ Pa/m, and an 80 mm spacing. (b) Model with 634 open square cracks with a side-length of 80 mm and a crack density of 0.1.

circumscribed circle, for which expressions are given in O'CONNELL and BUDIANSKY (1974). Figure 15b shows a model with a crack density of 0.1, with 538 square cracks, and an 80 mm side-length.

Figure 16 compares the results of three different fracture models for the wavepath from sensor 3 to sensor 10, indicating the effect on the *P*- and *S*-wave arrivals. The values of these arrivals and amplitudes are also compared in Table 3. The fractures should shift the waveform from the elastic case toward the recorded waveform. The

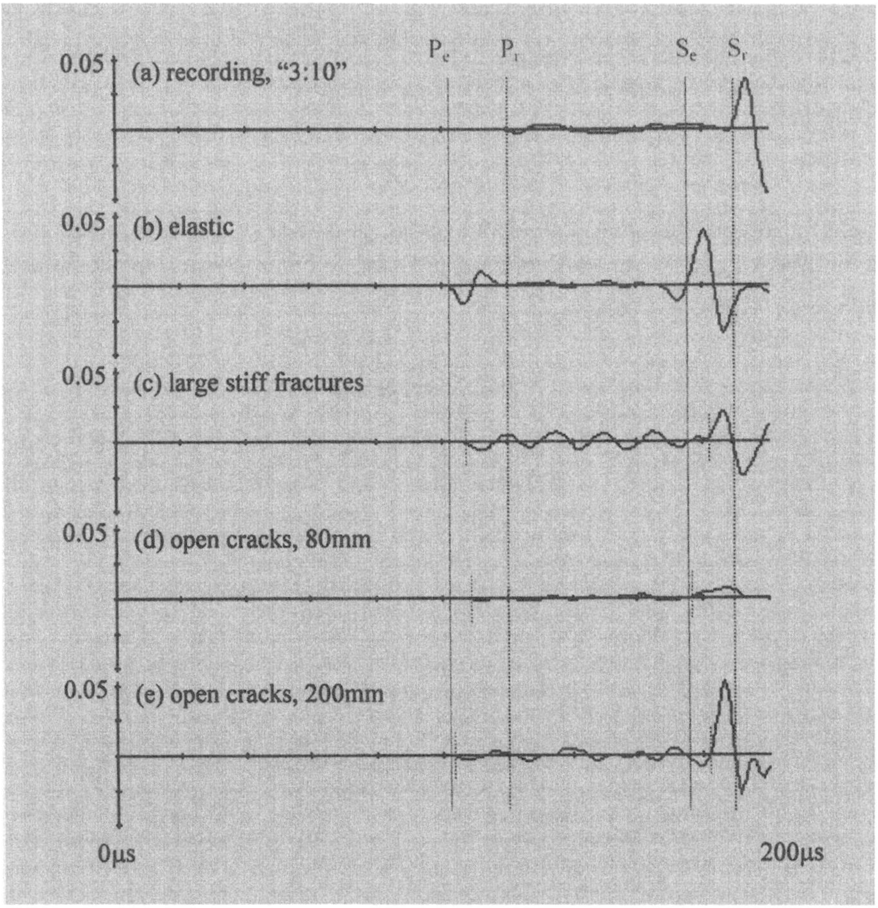

Figure 16
The effects of three different fracture sets on the modelled response for sensor 10, with the source at sensor 3. (a) and (b) are the recording and elastic model from Figure 13. (c) Model with large parallel fractures with a normal and shear fracture stiffness of 5*e*12 Pa/m and 2.5*e*12 Pa/m and an 80 mm spacing. (d) Model with 634 open square cracks with a side-length of 80 mm and a crack density of 0.1. (e) Model with 41 open square cracks with a side-length of 200 mm and a crack density of 0.1. In all cases fractures are parallel to the tunnel surface. The measured waveform is in volts, while modelled waveforms are velocity and have been scaled to the data. P_r, S_r, and P_e, S_e are the *P* and *S* arrivals of the measured waveform and elastic model, respectively.

Table 3

Effect of the different models of fracturing on the amplitude and time of first arrivals for P and S waves, for sensor 10 in the acoustic emission experiment. (The error is expressed as the difference between the modelled and measured values as a percentage of the measured value.)

Model	*P* wave		*S* wave		*P* wave		*S* wave	
	Time (μs)	Ampl. (mV)	Time (μs)	Ampl. (mV)	Time (%err)	Ampl. (%err)	Time (%err)	Ampl. (%err)
Experiment	117.9	−2.8	188.3	35.4				
Elast	100.0	−13.1	174.9	39.6	−15	370	−7	12
Large stiff fractures	103.8	−6.3	181.0	21.5	−12	125	−4	−40
Open 80 mm cracks	103.1	1.3	174.9	7.2	−13	−54	−7	−80
Open 200 mm cracks	100.8	3.0	181.7	52.2	−15	8	−4	47

first fracture model contains large fractures coupled by a fracture stiffness and spaced at 80 mm, and has the greatest effect on the waveform causing significant delays in both the *P* and the *S* arrival. Greater *P*-wave attenuation and delay is still required to match the experiment. The second model contains a random set of open 80 mm square cracks with a fracture density of 0.1, and also causes delayed arrivals, but excessively attenuates the waves. The third model contains a random set of open 200 mm square cracks and causes little delay in the arrivals however with very high attenuation. These observations were consistent when considering the other paths from Figure 14 which are oblique to the tunnel.

These results suggest that the fracture stiffness approach is more likely to account for both the amplitude and arrival effects, perhaps using more densely spaced fractures. The cases of open fractures with a particular density imply that the real openings are considerably smaller than those modelled, as there is a greater effect on attenuation than on arrival.

6. Conclusions

This work has presented a variety of different studies at different scales of the problem but with an overriding theme to provide a measure of how well current models of wave propagation in rock match the real behaviour of waves around deep excavations.

The first section shows some of the difficulties associated with modelling on the true problem-size scale. Although the study was based on a controlled event with an engineered source and *a priori* knowledge, lack of knowledge of the source became an overriding factor in interpreting the model behaviour. The main discrepancy was that a large shear and surface wave was generated in the model which was not present in the data. Lack of certainty in the complex blast source meant that the differences could not definitely be ascribed to the medium and the interaction with fracturing. Nevertheless important conclusions were noted in that the small calibration blast produced waves travelling close to the skin of the excavation, and modelling these motions proved successful with an elastic medium. Realising goals outlined in the introduction requires that once damaging waves have interacted with the near excavation, the models correctly predict the waves which are transmitted to other parts of the excavation. It is noteworthy then, that modelling the propagation of large amplitude waves along the excavation surface proved significantly less successful than modelling the propagation of small amplitude waves.

A number of conclusions can be made regarding the modelling of fractures. Earlier experimental and theoretical work validating the displacement discontinuity model of a fracture is supplemented by demonstrating numerically that the model not only produces similar frequency-dependent effects on wavespeed and attenuation, but that it can accurately reproduce the waveforms. The work highlights that for realistic wave propagation it is important that models of fractures include a fracture stiffness coupling the surfaces of the fracture, rather than considering fractures as simply open or closed. An important extension is the finding that where the local stress state can vary across a single fracture, that fracture must be modelled with a fracture stiffness that is stress-dependent. This has important implications for modelling the fracture zone around openings where the stress state is highly non-uniform.

Modelling the fractures in *in situ* rock show important preliminary findings. Although the modelling could not practically represent sufficiently small fractures, results did indicate that the sizes modelled were unrealistically large. Such models could give some upper bounds to estimates of fracture size. The fracture stiffness approach seems viable to model a distribution of micro-cracks for wavelengths greater than their separation. A two-part approach may be possible where the data is matched with a distribution of macro-cracks with fracture stiffness, and more detailed models relate the fracture stiffness to a distribution of micro-cracks.

Acknowledgements

I acknowledge Prof. L.J. Pyrak-Nolte for use of her published data, as well as for beneficial discussions. The modelling and developments in this paper have been performed while the first author pursues a Ph.D. degree at Liverpool University in the United Kingdom, under the supervision of Professor R.P. Young, on the subject of seismic wave interaction with underground openings. This work is supported under the project GAP 601, at the CSIR Division of Mining Technology, and is sponsored by SIMRAC (Safety in Mines Research Advisory Committee). This sponsorship is gratefully acknowledged.

Appendix A:

Brief Summary of the Models Used in Each Section

(i) Section 1

The size of the modelled region was $(x, y, z) = (280\text{ m}, 120\text{ m}, 360\text{ m})$. 1.55 million grid-points were used, with a 2 m spacing, and 4th-order spatial accuracy. The elastic parameters were $E = 80$ GPa, $v = 0.2$, $\rho = 2700\ \text{kg/m}^3$, $c_p = 5738$ m/s, and $c_s = 3514$ m/s. The event modelled was a vertical slip event in the footwall of the pillar. The source centre was 1 m into the pillar, 12 m below the plane of the stope, and 12 m back from the stope face. The region of slip was a square area 28 m by 28 m extending upwards to the plane of the stope and to the stope face. The rupture was propagated from the centre at 3000 m/s. The stress drop at a point was 9.35 MPa, and occurred over 2 ms. Further information pertinent to tabular mining layouts and associated rockbursts can be found in HANDLEY *et al.* (1997), MALAN (1999), or JAGER and RYDER (1999).

(ii) Section 3

The size of the modelled region was $(x, y, z) = (14\text{ m}, 12\text{ m}, 38\text{ m})$. 0.8 million grid-points were used, with a 0.2 m spacing and 4th order spatial accuracy. The elastic parameters were $E = 80$ GPa, $v = 0.2$, $\rho = 2700\ \text{kg/m}^3$, $c_p = 5738$ m/s, and $c_s = 3514$ m/s. The tunnel dimension was $(x, y, z) = (3.6\text{ m}, 3.6\text{ m}, 37\text{ m})$. Each blast hole was modelled as a line of pressure propagating at the velocity of detonation of the blast. The experimental blast comprised five blast holes varying in length from 4 m to 7 m, and with a velocity of detonations of 3600 m/s or 3800 m/s, and with the five detonations staggered by 71 μs. The position, dimension and blast parameters for each blast hole are given in HILDYARD and MILEV (1999), for both the main experiment and the calibration blast.

(iii) Section 4

The size of the unfractured model was $(x, y, z) = (90\text{ mm}, 100\text{ mm}, 90\text{ mm})$.The size of the fracture models was $(x, y, z) = (90\text{ mm}, 90\text{ mm}, 90\text{ mm})$. The elastic parameters were $E = 210$ GPa, $v = 0.3$, $\rho = 7800\ \text{kg/m}^3$, $c_p = 6020$ m/s, and $c_s = 3218$ m/s. The normal and shear fracture stiffness for uniform fracture stiffness models was $k_n = 6e13$ Pa/m and $k_s = 2e13$ Pa/m, respectively. Sources were applied over a 22 mm square region, and the received signal was averaged over a 22 mm square region.

- *P*-wave fracture models used 0.75 million grid-points with a 1 mm spacing and 4th order spatial accuracy. The source excited the vertical stress (σ_{yy}).

– S-wave fracture models used 5.96 million grid-points with a 0.5 mm spacing and 4th order spatial accuracy. The source excited shear stress, σ_{xy} or σ_{yz} depending on the polarisation considered.

Coupled models applied a stress of 68 MPa to the source area (a square region of 22 mm) on the y-z and x-z faces of the block, based on an experimental biaxial load of 30 kN. The stress-dependent stiffness was modelled as $k_n = a/(a - b(-u_n^{\mathrm{rel}}))^2$, where $-u_n^{\mathrm{rel}}$ is the relative displacement or closure of the two crack surfaces, $a = 1e{-}13$ m/Pa, $b = 4.38e{-}8$ Pa^{-1}, and with $k_s = k_n/3$.

(iv) Section 5

The size of the modelled region was $(x, y, z) = (800$ mm, 800 mm, 1200 mm). 6.25 million grid-points were used, with a 5 mm spacing and 4th order spatial accuracy. The elastic parameters were $E = 76.8$ GPa, $\nu = 0.245$, $\rho = 2630$ kg/m^3, $c_p = 5890$ m/s, and $c_s = 3425$ m/s. Sources and receivers covered a 10 mm square region, and were modelled as velocity normal to the sensor orientations (i.e. normal to the borehole, and facing inward toward the centre of the array (cf. Fig. 12)). The source wave shape is approximately a "full-wave" pulse over 15 μs, and is constructed from the function $-a\tau e^{-15\tau^2}$, where $a \approx 9.03$, $\tau = t/T$, $T = 15e{-}6$, and t is time in seconds. If the values in the figures are interpreted as 'mm/s', then the source has a maximum amplitude of 20 mm/s. The open crack models consisted of 41 200 mm × 200 mm randomly spaced cracks or 634 80 mm × 80 mm cracks in a volume of 770 mm × 770 mm × 870 mm, equating to a crack density of 0.1. The model with extensive stiff fractures contained 12 fractures with a spacing of 80 mm, and with a normal and shear fracture stiffness of $k_n = 5e12$ Pa/m and $k_s = 2.5e12$ Pa/m, respectively.

References

Adams, G. R. and Jager, A. J. (1980), *Petroscopic Observations of Rock Fracturing Ahead of Stope Faces in Deep-level Gold Mines*, J. South African Inst. Min. Metall. *80*(6), 204–209.

Bandis, S. C., Lumsden, A. C. and Barton, N. R. (1983), *Fundamentals of Rock Joint Deformation*, Int. J. of Rock Mech. *20*(6), 249–268.

Cai, J. G. and Zhao, J. (2000), *Effects of Multuiple Parallel Fractures on Apparent Attenuation of Stress Waves in Rock Masses*, Int. J. Rock Mech. and Mining Sci. *37*(4), 661–682.

Carlson, S. R. and Young, R. P. (1992), *Acoustic Emission and Ultrasonic Velocity Study of Excavation-induced Microcrack Damage in the Mine-by Tunnel at the Underground Research Laboratory*, Internal Report #RP015AECL, pp. 901–907.

Carlson, S. R. and Young, R. P. (1993), *Acoustic Emission and Ultrasonic Velocity Study of Excavation-induced Microcrack Damage at the Underground Research Laboratory*, Int. J. of Rock Mech. *30*(7) 901–907.

Crampin, S. (1981), *A review of wave motion in anisotropic and cracked elastic-media*. In *Wave Motion*, vol. 3, pp. 343–391.

Cundall, P. A. (1992), *Theoretical Basis of the Program WAVE*, Unpublished Internal Report, COMRO (now CSIR Division of Mining Technology, CSIR, South Africa), pp. 1–12.

DAEHNKE, A. (1997), *Stress Wave and Fracture Propagation in Rock*, Ph. D. Thesis, Vienna University of Technology.

Graves, R. W. (1996), *Simulating Seismic Wave Propagaion in 3D Elastic Media Using Staggered-grid Finite Differences,* Bull Seismol. Soc. Am. *86*(4), 1091–1106.

GU, B. L., NIHEI, K. T., and MYER L. R. (1996), *Numerical Simulation of Elastic Wave Propagation in Fractured Rock with the Boundary Integral Equation Method,* J. Geophys, Res. *101*(B7), 15,933–15,943.

HANDLEY, M. F., HILDYARD M. W., and SPOTTISWOODE, S. M. (1996), *The influence of deep mine stopes on seismic waves*, Proc. 2nd North American Rock Mechanics Symposium Montreal, pp. 499–506, 19–21 June, 1996.

HANDLEY, M. F., SELFE, D. A., VIEIRA, F. M. C. C., MACCELARI M. J., and DEDE, T. (1997), *Current Position of Strike Stabilising Pillar and Bracket Pillar Design Guidelines for Deep Level Tabular Orebodies*, J. South African Inst. of Mining and Metall. *97*(3), 103–118.

HILDYARD, M. W., DAEHNKE, A., and CUNDALL, P. A. (1995), WAVE: *A computer program for investigating elastodynamic issues in mining*, Proc. of the 35th US Symp. on Rock Mech., pp. 519–524

HILDYARD, M. W. and MILEV, A. M. (1999), *Modeling seismic wave interaction with a tunnel due to an artificial rockburst*, Proc. 2nd SARES (2nd South African Rock Engin. Symp.) (ed. Hagan, T. O.) pp. 273–280.

HUDSON J. A. (1981), *Wave Speeds and Attenuation of Elastic Waves in Material Containing Cracks*, Geophys. J. Roy. Astr. Soc. *64*, 133–150.

JAGER, A. J. and RYDER, J. A., *A Handbook on Rock Engineering Practice for Tabular Hard Rock Mines*, (published by The safety in Mines Research Advisory Committee (SIMRAC), Johannesburg, South Africa, 1999).

KOUZNIAK, N. and ROSSMANITH, H. P. (1998), *Supersonic Detonation in Rock Mass – Analytical Solutions and Validation of Numerical Models – Part 1: Stress Analysis*, FRAGBLAST (Journal) *2*(4), 449–486

LEVANDER, A. R. (1988), *4th-order Finite-difference P-SV Seismograms*, Geophys. *53* (11), 1425–1436.

LIU, E. R., HUDSON, J. A., and POINTER, T. (2000), *Equivalent Medium Representation of Fractured Rock*, J. Geoph. Res. *105*(B2), 2981–3000.

LYSMER, J. and KUHLEMEYER, R. (1969), *Finite Dynamic Model for Infinite Media*, J. Eng. Mech., ASCE *95* (EM4), 859–877.

MADARIAGA, R. (1976), *Dynamics of an Expanding Circular Fault*, By. Seismol. Soc. Am. *66*(3), 639–666.

MALAN, D. F. (1999) *Time-dependent Behaviour of Deep Level Tabular Excavations in Hard Rock*, Rock Mech. Rock Eng. *32*(2), 123–155.

MENDECKI, A. J. (1993), *Real time quantitative seismology in mines, keynote lecture* In Proc. 3rd Int. Symp. *Rockbursts and Seismicity in Mines* (ed. Young, R. P.), (Balkema, Rotterdam 1993) pp. 261–266.

MILEV, A. M., SPOTTISWOODE, S. M., HILDYARD M. W., RORKE, A. J., and FINNIE, G. J. (2000), *Simulated Rockburst – source design, seismic effect and damage*, Proc. of the 4th North American Rock Mech. Symp, July 31–August 3, 2000, Seattle. U.S.A.

MYER, L. R., HOPKINS, D., PETERSON, J. E., and COOK, N. G. W., *Seismic wave propagation across multiple fractures*. In *Fractured and Jointed Rock Masses* (eds. Myer, L. R., Cook, N. G. W., Goodman, R. E., and Tsang, P.), (Balkema, Rotterdam 1995) pp. 105–109.

NAPIER, J. A. L., DAEHNKE, A., DEDE, T., HILDYARD, M. W., KUIJPERS, J. S., MALAN, D. F., SELLERS E. J., and TURNER, P. A. (1997), *Quantification of Stope Fracture Zone Behaviour in Deep Level Gold Mines*, J. South African Inst. Min. Metall. *97*(3), 119–134.

O'CONNELL, R. J. and BUDIANSKY, B. (1974), *Seismic Velocities in Dry and Saturated Cracked Solids*, J. Geophys. Res. *79*, 5412–5426.

PYRAK-NOLTE, L. J., MYER, L. R., and COOK, N. G. W. (1990a), *Transmission of Seismic Waves Across Single Natural Fractures*, J. Geophys. Res. *95*(B6), 8617–8638.

PYRAK-NOLTE, L. J., MYER, L. R., and COOK, N. G. W. (1990b), *Anisotropy in Seismic Velocities and Amplitudes from Multiple Parallel Fractures*, J. Geophys. Res. *95*(B7), 11,345–11,358.

REDDY, N. and SPOTTISWOODE, S. M. (2000), *The Influence of Geology on a Simulated Rockburst Experiment*, J. South African Inst. of Min. Metall, submitted.

SAYERS, C. M. and KACHANOV, M. (1991), *A Simple Technique for Finding Effective Elastic-constants of Cracked Solids for Arbitrary Crack Orientation Statistics*, Int. J. Solids Struct. *27*(6), 671–680.

SCHOENBERG, M. (1980), *Elastic Wave Behaviour Across Linear Slip Interfaces*, J. Acoust. Soc. Am. *68*(5), 1516–1521.

VIRIEUX, J. (1986), *P-SV Wave Propagation in Heterogeneous Media – Velocity-stress Finite-difference Method*, Geophys. *51*(4), 889–901.

YOUNG, R. P., and COLLINS, D. S. (1997), *Acoustic Emission/Microseismicity Research at the Underground Research Laboratory, Canada* (1987–1997), Internal Report #RP037AECL, 1–126.

(Received July 14, 2000, Revised/accepted January 15, 2001)

Pure appl. geophys. 159 (2002) 277–307
0033–4553/02/030277–31 $ 1.50 + 0.20/0

Pure and Applied Geophysics

Tomographic Imaging of Thermally Induced Fractures in Granite Using Bayesian Inversion

WOJCIECH DĘBSKI[1] and R. PAUL YOUNG[2]

Abstract — The internal structure of rock samples studied in laboratory experiments can be described by a variety of physical parameters. Some of them, like the velocity of acoustic waves, enhanced velocity or quality factor can be reconstructed by means of ultrasonic tomography. This article presents the results of classical velocity tomography imaging, accompanied by the results of attenuation tomography and recently introduced enhanced velocity tomography obtained for a Lac Du Bonnet granite sample subjected to thermal stresses. To invert acoustic data recorded during six heating cycles, a Bayesian inversion scheme accompanied by a genetic algorithm optimization approach and the robust Cauchy norm have been used. To obtain the highest possible spatial resolution of images the inversion was performed in two steps. In the first step a crude parameterization of the sample was used. The result of this stage was next taken as an *a priori* model for a final inversion with refined parameterization. The choice of parameterization (cell sizes) and damping parameters at both stages was based on an analysis of the resolution operator. Both velocity and enhanced velocity tomography accurately imaged changes in the rock microstructure caused by thermal stresses. However, enhanced velocity tomography gave a much better spatial resolution than velocity tomography. On the other hand, attenuation tomography based on inversion of pulse rise times was able to image only a rough structure of the sample and it has difficulty with reasonable imaging of the crack formed in the sixth heating cycle.

Key words: Bayesian inversion, genetic algorithm, ultrasonic tomography, attenuation, thermal fracturing.

1. Introduction

Knowledge of the internal structure of rock samples, or more generally rock masses, is very important for both theoretical and practical applications. This includes information about the spatial distributions of mechanical, electrical, and other physical parameters and their change in time or in response to a change of external conditions.

One of the most effective methods used to accumulate the desired information concerning the mechanical properties of rocks is active acoustotic tomography. It

[1] Institute of Geophysics, Polish Academy of Sciences, ul. Księcia Janusza 64, 01-452 Warsaw, Poland. E-mail: debski@igf.edu.pl

[2] Department of Earth Sciences, Liverpool University, 4 Brownlow Street, Liverpool L69 7GP, UK.

assumes that waves used to probe a medium have short enough wavelengths in comparison to the heterogeneities of the imaged media and they propagate as rays (HANYGA and SEREDYŃSKA, 2000; ČERVENÝ *et al.*, 1997; AKI and RICHARDS, 1985). Different characteristics of recorded ultrasonic waveforms may be used to reconstruct the spatial distributions of various mechanical characteristics of rocks. The two most popular techniques, namely velocity and attenuation tomography, are based on recording two of the most obvious properties measured signals: energy propagation time and energy dissipation due to the non-elasticity of real materials. A new tomography imaging technique was put forward only recently by DĘBSKI and YOUNG (1999). It consists of a combination of both velocity and attenuation tomography in one scheme to maximize the advantage of the robustness of velocity and the sensitivity to the micro-structure of attenuation.

Active acoustic velocity tomography is used routinely to image the velocity distribution, subsequently interpreted in terms of rock micro-structure (KANEKO *et al.*, 1979; JANSEN *et al.*, 1993; YOUNG and MAXWELL, 1992; MAXWELL and YOUNG, 1996; FRIEDEL *et al.*, 1996). The results obtained in this way are known to be quite robust (see e.g., FALLS, 1993; JANSEN *et al.*, 1993).

Another promising application of the ultrasonic technique is attenuation tomography (GIBOWICZ and KIJKO, 1994; AKI and RICHARDS, 1985) which is used to image the acoustic attenuation structure of a rock described by the quality factor Q. However, this technique is considerably more difficult than velocity tomography. This is because attenuation of acoustic waves is a far more subtle effect than the variation of delays of energy arrival times due to velocity heterogeneities. There are many factors (see, for example, PETTITT, 1997; DĘBSKI, 1997b; SIATA, 1999; WATANABE and SASSA, 1996; FEUSTEL and YOUNG, 1993) that can easily disturb attenuation measurements, consequently it is difficult to obtain a reliable Q image. For this reason, a very high quality of data used for attenuation tomography must be ensured (WATANABE and SASSA, 1996; FEUSTEL and YOUNG, 1993). Nevertheless, the additional effort necessary to obtain an image of Q is worthwile because Q is regarded to be more sensitive to the microcrack structure than acoustic wave velocity. This was reported, for example, by KANEKO *et al.* (1979), MURPHY (1984), YOUNG and HILL (1986), KLIMENTOS and MCCANN (1990), BEST *et al.* (1994), PEACOCK *et al.* (1994), WATANABE and SASSA (1996), BEST and SAMS (1997), and SHATILO *et al.* (1998).

Many algorithms have been developed to date for processing tomography data. Ultrasonic tomography as a small-scale inverse problem offers a unique possiblity to test new inversion techniques. One such encouraging and powerful technique that allows the use of global optimization algorithms for example, is the Bayesian inverse theory (TARANTOLA, 1987; DĘBSKI, 1997a; DUIJNDAM, 1998). It consists in building the *a posteriori* probability density over space of all models, which describes the probability of each given model being the true one. The model for which the *a posteriori* probability attains the global maximum is called the *maximum likelihood solution* and is often regarded as the solution of the inverse problem (GOUVEIA and SCALE, 1998; SCALES and SMITH, 1995; TARANTOLA, 1987; JACKSON and MATSU'URA, 1985). Searching for the *maximum likelihood solution* in fact only reduces the Bayesian method to a classical-like optimization approach enhanced by the possibility of using different norms and new regularization schemes (PARKER, 1994; DUIJNDAM, 1988; TARANTOLA, 1987). Within the framework of this technique, the regularization of an inversion procedure is achieved by using *a priori* information represented by a probability distribution over the model space (TARANTOLA, 1987; PENDOCK, 1993; BARNES *et al.*, 1996; CASSIANI *et al.*, 1998; EPPSTEIN and DOUGHERTY, 1998).

A few questions arise in the context of the Bayesian inversion technique. The first one is the choice of the optimization method used to find the maximum likelihood model. In a classical approach, a version of the damped least-squares algorithm is most often chosen for tomography imaging because of the simplicity and high speed of calculations (LUTTER *et al.*, 1999; BOSHI and DZIEWONSKI, 1999; NOLET, 1987). However, the least-squares algorithm is known to be non-robust (KIJKO, 1994; MENKE, 1989; SCALES *et al.*, 1988; TARANTOLA, 1987; DĘBSKI, 1999) which limits its application only to cases when high quality data can be assured. This requirement is usually not a problem for velocity imaging but can rarely be fulfilled in an attenuation tomography study (SIATA, 1999; FEUSTEL, 1998; GAO, 1997; DĘBSKI, 1997b; WATANABE and SASSA, 1996). As our aim was to perform both velocity and attenuation tomography, we decided to use a robust method based on an evolutionary optimization strategy approach to search for maximum likelihood parameter distributions. This optimization method, when combined with the choice of a robust norm like the l_1 norm, Cauchy norm, etc. is expected to overcome the problems of the classical least-squares-based algorithms (KOPER *et al.*, 1999; MIETTINEN *et al.*, 1999; MICHALEWICZ, 1996; SEN and STOFFA, 1995; ZHOU *et al.*, 1995; BICHKAR *et al.*, 1998).

Another question within the context of Bayesian inversion techniques is the choice of a probability distribution describing *a priori* information. In the study presented here we performed data inversion in two steps, using a multi-scale technique (see, for example, JIN and BEYDOUN, 2000; BUNKS *et al.*, 1995; WILLIAMSON, 1990). In the first stage, when a crude parameterization of the sample was used, constant velocity, enhanced velocity, and Q distributions were taken as *a priori* models. In the next step the results of the previous inversion with a "crude" parameterization were used as an *a priori* model for a more detailed inversion with a finer grid and with an increased smoothing (damping) factor. Such a procedure, according to the Bayesian theory, allows more efficient use of information contained in data, provided it is accompanied by an appropriate change of weights assigned to *a priori* and experimental data (WANG and BRAILE, 1996; SCALES and SMITH, 1995).

The goal of this paper is twofold. Initially it presents the results of ultrasonic tomography of a Lac Du Bonnet granite sample subjected to induced damage by heating. The analysis covers images of velocity, enhanced velocity, and Q factor. Secondly, we present the method for a robust tomography imaging incorporating resolution analysis for an optimum parameterization of the imaged medium.

The paper starts with an introduction to the Bayesian inverse theory. Next, basic ideas, assumptions, and limitations of velocity tomography accompanied by a discussion of attenuation tomography based on inversion of pulse rise times and enhanced velocity tomography are presented. Thereafter, the experiment with a Lac Du Bonnet granite sample and the algorithm used for data inversion are described. The paper concludes with the discussion of the derived results.

2. Theory

Bayesian Inversion

The solution of the general inverse problem according to the Bayesian inverse theory consists in building the *a posteriori* probability distribution $\sigma(\mathbf{m})$ over the model space which describes the probability of a given model $(\mathbf{m})$ being the true one (Tarantola, 1987; Dębski, 1997a). It has been shown by Tarantola and Vallete (1982) that, in general, $\sigma(\mathbf{m})$ is the product of a distribution $f(\mathbf{m})$ describing *a priori* information by a *likelihood function* $L(\mathbf{m})$ which measures to what extent theoretical prediction fit the observed data:

$$\sigma(\mathbf{m}) = \text{const.} f(\mathbf{m}) L(\mathbf{m}) \tag{1}$$

where the constant represents a normalization of the probability density and the *likelihood function* $L(\mathbf{m})$ reads (see, for instance, Tarantola 1987; Dębski, 1997a; Duijndam, 1988)

$$L(\mathbf{m}) = \exp\{-\|\mathbf{t}^{obs} - \mathbf{t}^{th}(\mathbf{m})\|\} \ . \tag{2}$$

The symbol $\|\cdot\|$ stands for a norm which is employed to measure the "distance" between observed ($\mathbf{t}^{obs}$) and predicted ($\mathbf{t}^{th}$) data. The latter are solutions of the forward modeling problem

$$\mathbf{t}^{th} = \mathbf{G}(\mathbf{m}) \ . \tag{3}$$

The complete construction of $\sigma(\mathbf{m})$ is impossible in the case of a tomography study due to the large size of the model space. In such a situation, the classical solution of a tomography problem consists in finding a model for which $\sigma(\mathbf{m})$ reaches its maximum. The $\mathbf{m}^{ml}$ model found in this way is called the *maximum likelihood solution*.

In the present studies the Cauchy norm was assumed for both $f(\mathbf{m})$ and $L(\mathbf{m})$ (Dębski, 1997a)

$$\|\mathbf{t}^{obs} - \mathbf{t}^{th}\| = \sum_i \ln\left\{1 + \left(\frac{t_i^{obs} - t_i^{th}}{\sigma_D^i}\right)^2\right\} \tag{4}$$

with constant data $\sigma_D^i = \sigma_D$ and model $\sigma_M^i = \sigma_M$ variances. This norm is known to be considerably more robust than the l_2 (Gaussian) norm, so the inversion results are less sensitive to the presence of outliers in the data and to modeling errors (DĘBSKI, 1997a; NOBLE, 1992; TARANTOLA, 1987). $L(\mathbf{m})$ and $f(\mathbf{m})$ for this particular choice read:

$$L(\mathbf{m}) = \prod_i \frac{1}{1 + \left(\frac{t_i^{th}(\mathbf{m}) - t_i^{obs}}{\sigma_D}\right)^2} \tag{5}$$

$$f(\mathbf{m}) = \prod_j \frac{1}{1 + \left(\frac{m_j - m_j^{apr}}{\sigma_M}\right)^2} \ . \tag{6}$$

For this special choice, the $\mathbf{m}^{ml}$ solution can be found through a numerical search for the maximum of $\sigma(\mathbf{m})$.

To solve this optimization problem and to ensure that the maximum found is the global maximum, a version of the evolutionary strategy approach (enhancement of classical genetic algorithms) (MICHALEWICZ, 1996), described further, was applied.

It often happens in tomography imaging that there exists a null subspace of the forward modeling operators (ROWBOTHAM and PRATT, 1997; DEAL and NOLET, 1996) or, in other words, there exists a subset $\mathbf{m}^{nul}$ of the set of all $\mathbf{m}$ such that

$$\mathbf{G}(\mathbf{m}^{nul}) = 0 \tag{7}$$

This circumstance typically arises if there is a region of the imaged body poorly covered by rays with respect to the spatial parameterization used (see, for example, CHANG, 1996). In case of parameterization by division of the body into homogeneous cells, it may ensue when the cells are too small that some are probed by no rays. As it follows from eqs. (2), (3) and (7), the likelihood function is then independent of $\mathbf{m}^{nul}$ and the values assigned to $\mathbf{m}^{nul}$ by searching for the maximum of $\sigma(\mathbf{m})$ are determined by the form of the distribution $f(\mathbf{m})$. In our case, as it follows from equation (6), they will be determined by $\mathbf{m}^{apr}$ model.

The essential part of any inverse problem is the analysis of imaging errors. The Bayesian approach providing the solution of the inverse problem as the *a posteriori* probability distribution $\sigma(\mathbf{m})$ allows formation of any analysis of the imaging errors, correlation, resolution, etc. However, if the method is used only to search for $\mathbf{m}^{ml}$, an additional error analysis is required. The simplest one is based on the analysis of the properties of the *a posteriori* covariance matrix calculated assuming linearization of the forward problem around $\mathbf{m}^{ml}$

$$\mathbf{G}(\mathbf{m}) \approx \mathbf{G}(\mathbf{m}^{ml}) + \mathbf{A} \cdot (\mathbf{m} - \mathbf{m}^{ml}) \ . \tag{8}$$

Then the *a posteriori* covariance matrix may be approximated by the least-squares-based formula for a linear tomography in which, for constant C_D, the covariance matrix reads (DĘBSKI, 1997a; MENKE, 1989; TARANTOLA, 1987):

$$\mathbf{C}^{post} = C_D\left(\mathbf{A}^T\mathbf{A} + C_D\mathbf{C_M}^{-1}\right)^{-1} . \tag{9}$$

Within the same set of assumptions the resolution operator **R**, which by definition maps the true model $\mathbf{m}^{true}$ onto the estimated one ($\mathbf{m}^{ml}$) (BACKUS and GILBERT, 1967; MENKE, 1989):

$$\mathbf{m}^{ml} - \mathbf{m}^{apr} = \mathbf{R} \cdot (\mathbf{m}^{true} - \mathbf{m}^{apr}) \tag{10}$$

reads (MENKE, 1989; TARANTOLA, 1987)

$$\mathbf{R} = \mathbf{A}^T\mathbf{A} \cdot \left(\mathbf{A}^T\mathbf{A} + C_D\mathbf{C_M}^{-1}\right)^{-1} \tag{11}$$

The formula of eq. (9) may be used to estimate the imaging error even in the case of a sequential data inversion, as is the case, for example, with the multi-scaling technique (SCALES and SMITH, 1995). Thus, according to the Bayesian inverse theory, the *a posteriori* covariance matrix of the current inversion becomes the *a priori* one ($\mathbf{C_M}$) for the subsequent data inversion (SCALES and SMITH, 1995).

In the case of perfectly resolved parameters, the resolution operator **R** is the unit matrix (BACKUS and GILBERT, 1967; MENKE, 1989) and, according to eq. (10), $\mathbf{m}^{ml}$ is a good estimation of the sought $\mathbf{m}^{true}$ model. Otherwise the parameters are not resolved independently because the data supply insufficient information for the full estimation of $\mathbf{m}^{true}$. In this situation, according to Backus and Gilbert (1967), we can obtain only the filtered version of the sought $\mathbf{m}^{true}$ model.

Although the calculation of $\mathbf{C}^{post}$ and/or **R** is cumbersome generally (NOLET *et al.*, 1999; YAO *et al.*, 1999) due to the model and data space sizes, it is relatively easy to find them in small-scale inverse problems such as the one discussed in this paper.

Imaging Physical Properties

The methodology described above can be applied to obtain the spatial distribution of any physical parameter by inversion of an appropriate set of data. All we need is a way to calculate the predictions for a given model which may be compared with the measured data. As regards tomography imaging, this forward modeling task requires calculations of line integrals of a selected parameter along ray paths.

In this article we are dealing with three different parameters inverting three different sets of data read from recorded ultrasonic waveforms. They are as follows: the *P*-wave velocity obtained by inversion of the first arrival times of *P* waves, the enhanced velocity reconstructed from the pulse maximum arrival times, and the quality factor Q estimated by inversion of the pulse rise times. For the sake of completeness, below we describe the velocity tomography, attenuation tomography based on inversion of pulse rise times, and the newely introduced enhanced velocity tomography (DĘBSKI and YOUNG, 1999; DĘBSKI, 2001).

Velocity Tomography

Velocity tomography imaging belongs to the class of inverse problems as the specific case for which the forward modeling problem is determined by the line integral of slowness (reciprocity of velocity) (AKI and RICHARDS, 1985; MENKE, 1989; LEE and PEREYRA, 1993):

$$t^{th} = \int_{\text{ray}} m(x)dx \tag{12}$$

where $m(x)$ denotes slowness, t^{th} is the predicted travel time, and the integral is taken along a seismic ray path. The ray path depends on the slowness distribution, consequently the above relation between t^{th} and $m(x)$ is, in fact, nonlinear (IYER and HIRAHARA, 1993; NOLET, 1987; AKI and RICHARDS, 1985). However, in the case of tomography of local stuctures (for example in mines, rock sample, etc.) if the slowness distribution does not have a large gradient, this nonlinearity can be neglected with a slight loss of sharpness of tomograms (MAXWELL and YOUNG, 1993). The discretized form of eq. (12) reads:

$$t_i^{th} = \sum_j G_{ij}(\mathbf{m})m_j \tag{13}$$

where the G_{ij} element of the **G** operator is a fraction of ith ray paths in the jth cell. Neglecting the dependence of the ray on velocity leads to the following linear forward modeling equation:

$$t_i^{th} = \sum_j G_{ij}m_j \ . \tag{14}$$

Using eqs. (13) or (14) for travel-time calculations needed for tomography imaging rises two questions. The first one is when ray approximation may be used to solve tomography forward problem. The second question is when the straight-line ray-path approximation is justified.

The ray approximation assumes that the frequency of waves is infinitely high. In practice this means the wavelength must be smaller than the Fresnel volume (WIELANDT, 1987; ČERVENÝ, 1987; WATANABE and SASSA, 1993; HANYGA and SEREDYŃSKA, 2000). Velocity tomography imposes another bound on the wavelengths of waves used for tomography imaging. Interference and diffraction effects at velocity heterogeneities inside the imaged body are required to be negligible (NOLET, 1987; WIELANDT, 1987). As was found by WILLIAMSON (1991), this requirement implies that the smallest feature which can be resolved by ray-based tomography is of the order of the first Fresnel zone radius. To overcome the above limitations, more advanced forward modeling methods are necessary, such as, for example, the solutions of a full-wave equation (WATANABE, 1997; RIBODETTI and VIRIEUX, 1998), or the "fat rays" approach (WATANABE and SASSA, 1993; LOMAX, 1999; LEIDENFROST *et al.*, 1999; HUSEN and KISSLING, 2002).

The second question is connected with a straight ray-path assumption. The validity of the straight ray-path assumption depends crucially on the imaged structure. If the velocity gradients are small enough so that the departure of an actual ray from a straight line at the receiver position is smaller than a few wavelengths, the straight line approximation may be used (WILLIAMSON, 1991; MAXWELL and YOUNG, 1993). This is usually the case in tomography imaging of small objects like rock samples or pillars in mines (MAXWELL and YOUNG, 1993; WATANABE and SASSA, 1996). In such applications modeling errors introduced by the straight-line ray-path approximation only decrease the contrast of the images (MAXWELL and YOUNG, 1993).

The situation is more complicated if there is a region of high velocity contrast inside the imaged body. In this case the ray paths are bent by the strong heterogeneity, which has two implications. Initially it leads to an erroneous travel-time prediction for straight-line modeling due to the change of ray-path lengths. Secondly the actual, bent rays probe a different area than straight-line ones. Both effects may lead to serious modeling errors, especially in the case of attenuation tomography (DĘBSKI, 1997b).

Pulse Broadening Attenuation Tomography

One of the effects of the inelasticity of real materials is that acoustic, or more generally, seismic waves are attenuated when propagating through the medium. This loss of energy manifests itself not only in the decrease of the wave amplitude along the travel path but also in a change of the signal spectrum. Stronger damping of the high-frequency part of the signal spectrum can be seen as a broadening of the recorded pulse or equivalently as an increase of the pulse rise time. This broadening has been found to be related to the Q and v factors as follows (KAVETZKY *et al.*, 1990; GLADWIN and STACEY, 1974):

$$t_{rt} = t_o + c \int_{\text{ray}} \frac{ds}{Q(s)v(s)} \,, \tag{15}$$

where t_{rt} is the pulse rise time, t_o is the 'pure' source pulse rise time (this is what we should observe for a perfectly elastic medium and ideal source/receiver-medium coupling), v is the acoustic wave velocity, Q is the quality factor, and the integral is taken along the ray path. The constant c depends on the type of source (BLAIR and SPATHIS, 1984; GLADWIN and STACEY 1974; KAVETZKY *et al.*, 1990) and on how t_{rt} is defined. If the standard definition of t_{rt} as the time when the pulse amplitude changes from 10% to 90% of the pulse maximum amplitude is used, then $c \sim 0.5$ (BLAIR, 1990; KAVETZKY *et al.*, 1990).

The spatial distribution of Q can be derived from eq. (15) by any standard tomography inversion computer code, assuming the velocity distribution is already known (WATANABE and SASSA, 1993; DĘBSKI, 2002).

Enhanced Velocity Tomography

Both the velocity and attenuation tomography methods can be used separately to obtain independent information pertaining to the velocity and Q parameters. However, often neither v nor Q but some image of the internal rock microstructure such as crack distribution, porosity, etc. is desired. From this point of view it is reasonable to combine both velocity and attenuation tomography into one scheme to maximize the advantage of the robustness of velocity and the sensitivity to the microstructure of attenuation.

It has been proposed by DĘBSKI and YOUNG (1999) that the first peak maximum arrival time provides information which can be used for such an enhanced imaging of the rock microstructure. The first peak maximum arrival time t_{pt} is an algebraic sum of the pulse travel time and pulse rise time so that it contains information about both velocity and attenuation structure. It has been found (DĘBSKI and YOUNG, 1999) to be related to the parameter

$$\tilde{v} = v \frac{Q}{Q+c} \tag{16}$$

through the integral:

$$t_{pt} = t_o + \int_{\text{ray}} \frac{ds}{\tilde{v}(s)} \tag{17}$$

where the constants c and t_0 have been defined in eq. (15). The constant c links the pulse rise time to the integral of the inverse of the product of the velocity and Q (eq. (15)). As was mentioned in the previous subsection, the value of c depends on the experimental details and on how the pulse rise time is measured (BLAIR, 1990; KAVETZKY *et al.*, 1990). Equation (17) always remains valid, however, provided the relation between pulse rise time and Q_v has the form described by eq. (15) (DĘBSKI and YOUNG, 1999; DĘBSKI, 2002). The experimental setup and the method of pulse rise time measurement only influence the relation between $\tilde{v}$ and Q.

The relation between t_{pt} and $\tilde{v}$ described by eq. (17) has the same form as the relation between time and velocity (eq. (12)), so $\tilde{v}$ can be imaged by means of any standard tomography codes used for velocity tomography.

For highly elastic media ($Q/c \gg 1$) $\tilde{v} \approx v$ and images of $\tilde{v}$ and v will be similar. On the other hand, if $Q/c < 1$ then $\tilde{v} \propto vQ$ and imaging of $\tilde{v}$ corresponds to imaging of the Q distribution. Therefore $\tilde{v}$ effectively combines velocity and attenuation in a single tomography scheme.

There is another advantage of using $\tilde{v}$ rather than v which concerns the quality of the data used. The first arrival time data are picked from recorded waveforms. Because of noise it can be a significant problem to do this precisely to the required accuracy. Conversely, determining the maximum of the pulse is usually decidedly easier because of a larger signal-to-noise ratio.

Having reconstructed the v and $\tilde{v}$ images, one can construct an image of Q using the rewritten eq. (16):

$$Q = c\frac{\tilde{v}}{v - \tilde{v}} . \tag{18}$$

It is clear from the above formula that even if the v and $\tilde{v}$ distribution can be reasonably reconstructed, the Q distribution can potentially be numerically unstable due to the denominator structure in this formula. This is another manifestation of well-known difficulties with imaging of attennuation structure.

3. *Lac du Bonnet Granite Sample*

The procedure described above has been applied to study the structure of a Lac du Bonnet granite sample (JANSEN *et al.*, 1993; CARLSON *et al.*, 1993; MCKIRDY *et al.*, 1998). The 30-cm diameter and 20-cm high sample cylinder of medium grain size, of pink granite from the Underground Research Laboratory in Canada, was subjected to induced damage by heating. The heater was mounted in a central, 3-cm diameter hole drilled through the sample. The sample was heated in six cycles with an increased peak temperature in each cycle (JANSEN *et al.*, 1993; CARLSON *et al.*, 1993). When each peak temperature was achieved the sample was cooled to room temperature and then a tomography scan was performed. Collection of tomography data at room temperature ensured that the resulting images would only show structural changes due to the formation of new fractures in the sample, rather than changes due to thermal expansion effects. In the final heating cycle, when the temperature of the sample reached about 250 °C, thermal stresses resulted in a visible radial crack forming from the external surface to the inner hole of the sample at an angle of 65°, represented in Figure 1 by a white line.

Ultrasonic waveform data were acquired with 24 transducers mounted 15° apart along the cylinder's mid-plane as shown in Figure 1. Each transducer was sequentially switched to a pulse generator while those remaining were connected through amplifiers to 4 four-channel digital oscilloscopes operating with a 0.1 μs sampling period. The limited number of recording channels brought about acoustic waveforms that were recorded by only 16 transducers for each "shot". The six closest transducers and the one mounted opposite to the transducer generating acoustic

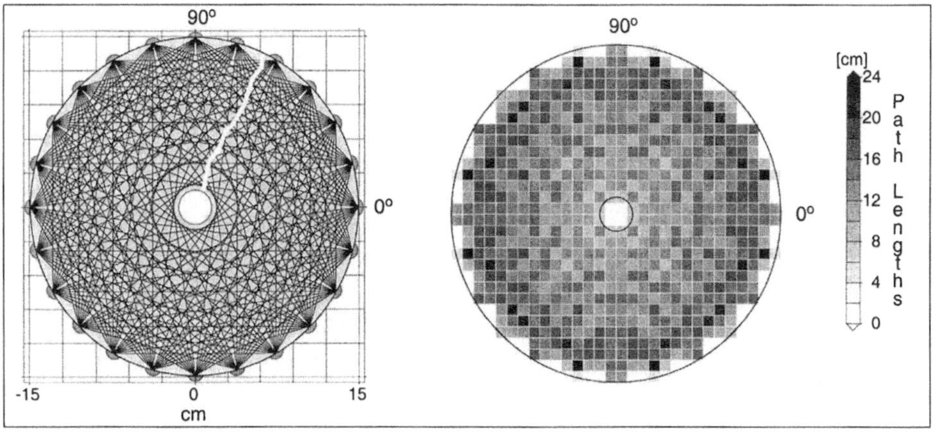

Figure 1
Ray paths (left) and sum of fractions of lengths of all rays passing through cells (right) used to parameterize the studied rock sample. Transducer positions and a sketch of the crack formed during the final heating cycle are marked on the left plot.

waves were not used. Altogether, 384 waveform traces were recorded. The acoustic pulses were generated by a Panametrics 5055R pulser which provided pulses with the rise time $t_0 = 0.3$ μs. The first arrival times, the first peak maximum arrival times, and the pulse rise times were carefully picked from the recorded waveforms and, after correction for apparent anisotropy, used for inversion for the v, $\tilde{v}$, and Q parameters, respectively.

Data

The Lac du Bonnet granite was reported to exhibit a few percent anisotropy (JANSEN *et al.*, 1993). To remove the influence of the anisotropy on final images, travel-time data were corrected for apparent anisotropy (Fig. 2).

The wave velocity was modeled using a function proposed by THOMSEN (1986) which indicated about 3–4% relative anisotropy. Travel times were corrected for this apparent anisotropy before inversion. This correction and isotropic inversion is conceptually identical to holding the orientation and relative anisotropy constant and inverting for heterogeneity in the parameters distribution along the fastest direction (MAXWELL and YOUNG, 1992). The corrected first arrival times and the pulse rise time data collected for all six scans are plotted as a function of ray-path lengths in Figure 3. The first arrival time data are plotted as reduced times, i.e.,

$$t^{red} = t^{obs} - l/v \tag{19}$$

where l is the ray-path length and the reduction velocity $v = 0.6$ cm/μs was taken.

A quick analysis of the data indicates that no signs of essential changes of the rock sample structure were observed during the first five heating cycles. Only a small

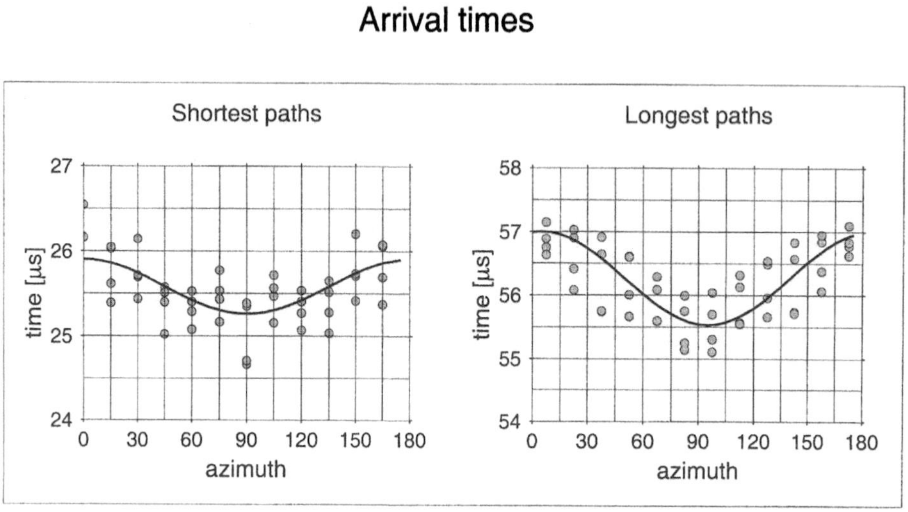

Figure 2
An example of arrival time data for the shortest (left) and the longest (right) ray paths plotted as a function of the ray paths azimuth. Solid line shows the best fit of the THOMSEN (1986) model with 4% anisotropy predicted.

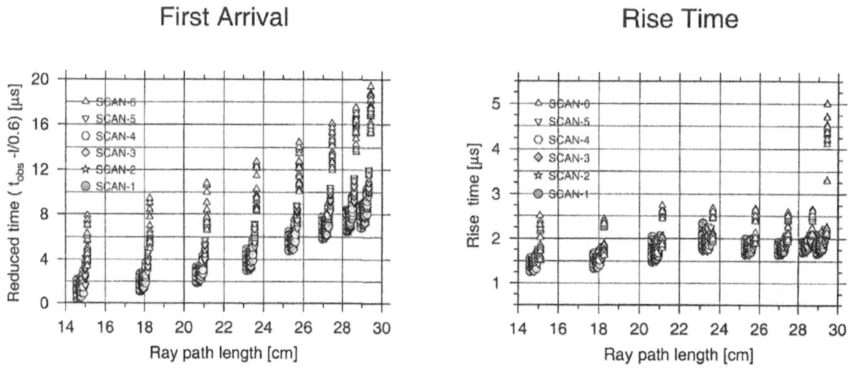

Figure 3
Reduced travel-time data used for imaging the velocity distribution (left) and the pulse rise time data used to reconstruct the Q distribution (right) for all six tomography scans. Note the essential increase of both travel times and pulse rise times measured in the six heating cycles along rays passing through the crack.

homogeneous increase of travel times was observed. The changes of pulse rise time were hardly observable during those scans. The data recorded in the sixth scan, after the forming of the crack, show an essential difference in comparison with the data obtained during previous tomography scans. Both the travel times and the pulse rise times increased significantly. Moreover, one can clearly observe a splitting between data recorded along the paths passing through the crack and others. The crack has significantly increased the travel times and pulse rise times of waves passing through its neighborhood, which implies a decrease of velocity and increase of attenuation (decrease of Q) in the area around the crack.

The recorded travel times for the shortest ray path (about 15 cm) were approximately 25 μs and for the longest – about 55 μs for first arrival times in the first scan. Thus the relative measurement errors due to the finite (0.1 μs sampling period range between 0.2–0.4%. A similar calculation for the pulse rise times gives an estimation of the errors at the 5% level. In reality the accuracy of the first arrival time picking was about 4 samples (0.4 μs) due to noise, decreasing the data picking accuracy to 1–2% in the case of the first arrival times and extending to 25% for rise time data. This accuracy of travel-time picking is acceptable for velocity tomography but must be essentially improved for pulse rise time data to make Q tomography possible. Improvement may be achieved by measuring the pulse rise time, not as the difference between first arrival and pulse maximum arrival times, but as the time when the pulse amplitude changes from 10% to 90% of the pulse maximum amplitude (BLAIR, 1990; KAVETZKY *et al.*, 1990). This method allowed measurement of the rise time with an accuracy of about 0.02 μs which gave relative errors of about 1–2%. The peak arrival times were measured with a simple interpolation with an accuracy of half of the sampling period, i.e., 0.05 μ, which gives a relative accuracy ranging 0.1–0.3%. In reality, as the information regarding hetrogeneities of the sample are provided not by the absolute arrival times or the pulse rise times but rather by the differences between them measured along different paths, the actual data accuracy is worse.

Algorithm

For the purpose of the current tomography study, the spatial distribution of the considered parameters, namely velocity v, enhanced velocity $\tilde{v}$, and quality factor Q were approximated by a set of square, homogeneous, equal size cells (see right plot Fig. 1).

The minimum possible size of the cells that could be used may be estimated by a length scale of spatial heterogeneities for which diffraction effects may still be disregarded (WILLIAMSON 1991). The transducer configuration used is very symmetric (see left plot in Fig. 1), so that according to WILLIAMSON'S (1991) suggestion, tomography resolution can be of the order of a single wavelength. The maximum of the frequency spectrum of recorded signals was about 0.5 MHz (PETTITT, 1997), so with the average velocity $v \approx 0.55$ cm/μs this gives the wavelength of about 1 cm. However, for such a detailed parameterization there are regions of the samples (see Fig. 1) poorly covered by rays. Moreover, the inverse problem becomes underdetermined as the number of parameters (720) in this case is almost twice larger than the number of data (384). In such a situation a simple data inversion using cells of 1 cm size could introduce large numerical instabilities to the final images. For this reason, to invert the collected data efficiently with the above spatial resolution, a complex multi-scale algorithm using gradually decreased cell sizes was used.

The idea (see, for example, JIN and BEYDOUN, 2000; WILLIAMSON, 1990) is to begin data inversion with a crude parameterization of the sample for which the spatial resolution determined by the block sizes is small. At this stage only the roughest features of parameter distribution can be recovered. This initial parameterization may be chosen in such a way that the resolution operator given by eq. (11) is almost diagonal. Then, all the parameters are well resolved (well constrained) and they are practically independent of the choice of the *a priori* model $\mathbf{m}^{apr}$ and weakly dependent on the damping parameter $\gamma = \sigma_D/\sigma_M$ ranging up to $\gamma \leq 1$ (see Fig. 4).

In the next step the result of the "crude" inversion is used as the *a priori* input model for the detailed imaging with decreasing cell size. Simultaneously the γ ratio may (should) be increased. The larger value of the damping factor γ implies that the final $\mathbf{m}^{ml}$ model is forced to be close to the *a priori* model $\mathbf{m}^{apr}$ which contains, by its construction, the roughest features of the sought parameter distribution. This way

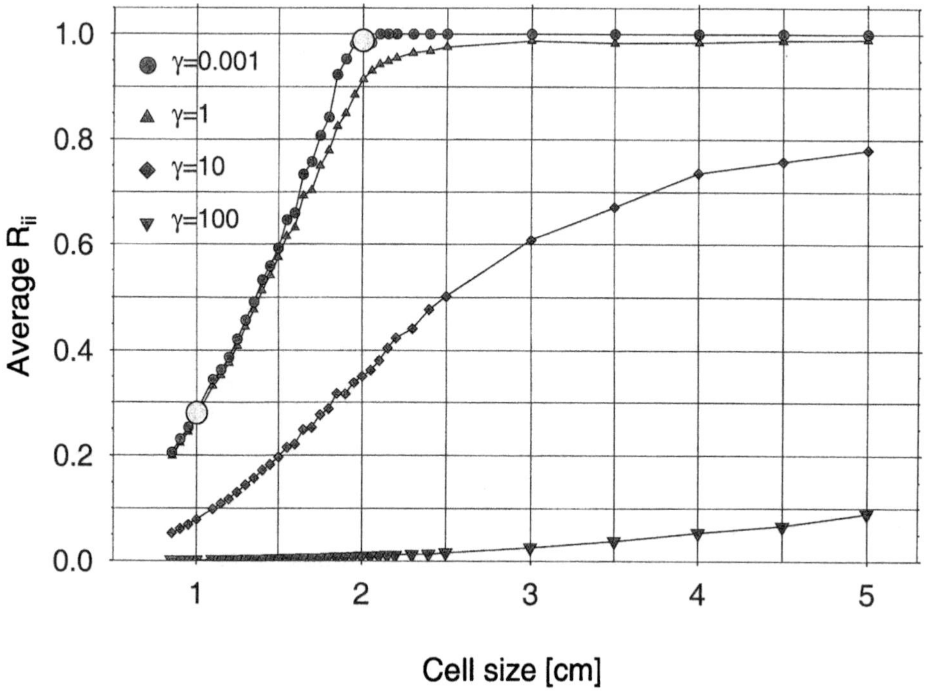

Figure 4
A few resolution curves describing the dependence of the average of diagonal elements of the resolution operator as a function of cell size for a few damping factors γ. The large open circles mark cell sizes, and γ chosen for the initial (right circle) and final (left circle) inversions.

the final image combines smoothness of the initial, well resolved but rough model with the high spatial resolution provided by inversion with the fine parameterization.

In the current study the cell size used for the initial inversion was determined by noticing when the average value of diagonal elements of the resolution operator **R** starts to fall (see Fig. 4). This happens for 2 cm cells and this value was chosen for the crude parameterization. The damping factor used at this stage was taken arbitrarily to be $\gamma = 10^{-3}$. The final data inversion was performed taking cells of the size of 1 cm, according to the previous estimation. The γ factor was estimated by the standard analysis of a trade-off between data and model variance as a function of γ (MAXWELL and YOUNG, 1992). As it follows from the trade-off curve which is shown in Figure 5, the optimum choice of γ for the final inversion is $\gamma \approx 0.5$.

Let us observe that, as it follows from eqs. (5) and (6), γ has a dimension of squared length (m^2). Therefore, its square root may be interpreted as the distance below which spatial variation of the parameters is smoothed due to the damping of the inversion procedure. A similar smoothing effect is introduced by a finite cell size. In our case both smoothing scales are close to each other.

To find the $\mathbf{m}^{ml}$ model for which the *a posteriori* probability density function described by eqs. (1), (5), and (6) attains its global maximum, the delta-coding

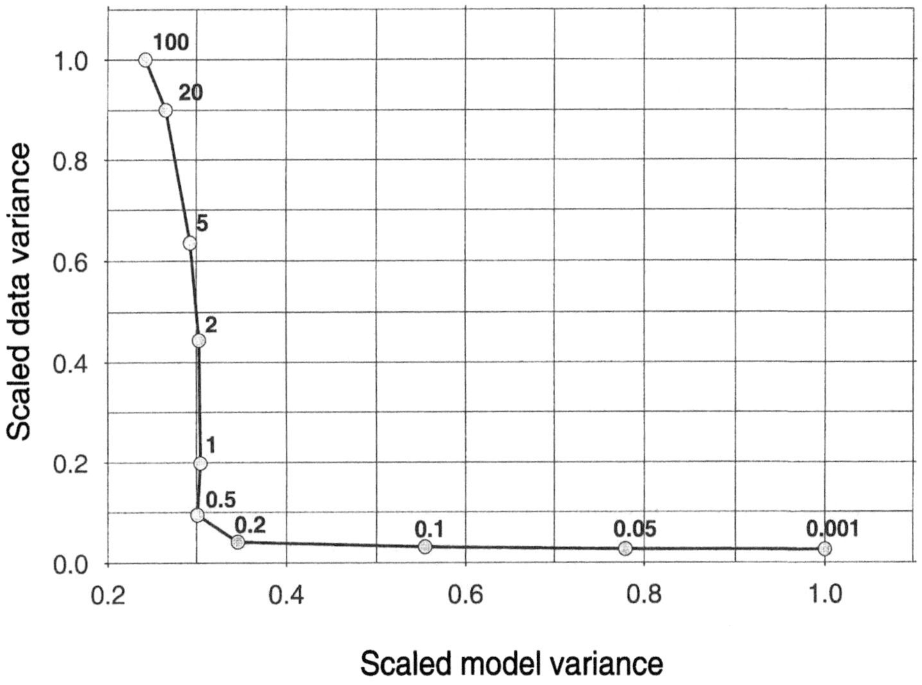

Figure 5
Trade-off curve: the correlation between data and model variances for different values of the damping factor γ.

version of the genetic algorithm (GA) (MICHALEWICZ, 1996; WHITLEY *et al.*, 1991) was used. This algorithm is based on the double application of the GA code. In the first stage GA is used to obtain the basic slowness (velocity) model by inspecting the full model space: all possible values of velocities assigned to each cell are randomly tested at this level. In the next step the model just found is taken as the base model, and only small perturbations in relation to it are sought (WHITLEY *et al.*, 1991). We decided to use this type of evolutionary strategy because classical genetic algorithms are known to cause problems with the precise estimation of the optimal solution when applied to numerical optimization problems (MICHALEWICZ, 1996).

The optimization code designed for this study uses the dynamic non-uniform mutation operator and the arithmetic crossover operator (MICHALEWICZ, 1996).

The estimation of imaging errors was made by calculation of the *a posteriori* covariance matrix according to eq. (9) using as σ_D the values estimated above: 0.4 μs, 0.05 μs, and 0.02 μs for the arrival times, peak arrival times and rise time, respectively. As the inversion consisted of two steps, calculation of the *a posteriori* covariance matrix was also a two-step procedure. Firstly, the *a posteriori* covariance matrix for the initial, crude inversion was calculated using as $(\mathbf{C}_\mathrm{M})$ a matrix proportional to the unit one with the proportionality constant $\sigma_M = 10^{-3}$. Next, the covariance matrix just calculated was taken as the *a priori* one $(\mathbf{C}_\mathrm{M})$ for the calculation of the final covariance matrix. The diagonal elements of the covariance matrices are shown in Figure 6. The spatial distribution of imaging errors essentially reflects the ray-path distribution: regions covered by more rays are more accurately resolved and *vice versa* (see Fig. 1).

4. *Results*

We start the discussion of the tomography imaging results with an analysis of the residuals for the *a posteriori* models. We only restrict our analysis to the first and last scans, as the results for the other scans are similar. The residuals calculated for these two scans are shown in Figure 7 for v, $\tilde{v}$, and Q.

It is clear that the distribution of the residuals calculated for the final *a posteriori* models are substantially narrower than those calculated for the initial *a priori* models for all three parameters. Here by an initial model we understand the model which is the result of the first step, "crude" inversion (see Fig. 8).

The visible reduction of residuals means that the found maximum likelihood models can reproduce the observational data quite well. We also observe that the histograms of the *a posteriori* residuals for v are wider than for $\tilde{v}$. This is caused by less accurate picking of first arrival times than peak arrival times, as was discussed. The side lobes that appear in the residual distribution of peak arrival times (middle line in Fig. 7) are typical round-off errors due to the finite discretization. In the case

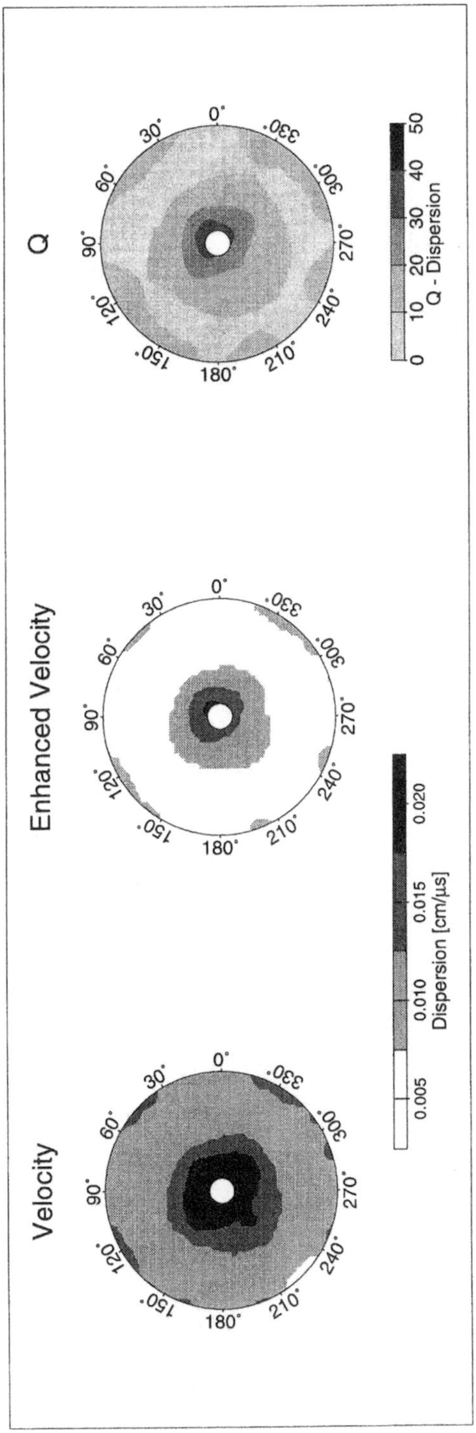

Figure 6

Spatial distribution of *a posteriori* imaging errors of velocity (left), enhanced velocity (center) and quality factor (right).

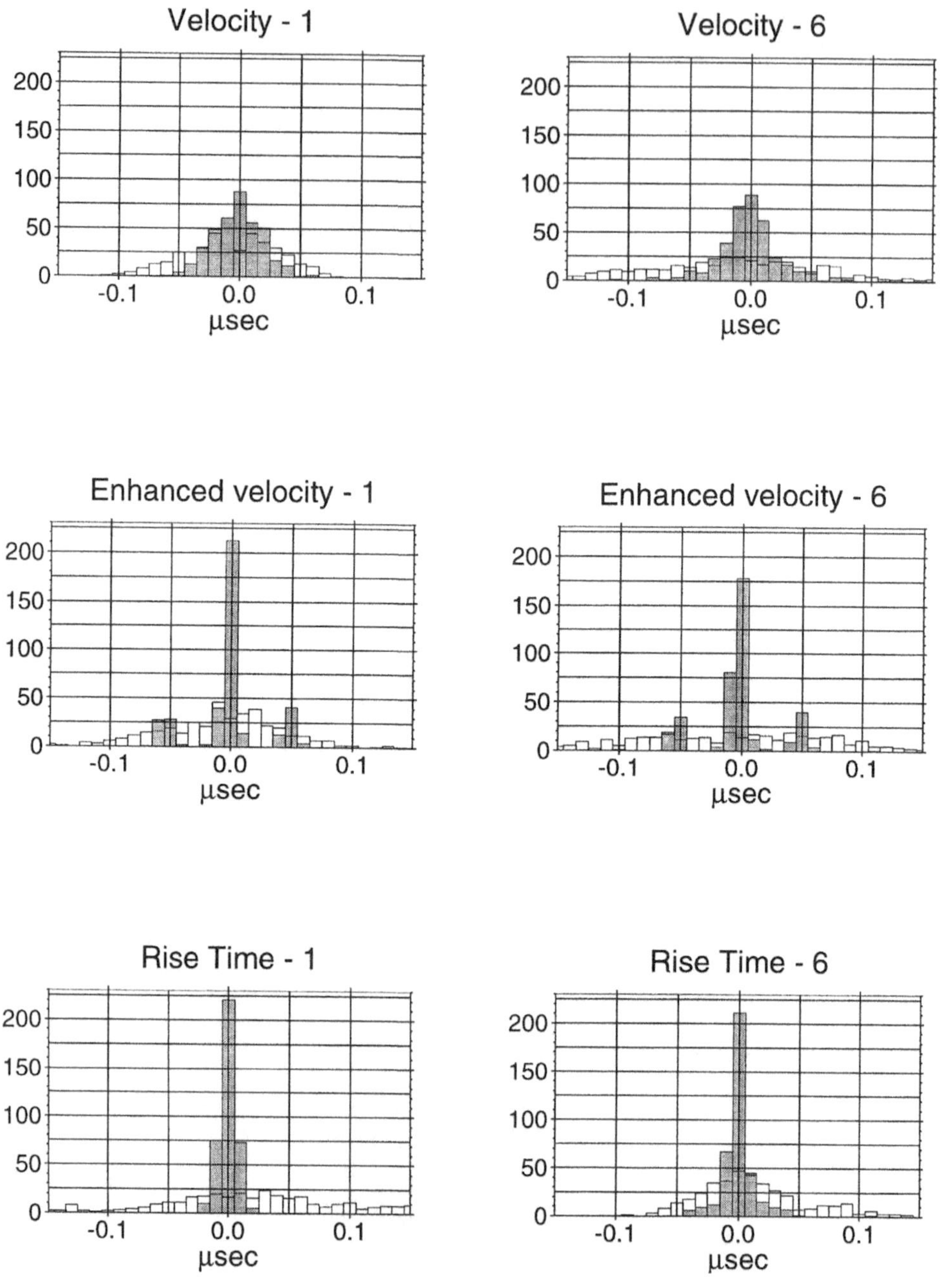

Figure 7
Histograms of the residuals of travel times (upper row), pulse travel times (middle row), and pulse rise times (lower row) for the *a priori* model used for the initial inversion (open bars) and the *a posteriori* models (closed bars) for the first (left column) and the last (right column) scans. Apparently more spread out *a posteriori* histograms of first arrival times than pick arrival times are caused by worse time picking accuracy (0.4 μs for t_{at} and 0.05 μs for t_{pt}). The side lobes seen at histograms of the *a posteriori* pick arrival times (middle row) are round-off errors due to the finite (0.05 μs) sampling interval. Significant reduction of *a posteriori* residuals with respect to a *priori* residuals shows that tomography could find the models which reproduce the observational data well.

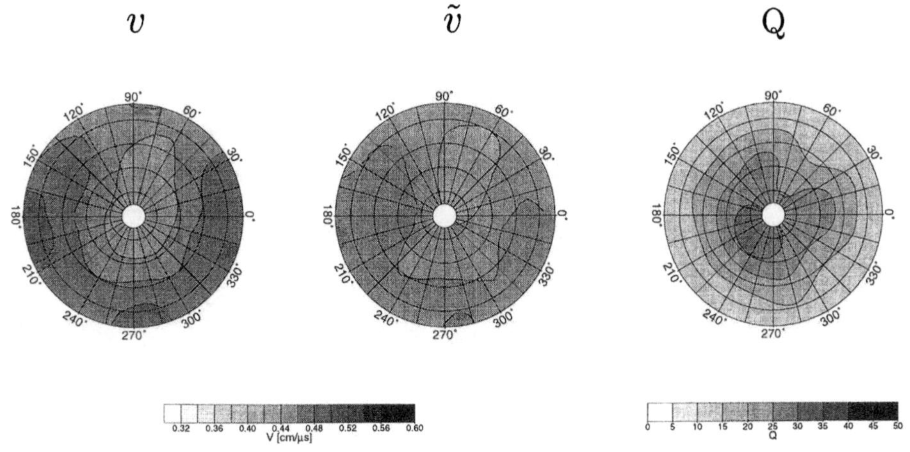

Figure 8
An example of *a priori* models used for the final, detailed inversion (sixth scan). These distributions were obtained by the initial inversion of the same data sets as used for the final tomography but with parametrization of v, $\tilde{v}$, and Q distributions by 2×2 cm cells.

of travel times (first line in Fig. 7) these finite sampling errors are masked by considerably increased time-picking uncertainties.

Figure 8 shows the images of velocity, enhanced velocity and Q factor obtained by initial,"crude" inversion for the first and last scans. Due to the large cell size (2 cm) these distributions are smooth with no spatial details. Nevertheless, they reveal the existence of a ring-type structure discussed below. Its existence was confirmed by the post-experiment measurements of the velocity along the cylinder axis at different distances from the symmetry axis, shown in Figure 13.

Figure 9 shows the final images of the velocity distributions recorded during all six heating cycles marked by the number of the consecutive heating cycle from 1 to 6.

The lower velocity zone in the central part of the sample, seen in plots 1–5, may be related to a fracturing of the sample during the drilling of the inner hole. The lower velocity zone observed around the external surface of the sample may have also been caused by the fracturing of the rock during the preparation of the sample. It is clear from this firgure that there was no essential change in the velocity distribution during the first three heating cycles. In the fourth and fifth cycles the overall diminishing of the velocity can be observed, although the shape of the velocity distribution has not changed. The tomogram obtained after the forming of the main crack shows further diminishing of the velocity in the external and central parts of the sample. In addition, the location of the induced crack has been correctly recovered, although the damaged area around it has been imaged with a rather low spatial resolution.

Figure 10 shows the images of the enhanced velocity distributions $\tilde{v}$, recorded during all six heating cycles.

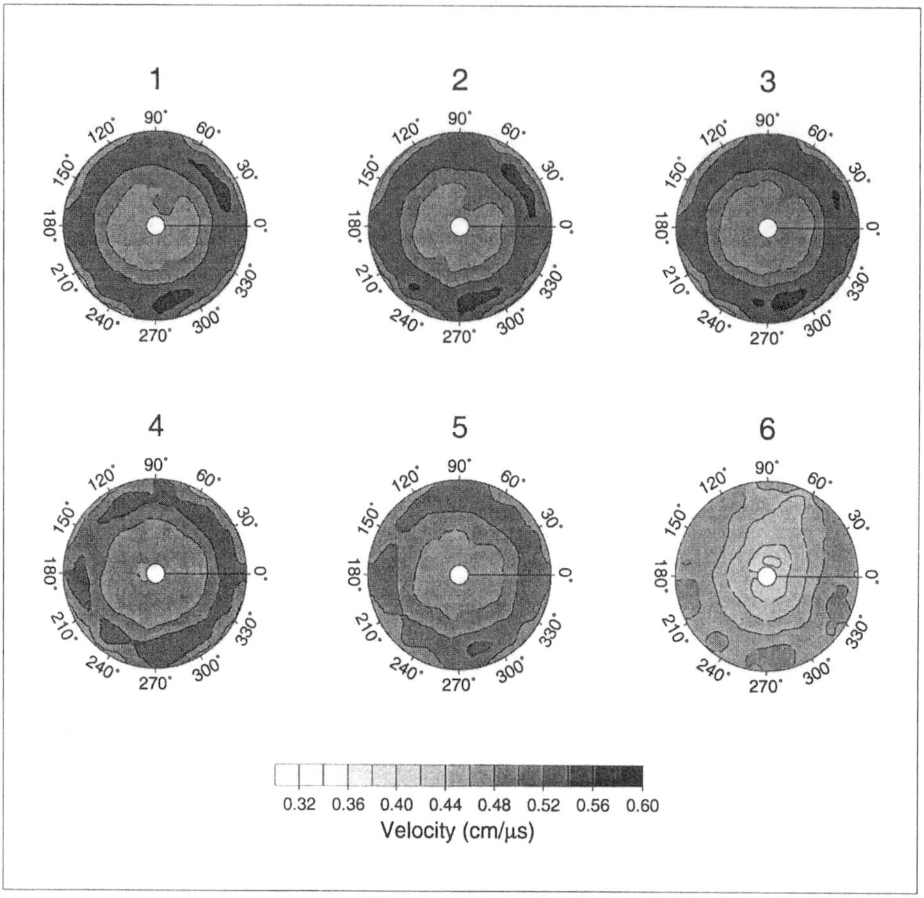

Figure 9
The velocity images obtained for all six scans. Note the progressive decrease of velocity seen mainly in the outer part of the sample after the second scan. Although the area of the main crack is well imaged in the sixth scan (plot 6), the achieved spatial resolution is moderate.

Similarly to the velocity images, the $\tilde{v}$ distribution displays an area of a lower $\tilde{v}$ around the inner hole and in the outermost part of the sample. However, diminishing of $\tilde{v}$ in subsequent heating cycles starts to be observed earlier than the changes in v. Already in the second cycle one can notice the decrease of $\tilde{v}$ around the inner hole while almost no change of v is observed. Also the amplitude of changes of $\tilde{v}$ in the subsequent 1–5 scans are larger than for v. The main difference between the images of $\tilde{v}$ and v is seen when comparing plot 6 in Figures 10 and 9. The distribution of $\tilde{v}$ shows vastly better spatial resolution than the v distribution. The main crack area at the angle of 65° is mapped by $\tilde{v}$ very precisely encircled by a narrow damage zone. Moreover, note the presence of a small area of very low $\tilde{v}$ around the inner hole at an angle of about 250°. The presence of this disturbance is discussed in the next section.

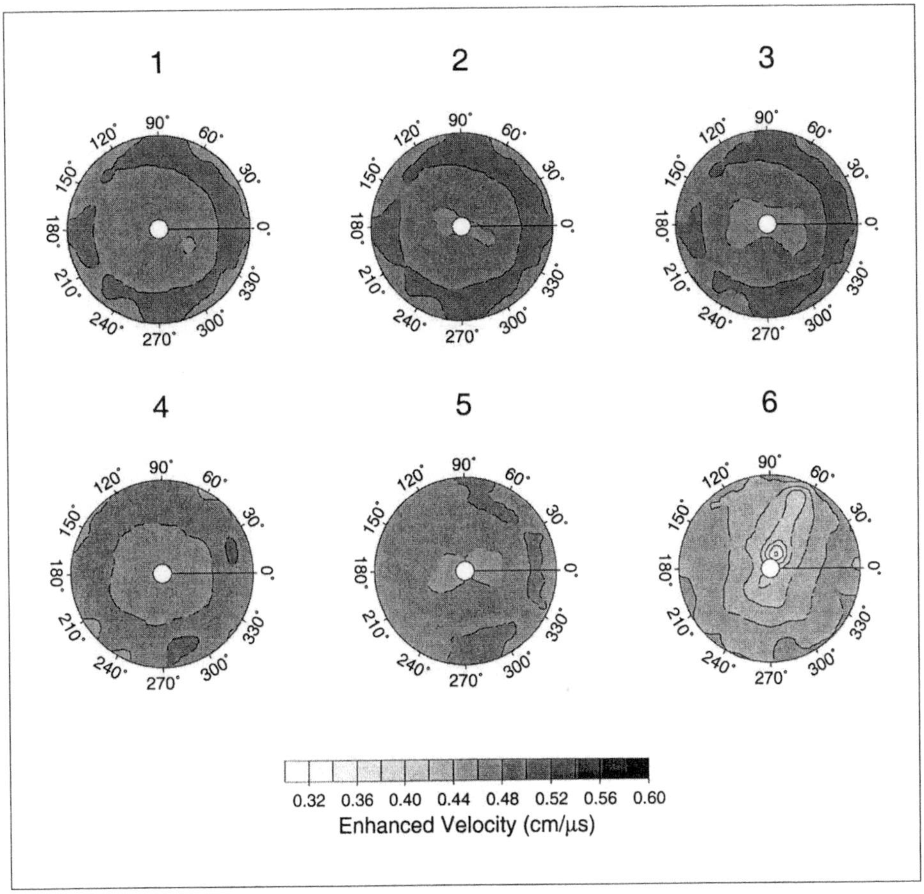

Figure 10
The enhanced velocity $\tilde{v}$ images obtained for all six scans. As in the case of the velocity images, the progressive decrease of $\tilde{v}$ noticed mainly in the outer part of the sample can be observed from the second scan. However, in comparison to the v image the main crack area is now imaged with considerably higher spatial resolution.

Finally, Figure 11 shows all six Q images.

Again one can observe the ring-type character of the Q distribution. The higher Q in the middle of the sample is bounded by a lower Q zone in the outer part of the sample. On the other hand, the damage area around the inner hole has hardly been imaged at all unlike the case of v and $\tilde{v}$. A barely noticeable outline is only visible in plot 1. In addition, the Q images seem to be more "noisy" than v and $\tilde{v}$. However, the most important thing is that the Q image merely suggests the existence of the main crack (plot 6). Only a sketchy outline of the crack in the inner and middle part of the sample is seen. This plot, similar to $\tilde{v}$, also shows the presence of a strong

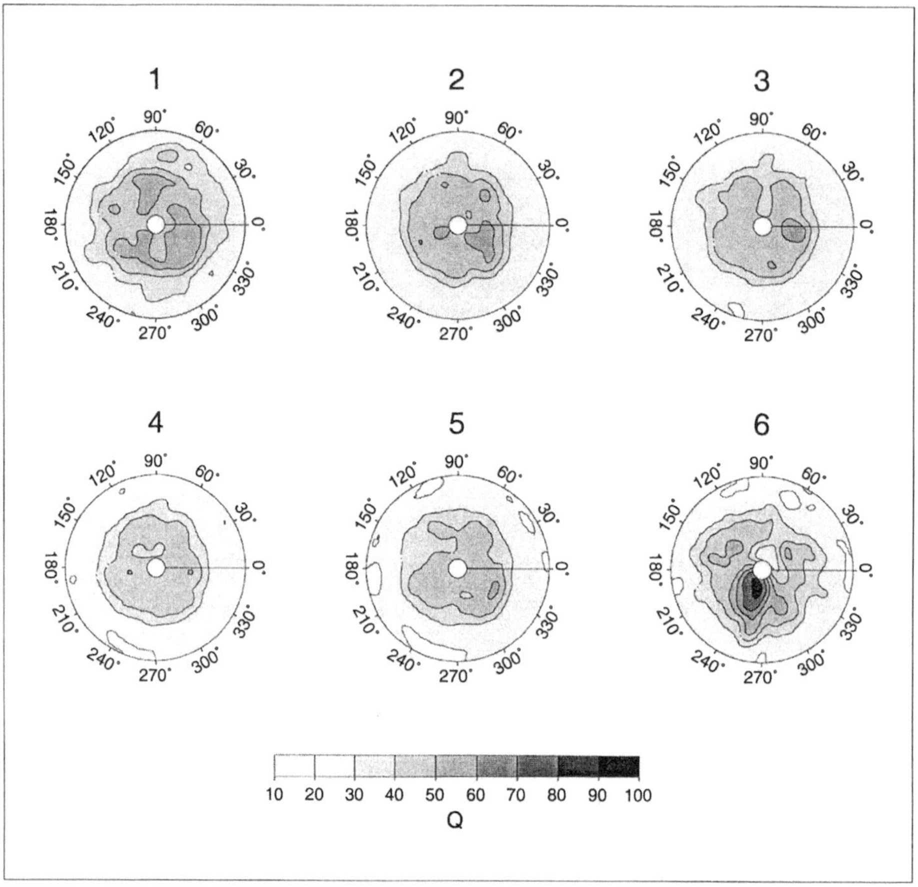

Figure 11
Images of the quality factor Q for all six scans obtained by inversion of pulse rise time data. No significant changes in Q distributions were observed during the first five scans. The main crack area formed in the final heating cycle was only sketched in plot 6, mainly in the inner part of the sample (see text for details).

heterogeneity at an angle of about 250° around the inner hole and in the middle part of the sample. This time however it is a region of increased Q values with respect to the background.

Figures 13 and 14 show the sections of v and $\tilde{v}$ radial (Fig. 13) and perpendicular to the main crack (Fig. 14) for the final, sixth tomography scan. The sketch of the location of the sections is shown in Figure 12.

In addition to the tomography scans, the velocity and enhanced velocity have been measured along the cylinder axis for the profiles A and B. The results of the measurements are shown in Figure 13. A variation of v and $\tilde{v}$ with the distance from the center similar to the one indicated by the tomography images is seen. This

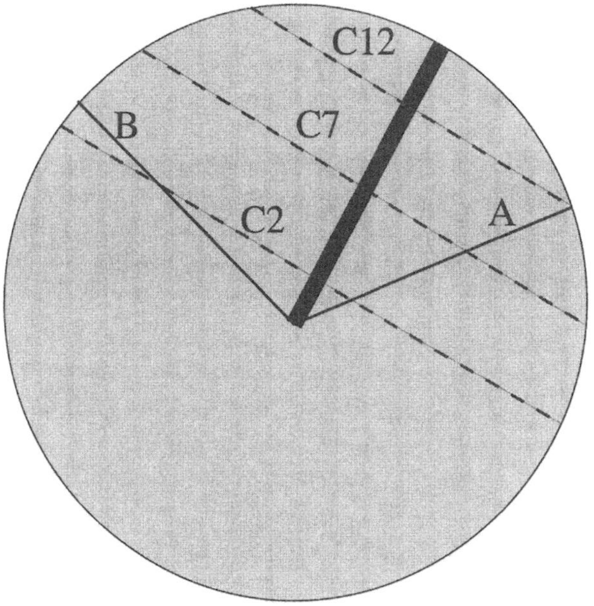

Figure 12
The scheme of the location of radial profiles A and B along which an additional measurement of velocity of the damaged sample was carried out. Additional profiles C2, C7 and C12 are perpendicular to the main crack. They pass at distances of 2 cm (C2), 7 cm (C7), and 12 cm (C12) from the sample symmetry axis.The thick radial line marks the crack position.

evidence strongly supports our conviction that the ring-type structures visible in the tomography images have a physical cause and are not a numerical artifact due to, for example, the non-homogeneous ray-path coverage.

In Figure 15 the sections of v and $\tilde{v}$ along the profile A for the first, third and fifth scans are shown. The dependence of v and $\tilde{v}$ on the distance from the sample center exhibits a decrease of the amplitude of its variations for subsequent scans. This effect is better visible in the case of $\tilde{v}$ than for v. If the observed flattening is assumed to be a consequence of changes in the rock microstructure then $\tilde{v}$ is more sensitive to the internal state of the imaged body.

5. *Discussion*

The results of ultrasonic tomography presented in this paper can be viewed in two different ways. From one point of view they provide information about the response of a rock sample to changes of its microstructure induced by heating. The observed decrease of v and $\tilde{v}$ suggests that each heating/cooling cycle made irreversible changes in the sample microstructure. This may be due to a release of pre-existing stresses

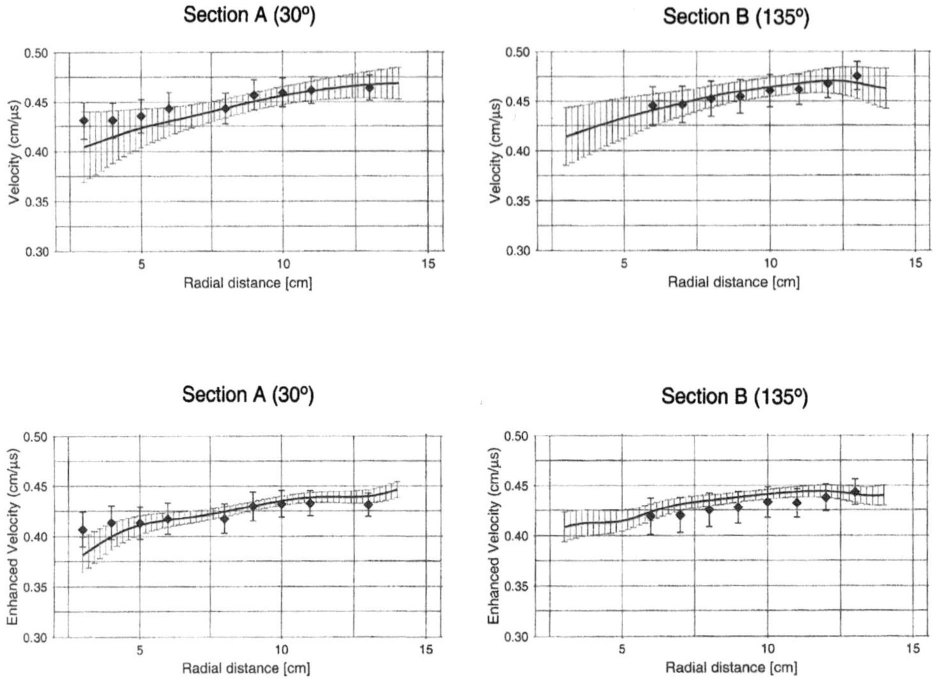

Figure 13
Sections of the velocity (upper row) and enhanced velocity (lower row) distribution along the radial profiles A (left column) and B (right column). The filled squares denote the velocity and enhanced velocity measured from top to bottom of the sample along profiles A and B.

caused by sample preparation, but also to a more fundamental change in the sample microstructure such as opening or closing microcracks, creating new cracks or grain composites, etc. The observed decrease of v or $\tilde{v}$ in the subsequent tomography scans apparently depends on the distance from the heater (see Fig. 15). This change and its spatial characteristics must be explained by the rock-mechanics theory. Given the results presented above, there is an opportunity not only to describe this effect qualitatively but also to try to build a model which could interpret the observed changes quantitatively.

On the other hand the presented results can be regarded as another step in the application of the tomography imaging technique to monitor rock-mass changes. The results obtained confirmed the usefulness of active velocity and enhanced velocity tomography for imaging changes of internal rock structure caused by changes of external conditions (MAXWELL and YOUNG, 1992, 1996). From this point of view enhanced velocity tomography seems very promising because of its robustness and sensitivity to rock microstructure larger than that offered by classical velocity tomography.

Contrary to velocity and enhanced velocity tomography, attenuation tomography results are not definitively conclusive. From one side the obtained images of Q

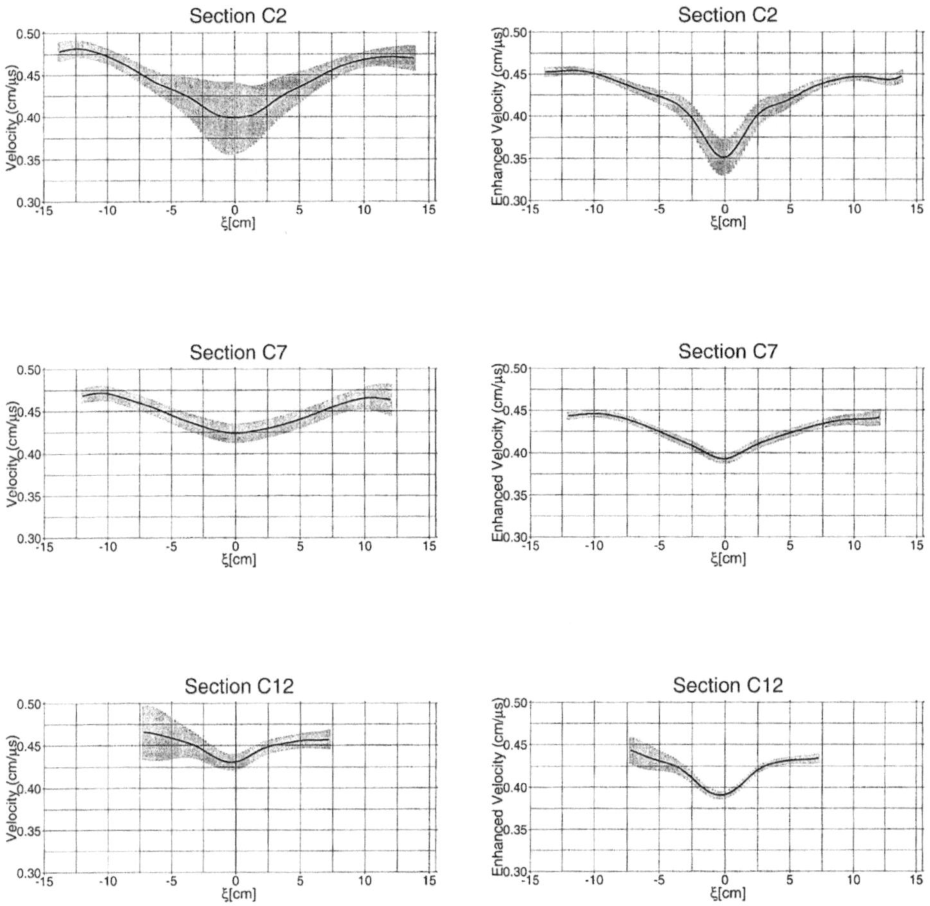

Figure 14

Sections of the velocity (left column) and enhanced velocity (right column) distributions for profiles C2 (upper row), C7 (middle row) and C12 (lower row) perpendicular to the main crack. The shaded area marks the imaging errors. Note the more precise mapping of the crack damage zone, especially in the inner and outermost parts of the sample (profiles C2 and C12) by the enhanced velocity than by the velocity.

exhibit a ring-type structure similar to that obtained for v and $\tilde{v}$ distributions. The most external part of the sample is imaged in a coherent way by Q and v or $\tilde{v}$. This part of the sample subjected to microstructure damage during sample preparation exhibits lower $v/\tilde{v}$ and, as expected, lower Q. Thus all images are consistent with each other and consistent with our expectations. However, this ends the coincidence between $v/\tilde{v}$ Q distribution. The Q images do not exhibit, for example, significant differentiation in the inner and central parts, whereas v and $\tilde{v}$ clearly indicate the existence of an area of lower $v/\tilde{v}$ in the innermost part of the sample. However, the most important matter is that Q distribution for the final, sixth scan only sketches the main crack damage zone. There are two most probable reasons for that. First, the pulse rise time data used for retrieving Q distribution are subjected to a number of

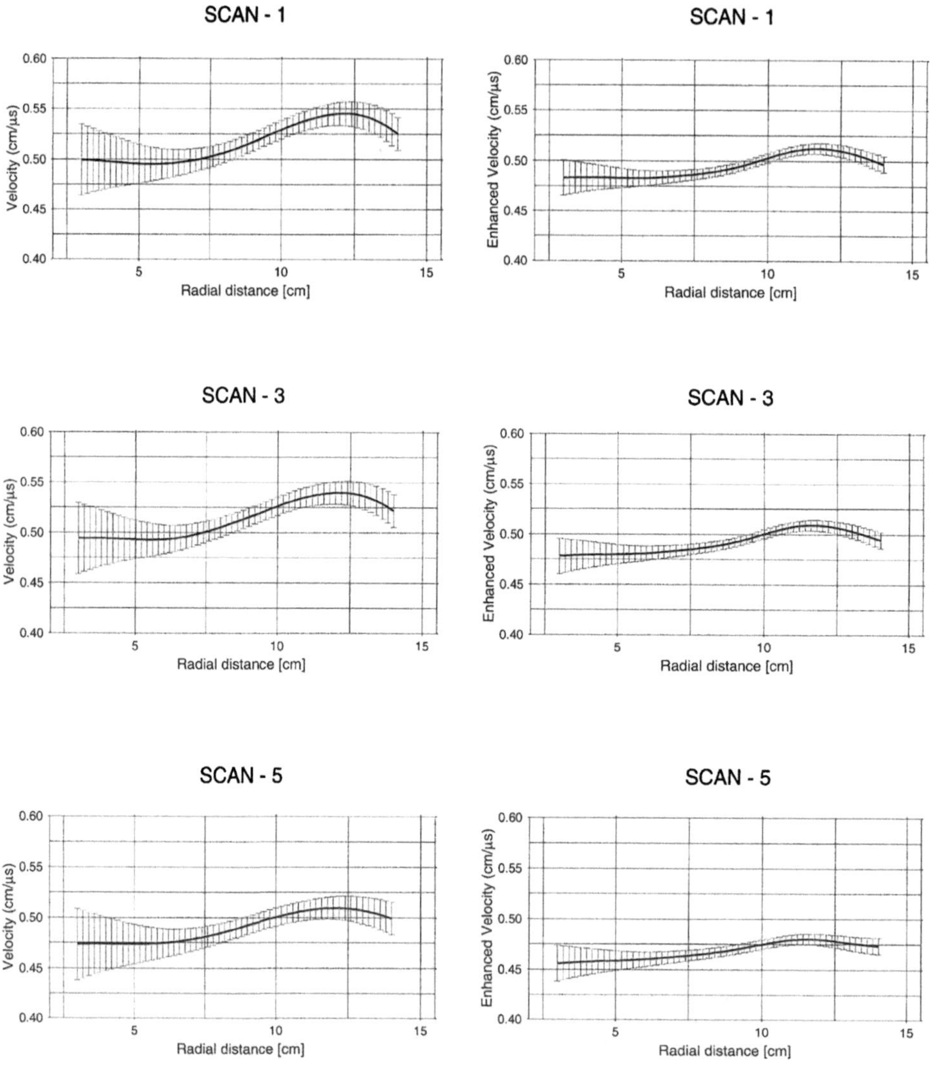

Figure 15

Radial sections of the velocity distributions v (left column) and enhanced velocity $\tilde{v}$ (right column) along profile A for the first scan (upper row), third scan (middle row) and fifth scan (lower row). Both v and $\tilde{v}$ sections exhibit a similar ring-type structure (see text for discussion), however due to lower imaging errors the $\tilde{v}$ distributions provide more accurate images.

disturbing factors, such as different coupling of transducers with the rock surface, different characteristics of each transducer, complicated angular-dependent response of transducers, etc. (DĘBSKI, 1997a; PETTITT, 1997). These factors may disturb or even totally mask the subtle changes of pulse rise times caused by spatial heterogeneities of Q. The other cause of unsatisfactory imaging of the main crack may be the assumption of the straight line wave propagation paths. Formation of the crack has created an area with a very large velocity gradient (see Fig. 10) where large ray bending is expected. Ignoring this effect and inverting the data on the assumption of straight line ray paths obviously creates major modeling errors. Solving the problem of the unsatisfactory recovering of the main damage zone requires further studies. Firstly, repeating the inversion of rise time data using a full nonlinear tomography approach is necessary. Secondly, an attempt has to be made to use other than rise-time data for gathering information about Q.

Finally we discuss the presence of an anomaly that appears around the inner hole at an angle of 250°. This disturbance is best seen in the $\tilde{v}$ distribution as a small region, slightly biased towards smaller angles, as the prolongation of the main crack damage zone beyond the inner hole. The velocity distribution resolves this area poorly and also exhibits a smaller velocity in this part of the sample. On the other hand, the Q distribution shows high Q disturbance in this area. The importance of possible anomalies in this region derives from the fact of a high stress concentration in this region, as was found by McKirdy *et al.* (1998). As a result of a detailed inspection of the top surface of the sample in that region, a small (about 0.5 mm length) secondary microcrack has been found. Its location coincides well with the small $\tilde{v}$ anomaly at the 250° angle. The bottom part of the sample was damaged around the inner hole by drilling, so no attempt was made to look for possible cracks at the bottom plane of the sample. The $\tilde{v}$ image suggests that the secondary microcrack visible at the top may continue to the tomography plane. The interpretation of the positive Q anomaly in this region is ambiguous, however. From one point of view, stress concentration might have caused a compaction of the granite material which would lead to an increase of Q as suggested by the obtained result (Fig. 10). On the other hand, because inversion of pulse rise times was not unable to determine the main damage zone reliably, it is probable that the considered anomaly is a numerical artifact which by chance coincides with the location of the microcrack. More precise attenuation imaging is necessary to clarify this point.

Acknowledgments

The authors would like to thank Gerhard Pratt and an anonymous reviewer for their valuable comments that significantly improved the original manuscript. We are also indebted to S. J. Gibowicz for numerous discussions. This work was founded in part by the Polish State Committee for Scientific Research under contract No. 9-T12B-034-18. All plots were prepared using the GMT software (Wessel and Smith, 1998).

References

Aki, K., and Richards, P. *Quantitative Seismology* (San Francisco, Freeman and Co.1985).

Backus, G., and Gilbert, F. (1967), *The Resolving Power of Gross Earth Data*, Geophys. J. R. Astr. Soc. *13*, 247–276.

BARNES, C., CHARARA, M., and TARANTOLA, A., *Geological information and the inversion of seismic data.* In *Inverse Methods Interdisciplinary Elements of Methodology, Computation, and Application*, vol. 63 of *Lecture Notes in Earth Sciences* (eds. Jacobsen, B. H., Moosegard, K., and Sibani, P.) (Berlin, Springer-Verlag 1996) pp. 98–113.

BEST, A. I., MCCANN, C., and SOTHCOTT, J. (1994), *The Relationships between the Velocities, Attenuations, and Petrophysical Properties of Reservoir Sedimentary Rocks*, Geophys. Prosp. *42*, 151–178.

BEST, A. I., and SAMS, M. S. (1997), *Compressional Wave Velocity and Attenuation at Ultrasonic Frequencies in Near-surface Sedimentary Rocks*, Geophys. Prosp. *45*, 327–344.

BICHKAR, R., SINGH, S. K., and RAY, A. K. (1998), *Genetic Algorithmic Approach to the Detection of Subsurface Voids in Cross-hole Seismic Tomography*, Pattern Recogn. Lett. *19*(5–6), 527–536.

BLAIR, D. P. (1990), *A Direct Comparison Between Vibrational Resonance and Pulse Transmission Data for Assesment of Seismic Attenuation in Rock*, Geophysics *55*(1), 51–60.

BLAIR, D. P., and SPATHIS, A. T. (1984), *Seismic Source Influence in Pulse Attenuation Studies*, J. Geophys. Res. *84*, 9253–9258.

BOSHI, L., and DZIEWONSKI, A. M. (1999), *High- and Low-resolution Images of the Earth's Mantle: Implications of Different Approaches to Tomographic Modeling*, J. Geophys. Res. *104*(B11), 25,567–25,594.

BUNKS, C., SALECK, F. M., ZALESKI, S., and CHAVENT, G. (1995), *Multiscale Seismic Waveform Inversion*, Geophysics *60*(5), 1457–1473.

CARLSON, S. R., JANSEN, D. P., and YOUNG, R. P. (1993), *Thermally- induced Fracturing of Lac du Bonnet Granite*, Technical Report RP020AECL, Department of Geological Sciences, Queen's University, Kingston, Canada.

CASSIANI, G., BOHM, G., and VESNAVER, A. (1998), *A Geostatistical Framework for Incorporating Seismic Tomography Auxiliary Data into Hydraulic Conductivity Estimation*, J. Hydrol. *206*(1), 58–74.

ČERVENÝ, V. (1987) *Raytracing algorithms in three-dimensional laterally varying layered structures.* In *Seismic Tomography with Application in Global Seismology and Exploration Geophysics* (ed. Nolet G.), (Dordrecht, D. Reidel Publishing Company 1987) pp. 99–135.

ČERVENÝ, V., MOLOTKOV, I. A., and PŠENČIK, I. *Ray Method in Seismology* (Praha, Univerzita Karlova 1997).

CHANG, X. (1996), *Relationship Between Ray Distribution and Reconstructed Velocity Image in Reflection Tomography*, J. Appl. Geophys. *35*(2), 145–150.

DĘBSKI, W. (1997a), *The Probabilistic Formulation of the Inverse Theory with Application to the Selected Seismological Problems*, Ph.D. Thesis, IGF PAN. Publs. Inst. Geophys. Pol. Acad. Sc. *B-19 (293)*.

DĘBSKI, W. *Study of the image reconstruction accuracy of active amplitude tomography.* In *Rockburst and Seismicity in Mines Proceedings* (eds. Gibowicz, S. J., and Lasocki, S.) Postbus 1675, (Rotterdam, Netherland, Balkema 1997b) pp. 141–144.

DĘBSKI, W. (1999), *Preliminary Results of P-wave Tomography Determined from Selected Aftershocks of the Egion 1995 Earthquake, Gulf of Corinth, Greece*, Acta Geophys. Pol. *XLVII*(2), 165–184.

DĘBSKI, W. (2002), *Imaging rock structure using acoustic waves: methods and algorithms.* In *Seismogenic Process Monitoring* (ed. Osagawara, H.) (Balkema. in press).

DĘBSKI, W., and YOUNG, R. P. (1999), *Enhanced Velocity Tomography: Practical Method of Combining Velocity and Attenuation Parameters*, Geophys. Res. Lett. *26*(21), 3253–3256.

DEAL, M., and NOLET, G. (1996), *Nullspace Shuttles*, Geophys. J. Int. *124*, 372–380.

DUIJNDAM, A. (1988), *Bayesian Estimation in Seismic Inversion. Part I: Principles*, Geophys. Prosp. *36*, 878–898.

EPPSTEIN, M. J., and DOUGHERTY, D. E. (1998), *Optimal 3-D Travel-time Tomography*, Geophys. *63*(3), 1053–1061.

FALLS, S. D. (1993), *Ultrasonic Imaging and Acoustic Emission Studies of microcrack Development in Lac du Bonnet Granite*, Ph.D. Thesis, Queen's University, Ontario, Canada.

FEUSTEL, A. J. (1998), *Seismic Attenuation in Underground Mines: A Comparative Evaluation of Methods and Results*, Tectonophys. *289*, 31–49.

FEUSTEL, A. J., and YOUNG R. P. (1993), *Attenuation Analysis at the AECL Underground Research Laboratory Using the Spectral Ratio Method; Preliminary Results*, Technical Report RP022AECL, Atomic Energy of Canada, Ltd.

FRIEDEL, M., JACKON, M. J., WILIAMS, E. M., OLSON, M. S., and WESTMAN, E. (1996), *Tomographic Imaging of Coal Pillar Conditions: Observations and Implications*, Int. J. Rock Mech. Min. Sci. and Geomech. Abstr. *33*(3), 279–290.

GAO, S. (1997). *A Bayesian Nonlinear Inversion of Seismic Body-wave Attenuation Factors*, Bull. Seismol. Soc. Am. *87*(4), 961–970.

GIBOWICZ, S. J., and KIJKO, A. *An Introduction to Mining Seismology* (San Diego, Academic Press 1994).

GLADWIN, M. T., and STACEY, F. D. (1974), *Anelastic Degradation of Acoustic Pulse in Rock*, Phys. Earth Planet. Int. *8*, 332–336.

GOUVEIA, W. P., and SCALE, J. (1998), *Bayesian Seismic Waveform Inversion: Parameter Estimation and Uncertainty Analysis*, J. Geophys. Res. *103*(B2), 2759–2779.

HANYGA, A., and SEREDYŃSKA, M. (2000), *Raytracing in Elastic and Viscoelastic Media*, Pure appl. geophys. *157*, 679–717.

HUSEN, S., and KISSLING, E. (2001), *Local Earthquake Tomography Between Rays and Waves: Fat Ray Tomography*, Phys. Earth Planet. Int., in press.

IYER, H., and HIRAHARA, K. *Seismic Tomography, Theory and Practice* (London, Chapman and Hall 1993).

JACKSON, D. D., and MATSU'URA, M. (1985), *A Bayesian Approach to Nonlinear Inversion*, J. Geophys. Res. *90*(B1), 581–591.

JANSEN, D. P., CARLSTON, S. R., YOUNG, R. P., and HUTCHINS, D. A. (1993), *Ultrasonic Imaging and Acoustic Emission Monitoring of Thermally Induced Microcracks in Lac du Bonnet Granite*, J. Geophys. Res. *98*(B12), 22,231–22,243.

JIN, S., and BEYDOUN, W. (2000), *2D Multiscale Non-linear Velocity Inversion*, Geophys. Prosp. *48*(1), 163–180.

KANEKO, K., INOUE, I., SASSA, K., and ITO, I. (1979), *Monitoring stability of rock structures by means of acoustic wave attenuation.* In *Proc. 4th Congress of ISRM*, pp. 287–292.

KAVETZKY, A., CHITOMBO, G. P. F., MCKENZIE, C. K., and YANG, R. L. (1990), *A Model of Acoustic Pulse Propogation and its Application to Determine Q for Rock Mass* Int. J. Rock Mech. Min. Sci and Geomech. Abstr. *27*(1), 33–41.

KIJKO, A. (1994, April), *Seismological Outliers: L1 or Adaptive lp Norm Application*, Bull. Seismol. Soc. Am. *84*(2), 473–477.

KLIMENTOS, T., and MCCANN, C. (1990), *Relationship among Compressional Wave Attenuation, Porosity, Clay Content, and Permeability in Sandstones*, Geophys. *55*, 998–1014.

KOPER, K. D., WYESSION, M. E., and WIENS, D. A. (1999), *Multimodal Function Optimization with a Niching Genetic Algorithm: A Seismological Example*, Bull. Seismol. Soc. Am. *89*(4), 978–988.

LEE, W. H. K., and PEREYRA, V., *Mathematical introduction to seismic tomography*. In *Seismic Tomography, Theory and Practice*, (eds. Iyer, H. M., and Hirahara, K.), (London, Chapman and Hall. 1993) pp. 9–20.

LEIDENFROST, A., ETTRICH, N., GAJEWSKI, D., and KOSLOFF, D. (1999), *Comparison of Six Different Methods for Calculating Travel-times*, Geophys. Prosp. *47*, 269–297.

LOMAX, A. (1999), *Path-summation Waveforms*, Geophys. J. Int. *138*, 702–716.

LUTTER, W. J., FUIS, G. S., THURBER, C. H., and MURPHY, J. (1999), *Tomographic Images of the Upper Crust from the Los Angeles Basin to the Mojave Desert, California: Results from the Los Angeles Region Seismic Experiment*, J. Geophys. Res. *104*(B11), 25,543–25,565.

MAXWELL, S. C., and YOUNG, R. P. (1992), *Sequential Velocity Imaging and Microseismic Monitoring of Mining-induced Stress Change*, Pure appl. geophys. *139*(3/4), 421–447.

MAXWELL, S. C., and YOUNG, R. P. (1993), *A Comparison Between Controlled Source and Passive Source Seismic Velocity Images*, Bull. Seismol. Soc. Am. *83*(6), 1813–1834.

MAXWELL, S. C., and YOUNG, R. P. (1996), *Seismic Imaging of Rock Mass Responses to Excavation*, Int. J. Rock Mech. Min. Sci. and Geomech. Abstr. *33*(7), 713–724.

MCKIRDY, D. M., PETTIT, W. S., and YOUNG, R. P. (1998), *Thermo-elastic Modelling of Fracturing in Granite*, presented at GEOSCIENCE-98, Keele, U.K.

MENKE, W. *Geophysical Data Analysis: Discrete Inverse Theory*. International Geophysics Series (San Diego, Academic Press 1989).

MICHALEWICZ, Z. *Genetic Algorithms + Data Structures = Evolution Programs* (Berlin: Springer-Verlag 1996).

MIETTINEN, K., MÄKELÄ, M. M., NEITTAANMÄKI, P., and PÉRIAUX, *Evolutionary Algorithms in Engineering and Computer Science* (Chichester, John Wiley and Sons Ltd 1999).

MURPHY, W. F. (1984), *Acoustic Measures at Partial Gas Saturation in Tight Sandstones*, J. Geophys. Res. *89*, 11,549–11,559.

NOBLE, M. (1992), *Nonlinear Inversion of Seismic Data.* Ph.D. Thesis, Inst. de Physique du Globe de Paris.

NOLET, G. *Seismic Tomography* (Dordrecht, D. Reidel Publishing Company 1987).

NOLET, G., MONTELLI, R., and VIRIEUX, J. (1999), *Explicit, Approximate Expressions for the Resolution and a posteriori Covariance of Massive Tomographic Systems*, Geophys. J. Int. *138*(1), 36–44.

PARKER, R. L. *Geophysical Inverse Theory* (New Jersey, Princeton University Press 1994).

PEACOCK, S., MCCANN, C., SOTHCOTT, J., and ASTIN, T. R. (1994), *Experimental Measurement of Seismic Attenuation in Microfractured Sedimentary Rock*, Geophysics *59*, 1342–1351.

PENDOCK, N. (1993), *Bayesian image reconstruction*, In *Proceedings of the 8th Scandinavian Conference on Image Analysis.*, (NOBIM-Norwegian Soc. Image Process and Pattern Recognition) vol. 1, pp. 573–578.

PETTIT, W. (1997), *The Evaluation of a Transducer's Azimuthal Response Using an Aluminium Half-cylinder*, Technical Report, Keele University.

RIBODETTI, A., and VIRIEUX, J. (1998), *Asymptotic Theory for Imaging the Attenuation Factor Q*, Geophysics *63*(5), 1767–1778.

ROWBOTHAM, P. S., and PRATT, R. G. (1997), *Improved Inversion Through Use of the Null Space* Geophysics *62*(3), 869–883.

SCALES, J., GERSZTENKORN, A., and TREITEL S. (1988), *Fast l_p Solution of Large, Sparse, Linear Systems: Application to Seismic Travel Time Tomography*, J. Comp. Phys. *75*, 314–333.

SCALES, J., and SMITH, M. (1995), *Introductory Geophysical Inverse Theory*, Technical Report, Center for Wave Phenomena, Colorado School of Mines, Golden, Colorado.

SEN, M., and STOFFA, P. L., *Global Optimization Methods in Geophysical Inversion* (Amsterdam, Elsevier Science 1995).

SHATILO, A. P., SONDERGELD, C., and RAI, C. S. (1998), *Ultrasonic Attenuation in Glenn Pool Rocks, Northeastern Oklahoma*, Geophysics *63*(2), 465–478.

SIATA, R. (1999), *Accuracy and Effectivity of Delimination Methods of Field of Seismic Waves Attenuation in Coal Seam (Practical Example).* In *Acta Montana.*

TARANTOLA, A. *Inverse Problem Theory: Methods for Data Fitting and Model Parameter Estimation* (Amsterdam, Elsevier 1987).

TARANTOLA, A., and VALLETE, B. (1982), *Inverse Problems = Quest for Information*, J. Geophys. *50*, 159–170.

THOMSEN, L. (1986), *Weak Elastic Anisotropy*, Geophysics *51*, 1954–1966.

WANG, B., and BRAILE, L. W. (1996), *A multi-scale approach for seismic tomography.* In *International Conference on Image Processing*, (ed. I. S. P. Soc.), vol. 2, pp. 713–716.

WATANABE, T. (1997), *Full-waveform seismic tomography based on the acoustic and elastic wave-equation.* In *Proceedings of ISEEG'97 Conference*, ISEEG, pp. 93–98.

WATANABE, T., and SASSA, K. (1996), *Seismic attenuation tomography by use of amplitude attenuation and pulse broadening.* In *Expanded Abstract of 63rd SEG Meeting*, pp. 33–36.

WATANABE, T., and SASSA, K. (1996), *Seismic Attenuation Tomography and its Application to Rock-mass Evaluation*, Int. J. Rock Mech. Min. Sci. and Geomech. Abstr. 33(5), 467–477.

WESSEL, P., and SMITH, W. H. F. (1998), *New, Improved Version of the Generic Mapping Tools Released*, EOS, Trans. AGU *79*, 579.

WHITLEY, D., MATHIAS, K., and FITZHORN, P. (1991), *Delta coding: An iterative search strategy for genetic algorithms.* In *Proc. of the Fourth International Conference on Genetic Algorithms.*

WIELANDT, E. (1987), *On the validity of the ray approximation for interpreting delay times.* In *Seismic Tomography with Application in Global Seismology and Exploration Geophysics* (ed. Nolet, G.) (Dordrecht: D. Reidel Publishing Company) pp. 85–99.

WILLIAMSON, P. R. (1990), *Tomographic Inversion in Reflection Seismology*, Geophys. J. Int. *100*, 255–274.

WILLIAMSON, P. R. (1991), *A Guide to the Limits of Resolution Imposed by Scattering in Ray Tomography*, Geophys. *56*(2), 202–207.

YAO, Z. S., ROBERTS, R. G., and TRYGGVASON, A. (1999), *Calculating Resolution and Covariance Matrices for Seismic Tomography with the LSQR Method*, Geophys. J. Int. *138*, 886–894.
YOUNG, R. P., and HILL, J. J. (1986), *Seismic Attenuation Spectra in Rock Mass Characterization; a Case Study in Open-pit Mining*, Geophys. *51*, 302–332.
YOUNG, R. P., and MAXWELL, S. C. (1992), *Seismic Characterization of a Highly Stressed Rock Mass Using Tomographic Imaging and Induced Seismicity*, J. Geophys. Res. *97*(B9), 12,361–12373.
ZHOU, R., TAJIMA, F., and STOFFA, P. L. (1995), *Application of Genetic Algorithms to Constrain Near-source Velocity Structure for 1989 Sihuan Earthquakes*, Bull. Seismol. Soc. Am. *85*(2), 590–605.

(Received June 9, 2000, revised/accepted January 4, 2001)

Pure appl. geophys. 159 (2002) 309–343
0033–4553/02/030309–35 \$ 1.50 + 0.20/0

Pure and Applied Geophysics

A Simple Analogue Experiment to Account for Power-law and Exponential Decays of Earthquake Sequences

HIROSHI OGASAWARA[1]

Abstract — An event rate decays either exponentially in earthquake swarms, or according to the power law in aftershock sequences, suggesting different physics for these decays. In order to investigate the physics, I measured decays in the rate of acoustic emissions (AE) that occur to relax thermal stress after turning off a home cooking apparatus. Such AE are analogous to an earthquake induced by fluid intrusion because thermo-elasticity was precisely analogous to poro-elasticity. To consider what controlled the decays, different heating rates and durations were specified in several experimental runs for which the mechanical and thermal properties were identical. Temperature decayed exponentially with the same time constant. However, both a power-law decay with a p value roughly equal to two and an exponential decay were observed. To ascertain what was occurring at the AE source, the stress and frictional strength at the AE source were numerically estimated. The stress at the AE source decayed exponentially for both power-law decay and exponential decay. In power-law decay, strength was initially very low and recovered immediately and significantly. However, in exponential decay, the change in strength during the initial stage was not as large as that in power-law decay. It was demonstrated that the time constant of exponential event-rate decay was identical to that of temperature decay if strength is constant.

Key words: Strength, stress, power-law decay, exponential decay, thermoelasticity, poro-elasticity.

1. Introduction

Decay for an aftershock event rate $n(t)$ fits the power law well, meaning Omori's original formula as modified by UTSU (1957, 1961):

$$n(t) = \xi(t + c)^{-p} \tag{1}$$

where t is time, and ξ, c and p are empirically-determined constants. There have been numerous discussions concerning what those empirical constants mean physically. For example, UTSU *et al.* (1995) compiled the parameter p for more than 200 aftershock sequences that had been published in over 50 papers, noting in summary that $0.6 < p < 2.5$, with a median of about 1.1. These authors also reviewed papers that mentioned causes of variation in p (e.g., MOGI, 1962a, 1967a; MIKUMO and

[1] Faculty of Science and Engineering, Ritsumeikan University, 1-1-1 Noji Higashi, Kusatsu, 525–8577, Japan. E-mail: ogasawar@se.ritsumei.ac.jp

MIYATAKE, 1979; KISSLINGER and JONES, 1991; CREAMER and KISSLINGER, 1993) and concluded: "The variability of p value may be related to tectonic conditions of a region such as structural heterogeneity, stress and temperature, nonetheless it is not clear which factor is the most significant in controlling a p value."

With respect to earthquake swarms, MOGI (1967b) investigated decays in event rates for 30 earthquake swarms in Japan. He found that an event rate often decays exponentially following a large peak in the rate:

$$n(t) = \eta \exp(-qt) \tag{2}$$

where η and q are empirically-determined constants. An exponential decay has also been documented in other earthquake swarms: e.g., nine earthquake swarms on the Izu Peninsula, Japan (1978–1984) by MOGI (1991); a Hakodate earthquake swarm, Japan (1980–1981) by TSUKUDA (1991) and an Iriomote Island earthquake swarm, Japan by TSUKUDA (1993).

Three large peaks in the event rate were observed in the Matsushiro earthquake swarm, Japan, in 1965–1967. MOGI (1987) documented that the event rate decayed exponentially after the first and second peaks. After the third peak, however, a substantial amount of water seepage was followed by an earthquake swarm that did not decay exponentially (Fig. 1). For some 20 earthquake swarms (1978–1989) off the eastern Izu peninsula, Japan, MOGI (1991) pointed out the tendency for swarms before 1985 to decay exponentially, while those after 1984 did not. TSUKUDA (1991,

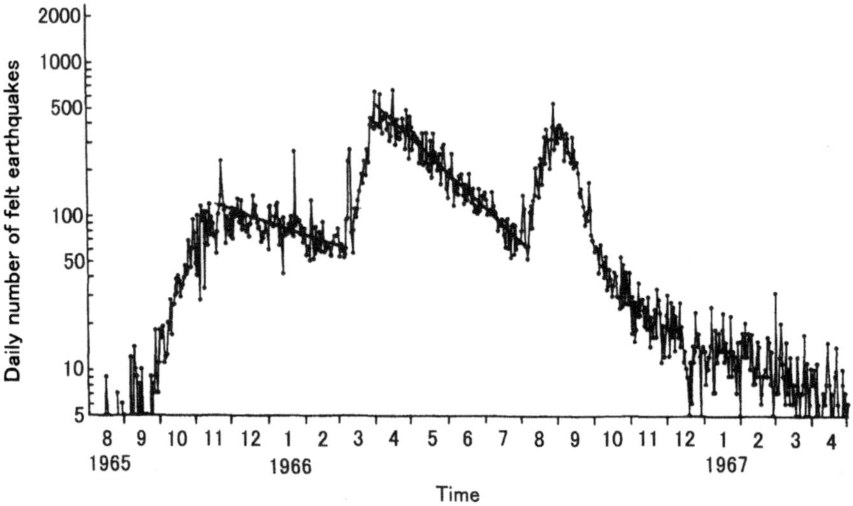

Figure 1

An example of exponential and power-law decays in an earthquake swarm (Matsushiro earthquake swarm, 1965–1967; after MOGI, 1987). A logarithm of the daily number of earthquakes felt at the Matsushiro observatory, Japan Meteorological Agency, is plotted versus time. After two peaks (November 1965 and March 1966), the event rate decayed exponentially but not after the peak of August 1997.

1993) fitted the modified Omori's law to the non-exponential decay swarms and found that p was roughly equal to two.

MOGI (1962b) conducted laboratory rock fracture experiments and observed exponential decay in AE bursts under constant stress. As a result, he suggested that exponential decay in earthquake swarms occurred under constant stress or constant pore pressure (MOGI, 1987), as observed in his earlier experiment (MOGI, 1962b). He also suggested that non-exponential decay was associated with a significant decrease in pore pressure for the Matsushiro earthquake swarm. However, HIRATA (1987) found both an exponential and a power-law decay for AE bursts during a creep test while applying constant stress to a dry rock specimen, although the difference in the decays was minor. He found that decay in AE bursts was initially exponential, and gradually shifted to a power-law decay, followed by failure of the specimen. He briefly mentioned that the change in fracture condition is the cause of this shift, but he did not elaborate on this because there was no available information on stress or strength in the earthquake source area. MOGI's (1962b) conclusion constrains the loading condition external to an earthquake source area, however it does not constrain the conditions of either stress or strength in the source area.

Reviewing the works mentioned above, and others, KOYAMA (1997) stated, "We do not know any physical relationship between an exponential time-decay of earthquake swarms and a power-law decay of aftershocks, although some studies derive such a relation based on physically-unproven *a priori* assumption."

In order to answer the questions posed by UTSU *et al.* (1995) and KOYAMA (1997), one must investigate event-rate decay that occurs in an environment with identical mechanical characteristics. However, there have been few such attempts to do this, as relevant as such work is.

OGASAWARA (1990a) demonstrated that Acoustic Emission (AE) frequently occurs after turning off a heating or a cooking apparatus such as a gas heater or an electric cooking pot at home, or a motor vehicle engine. Such AE activity decays according to Omori's modified power law, obeys the Gutenberg-Richter's law, and is reproducible. In this report, I will first demonstrate that investigating AE in this way will contribute to an understanding of earthquake physics.

OGASAWARA (1990a) used a gas heater but it contains many potential AE sources, which made further investigations inconclusive. He later found an electric cooking pot to be a suitable and very simple structure (OGASAWARA, 1990b) in which the AE source was confined to a limited area and AE again resembled those of aftershocks. OGASAWARA *et al.* (1992) changed both the rate and duration of heating in the rice cooker experiment and found different decays. In this paper I describe in detail the experiment of OGASAWARA *et al.* (1992) as one, which allowed for differences in decay under different tectonic conditions, although in materials with the same properties in a seismogenic area.

I analyzed the data of OGASAWARA *et al.* (1992) using the maximum likelihood method (OGATA, 1983; MATSU'URA, 1993). I investigated Akaike's Information

Criterion (AIC; AKAIKE, 1974) to see whether a power-law or an exponential decay better fits the observed decay in an event rate. I also carried out a numerical analysis to estimate strength and shear stress in a seismogenic area where a power-law or an exponential decay is observed. I considered what controls might be on the decay of an AE rate.

2. *Why a Rice Cooker?*

a) Analogy of Seismicity by Pore Pressure Change

Temporal change in the rate of earthquake occurrences is dependent upon changes in shear stress along a source fault and frictional strength, F. F is represented as:

$$F = \mu(\sigma_n - P) \tag{3}$$

where μ is a friction coefficient, σ_n the normal stress on a source fault and P pore pressure (e.g., JAEGER and COOK, 1979). For example, both recovery in friction and relaxation in stress are necessary to account for an aftershock sequence (e.g., DIETERICH, 1972a; MIKUMO and MIYATAKE, 1979).

The empirical constitutive law of friction (e.g., DIETERICH, 1972b, 1978; RUINA 1983) describes well a change in μ and can account for a variety of seismicities (e.g., DIETERICH, 1994; SCHOLZ, 1998). However, a major change in P will necessarily result in a considerably larger change in F. A large change in P can, therefore, play a dominant role in various aspects of seismicity, such as creep on the San Andreas fault (e.g., IRWIN and BARNES, 1975), earthquake swarms (e.g., OHTAKE, 1976), in fluid injection experiments (e.g., EVANS, 1966; HEALY *et al.*, 1968), and in seismicity induced by impoundment of water reservoirs (e.g., GUPTA and RASTOGI, 1976; SIMPSON *et al.*, 1988). Thus, an investigation of seismicity induced by changes in P is inherently important. Many laboratory experiments have been carried out to study the relationship between AE activity and pore pressure. The stress or pore pressure on a source fault, which has usually been left unstudied, is central to the present study.

As RICE and CLEARY (1976) state, a linear formation of the poro-elastic effect is precisely analogous to the theory of linear thermoelasticity. By investigating AE caused by thermal stress, therefore, I gain insights into seismicity induced by pore pressure changes.

b) A Rice Cooker and the AE Mechanism

Imagine two plates with an initial length L_0, elastic moduli E_1 and E_2, and thermal expansion coeffcients α_1 and α_2 as shown in Figure 2. I assume that the left edges of the plate are fixed. To avoid a bending effect, vertical deformation is

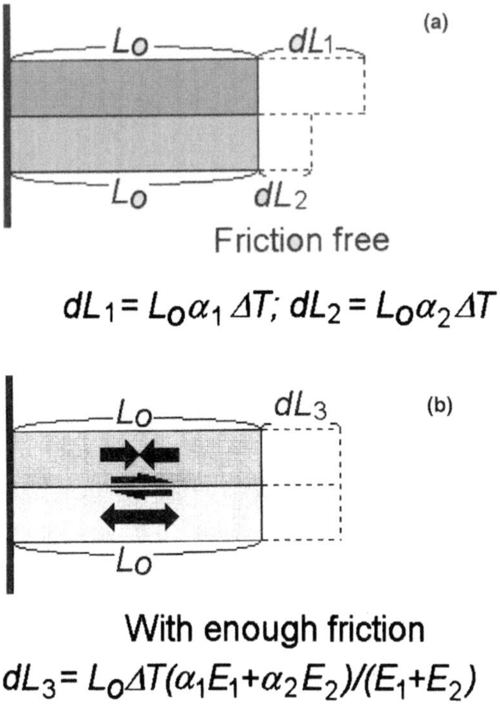

Figure 2
Schematic diagram showing the generation of shear stress along the interface of two materials with different properties for a temperature increase ΔT. Left edges are fixed. Vertical displacement is constrained to prevent bending. (a): Friction free. (b): with enough friction.

constrained. If the plates are free to expand, each plate will expand without stress, as shown by the dashed line in Figure 2a that is associated with an increase in temperature ΔT. However, if friction between the plates is extensive enough to keep them in contact (Fig. 2b), then compression and tensile stresses equal to $E_1E_2|\alpha_1\alpha_2|\Delta T/(E_1+E_2)$ will arise in the plates, and induce shear stress at the interface to counterbalance them (e.g., TIMOSHENKO and GOODIER, 1970). If the shear stress exceeds the frictional strength, AE will occur at the interface to release the stress.

Figure 3 shows the electric rice cooker used in the experiment by OGASAWARA *et al.* (1992). A rice cooker is used to cook rice for oriental dishes. It consists of a heating component, a pot, and a body structure to support them. In the heating component, a heater with a rounded U shape is clamped to the body by bolts A and B. Bolt A plays a key role in AE activity because no AE occurs if bolt A is removed. In contrast, no change in AE activity is seen when bolt B is taken off. Consequently, the AE source is limited to the area clamped by bolt A.

Figure 4 schematically represents the structure around bolt A. The heater is sandwiched by a steel plate and a porcelain insulator, and is clamped to an

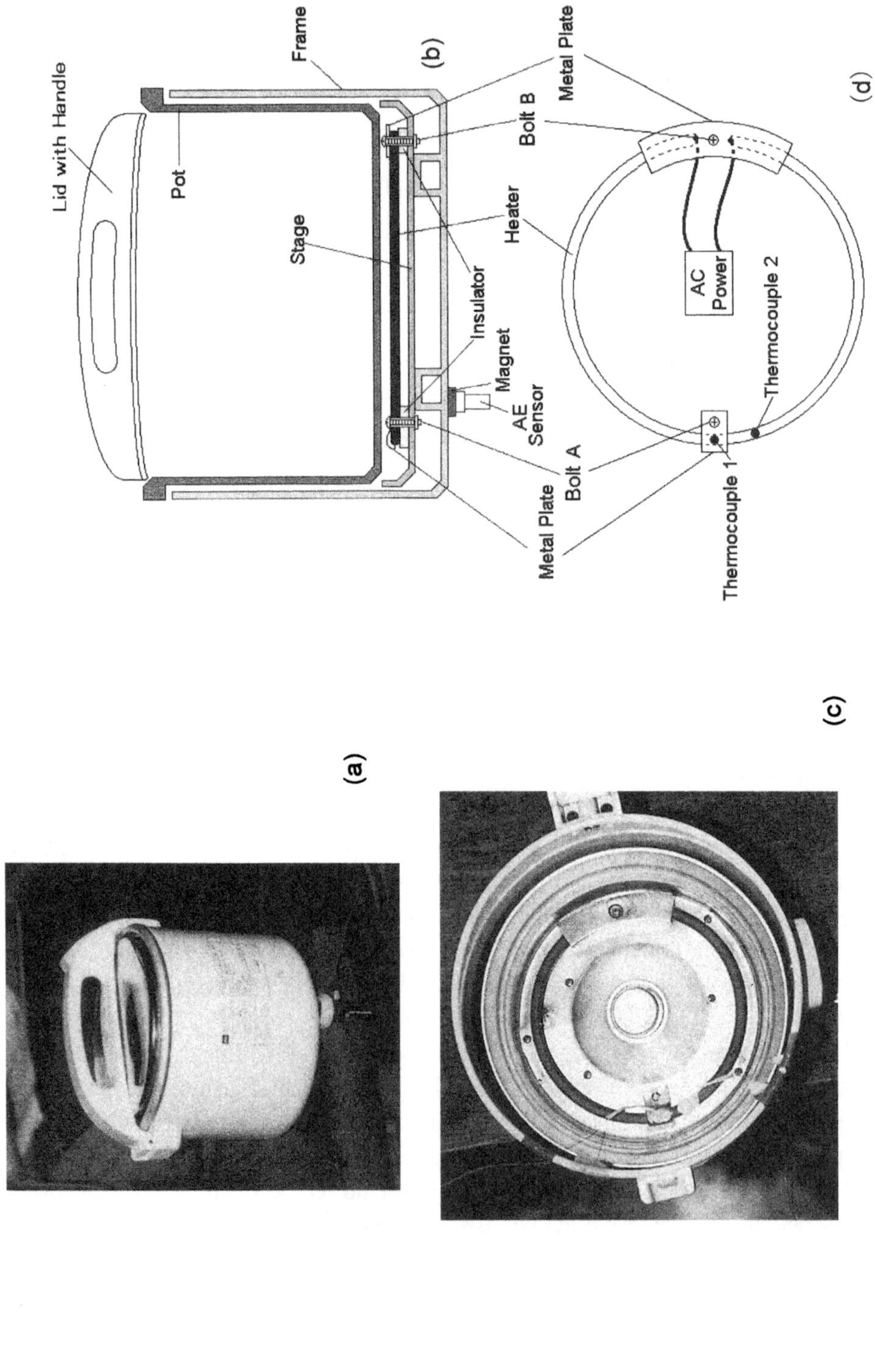
(a)
(b)
(c)
(d)
Lid with Handle
Pot
Frame
Stage
Heater
Insulator
Magnet
AE Sensor
Bolt A
Bolt B
Metal Plate
Metal Plate
AC Power
Thermocouple 1
Thermocouple 2

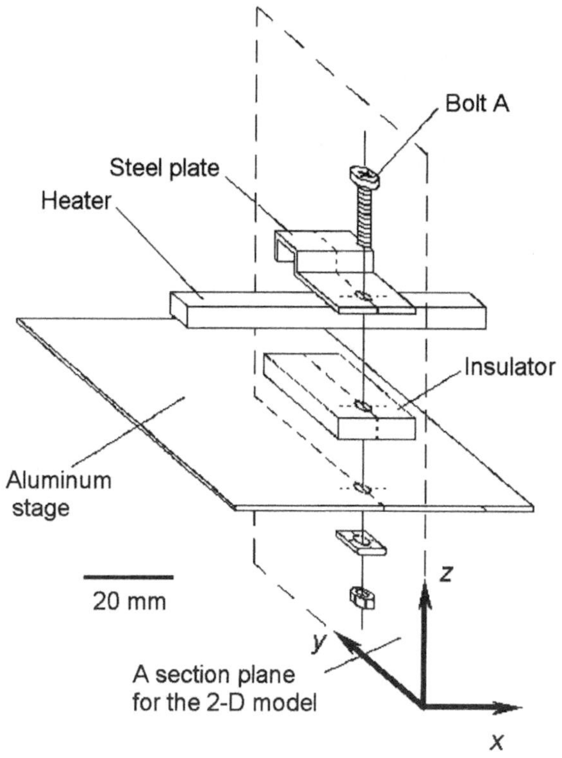

Figure 4
Schematic diagram representing the configuration of the core parts of the rice cooker shown in Figure 3. As the coefficient of thermal expansion and elastic modulus are not the same for different parts, AE will occur to release thermal shear stress on interfaces of the parts during heating or cooling. With a change in temperature, bolt A expands or shrinks, leading to changes in frictional strength on the interfaces.

aluminium stage by bolt A. If one assumes physical properties for each part as shown in Table 1 (data after BAUCCIO, 1994), and increases the temperature by 100 °C, the thermal stress shown in Table 2 is expectable. Stress exceeding 10 MPa can be expected at the interfaces between the steel plate and the porcelain, between the heater (MgO) and the porcelain, and between the porcelain and the aluminium stage, making them candidates for acoustic emissions. It is easy to change the temperature by 100 °C in the rice cooker. Stress in the sources areas of natural earthquakes

◄

Figure 3
The electric rice cooker used in the experiment. (a) Side view. (b) Structure of the rice cooker (side view). (c) Top view of the core elements of the rice cooker when the lid and pot are taken off. (d) Schematic diagram showing the configuration of the heater, steel plate, and thermocouples (top view).

Table 1

Physical properties of familiar materials (after BAUCCIO, 1994)

	Thermal cond. W/(mK) @RT	Thermal expns. $\times 10^{-6}$/K	Specific heat kJ/(kgK)	Density kg/m^3	Young's modulus GPa	Thermal stress MPa/K	Thermal diffusivity $\times 10^{-6}$ m^2/s	Emissivity
			Ceramics					
MgO	59	13.5	1.17	3600	210	2.84	4.07	0.9
			Whiteware and clay based products					
Earthenware	1.26	7.3–8.3	1*	2200	55	0.41–0.46	0.57	
Hard porcelain	1*	3.5–4.5	1*	2400	75			0.92
Bone china	1.26	8.4	1*	2700	96	0.81	0.47	
Hotel china		7.3–8.3	1*	2600	82	0.60–0.68		
			Metal					
Aluminium	160	22.4	0.96	2700	70	1.57	61.73	0.09
Steel	38	14	0.45	7800	200	2.80	10.83	0.31

*: assumed

Table 2

Thermal stress in MPa per 100 °K change for the rice cooker parts that are in tight contact

	Porcelain	Steel	Aluminium
MgO	52.5	5.1	
Porcelain		54.5	
Steel			43.6

several kilometers deep may be the same as shown in Table 2. Therefore, the rice cooker experiment is one of the easiest analogues to infer the activity at the source areas of natural earthquakes.

The frictional strength F for the rice cooker is represented as:

$$F = \mu(\sigma_b - B) \tag{4}$$

where σ_b is a constant representing the stress to clamp bolt A (Fig. 4) before the experiment. B is a variable that represents expansion of bolt A during the experiment; the steel bolt will expand more than the porcelain insulator (Fig. 4). With an increase in temperature, the bolt will loosen, leading to a decrease in F, whereas a decrease in temperature conversely leads to an increase in F. It should be noted that Eq. (4) is identical to Eq. (3), the fracture criteria for an earthquake. Temperature plays the same role as pore pressure. Therefore, the rice cooker experiment can be regarded as mimicking an earthquake generation experiment under a constant normal stress with varying pore pressure.

Phenomena observed in the rice cooker experiment occur in a time scale on the order of 100 s, a dimension of about 1 cm, and a thermal diffusivity of 10^{-6} m^2/s

Table 3

General rock fracture categories and transmissivity of the water (after PUSCH*, 1995)*

Length (m)	Spacing (m)	Width (m)	Gauge thick (m)	Transmissivity (m^2/s)
100–1000	10–100	1–10	< 0.01	10^{-9}–10^{-6}
1000–10000	100–1000	10–100	0.1	10^{-7}–10^{-4}
> 10^4	> 10^3	> 10^2	1	10^{-5}–10^{-2}

(BAUCCIO, 1994). Poroelasticity is precisely analogous to thermo-elasticity (RICE and CLEARY, 1976). Both temperature and pore pressure behave according to the diffusion equation. Pore water diffusion is a fissure zone longer than 10 km and wider than 100 m has a diffusivity of about 10^{-2} m^2/s (Table 3; e.g., PUSCH, 1995). Taking scaling similarity into account, a phenomenon seen in a 1-cm dimension and a 100-s duration with a 10^{-6} m^2/s diffusivity corresponds to a phenomenon in a 100-m dimension and a 10-day duration with a 10^{-2} m^2/s diffusivity. Therefore, the phenomenon observed in the rice cooker experiment is comparable to an earthquake swarm occurring in such a fissure zone on such a time scale.

Since stress in the rice cooker changes with temperature, it changed independent of AE in the rice cooker experiment. This situation is not usually expected in an aftershock process. However, significant crustal movements seemingly independent of earthquake occurrences are often observed in earthquake swarms. For instance, in a swarm off the eastern Izu peninsula, Japan, YAMAMOTO *et al.* (1991) observed significant crustal movement that is explainable not by the cumulative moment of earthquakes but by the opening of vertical cracks. The rice cooker experiment contributes insights into such seismicity.

As described in detail above, temperature is equivalent to pore pressure. Frictional strength is described by the same equation as that for natural earthquakes. Therefore, by using the rice cooker I can investigate seismicity associated with changes in shear stress and frictional strength by analogy with pore pressure changes.

3. The Experiment of OGASAWARA et al. *(1992)*

a) Experimental Methods and Observation System

The lid and the pot were taken off as shown in Figure 3c. An AE sensor (an accelerometer with a sensitivity of 100 m V/G and a frequency range extending to 10 kHz) was attached with a magnet to the bottom of the frame, to be away from high temperatures. Signals were amplified by 1000 times and recorded by the audio channels of a consumer video recorder (Fig. 5). In order to monitor the temperature

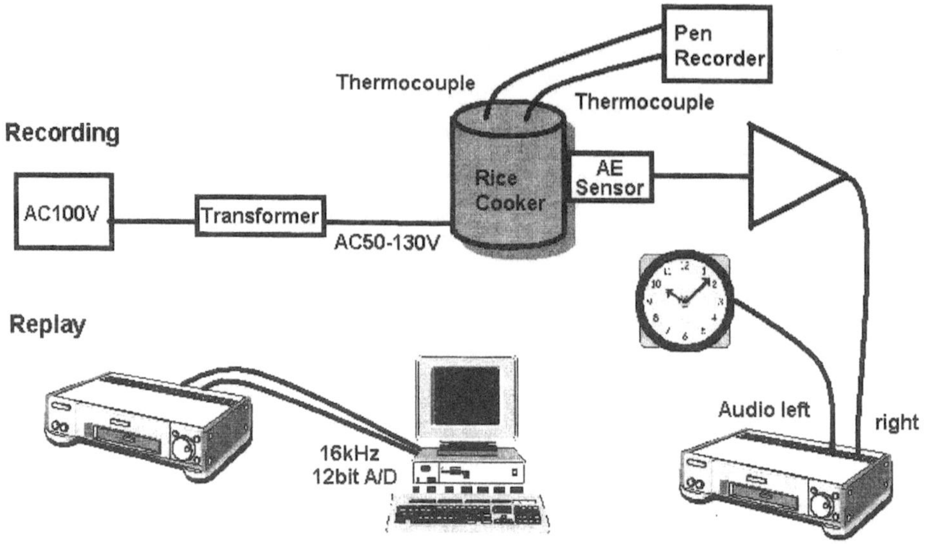

Figure 5
Schematic diagram of the data acquisition system. A transformer controls AC voltage supplied to the rice cooker.

in the AE source area, thermocouples were installed on the steel plate and the heater using alumina cement. Their output was recorded by a pen recorder. After heating at different rates and durations, as described in the next section, the switch was turned off and AE activity was recorded until it ceased.

The accelerograms were replayed and digitized with a 12-bit 16.1-kHz A/D conversion and stored on the hard disk of a PC. Figure 6 (upper) shows the accelerograms for 30 sec after the switch was turned off in experimental run 1, as described in the next section; a close-up view of the first event is also shown in the inset. To read the maximum amplitude and occurrence time, each event was enlarged as shown in Figure 6 (lower). Suspicious accelerograms were replayed on a speaker and were listened to in order to check whether they were man-made noise or AE activity. I counted events with an amplitude larger than 45 mV and recorded their occurrence times and maximum amplitudes. At about 8.02 s in Figure 6 (lower), a small event with a maximum amplitude of $\sim$20 mV may be hidden. However, there was slight possibility of overlooking events with amplitudes exceeding 45 mV.

b) Experimental Conditions

In order to determine what controlled the AE activity, experiments were carried out under four different combinations of different rates and durations of heating. I changed the heating rate by controlling AC voltage supplied to the rice cooker with a

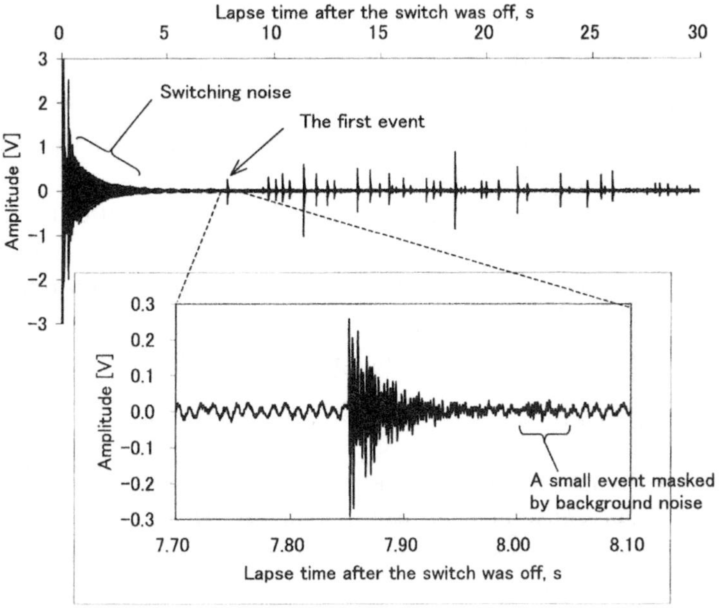

Figure 6
An example of an accelerogram for 30 seconds in run 1. The large acceleration until 4 s is noise generated when the switch was turned off. The first event is enlarged in the lower part of the figure. A small event masked by background noise that was not analyzed is also indicated in the lower figure.

Table 4

Experimental conditions and summary of the results

Runs	Run 1	Run 2	Run 3	Run 4
Supplied power, V	50	100	130	100
Heated period, s	600	68	37	128
Temp. when off, °C	225	225	225	400
Max. Temp., °C	225	251	256	404
Max. Temp. diff., °C	25	55	59	55
Number of AE	115	304	394	628
Decay of AE activity	power law	Exponent	exponent	power law
q value, s^{-1}	0.0155	0.0283	0.0229	0.0114
c value, s	69.4	111.1	151.6	30.6
p value	2.94	4.92	5.20	1.76

transformer (Fig. 5). As listed in Table 4, in runs 1, 2 and 3, the rice cooker was turned off when the temperature of the heater reached 225 °C, when the AC voltage was 50, 100 and 130 V, respectively, with the heating rates and durations being different. The AC voltage in run 4 was 100 V, which was the same as in run 2, however the rice cooker was turned off at 400 °C, which was decidedly hotter than in run 2.

c) Temperature Change

Figure 7 shows a logarithm of temperature plotted versus lapse time after the rice cooker was switched off. As can be seen in run 1, the switch was turned off when the temperature nearly reached a stationary state. In runs 2 and 3 the switch was turned off before reaching a stationary state.

In runs 1 and 4 the temperature started to decrease soon after the switch was turned off, whereas in runs 2 and 3 it continued increasing for a while after the switch was turned off. However, it is significant that in all runs the temperature decayed exponentially with a quite similar time constant of about $-0.004\ s^{-1}$ (compare the solid guide line in Fig. 7a) after the switch was turned off.

d) The Role of the Bolt

In Figure 8 the maximum amplitude of AE, A_{max}, the cumulative amplitude, ΣA_{max}, and the cumulative number of AE events, N, are plotted versus lapse time after the cooker was switched off. When the switch was turned off, the bolt was expectably the loosest, because it had thermally expanded the most, resulting in low AE activity. However, the bolt became tight with decreasing temperature. A high stress rate will lead to a substantial number of smaller events. However, larger events will not occur until the bolt is sufficiently tight, as generally summarized in Table 5. In sum, the role of the bolt is central to controlling the manner of AE activity.

OGASAWARA *et al.* (1998) conducted an experiment in which the parts shown in Figure 4 were cooled with a fan, but only during heating periods. They detected AE that began soon after turn-off, once again suggesting that the bolt controlled the AE activity.

Therefore, there is abundant qualitative confirmation that AE activity is controlled by not only decay in stress rate, but also by the recovery in frictional strength due to contraction of the bolt.

e) Exponential or Power-law Decay?

In Figure 8 one sees that the cumulative number of AE and the cumulative maximum amplitude did not increase after $t = 100$ s in run 2 or $t = 200$ s in run 3, whereas both continued to increase in runs 1 and 4. Differences in decay were therefore thoroughly investigated using the maximum likelihood method (OGATA, 1983). I used computer code from MATSU'URA (1993), which is essentially the same as is included in IASPEI Software Library Volume 6. I applied a power-law decay represented by Eq. (1) and an exponential decay represented by Eq. (2) to the observed number of AE, $N(t)$. Akaike's Information Criterion (*AIC*; AKAIKE, 1974) is introudced here to compare the fit:

$$AIC = -2 \max \ln(\text{likelihood}) + 2(\text{number of adjusted parameters}) , \quad (5)$$

refer to OGATA (1983) for more detail on likelihood.

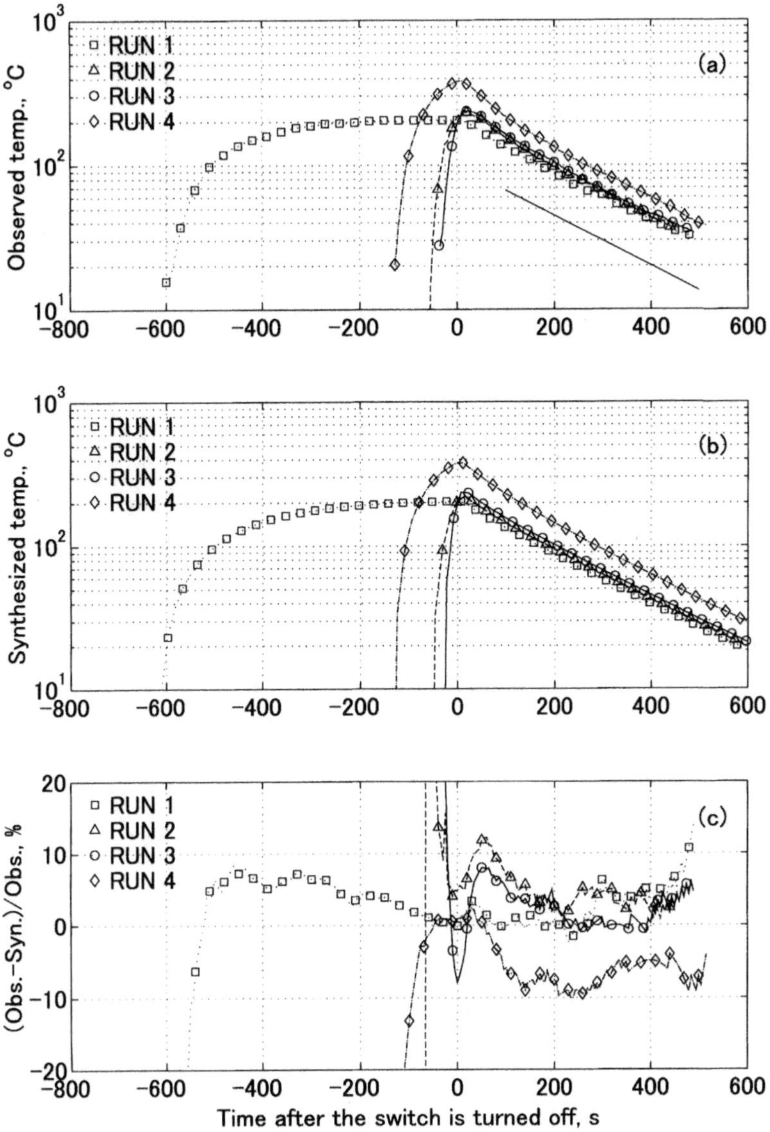

Figure 7

Temporal change in temperature (log scale) versus lapse time since the switch was turned off. Observed and synthesized temperature changes referred to room temperature are shown in (a) and (b), respectively. In Figure 7a, an exponential decay with a time constant of $q = -0.004\ s^{-1}$ is shown as a guide by the solid line without symbols. Difference between the observed and synthesized temperature normalized to the observed one is shown in (c). In run 1, heating took place for the longest time period. In runs 1, 2 and 3 the switch was turned off when the temperature was 225 °C, whereas 50 V, 100 V and 130 V were supplied, respectively. 100 V was supplied in run 2 and run 4 but the switch was turned off at 400 °C in run 4. Note that the temperature decays exponentially with the same time constant in all four runs.

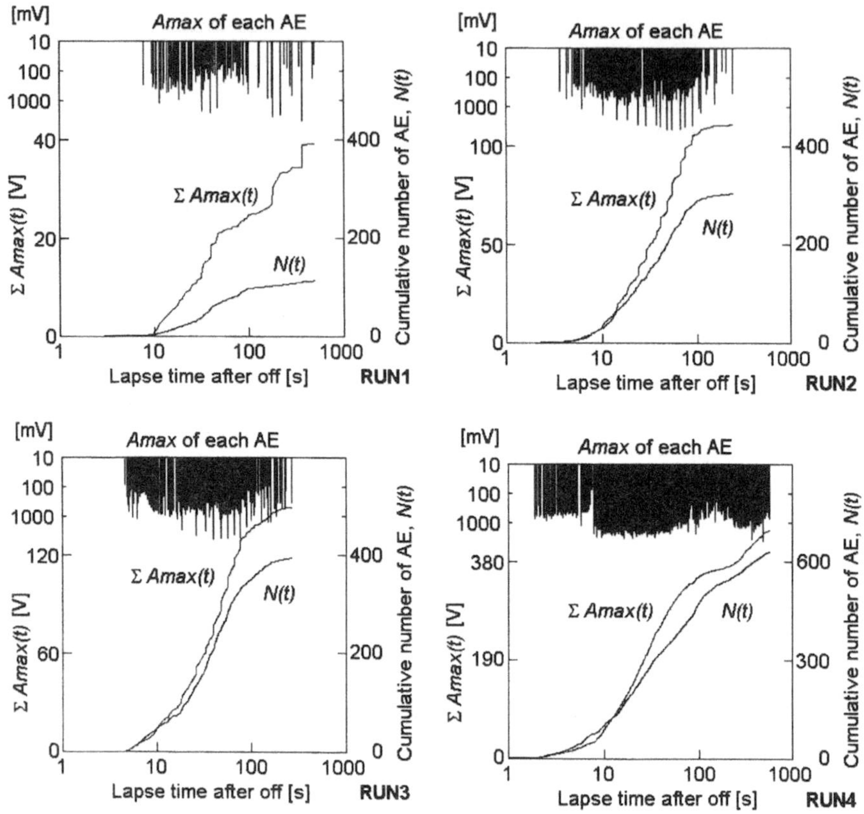

Figure 8
Maximum amplitude for each event (bar graph), cumulative maximum amplitude (ΣA_{max}) and cumulative number of AE (N) after turning off the rice cooker.

Tables 4 and 5 show the best-fit parameters in Eqs. (1) and (2) with *AIC*. The differences in *AIC* between power-law and exponential decays are 11 at its minimum and 158 at its maximum. In a release note with MATSU'URA'S (1993) computer code, she notes that the difference is significant if the difference in AIC is larger than three. As a result, the difference in the fit of models is significant in our analysis.

Table 5

Relating the characteristics of AE activity to thermal stress conditions

Temperature	Contact	Friction	Shear stress	Shrinking rate	AE rate	AE magnitude
High	Very loose	Very small	Very small	Fastest	No	–
	Loose	Small	Small	Fast	Frequent	Small only
	Tight	Medium	Medium	Slow	Most frequent	Large
	Tighter	Large	Large	Slowest	Seldom	Large
Low	Tightest	Largest	Largest	No	No	–

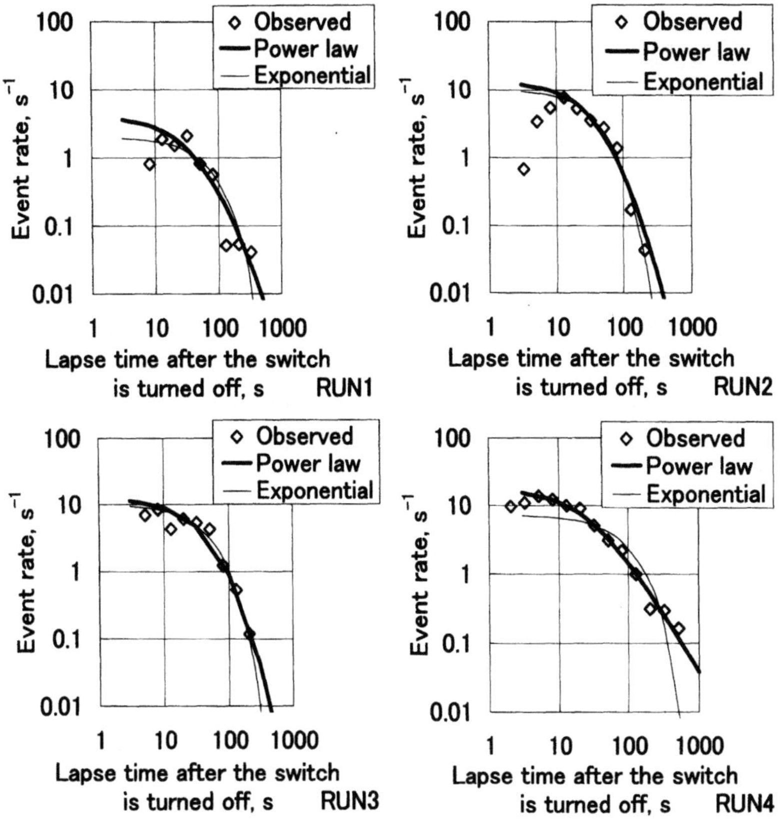

Figure 9
Log event rate AE versus log time after the switch was turned off. Diamonds show the observed event rate. The best-fit power-law and exponential decays are also shown.

Figure 9 shows the logarithms of the AE rate plotted versus the logarithms of lapse time after being turned off. In run 4, a power-law decay (the thick line in Fig. 9) clearly fits decidedly better than does an exponential decay (the thin line in Fig. 9). The p value for run 4 is 1.76, being nearly equal to two that TSUKUDA (1991, 1993) pointed out for decays in the final stages in earthquake swarms.

The differences in runs 1, 2 and 3 are somewhat subtle in Figure 9. However, in Figure 10 there is clearly a significant difference in residuals in the cumulative number of AE, $N(t)$, from the best-fit models. In Figure 10, run 4 again decays beautifully according to the power law. Run 1 also fits the power law better, although less clearly than does run 4. Because significantly smaller residuals are seen for the exponential decay in runs 2 and 3 (Fig. 10), exponential decay fits them better.

In analyzing natural earthquake swarms (e.g., MOGI, 1967b; TSUKUDA, 1993), one does not always know when "*the switch is off*," and in such cases, one usually fits decay models to the data only after seismicity starts to decay. With this in mind, one

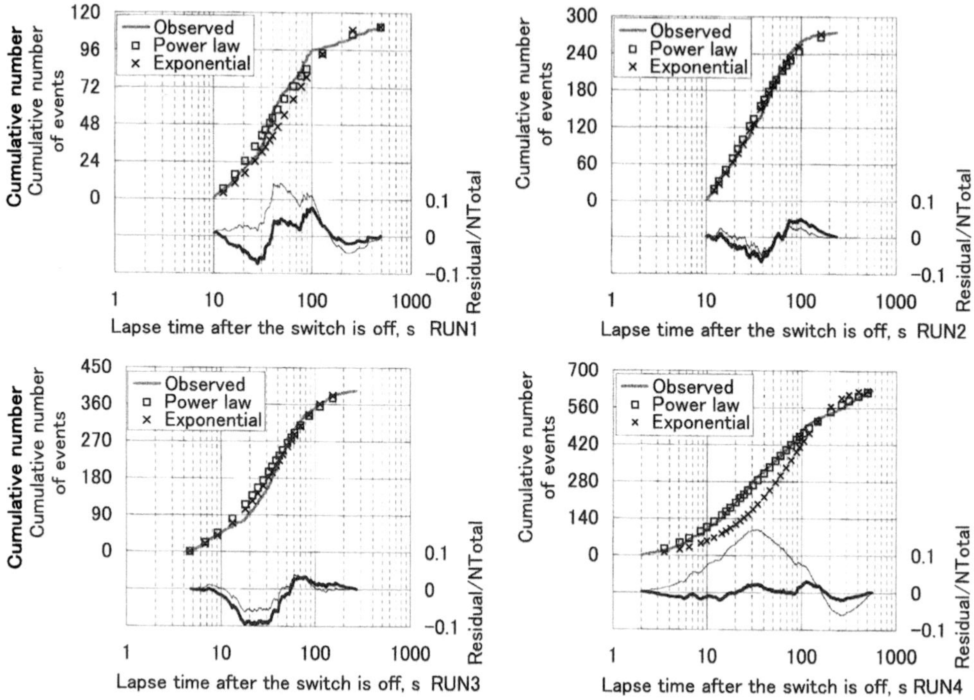

Figure 10
Cumulative number and residual versus log time after the switch was off. The thick gray line shows the observed numbers. The thick and thin black lines respectively represent the residuals for power-law and exponential decay, normalized by the total observed number of AE.

should analyze the present data in a manner similar to the standard analysis for natural earthquake swarms. Throughout this paper I will discuss use of the parameters obtained by fitting the models to the data after the activity started to decay. In order to check the influence of data selection on the fit, I fitted Eqs. (1) and (2) to all the data after the switch was off, as well as to the data after the decay started. Because smaller *AIC* means a better fit (AKAIKE, 1974), power-law decay fits better for runs 1 and 4 independent of data selection, whereas exponential decay fits better for runs 2 and 3, as summarized in Table 6.

It should be noted that both power-law and exponential decay were seen, even thogh temperature decays exponentially with the same time constant. To account for decay in AE activity, something other than the temperature must be taken into account.

f) Magnitude-frequency Relationship

Figure 11 shows the relationship between maximum amplitude and an AE rate. The source area is narrow compared to the hypocentral distance (Fig. 3), and the

Table 6

Comparison of AIC

	Run 1		Run 2		Run 3		Run 4	
Data	All	After peak	All	After peak	All	After peak	All	After peak
AE number	114	111	304	275	394	394	628	626
Exponentia	326	326	−49	−31	−153	−153	51	34
Power law	321	310	−32	−20	−132	−132	-81	−124

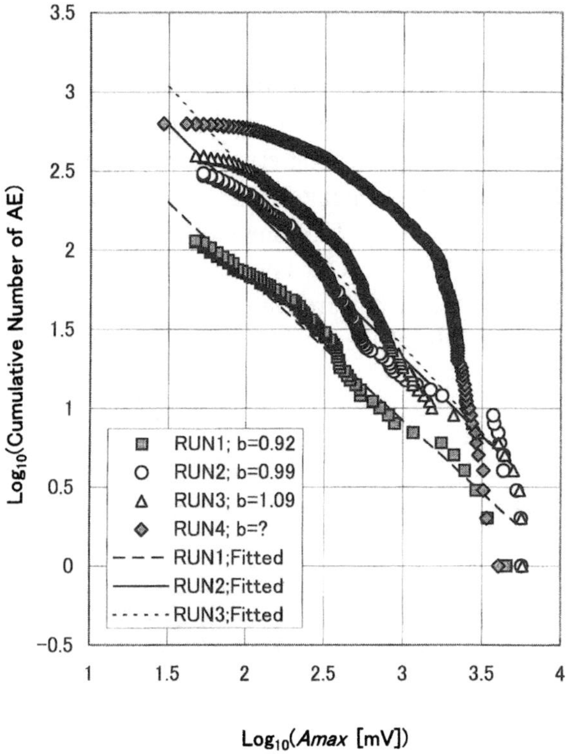

Figure 11
Cumulative number of AE versus maximum amplitude on a log-log scale.

hypocentral distance can be regarded as nearly the same for all events. Therefore, Figure 11 can be regarded as a magnitude-frequency relationship.

In Figure 11 runs 1, 2 and 3 seem to be nearly straight, except for the large events. Runs 1, 2 and 3 have respective *b* values of 0.92 ± 0.03, 0.99 ± 0.04, and 1.09 ± 0.05. Run 4 is not straight at all and I cannot deduce any meaningful *b* value, but it is obvious that larger events occurred more frequently than smaller events. Therefore, the runs which decay according to the power law have smaller *b* values or a tendency to include larger events.

4. Numerical Analysis to Learn What is Taking Place in the AE Source Area

Thus far, the stress or strength at the AE source has not been addressed at length. In the following sections, therefore, I describe a quasi-three-dimensional numerical analysis to determine the temperature distribution in all core elements. Based on the stress and the strength thus obtained, I discuss what controls decay in an event rate.

a) Methods

Temperature change was numerically estimated for the model shown in Figure 12, the shape of which is identical to the *yz* section of the core elements around the AE source (Fig. 4). I assumed that no heat was transferred along the *x* axis, thus temperature was independent of the *x* coordinate.

The properties of the heat-generating element were assumed to be the same as those of the MgO ceramics surrounding it. The steel bolt was modeled as a vertical bar united with a folded steel plate. To represent the alumina cement used to attach thermocouples in the experiment of OGASAWARA *et al.* (1992), a layer with the same properties as those of porcelain was present on the top. Table 1 shows the physical properties assumed for all the elements (BAUCCIO, 1994).

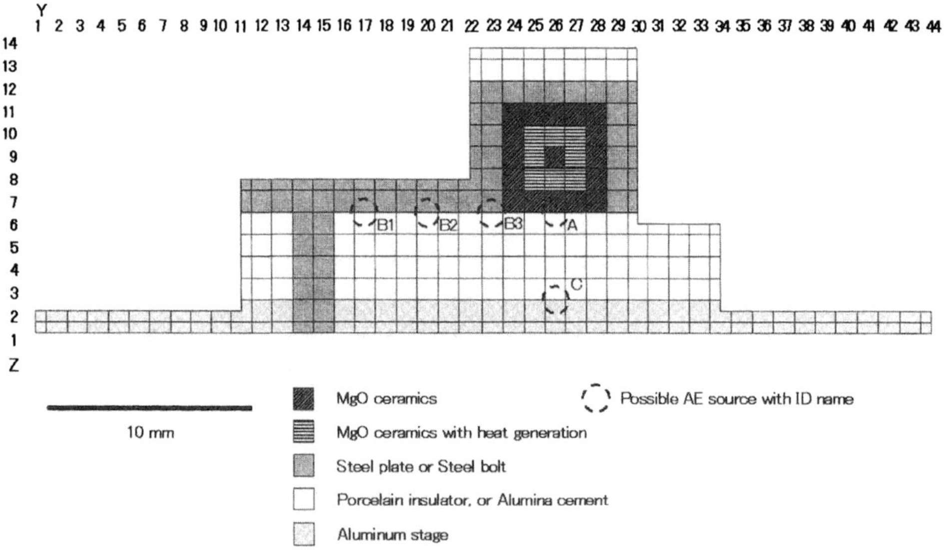

Figure 12

A two-dimensional model of temperature changes round the AE source. The model is nearly identical to the *yz* section in Figure 4. A, B1, B2, B3 and C are the candidates for the location of AE activity.

Using a conventional finite-difference approximation (e.g., CHAPMAN, 1987), I solved the diffusion equation:

$$dT/dt = \kappa_n/\rho_n\alpha_n(d^2T/dy^2 + dT^2/dz^2) + Q \tag{6}$$

where T and t are absolute temperature and time, respectively; κ_n, ρ_n, α_n, and Q are thermal diffusion coefficient, density, specific heat capacity and generated heat, respectively, with suffix n being the identification number of the materials. These properties were assumed to be constant, regardless of temperature. The specific heat of porcelain was assumed to be 1 kJ/(kgK).

While the switch was on, heat Q was generated in the heater. The heater used in the experiment by OGASAWARA *et al.* (1992) operated at a power of 400 W at 100 V; the voltage supplied for runs 1, 2 and 3 was 50 V, 100 V and 130 V, respectively (Table 4). Therefore, the power supplied for runs 1 and 3 was respectively $0.25 \times Q_2$ and $1.69 \times Q_2$, where Q_2 is the power supplied in run 2.

I also assumed a room temperature of 20 °C (293°K), with heat loss by convection and radiation as follows:

$$\text{Heat loss by convection} = h(T - 293) \tag{7}$$

$$\text{Heat loss by radiation} = \varepsilon_n\sigma(T^4 - 293^4) \tag{8}$$

where h is a coefficient for free convection in air, which is normally between 5 and 25 $\mathrm{Wm^{-2}\,K^{-1}}$ (e.g., CHAPMAN, 1987). I assumed $h = 20\ \mathrm{Wm^{-2}\,K^{-1}}$ to obtain the simulated temperature change that best fits the temperature change observed by OGASAWARA *et al.* (1992). ε_n in Eq. (8) is emissivity, generally being much less than 1 and dependent upon surface conditions (see CHAPMAN, 1987). σ is the Stefan-Boltzmann coefficient of $5.67 \times 10^{-8}\ \mathrm{Wm^{-2}\,K^{-4}}$. As to $h = 20\ \mathrm{Wm^{-2}\,K^{-1}}$ and $\varepsilon_n = 1$, and convection is dominant for $T < 300$ °C. If $T > 300$ °C, then radiation is greater than convection; for instance, at $T = 400$ °C, radiation is about 1.5 times greater than convection. Most temperature ranges in the experiments of OGASAWARA *et al.* (1992) were lower than 300 °C, implying that convection was dominant. This is why T decayed exponentially in the experiment after the unit was turned off (Fig. 7a):

$$C_t\, dT/dt = -h(T - 293^\circ\mathrm{K}) \ , \tag{9}$$

where C_t is a thermal capacity.

b) Space and Time Variations of Temperature

Initially, I did not use the x dimension effect to estimate temperature. Immediately after the switch was turned off, a rapid decrease in temperature was calculated, whereas a slight increase in temperature was observed at $t = 0$–20 s in runs 2 and 3 (Fig. 7a).

In the pure two-dimensional model that I used in the initial estimation, the length of the steel plate, the heater, the insulator and the stage were assumed to be comparable in the x direction. However, as seen in Figure 4, the length of the steel plate and the insulator in the x direction, is considerably smaller than that of the heater and the stage, by at least a factor of 20. Therefore, the thermal capacity of the heater has been underestimated, resulting in the rapid decrease in temperature noted above. Although three-dimensional analysis can solve the problem of underestimation in the thermal capacity of the heater, it would be time-consuming. Therefore, in order to take the quasi-three-dimensional effect into account, I multiplied the thermal capacity and the heat loss of the heater by a factor of 20 and carried out a two-dimensional analysis. Figure 7 shows the temperature thus obtained at the place where the thermocouple was installed to monitor temperature in the experiment of OGASAWARA *et al.* (1992). As can be seen, the synthesized temperature (Fig. 7b) is nearly identical to the observed temperature (Fig. 7a). Differences between the observed and synthesized temperature are within 10% for the period from ~10 s before the switch was turned off (see Fig. 7c).

In section 3, temperature data only on the surface were available, and discussion here has also been limited. However, one can now see the temperature distribution and its temporal variation everywhere for all runs. Let us look carefully at the temperature in the yz section in several stages for all the runs shown in Figure 13.

In run 1, 50 V was supplied for the longest period, until the temperature of the plate surface reached 225 °C and a power-law decay of $p = 2.94$ was observed (Table 4). The synthesized temperatures at $t = 0$ s are about 200 °C most places in the yz section. It later cooled down homogeneously as well.

In runs 2 and 3, 100 V and 130 V were supplied, respectively, for a short period until 225 °C was reached, and the exponential decays were observed in the event rate (Table 4). Low temperature around the bolt (indicated by an arrow in Fig. 13) is an important characteristic in the temperature distribution of runs 2 and 3. Until $t = 45$ s, the temperature around the bolt was cooler than in run 1.

In run 4, 100 V was supplied for a long period to reach 400 °C than in run 2, and the power-law decay with $p = 1.76$ was observed (Table 4). The most salient characteristic is that the temperature around the bolt was the highest among all runs. The distribution was not very homogeneous until $t = 100$ s, but the later temperatures were distributed homogeneously. The temperature distribution at $t = 200$ s in run 4 is nearly the same as that at $t = 100$ s in run 1. At $t > 200$ s in run 4, decay occurs as it does at $t > 100$ s in run 1.

As mentioned above, one sees fundamental differences in the temperature distribution in the four runs regardless of the fact that temperature decays exponentially with time for all runs. Using these temperature data, I will next

▶

Figure 13
Space and time change for the synthesized temperature. Snapshots of temperature distribution on yz sections are shown with time advancing downward. Experimental runs from 1 to 4 are respectively shown from the leftmost to rightmost columns. The location of the bolt ($y = 14$–15) is pointed by an arrow.

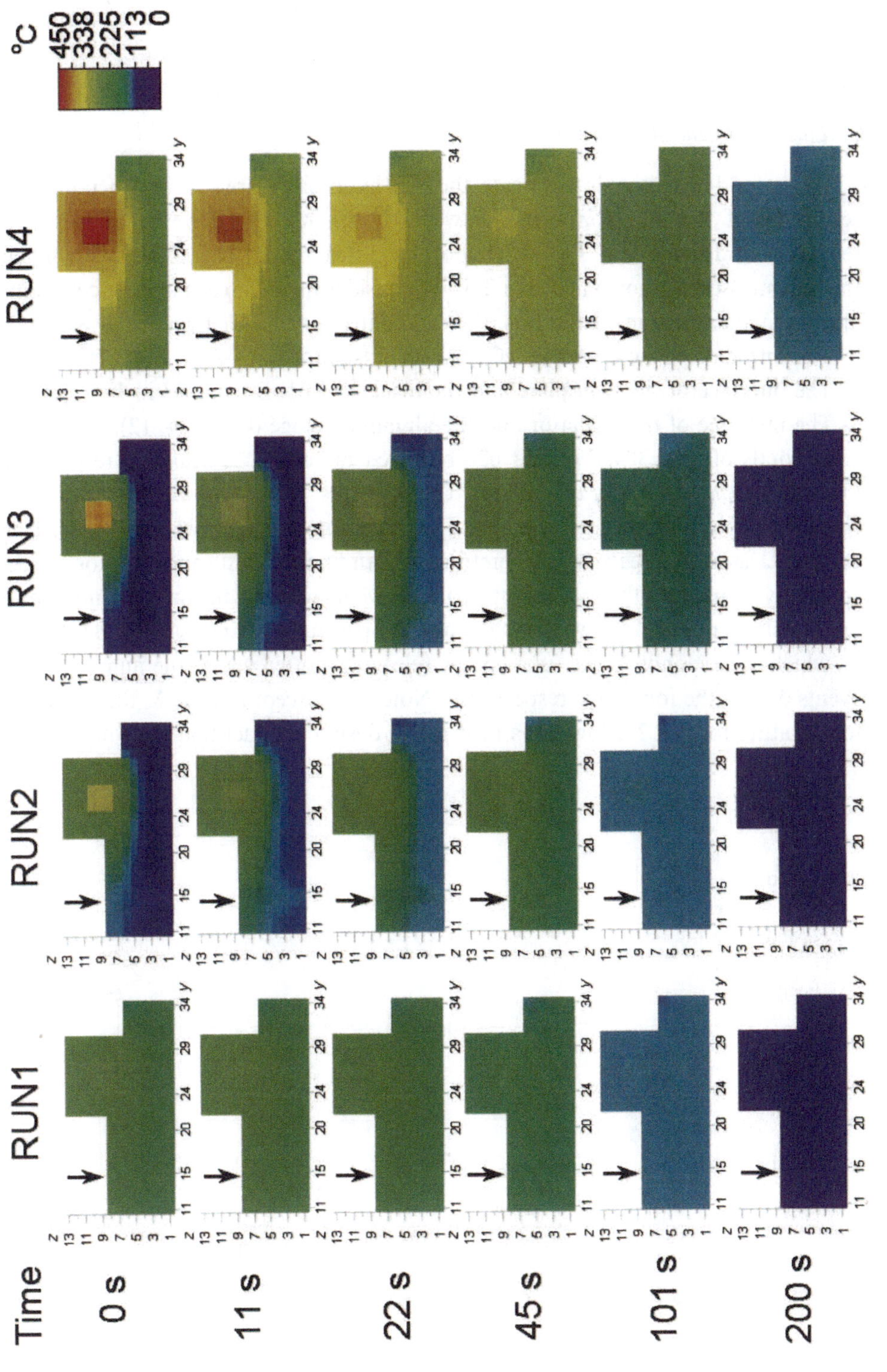
Time
RUN1
RUN2
RUN3
RUN4
0 s
11 s
22 s
45 s
101 s
200 s
°C
450
338
225
113
0

discuss the location of the source and the cause of the difference that can account for decay of the event rate.

c) Location of the AE Source

Thus far I was unable to locate the AE hypocenter area owing to poor time resolution of the AE monitoring, however it can now be done by allowing for temperature distribution. The AE hypocenter area should be a place with higher thermal stresses, as shown in Table 2. If we consider the difference in elastic modulus and thermal expansion coefficients, the candidates for the AE hypocenter are:

- The interface of the heater (MgO) and the insulator (area A in Fig. 12),
- The interface of the steel plate and the insulator (B1, B2 and B3 in Fig. 12),
- The interface of the insulator and the aluminum stage (C in Fig. 12).

Even totals of 114, 304, 394 and 628 occurred in runs 1, 2, 3 and 4, respectively (Table 4; OGASAWARA *et al.*, 1992). These events totals can be thought of as fractions of the total energy released. The maximum temperature reflects the total energy released from each site. Therefore, we can find the best candidate for the AE source by comparing the total number of AE events with the maximum temperature at each site. Figure 14 shows the maximum temperatures in areas A, B1, B2, B3 and C, which corresponds to the total energy released, plotted versus the number of AE events during the four runs, respectively. Note that, except in area A, the maximum temperature for run 2 is lower than for run 1, despite the fact that the total number

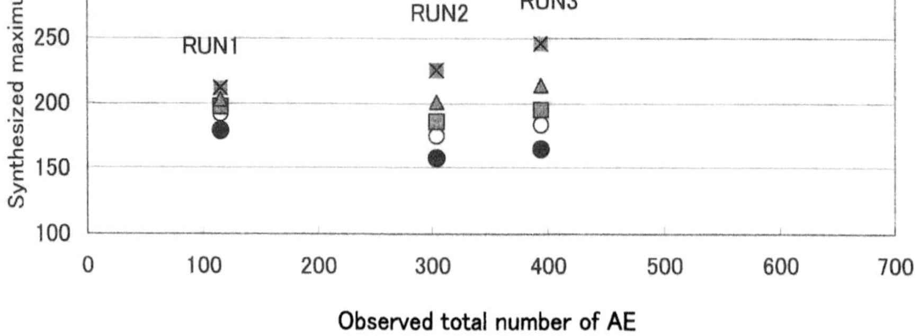

Figure 14
The correlation between the total number of AE and the maximum temperature at the candidate locations of AE activity shown in Figure 12.

of AE events tripled from run 1 to run 2. Even though the total number of AE events increased by four times from run 1 to run 3, the maximum temperature in run 3 is either lower than in run 1, or does not significantly increase, except in area A.

Moreover, as can be seen in Figure 13, with increasing distance from the heater, the temperature change showed significant delays in runs 2, 3 and 4, compared to area A. For example, in run 3 the temperature peaked at $t = 4$–7 s in area A, and at $t = 10$–14 s, 15–26 s, 28–39 s, and 54–71 s in areas B1, B2, B3, and C, respectively. With increasing distance, temperature rates around the peak temperatures are quite slow at B1, B2, B3, and C, compared with area A. However, this cannot explain the initial rapid change in event rate for each run. An additional factor is that the surfaces of the heater and the insulator are relatively rough, whereas the surfaces of the steel plate and the aluminum stage are relatively smooth. In sum, I suggest that area A, the interface of the heater and the insulator, is the main hypocenter area.

d) Change in Frictional Strength and Shear Stress at the AE Source

We next investigate strength and stress in area A, beginning with frictional strength at the AE source. A steel bolt expands several times more than porcelain (see Table 1). Therefore, the higher the temperature is, the looser will be bolt A and the lower the frictional strength. The temperatures averaged over the length of bolt A were 317, 205, 137 and 112 °C for runs 4, 1, 2 and 3, respectively, being significantly different for different runs. For example, AE in run 4 began with the lowest frictional strength, whereas AE in run 3 began with the highest strength.

Figure 15 shows temporal change in the expansion of bolt A. It should be noted that, at $t = 0$ s, the expansion of bolt A in run 4 is about 1.8 times greater than that in run 1, keeping in mind that the vertical axis in Figure 15 is a logarithmic scale and a given change rate in expansion is larger for a greater magnitude of expansion. Therefore, as shown, the rate in shrinkage (strength recovery) is the highest in run 4.

In runs 2 and 3, where exponential event-rate decay was observed, the expansion of bolt A was only 50–60% of that observed for run 1 at $t = 0$ s. As seen in Figure 15, at $t = 0$–40 s, bolt A continues shrinking during run 1, while it continues expanding in runs 2 and 3. It is only after $t = 40$–80 s that the expansions of the bolt in runs 1–3 are similar, and that they shrink with time in a similar manner. After $t = \sim 100$ s, bolt shrinks exponentially in all runs with a time constant of about $-0.004\ s^{-1}$ (compare the gray thick line in Fig. 15).

Next, we examine the temporal change of temperature in area A (Fig. 16). In Figure 16 (lower) there are differences in exponential decay prior to $t = 10$ s. However, at $t > 10$ s, there is no significant difference in exponential decay for all runs. The temperatures in area A in all runs seem to decay exponentially with the same time constant of about $-0.004\ s^{-1}$, keeping in mind that the temperature in

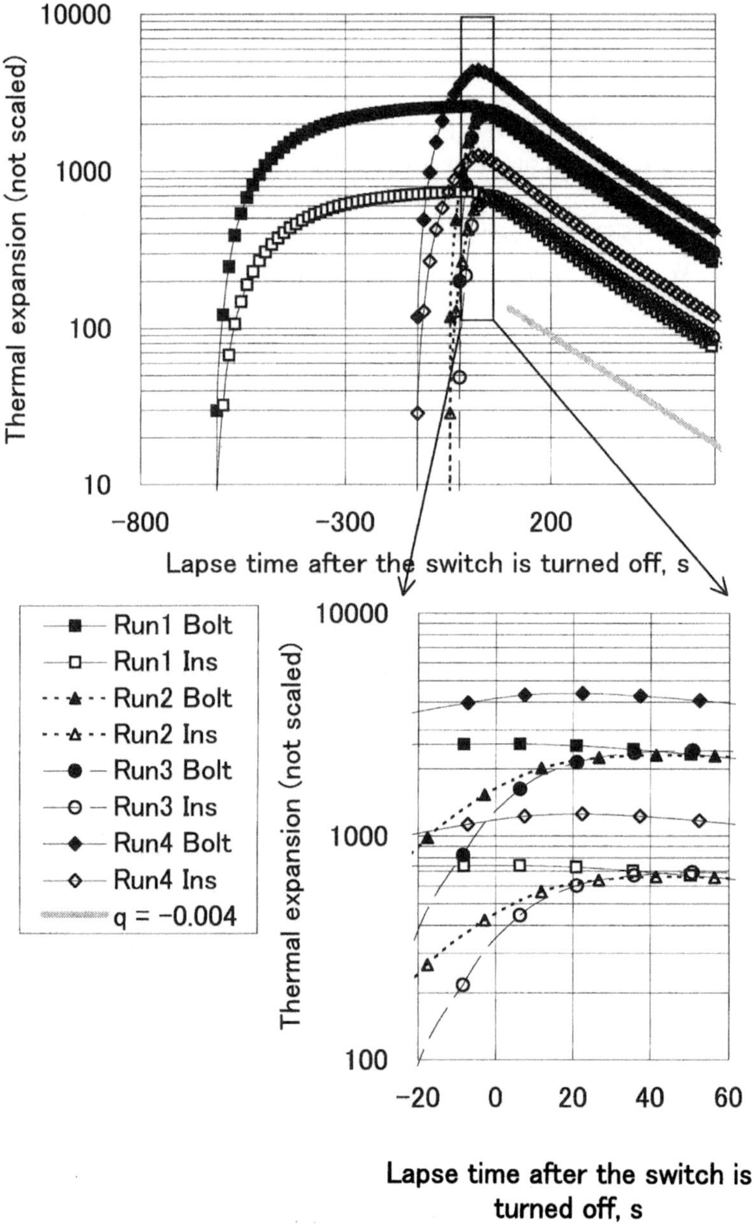

Figure 15

The thermal expansion of bolt A and the insulator (Ins) near the bolt. The expansion of bolt A is inversely proportional to the frictional strength in the AE source area. The strength for runs 1 and 4, when a power-law decay prevails, is low soon after turning the switch. In contrast, the strength for runs 2 and 3, when an exponential decay prevails, is high soon after the switch was turned off, but decreases with time until $t = 60$ s, and then increases again. An exponential decay with a time constant of $q = -0.004\ s^{-1}$ is shown as a guide by the gray thick line.

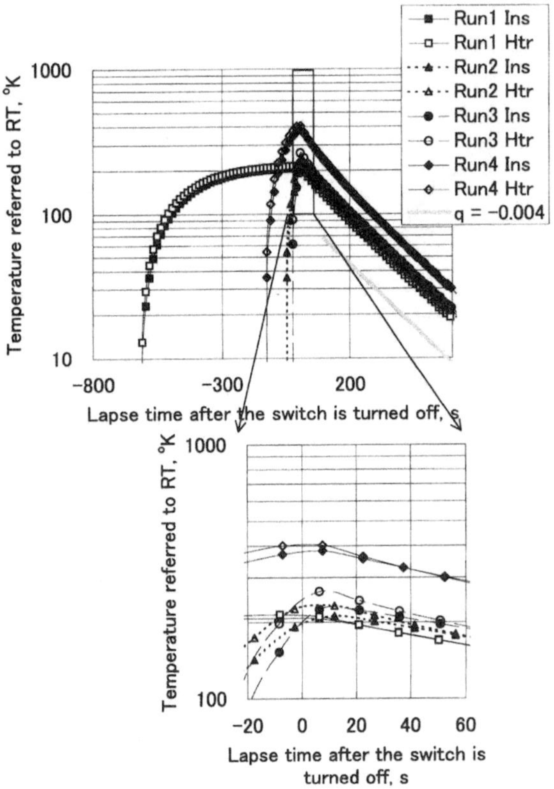

Figure 16
The synthesized temperature at the insulator (Ins) and the heater (Htr) at point A (see the location in Fig. 12), the most probable source area of AE activity. Note that the temperature will decay exponentially for all runs after the switch is turned off. An exponential decay with a time constant of $q = -0.004\ s^{-1}$ is shown as a guide by the gray thick line.

area A corresponds to shear stress in this area. Therefore, shear stress in area A is no indication of the difference in decay of an event rate observed in the experiment.

5. *Discussion*

a) Conditions for Non-exponential Decay

I will discuss how the conditions of non-exponential decay are satisfied by the rice cooker experiment.

Figure 17 shows the theoretical exponential decay with nearly the same decay constant in runs 2 and 3, and a theoretical power-law decay with nearly the same

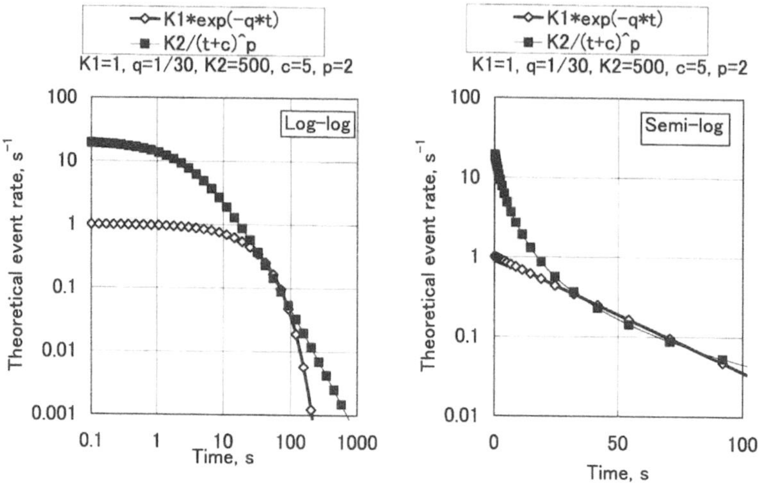

Figure 17
Comparison between theoretical exponential and power-law decays in event rate. In order to consider the difference between these decays, parameters nearly equal to those in the rice cooker experiment are assumed. From $t =$ several tens of seconds to $t =$ a few hundred seconds, the difference can be very small between the power-law and exponential decays. However, the event rate is initially much higher for power-law decays than for exponential decays.

decay constant as in run 4. In the earlier half-period ($t <$ a few tens of seconds), the event rate for power-law decay is significantly higher and the decay is faster. However, when power-law decay prevailed in runs 1 and 4 (Fig. 15, lower), frictional strength began at the lower level (larger expansion) of $t = 0$ s and recovered over time. When exponential decay prevailed, in runs 2 and 3, strength began at the higher level (smaller expansion) of $t = 0$ s and did not begin to recover until $t = 40$–80 s. These relationships may possibly explain the significant differences in the theoretical power-law and exponential decays during the initial half-period.

In the later period ($t \sim$ several tens of seconds), there are no significant characteristics in the observed stress and strength in Figures 15 and 16, which corresponds well with their being no significant difference in the theoretical power-law and exponential decays in Figure 17. Nonetheless, clear difference is seen at the final stage in Figure 17 ($t >$ 100 s). However, the event rate may be too low to be detected in the rice cooker experiment, or may be masked by background seismicity in the case of natural seismicity.

We can more rigorously evaluate whether the observed characteristics in stress and strength can account for the exponential and non-exponential decay during the initial half-period.

An earthquake will occur when shear stress $\tau(t)$ on a source fault exceeds frictional strength $F(t)$. Immediately after the earthquake, the stress $\tau(t)$ will decrease to some level τ_2 and start to recover. If the event rate is high enough, and the change in $F(t)$ is negligibly small, as shown in Figure 18, $\tau(t)$ approximates a linear change with time, so the recurrence period of events P_r will be:

$$P_r = (F(t) - \tau_2)/(d\tau(t)/dt) . \tag{10}$$

This situation is the same as the time-predictive model (SHIMAZAKI and NAKATA, 1980). As an event rate $n(t)$ is the reciprocal of a recurrence period P_r,

$$n(t) = 1/P_r = (d\tau(t)/dt)/(F(t) - \tau_2) . \tag{11}$$

$\tau(t)$ is proportional to temperature $T(t)$, and $T(t)$ decayed exponentially with a time constant $q = \sim -0.004\ s^{-1}$ in the experiment, as shown in Figure 7 or 16. Therefore, I substituted $\tau(t) = \beta\ \exp(-qt)$ in Eq. (11), where β is a constant. The natural logarithms of an event rate will be:

$$\ln n(t) = \ln \beta - qt - \ln(F(t) - \tau_2) . \tag{12}$$

If the third term on the right-hand side of Eq. (12) changes only slightly or not at all, then the event rate will decay exponentially with a time constant of q. For the electric rice cooker, q is proportional to the convection constant h and inversely proportional to a thermal capacity C_t, as described in Eq. (9).

However, it should be noted that, if $F(t) - \tau_2 \ll 1$, then the third term on the right-hand side of Eq. (12) will be very large. In other words, the event rate will be

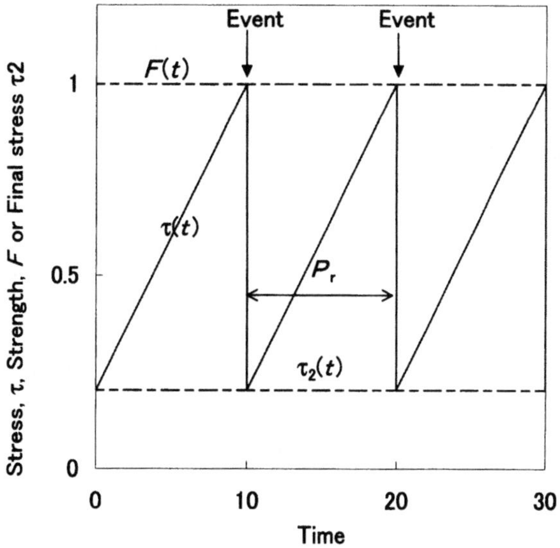

Figure 18
Approximate relationship between time derivatives of stress, $\tau(t)$, frictional strength $F(t)$, final stress $\tau_2(t)$ and recurrence period P. If the event rate is high enough to regard $F(t)$ as constant, then $P = (F(t) - \tau_2)/d\tau/dt)$.

considerably larger than occurs in exponential decay. If $F(t) - \tau_2$ is initially very small and recovers significantly, then $n(t)$ decays quite differently from exponential decay. This is a very simple, interesting inherent relationship between event rate, stress and strength.

As discussed above, the results of the rice cooker experiment explain under what conditions an event rate decays non-exponentially. However, this does not necessarily tell us under what conditions an event rate decays according to the power law, which requires further investigation.

b) What Can Account for Power-law Decay?

Now, using data from the present numerical analysis, we find out what controls decay. I have calculated $d\tau(t)/dt$ using the data shown in Figure 16; for simplicity, let τ_2 in Eq. (11) equal zero. $F(t)$ in Eq. (11) can be obtained by Eq. (4) but σ_b was not measured in experiment of OGASAWARA *et al.* (1992). Thus, we assume several cases of σ_b to determine frictional strength when the switch was off, F_{off} that best fits the observed data. For example, if 95% of σ_b is relaxed when the switch was off, then F_{off} is 0.05 $\mu\sigma_b$; if 76, 64, 48 and 24% of σ_b are relaxed, then $F_{\text{off}} = 0.24\ \mu\sigma_b$, 0.36 $\mu\sigma_b$, 0.52 $\mu\sigma_b$ and 0.76 $\mu\sigma_b$, respectively.

Figure 19 shows $n(t)$ that was synthesized in run 4 using the data shown in Figures 15 and 16, and assuming various magnitudes of F_{off}. What is striking is that, in the case of the lowest frictional strength ($F_{\text{off}} = 0.05\ \mu\sigma_b$ see 95% relaxed in Fig. 19), the estimated $n(t)$ nearly coincides with the experimental results (the squares in Fig. 19), and with decay according to the power law with $p = 1.76$. Conversely, in Figure 19 (right), $n(t)$ in the case of higher frictional strength (smaller relaxation) is nearly straight in a semi-log plot. Namely, power-law decay is transformed into exponential decay with decreasing amounts of relaxation in σ_b (with increase in F_{off}). The time constant of the exponential decay is about $-0.004\ \text{s}^{-1}$, the same as those for temperature change (compare Fig. 19 with Figs. 7a, 15 and 16). This is also very interesting.

Figure 20 shows the synthesized $n(t)$ for runs 1, 2 and 3, assuming the same σ_b as best fits run 4. Exponential decays were successfully reproduced for runs 2 and 3. Again, the time constant is nearly equal to $-0.004\ \text{s}^{-1}$.

The synthesized event rate for run 1 does not seem to reflect power-law decay in Figure 20. However, this may reflect the fact that the difference in the fitting of exponential decay and power-law decay originally was not large for run 1 (see Figs. 9 and 10 or Table 6).

Figure 21 represents the temporal change in friction for all runs, using the assumption σ_b that satisfies $F_{\text{off}} = 0.05\ \mu\sigma_b$ in run 4. As is shown, the frictional strength in run 4 is initially very low but recovers considerably. Compared with run 4, initial frictional strengths in runs 1, 2 and 3 are not as low on the log scale. Although both stress and strength change exponentially with time, an event rate is

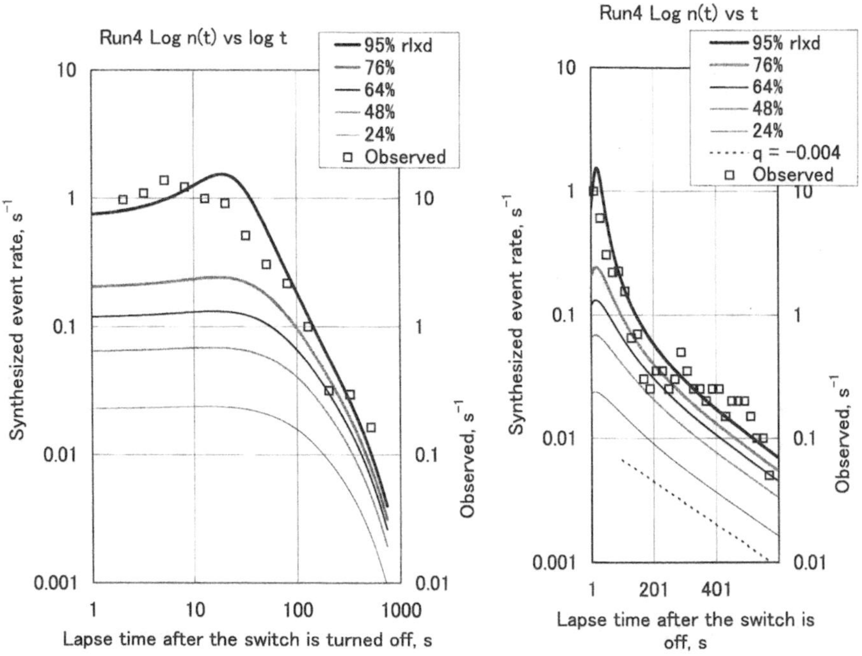

Figure 19
The manner of decay in event rate for run 4 is dependent upon the frictional strength. (Left) Log-log plot of the synthesized event rate versus lapse time, (right) a semi-log plot. The event rate is synthesized by Eq. (8) and by using the synthesized expansion of bolt A and temperature shown in Figures 15 and 16. For comparison, the event rate actually observed in laboratory experiments by OGASAWARA *et al.* (1992) is also shown by squares. Note that power-law decay can be seen for the case where 95% of σ_b is relaxed when the switch was turned off (95% rlxd), whereas exponential decay occurs for the case where the smaller amount of σ_b is relaxed. An exponential decay with a time constant of $q = -0.004$ s^{-1} is shown as a guide by the dotted line in the semi-log plot (right).

dependent upon a rate of stress, not upon its magnitude. On the other hand, the event rate is dependent upon the magnitude of strength. This is why decay in an event rate appears to be both exponential and power-law dependent for a given magnitude of strength.

Equation (12) tells us that $n(t)$ decays exponentially with the time constant q, if $(F(t) - \tau_2)$ is constant. In Figure 20, $F(t)$ becomes a constant with increasing time. That is why $n(t)$ decays exponentially with the time constant $q = -0.004$ s^{-1} for larger t in Figures 19 and 20.

Therefore, it appears that we have successfully accounted for the fundamental characteristics of the event-rate decay observed in the rice cooker experiment.

c) Problem

In spite of the foregoing discussion, there are only minor differences in the event rate estimated for runs 1, 2 and 3 (Fig. 20), whereas there are broad differences in the

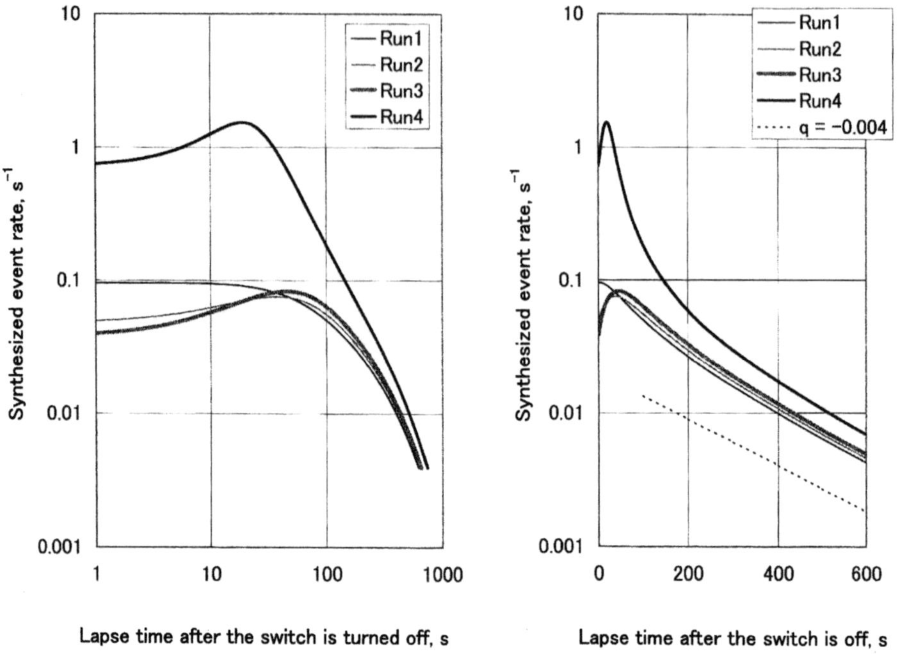

Figure 20

The event rate synthesized by Eq. (8) and the data shown in Figures 15 and 16 for runs 1, 2, 3 and 4, assuming the same pre-stress σ_b as that which fits run 4 the best. An exponential decay with a time constant of $q = -0.004\ \mathrm{s}^{-1}$ is shown as a guide by the dotted line in the semi-log plot (right).

actual event rate for those runs (Fig. 9). There were larger b values observed in runs 2 and 3 (Table 4). To some extent, this may explain the larger event numbers in runs 2 and 3, since I assumed the same b value for all runs to synthesize event numbers. However, the difference is too large to be explained by differences in b value. The low synthesized event rate for runs 2 and 3 is probably caused by the limit of the present quasi-three-dimensional analysis. To realize the actual temporal change in temperature we must multiply the heat capacity of the heater by twenty times. At the same time, we must assume that heat was lost by convection and radiation to the surrounding air from the surface of the heater. In fact, we must assume that heat loss and the resulting lower temperature in the AE area produced the lower synthesized event rate.

d) Is the Rice Cooker Experiment Analogous to an Earthquake Swarm?

Thus far the discussions herein have been developed with the assumption that both shear stress and strength along a source fault change exponentially with time. It is tempting to think that one might assume any functions for stress or strength change, and mathematically combine any of these. If this were the case, it might seem useless to discuss only cases where both stress and strength change exponentially over time. However, this view would be incorrect.

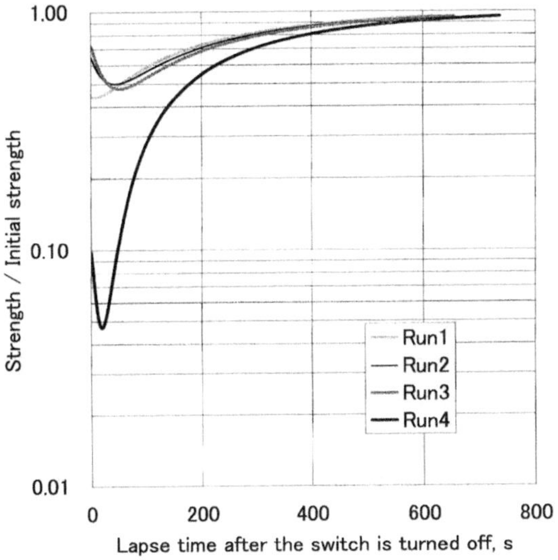

Figure 21
The temporal change in strength for four runs estimated with σ_b that satisfies $F_{\text{off}} = 0.05\ \mu\sigma_b$. Note that strength is plotted on a log scale. In run 4, a power-law decay in an event rate is observed and the strength is initially very low and recovers substantially with time.

It has often been noted for earthquake swarms that seismic events are located around tensile cracks that strike parallel to the maximum horizontal stress (e.g., HILL, 1977). Pore pressure opens the tensile cracks, leading to a change in stress state around these cracks. The stress rate on the earthquake source cracks, is proportional to the pore-pressure rate in the tensile cracks. On the other hand, pore fluid can percolate into earthquake sources, thereby reducing strength.

Permeability in the cracks or the earthquake source is considerably larger than in country rock. As a result, pore pressure will decay exponentially, largely as water leaks from holes in a bucket. Therefore, it is natural to assume that stress and strength at an earthquake source change exponentially with time for earthquake swarms.

The linear development of a pore-elastic effect is precisely analogous to the theory of linear thermo-elasticity, as RICE and CLEARY (1976) have noted. Therefore, temperature in the rice cooker experiment may be regarded as the analogue of pore pressure in an earthquake swarm. The rice cooker experiment therefore is a good analogy of earthquake swarm driven by fluid activity. For example, we have demonstrated that the time constant q of an AE event rate is identical to that of a decrease in temperature if strength is constant. It is therefore suggested, by analogy, that the time constant q of exponential event-rate decay in earthquake swarms is identical to that of pore-pressure decrease. The q of temperature decrease for the rice cooker experiment is inversely proportional to thermal capacity C_t and proportional to a convection coefficient h, as described in Eq. (9). Therefore, by analogy, the q of earthquake swarms may constraint pore-fluid leakage or pore-fluid capacity.

6. Concluding Remarks

Consequently, we have successfully demonstrated that an event rate *prima facie* decays either exponentially or according to a power law, depending upon a magnitude of strength, or how high the pore pressure is at an earthquake source. This is true even though the physical properties are the same, and both stress and strength change exponentially with the same time constant.

The observations made thus far in the rice cooker experiment can be used to interpret decays in earthquake swarms or phenomena observed in relevant laboratory rock fracture experiments, as follows:

1) The condition of power-law decay is initially quite low strength followed immediately by considerably recovery in strength in the seismic source area. Such a condition prevails when fluid intrudes a seismic area enough to considerably reduce strength. In such a case, the pore pressure in the shear crack will rapidly decrease immediately after the termination of fluid intrusion. The condition of exponential decay is an initial high strength or a low strength rate. This condition prevails when an earthquake source is isolated from pore fluid, or if percolation of pore fluid is hard around the earthquake source.

2) Confining pressure increase with depth. Therefore, a given pore pressure will decrease in strength more significantly at shallower depths. As noted in section 1, MOGI (1991) pointed out for about 20 earthquake swarms off the eastern Izu peninsula, Japan, that swarms before 1985 tended to decay exponentially, while those after 1984 tended to not decay exponentially. He also noted that the latter swarms were shallower than the former swarms. These observations fit the interpretation that, as swarms become shallower, pore pressure makes strength decrease considerably, leading to a power-law decay. With increased depth, permeability decreases and an earthquake source tends to be isolated from pore fluids, which can lead to exponential decay in an event rate.

3) During the initial stage of an earthquake swarm, pore fluid will be unable to reach earthquake sources to sufficiently reduce strength. A seismic burst following every intermission of fluid intrusion during the initial stage of earthquake swarm, therefore, may exhibit an exponential decay in event rate. MATSU'URA (1994, personal communication) investigated the earthquake swarms off Izu peninsula, Japan. She noted a tendency for exponential decay to not imply the termination of a swarm, even though the power law suggested a termination. The conclusion I reached in the present study can also account for this tendency.

4) HIRATA (1987) investigated the temporal change in decay in AE event rates observed in the bursts during a laboratory rock-creep test. He found that the AE event rate decayed exponentially in earlier bursts, however that it followed the power law for bursts near the main failure of a specimen. He speculated that a change in the fracture condition caused a change in decay. However, I infer that, just prior to final failure of the specimen, AE bursts occurred in an area with lesser strength, which was the initiation area for the final failure.

Acknowledgments

First, I would like to thank Professor Kazuo Mino, who has encouraged me and shared many in-depth discussions. The experiments were carried out in experimental classes on geophysics at the Undergraduate School of Liberal Arts, Kyoto University. I thank the more than 20 students who have conducted experiments over the years. I also thank Professors Shogo Matsushima, Satoshi Sakai and Takahiro Ohkura, and Dr. Naoto Ishikawa, who kindly assisted with the experiment. Dr. Ritsuko S. Matsu'ura kindly provided the computer program AFT12.FOR, used for determining p values. I am grateful for the helpful suggestions and comments of Dr. C. Trifu and three anonymous referees, which significantly improved the manuscript.

REFERENCES

AKAIKE, H. (1974), *A New Look at the Statistical Model Identification*, IEEE Trans. Autom. Control *19*, 716–723.

BAUCCIO, M. L. (ed.), *ASM Engineering Materials Reference Book*, 2nd edn. (ASM International, Materials Park 1994).

CARSLAW, H. S. and JAEGER, J. C., *Conduction of Heat in Solids*, 2nd edn. (Oxford University Press, London 1959).

CHAPMAN, A. J. *Fundamentals of Heat Transfer* (Macmillan Pub., New York 1987).

CREAMER, F. H., and KISSLINGER, C. (1993), *The Relation Between Temperature and the Omori Decay Parameter for Aftershock Sequences near Japan*, EOS, *74*, Suppl., 417.

DIETERICH, J. (1972a), *Time-dependent Friction as a Possible Mechanism for Aftershocks*, J. Geophys. Res. *77*, 3771–3781.

DIETERICH, J. (1972b), *Time-dependent Friction in Rocks*, J. Geophys. Res. *77*, 3690–3697.

DIETERICH, J. (1978), *Time-dependent Friction and the Mechanics of Stick-slip*, Pure appl. geophys *116*, 790–806.

DIETERICH, J. (1994), *A constitutive Law for Rate of Earthquake Production and its Application to Earthquake Clustering*, J. Geophys. Res. *99*, 2601–2618.

EVANS, D. M. (1966), *Man-made Earthquakes in Denver*, Geotimes *10*, 11–17.

GUPTA, H. K. and RASTOGI, B. K., *Dams and Earthquakes, Developments in Geotechnical Engineering 11* (Elsevier Sceintific Publishing Co., Amsterdam, 1976).

HEALY, J. H., RUBEY, W. W., GRIGGS, D. T., and RALEIGH, C. B. (1968), *The Denver Earthquakes*, Science *161*, 1301–1310.

HILL, D. P. (1977), *A Model for Earthquake Swarms*, J. Geophys. Res. *82*, 1347–1352.

HIRATA, T. (1987), *Omori's Power-law Aftershock Sequences of Microfracturing in Rock Fracture Experiment*, J. Geophys. Res. *92*, 6215–6221.

IRWIN, W. P. and BARNES, I. (1975), *Effects of Geological Structure and Metamorphic Fluids on Seismic Behaviour of the San Andreas Fault System in Central and Northern California.* Geology *3*, 713–716.

JAEGER, J. C. and COOK, N. G. W., *Fundamentals of Rock Mechanics* (3rd edn), (Chapman and Hall 1979), 593 pp.

KISSLINGER, C. and JONES, L. M. (1991), *Properties of Aftershocks in Southern California*, J. Geophys. Res. *96*, 11,947–11,958.

KOYAMA, J., *The Complex Faulting Process of Earthquakes* (Kluwer Academic Publishers, Dordrecht, 1997).

MATSU'URA, R. S. (1993), *Median Values of Parameters in the Modified Omori Formula - for Main Shocks in and Near Japan of M6.0 and Larger (1969–1991)*, Abstracts, 1993 Japan Earth and Planetary Science Join Meeting, 224.

MIKUMO, T. and MIYATAKE, T. (1979), *Earthquake Sequences on a Frictional Fault Model with Non-uniform Strengths and Relaxation Times*, Geophys. J. R. astr. Soc. *59*, 497–522.

MOGI, K. (1962a), *On the Time Distribution of Aftershocks Accompanying the recent Major Earthquakes in and Near, Japan*, Bull. Earthq. Res. Inst., Univ. Tokyo *40*, 175–182.

MOGI, K. (1962b), *Study of Elastic Shocks Caused by the Fracture of Heterogeneous Materials and its Relations to Earthquake Phenomena*, Bull, Earthq. Res. Inst., Univ. Tokyo *40*, 125–173.

MOGI, K. (1967a), *Earthquakes and Fractures*, Tectonophysics *5*, 35–55.

MOGI, K. (1967b), *Regional Variation of Aftershock Activity*, Bull Earthq. Res. Inst., Univ. Tokyo *45*, 711–726.

MOGI, K. (1987), *A Generation Mechanism of Matsushiro Earthquake Swarm and its Relation to the Niigata Earthquake*, Abst. Seimol. Soc. Jpn., no. 2, 97 (in Japanese).

MOGI, K. (1991), *Recent Irreversible Crustal Activity in East Izu Peninsula*, Chikyu Monthly *13*, 55–62 (in Japanese).

OGASAWARA, H. (1990a), *A Preliminary Experimental Study on the Aftershocks of Earthquakes-observation of the Acoustic Emissions Occurring after Turning out a Gas Stove*, Zisin *2*, 43, 69–76 (in Japanese with English abstract and figure captions).

OGASAWARA. H. (1990b), *Comparison of aftershocks of earthquake and AE that occurs after turning out gas stove and so on.* In *Progress in Acoustic Emission V* (Yamaguchi, K. Takahashi, K. and Nitisuma, H. eds.) (The Japanese Society for Non-Destrutive Inspection 1990b) pp. 266–272.

OGASAWARA, H., SAWANO, H., MIYAKE, K., OHIGASI, A., OZAKI, H., and NISHIIKE, Y. (1992), *Comparison of Aftershocks with AE Emitted from Electric or Gas Apparatus in our Neighbors (3) — Difference Between Activities of AE in Different Electric Source Voltage*, Abstract, 1992 Japan Earth and Planetary Science Joint Meeting, 58.

OGASAWARA, H., TANIKAWA, W., KURAHASHI, K., SUGITO, N., TSUDA, R., and YOSHIDA, K. (1998), *An Experiment on Two Modes of Aftershocks Using an Electric Rice Cooker and a Fan*, Abstract 1998 Japan Earth and Planetary Science Joint Meeting, 296.

OGATA, Y. (1983), *Estimation of the Parameters in the Modified Omori Formula for Aftershock Frequencies by the Maximum Likelihood Procedure*, J. Phys. Earth *31*, 115–124.

OHTAKE, M. (1976), *A Decade since Matsuhiro Earthquake Swarm*, Kagaku *46*, 306–313, (in Japanese).

PUSCH, R., *Rock Mechanics on a Geological Base* (Elseiver, Amsterdam, 1995).

RICE, J. R. and CLEARY, M. P. (1976), *Some Basic Stress Diffusion Solutions for Fluid Saturated Elastic Porous Media with Compressible Constituents*, Rev. Geophys. Space Phys. *14*, 227–241.

RUINA, A. L. (1983), *Slip Instability and State Variable Friction Laws, J. Geophys. Res. 88*, 10,359–10,370.

SCHOLZ, C. H. (1998), *Earthquake and Friction Laws, Nature 391*, 37–42.

SHIMAZAKI, K. and NAKATA, T. (1980), *Time-predictable Model for Large Earthquakes*, Geophys. Res. Lett. *7*, 279–282.

SIMPSON, D. W, LEITH, W. W., and SCHOLZ, C. H. (1988), *Two Types of Reservoir-induced Seismicity*, Bull. Seismol. Soc. Am. *78*, 2025–2040.

TIMOSHENKO, S. P., and GOODIER, J. N., *Theory of Elasticity, 3rd edn.* (International Student Edition, McGraw-Hill, Tokyo 1970).

TSUKUDA, T. (1990), *Some Characteristics of the Seismicity in the Northern Fossa Magma, Central Japan*, Earth Science Mag. *99*(1), 32–42 (in Japanese).

TSUKUDA, T. (1991), *Earthquake Swarm*, The Earth Monthly *13*, 673–678.

TSUKUDA, T. (1993), *West Iriomote Island Earthquake Swarm Viewed as Time Series*, The Earth Monthly *15*(4), 218–224, (in Japanese).

TURCOTTE, D. L. and SCHUBERT, G., *Geodynamics: Application of Continuum Physics to Geological Problems* (John Wiley and Sons, New York 1982).

UTSU, T. (1957), *Magnitudes of Earthquakes and Occurrence of their Aftershocks*, Zisin *2*, *10*, 25–45 (in Japanese with English abstract).

UTSU, T. (1961), *A Statistical Study on the Occurrence of Aftershocks*, Geophys. Mag. *30*, 521–605.

UTSU, T., OGATA, Y., and MATSU'URA, R. S. (1995), *The Centenary of the Omori Formula for a Decay Law of Aftershock Activity*, J. Phys. Earth *43*, 1–33.

YAMAMOTO, E., OKADA, Y., and OHKUBO, T. (1991), *Ground Tilt Changes Preceding the 1989 Submarine Eruption off Ito, Izu Peninsula*, J. Phys. Earth *39*, 165–176.

(Received April 20, 2000, revised January 18, 2001, accepted January 20, 2001)

Pure appl. geophys. 159 (2002) 345–369
0033–4553/02/030345–25 $ 1.50 + 0.20/0

Pure and Applied Geophysics

Induced Microearthquake Patterns in Hydrocarbon and Geothermal Reservoirs: Six Case Studies

W. SCOTT PHILLIPS,[1] JAMES T. RUTLEDGE,[1] LEIGH S. HOUSE,[1] and MICHAEL C. FEHLER[1]

Abstract—The injection or production of fluids can induce microseismic events in hydrocarbon and geothermal reservoirs. By deploying sensors downhole, data sets have been collected that consist of a few hundred to well over 10,000 induced events. We find that most induced events cluster into well-defined geometrical patterns. In many cases, we must apply high-precision, relative location techniques to observe these patterns. At three sedimentary sites, thin horizontal strands of activity are commonly found within the location patterns. We believe this reflects fracture containment between stratigraphic layers of differing mechanical properties or states of stress. At a massive carbonate and two crystalline sites, combinations of linear and planar features indicate networks of intersecting fractures and allow us to infer positions of aseismic fractures through their influence on the location patterns. In addition, the fine-scale seismicity patterns often evolve systematically with time. At sedimentary sites, migration of seismicity toward the injection point has been observed and may result from slip-induced stress along fractures that initially have little resolved shear. In such cases, triggering events may be critical to generate high levels of seismic activity. At one crystalline site, the early occurrence of linear features that traverse planes of activity indicate permeable zones and possible flow paths within fractures. We hope the continued development of microseismic techniques and refinement of conceptual models will further increase our understanding of fluid behavior and lead to improved resource management in fractured reservoirs.

Key words: Induced microseismicity, geothermal, oil and gas, fluid flow, location.

Introduction

Microseismic monitoring has been carried out during the hydraulic stimulation of experimental hot-dry-rock geothermal reservoirs since the mid-1970s (PEARSON, 1981; ALBRIGHT and PEARSON, 1982). Mapping the cloud of microseismicity proved valuable in estimating the rough extent and orientation of a stimulation. For example, a 1983 stimulation (21,000 m^3) at Fenton Hill, New Mexico yielded over 11,000 locatable seismic events. The mapped event cloud formed a 0.3-km^3 tabular volume surrounding the injection point; however, the cloud did not extend to a nearby well, located 200 m distant (HOUSE, 1987). Neither was an intended fluid connection observed in that well. Later, a sidetracking operation was performed to

[1] Seismic Research Center, Los Alamos National Laboratory, Los Alamos, NM, USA.

drill into the seismic cloud, where a flow connection was established between the wells, allowing circulation of the man-made geothermal reservoir.

Although microseismic clouds have played a key role in the development of hot-dry-rock geothermal reservoirs, these results produced little information concerning the intricacies of the fluid-flow network. Recently, high-precision relative location techniques have been successfully applied to geothermally induced events, offering the ability to observe patterns at unprecedented levels of spatial detail (PHILLIPS *et al.*, 1997; TEZUKA and NIITSUMA, 1997; LI *et al.*, 1998; GAUCHER *et al.*, 1998; LEES, 1998). To make practical use of these results, we must ascertain the detailed relationships between seismicity, fracture geometry, stress and fluid flow. We generally rely on simple interpretive models, such as that in which increased pore pressure reduces normal stress and causes slip (PEARSON, 1981). Thus the presence of seismicity at a given location is taken to indicate a pressure connection, although not necessarily a high-permeability connection, between that location and the injection well. Furthermore, this model predicts little shear-slip seismicity along fractures that open in tensile mode, potentially the most conductive flow paths. As experience is gained in interpreting high-precision location patterns, we hope to better use the patterns to infer locations of high-flow paths, even if those paths are seismically inactive.

The successful application of microseismic observation techniques in hot-dry-rock geothermal reservoirs is partially attributed to the simple and efficient propagation paths through the host crystalline rock. Microseismic techniques might be equally useful in sedimentary, hydrocarbon reservoirs if sufficient numbers of events can be observed. Of course the more complex and less efficient propagation through sedimentary rock should degrade waveforms; however, successful experiments have been run for years (DOBECKI, 1983; SARDA *et al.*, 1988; FIX *et al.*, 1989; VINEGAR *et al.*, 1991; WARPINSKI *et al.*, 1996; ZHU *et al.*, 1996; MAXWELL *et al.*, 1998; DYER *et al.*, 1999) and a number of recent experiments have collected data sets of sizes similar to geothermal data sets, given the volumes of injected fluid (Fig. 1). These encouraging results should lend impetus to studies relating high-precision seismicity patterns, fracture systems and fluid flow in hydrocarbon reservoirs.

In this paper we review location results from six case studies: four in sedimentary and two in crystalline environments. In three cases absolute location techniques are sufficient to resolve simple patterns; in the remaining cases, high-precision relative location must be applied to resolve the initial cloud-like patterns into point clusters, line segments, planar patches or combinations thereof. We speculate on the mechanisms that might produce the observed patterns in different environments, with the goal of better understanding fluid flow in fracture systems.

Microseismic Data and Location Methods

Because of their small size, induced microearthquakes must be monitored using borehole instruments. In the studies presented here, the larger induced events

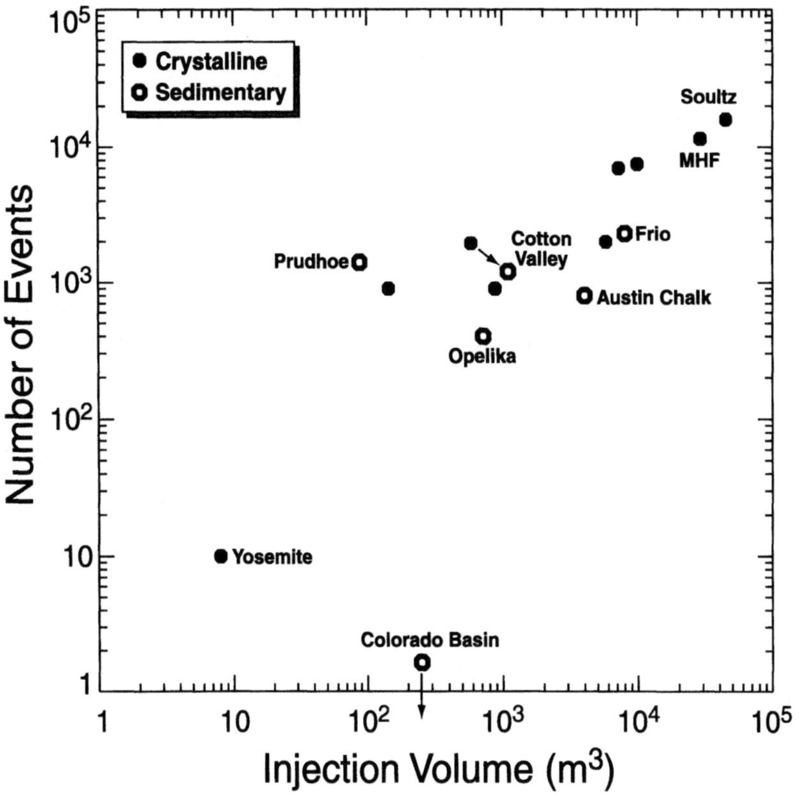

Figure 1
Number of recorded events versus injection volume from a sampling of crystalline and sedimentary reservoir experiments. Crystalline rock experiments are represented by filled circles and include many Fenton Hill injections. Sedimentary experiments are represented by open circles. The Colorado basin experiment produced no observable events, as indicated by an arrow. Results from a Cotton Valley injection coincide with a Fenton Hill experiment, also indicated by an arrow.

approach a magnitude of +1, while most events fall in the magnitude −4 to −2 range, varying somewhat with experiment. Large shear waves and the presence of both compressional and dilatational first arrivals indicate that most induced events have dominantly shear-slip mechanisms. Tensile mechanisms have been observed during some stimulations; however these events produce relatively little seismic energy and display a distinctive spectral peaking that reflects modes of the fluid filled crack (BAME and FEHLER, 1986; FERRAZZINI *et al.*, 1990). From spectral studies, Fenton Hill shear-slip events, for example, have moments 10^{13} to 10^{18} dyne-cm, source radii 1–100 m and Brune stress drops 0.1 to 20 bars (FEHLER and PHILLIPS, 1991). Furthermore, FEHLER and PHILLIPS (1991) found that small events dominated the energy release at Fenton Hill (high Gutenberg-Richter *b* value). This behavior is similar to what is often observed in earthquake swarms and volcanic sequences and is consistent with pronounced small-scale heterogeneity in permeability and pore pressure along fracture surfaces (SYKES, 1970). Thus, induced failure often occurs as a sequence of many small events rather than a few large events, allowing the use of event location patterns to delineate the geometry of slipping joints or fractures.

The requirement for downhole recording imposes severe restrictions on monitoring. Success has been achieved with only one geophone; however, three or more geophones in two or more wells are required to perform high-precision location (two stations in different wells can be sufficient if reflected arrivals are observed, as a later example will show). The sparse monitoring networks make it necessary to use shear waves, and occasionally reflected phases and particle motion (hodogram) data, to locate events. In addition, coupling effects are difficult to control in wellbores and we often see spectral peaks in the data (monochromatic waveforms), perhaps caused by a poor bond between the well casing and the formation. In recent years, some microseismic experiments have been run using multi-level geophone strings that are cemented into wells expressly drilled for or sacrificed to monitoring (e.g., GIBOWICZ *et al.*, 1991; KECK and WITHERS, 1994). This type of deployment can give excellent depth constraint on locations and minimizes coupling related effects, but is considerably more expensive than temporary deployments in existing wells.

Although the downhole recording environment has drawbacks, avoiding the scattering and low attenuation of near-surface materials can yield simple and easily interpreted waveforms. Furthermore, we generally observe that waveforms from nearby events are very similar to one another, likely the result of common paths and source orientations (MORIYA *et al.*, 1994; ROFF *et al.*, 1996). Relative arrival times of similar waveforms can be precisely determined, which, in turn, allows high precision relative locations to be calculated. Figure 2 shows vertical component seismograms from two areas of the Fenton Hill reservoir. Consistent *P*- and *S*-wave arrival times can be chosen by marking the same point in the waveform for all events. Conversely, timing the events in order of occurrence as is normally done during routine processing, would result in inconsistent picks because the similarities between waveforms would not be noticed. This is especially apparent for the *S* waves from the second group shown in Figure 2.

For high-precision, relative location work, we determine arrival times manually using a procedure that approximates the precision of automatic cross-correlation schemes through choosing waveform peaks rather than first breaks. In tests on Fenton Hill data, applying a cross-correlation method (after VANDECAR and CROSSON, 1990) produced virtually no change in the relative location pattern, compared to results determined originally by visual cross correlation. When picking times manually, looking at larger events first helps to identify waveform shapes that are more difficult to find in poorer signal-to-noise traces. We also experiment with filters to enhance the visual similarity of waveforms. With events that display poor waveform correlation, the peak following the first break is chosen irrespective of polarity. Poor correlation is common and can most often be attributed to near nodal takeoff angles. Automatic, high-precision timing methods have been investigated

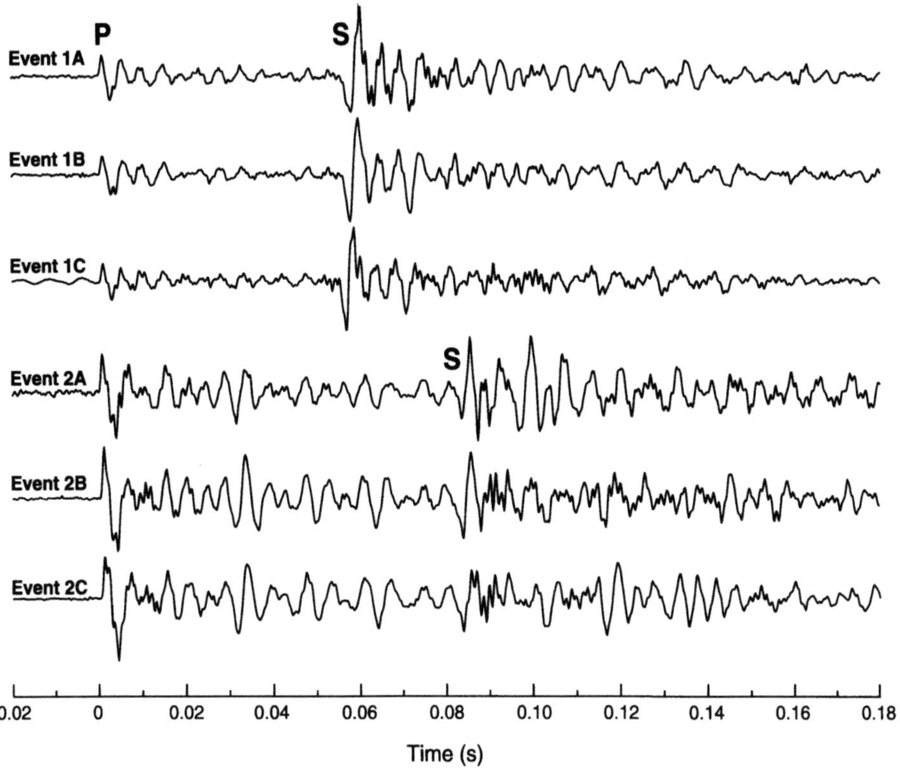

Figure 2

Vertical component seismograms generated by six events during the MHF injection at Fenton Hill. The top three events occur in the cluster shown in Figure 8. The bottom three occur in a separate cluster, as described in PHILLIPS *et al.* (1997).

with downhole data sets (starting with initial arrivals rather than high-precision arrivals as for the test mentioned above); however, while results to date are promising, direct comparisons reveal that the automatic results provide less interpretable detail than can be obtained manually. Possible reasons and implications of this will be discussed later.

Calibration is a critical component of microseismic location. This involves determining velocity structures, station corrections and geophone orientations if particle motion information is to be used in location. Calibration is less important for relative location studies because path effects can be assumed to be similar. However, the absolute locations of the resulting event patterns are likely to be more uncertain than are their relative locations.

To calculate relative locations, we use the master-event technique (EVERNDEN, 1969). This only requires a generic event location code. An event with the largest number of high-quality arrivals is chosen as the master. The hypocenter of this event is fixed at some initial location and station corrections are set equal to the arrival-time residuals for each station and phase type. These station corrections are then applied to the remaining events to obtain locations relative to the master event. Next, sizes, shapes and orientations of the location error ellipsoids are examined to eliminate systematic error as the source of any patterns that may be observed. We approximate errors by assuming the locations are perfectly calibrated absolute locations. First, relative data errors for each station and phase type are estimated based on the frequency content and degree of similarity of the filtered waveforms used to determine arrival times. After computing locations using weights based on the relative errors, a constant multiplier is applied to produce absolute arrival-time errors that best match rms residuals for the full data set. We have assumed that the frequency content of the seismograms more directly controls relative arrival error than do signal-to-noise effects. This is a valid assumption because poor signal-to-noise waveforms that display little similarity with the rest of the data are eliminated from the study.

Results

We present location patterns obtained from six experiment sites. Four are sedimentary, the other two are crystalline environments. In five cases, microseismic events were induced by fluid injection; in one case (Clinton County, Kentucky), events were induced by oil production. In three of the six cases precise relative locations were necessary to observe detailed patterns in the microseismicity. Experimental details, including methods of velocity model calibration, are summarized in Table 1.

Austin Chalk, Giddings Field, Texas

The Austin chalk is a fractured limestone with a matrix porosity of 10% and matrix permeability of 0.01 to 0.1 millidarcys. Fractures result from the bending of the brittle limestone over a deeper and older Jurassic shelf margin, which trends northeast-southwest, roughly parallel to the Gulf coast of Texas. These fractures enable the Austin chalk to produce oil at relatively high rates (up to 160 m^3 per day) and reach single well maximum cumulative production approaching 80,000 m^3 of oil. Hydraulic stimulation is used to complete new wells and to enhance production from older wells. During stimulation, the water forced into untapped areas of the chalk is thought to replace, and thus mobilize hydrocarbons residing in small cracks through imbibition.

To study the stimulated fracture systems, we monitored Austin chalk microseismicity during two 4000 m^3 stimulations at depths greater than 2000 m in the Giddings field (PHILLIPS *et al.*, 1998). We deployed a single three-component geophone at one site and two geophones in separate wells at a second site. Over 480 and 770 microearthquakes were collected during the two stimulations. Many

Table 1

Microseismic experiment descriptions

Site	Activity	Generic rock type	Downhole sensors	Wells	Velocity calibration	Location procedure	Location error (m)	Reference for results presented
Austin Chalk	Stimulation	Clastic/lime sediment	2, plus reflection	2	Layered (ABC)	Absolute	10	PHILLIPS *et al.* (1998)
Lower Frio	Stimulation	Clastic sediment	50	2	Homogeneous (AB)	Absolute	2	HOUSE and FLORES (2001)
Cotton Valley	Stimulation	Clastic sediment	6	2	Homogeneous (ABC)	Relative	2	None
Clinton Co.	Production	Massive limestone	3	2	Layered (ABC)	Absolute	4	RUTLEDGE *et al.* (1998b)
Fenton Hill	Stimulation	Crystalline	4	4	Homogeneous (AB)	Relative	3	PHILLIPS *et al.* (1997)
Soultz	Stimulation	Crystalline	4	4	Homogeneous (AB)	Relative	4	PHILLIPS (2000)

A reflected phase can be thought of as an additional sensor at depth. Letters represent velocity calibration methods as follows: A) shots; B) well logs; C) hypocenter-velocity inversion. Location errors are average, $1-\sigma$, major axis lengths; errors for absolute procedures do not include model errors.

seismograms included an *S*-wave reflection off a high contrast interface below the Austin chalk.

At the two-geophone Giddings site, arrival times from *P*, *S* and reflected phases allowed events to be located in three dimensions without having to rely on particle motion (hodogram) constraints. Results of a hypocenter-velocity inversion applied to 90 high-quality events aligned with the expected fracture trend and extended over 800 m from the injection point (Fig. 3). The reflection data allowed event depths to be constrained to a narrow range along the base of the Austin chalk. In this area,

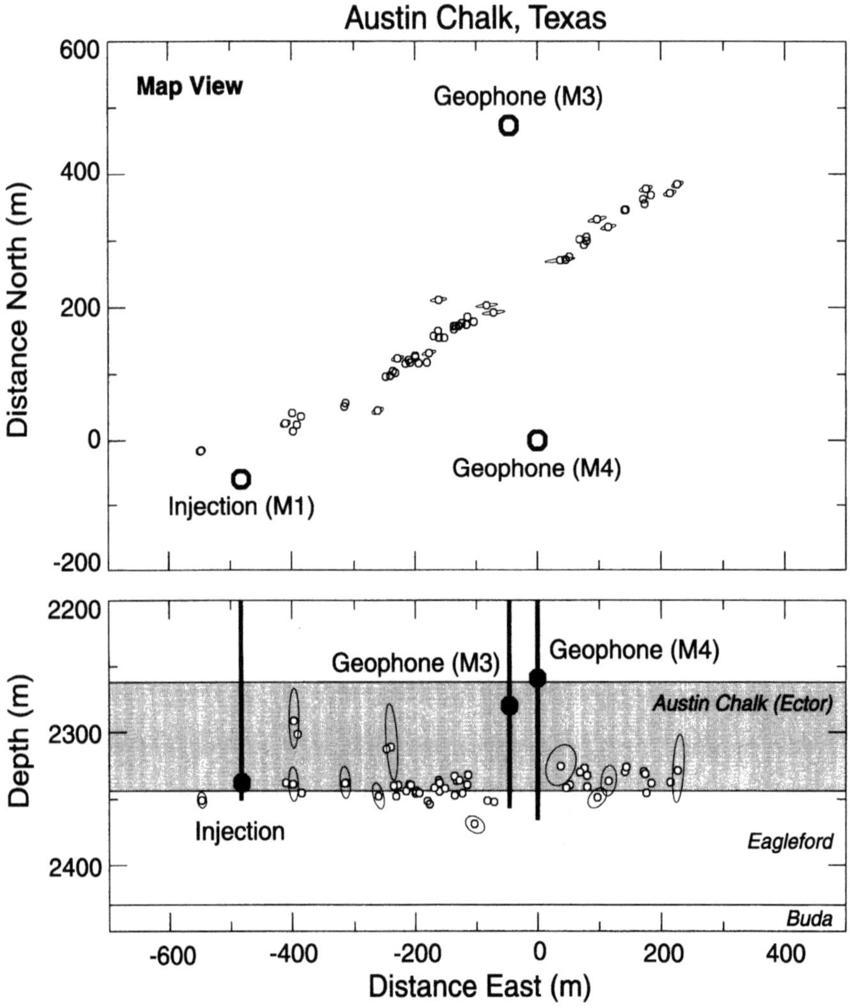

Figure 3

Map and cross-section views of events located in the Austin chalk at the Matcek site, Giddings field, Texas. Monitor and injection wells are shown as open circles in the map view and depths are indicated by solid circles in the cross section. Event locations are indicated by the small open circles. Error ellipses for representative events are included. The cross section is vertically exaggerated by a factor of two.

reservoir engineers rely on a thin, ductile layer of volcanic ash embedded within the Austin chalk to contain the stimulation within the deeper, producing portion of the Austin chalk. The Eagleford shale limits the stimulation at the base of the Austin chalk. The tight depth distribution of microearthquake locations demonstrates that containment was effective at this site.

To increase the number of locatable events, we fixed event depths to injection depths and obtained densely populated seismic zones that aligned with the expected fracture trend at both sites (see PHILLIPS *et al.*, 1998). However, the seismic zone at the first site was five times wider (horizontally) than the zone at the second site shown in Figure 3. There was a large increase in the oil production rate from the first site immediately following stimulation, but little increase from the second. This suggests that the width of the microseismic pattern reflects the degree to which ancillary fractures are accessed, and thus indicates the lateral extent and effectiveness of the stimulation.

Frio Formation, Texas

In October 1993, an injection experiment was performed in Jasper County, near Beaumont, Texas (KECK and WITHERS, 1994). The purpose of this experiment was to investigate the safe use of hydraulic fracturing for solid waste disposal. To accomplish this, an injection well and two monitoring wells were drilled to depths of 1460 m. Fluids (8000 m^3) and solids were injected into the highly permeable and unconsolidated Lower Frio sandstone over a depth interval of 1349 to 1407 m. The Lower Frio was bounded above and below by relatively impermeable shales.

Induced microearthquakes were continuously monitored to determine the lateral extent and the height of the fracture system that was produced (WITHERS and RIEVEN, 1996; RIEVEN, 1999). Each monitor well contained 25 three-component geophone packages at 9 m (30 ft) intervals that were cemented in. Nearly 2900 seismic events were recorded during the entire experiment. Most were too small to be located reliably. A total of 54 well-recorded microearthquakes was chosen for further study, based on numbers of *P*- and *S*-wave arrivals (HOUSE *et al.*, 1996; HOUSE and FLORES, 2001).

Epicenters of the 54 high-quality events, located in an absolute sense using *P*- and *S*-arrival times, extend about 200 m southwest from the injection well (Fig. 4). The near-injection seismicity scatters around the injection well, while a distant cluster falls along a narrow, horizontal trend, aligned toward the injection borehole. The tight hypocentral distribution of the distant cluster is surprising, because one would expect that the unconsolidated nature of the Frio Formation would not support the development of such a well-defined zone of deformation (HOUSE *et al.*, 1996). WITHERS and RIEVEN (1996) report a temporal migration towards the injection well within this group of events, and similar behavior for a second cluster recorded during a later injection.

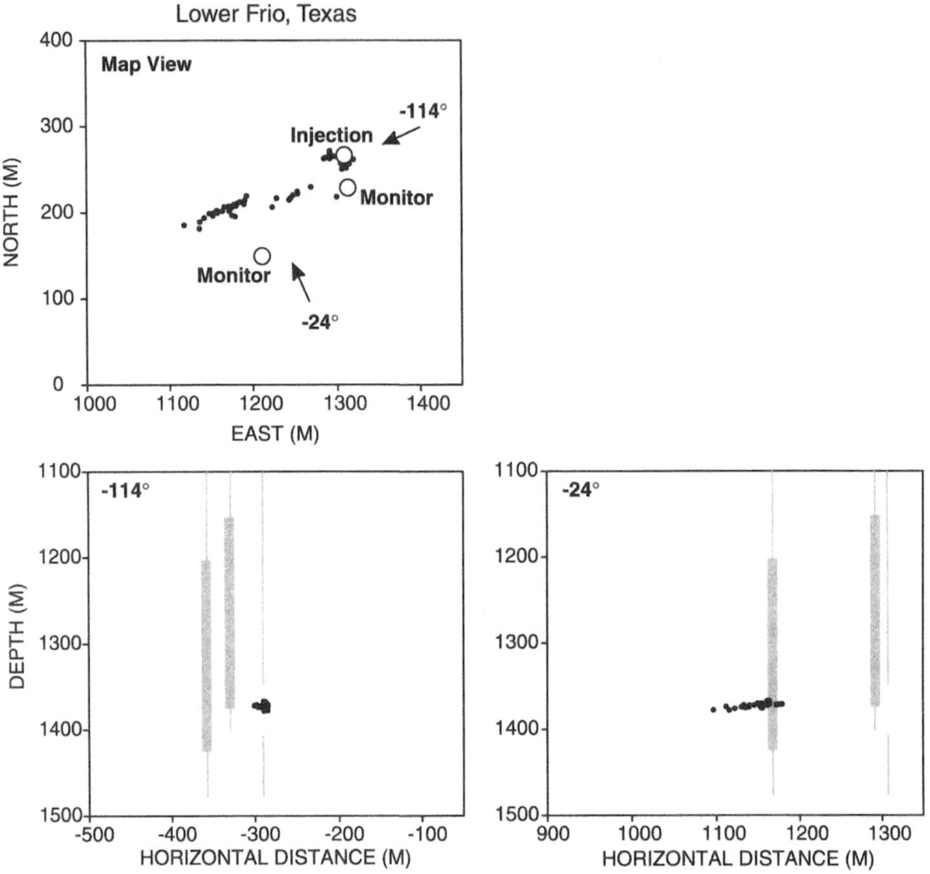

Figure 4
Map and cross-section views of events located in the lower Frio during the waste injection experiment. Monitor and injection wells are indicated by open circles in the map view. Arrows show cross-section view directions. In cross-section views, wells are indicated by light gray lines; thick lines represent locations of geophone strings. The gap in the injection well represents the injection interval. Only events from the linear cluster located southwest of the injection well are included in the cross-section views.

Cotton Valley, Texas

In May, 1997, a consortium of operators and service companies conducted an extensive hydraulic fracture imaging demonstration in the Cotton Valley tight gas sands of East Texas (WALKER, 1997). Well completions and recompletions in this type of reservoir account for many of the hydraulic fracture operations conducted in the United States and, therefore, constitute an important potential application of microseismic techniques.

At this site, two 48-level, 3-component geophone arrays were cemented behind the casing of two offset monitor wells. We analyzed microseismicity recorded during the second of three injection stages, a 1100 m^3 injection at depths of 2757 m to 2838 m. Nearly 1200 events were detected and 290 were initially located (RUTLEDGE *et al.*, 1998a). The locations show a diffuse, tabular feature striking N70°E and dipping 45°SE.

A tight cluster of 161 events located 120 m east of the treatment well generated similar *P* and *S* waveforms. For this cluster, we manually determined relative arrivals to sub-sample precision using data from 6 stations representative of both monitor well arrays. After applying the master-event location technique to the *P*- and *S*-wave data, relative locations collapse into a thin horizontal strand, 40-m long and less than 3-m thick (Fig. 5). The events separate into two main groups, distinguished by a change in *S* waveform character at one station. The change could result from the variation in takeoff angles from one end of the feature to the other, a near-source

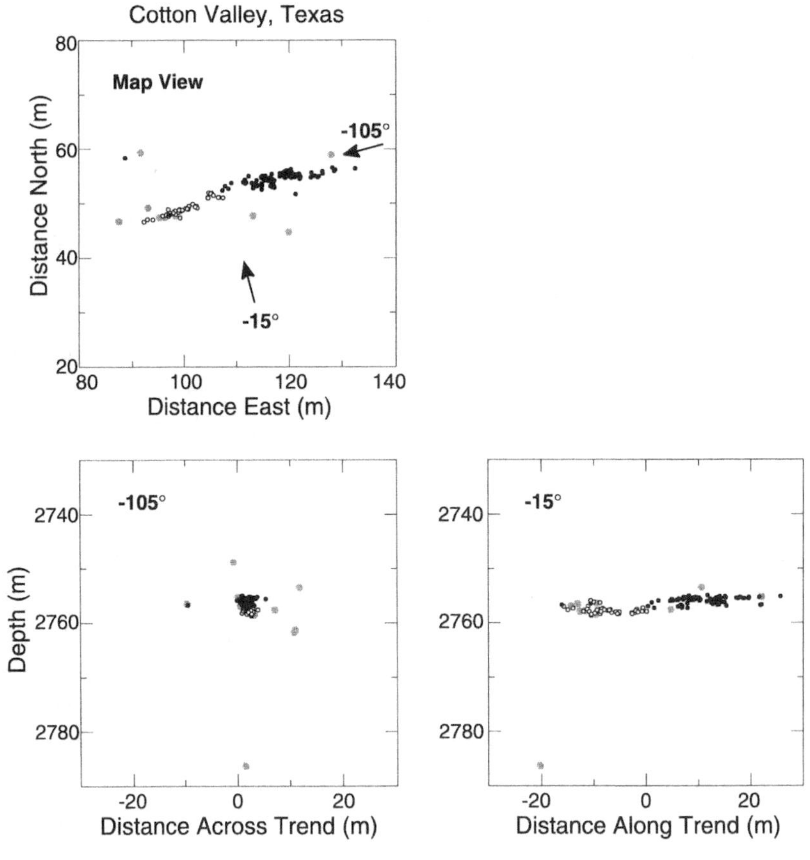

Figure 5
Map and cross-section views of a relocated cluster of Cotton Valley events. Event groups based on waveform type are represented by different symbols (filled circles, open circles, gray filled circles). Map view arrows give along and across trend cross-section view directions.

path effect, or a change in source mechanism. A third, minor group generated waveforms dissimilar to the first two. The linear feature is especially well defined along its western half and breaks into small subclusters in the eastern half. A slight bend can be seen toward the eastern end. Location error ellipsoids are oriented along the trend, as expected for the station geometry, but are of magnitude less than 2 m.

Fracture growth patterns are revealed by the remarkable temporal behavior of the relocated events (Fig. 6). The seismicity begins at the eastern end, near the bend (Fig. 5), then migrates westward toward the injection well. In the eastern half of the cluster we observe subclusters that rupture repeatedly. Within each subcluster, similar sized events occur and also appear to migrate westward. The subclusters may represent discrete, isolated fractures separated by a few meters, with event sizes within each subcluster limited by fracture dimension. These patterns would be obscured if location errors were more than a few meters, which further supports the resolving power of the relative mapping method. A third set of events, distinguished by waveform types that were less similar to the two main groups (gray filled circles), occurred when the injection well was shut in.

Clinton County, Kentucky

In Clinton County, Kentucky oil is produced from low porosity carbonate rocks of Ordovician age, at depths from 230 to 730 m. A preponderance of isolated, high volume wells suggests the existence of isolated zones of fracture storage and high permeability. Initial production rates as high as 64 m^3 per hour and cumulative production of 16,000 m^3 from a single well have been reported (HAMILTON-SMITH *et al.*, 1990). However, fracture orientations are generally unknown and are often assumed to be vertical. We deployed geophones near high volume producing wells at a number of Clinton Co. sites with the goal of delineating the reservoir fracture systems (RUTLEDGE *et al.*, 1998b).

Three geophones were deployed in two existing wells at depths between 240 and 430 m at the Tallent site in Clinton Co. and monitoring started six weeks after a nearby well began production. Over 3200 events were recorded during a period of 28 weeks. The event rate was clearly associated with the production rate, but with a lag of two to three weeks (RUTLEDGE *et al.*, 1998b). Events may have been triggered by tiny increases in horizontal compressive stress associated with the depletion of an adjacent fracture, or by small pore pressure changes associated with the replacement of oil by brine (RUTLEDGE *et al.*, 1998b).

The Tallent site was calibrated using shot and well log data and a hypocenter-velocity inversion applied to a high-quality subset of events. Following this, events were located using absolute location techniques; over 1200 of the best quality locations are shown in Figure 7. Due to the high data quality and the similar distributions of measured arrivals from event to event, the absolute location patterns are very close to those that would be obtained using relative location techniques.

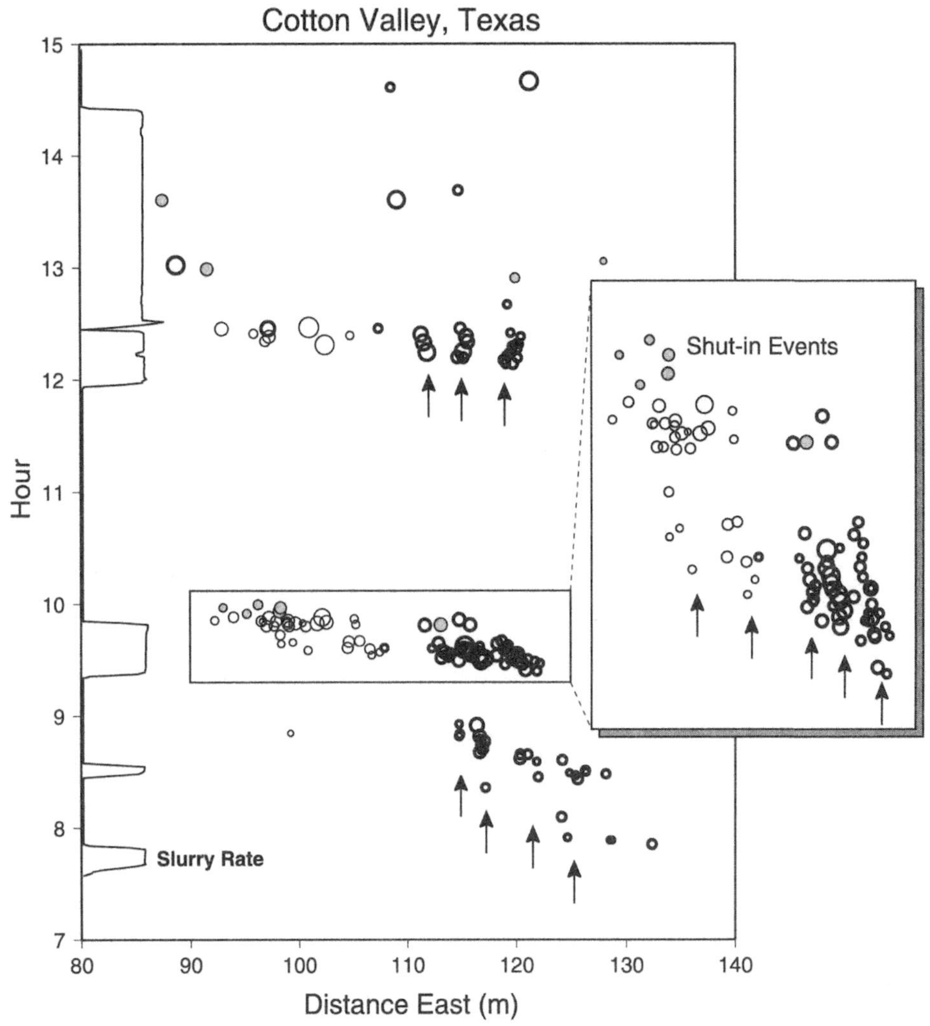

Figure 6
Spatio-temporal behavior of the Cotton Valley cluster. Events are represented by circles whose diameters are proportional to peak log trace amplitude. Line weights represent event groups based on waveform type that are shown in Figure 5. The slurry (injection) rate is included on the left (peak rate 105 l/s). Activity associated with the third injection pulse is scaled up in the inset plot. Arrows indicate subclusters of events discussed in the text.

Results show three fracture planes striking N65°E to N82°E and dipping 15° to 20° to the NW or to the SE that surround, but do not intersect, the producing interval. Relatively few events populate the deepest fracture (200 events total; 80 high-quality locations are shown in Fig. 7), but unique, *S*-nodal (large *P*, small *S*) waveforms observed at one geophone distinguish these events from those of the main group. The deep event group forms an elongate planar pattern that intersects the main fracture along its northern, well-defined edge.

The seismically active fractures can be extrapolated to old production intervals in wells from which 725 m^3 of oil had been extracted in the nine months preceding monitoring. Two wells were later drilled into the main mapped fracture and

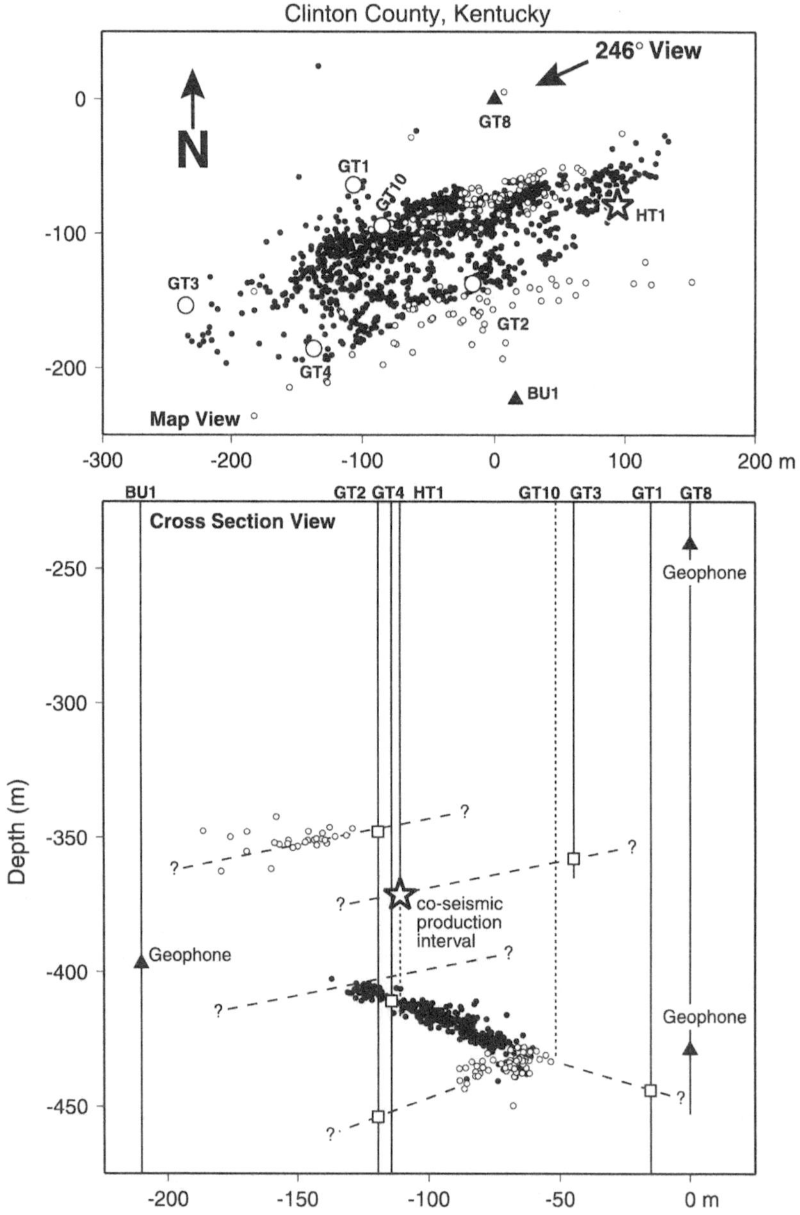

produced brine. Thus the microearthquake locations defined fractures that had contained oil, but were drained and subsequently recovered to hydrostatic pressure via brine invasion. If knowledge of the fracture orientation can be extrapolated to other sites in this region, these results should aid field development by improving drilling strategies (horizontal drilling will help little, for example), placement of offset wells and abandonment or pressure maintenance operations to avoid premature contamination of producing wells by brines.

Fenton Hill, New Mexico

In December 1983, a massive hydraulic fracturing (MHF) experiment was carried out at the hot-dry-rock geothermal site at Fenton Hill, New Mexico. Over 21,000 m^3 of water were injected in 61 hours. The injection point was 3460 m deep in predominantly granitic, Precambrian basement under the southwest flank of the Jemez caldera. The purpose of the hydraulic fracturing was to create a reservoir through which fluids could be circulated to extract geothermal heat (DREESEN and NICHOLSON, 1985; FRANKE and NUNZ, 1985).

The MHF experiment was instrumented with both shallow and deep borehole sensors to monitor accompanying seismic activity (HOUSE, 1987). Two vertical-component sensors were sited near the top of the basement at depths greater than 500 m and two pressure- and temperature-hardened, triaxial sensors were sited within the basement at depths of 2855 m and 3300 m. Over 11,000 locatable microearthquakes were recorded. They filled a tabular volume, 1 km by 1 km by 300 m in size, striking N10°W and dipping 65° to the east.

Precise relative location techniques were applied to five clusters within the microseismic cloud (PHILLIPS *et al.*, 1997). In all cases, initially scattered hypocenter patterns tightened dramatically after repicking and relocation. The example shown in Figure 8 is small in extent, however otherwise typical of the five clusters. Events fall onto an isolated planar patch about 50 m wide. In the face-on view (bottom right in Fig. 8), sharp, straight edges bound the planar patch, roughly forming a parallelogram. This bounding edge pattern was seen in four of the five clusters studied; in the fifth cluster, events fell onto a single, isolated linear segment. The edges of the planar

◄

Figure 7

Map (top) and cross-section (bottom) views of seismicity recorded at the Tallent site, Clinton County, Kentucky. In the map view, the current producing well is represented by a star, monitor wells by triangles and remaining production wells by open circles. Event locations are represented by dots with black fill indicating the main fracture and white fill indicating the two smaller fractures. An arrow indicates the cross-section view direction. In the cross-section view, wells existing prior to geophone deployment are indicated by solid lines, wells drilled following deployment by dotted lines. Old production intervals are represented by squares, the current producing interval by a star and geophone positions by triangles. Possible aseismic fractures are indicated by dashed lines. Only seismicity that occurred east of −20 m is included in the cross-section plot.

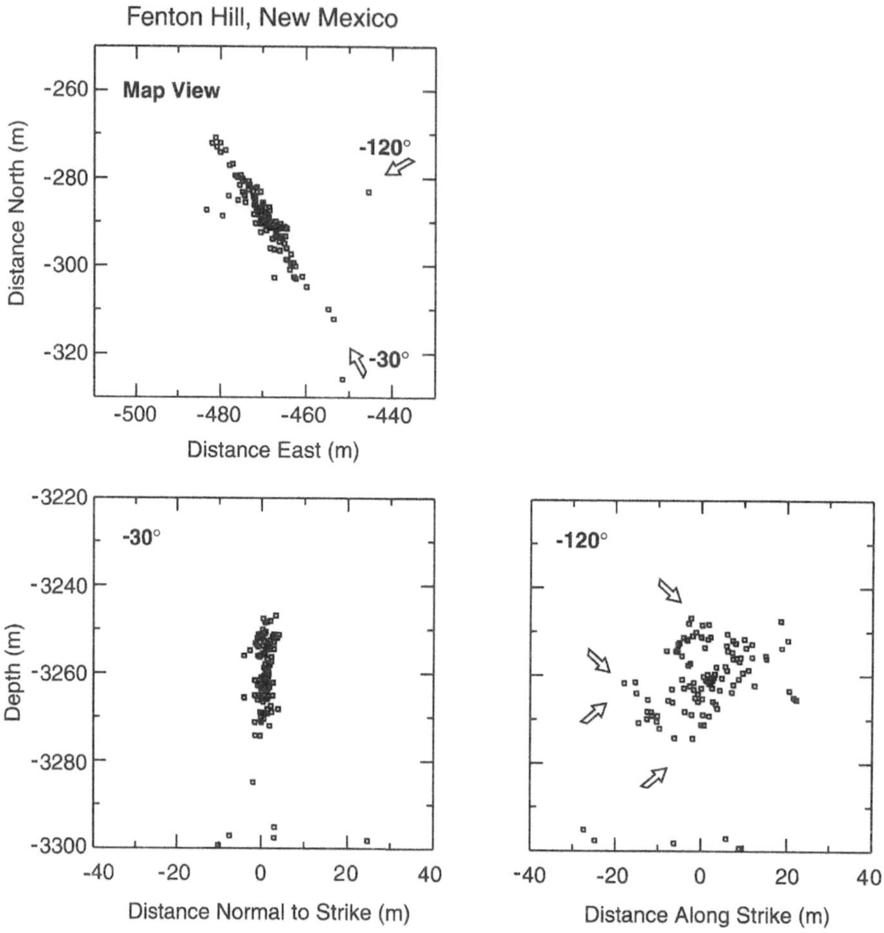

Figure 8
Map and cross-section views of a relocated Fenton Hill cluster. Arrows indicate cross-section view angles (map view) and edges discussed in the text (broadside view).

clusters likely result from truncation of the slipping surface by aseismic joints. Thus the seismicity allows some constraint on position and orientation of the aseismic joints, which may form part of the fluid-flow network. The isolated linear segment is thought to occur along the intersection of two otherwise aseismic joints. This conclusion had been drawn earlier by ROFF *et al.* (1996) after examining many linear and planar features in the original, but lower quality location set.

Soultz-sous-Forêts, France

The Soultz-sous-Forêts geothermal site is located in the Alsace region of France, along the western edge of the Rhine Graben at a point where especially high subsurface temperature gradients can be found. At Soultz, the reservoir region lies in granitic rock at depths of 2 to 4 km. Temperatures reach 168° at a depth of 3.8 km. In September 1993, 44,000 m^3 of water was injected at a depth of 2850 m in the first stage of reservoir creation (JUNG *et al.*, 1996).

During injection, seismic activity was recorded by three, four-component borehole sensors (Galperin orientations with one vertical component) and one borehole hydrophone deployed at depths between 1.3 and 2.1 km. Over 16,000 events were recorded. The seismic cloud is tabular in form, with dimensions of about 1.5 by 1.2 by 0.5 km, strikes NNW and dips nearly vertically (JONES *et al.*, 1995). The overall shape is similar to that observed for the MHF injection at Fenton Hill.

A cluster analysis based on spatial locations identified numerous concentrations of microearthquake activity throughout the September 1993 cloud of seismicity. Two of these clusters were chosen for a precise relative location study (PHILLIPS, 2000). Relative *P*- and *S*-wave arrival times were determined manually, imitating cross-correlation picking when possible, and choosing first break peaks for data that were inconsistent or nodal. For the deeper of the two clusters, events were grouped based on *S*-waveform shapes at one station, and these groups became spatially distinct after location.

Following relative relocation, the clusters collapsed into well-defined patterns (Fig. 9). Both clusters consist of two planar patches that intersect and terminate against one another, as seen in views along the trends of the planar intersections (Fig. 9, top). The slip inferred from focal mechanisms causes a mass deficit on the inside corner of the planar intersections and acts to decrease normal stress, perhaps resulting in further slip. This effect is most obvious for the deep cluster and may be the mechanism responsible for the high levels of activity. Linear segments within the seismicity become apparent in the broadside views, (linear features are indicated by arrows in Fig. 9, bottom, with annotations representing similarly oriented groups), possibly indicating cross-cutting fractures. The linear features are more apparent in the shallow cluster (arrows a and b).

Both clusters display interesting temporal behavior. Events are coded by the order in which they occur in the face-on views (Fig. 9, bottom). In the shallow cluster, most of the earliest events fall along the two linear segments identified above. Activity began along these segments and spread into the planar patches, (especially into the lower left area, Fig. 9, bottom left), as the pressurization continued. In the deep cluster, activity began on plane 4 along linear segments 4a and 4c and tailed off with a shift of activity to plane 3. Activity continued along two parallel linear segments, both labeled 3a, and spread in between with time. The early activity along linear segments indicates their association with permeable zones within the fracture. Observations of turbulent flow in this reservoir lead KOHL *et al.* (1997) to suggest that flow is confined to narrow channels, consistent with our interpretation if our seismic results are representative of the reservoir as a whole. If one may extrapolate the seismic results, linear segments could indicate important components of the

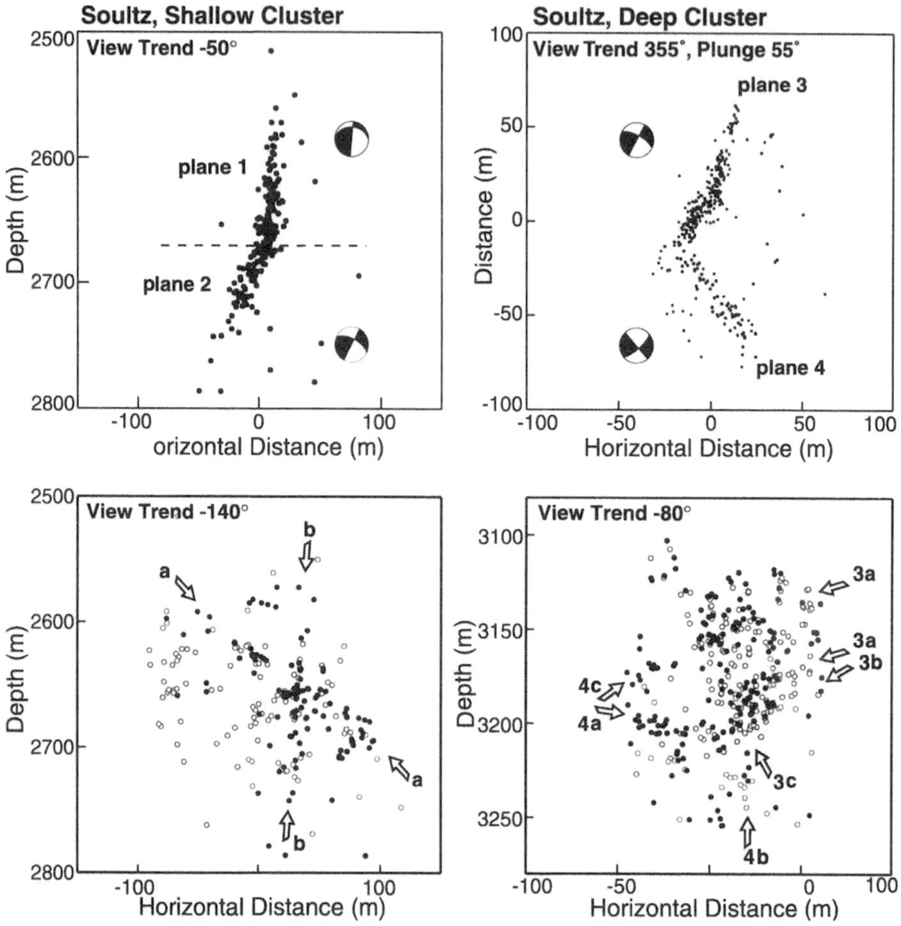

Figure 9

Cross-section views of two relocated Soultz clusters. The shallow cluster is shown on the left, the deep cluster on the right. Top views are along the trends of the fracture intersection axes. Bottom views are broadside to plane 1 and plane 3. Projected focal mechanisms are indicated in the along axis views. A dashed line divides shallow cluster planes based on *P*-polarity patterns. In the broadside views, black filled dots represent the earlier 50% of the events that occurred in each cluster, open dots represent the later 50%. Various linear features discussed in the text are indicated by the annotated arrows.

fluid-flow network at Soultz, and mapping their distribution should lead to important advances in understanding reservoir behavior.

Discussion

In a massive, competent limestone and in crystalline environments we observe seismicity that organizes into combinations of linear and planar features. At Fenton

Hill, straight, nearly parallel edges bound planar patches of activity (Fig. 8) and have been interpreted as representing aseismic joint sets that truncate the seismically active joints. The planar patches are isolated from other seismicity, however, in one cluster, a unique family of events falls along one edge (PHILLIPS *et al.*, 1997), possibly representing slip on the truncating joint. One or more of the aseismic, truncating joints could have acted as pressure conduits to the otherwise isolated patches. Thus these aseismic joints could be part of the fluid network and their positions and orientations can be partially constrained by patterns in the seismically active joints. Results from Soultz and Clinton Co., Kentucky support the truncation model, although, at these sites the truncating features are seismically active. In both cases, planar zones of seismicity truncate each other and form "V-shaped" patterns when viewed along the intersection trend. At Soultz we observe additional linear features within the seismically active fractures. These could be traces of intersecting, aseismic joints, but could also mark changes in rock composition, such as small dikes, but we can only speculate. Either way, the linear segments represent permeable channels within fractures because they are observed to develop early in the injection. Thus, precise event locations allow us to do more than simply ascertain the position, orientation and extent of active fractures; they also allow us to make inferences as to fluid-flow paths within fractures, as well as constrain positions of aseismic fractures that also may be permeable.

In clastic sedimentary environments we observe a strong tendency for induced microseismicity to organize into thin, roughly horizontal strands of activity. This behavior is especially surprising in the Lower Frio, given the unconsolidated nature of these sediments, as noted by HOUSE and FLORES (2001). From studies in crystalline environments, linear features are thought to occur along intersections between otherwise aseismic fractures; however, this mechanism gives rise to a multitude of possible linear orientations. The near horizontal orientations found consistently at the clastic sedimentary sites leads us to suspect that stratigraphy, instead, controls the behavior. Stratigraphic changes can limit fracture growth through an increase in ductility or an increase or decrease in mechanical strength and supported stress at a layer boundary. The observed confinement of microseismic events between the tuff and shale layers within and below the Austin chalk, respectively, support this interpretation. Similar effects have been demonstrated in fracture mine back experiments (WARPINSKI and TEUFEL, 1987). We propose that this also controls the linear patterns observed in the Cotton Valley and lower Frio studies. We look forward to refining these ideas as additional field information, including comparison with well log stratigraphy, fracture and flow data, is included in these studies.

Strong spatio-temporal relationships have also been observed at sedimentary sites. The observed patterns lend credibility to the estimated location accuracy and may yield important insight into microseismic generation. One of the more intriguing temporal behaviors is the small-scale systematic migration of events within a cluster towards the injection well at Cotton Valley (Fig. 6). WITHERS and RIEVEN (1996) also report migration of seismicity towards the injection well within two, separate clusters of activity in the Lower Frio. In general, large-scale clouds of activity grow with time. We might expect migration of cluster events, if it were due to fracture extension, to evolve away from the active well with time as well. Such behavior is observed, but only during early stages of injection in the Lower Frio (RIEVEN, 1999). A possible interpretation of the reverse migration within clusters is that the inflating fracture is oriented normal to the minimum principal stress and supports little resolved shear, in which case initial slip requires a particular mechanical condition that occurs infrequently along its length. (Alternatively, WITHERS and RIEVEN (1996) observe that events increase in size with distance from the injection point during pressurization of the Lower Frio and propose that this results from an increasing accumulation of energy at the fracture tip with time.) In either case, once slip has initiated at a distant point, local stresses are altered in both directions along the fracture, increasing the potential for slip. Migration towards the well may occur preferentially because pore pressure is higher in that direction. Thus, because reverse migration is commonly observed within clusters, slip-induced stresses can be responsible for significant fractions of the seismicity along inflating fractures oriented perpendicular to the minimum stress.

The effects of seismic slip on the local stress field may also be important at the Soultz site, where we observe concentrations of seismicity where fractures intersect and truncate one another (Fig. 9). Focal mechanisms show that the seismic slip would cause a mass deficit at the inside corners of fracture intersections, which means slip on either fracture would reduce normal stresses on the other fracture and would facilitate additional slip events. This mechanism could also explain the high level of activity in the Cotton Valley cluster, where the initial slip occurs near a slight bend at the eastern end of the linear segment, although further study is needed to confirm this interpretation. Thus, slip-induced stress could cause the systematic migration discussed previously, as well as the clustering patterns we observe. It could ultimately affect the flow network and should be an important focus for future, quantitative work.

Significant progress has been made recently in the development of methods that rely on routinely determined event hypocenters to find details in clouds of seismicity. The three-point method (FEHLER *et al.*, 1987) finds planar orientations by examining all combinations of three hypocenters simultaneously. The orientations of Fenton Hill planes found with this technique are consistent with known stress fields. The collapsing method (JONES and STEWART, 1997) shifts hypocenters toward the center of mass of nearby events and is able to resolve numerous geometrical features that are otherwise obscured by location error. This method has been generalized into a joint hypocenter (JHD) method (FEHLER *et al.*, 2000). Another method that identifies clusters of events based on *S*-to-*P* amplitude ratios found many small linear and planar clusters that matched orientations obtained by applying the three-point method to a Fenton Hill data set (ROFF *et al.*, 1996). Of these methods, collapsing is the easiest to apply. The positions and orientations of planar features obtained using the collapsing method have been shown to agree with those obtained by precise arrival time repicking (FEHLER *et al.*, 2000); however, repicking reveals an extra level of complexity, such as the bounding edges and linear features within planes, upon which much of our interpretation is based.

Precise manual repicking of arrival times requires painstaking attention to detail and, depending on the data set, can take considerable time. As many as 1000 events can be reasonably processed in this manner. However, for large data sets, such as the 16,000 events in a single injection at Soultz, manual methods become impractical. Automated methods must be investigated for this reason. Automated, cross correlation repicking methods have been successfully applied to natural earthquake data sets (DODGE *et al.*, 1995; SHEARER, 1997; RUBIN *et al.*, 1999), but have been less successful with induced data sets. Induced events from the same geometrical feature often produce waveforms that correlate poorly at one or more stations, even though waveforms at other stations might display high levels of similarity. This primarily results from near nodal takeoff angles to those stations (RUTLEDGE *et al.*, 1998b; PHILLIPS, 2000), but perhaps also from near source path variation or varying slip directions along the fracture due to stress heterogeneity. The poorly correlated waveform arrivals are necessary to constrain hypocenters and cannot be discarded, as can be done in natural earthquake studies for which many more stations are generally available. Furthermore, downhole site conditions can yield band-limited seismograms, which causes difficulty in identifying the correct cycle when aligning waveforms (cycle skipping). Manual picking is better able to identify consistent nuances in such waveforms that allow their alignment. Because the eye is able to identify these nuances, it should be feasible to modify automatic relative picking methods to do the same. Automatic methods are currently being tested on the Soultz deep cluster data set shown in Figure 9 and are being compared to results from our manual repicking (ASTER and ROWE, 2000). This work shows that relative event locations obtained automatically reproduce the intersecting planes shown in Figure 9, but must be developed further to obtain the finer scale features such as the linear segments that we find important for interpretation. However, the automated methods show much promise and, with further development, should allow large data sets to be precisely located in a timely fashion.

Conclusions

In our experience with induced events in geothermal and hydrocarbon reservoirs, seismicity patterns are rarely diffuse, but are composed of distinct geometrical features. In many cases, precise relative location must be applied to observe these patterns. When few stations are available to constrain locations, arrival time error leaks into the event locations and causes scatter. Thus, while precise repicking shifts arrival times only a fraction of the dominant period, it can improve location patterns significantly. The large number of small events (high Gutenberg-Richter *b* value) that are generally observed during induced experiments (e.g., FEHLER and PHILLIPS, 1991) means that abundant events are available to delineate slip surfaces and boundaries well. We suggest that routinely determined location patterns and subsequent interpretations will be considerably improved by carrying out precise relative locations with any induced data set that appears diffuse. Furthermore, interpretations crafted to be consistent with diffuse seismicity should be viewed with this possibility in mind.

We find that horizontal, linear features dominate the seismicity patterns in clastic sedimentary environments common to oil and gas reservoirs. Well-defined temporal migration within clusters, most often towards the active well, also is observed at these sites. We believe that the linear patterns result from stratigraphic variations that confine the stimulated fractures, and that the migrating seismicity results from slip-induced stress concentrations that are initiated by triggering events. A fracture oriented orthogonal to the least principal stress, experiencing little resolved shear, may require a triggering event to produce significant amounts of seismicity. Such a triggering event may occur where conditions are locally more favorable for slip, such as at a bend in the fracture.

In crystalline rock and massive, brittle sediments, the location patterns indicate interacting networks of fractures. Not only can active fractures be identified and mapped, but the positions and orientations of aseismic fractures also can be inferred through the imprints they leave in the seismicity patterns, namely, the straight edges bounding the seismically active planar patches or the linear segments that cut though them. Aseismic fractures could be important components of the fluid-flow network. Furthermore, linear zones within seismically active fractures are observed to develop early and could mark permeable channels along fracture intersections at the Soultz geothermal site. Therefore, in addition to constraining the geometry of the fracture network, seismicity can also be used to infer zones of higher permeability.

The mapping of large data sets, such as Fenton Hill and Soultz, to the level of precision demonstrated here will rely on the development of automatic methods. Waveform cross-correlation methods have shown much success with natural earthquake data sets. However, induced seismicity data sets present more difficulty because of the small number of stations that are available. This means that all arrivals must be relied upon to constrain event locations, even if nodal takeoff angles produce poorly correlated waveforms or if wellbore site conditions produce band-limited waveforms at particular stations, as is commonly observed. Manually determined arrival times can be used to guide the further development of automatic methods that will more reliably align such waveforms (ASTER and ROWE, 2000).

Acknowledgments

We are grateful to Professor Hiroaki Niitsuma and members of the MTC project for their advice and encouragement. We also thank Dave Anderson, Tom Fairbanks, Jim Albright and Bob Hanold for their operational and technical support. Comments from two anonymous reviewers greatly helped us improve the manuscript. This study was partially supported by the DOE Oil Recovery Technology Partnership and the DOE Office of Basic Energy Sciences.

References

ALBRIGHT, J. N., and PEARSON, C. F. (1982), *Acoustic Emissions as a Tool for Hydraulic Fracture Location: Experience at the Fenton Hill Hot Dry Rock Site*, J. Soc. Pet. Eng. *22*, 523–530.

ASTER, R. C., and ROWE, C. A. (2000), *Automatic phase pick refinement and similar event association in large seismic datasets.* In *Advances in Seismic Event Location*, (C. Thurber and N. Rabinowitz, eds) *in press.*

BAME, D. and FEHLER, M. C. (1986), *Observations of Long-period Earthquakes Accompanying Hydraulic Fracturing*, Geoph. Res. Lett. *13*, 149–152.

DOBECKI, T. L. (1983), *Hydraulic Fracture Orientation by Use of Passive Borehole Seismics*, paper SPE 12110, presented at the 1983 SPE Annual Technical Conference and Exhibition, San Francisco.

DODGE, D. A., BEROZA, G. C., and ELLSWORTH, W. L. (1995), *Foreshock Sequence of the 1992 Landers, California, Earthquake and its Implications for Earthquake Nucleation*, J. Geophys. Res. *100*, 9865–9880.

DREESEN, D. S., and NICHOLSON, R. W. (1985), *Well Completion and Operations for MHF of Fenton Hill Well EE-2*, Trans. Geotherm. Res. Coun. *9, Pt II*, 105–110.

DYER, B. C., JONES, R. H., COWLES, J. F., BARKVED, O., and FOLSTAD, P. G. (1999), *Microseismic Survey of a North Sea Reservoir*, World Oil, *March*, 74–78.

EVERNDEN, J. F. (1969), *Identification of Earthquakes and Explosions by Use of Teleseismic Data*, J. Geophys. Res. *74*, 3828–3856.

FEHLER, M. C., and PHILLIPS, W. S. (1991), *Simultaneous Inversion for Q and Source Parameters of Microearthquakes Accompanying Hydraulic Fracturing in Granitic Rock*, Bull. Seismol. Soc. Am. *81*, 553–575.

FEHLER, M. C., HOUSE, L. S., and KAIEDA, H. (1987), *Determining Planes Along Which Earthquakes Occur: Method and Application to Earthquakes Accompanying Hydraulic Fracturing*, J. Geophys. Res. *92*, 9407–9414.

FEHLER, M. C., PHILLIPS, W. S., HOUSE, L. S., JONES, R. H., ASTER, R., and ROWE, C. A. (2000), *A Method for Improving Relative Earthquake Locations*, Bull. Seismol. Soc. Am., *in press.*

FERRAZZINI, V., CHOUET, B., FEHLER, M. C., and AKI, K. (1990), *Quantitative Analysis of Long Period Events Recorded During Hydrofracture Experiments at Fenton Hill, New Mexico*, J. Geophys. Res. *95*, 21,871–21,884.

FIX, J. E., ADAIR, R. G., FISHER, T., MAHRER, K., MULCAHY, C., MYERS, B., SWANSON, J., and WOERPEL, J. C. (1989), *Development of Microseismic Methods to Determine Hydraulic Fracture Dimensions*, Gas Res. Inst., Tech. Rep. No. *89*-0116.

FRANKE, P. R., and NUNZ, G. J. (1985), *Recent Developments in the Hot Dry Rock Geothermal Energy Program*, Geothermal Resources Council Annual Meeting, Preprints, *1–4.*

GAUCHER, E., CORNET, F. H., and BERNARD, P. (1998), *Induced Seismicity Analysis for Structure Identification and Stress Field Determination*, Paper SPE 47324, Proc. SPE/ISRM, Trondhiem, Norway.

GIBOWICZ, S. J., YOUNG, R. P., TALEBI, S., and RAWLENCE, D. J. (1991), *Source Parameters of Seismic Events at the Underground Research Laboratory in Manitoba, Canada: Scaling Relations for Events with Moment Magnitude Smaller than –2*, Bull. Seismol. Soc. Am. *81*, 1157–1182.

HAMILTON-SMITH, T., NUTTAL, B. C., GOODING, P. J., WALKER, D., and DRAHOVZAL, J. A. (1990), *High Volume Oil Discovery in Clinton County, Kentucky*, Kentucky Gol. Surv., Ser. 11, Inf. Circ. 33.

HOUSE, L. S., and JENSEN, B. (1987), *Focal Mechanisms of Microearthquakes Induced by Hydraulic Injection in Crystalline Rock*, EOS Trans. *68*, 1346.

HOUSE, L. S. (1987), *Locating Microearthquakes Induced by Hydraulic Fracturing in Crystalline Rock*, Geophys. Res. Lett. *14*, 919–921.

HOUSE, L. S., FLORES, R., and WITHERS, R. (1996), *Microearthquakes Induced by a Hydraulic Injection in Sedimentary Rock, East Texas*, SEG 66th Ann. Meeting, pp. 110–113.

HOUSE, L. S., and FLORES, R. (2001), *Seismological Studies of a Fluid Injection in Sedimentary Rocks, East Texas*, Pure appl. geophys, in press.

JONES, R. H., and STEWART, R. C. (1997), *A Method for Determining Significant Structures in a Cloud of Earthquakes*, J. Geophys. Res. *102*, 8245–8254.

JONES, R. H., BEAUCE, A., JUPE, A., FABRIOL, H., and DYER, B. C. (1995), *Imaging Induced Microseismicity During the 1993 Injection Tests at Soultz-sous-Forets, France*, Proc. World Geotherm. Cong., Florence, 2665–2669.

JUNG, R., RUMMEL, F., JUPE, A., BERTOZZI, A., HEINEMANN, B., and WALLROTH, T. (1996), *Large-scale Hydraulic Injections in the Granitic Basement in the European HDR Programme at Soultz, France*, Proc. 3rd Int. HDR Forum, Santa Fe, 75–76.

KECK, R. G., and WITHERS, R. J. (1994), *A Field Demonstration of Hydraulic Fracturing for Solid Waste Injection with Real-time Passive Seismic Monitoring*, Paper SPE 28495, SPE Annual Technical Conference and Exhibition, New Orleans.

KOHL, T., EVANS, K. F., HOPKIRK, R. J., JUNG, R., and RYBACH, L. (1997), *Observation and Simulation of non-Darcian Flow Transients in Fractured Rock*, Water Resour. Res. *33*, 407–418.

LEES, J. M. (1998), *Multiplet Analysis at Coso Geothermal*, Bull. Seismol. Soc. Am. *88*, 1127–1143.

LI, Y. P., CHENG, C. H., and TOKSOZ, M. N. (1998), *Seismic Monitoring of the Growth of a Hydraulic Fracture Zone at Fenton Hill, New Mexico*, Geophysics *63*, 120–131.

MAXWELL, S. C., YOUNG, R. P., BOSSU, R., JUPE, A., and DANGERFIELD, J. (1998), *Microseismic Logging of the Ekofisk Reservoir*, Paper SPE 47276, Proc. SPE/ISRM, Trondhiem, Norway.

MORIYA, H., NAGANO, K., and NIITSUMA, H. (1994), *Precise Source Location of AE Doublets by Spectral Matrix Analysis of Triaxial Hodogram*, Geophysics *59*, 36–45.

PEARSON, C. (1981), *The Relationship Between Microseismicity and High Pore Pressures During Hydraulic Stimulation Experiments in Low Permeability Granitic Rocks*, J. Geophys. Res. *86*, 7855–7864.

PHILLIPS, W. S., HOUSE, L. S., and FEHLER, M. C. (1997), *Detailed Joint Structure in a Geothermal Reservoir from Studies of Induced Microearthquake Clusters*, J. Geophys. Res. *102*, 11,745–11,763.

PHILLIPS, W. S., FAIRBANKS, T. D., RUTLEDGE, J. T., and ANDERSON, D. W. (1998), *Induced Microearthquake Patterns and Oil-producing Fracture Systems in the Austin Chalk*, Tectonophysics *289*, 153–169.

PHILLIPS, W. S. (2000), *Precise Microearthquake Locations and Fluid Flow in the Geothermal Reservoir at Soultz-sous-Forêts, France*, Bull. Seismol. Soc. Am. *90*, 212–228.

RIEVEN, S. A. (1999), *Analysis and Interpretation of Clustered Microseismicity at Geothermal and Petroleum Reservoirs*, Ph.D. Thesis, MIT, Cambridge, 410 pp.

ROFF, A., PHILLIPS, W. S., and BROWN, D. W. (1996), *Joint Structures Determined by Clustering Microearthquakes Using Waveform Amplitude Ratios*, Int. J. Rock Mech. Geomech. Abs. *33*, 627–639.

RUBIN, A. M., GILLARD, D., and GOT, J.-L. (1999), *Streaks of Microearthquakes Along Creeping Faults*, Nature *400*, 635–641.

RUTLEDGE, J. T., PHILLIPS, W. S., HOUSE, L. S., and ZINNO, R. J. (1998a), *Microseismic Mapping of a Cotton Valley Hydraulic Fracture Using Decimated Downhole Arrays*, Proc. 68th Ann. Mtg., SEG, 338–341.

RUTLEDGE, J. T., PHILLIPS, W. S., and SCHUESSLER, B. K. (1998b), *Reservoir Characterization Using Oil-production-induced Microseismicity, Clinton County, Kentucky*, Tectonophysics *289*, 129–152.

SARDA, J.-P., PERREAU, P. J., and DEFLANDRE, J.-P. (1988), *Acoustic Emission Interpretation for Estimating Hydraulic Fracture Extent*, Paper SPE 17723, SPE Gas Technology Symposium, Dallas.

SHEARER, P. M. (1997), *Improving Local Earthquake Locations Using the L1 Norm and Waveform Cross Correlation: Application to the Whittier Narrows, California, Aftershock Sequence*, J. Geophys. Res. *102*, 8269–8284.

SYKES, L. R. (1970), *Earthquake Swarms and Sea-floor Spreading*, J. Geophys. Res. *75*, 6598–6611.

TEZUKA, K., and NIITSUMA, H. (1997), *Integrated Interpretation of Microseismic Clusters and Fracture System in a Hot Dry Rock Artificial Reservoir*, Proc. 67th Ann. Mtg., SEG, 657–660.

VANDECAR, J. C., and CROSSON, R. S. (1990), *Determination of Teleseismic Relative Phase Arrival Times Using Multi-channel Cross-correlation and Least-squares*, Bull. Seismol. Soc. Am. *80*, 150–169.

VINEGAR, H. J., WILLS, P. B., DEMARTINI, D. C., SHYLAPOBERSKY, J., DEEG, W. F., ADAIR, R. G., WOERPEL, J. C., FIX, J. E., and SORRELLS, G. G. (1991), *Active and Passive Seismic Imaging of a Hydraulic Fracture in Diatomite*, Paper SPE 22756, SPE Annual Technical Conference and Exhibition, Dallas.

WALKER, R. N. (1997), *Cotton Valley Hydraulic Fracture Imaging Project*, SPE Paper 38577, SPE Annual Technical Conference and Exhibition, San Antonio.

WARPINSKI, N. R., and TEUFEL, L. W. (1987), *Influence of Geologic Discontinuities on Hydraulic Fracture Propagation*, J. Pet. Tech. *39*, 209.

WARPINSKI, N. R., WRIGHT, T. B., UHL, J. E., ENGLER, B. P., DROZDA, P. M., PETERSON and BRANAGAN, P. T. (1996), *Microseismic Monitoring of the B-sand Hydraulic Fracture Experiment at the DOE/GRI Multi-site Project*, Paper SPE 36450, SPE Annual Technical Conference and Exhibition, Denver.

WITHERS, R. J., and RIEVEN, S. A. (1996), *Fracture Development During Cuttings Injection Determined by Passive Seismic Monitoring*, SEG 66th Ann. Meeting, 106–109.

ZHU, X., GIBSON, J., RAVINDRAN, N., ZINNO, R., and SIXTA, D. (1996), *Seismic Imaging of Hydraulic Fractures in Carthage Tight Sands: A pilot Study*, The Leading Edge, *15*, 218–224.

(Received August 10, 2000, revised/accepted January 29, 2001)

Pure appl. geophys. 159 (2002) 371–401
0033 – 4553/02/030371 – 31 $ 1.50 + 0.20/0

Pure and Applied Geophysics

Seismological Studies of a Fluid Injection in Sedimentary Rocks, East Texas

LEIGH S. HOUSE[1] and RODERICK FLORES[1,2]

Abstract—Induced microseismicity data from a large volume fluid injection into sedimentary rock was analyzed to study the fracture system, fluid pathways, and state of stress in the lower Frio formation in east Texas. Seismicity data are from two arrays of 25 3-component geophone packages sited in two monitoring boreholes. From a total of 2,894 event triggers, a subset of 54 microearthquakes was chosen for their high quality seismograms and clear *P* and *S* arrivals. Arrival times were picked with a precision of 0.5 to 1.0 ms, and microearthquakes were located with hypocentral uncertainties estimated as less than 10–20 m. Hypocenters farthest from the injection well define a nearly horizontal tube of seismicity approximately aligned in the direction of the injection well. A simultaneous inversion of arrival times for transverse isotropic velocity structure and hypocenters yielded *P*-wave anisotropy of −14% and *S*-wave anisotropy of −2%. Thus, velocities along vertical ray paths are higher than those along horizontal paths, probably because of lithologic differences. Single-event focal mechanisms were determined for 47 events, and many of them are normal fault type. The minimum principal stress derived from the focal mechanisms is nearly horizontal and trends approximately north-south, consistent with the regional stress state. An imaging analysis of the seismograms shows the presence of strong seismic scatterers at positions that correlate with boundaries seen in the hypocenters; both features probably result from a similar set of heterogeneities. This study demonstrates the abundance of information that can be extracted from induced seismicity data and underscores the value of induced seismicity monitoring for studying the fluid and fracture systems created by fluid injections.

Key words: Induced seismicity, hydraulic fracturing, seismicity monitoring, fault plane solutions, seismic imaging.

Introduction

Many researchers have studied seismicity induced by fluid injections in crystalline rock (BAME and FEHLER, 1986; HOUSE, 1987; FEHLER and PHILLIPS, 1991; BLOCK *et al.*, 1994; CORNET and YIN, 1995; JONES *et al.*, 1995; SASAKI, 1998; PHILLIPS, 2000; PHILLIPS *et al.*, 2002). A few have studied seismicity induced by fluid injections in sedimentary rocks (VINEGAR *et al.*, 1992; PHILLIPS *et al.*, 1998; RUTLEDGE *et al.*, 1998). Collectively, these studies have shown that the induced seismicity data provide

[1] Los Alamos Seismic Research Center, Los Alamos National Laboratory, Los Alamos, NM, 87545 U.S.A. E-mail: house@lanl.gov
[2] Now at Truchas Consulting, LLC, Albuquerque, NM.

highly detailed information about the fracture systems that no other geophysical methods can provide (e.g., PHILLIPS *et al.*, 2002).

An injection experiment into a sedimentary rock formation, termed the "Deep Well Treatment and Injection" (DWTI) project, was done by ARCO in October 1993 near Beaumont, in eastern Texas (see Fig. 1). The purpose of the experiment was to demonstrate the use of hydraulic fracturing to dispose of solid wastes. KECK and WITHERS (1994) describe the experiment and report results of mapping about two hundred of the induced microearthquakes (see also WITHERS and RIEVEN, 1996). To carry out the experiment, an injection well and two seismicity monitoring wells were drilled to total depths of about 1460 m. Fluids were injected into the Lower Frio sandstone, an unconsolidated sand that has a porosity of about 35% and a permeability of about $2.0 \cdot 10^{-12}$ m^2 in the injection zone (ATLANTIC RICHFIELD CORP., 1994). Relatively impermeable shales above and below the Lower Frio were expected to confine the vertical extent of the fracture system. The overlying shale is about 40 m thick, the underlying shale is about 450 m thick (ATLANTIC RICHFIELD CORP., 1994).

The Frio formation is a progradational wedge of coastal plain sediments of Oligocene-Miocene age located at the northern edge of the Gulf Coast Tertiary basin (GALLOWAY *et al.*, 1982). It has been a prolific hydrocarbon reservoir for more than 50 years. It consists of interfingered marine and nonmarine sands and shales (GALLOWAY *et al.*, 1982). Gamma ray logs from the three DWTI boreholes (Fig. 2a) show the character of the Frio formation, with its two distinct sand layers and

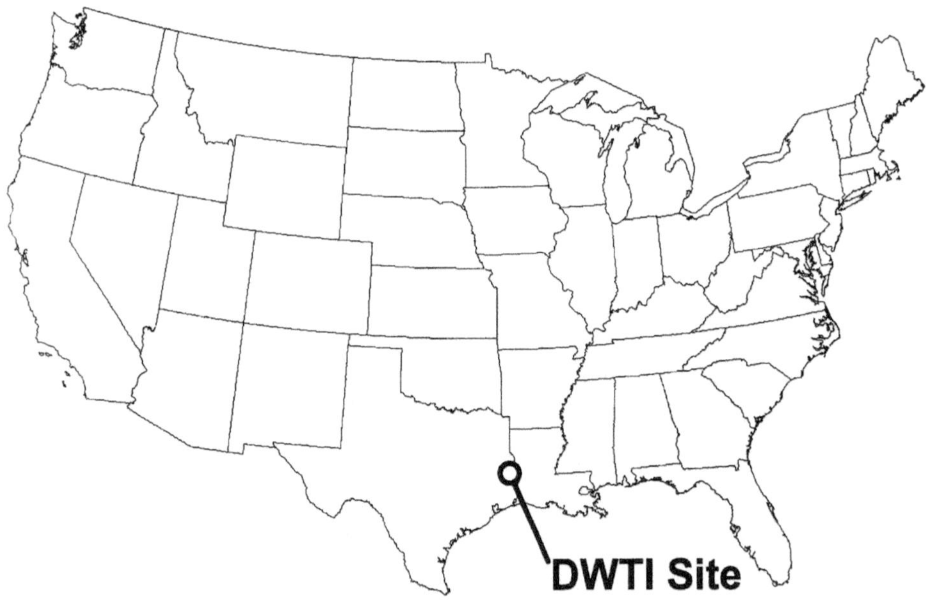

Figure 1
Map of the U.S. showing the location of the DWTI experiment.

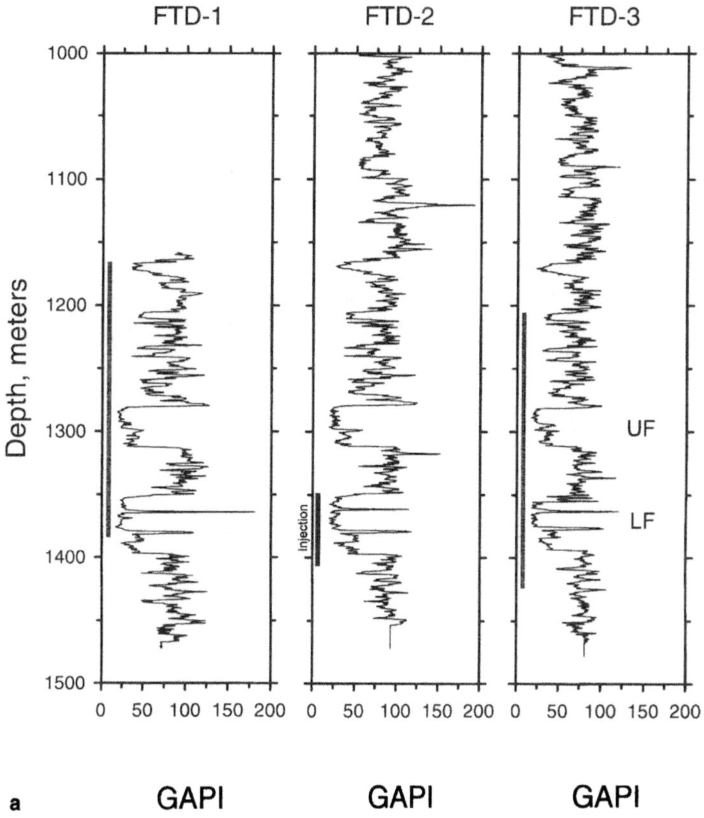

Figure 2

Logs from the three DWTI wells. (a) Gamma logs from all three DWTI wells. Gray bars in FTD-1 and FTD-3 plots are the depths of the geophone packages. Black bar in FTD-2 plot is the extent of the injection (perforated) interval. The distinct sands of the Upper and Lower Frio formations (UF and LF) and the shales that bound them are readily identifiable. Two shale interbeds in the Lower Frio provide a distinct signature to correlate between the wells. (b) Velocity logs for the injection well FTD-2. Bars show depths of the Upper Frio (UF), the Lower Frio (LF), and the injection interval.

bounding shales. Correlation of the logs from the three wells indicates that the Frio formation is essentially flat-lying in the area of the DWTI experiment. Figure 2b shows velocity logs from the injection well, FTD-2. Neither of the sand intervals of the Frio formation (Upper Frio or Lower Frio) is readily discernable in the *P*-wave velocity logs. The Lower Frio is seen to have a higher *S*-wave velocity than the adjacent shales. A depth interval of about 110 m of the Frio formation and adjacent shales was cored while drilling well FTD-1 with about 60% recovery (KECK and WITHERS, 1994). Core from the Lower Frio was friable at surface conditions (R. Withers, personal communication, 1994).

The DWTI experiment was conducted as a series of four distinct injection cycles. The first, cycle 1, was a small volume injection as a test of fracture closure pressure and leak-off behavior. Sand was added to the injected fluids in the three main

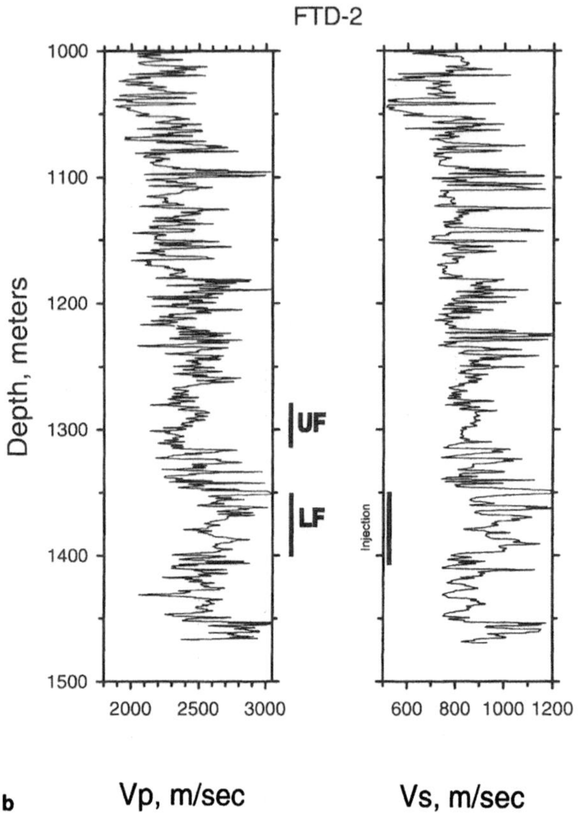

Figure 2b

injection cycles (for details of the injections, see Table 1 and Fig. 3). The injection rate was kept nearly constant during each of the injection cycles. The downhole pressure in the injection well varied between 20.7 and 24.1 MPa during the injection cycles, and slowly dropped between cycles while it was shut in. Fluid density was increased slowly during injection cycles two to four.

Table 1

Injection parameters and microearthquakes detected during the DWTI

	Cycle 1	Cycle 2	Cycle 3	Cycle 4
Liquid injected (Bentonite slurry; m^3)	240	2,465	2,226	2,385
Sand injected (kg)	0	302,040	498,870	634,920
Events detected	157	1,293	881	563
Events with peak amplitude ≧ 150 counts at FTD-3 geophone	0	70	22	10
Events with peak amplitude ≧ 500 counts at FTD-3 geophone	2	22	8	2
Events located for this study	2	46	6	0

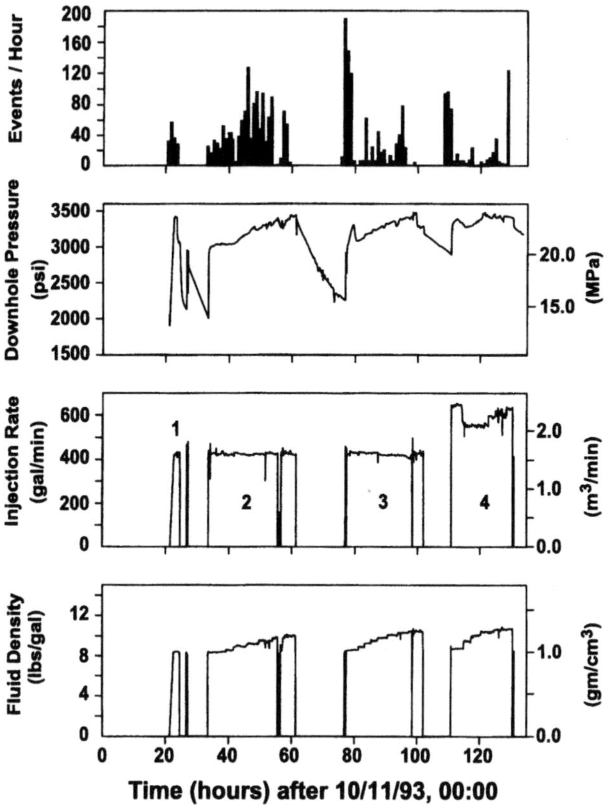

Figure 3

Time history of the injections and induced seismicity. From top to bottom, the seismicity rate, downhole injection pressure, injection rate, and density of the injected fluid are shown. The injection was done in four distinct cycles on the injection rate plot. Solids (sand) were added to the injected fluid in cycles 2, 3, and 4. Event times and fluid data samples are referenced to different clocks that were not synchronized.

Monitoring of Induced Seismicity

Induced microearthquakes were monitored by geophone packages in two monitoring wells, FTD-1 and FTD-3. An azimuth of about N40°E was anticipated for the DWTI fracture system from regional stresses, borehole breakouts, and measurements on oriented core samples (KECK and WITHERS, 1994). The monitor wells were positioned slightly offset laterally from the expected fracture azimuth. Monitor well FTD-1 was 24.4 m from the injection well (FTD-2) at an offset of about 18 m from the expected fracture trend. Monitor well FTD-3 was 155.5 m from the injection well with a lateral offset of about 27 m from the expected fracture trend. The distances of the monitoring wells from the injection well were chosen to help assure that induced seismicity would be recorded by at least one of them. Each monitor well contained 25 geophone packages cemented in at depth intervals of 9.1 m. Each package contained three orthogonal 30 Hz geophones. With a recording capacity of only 96 input channels, however, the two horizontal components could be recorded from only every other geophone package; vertical components were recorded from every geophone package. The recording system acquired 16-bit data samples at a rate of 2,000 samples per second.

A total of 2,894 event triggers were recorded during the four injection cycles (see Table 1). All but four events in the first cycle were very small, with peak amplitudes less than 150 digital counts at a deep, quiet reference geophone in the close in monitoring well, FTD-1. Events this small were difficult to locate because they had few clear arrivals. Based on *S*-*P*-arrival times, the fracture extended to about 150 m from the injection well during the first cycle (KECK and WITHERS, 1994). In contrast, the second cycle had more than 70 events with amplitudes larger than 150 counts, including 22 larger than 500 counts at the FTD-1 reference geophone. The fracture appears to have grown to at least 200 m long during this cycle. Early in the cycle the event rate was low (Fig. 3, top) and few events of significant size were detected. After about an hour of sand injection, several large events were detected near the distant monitor well, FTD-3. The remainder of this cycle is characterized by swarms of events separated by several minutes of quiescence (KECK and WITHERS, 1994). The third injection cycle produced events with smaller maximum amplitudes. Only 22 had peak amplitudes larger than 150 counts and only 8 were larger than 500 counts. The fracture appears to have grown to about 365 m during this phase (KECK and WITHERS, 1994). The fourth cycle had only 10 events larger than 150 counts in peak amplitudes and only 2 larger than 500 counts. The fracture length was not estimated for this cycle, although it may have continued growing (WITHERS and RIEVEN, 1996).

Although we did not determine magnitudes of the events, we estimated event magnitudes as between about −2 and 2. The lower magnitude is estimated from experience with other hydraulic injections. The upper magnitude is an upper bound estimated from the fact that none of the events were felt at the surface. FEHLER and PHILLIPS (1991) studied microearthquakes induced by hydraulic fracturing in granitic rock and estimated source dimensions of between a few meters to a few tens of meters. We assume that DWTI microearthquakes are generally comparable in size.

Selecting Microearthquakes to Study

We chose a subset of 54 microearthquakes from the nearly 3,000 events recorded during the experiment. Generally, these were among the larger events with higher signal-to-noise ratios. Most of the events recorded were small compared to the background noise, and few arrival times could be picked reliably. We discarded all of the events from the final injection cycle (cycle 4) because they were small and had limited downhole station coverage. Events from the first three cycles were given quality ratings from visual inspection. Approximately 1,800 were discarded because they lacked at least several *S*-wave arrivals at both borehole arrays as well as at least one P-wave arrival. The remaining events were ranked according to peak amplitude (in digital counts) at any geophone. In addition, we carefully looked for any events whose S-P times indicated that their epicenters were within a gap in the seismicity. A final step was to select events to fill time gaps in the initial data set. A total of 54 microearthquakes were chosen for study.

Orienting the Geophone Packages

Determining the orientations of the horizontal geophones was essential for rotating the seismograms into P, S_V and S_H components. Orientations were mainly determined from recordings of four primacord shots set off within the injection well, FTD-2. The shots were of varying sizes (0.9 m to 2.74 m) and depths (1345 m to 1433 m). The first (and largest) shot was recorded by all 25 packages in the FTD-1 monitor well and by 7 packages in the more distant FTD-3 well. The remaining three shots were recorded by all 25 packages in FTD-3 and by 7 packages in FTD-1. Because of these lop-sided recordings, orientations of the geophone packages in the close-in borehole, FTD-1, are more uncertain than those in FTD-3. To better determine orientations of the geophones in FTD-1, we used the P-wave particle motions from preliminary hypocenters of the 54 events.

Particle motions from a few geophones near the top of both monitoring wells reflect apparent shifts during the experiment. Particle motions from the primacord shots and the early microearthquakes were consistent with those of the other geophone packages. Partially through injection cycle 2, particle motions were inconsistent. It was difficult to determine exactly when the anomalous particle motions first appeared, however those from event 47 (12 Oct. 1993, 03:49; cycle 1) are correct, while those from event 358 (12 Oct. 1993, 21:35; cycle 2) are shifted. The shifts seem larger from geophones in FTD-1. These apparent changes in the particle motions are unexplained. The geophones were cemented in, although perhaps the quality of the cement was poorer at shallow depths. W. S. Phillips (personal communication, 1995) identified a similar apparent shift of particle motions in data from a 3-component geophone package that monitored a fluid injection at Fenton Hill, New Mexico.

Homogeneous Velocity Structure and Event Locations

Hypocenters were computed using an isotropic half-space velocity model with V_P of 2.47 m/ms and V_S of 1.12 m/ms. These velocities were obtained from travel times of the primacord shots, and approximately agree with velocities of the Lower Frio formation estimated from sonic logs (Fig. 2b).

Seismograms were rotated into P, S_V, and S_H components and corresponding P, S_V, and S_H arrival times were picked from them. The rest of the geophone packages had only vertical component data, and only *P* arrivals were picked from them. An average of 40 *P* and 58 *S* arrivals were picked for each event. The geophones in the two monitoring wells define a single plane of station coverage, and from arrival times alone, hypocenters could lie on either side of that plane. We assumed the events were on the same side of the plane of station coverage as the injection well and verified this assumption from *P*-wave particle motions of several events. The trial epicenter for all the events was then chosen on the north-west side of the plane of the two wells. The trial depth of 1355 m was chosen within the injection interval.

Epicenters extend southwest from the injection well (FTD-2) 200 m (Fig. 4) and define three distinct zones of seismicity (top, Fig. 4). The near zone of events is scattered around the injection well. A few of these occurred early – in injection cycles 1 and the first half of cycle 2 (Fig. 3). Most occurred later – in the second half of cycle 2 and in cycle 3. The far zone of seismicity lies in a narrow zone aligned toward the injection borehole. These events all occurred during injection cycle 2. A middle zone of seismicity lies between the near and far zones, and is separate from them. These middle zone events generally had relatively low amplitudes which made them difficult to locate, and all but one occurred during injection cycle 2. The separation of the events into three distinct zones suggests that the fluid pathways produced by the hydraulic fracturing were discontinuous, perhaps because of discontinuities in the Frio formation. The relatively small number of events in the middle zone suggests that this may be a region of less competent material than the areas to the northeast and southwest. Since there is no obvious indication of discontinuous characteristics in this area in either the well logs or nearby reflection profiles, we cannot confirm that interpretation.

Hypocenters are confined to a narrow range of depths, between about 1355 and 1390 m (Fig. 4), and all lie within the lower Frio formation. With the large vertical extent of the geophone arrays, event depths should be well determined. The far cluster is characterized by a particularly tight depth distribution, and, considering their narrow epicentral trend, the microearthquakes in this zone lie in a tube-like volume. Such a tight event distribution seems surprising because the Frio formation has high permeability. PHILLIPS *et al.* (2002) discuss similarly tight distributions of event hypocenters from several fluid injections and interprets the features to be the intersection of two differently oriented joints or fractures.

Inversion for Transverse Isotropic Elastic Moduli

Geological formations with well-developed horizontally stratified sediments, such as the Frio formation, propagate seismic waves at velocities that vary with incidence angle (LEVIN, 1979). This state is termed transverse-isotropy (T-I) or hexagonal symmetry. It is the simplest case of anisotropy that has general applicability in geophysics. It is characterized by two directions, typically both horizontal, in which

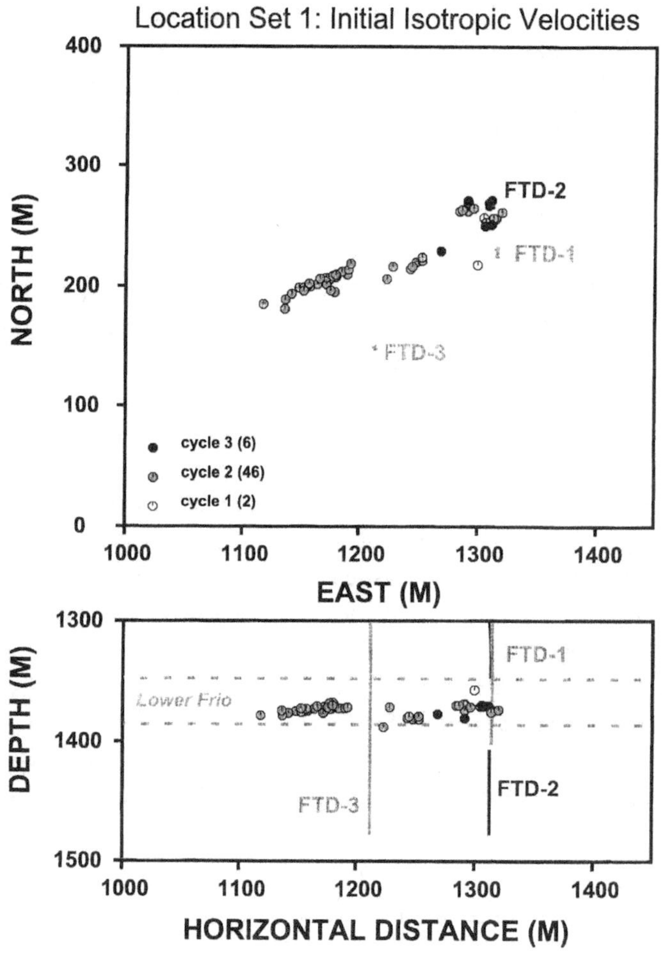

Figure 4
Microearthquake hypocenter set 1, shown in map view (top) and west to east cross section (bottom). These were located using the initial homogeneous velocity structure. Wells FTD-1 and FTD-3 were for monitoring, well FTD-2 for injection; the injection interval is the gap in the trace for FTD-2 in the cross section. Distances are in meters.

elastic moduli are the same, and a third direction, typically vertical, in which moduli are different. Generally velocities increase with depth, and as a result, horizontally travelling waves will propagate faster than vertically travelling waves.

Assuming that anisotropy of the Frio formation was weak (i.e., less than 10–20%), we used the relations presented by THOMSEN (1986) to formulate a simultaneous inversion for T-I elastic moduli and event hypocenters. There are five independent elastic moduli (C_{11}, C_{13}, C_{33}, C_{44}, and C_{66}). P- and S_V-wave propagation are described by constants C_{11}, C_{13}, C_{33}, and C_{44}, and S_H-wave propagation is described by just C_{44} and C_{66}. Although the geophone coverage was dense vertically in the two wells, lateral ray sampling of the rock volume was limited. Thus, we inverted for a 1-D (half-space) T-I structure, rather than 2-D or 3-D T-I structure.

Vertical P-wave and S_H-wave velocities are (respectively):

$$\alpha_0 = \sqrt{C_{33}/\rho}; \quad \text{and} \quad \beta_0 = \sqrt{C_{44}/\rho}. \tag{1}$$

Following THOMSEN (1986), the non-dimensional anisotropy parameters ε and γ represent the percent anisotropy (the fractional difference between horizontal and vertical velocities) for P and S-waves, respectively:

$$\varepsilon = \frac{C_{11} - C_{33}}{2C_{33}}; \quad \text{and} \quad \gamma = \frac{C_{66} - C_{44}}{2C_{44}}. \tag{2}$$

We used P- and S-arrival times for all 54 events in the inversion. The hypocenters from the homogeneous velocity structure (Fig. 4) were used as initial hypocenters for the inversion. Initial values of the elastic constants for the inversion were computed from the homogeneous P and S velocities. Two densities (2.1 g/cm^3 and 2.5 g/cm^3) were used in two separate inversions to test the effect that density has on the moduli. The lower value (2.1 g/cm^3) is probably closer to the actual density of the Frio formation at the depths of the DWTI (CLARK, 1996).

Results of the inversions are summarized in Table 2. The hypocenters and anisotropic parameters are quite similar for both densities, which confirms that the

Table 2a

Transverse isotropy parameters, density of 2.1 g/cm^3

Parameters	All events (54)	Far zone (28)	Near zone (17)
Elastic moduli from inversion			
C_{11}	8.7891	8.7986	9.0367
C_{13}	6.2635	6.3217	6.3846
C_{33}	12.2410	12.0899	12.3332
C_{44}	2.1715	2.2022	2.1875
C_{66}	2.0635	2.0722	2.1146
Calculated parameters			
ε (P anisotropy)	−0.1410	−0.1361	−0.1336
γ (S anisotropy)	−0.0249	−0.0295	−0.0167
V_{P_Z} (m/ms)	2.41	2.39	2.42
V_{S_Z} (m/ms)	1.01	1.02	1.02
Poisson's ratio$_Z$	0.39	0.38	0.39

Three inversions were carried out for each of three event sets: all 54 events, the 28 in the far event group, and the 17 in the near group. The central event group, with only seven events, had insufficient data for a separate inversion. Elastic moduli from the far group are similar to those from all 54 events. Moduli from the near group are generally larger than those from the far group, suggesting the presence of a geological or structural discontinuity between the far and near zones.

Table 2b

Transverse isotropy parameters, density of 2.5 g/cm³

Parameter	All events (54)	Far zone (28)	Near zone (17)
Elastic moduli from inversion			
C_{11}	10.4592	10.5697	10.6874
C_{13}	7.3589	7.4476	7.6297
C_{33}	14.3919	14.5114	14.5806
C_{44}	2.6029	2.5845	2.6608
C_{66}	2.4888	2.4909	2.5176
Calculated parameters			
ε (*P* anisotropy)	−0.1366	−0.1358	−0.1335
γ (*S* anisotropy)	−0.0219	−0.0181	−0.0269
V_{P_Z} (m/ms)	2.40	2.41	2.41
V_{S_Z} (m/ms)	1.02	1.01	1.03
Poisson's ratio$_Z$	0.39	0.39	0.39

The events were separated into the same groups as Table 2a; a higher bulk density was chosen, about the highest that is reasonable for the Frio formation. Results are similar to those of Table 2a. Although the elastic moduli are different from those for a density of 2.1 g/cm³, the calculated parameters are similar.

inversion results do not depend strongly on density. The average vertical *P*-wave velocity from the two inversions is 2.40 m/ms and the average S_V velocity is 1.02 m/ms. Results from the inversion yield a *P*-wave anisotropy (ε) of about −14% and *S*-wave anisotropy (γ) of about −2%, indicating that the horizontal velocities are slightly lower than the vertical velocities. Those results counter what we expected, since in a typical T-I formation, the horizontal velocity exceeds the vertical.

We did not incorporate station corrections in the T-I inversion. Our resources were limited, and we could not adequately explore possible trade-offs between transverse-isotropy and station corrections. Hypocenters from the T-I inversion follow the same general patterns as those from the homogeneous velocity model, although some details differ (compare Figs. 4 and 5). The near zone of hypocenters from the T-I inversion lies to the southwest and slightly farther away from the injection well, compared to those from the homogeneous velocity model. Events in the central zone of Figure 5 are farther from the remaining epicenters than those of Figure 4. Epicenters in the far zone of the figures plot are less linear than those from the homogeneous velocity structure (Fig. 4), and trend rotationally clockwise into a nearly east-west trend compared to the more southwesterly trend of the original epicenters. That east-west tendency is oblique toward the injection well. Collectively, these features emphasize the likelihood of structural discontinuities between the injection wellbore and the far zone, which was suggested previously. The depth distribution of the T-I hypocenters is even narrower and slightly shallower than that of the homogeneous hypocenters (Fig. 5). The far zone of the hypocenters is nearly horizontal.

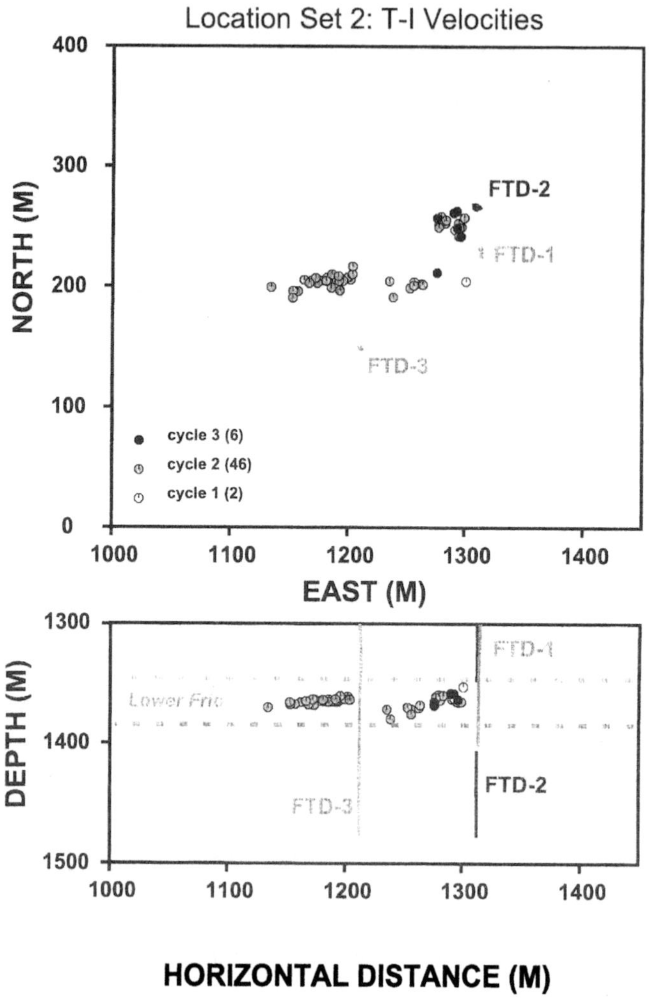

Figure 5
Microearthquake hypocenter set 2. These were located using a halfspace structure with velocities from the transverse isotropy inversion with density of 2.1 g/cm^3. Details of the plot are similar to Figure 4.

Event Relocations Using Homogeneous Structure With Velocities from the T-I Inversion

To explore the effects of differing velocities on the hypocenters, a third set of locations for the 54 events was calculated with a purely isotropic velocity structure using the vertical velocities, α_0 and β_0 (also termed V_{P_Z} and V_{S_Z}) averaged from the two T-I inversions (both densities). The hypocenters obtained are surprisingly different from the original ones, given the slight changes in velocities: V_P of 2.40 rather than 2.47 m/ms and V_S of 1.02 rather than 1.12 m/ms (compare Figs. 4, 5, and 6). The epicenters in Figure 6 are distributed in three distinct zones. The near zone is

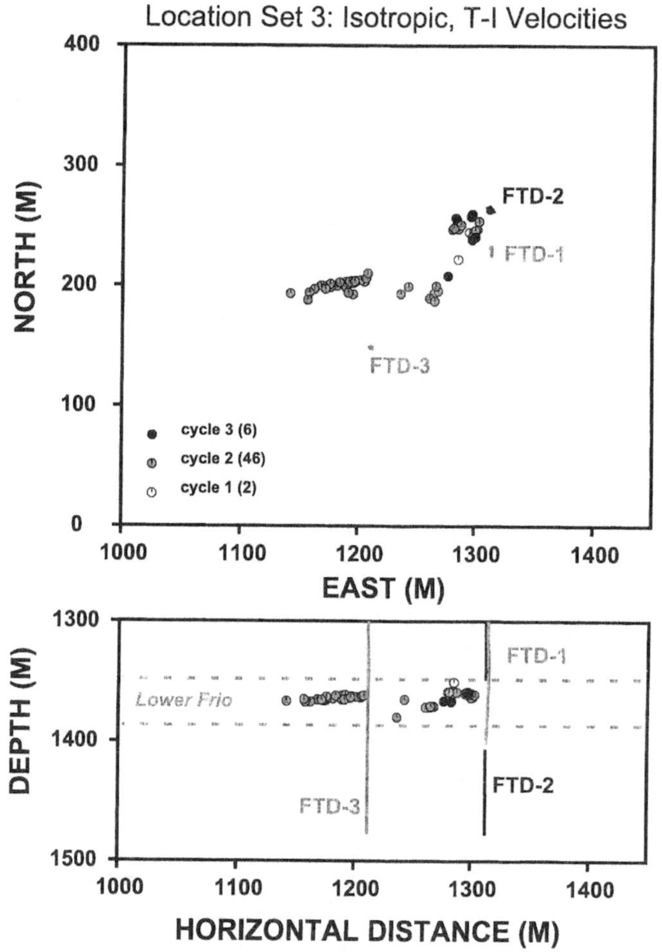

Figure 6
Microearthquake hypocenter set 3. These were located using a homogeneous halfspace velocity structure using the vertical velocities from the transverse-isotropy inversions (V_{P_Z} and V_{SH_Z}). Details of the plot are similar to Figure 4.

similar to the T-I case (Fig. 5). The central zone is positioned closer to the line between the monitor wells. It also inclines differently overall from the trends of both the original and full T-I velocity locations. The far cluster has a trend that is oriented slightly clockwise from the direction to the injection well and aligns directly with monitor well FTD-1. The depths are nearly as tightly clustered as those from the full T-I velocities.

Comparing the root-mean-square (RMS) error of the travel times for the different velocity models provides a simple measure of the quality of fit (Table 3). The RMS for the T-I inversion is approximately 30% lower than for the purely isotropic case.

Table 3

Velocity models and root-mean-square residuals

Arrival type	Pure isotropy		T-I inversion						T-I V_Z, average of $\rho = \{2.1, 2.5\}$	
	V (m/ms)	RMS (ms)	$\rho = 2.1$			$\rho = 2.5$			V (m/ms)	RMS (ms)
			Vz (m/ms)	V_{XY} (m/ms)	RMS (ms)	Vz (m/ms)	V_{XY} (m/ms)	RMS (ms)		
P only	2.47	2.78	2.41	2.07	2.10	2.40	2.07	2.08	2.40	2.76
S_H only	1.12	3.99	1.02	0.99	2.91	1.02	1.00	2.91	1.02	3.64
Both (P & S_H)		3.60			2.58			2.58		3.36

Velocity models and root-mean-square (RMS) residuals, in milliseconds, from calculating hypocenters. Results are averages for all 54 events. Velocities and RMS values are tallied for just P waves (P), for just S_H waves (S_H) and for both. The pure isotropy velocity model is a homogeneous halfspace model with velocities obtained from primacord shots. The T-I (transverse-isotropy) Inversion results were from a simultaneous inversion for elastic moduli and hypocenters. Vz are the vertical velocities, V_{XY} are the horizontal velocities. The preferred bulk density of 2.1 g/cm^3 and a comparison value of 2.5 g/cm^3 were used for the T-I inversions. The T-I Vz velocity model is a homogeneous halfspace model with velocities obtained from the averages of the vertical velocities (Vz) from the two T-I inversions.

The similar RMS values from the two T-I inversions confirms that the results are essentially independent of density. Using the isotropic model with the vertical velocities from the T-I inversion (Location Set 3) results in about a 10% reduction of RMS for S waves compared to the original homogeneous velocities, but no change for the P waves. The RMS error for P arrivals is lower for the T-I inversion (by about 25%), but about the same for the purely isotropic initial velocity model and the isotropic model using vertical velocity from the T-I inversion.

Focal Mechanisms and Stresses

Even with only two monitoring wells, the large vertical extent of the geophones provided coverage of much of the focal sphere of the earthquakes. To simplify the analysis and to obtain results within the limited resources available for this study, we assumed double-couple earthquake mechanisms.

P-wave first motions were read from as many geophone packages as possible for all 54 events. Most waveforms have substantial signal-to-noise ratio, and first motions were easily picked. The close vertical positions of the geophones allowed correlating first motions from nearby geophones, and it was possible to constrain some nodal planes from systematic differences in first motions. We used the third set of hypocenters and the corresponding velocity structure to compute the take off angles of rays from earthquakes to stations. Double-couple focal mechanisms were obtained with the program FPFIT (REASENBURG and OPPENHEIMER, 1985). Figure 7 presents an example focal mechanism. One nodal plane is tightly constrained, the other somewhat less so. A total of 47 events have reasonably well-constrained mechanisms; the remaining 7 were discarded because they had too many discrepant first motions or their mechanisms were inadequately constrained.

Focal mechanisms for all 47 events are shown in Appendix A, divided by event groups. FPFIT found multiple mechanisms for 15 events and the multiple solutions are plotted in Appendix A. We chose preferred mechanisms after reviewing waveforms; these are identified with a *P* prefix on the event number. Most mechanisms are fairly well constrained, although with only two monitoring wells, a few are not, for example, events 650, 720, and 990 in the distant event group (Appendix A). Nevertheless, one nodal plane must have shallow dip and the other must be nearly vertical. The differing mechanisms obtained from nearly identical patterns of first motions (e.g., events 650 and 990) illustrate the range of strike that is compatible with the first motions. Although perhaps these mechanisms should have been removed from the data set, they nevertheless provide information that is not seen in the remaining mechanisms. FPFIT calculates uncertainty in the strike, dip,

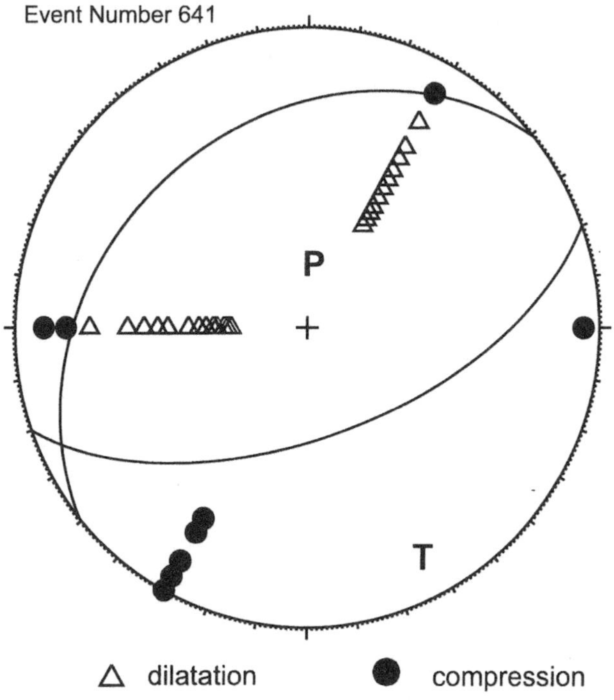

Figure 7
Example of single-event focal mechanism. This is an event from the middle group. There are 33 *P*-wave first motions for this normal faulting type event. FPFIT found multiple solutions for this event; we chose this one as most consistent with its clear first motions.

and rake of each mechanism. The average of the strike uncertainty is about 20° and the average in the dip is about 15°. Several mechanisms contain many first motions that are near a nodal plane (for example, events 556, 650, and 1151 of the far group). These mechanisms could be dismissed as resulting from indeterminate first motions. However, the ability to correlate waveforms among many closely spaced geophones provides more reliable first motion picks and these mechanisms are generally moderately well constrained.

A total of 21 of the 47 mechanisms shown in Appendix A are predominantly normal faulting, more than any other type of focal mechanism. Most of the events in the near and middle groups are this type (16 of 25). The next most common mechanism has one nearly horizontal nodal plane, with the other nearly vertical (events 47, 966, 1135, 1256, 1296, and 2317). The nearly vertical plane generally has a strike of NW-SE, nearly perpendicular to the strike of the hypocentral distribution. In the far event group, 12 of the 26 events have mechanisms with one nearly vertical nodal plane and the other nearly horizontal. The strike of the nearly vertical plane is generally NE-SW for seven of the events (467, 581, 713, 720, 769, 808, and 990), and NW-SE for five (events 530, 556, 636, 650, and 1151). Five mechanisms are mainly normal faulting, (events 625, 673, 734, 762, and 1008). Four of the remaining mechanisms are mainly reverse faulting (events 501, 515, 962, and 1012).

Figure 8 is a plot of the *P* and *T* axes of mechanisms from the near and far event groups. There are 18 focal mechanisms in the near (east) group and 22 in the far

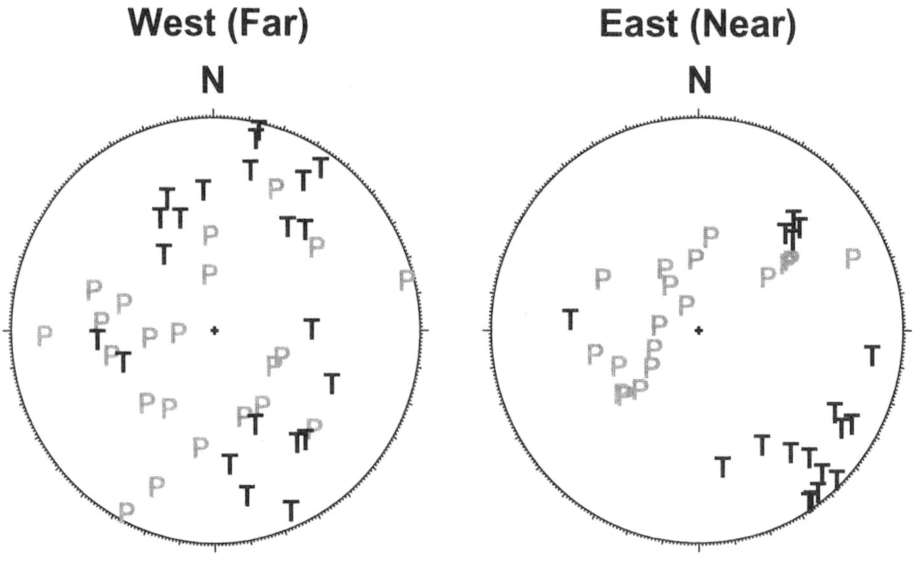

Figure 8
Plots of the *P* and *T* axes of the focal mechanisms from events in the west (far; left plot) and east (near; right plot) seismicity groups. Mechanisms from the west event group have a larger diversity of *P* and *T* axes than those from the east event group.

(west) group. Since the middle event group had only 7 mechanisms, we did not plot them separately. Mechanisms from the near event group have *T* axes that are nearly horizontal or gently plunging, and trend generally SE or NW. *P* axes plunge more steeply and lie in a band that extends NE to SW. Events in the far event group have greater variation of individual mechanisms, with more overlap between the distribution of the *T* and *P* axes. This suggests a more variable stress field or that the difference in the magnitudes of the principal stresses is small. Alternatively, there may be structural discontinuities or heterogeneities in the physical properties of the Frio formation, perhaps related to the depositional process.

The injection site lies within the Gulf Coast stress province (ZOBACK and ZOBACK, 1980), which has a predominantly normal faulting stress field, with the least principal horizontal stress trending approximately N-S. Focal mechanisms from the near and middle event groups are generally consistent with that regional stress field, although the grouping of the *T* axes suggests that the least principal horizontal stress trends NW-SE rather than N-S (Fig. 8).

To better study stresses, all 47 focal mechanisms were used to solve for a single uniform stress field employing the grid-search technique of MICHAEL (1987). The resulting stress field is shown in Figure 9, along with the 95% confidence intervals obtained using the bootstrap technique of MICHAEL (1987). The minimum principal stress, σ_3, is nearly horizontal and trends about N-S, in good agreement with the NNW-SSE to N-S trend of the least principal horizontal stress for this portion of the Gulf Coast stress province (ZOBACK and ZOBACK, 1980). The maximum and intermediate principal stresses, σ_1 and σ_2, respectively, both plunge about 45° along a trend that is nearly east-west. This stress state would favor oblique-normal faulting. The 95% confidence limits (Fig. 9) on the orientations of the principle stresses are moderately large, a result of the diversity of the individual mechanisms. The ratio of principal stress magnitudes, ϕ $\left(\phi = \left[\frac{(\sigma_2-\sigma_3)}{(\sigma_1-\sigma_3)}\right]; \text{MICHAEL, 1987}\right)$, from the stress analysis is about 0.8. The stress ratio is generally not well determined solely from focal mechanisms, however a value of 0.8 suggests that σ_2 has similar magnitude as σ_1, consistent with their overlapping confidence intervals in Figure 9. This would be a nearly uni-axial stress state. To review our choice of using all focal mechanisms to solve for a single stress field, we performed a separate stress analysis using only the 28 events in the western group. The stress field from these events closely corresponds to the stress field from all 47 events (Fig. 9), although it is rotated CCW by about 30°. The 95% confidence intervals in the principal stress axes from the full event set include the principal stress axes from the 28 event subset.

Imaging With Waveforms of Induced Microearthquakes

The presence of secondary arrivals in the waveforms of many of the microearthquakes suggested that they contained additional information regarding significant

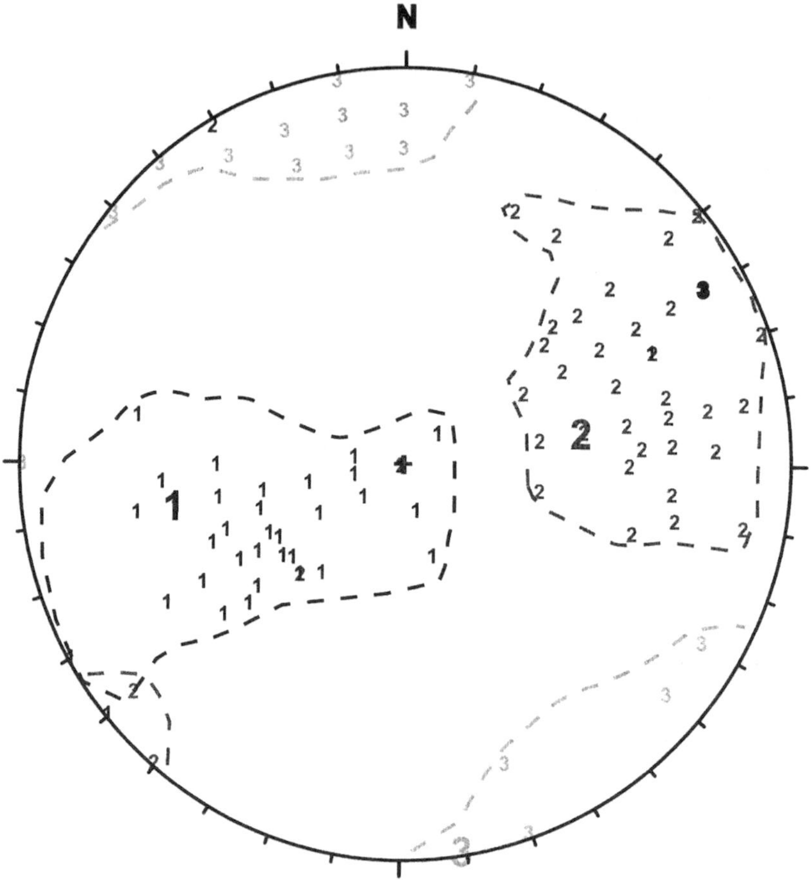

Figure 9
The stress field derived from all 47 focal mechanisms using the grid search method of MICHAEL (1987). The large numbers (1, 2, 3) show the positions of the maximum, intermediate, and minimum principal stresses, respectively. The dashed lines approximately bound the 95% confidence limits in the stress field. Small numbers show possible positions of the principal stresses computed using the bootstrap method described by MICHAEL (1987).

discontinuities near the fracture system. Secondary seismic arrivals are recorded after the direct *P* and *S* arrivals and are distinct from the *P* or *S* coda. They result from scattering of *P*- or *S*-wave energy by discontinuities in the rock. The Frio formation and the adjacent shales are relatively uniform laterally, and the strongest discontinuities might be fluid pathways created by the injection itself.

Figure 10 shows secondary arrivals in seismograms from an example event. The waveforms plotted are the rotated *P*, S_V and S_h components from six 3-component geophone packages. Prominent secondary arrivals are noted on the S_h component traces.

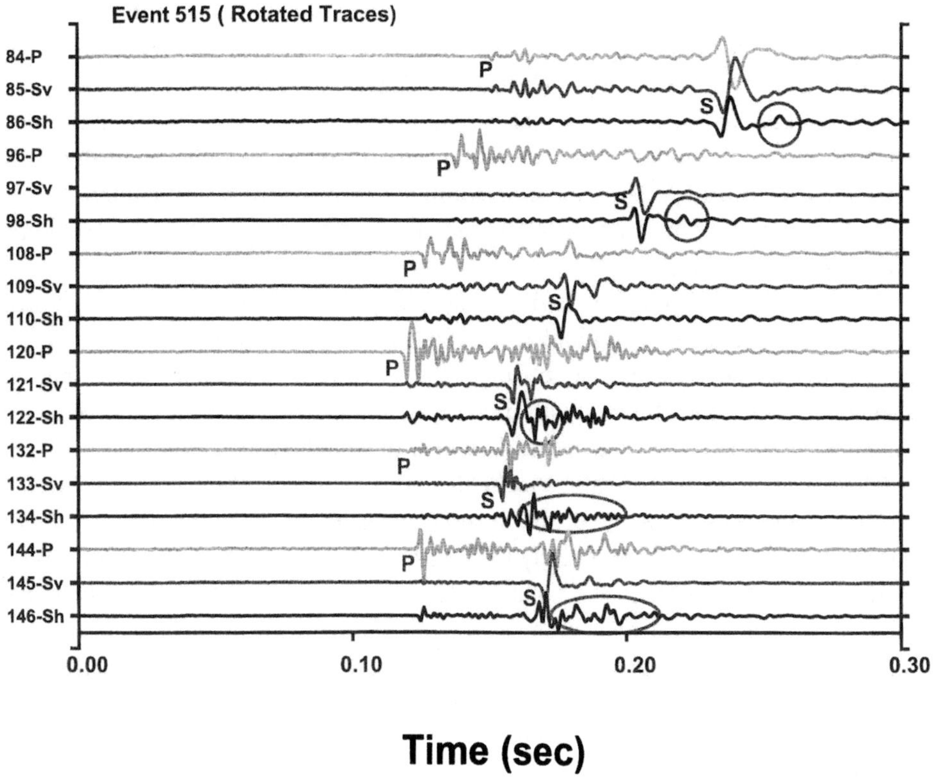

Figure 10
Example of seismograms showing secondary arrivals (circled) identified after the S arrivals on many traces. Traces are rotated P, S_V and S_H components from six geophone packages in the farther monitoring borehole, FTD-3. Traces are from one of the earthquakes used in the diffraction stacking.

To exploit the information contained in the secondary arrivals, a diffraction stacking analysis was used. FLORES *et al.* (1995) and HOUSE *et al.* (1997) summarized the method and presented results from an analysis of this data set and another from an injection into crystalline rock. Conceptually, the diffraction stack method is similar to the Kirchhoff migration method used with seismic reflection data. In this implementation we divided the rock into cubes 5 m on a side, which is about 1/2 of a wavelength (the predominant frequency is about 170 Hz, and V_S is 1.02 km/s). Travel times are computed for the direct arrival, as well as for secondary arrivals, taking each of the cubes to be a possible scatterer. The difference in travel time between the direct ray path and the path including the scatterer provides the time delay of a secondary arrival after the direct arrival. The data sample at that time is added to the stack value of the cube. The stack value of each cube is normalized by the number of seismograms that were summed into the cube (number of "hits"). The resulting 3-D volume of normalized stack values is the end product of the analysis.

As with Kirchhoff migration, the reliability of the travel times is crucial for properly imaging the scatterers. Proper imaging also depends on the summation of secondary arrivals from sources and receivers that are well distributed in space. Unfortunately, earthquakes are not uniformly distributed in space, nor are the geophone monitoring packages. Thus, the diffraction ellipsoids will not be evenly illuminated and artifacts will remain. With a homogeneous velocity structure, these artifacts will be ellipsoidal. Thus, scattering signatures that are ellipsoidal should be interpreted skeptically. The spatial resolution of the method is limited by the frequency content of the seismograms and the spatial coverage of the earthquake sources and the receivers.

The diffraction stacking method used here is similar to methods used by LYNNES and LAY (1989) and SPUDICH and IIDA (1993) to study scatterers in 2-D. It is also similar to the method that SOMA and NIITSUMA (1997) used to study induced seismicity related to geothermal injections, although their method requires secondary arrivals to have linear particle motions, which our method does not.

There are significant benefits to using the induced microearthquakes as sources to interrogate the rock mass. First, microearthquakes are considerably more energetic than man-made sources, particularly for shear waves. Second, microearthquakes are distributed throughout the rock volume being studied, whereas artificial sources can only be placed in nearby boreholes or at the surface. Drawbacks to using microearthquakes as sources are that neither origin times nor hypocenters are known beforehand. Both must be calculated and hold uncertainties because of errors in the arrival times and the velocity structure. These uncertainties may produce spatial smearing.

The seismograms that are used must be carefully chosen to achieve favorable results from the method. The events must be well located and have seismograms with good signal-to-noise. We did not set quantitative limits for either of these, but since we had carefully picked arrival times, we assumed the events were well located. We used earthquake hypocenters computed from the third location set (Fig. 6). Although we judged them to be well located, the hypocentral uncertainties may still be 10 m or so. We estimated the signal-to-noise ratio from a visual inspection of the seismograms. To compensate for differing event sizes, we normalized the trace amplitudes according to the rms value of the trace. To avoid having to correct for differences in focal mechanisms or source-time functions we selected seismograms with similar direct S waveforms. We further simplified the analysis by using only S_h component seismograms, therefore we only had to consider S-wave propagation for both direct and scattered waves. Finally, we selected earthquake hypocenters distributed as uniformly as possible to minimize the artifacts from concentrations or holes in the spatial distribution of sources.

The data used in the diffraction stacking analysis do not fully satisfy the rather stringent criteria described. The need to choose seismograms with similar waveforms at the same set of receivers resulted in selecting 20 earthquakes in the western (far) group of earthquakes (see Fig. 11). These hypocenters lie in an E-W line of seismicity, and provide sparse N-S or vertical spread. To improve that limited

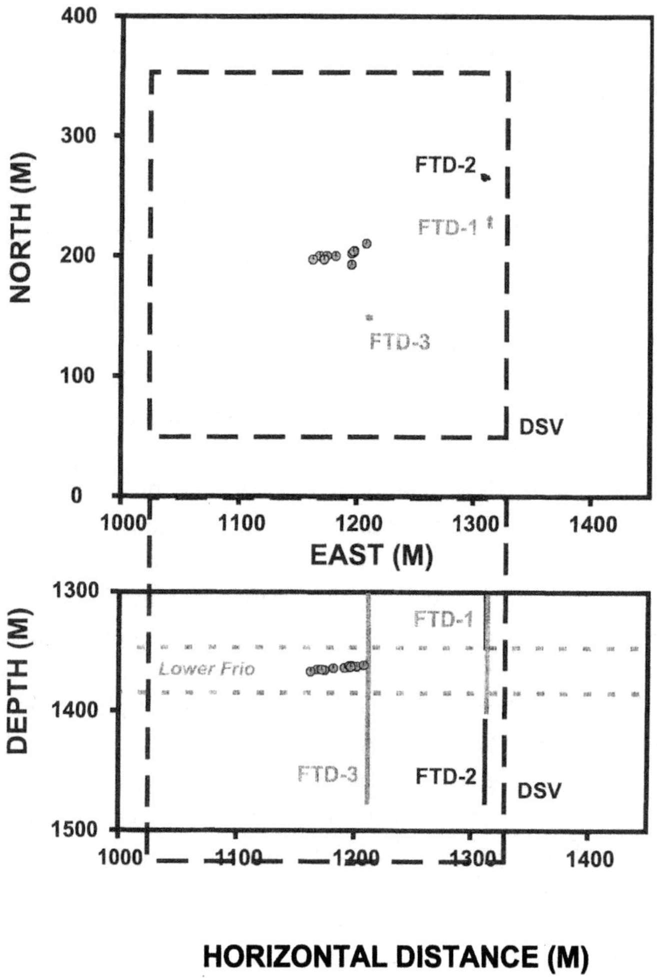

Figure 11
Plot of the 20 earthquakes used in the diffraction stack analysis. The dimensions of the diffraction stack volume are shown (labelled "DSV"). The DSV extends above and below the depths plotted in the cross section. Hypocenters are from location set 3, shown in Figure 6.

coverage, we used seismograms from 10 geophone packages spread over a depth range of 165 m in the monitor borehole, FTD-3. The depth extent of the geophone packages is orthogonal to the E-W extent of the hypocenters. Although we wanted to include seismograms from the other monitoring borehole, FTD-1, those waveforms were inadequate quality. Figure 11 shows the hypocenters of the earthquakes and the extent of the stack volume used.

After tests of the method, we found that using the amplitude envelopes of the seismograms produced better results than using the seismograms themselves. The fairly poor results from using the seismograms presumably was because of uncertainty in the hypocenters and travel times. Since scattering cubes are small, relatively minor errors in hypocenters or travel times could place stack values in with the wrong cube, resulting in destructive rather than constructive interference of the waveforms. Using amplitude envelopes produced more coherent and interpretable results.

Results are shown in Figure 12. The figure shows three 2-D sections from the diffraction stack cube, along with the hypocenters of all 54 earthquakes. The positions of the three sections were chosen to show some of the interesting features in the stack cube. Figure 12A is a plan view (horizontal section). The plot shows that much of the rock volume is relatively homogeneous, with stack values that plot as dark blue (background). There is an area of high scattering near the middle of the plot (light blues, yellows, and reds), which includes circular or elliptical shaped features that may be partly artifacts. Nevertheless, the area of highest scattering (yellows and reds) underlies the earthquakes of the eastern half of the far event group. Near the western end, several epicenters hook around to the south, approximately following the trend of a curved area of intermediate scattering (middle blues). The areas of the middle and near injection event groups have low scattering and presumably are relatively homogeneous rock. An arc of larger scattering passes through the injection well. There is a scattering feature within this arc although the spatial coverage of sources and receivers is not adequate for actually resolving it. Perhaps the scattering feature is associated with the injection well.

Figure 12B is a vertical section looking north. The section plane is at a north-south coordinate of 185 m, about the position of the southwestern most event of the far event group. In the middle of the plot the highest scattering (yellows and reds) defines a line that dips to the east. It starts below the middle of the far event group and terminates at about the position of monitoring well FTD-3 and the eastern end of the earthquakes in the far event group. The vertical extent of the large scattering area coincides with the injection interval, which is indicated by the gap in the trace of well FTD-2.

Figure 12C is a vertical section looking west. In this view, the far event group lies within an area of high scattering, and the middle event group lies in an area of low scattering. The area of highest scattering lies 10 to 15 m below the hypocenters of the western group of events. There is also an area of low scattering that extends to the south and down from the events of the far event group. Although roughly accurate, this is probably a real feature since the center of the arc is quite a bit below the events in the far cluster which were the sources used for these images.

Discussion

Seismic monitoring of this fluid injection provided the opportunity to study induced seismicity in an unconsolidated, laterally continuous sedimentary environment. The vertical extent of seismic monitoring allowed calculation of high precision earthquake hypocenters, inversion for transverse-isotropic velocity structure, single-event focal mechanisms, and imaging of seismic scatterers. One factor common to

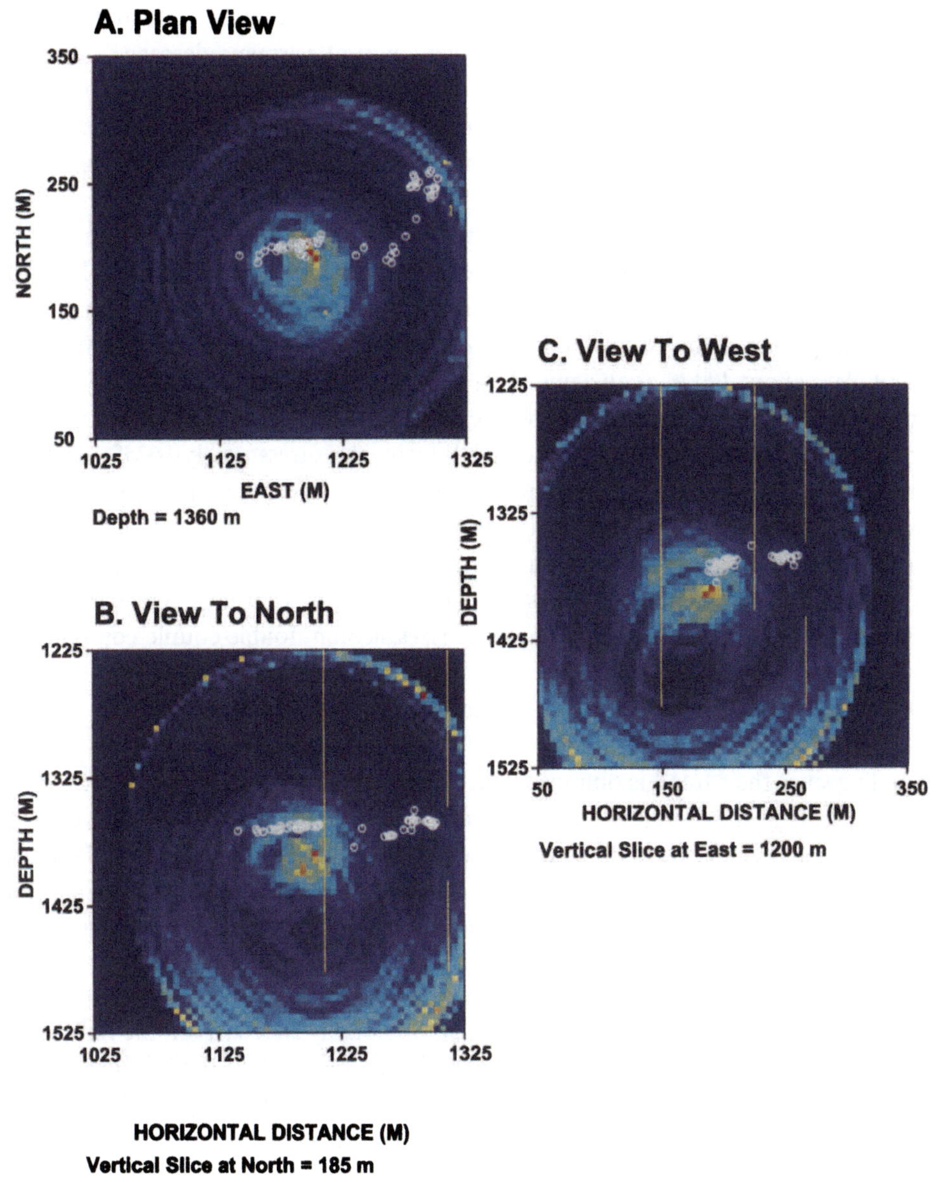

Figure 12
2-D slices from the diffraction stack cube, with hypocenters of all 54 earthquakes (white circles). Hypocenters are from location set 3, shown in Figure 6. Yellow lines are wells; the injection interval in well FTD-2 is the open portion. Monitoring well FTD-1 was omitted from plot B because it plots on top of injection well FTD-2. Plot A: horizontal slice (plan view) at a depth of 1360 m. Plot B: vertical slice, looking north, at 185 m north coordinate. Plot C: vertical slice, looking west, at an east coordinate of 1200 m.

results of all these studies is the heterogeneity that was seen. Although the Frio sandstone is presumed to be laterally continuous on the scale of km, the seismicity data imply that it is not on a scale of a few 10's of m.

Distribution of seismicity. The hypocenters plotted in Figures 4, 5, and 6 divide into three groups of events separated by two zones of seismic quiescence. WITHERS and RIEVEN (1996) located a larger set of 200 microearthquakes from this injection, and their plots show a hypocenter distribution similar to that shown in Figures 4, 5, and 6, although they found two additional groups of seismicity. Each of these are separated from the three groups seen in Figure 4 by a gap in seismicity. One of these event groups extends approximately 150 m farther to the southwest from the injection zone than the "far" group of events in Figure 4, along nearly the same trend. The other group extends almost 150 m north from the injection well. Including the event group in the southwest, the total extent of seismicity from the injection borehole is about 400 m at an azimuth of about 246°. The trend of seismicity is at an angle of about 77° (nearly perpendicular) to the azimuth of the minimum compressive principal stress (σ_3) derived from the stress analysis of the focal mechanisms (169°). The induced microearthquakes align approximately along the expected direction of the fracture opening. Assuming they are double-couple mechanisms, the 47 focal mechanisms are fairly well constrained. It could be worthwhile to obtain moment tensor solutions for these events. Since they were induced by fluid injections some may have sizeable non-double-couple components. MILLER *et al.* (1998) note a number of studies that have reported non-double-couple mechanisms, many of which are in volcanic or geothermal areas.

The events in the north-trending event group of WITHERS and RIEVEN (1996) are puzzling since their trend is oblique to the trend of the rest of the hypocenters. None of the events in this group passed the screening criteria used to select the events for this study, and all occurred in the last injection phase (cycle 4). WITHERS and RIEVEN (1996) attribute the occurrence of these events to a build up of fluid pressure near the injection well, perhaps because of the large volume of fluid that had been injected by that time (about 7,600 m^3).

The spatial extent of the seismicity seen in the hypocentral plots (Figures 4, 5, and 6) defines a fracture half length of about 250 m, notably shorter than the half length of about 400 m computed from fracture modeling done before the injection (KECK and WITHERS, 1994). The discrepancy can be resolved by including the hypocenters determined by WITHERS and RIEVEN (1996) that extend to the SW of the events studied here. The 400 m extent of that seismicity from the injection well is in excellent agreement with the computed fracture length. The expected height of the fracture system was about the full 45 m thickness of the Frio formation. The depth range of the hypocenters is close to that, although most events occurred in a narrower depth range, about 25 m (see Figs. 4, 5, and 6).

The hypocenters in the far (west) event group are confined to a narrow, tube-like distribution, which is surprising in a uniform, unconsolidated, high permeability sandstone. PHILLIPS *et al.* (2002) interpreted such 'tubes' of seismicity as indicating the intersections of preexisting joints of different orientations. The joints themselves may have little or no seismicity. But intersections of joints could have significant conductivity and therefore be seismically active. Since the Frio formation has high permeability, we expected to see a diffuse distribution of hypocenters. WITHERS and RIEVEN (1996) postulate that the injected fluids, which included bentonite clay, may have produced a mud cake around the fracture system that would limit the leakoff of the fluids. Since fluid pressure would decrease with distance from the injection well, a mud cake effect might be especially effective in confining the spatial extent of fluids at extended distances from the injection well.

All the seismicity studied here is positioned on one lobe of the fracture system created by the injection. The absence of hypocenters to the NE of the injection well is notable, and may be an artifact of the station geometry. Our earthquake selection criteria were rigorous, and required that events were recorded with good signal-to-noise by geophone packages in both monitoring boreholes. Events comparable in size to those we studied could have occurred to the north-east of the injection well and not have been recorded at FTD-3, the more distant monitoring borehole. WITHERS and RIEVEN (1996) used less restrictive selection criteria and located nearly four times as many events, and yet none of them were located to the NE of the injection well. The fracture modeling assumed the fracture would have a second, symmetric side or "wing" (KECK and WITHERS, 1994). With the high permeability of the Frio formation and the lack of an obvious discontinuity, such as a fault, to prevent the fracture system from extending to the NE, the lack of seismicity to the NE of the injection well is problematic. We assert that the fracture probably extended to the NE and that the induced seismicity was too poorly recorded for the events to be locatable.

Hypocentral Differences. The differences in the three velocity models used to locate the earthquakes are slight, yet the hypocenters are surprisingly different (compare Figs. 4, 5, and 6). Each hypocenter plot contains features that look interesting, and it is difficult to choose one location set as generally better than the others, based only on the hypocenters. The hypocenter set calculated with the Transverse-Isotropic velocities has the minimum RMS (root-mean-square) travel time residuals (see Table 3). The RMS of that location set is about a full msec lower than the RMS of both the other two location sets. Yet these hypocenters (Fig. 5) show features that look more diffuse and less clearly focused than comparable features in Figures 4 or 6. In addition, the strike of the far group of events, in both Figures 5 and 6, is rotated clockwise from the direction to the injection well, FTD-2. Nevertheless, if structural effects or the variation of rock properties control the distribution of hypocenters, the trend of the seismicity in the far event group could actually be different from the direction to the injection well. Subsequently, details of the hypocentral distribution may not provide a meaningful basis from which to judge the quality of the hypocenters. We assume that the locations from the full T-I velocity structure are the best location set because these resulted in the lowest RMS.

Structural or Stratigraphic Discontinuities. As noted earlier, the group of hypocenters that lies near the injection well is separated from the middle group of hypocenters by a gap of 25–30 m. The middle group of events is also separated by a gap of 25–30 m from the far group of seismicity. If the Frio formation were laterally uniform, we would expect the microearthquake hypocenters to be more evenly distributed, without such obvious concentrations and gaps. We interpret the gaps in seismicity to result from structural or stratigraphic discontinuities in rock properties. That interpretation is consistent with the positions of some of the prominent scatterers seen in the diffraction stacking results in Figure 12. In particular, the eastern and western edges of the far event group coincide with the edges of an area of strong scatterers in the middle of the image volume (light blues, yellows, and reds). The lack of strong scatterers between the far event cluster and the injection well implies that the rock is fairly uniform, although the gap in the hypocenter distribution suggests that there is a discontinuity between the near and middle event groups. Moderate scattering (light blue) near the injection well (Fig. 12A) contrasts with low scattering (dark blue) between the injection well and the far event group and may indicate that some of the injected fluid stayed near the injection well. Considering its 3-D extent, the arc shape of this feature is probably an artifact and indicates that the spatial distribution of the data (hypocenters or stations) was not adequate to fully image scatterers situated there. Interpreting the lack of strong scatterers between the injection well and the far group of events to indicate that the rock is uniform would be consistent with the diffuse spatial character of the hypocenters in the near and middle event groups. On the other hand, the strong scattering near the far event group implies that the rock has strong discontinuities (either structural or stratigraphic) in it.

T-I Velocities. The negative *P*-wave anisotropy (−14%) was a surprising result, since it means that *P* waves travel faster vertically than horizontally. In a horizontally layered formation with velocities that increase with depth, rays along horizontal paths should sample higher velocities, and thus travel faster than rays along vertical paths. In addition, there is little, if any, *S*-wave anisotropy. A simple explanation could be that the anisotropy of *P* velocities results from the volume of fluid injected, which would increase the porosity of the Frio formation. Using a relation developed from log measurements in the Frio formation (CASTAGNA *et al.*, 1985), a 4% increase in porosity would decrease V_p by about 14%. However that increase in porosity would also decrease V_s by about 20%. Other possible explanations are that the anisotropy resulted from artifacts in the data and data analysis or differences in the rock types sampled by horizontal and vertical ray paths.

For the anisotropy to result from artifacts in the data and data analysis would require that the seismic ray coverage was inadequate or that there was a systematic bias or mistiming of the arrivals. But the large vertical extent of the monitoring packages in the two boreholes resulted in substantial vertical ray coverage, which should be adequate to fully constrain the 1-D anisotropy inversion. Furthermore, since one person picked the arrivals for all the events, and the picks were made from high quality seismograms, it is unlikely that there was any systematic mispicking of either the *P*- or the *S*-wave arrival times.

The remaining possibility is that the P-wave anisotropy results from differences in the lithologies sampled by rays travelling vertically compared to those travelling horizontally. Vertical travelling rays would sample both predominantly sandy and predominantly shaley layers of the Frio formation, whereas horizontally travelling rays would primarily sample sandy layers. CASTAGNA *et al.* (1985) show V_p/V_s values in Gulf Coast sands that are typically some 10% lower than V_p/V_s in shales. From the results of the T-I inversion (all events) in Table 2a, the vertical V_p/V_s is 2.37, and the horizontal is (2.07/0.99) or 2.09, about 12% lower than the vertical V_p/V_s. Thus, if V_s were constant, V_p would decrease by about −12%. That is about the right value of anisotropy and it suggests that the V_p anisotropy results from the differing lithologies.

Conclusions

Extensive monitoring of seismicity induced by a hydraulic injection into the highly permeable Frio formation of east Texas produced a large, high quality microearthquake data set. Most of the induced microearthquakes were small, although 54 were large enough to be well recorded at both of the two monitoring wells and were selected for study.

These microearthquakes exhibit two different types of spatial character. Events near the injection borehole have scattered epicenters with no clear trend. The epicenters of events far from the borehole align either NE-SW or E-W, depending on the velocity structure used to locate them. Between the near and far groups of microearthquakes is a poorly defined middle group, in which few events occurred. The relatively nonseismic behavior in the middle zone suggests the presence of a discontinuity in the Frio formation within this zone. However, neither the well logs nor nearby seismic reflection profiles provide any clear indication of such a discontinuity.

The large depth extent of the monitoring geophone packages and the near horizontal layering of the sediments allowed for invertion of arrival times for transverse isotropic elastic moduli. A simultaneous inversion for these moduli and earthquake hypocenters produced surprising results. The P-wave anisotropy of −13.7% far exceeds the S-wave anisotropy of −2.2%. Also, both the P-wave velocity and the S-wave velocity are higher for vertical propagation than horizontal propagation. Separate inversions of data from events in the far group (west) and in the near group (east), yielded P-wave anisotropies of −13.6% for the far group, and −13.3% for the near group. The difference in S-wave anisotropies is reversed, with −1.8% in the far zone and −2.7% in the near zone. These differences suggest the presence of a discontinuity within the middle zone.

Single-event focal mechanisms were determined for 47 of the events. These microearthquakes show large components of shear slip (double-couple). The mechanisms have a large range of types and orientations, which is surprising since the Frio formation is unconsolidated, and should not have obvious pre-existing faults or fractures.

The focal mechanism data were used to solve for a single uniform stress field. The least principal stress is nearly horizontal and trends N-S, which agrees well with published results for the Gulf Coast. The maximum and intermediate stresses both plunge about 45°, which is a stress regime conducive to normal to oblique slip.

Acknowledgments

We thank ARCO for providing copies of the microearthquake data. We appreciated discussions with Robert Withers, Shirley Rieven, Mike Fehler, and Scott Phillips. We are also grateful for the advice and information provided by Richard Keck, Robert Dart, and Dennis Corrigan. Tom Fairbanks provided assistance with the field data collection. Andy Michael supplied a copy of his stress analysis programs and valuable advice. We thank Jim Rutledge and two anonymous reviewers for their reviews of the manuscript and helpful comments. We also thank Cezar Trifu for his patience and skill in handling this manuscript. Analysis of this data set was supported by the U.S. Department of Energy, Office of Basic Energy Sciences (OBES), through contract W-7405-ENG-36.

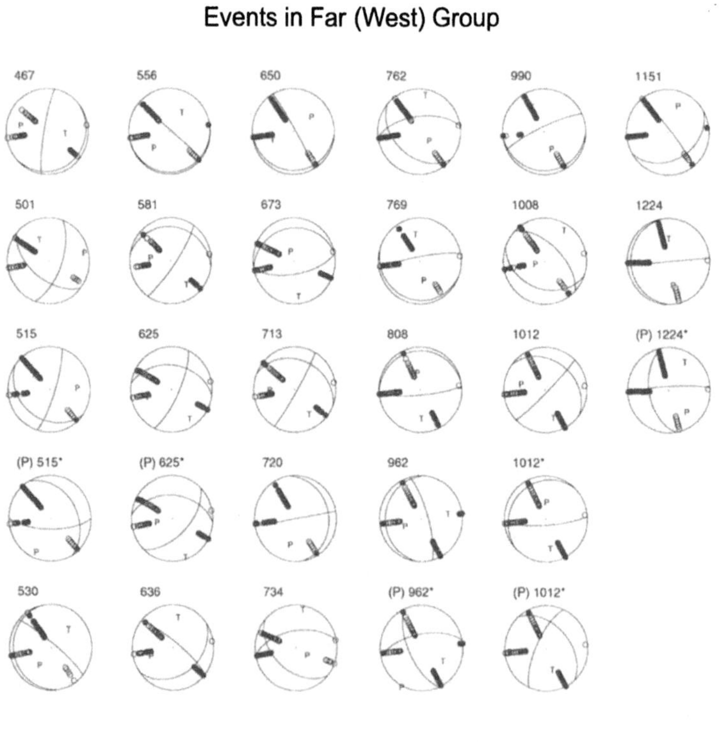

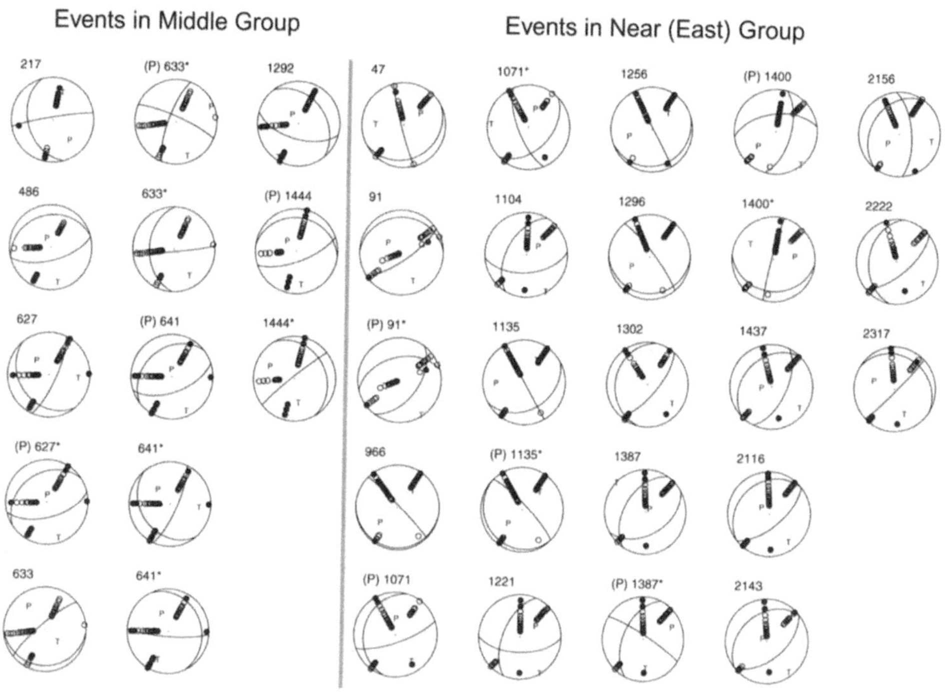

Appendix A. Single-Earthquake Focal Mechanisms

Focal mechanisms for 47 events, obtained from *P* wave first motions of individual microearthquakes using the program FPFIT (REASENBURG and OPPENHEIMER, 1985). Focal mechanisms from seven events were rejected because of excessive numbers of discrepant first motions or lack of constraint. Focal mechanisms are separated into the same three event groups as the hypocenters: west is the far event group – farthest from the injection well and east is the near event group. Numbers above focal mechanisms are event ID numbers. Multiple solutions were found for several events and are indicated by asterisks after the event number; the solution we preferred is noted with a "*P*" prefix on the event number. Filled circles are compressions, open circles are dilatations.

REFERENCES

ATLANTIC RICHFIELD CORP. (1994), *Deep Well Treatment and Injection Program: Fracture Technology Field Demonstration Project* (The Atlantic Richfield Company, Los Angeles, CA).

BAME, D. and FEHLER, M. C. (1986), *Observation of Long-period Earthquakes Accompanying Hydraulic Fracturing*, Geophys. Res. Lett. *13*, 149–152.

BLOCK, L., CHENG, C. H., FEHLER, M. C., and PHILLIPS, W. S. (1994), *Seismic Imaging Using Microearthquakes Induced by Hydraulic Fracturing*, Geophysics *59*, 102–112.

CASTAGNA, J. P., BATZLE, M. L., and EASTWOOD, R. L. (1985), *Relationship Between Compressional Wave and Shear-wave Velocities in Clastic Silicate Rocks*, Geophysics *50*, 571–581.

CLARK, S. P., Jr. (ed.), *Handbook of Physical Constants* (Geological Society of America, Memoir 97, Penrose, CO 1966).

CORNET, F. H. and YIN, J. (1995), *Analysis of Induced Seismicity for Stress Field Determination and Pore Pressure Mapping*, Pure appl. geophys. *145*, 677–700.

FEHLER, M. C. and PHILLIPS, W. S. (1991), *Simultaneous Inversion for Q and Source Parameters of Microearthquakes Accompanying Hydraulic Fracturing in Granite Rock*, Bull. Seismol. Soc. Am. *81*, 553–575.

FLORES, R. A., HOUSE, L. S., and WITHERS, R. J. (1995), *Microseismic Studies of a Hydraulic Injection into Sedimentary Rock, East Texas*, EOS (Trans. AGU), *76*, #46 (supplement), F601.

GALLOWAY, W. E., HOBDAY, D. K., and MAGARA, K. (1982), *Frio Formation of Texas Gulf Coastal Plain: Depositional Systems, Structural Framework, and Hydrocarbon Distribution*, AAPG Bulletin *66*, 649–688.

HOUSE, L. (1987), *Locating Microearthquakes Induced by Hydraulic Fracturing in Crystalline Rock*, Geophys. Res. Lett. *14*, 919–921.

HOUSE, L., FLORES, R., and WITHERS, R. (1996), *Microearthquakes Induced by a Hydraulic Injection in Sedimentary Rock, East Texas*, Soc. Expl. Geophys., Exp. Abstr. 66th Ann. Tech. Mtg., 110–113.

HOUSE, L., PHILLIPS, W. S., FEHLER, M., and RUTLEDGE, J. (1997), *Using Waveforms of Induced Microearthquakes to Image Hydraulic Fracture Systems*, Soc. Expl. Geophys., Exp. Abstr. 67th Ann. Tech. Mtg. 234–237.

JONES, R. H., BEAUCE, A., JUPE, A., FABRIO, H., and DYER, B. C. (1995), *Imaging Induced Microseismicity During the 1993 Injection Tests at Soultz-sous-Forets, France*, Proc. World Geotherm. Cong. Florence, Italy, 2665–2669.

KECK, R. G. and WITHERS, R. J. (1994), *A Field Demonstration of Hydraulic Fracturing for Solids Injection with Real-time Passive Seismic Monitoring*, SPE Paper 28495, Soc. Petr. Eng., 1994 Ann. Meet. Exp. Abstr., 319–334.

LEVIN, F. K. (1979), *Seismic Velocities in Transversely Isotropic Media*, Geophysics *44*, 918–936.

LYNNES, C. S. and LAY, T. (1989), *Inversion of P Coda for Isotropic Scatterers at the Yucca Flat Test Site*, Bull. Seismol. Soc. Am. *79*, 790–804.

MICHAEL, A. (1987), *Use of Focal Mechanisms to Determine Stress: A Control Study*, J. Geophys. Res. *92*, 357–368.

MILLER, A. D., FOULGER, G. R., and JULIAN, B. R. (1998), *Non-double-couple Earthquakes. 2 Observations*, Rev. Geophys. *36*, 551–568.

PHILLIPS, W. S., FAIRBANKS, T. D., RUTLEDGE, J. T., and ANDERSON, D. W. (1998), *Induced Microearthquake Patterns and Oil-producing Fracture Systems in the Austin Chalk*, Tectonophysics *289*, 153–169.

PHILLIPS, W. S. (2000), *Precise Microearthquake Locations and Fluid Flow in the Geothermal Reservoir at Soultz-sous-Forets, France*, Bull. Seismol. Soc. Am. *90*, 212–228.

PHILLIPS, W. S., RUTLEDGE, J. T., HOUSE, L. S., and FEHLER, M. C. (2002), *Induced Microearthquake Patterns in Hydrocarbon and Geothermal Reservoirs*, 345–369. Pure appl. geophys. *159*

REASENBURG, P. and OPPENHEIMER, D. H. (1985), *FPFIT, FPPLOT, and FPPAGE: Fortran Computer Programs for Calculating and Displaying Earthquake Fault Plane Solutions*, Open File Report 85-0739, U.S. Geol. Surv., Menlo Park, CA.

RUTLEDGE, J. T., PHILLIPS, W. S., HOUSE, L. S., and ZINNO, R. J. (1998), *Microseismic Mapping of a Cotton Valley Hydraulic Fracture Using Decimated Downhole Arrays*, Soc. Expl. Geophys., Exp. Abstr. 68th Ann. Tech. Mtg., 338–341.

SASAKI, S. (1998), *Characteristics of Microseismic Events Induced During Hydraulic Fracturing Experiments at the Hijiori Hot Dry Rock Geothermal Energy Site, Japan*, Tectonophysics *289*, 171–188.

SOMA, N. and NIITSUMA, H. (1997), *Identification of Structures Within the Deep Geothermal Reservoir of the Kakkonda Field, Japan, by a Reflection Method Using Acoustic Emission as a Wave Source*, Geothermics *26*, 43–64.

SPUDICH, P. and IIDA, M. (1993), *The Seismic Coda, Site Effects, and Scattering in Alluvial Basins Studied Using Aftershocks of the 1986 North Palm Springs, California, Earthquake as a Source Array*, Bull. Seismol. Soc. Am. *83*, 1721–1743.

THOMSEN, L. (1986), *Weak Elastic Anisotropy*, Geophysics *51*, 1954–1966.

VINEGAR, H. J., WILLS, P. B., DEMARTINI, D. C., SHLYAPOBERSKY, J., DEEG, W. F. J., ADAIR, R. G., WOERPEL, J. C., FIX, J. E., and SORRELLS, G. G. (1992), *Active and Passive Seismic Imaging of a Hydraulic Fracture in Diatomite*, J. Pet. Tech. *44*, 28–34 and 88–90.

WITHERS, R. J. and RIEVEN, S. (1996), *Fracture Development During Cuttings Injection Determined by Passive Seismic Monitoring*, Soc. Expl. Geophys., Exp. Abstr. 66th Ann. Tech. Mtg., 106–109.

ZOBACK, M. L. and ZOBACK, M. (1980), *State of Stress in the Conterminous United States*, J. Geophys. Res. *85*, 6113–6156.

(Received October 2, 2000, revised / accepted January 10, 2001)

Pure appl. geophys. 159 (2002) 403–420
0033–4553/02/030403–18 $ 1.50 + 0.20/0

Pure and Applied Geophysics

Production-induced Normal Faulting in the Valhall and Ekofisk Oil Fields

MARK D. ZOBACK[1] and JENS C. ZINKE[1]

Abstract—*In situ* stress and pore pressure data from the Valhall and Ekofisk oil reservoirs indicate that at the onset of production in both fields an incipient state of normal faulting existed in the crest of the anticlinal structures. In contrast, on the flanks of the structures the initial least principal stress values indicate an almost isotropic state of stress. Oil production from both fields caused marked pore pressure reductions as well as poroelastic reductions of the least principal stress in both the crest and flanks of the two structures. We demonstrate that as a result of production-induced pore pressure and stress changes, normal faulting appears to have spread out from the crests of the structures on to the flanks. Further evidence of a normal faulting stress state at Valhall has been found using data from a passive seismic monitoring experiment. Numerous microearthquakes were recorded during a six week monitoring period that are located at the very top of the reservoir or in the shale caprock immediately above it. An inverse/composite focal plane mechanism of these microearthquakes is consistent with a normal faulting stress regime.

Key words: Production-induced normal faulting.

Introduction

In this study we utilize *in situ* stress and pore pressure data from the Valhall and the Ekofisk fields in the Central Graben area of the North Sea to investigate active faulting within the reservoirs and cap rocks. Active faulting in oil and gas reservoirs is of importance for a variety of reasons. From an engineering perspective, slip on active faults appears to be the cause of sheared casings of production wells in some fields (MAURY *et al.*, 1992) and, in others, shear slip on pre-existing faults and bedding planes appears to be a serious source of wellbore instability during drilling (e.g., WILLSON *et al.*, 1998). In many low permeability reservoirs, critically-stressed faults (i.e., faults that are active in the present stress field) contribute importantly to overall reservoir permeability (FINKBEINER *et al.*, 1998; DOHLAKIA *et al.*, 1998). In other cases, re-activation of reservoir bounding

[1] Department of Geophysics, Stanford University, Stanford, CA 94305, USA.
E-mail: zoback@pangea.stanford.edu.

faults can cause a loss of seal capacity and leakage to occur (e.g., WIPRUT and ZOBACK, 1999) and slip on active faults may control the vertical extent of oil and gas column that a fault-bounded reservoir can contain (e.g., FINKBEINER *et al.*, 2000).

There have been a number of case studies in which both fluid withdrawal and fluid injection appear to have induced active faulting in oil and gas reservoirs (review by GRASSO, 1992). Since the classic study of injection-induced earthquakes in the Rangely oil field in Colorado (RALEIGH *et al.*, 1972), induced faulting in oil and gas reservoirs is usually associated with pore pressure increases due to water-flooding or hydraulic fracturing. In this case increases in pore pressure causes slip along pre-existing faults by reduction of the effective normal stress on the fault plane. The Coulomb criterion can be written in terms of effective stresses as:

$$\tau = \mu(S_n - P_p) \tag{1}$$

where τ is the shear stress on the fault, S_n is the normal stress, P_p is the pore pressure and μ is the coefficient of friction (cf., JAEGER and COOK, 1971). Thus, by raising the pore pressure, the effective normal stress on pre-existing faults $(S_n - P_p)$ is reduced, making it possible for faulting to occur at a level of shear stress at which fault slip would normally not occur.

A second mechanism of induced faulting in the vicinity of oil and gas reservoirs is poroelastic stress changes in the medium surrounding a compacting reservoir (SEGALL, 1985, 1989, 1992). In this case, faulting is induced by the superposition of pre-existing stresses with the stresses caused by pore pressure decreases in the reservoir. Application of this theory to the Lacq field in France (FEIGNIER and GRASSO, 1990) demonstrates a reasonably good correlation between the theoretically expected location of seismicity and that observed as long as the pre-existing stress state in the vicinity of the reservoir is characterized by normal faulting with a vertical maximum principal stress (SEGALL *et al.*, 1994).

In the present study, we discuss another mechanism by which normal faulting within reservoirs might be induced by poroelastic stress changes. We demonstrate that the stress path associated with reservoir production in the Valhall and the Ekofisk fields is such that normal faulting is enhanced in the reservoirs by the poroelastic decreases in the least horizontal stress decreases accompanying production-induced pore pressure decline. While this may seem counterintuitive in light of the conventional relationship between pore pressure and failure expressed in equation (1), we demonstrate that this is theoretically correct in terms of the Coulomb criteria for the observed stress path in the two oil fields. By stress path we mean the change in horizontal stress magnitudes in the reservoir accompanying changes of pore pressure with production. Similar processes may be active in other oil and gas fields where normal faulting within reservoirs appears to have been induced by hydrocarbon production (cf., DOSER *et al.*, 1991).

The Valhall and Ekofisk Oil Fields

The Valhall Field and the Ekofisk Field are located within the Central Graben in the southern part of the Norwegian North Sea (see Fig. 1a). The Valhall structure trends NW-SE and is an elongated anticline along a structure locally known as the Lindesnes Ridge (MUNNS, 1985). The reservoir is at a depth of approximately 2400 m subsea and consists of two late Cretaceous oil bearing formations: the Tor formation and the underlaying Hod formation which are overlain by Paleocene and Eocene age shale cap rock (Fig. 1b). Both formations are soft chalk facies with a primary due to formation overpressure inhibiting mechanical compaction of the chalk. The discovery well was drilled in 1975 and the field went into production in October 1982.

The Ekofisk structure is an elongated anticlinal structure with a principal fold axis in an approximately, north-south direction (VAN DEN BARK and THOMAS, 1981). The reservoir depth is approximately 2900 m subsea and consists of two fractured chalk intervals: the Ekofisk formation (Danian) and the Tor formation (Maastrichtian). A relatively impermeable layer of argillaceous, siliceous, and cherty chalk separates the two oil bearing horizons. Despite the high porosity of the chalks (up to 51%), matrix permeability is low (about 1 md, THOMAS *et al.*, 1987) and reservoir permeability (~150 md) appears to be controlled by an extensive natural fracture system (BROWN, 1987).

In the case of both of these fields, there has been significant concern about active faulting. There have been numerous occurrences of casing failures in both fields which have been interpreted as the result of shear along active faults. This is especially clear for casing failures in the shale caprock at Ekofisk (SCHWALL and DENNEY, 1994). Moreover, there is appreciable gas leakage through the shale cap rocks in both fields which may be exacerbated by flow through faults (MUNNS, 1985). Finally, it is well known that since Ekofisk went into production in 1971, appreciable sea-floor subsidence has occurred (e.g., NAGEL, 1998; RHETT, 1998), presumably due to compaction within the reservoir. Interestingly, even though full-field water flooding operations began at Ekofisk in 1987, subsidence has continued at a near constant rate of 38 cm/year. A limited amount of subsidence has occurred at Valhall (PATTILLO *et al.*, 1998). If faulting is being exacerbated by oil production, slip on active faults may be contributing to the reservoir compaction causing sea floor subsidence.

Least Principal Stress and Pore Pressure Analysis

In this study we focus on the Valhall field because we had direct access to least principal stress and pore pressure data. We have been constrained to use only published data for the Ekofisk field. One important difference between the two fields is that there has been a major water flooding program at Ekofisk since 1987 to

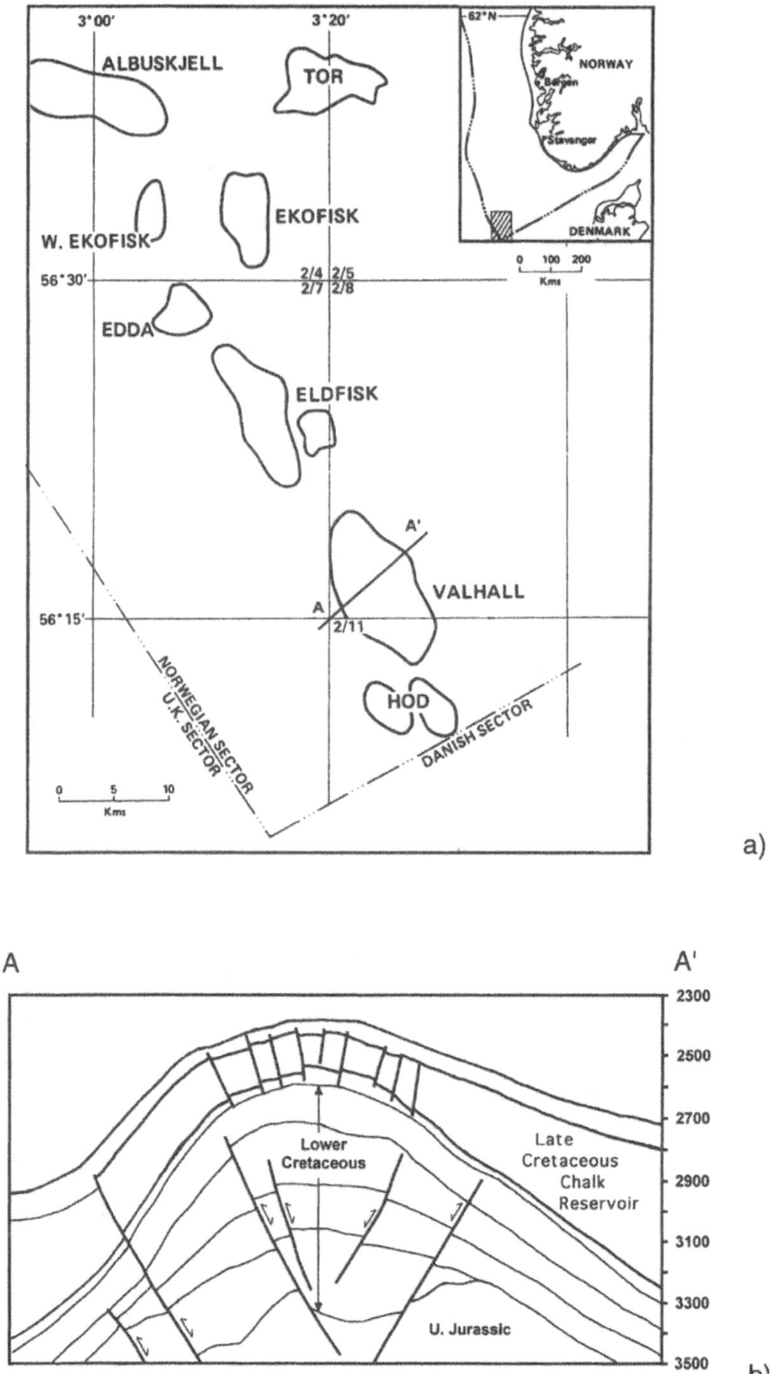

Figure 1
Location of the Valhall and Ekofisk oil fields are shown in a map of the southern Norwegian North Sea (a). A profile along line AA' illustrates the dome structure of the Valhall oil field (b).

maintain reservoir pressure and to attempt (unsuccessfully) to control subsidence. No water flooding has been carried out to date at Valhall. Because of this, we utilize stress and pressure data from Ekofisk acquired prior to the initiation of water flood operations. In fact, these data are similar to those at Valhall and lead to similar conclusions regarding induced faulting within the reservoir.

The data available from Valhall include conventional well logs, leak-off tests (LOTs) and minifracs to constrain the magnitude of the least principal stress, drillstem tests (DSTs) and remote formation tests (RTFs) to constrain pore pressure. Unfortunately, the original data from the pressure tests were not available to us. Data available from a passive seismic monitoring experiment consisted of microearthquake seismograms from six, three component seismometers deployed for a six week period in an observation well. These data are discussed below.

In Figure 2a we show a depth profile of all available pore pressure data at Valhall, and in Figure 2b we show all available estimates of the least principal stress. The vertical stress shown in Figure 2 was derived by integration of density logs. Leak-off tests (LOTs) and minifracs were used to constrain the least principal stress along with several formation integrity stresses (FITs). Remote formation test (RFT)

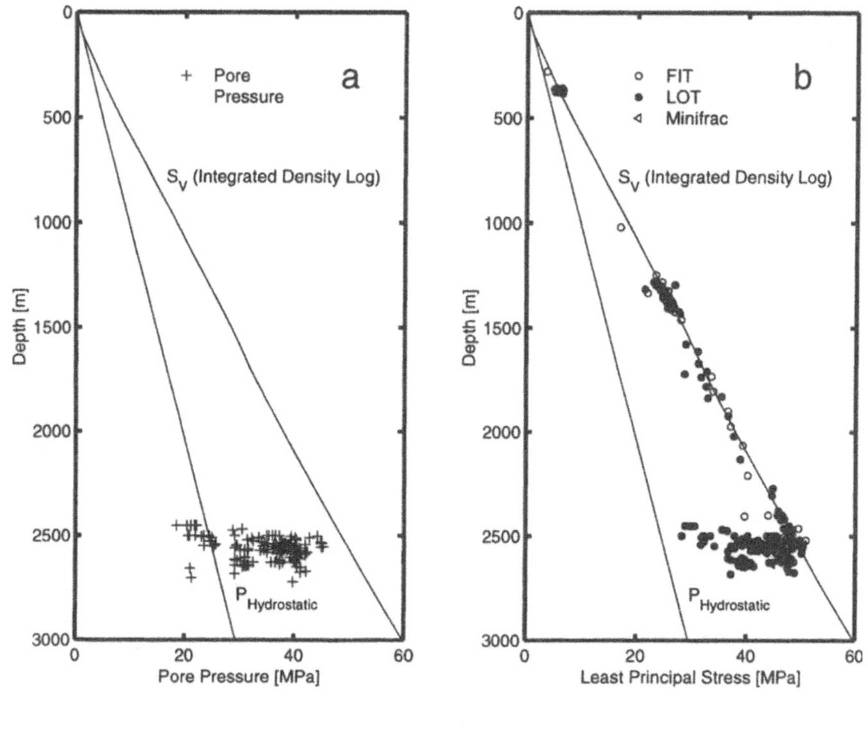

Figure 2

Depth profile of pore pressure (a) and least principal stress (b) at Valhall. LOT data appear to indicate that $S_3 \sim S_V$ in the overburden. In the reservoir, considerable scatter is seen in S_3 and in the pore pressure measurements, in part resulting from depletion in some parts of the reservoir.

data were principally used to constrain pore pressures. Pore pressure measurements are mainly limited to the reservoir ($\gtrsim$2400 m) and show considerable scatter, in part resulting from depletion in some parts of the reservoir. While the LOT data appear to indicate that $S_3 < S_v$ in the overburden ($\lesssim$2400 m), considerable scatter is seen in S_3 at the reservoir level. We interpret the scatter in S_3 values in the reservoir as being the result of production and accompanying poroelastic effects, as discussed below.

In Figure 3 we present the data from the reservoir section of the Valhall field (principally from the Tor formation) to evaluate the magnitude of the least principal stress with respect to pore pressure and position within the reservoir. To separate measurements with respect to their position in the crest or flank, we used the 2450 m depth contour on the top Tor structure map. By using colored symbols, we indicate the time that a given observation was made.

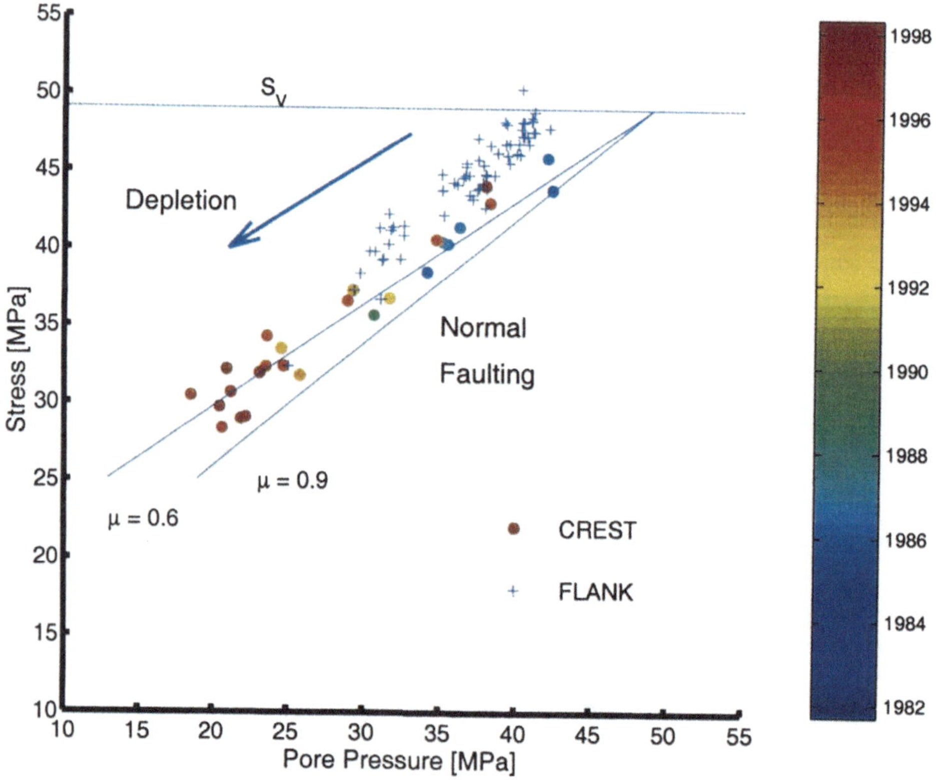

Figure 3
Least principal stresses versus pore pressure are shown for the Tor reservoir at Valhall. A horizontal line denotes the vertical stress S_V, and the inclined straight lines correspond to the Coulomb failure criterion for a coefficient of friction of 0.6 and 0.9 respectively. +'s denote values obtained for the flanks, and circles show values obtained for the crest. The color of the circles indicates the time that a given observation was made. As depletion progressed, pore pressure and stress dropped dramatically.

Several trends are apparent in these data. First, there is a clear reduction of pore pressure and S_3 with time that shows the overall effects of depletion. It is clear that both on the crest of the structure and on the flanks, pore pressures and least principal stresses in the reservoir were quite high in the early 1980s. As depletion occurred, pressure and stress dropped dramatically. On the crest of the structure, the change of the least principal stress with production ($\Delta S_3/\Delta P_p$) is about 0.7. The horizontal straight line in Figure 3 represents the vertical stress which is not expected to change with depletion for reservoirs which are laterally extensive with respect to their thickness (SEGALL and FITZGERALD, 1996). The inclined straight lines in Figure 4 correspond to the Coulomb failure condition for normal faulting utilizing coefficients of friction, μ, of 0.6–0.9, which are frequently measured in the laboratory for a wide variety of rocks (BYERLEE, 1978). The Coulomb failure condition is described by the following

$$(S_1 - P_p)/(S_3 - P_p) = ((\mu^2 + 1)^{1/2} + \mu)^2 \qquad (2)$$

(JAEGER and COOK, 1971) where $S_1 = S_v$ and $S_3 = S_{h\min}$. TOWNEND and ZOBACK (2000) review numerous cases where the Coulomb criterion and laboratory-derived coefficients of friction accurately describe the *in situ* stress state.

Note that the data presented in Figure 3, under initial, undepleted conditions, show that the stress and pore pressure in the crest of the structure was in a normal faulting stress state, just as illustrated in the simplified sketch in Figure 4. As depletion occurred, the stress path, or the change of horizontal stress magnitudes due to reduction in the pore pressure, was such that the crest remained in a normal faulting stress state even though pore pressure was decreasing with time. Active normal faulting was the natural state and as production occurred, the stress path was such that normal faulting continued. The intermediate principal stress, S_2, in normal faulting areas corresponds to the maximum horizontal stress, $S_{H\max}$. It is expected that both horizontal stresses were affected more or less equally by depletion. While we have no direct estimates of $S_{H\max}$ magnitudes at Valhall, utilization of a variety of techniques indicates very little difference between the two horizontal stresses (KRISTIANSEN, 1998).

Figure 3 also demonstrates that stress magnitudes on the flanks of the Valhall structure were initially appreciably higher than on the crest. In fact, both pore pressures and least principal stress values were initially close to the overburden stress (well above the values observed in the crest) and certainly not consistent with normal faulting. In fact, if the maximum horizontal stress is intermediate in magnitude between the least horizontal stress and the vertical stress (as appears to be the case in the crest), the initial stress state on the flanks was almost isotropic. What is quite interesting, however, is that despite the nearly isotropic initial stress state, the stress path accompanying production on the flanks of the reservoir is such that once depletion has reduced pore pressure to about 30 MPa, a normal faulting stress state

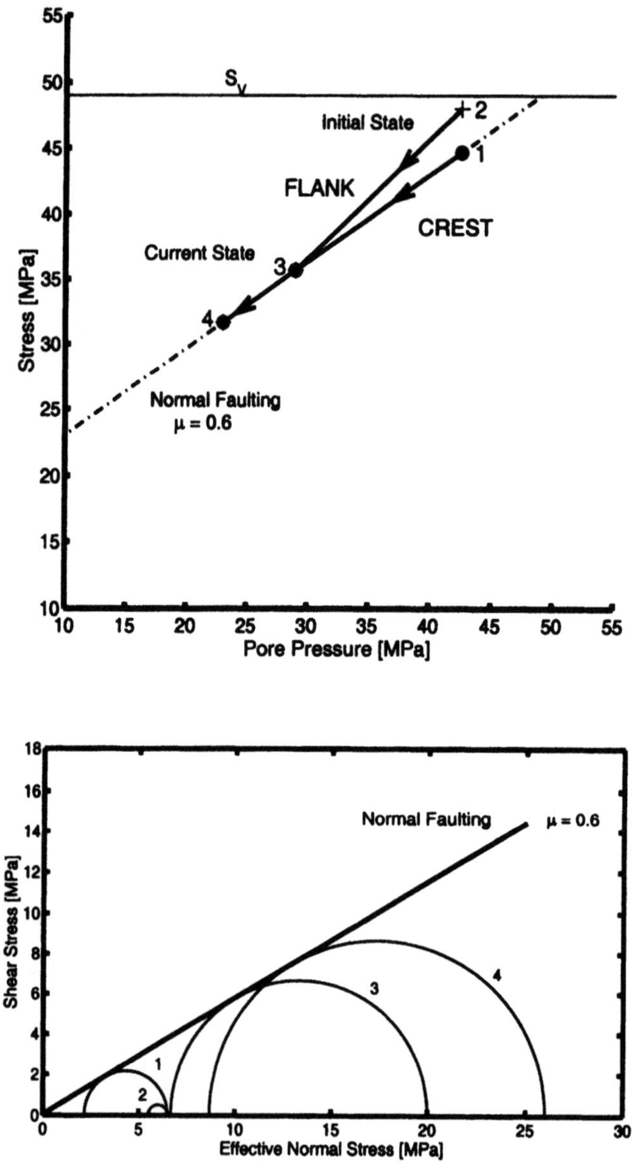

Figure 4
(a) Schematic relationship between initial stress states on the flank and crest of the Valhall structure (states 1 and 2, respectively) and their evolution (stress path) as the reservoir was depleted. A coefficient of friction of 0.6 was used to define a single failure line for simplicity. Note that the flank of the reservoir was not initially in a state of active normal faulting, the stress path was such that the normal faulting was induced by reservoir depletion. (b) Mohr-circle representation of the stress states shown in (a).

is encountered. Thus, depletion of the Tor formation appears to have induced normal faulting on the flanks of the reservoir. As normal faulting had already been occurring on the crest, it appears that it has spread outward from the crest of the structure onto the flanks as production and depletion have taken place.

The evolution of stress and pore pressure at Valhall are shown in a simplified sketch and accompanying Mohr diagrams in Figure 4. Initial stress conditions in the crest are illustrated by state 1 and evolve over time to states 3 and 4 as pore pressure is reduced and the least principal stress decreases due to poroelastic effects. Because the vertical stress does not change during depletion for laterally extensive reservoirs (SEGALL and FITZGERALD, 1996), the stress path is such that normal faulting is expected for each stress state (as illustrated by Mohr circles 1, 3 and 4).

The stress path and onset of faulting on the flanks of the Valhall reservoir are shown by states 2, 3 and 4 in Figure 4. The initial pore pressure and stress on the flanks (state 2) is certainly not consistent with normal faulting as the least principal stress is nearly equal to the vertical stress (note the extremely small Mohr circle). However, as pore pressure and stress change with production, the least horizontal effective stress increases slightly whereas the vertical effective stress increases considerably. Thus the stress path eventually intersects the normal faulting failure line (state 3) and with subsequent decreases in pore pressure, the least principal stress must decrease along this line because it cannot decrease more rapidly without exceeding the frictional strength of the reservoir (i.e., the Mohr circle cannot exceed the failure surface).

Poroelastic theory can be used to predict the magnitude of stress changes with depletion. For an isotropic, porous and elastic reservoir that is laterally extensive with respect to its thickness (i.e., a ratio of "breadth" to thickness greater than 20:1), the following expression (derived for an infinite, horizontal reservoir of finite thickness) is applicable (SEGALL and FITZGERALD, 1996).

$$\Delta S_h/\Delta P_p = \alpha\ (1-2v)/(1-v)\ , \tag{3}$$

where ΔS_h is the change in horizontal stress with a change in pore pressure (ΔP_p), v is Poisson's ratio and α is Biot's coefficient. $\alpha = 1 - K_b/K_g$, where K_b is the bulk modulus of the bulk rock and K_g is the bulk modulus of the mineral grains. It is clear from the observations in Figure 4 that on the flanks of the reservoir, $\Delta S_h/\Delta P_p$ is about 0.9. While we could attempt to use laboratory measurements of elastic modulii to compare this value with that expected value using equation (3), this is basically unnecessary. First, it is quite possible that the weak chalk reservoir is not deforming elastically. Second, because we have an observed stress path value of 0.88 on the flanks of the reservoir, we can consider it as being diagnostic of this reservoir under *in situ* conditions.

It is noteworthy that a nearly identical process is seen in the chalk reservoir of the Ekofisk field (Fig. 5), based on data from TEUFEL *et al.* (1991) which come from the time period prior to the initiation of water flooding operations. Because the Ekofisk structure has an oil column exceeding 300 m and we do not know the depths of the measurements reported by TEUFEL *et al.* (1991), which could come from a relatively large depth range, we have transformed pore pressure and least principal stresses into

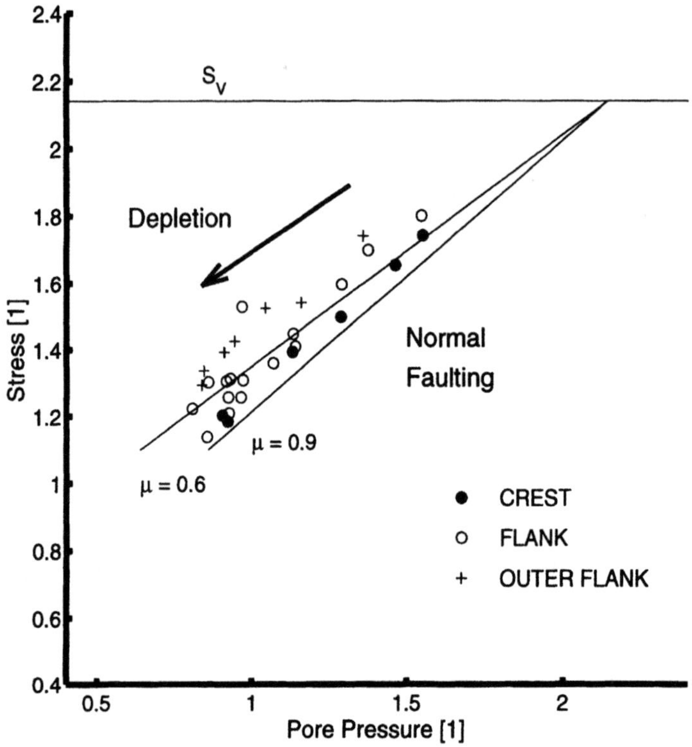

Figure 5
Least principal stresses versus pore pressure in terms of specific gravity are shown for the Ekofisk reservoir at Ekofisk (data from TEUFEL *et al.*, 1991). Filled circles, open circles, and crests denote values obtained for the crest, the flank, and the outer flank, respectively.

equivalent pressure gradients assuming an average depth of 2929 m, 3000 m, and 3075 m for the crest, flank, and outer flank of the structure, respectively. The vertical stress was obtained from integration density logs from several exploration wells.

Note in Figure 5 that the stress and pore pressure data from the Ekofisk reservoir demonstrate that just as at Valhall, under initial conditions, the crest of the structure was in a state of normal faulting but the flank and outer flank were not. Normal faulting would have been expected to continue as depletion and poroelastic stress changes occurred. As reservoir pressures decreased throughout, normal faulting was expected to spread out onto the flanks of the reservoir. The outer flank of the reservoir was almost in a state of normal faulting at the time of the last set of measurements.

TEUFEL *et al.* (1991) also discussed the data shown in Figure 5 in the context of faulting by relating the stress and pore pressure measurements to strength measurements made on Ekofisk core. However, our interpretation of incipient normal faulting in the crest of the structure and induced normal faulting on the flanks due to the poroelastic stress path expands upon their general interpretation.

Microearthquakes at Valhall and Ekofisk

Microseismic monitoring experiments were performed at both Valhall and Ekofisk. At Valhall, an array of six, three-component seismometers were deployed between June 1 and July 27, 1998 in a vertical section of one of the wells near the crest of the structure (Fig. 6). The seismic array was deployed about 300 m above the reservoir (Fig. 6). The spacing between the geophones was 20 m and the recorded seismic signals were digitized at 1 kHz. The locations of 328 microseismic events recorded during the ~7 week monitoring period are shown in Figure 6. Note that the majority of events occurred about 200 m to the west of monitoring well and that microearthquakes are occurring either at the very top of the reservoir or in the Paleocene shale cap rocks that overlay the Tor reservoir (MAXWELL, 2000).

To assess the state of stress in the Valhall cap rock associated with the microearthquakes, we could attempt to compute a focal mechanism for the events shown in Figure 6. However, because we have many earthquakes but effectively only a single recording site (i.e., the spacing of the sensors is much less than the path length), we computed a composite/inverse focal plane. To do this, we had to assume that the earthquakes were all occurring within a uniform stress field and all result from the same style of faulting (i.e., all would have the same focal plane mechanism had it been possible to analyze the events separately).

To obtain the inverse/composite focal mechanism, we positioned each microearthquake with respect to the focal sphere centered on each seismometer so that the ~300 events are assigned to different positions on the composite focal sphere. From the radiation patterns of a double-couple earthquake source (AKI and RICHARDS, 1980), it is known that the sign of the *P*-wave first motion is different in the four quadrants. In the focal plane mechanism shown in Figure 7, the filled symbols denote events with positive displacement, whereas open symbols denote events with negative displacement. Unfortunately, while the *P*-wave polarities suggest a normal fault mechanism (dilatations near the center of the focal sphere) the data are not sufficient to define the focal planes because of the limited distribution of microearthquakes.

In an attempt to overcome this limitation, we have also used *S*-to-*P* amplitude ratios because the position on the focal sphere where there is maximum *P*-wave amplitude (45° from the fault plane or auxiliary plane) coincides with the minimum of the *S*-wave amplitude. Similarly, we expect large S/P amplitude ratios (large symbols) along both the fault plane and the auxiliary plane, where the sign of the *P* wave changes from compressional to dilatational. We limited this analysis to waveforms with signal-to-noise ratios > 3. By combining the *P*-wave polarity and the S/P ratio data we define the focal mechanism shown in Figure 7. While there is considerable scatter in the mechanism due to the assumptions about the uniformity of the stress field and similarity of focal plane mechanisms that we have been required to make, the focal plane mechanism suggests normal faulting along NE-SW trending planes.

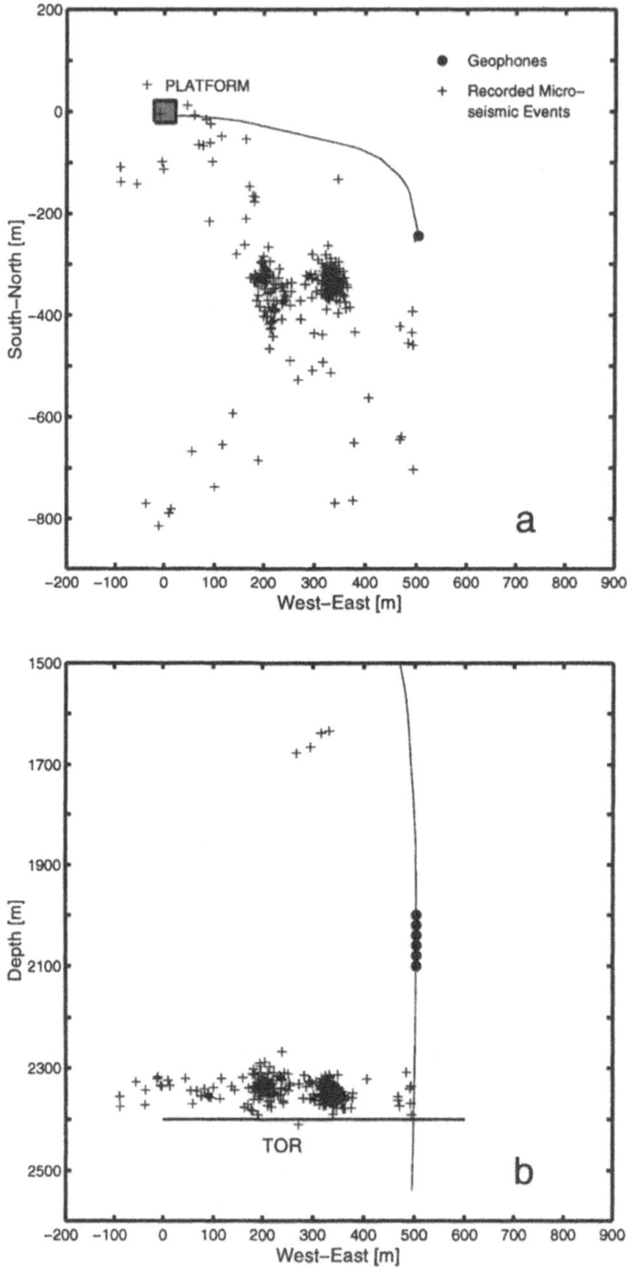

Figure 6
Well path with geophones and locations of the 328 at Valhall recorded microseismic events are shown in map view (a), and in a vertical west-east section (b). It is evident that the microseismic activity is concentrated in a narrow layer above the reservoir.

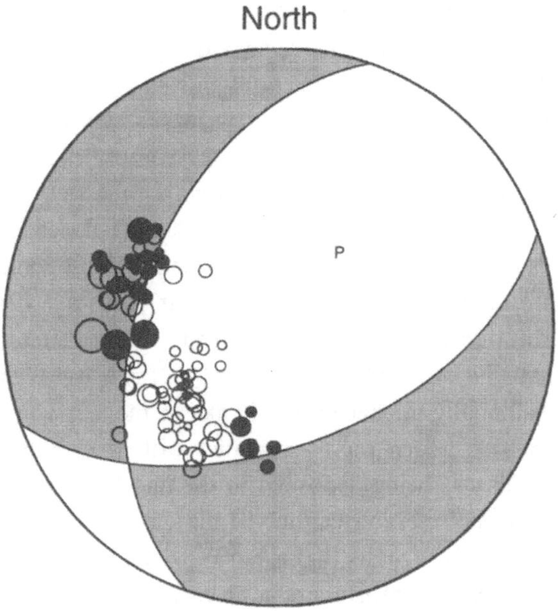

Figure 7
Inverse/composite focal mechanism for microearthquakes recorded at Valhall using the technique described in the text. The focal mechanism is an attempt to best fit both the polarity data (where open circles denote dilatational and filled circles denote compressional P-wave first motions) and S-to-P amplitude ratio (indicated by the size of the symbols). While there is considerable uncertainty in the focal plane mechanism, it is certainly suggestive of normal faulting.

Above a compacting reservoir we expect to observe an increase of horizontal compressive stresses and potentially, induced reverse faulting (see SEGALL, 1989). Because the earthquakes are normal faulting events, not thrust events, it suggests that the microearthquakes are not induced by the elastic stress changes due to compaction. Rather, they appear to be induced by normal faults propagating up into the cap rock from the reservoir. The least principal stress above the Valhall reservoir is either nearly isotropic or somewhat compressional because the vertical stress above the reservoir is equal to the least principal stress (Figs. 2, 3). While this might seem to suggest that production might induce reverse faulting there, the compressive stress perturbation above the reservoir is extremely small. As given by SEGALL and FITZGERALD (1996), immediately above a flat-lying ellipsoidal reservoir, the stress perturbation is given by

$$\Delta S_h/\Delta P_p = \alpha\{(1-2\nu)/(1-\nu)\}\{\pi/4\}\{A_3/A_1\} , \tag{4}$$

where A_3/A_1 is the ratio of the height of the reservoir, A_3, to its lateral extent, A_1. As A_3/A_1 is approximately 0.03 for Valhall, comparison of equations (3) and (4) demonstrates that the increase in horizontal stress with pore pressure changes due to production above the reservoir is only about 1% of the decrease in horizontal stress with pore pressure changes within it. In other words, the increase in horizontal compressive stress would be ~0.2 MPa above the reservoir as opposed to ~20 MPa, within it. Thus, the only way in which reverse faulting could have been induced above the reservoir would have been if a reverse faulting stress regime existed there initially. The focal plane mechanism of the induced earthquakes indicates that this clearly was not the case. This suggests that the normal faulting above the reservoir was caused by the upward propagation of normal faults from within the reservoir.

It is important to note that the NE trend of the fault planes implied by the focal plane mechanism is not consistent with the NW trending fault plane orientations that appear in structure contour maps of the reservoir (MUNNS, 1985). One possible interpretation of this discrepancy is that the current state of stress in the reservoir is not the same as that which caused the movement on the NW trending faults. In fact, as the chalk is substantially thicker on the flanks of the reservoir (Fig. 1b) it seems clear that much of the deformation associated with the anticlinal structure was occurring as the chalk was being deposited in the late Cretaceous. In addition, the faults mapped within the reservoir cannot be mapped up into the Tertiary section (MUNNS, 1985). All of this suggests that the NW-trending faults in the reservoir are no longer seismically active. Unfortunately, no clear trend is seen in the distribution of the microearthquake events (Fig. 6a).

Microseismic recording was carried out at Ekofisk for 18 days in April, 1997 with the same 6-level seismometer system utilized at Valhall. In this case, the seismometers were deployed at the same depth as the reservoir in a well located in the crest of the structure (MAXWELL *et al.*, 1998). While the events near the wellbore could be located with relatively high accuracy, the depth and location of the events further away were more uncertain. In a detailed analysis of the microearthquake locations, MAXWELL *et al.* (1998) report that the majority of the events appear to occur in the upper part of the reservoir, apparently in low porosity, stiffer layers. They also argue that the well-located events cluster along discrete NE-trending lineations subparallel to mapped faults in the reservoir, obviously suggesting that water injection is triggering slip on these pre-existing faults. This may well be the case. However, as the pressure and stress data presented in Figure 5 show that the crest of the structure was in an incipient state of normal faulting even prior to the onset of water injection, this relationship could be somewhat more complex. It is interesting that the NE trend of the microseismic lineations and pre-existing faults in Ekofisk are subparallel to the fault planes implied by the focal mechanism derived for Valhall.

Discussion

Sheared Wells, Gas Leakage, Reservoir Permeability and Subsidence

The observations presented above have a variety of geomechanical implications. First, it is obvious that both the active normal faulting in the crest of the structures and the induced faulting on the flanks may be responsible for some of the problems encountered with sheared well casings. It would be interesting to know if these problems "spread out" onto the flanks of the structures as the normal faulting stress state was reached. It is important to note that once the faulting process begins in part of the reservoir, water injection and pressure maintenance programs (such as implemented in Ekofisk) would not cause faulting to stop. In fact, once the pore pressure and least principal stress have "hit" the failure line (Figs. 3–5), faulting would be expected to continue unless water flooding actually reversed the pore pressure decline. Unfortunately, if the water injection caused local increases in pore pressure in the vicinity of faults it would likely trigger fault slip in terms of the conventional induced seismicity mechanism described at the beginning of this paper. This, of course, may also be directly relevant to compaction and subsidence at Ekofisk (c.f., NAGEL, 1998) as it is possible that some of the compaction and subsidence that has occurred is associated with slip on normal faults. One observation that may support this suggestion is that the initiation of a water injection and pressure maintenance program at Ekofisk did not cause subsidence to stop, or even slow down (NAGEL, 1998). Because of the arguments above, this is what would be expected in the case of slip on active faults. If reservoir compaction was the result of only the net increase in effective stress, the compaction should have stopped, or at least slowed down, once reservoir pressures stabilized.

With respect to reservoir permeability and gas leakage, active faulting in the reservoir is clearly a good news/bad news situation. The good news, of course, is that active faulting in the reservoirs is capable of increasing matrix permeability. BARTON *et al.* (1995), DHOLAKIA *et al.* (1998), CHANCHANI *et al.* (2001), and FINKBEINER *et al.* (1998) describe situations in which relatively small scale critically-stressed faults in the current stress field result in dramatic increases in reservoir permeability. It is quite interesting that despite the reservoir compaction accompanying depletion, reservoir productivity has remained steady, or has slightly increased, since production began (SULAK, 1991). In light of this observation, it is reasonable to assume that active shear faults may be enhancing the low matrix permeability (BROWN, 1987) and counteracting the permeability reductions expected to accompany compaction.

The bad news, of course, is that as there has been appreciable gas leakage through the shale cap rocks in both fields, which may have been exacerbated by enhanced fluid flow through faults propagating up into the cap rock from the reservoir. If the microseismicity is in the shale caprock at Valhall (Fig. 6), it is presumably due to the upward propagation of faults activated by fluid production in the reservoir. There also appears to be some faulting in the caprock at Ekofisk (MAXWELL *et al.*, 1998). Because of the water injection program at Ekofisk and the large strains accumulated over ~20 years of production (as well as the uncertainties in determination of the exact depth of the microearthquakes which are distant from the observation well), it is difficult to interpret the microearthquakes at Ekofisk with confidence. Nonetheless, in light of our findings for both Ekofisk and Valhall, it is interesting to speculate if there has been increased leakage caused by the initiation of faulting on the flanks of the reservoirs.

Conclusions

There are significant similarities in both the initial stress fields of the Valhall and Ekofisk fields and their evolution with time. In both fields, we found that under initial, undepleted conditions, the crest of the structures was in a state of active normal faulting. As depletion occurred, the stress path was such that the crest remained in a normal faulting stress state which appears to have spread outward to the flanks of the reservoirs. The locations of microseismic events reveal that normal faulting can be observed in and above the productive horizons. Whereas the normal faulting appears to be aseismic in the weak chalks of the reservoir, it is clearly seismic in the overburden. The significance of this as a possible mechanism responsible for problems with sheared well casings, reservoir permeability, cap rock integrity and subsidence is currently unknown.

Acknowledgements

We thank BP-Amoco for providing the data used in this study. Special thanks are due to L. Thomsen, M. Mueller, O. Barkved, T. G. Kristiansen, and P. G. Folstad from BP-Amoco for their assistance in data collection and constructive discussions. We thank S. C. Maxwell for constructive comments on the text. CSM Associates processed the seismic data of the passive seismic monitoring experiment and provided computed event locations. The views expressed in this paper are those of the authors and do not necessarily represent those of BP-Amoco or their co-ventures. Financial support for this project was provided by the Stanford Rock and Borehole Geophysics project.

References

Aki, K., and Richards, P. G., (*Quantitative Seismology, Theory and Methods*), vol. 1, (Freeman, San Fr ancisco, 1980), pp. 557.

Barton, C. A., Zoback, M.D., and Moos, D. (1995), *Fluid Flow Along Potentially Active Faults in Crystalline Rock*, Geology *23*, 683–686.

Byerlee, J. D. (1978), *Friction of Rock*, Pure appl. geophys. *116*, 615–626,

Brown, D. (1987), *The flow of water and displacement of hydrocarbons in fractured chalk reservoirs*. In *Fluid Flow in Sedimentary Basins and Aquifers* (eds. J. C. Goff and B. P. Williams) (Geological Society London) Special Publication *34*, 201–218.

Dholakia, S. K., Aydin, A., Pollard, D., and M. D. Zoback, (1998), *Development of Fault-controlled Hydrocarbon Migration Pathways in the Monterey Formation, California*, AAPG Bull. *82*, 1551–1574.

CHANCHANI, S. K., ZOBACK, M. D., and BARTON, C. (2000), *A Case Study of Hydrocarbon Transport Along Active Faults and Production-related Stress Changes in the Monterey Formation, California*, Geophys. Res. Lett., in press.

DOSER, D. I., BAKER, M. R., and MASON, D. B. (1991), *Seismicity in the War-Wink Gas Field, Delaware Basin, West Texas, and its Relationship to Petroleum Production*, Bull. Seismol. Soc. Am. *81*, 971–986.

FEIGNIER, B., and GRASSO, J. R. (1990), *Seismicity Induced by Gas Production: I. Correlation of Focal Mechanisms and Dome Structure*, Pure appl. geophys. *134*, 405–426.

FINKBEINER, T., BARTON, C. A., and ZOBACK, M. D. (1998), *Relationship Between in situ Stress, Fractures and Faults, and Fluid Flow in the Monterey Formation, Santa Maria Basin, California*, AAPG Bull. *81*, 1975–1999.

FINKBEINER, T., ZOBACK, M. D., Stump, B., and FLEMINGS, P. (2000), *Stress, Pore Pressure and Dynamically-constrained Hydrocarbon Column Heights in the South Eugene Island 330 Field, Gulf of Mexico*, AAPG Bull., (in press).

GRASSO, J. R. (1992), *Mechanics of Seismic Instabilities Induced by the Recovery of Hydrocarbons*, Pure appl. geophys. *139*, 507–533.

JAEGER, J. C., and COOK, N. G. W., *Fundamentals of Rock Mechanics* (Chapman and Hall, New York 1971).

KRISTIANSEN, G. (1998), *Geomechanical Characterization of the Overburden above the Compacting Chalk Reservoir at Valhall*, Eurock '98, SPE/ISRM Rock Mechanics in Petroleum Engineering, The Norwegian University of Science and Technology, Trondheim, Norway, 193–202.

MAXWELL, S. C., YOUNG, R. P., BOSSU, R., JUPE, A., and DANGERFIELD, J. (1998), *Microseismic Logging of the Ekofisk Reservoir*, Eurock '98, SPE/ISRM Rock Mechanics in Petroleum Engineering, The Norwegian University of Science and Technology, Trondheim, Norway, 387–393.

MAXWELL, S. C. (2000), *Comparison of Production-induced Microseismicity from Valhall and Ekofisk*, Paper presented at the Passive Seismic Method in E&P of Oil and Gas Workshop, 62nd EAGE Conference, May, 2000.

MAURY, V. M. R., GRASSO, J. R., and WITTLINGER, G. (1992), *Monitoring of Subsidence and Induced Seismicity in the Larq Gas Field (France): The Consequences on Gas Production and Field Operation*, Engineering Geology *32*, 123.

MUNNS, J. W. (1985), *The Valhall Field: A Geological Overview*, Marine and Petroleum Geology, *2*, 23–43.

NAGEL, N. B. (1998), *Ekofisk Field Overburden Modelling*, Eurock '98, SPE/ISRM Rock Mechanics in Petroleum Engineering, The Norwegian University of Science and Technology, Trondheim, Norway, 177–186.

PATTILLO, P. D., KRISTIANSEN, T. G., SUND, G. V., and KJELSTADLI, R. M. (1998), *Reservoir Compaction and Seafloor Subsidence at Valhall*, Eurock '98, SPE/ISRM Rock Mechanics in Petroleum Engineering, The Norwegian University of Science and Technology, Trondheim, Norway, 377–386.

RALEIGH, C. B., HEALY, J. H., and BREDEHOEFT, J. D. (1972), *An experiment in earthquake control at Rangely, Colorado*, Science *191*, 1230–1237.

RHETT, D. W. (1998), *Ekofisk Revisited: A New Model of Ekofisk Reservoir Geomechanical Behavior*, Eurock '98, SPE/ISRM Rock Mechanics in Petroleum Engineering, The Norwegian University of Science and Technology, Trondheim, Norway, 367–375.

SCHWALL, G. H., and DENNEY, C. A. (1994), *Subsidence Induced Casing Deformation Mechanisms in the Ekofisk Field*, Eurock '94, SPE/ISRM Rock Mechanics in Petroleum Engineering, Balkema, Delft, Netherlands, 507–515.

SEGALL, P. (1985), *Stress and Subsidence Resulting from Subsurface Fluid Withdrawal in the Epicentral Region of the 1983 Coalinga Earthquake*, J. Geophys. Res. *90*, 6801–6816.

SEGALL, P. (1989), *Earthquakes Triggered by Fluid Extraction*, Geology *17*, 942.

SEGALL, P. (1992), *Induced Stresses Due to Fluid Extraction from Axisymmetric Reservoirs*, Pure appl. geophys. *139*, 535.

SEGALL, P., GRASSO, J. R., and MOSSOP, A. (1994). *Poroelastic Stressing and Induced Seismicity Near the Lacq Gas Field, Southwestern France*, J. Geophys. Res. *99*, 15,423–15,438.

SEGALL, P., and FITZGERALD, S. D. (1996), *A note on Induced Stress Changes in Hydrocarbon and Geothermal Reservoirs*, Tectonophysics *289*, 117–128.

SULAK, R. M. (1991), *Ekofisk Field: The First 20 Years*, J. Petrol. Tech., 1265–1271.

Teufel, L. W., Rhett, D. W., and H. P. Farrell, *Effect of reservoir depletion and pore pressure drawndown on in situ stress and deformation in the Ekofisk field, North Sea.* In *Rock mechanics as a Multidisciplinary Science* (ed. J. C. Roegiers) (Balkena, Rotterdam 1991), pp. 63–72.

Thomas, L., Dixon, T., Evans, C., and Vienot, M. (1987), *Ekofisk Pilot Waterflood*, J. Petro. Tech. *39*, 221.

Townend, J., and Zoback, M. D. (2000), *How Faulting Keeps the Crust Strong*, Geology *28*, 399–402.

Willson, S., Last, N. C., Zoback, M. D., and Moos, D. (1998), *Drilling in South America: A Wellbore Stability Approach for Complex Geologic Conditions*, SPE 53940, in 1999 SPE Latin American and Caribbean Petroleum Engineering Conference, Caracas, Venezuela *21–23* April, 1999.

Wiprut, D., and Zoback, M. D. (1999), *Fault Reactivation and Fluid Flow Along a Previously Dormant Normal Fault in the Norwegian North Sea*, Geology *28*, 595–598.

(Received July 18, 2000, revised/accepted December 21, 2000)

Pure appl. geophys. 159 (2002) 421–447
0033–4553/02/030421–27 $ 1.50 + 0.20/0

Pure and Applied Geophysics

An Application of Induced Seismicity Data Analysis for Detection of Spatial Structures and Temporal Regimes of Deformation Processes in Hydrocarbon Fields

SERGEJ BORISOVICH TURUNTAEV[1] and OLGA ANDREEVNA RAZUMNAYA[1]

Abstract—An application of induced seismicity data to detection of deformation processes is considered. Methodology of the data interpretation is based on the conception of rock mass as a block medium adopting energy from internal and external sources and forming dissipative structures. The ideas from the stability theory of nonlinear systems are used. A spatial and temporal distribution of seismicity is analyzed to discover a regular, predictable component of seismic regime. Various methods were used for the analysis of seismic events listed in the seismic catalog in the region of Romashkino oil-field (Russia). Quasi-harmonious oscillations of seismic activity were found. They were synchronized in time with change of injection effectiveness. Spatial structures of seismicity were detected, and it was shown that these structures are in good agreement with tectonic faults. Data analysis shows correlation between seismic activity and imbalance of injected and extracted fluids. The obtained results show that water-injection effectiveness decreases during periods of increased seismic activity and increases during periods of low seismic activity.

Key words: Induced seismicity, oil recovery, seismic regime, methods of interpretation.

Introduction

Induced and triggered seismicity was observed during water reservoirs filling, development of mineral and hydrocarbon deposits, during execution of underground nuclear explosions and during carrying out of big-scaled construction works (NIKOLAEV, 1977; GAUPTA and RASTOGY, 1979; PASECHNIK, 1977; SIMPSON, 1986; NICHOLSON and WESSON, 1991; MILNE and BERRY, 1976; GRASSO, 1992). The seismic events which are directly caused by action on rock mass are usually called "induced" or "man-made" seismicity. Hypocenters of this type of earthquakes are located within or near the rock mass under action. Energies are usually not high (local magnitude of events is less than 3 on the Richter scale). If an action on a rock mass caused a disturbance of natural deformation regime in the rock mass, seismicity can be triggered. The source of triggered seismic energy is an energy of tectonic processes

[1] Institute for Dynamics of Geosphere Russian Academy of Sciences, Leninsky prosp., d.38, k.6, 117979, Moscow, Russia. E-mail: turunt@postman.ru

related to some stress readjustment (Simpson, 1986). In this case even weak actions can lead to an appearance of strong seismicity (Sadovsky *et al.*, 1988; Rodionov *et al.*, 1989). Theoretically earthquake energy can reach a maximum for natural seismicity and it depends on tectonic stresses in the region. Hypocenters are usually close to a man-made object, but can also appear at a distance of several kilometers from that object.

During a long-term development of hydrocarbon deposits, conditions of stability in reservoir can be disturbed. This can cause significant changes in stresses within the reservoir and the surrounding rocks. Injected fluid can propagate into cracks, cause an increase in pressure in pores and fractures and can serve as a sort of lubricant in the fractured zones. Earthquakes induced by filtration processes, as well as other tectonic and man-made earthquakes, are caused by rock blocks movement along joints. Pore-elastic forces can directly trigger a displacement along a fault in surrounding rock mass (pore-elastic forces factor). Pore pressure can be transferred from an injection zone into an earthquake zone through a fault or another permeable structure. Fluid migration in this case can be negligible (hydrostatic factor). Finally, fluids can migrate from injection zones to zones of earthquake preparation (hydrodynamic factor).

Hydrocarbon field development always has changes in induced seismic background which causes microseismic event occurrences. The energy of a seismic event depends on reservoir rocks properties, properties of the surrounding rocks, their stress state, the medium heterogeneity level and on deposit development rate and intensity.

It is natural to consider triggered and induced seismicity as one of the hazardous consequences of hydrocarbon field development. Examples of such hazardous and even catastrophic after-effects are well known. Fortunately those catastrophic events are extremely rare as concerns hydrocarbon deposit development. Generally weak seismic events are induced, and they can only be recorded with the help of sensitive seismometer network. These weak seismic events, induced as well as caused by natural deformation processes, carry important information pertaining to the location of active structures of rock massif, containing hydrocarbon deposit. They also carry information about changes in the stress-strain state of the rock massif with respect to time. With the help of interpretation of data on weak seismicity and microseismicity we can reveal faults that are currently active, find the location of the front of oil by water or gas displacement and estimate time variations of reservoir permeability and porosity. Deformation processes in the regions of hydrocarbon deposits, recorded by a number of investigators, are accompanied with changes in reservoir properties of the deposit. These processes have a certain duration that allows the use of information regarding changes of the reservoir stress-strain state in a time when scheduling measures for hydrocarbon recovery increase.

This paper presents a study of temporal and spatial characteristic features of the seismicity in the region of Romashkino oil field (Tatarstan, Russia). The main purpose of this study is an application of seismic data analysis to reveal regular (predictable) component of seismic regime and detection of active tectonic structures.

Data Characteristics

According to the map of seismic zoning, the southeastern part of Tatarstan, where the Romashkino oil field is located, is considered to be a seismically quiet area. In 1982–83 in the region of Almetjevsk city seismic events with intensity III-IV were noticed by citizens. A local seismic network was installed in 1985. From 1986 to 1992 this seismic network recorded 391 local earthquakes with magnitude $0.1 \leq M \leq 4$ on the Richter scale and with hypocenter depths ranging from 3 to 8 km (ISKHAKOV, 1988, 1992). An increase in seismic activity was recorded at the end of 1986–beginning 1987, in the middle of 1988, and at the end of 1991. The earthquake on September 29, 1986 reached intensity VI in the epicenter (magnitude $M = 3.8$, hypocenter depth 5–10 km) and that on September 19, 1988 registered intensity up to V. On October 28, 1991 an earthquake with $M = 4$, intensity in epicenter VI and a hypocenter depth of 6 km (ISKHAKOV *et al.*, 1992) occurred in the region of Almetjevsk.

Romashkino oil field is Russia's largest oil field. Its maximum dimension is about 70 km. Amplitude of the structure crest high of Devon sediments is 50–60 m, reservoir depth is 1600–1800 m (geological conditions of oil and gas accumulation and location, 1982). The deposit is a multilayer, oil-bearing formation formed by quartz sandstone with a thickness between 10 and 30 m and carbonates. The permeability of the sandstone layers ranges from 200 to 420 mD, the porosity is between 18.8 and 20.4%, oil-saturation is 69.4–90.5%. The initial reservoir pressure is 160 to 180 atm. The density of the oil is 0.8–0.85 g/cm^3, and the viscosity varies from 2.3 to 4 cPz on average.

Geology explorations in this region started in 1933. In 1947 an exploration deep drilling operation began. In July, 1948 oil was produced for the first time (MUSLIMOV, 1979). Water injections in the oil field started in 1954. In the first few years it did not compensate fluid extraction. In 1958, for the first time, the volumes of injected fluid exceeded the volumes of extracted fluid. In 1963 the difference between total volumes of injected and produced fluids was eliminated. In 1975 176.6×10^6 m^3 of water was injected. According to the state on January 1, 1976, the volume of injected fluid was 2.13×10^9 m^3, which was at that time equal to 104.7% of extracted fluid in reservoir conditions (MUSLIMOV, 1979). According to the General Schedule of Romashkino oil field development (1974–76), the water injection was thought to be at the top pressure of 200–250 atm. In some case the pressure was actually higher. For exploitation convenience Romashkino oil field is divided into more than 20 areas.

Figure 1 shows a map of the seismic activity in the region of Romashkino oil field. As a measure of seismic activity the following convention was used: a sum of cubic roots from the energies of earthquakes which took place within the area of per sq. km for the period of observations (1986–92). Most earthquake epicenters are situated along Altunino–Shunaksky depression structure, seismic activity distribution in the area does correspond to tectonic faults, revealed by BELOUSOV *et al.* (1994).

In Figure 2 a frequency-magnitude plot is shown (the relationship between a logarithm of the number of seismic events N and magnitude M). The magnitude of seismic events was calculated according to energy class K:

$$M = 0.67K - 3.2 \ , \tag{1}$$

obtained from Gutenberg–Richter law (GUTENBERG and RICHTER, 1956). Energy class is usually used in Russia to characterize seismic events. In Directions for data acquisition and data processing at seismic stations of the USSR united seismic network, 1981, it is defined as $K = \log_{10} E$, where E (in Joules) is the energy of seismic waves at a reference distance R_0 from the seismic source. In practice, amplitude to period ratio was used to calculate energy class with the help of correlation curves, plotted in preliminary investigations (ISKHAKOV, 1988). Only data

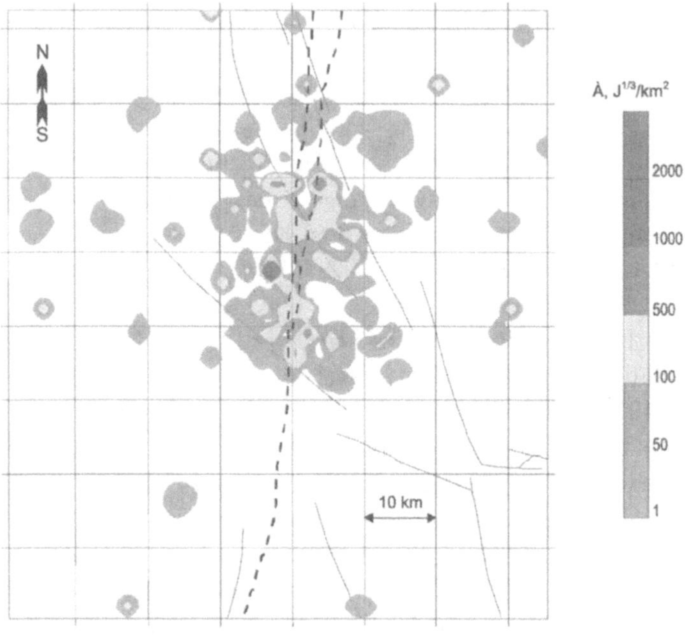

Figure 1
Distribution of seismic events energy release density in the region of the Romashkino oil-field (in cubic roots of energies per square km). Tectonic faults (after BELOUSOV *et al.*, 1994) and Altunino–Shunaksky depression structure are shown by solid and dashed lines, respectively.

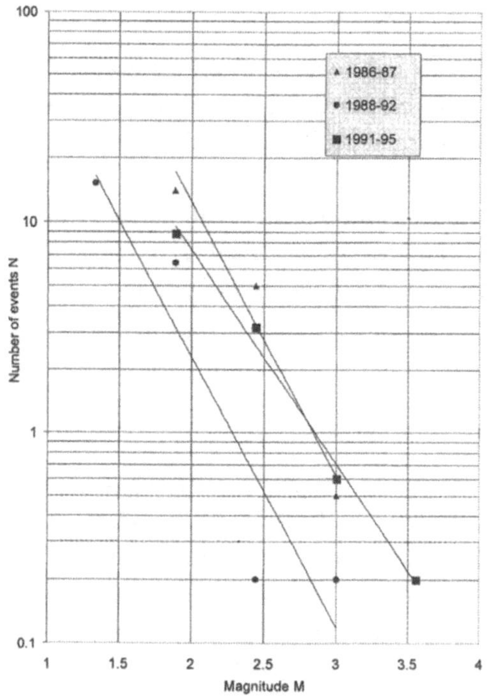

Figure 2
The relationship between a logarithm of the number of seismic events N and magnitude M.

from catalogs of instrumentally recorded seismic events were used (data from 1986 to 1995) for calculations, remote events with an epicenter distance more than 70 km were removed from consideration. Because of different representations of the seismic event catalogs and different methods of seismogram interpretation in the different periods of observations, the frequency-magnitude relations were plotted separately for the three different time intervals (1986–87, 1988–92, 1992–95), and an annual average number of events for these time periods was computed. It is well known that the relationship between a logarithm of the number of seismic events and their magnitude should be linear; a distinction from the direct line in the left part of the plot (weak seismic events) proves that sensitivity of the seismic network is not enough to record weak events. A distinction from the direct line in the right part of the plot (seismic event with high energies) is considered as a proof of insufficient duration of the observations. The linear part of the studied relationship corresponds to the magnitude range of seismic events that is representative for this particular seismic network recording. One can see that for the initial period of observations only the records of events with magnitude $M \geq 2$ can be considered representative, later in this study the events with $M = 1.5$ become representative too. The magnitude-frequency relationships for the above periods appears as:

$$\lg \bar{N} = 2.94 - 1.29M \quad \lg \bar{N} = 3.7 - 1.3M \quad \lg \bar{N} = 2.9 - 1.02M \ , \tag{2}$$

where $\bar{N}$ is an average annual number of seismic events. The slopes of these plots considerably exceed the value for natural seismicity 0.75 (SADOVSKY and PISARENKO, 1991) and reach values normal for induced and triggered seismicity (MIRZOEV *et al.*, 1987; TURUNTAEV, 1993).

We should nevertheless point out that both periods with the same representativities of observations and the overall number of recorded events, which can be classified as representative, do not allow consideration of the obtained result as final.

Seismic Activity Variations in Time

Seismic activity is one of the most often used characteristics of seismicity (PONOMARYOV and TEJTELBAUM, 1974). It allows transfer of a discrete representation of seismic events (which can be described with three spatial coordinates, time and energy) to a continuous one. As a measure of activity, a number of events taking place in a volume of a certain size during a certain time period or a sum of energies of these events raised in some power is generally used. It is shown by (SADOVSKY and PISARENKO, 1991) that the most preferable power is $\gamma/2$, where γ is a slope of frequency – log of energy relation. For the studied region $\gamma = 0.6$–0.7. This is the first reason for adopting the sum of cubic roots from the energies of events as a measure of seismic activity. The other reason is to assign physical meaning to the value, used as a characteristic of the seismic activity. To decrease an influence of an arbitrary choice in the initial location of the area and in the time interval, an overlap of areas and an overlap of time intervals are used. The values of these overlaps depend on the smoothness of the obtained distributions of an activity and also on the studied problem. Temporal and spatial components of the activity change are considered here separately.

Monthly seismic activity was calculated, which means a sum of cubic roots from the energies of events that took place during one month. The obtained temporal series were normalized on average value for this time period. Variation of seismic activity in the area of Romashkino oil field in 1986–1995 obtained with the help of this method is shown in Figure 3a.

Let us consider the temporal function of activity $A(t)$ as a sum of three components

$$A(t) = A_p(t) + A_r(t) + A_t(t) \ , \tag{3}$$

where A_p is high frequency random oscillations of activity, A_r is a regular component, A_t is slow variations, a trend. Let us try to find out if such a procedure would allow the display of a regular component in the behavior of the function $A(t)$. For this purpose let us convert coordinates $(A(t), t)$ to phase coordinates

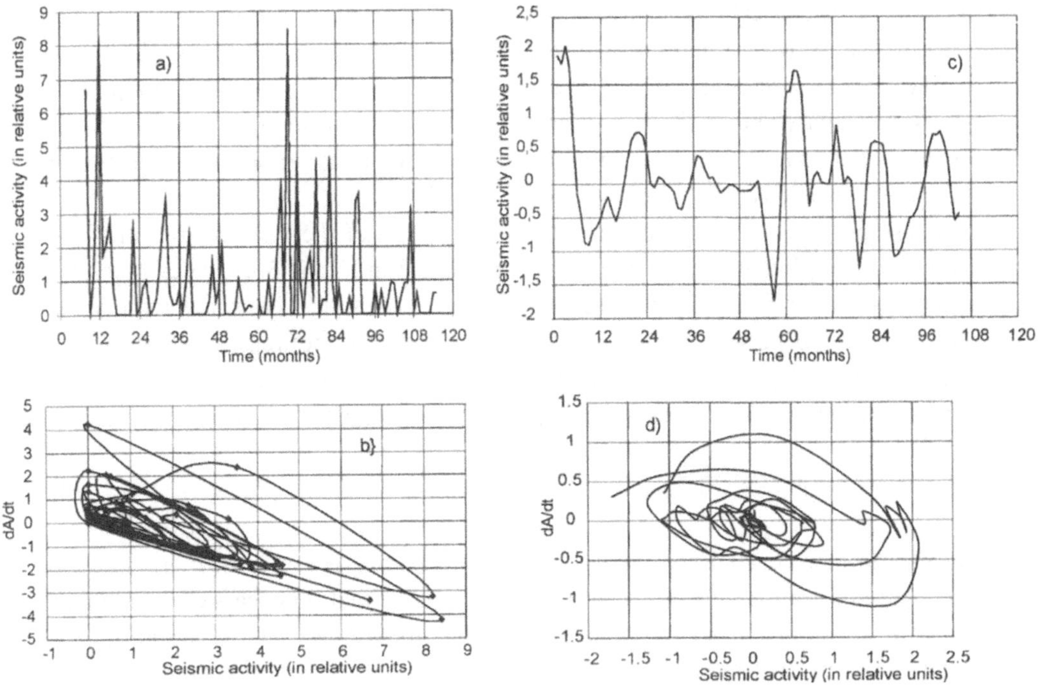

Figure 3
A seismic activity variation in the region of the Romashkino oil field: a – monthly seismic activity (a sum of cubic roots from the energies of events that took place during a month); b – monthly seismic activity change in phase coordinates; c – regular component of seismic activity variations; d – phase portrait after smoothing and trend removing.

$(A(t), dA(t)/dt)$, that means to coordinates activity – rate of activity variation. A point in such coordinates defines a state of seismic process at some instant of time and velocity of changes of this state. Expressing seismic activity in terms of phase coordinates is useful for several purposes:

1. Two characteristics of seismic process (its activity and rate of activity variation) are considered and transformed as independent values.

2. Phase portrait is more "sensitive" to procedures such as smoothing and trend removing, what simplifies choice of time period for calculation of activity, smoothing type, coordinate transformation.

3. Standard procedure of Fourier analysis is not effective if applied to quasi-harmonical oscillations with changing frequency and amplitude. In phase coordinates an increase of an amplitude of oscillations up to a constant value will look like outer limit cycle; a decrease of the amplitude to zero will look like the point type attractor, and so on.

4. After selection of procedures which transform the phase portrait to a form allowing a mathematical description, one can carry out reverse transformation and

obtain a mathematical description of the regular component of the original seismic process; that allows estimation of further development of the seismicity. Statistical significance of the obtained prognosis depends on the value of the chaotic component of seismic activity and of the rate of activity variation. Furthermore it also depends on the possibility to recognize bifurcation points of phase trajectories (points of changes of seismic regime type).

Phase portrait of the seismic process in the region of the Romashkino oil field, described by a plot shown in Figure 3a, is presented in Figure 3b. One can see that an activity state trajectory in phase coordinates looks chaotic. Let us now remove the random component with the help of moving window smoothing with half-a-year window with a step equal to one month; and let us remove a trend aided by linear transformation of coordinates according to the following formulas:

$$A' = A - 1.71 + 0.02t \quad \left(\frac{\partial A}{\partial t}\right)' = \frac{\partial A}{\partial t} - 0.25 + 0.005t \ . \tag{4}$$

The curve obtained after smoothing and trend removing is shown in Figure 3c, the final phase portrait is shown in Figure 3d. One can see that it appears close to a pair of cycles: trajectory starts at some initial point, then it "twists" as a spiral; at a certain moment the trajectory comes back to the external part of the spiral and then "twists" again. At the same time the trajectory never leaves a certain area.

The obtained phase portrait is very close to the limit cycle, which can be found as a solution of Duffing equation (HAKEN, 1983):

$$\dot{q}_2 + kq_2 + q_1^3 = C \quad q_2 = \dot{q}_1 \tag{5}$$

with parameters $k = 0.35$, $C = 6.6$. This equation describes behavior of an oscillator under action of an external force.

An outward motion of a spiral trajectory corresponds, in general, to an increase of an amplitude of seismic process activity oscillations while an inward motion corresponds to a decrease of amplitude oscillations. Oscillations of seismic process activity are not directly sinusoidal; their period tends to oscillate with an average value close to 12 months.

It is interesting to compare two cycles of seismic activity variations in the region of the Romashkino oil field. Both cycles start with the strongest earthquake for this region (during the period of observations) with intensity VI, each cycle lasts for about 5 years. In Figure 4 the two cycles of activity variations (1986–90 and 1991–95) are superimposed so that their first maximums coincide. It can be seen as an intriguing qualitative coincidence of the curves. That is evidence of some kind of regularity in seismic activity oscillations.

Let us point out that phase trajectories of the determining processes should never intersect each other. Intersections taking place in the studied case are probably caused by the projection of three- and more-dimensional attractors to a plane. A consideration of spatial movements of seismic activity centers can allow avoidance of

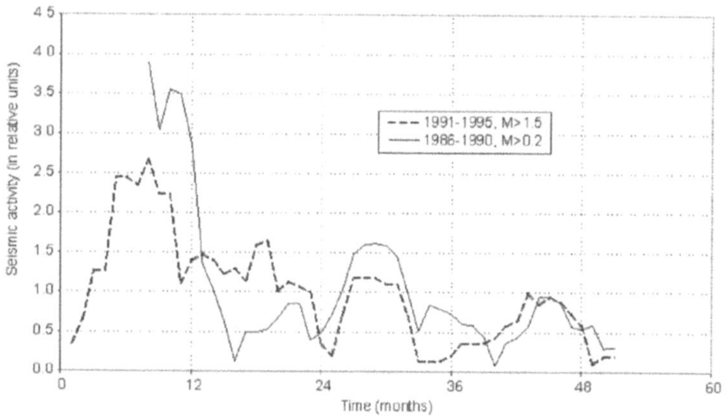

Figure 4
A comparison of two cycles of seismic activity in the region of the Romashkino oil field.

the violation of the determination in the points of intersection of phase trajectories (TURUNTAEV, 1994).

Trajectories of seismic activity variations are always a realization of a stochastic process. Accumulation of data will allow to detect more accurately a primary component of the seismic process and to determine an equation which will satisfactorily describe obtained trajectories in phase coordinates. Besides prognosis information, an existence of a regular component of the sequence of seismic events can characterize an energy state of the rock massif. It seems possible, that when a dissipation rate of energies accumulated in rock massif and originating from outside reaches certain values, an energy release due to seismic processes will become structured in space and time. In a rock massif subjected to oil production, along with a decrease of reserves of tectonic deformations energy, the possibility of release of this energy increases due to the continuation of an increase in the amount of extracted and injected fluids. An existence of quasi-periodical oscillations of seismic activity proves that energy flow (revealing itself in seismic events) is big enough. One can presume, that the regime of seismic events will be sensitive to the value of production and injection and to exploitation regimes. This will allow the seismicity to be affected by scheduling an exploitation regime according to revealed periods of seismic activity.

A Relation Between Spatial Structures of the Induced Seismicity and Tectonic Structures

The goal of this section is to create a method of spatial groups of seismic events detection and to analyze a spatial and temporal distribution of hypocenters of grouped seismic events. The main distinction of this work from a more common grouping procedure is an assumption of a relation between time and places of occurrence of grouped events, i.e., that the appearance of later events in a spatial group is caused by changes in stress field near epicentral zones of the previous events.

Seismic Events which are Close in Space to Each Other

Earthquakes located close to each other were chosen from the analyzed catalog of local events. Energy criteria were adopted as criteria of the closeness of seismic events. Closeness estimation procedure consisted of a comparison of the distance between a pair of event hypocenters with the size of effective stresses change zone that occurred as a result of the first event in the pair. Radius of influence zone R_{infl} is treated as proportional to the fault length L:

$$R_{infl} = A \times L \ .$$

The fault length was found according to an energy E of an earthquake:

$$L[m] = 0.4 * E^{1/3}[J] \ . \qquad (6)$$

Coefficient 0.4 was found from the following empiric formulas: $\lg L = 3.2 + 0.5M$, $\lg E = 11.8 + 1.5M$, where M is a magnitude of an earthquake. Dispersion of the coefficients of the above relations according to (KASAHARA, 1981; SADOVSKY *et al.*, 1987) gave an estimation for $E/V = 10^{1.2} \div 10^{2.3}$, where V is an earthquake zone volume in m^3. According to various authors' estimations (see, for example, the review in GORBUNOVA and KAL'MET'EVA, 1988), the coefficient A in (6) can range from 3 to 8.

The algorithm chosen to form groups of seismic events is now described. The first event in the analyzed catalog was considered, and a search for the next closest event (within the influence zone of the first event) was carried out. After such an event was detected, a zone of influence was considered as a sum of the zones of influences of the first and second events, and the search for events enclosed in the described cumulative zone was carried out and so on. A set of events formed the group of the event, which is first in the catalog, and will be called 'first event group'. Then the second event group was formed and so on. Aftershocks of the earthquake on September 19, 1988 were removed from the sample. That is how the sample consisting of events close in the described sense was created. Figure 5 shows the location of all the grouped events, beginning with the first one, with coefficients of influence equal to 5. Every first event in a group is shown with a different color which corresponds to a succession of grouped events appearance. Location of the main tectonic faults is also shown.

One can see that epicenters of grouped events are related to the faults, while all the seismic events are spread over a relatively large area (compare Fig. 1 with Fig. 5). There are different interpretations of the tectonic structures in the region of Romashkino oil field however Altunino–Shunaksky depression structure (dashed

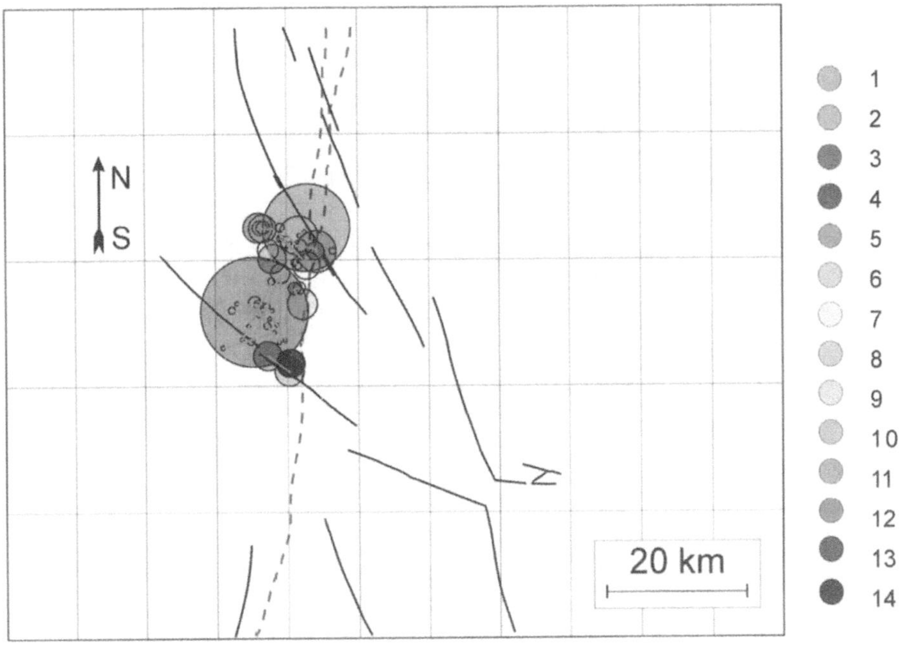

Figure 5
A map of influence zones of grouped events in 1986–1992. The centers of circles correspond to locations of epicenters. The diameters of circles are equal to dimensions of zones of seismic event influence defined by the length of fault with coefficient 5. The first event in each group has a different pattern. Location of main tectonic structures is shown.

lines), which separates Romashkino and Novoelokhovskoje oil fields, is always pointed out as a main linear structure. This depression is related with tectonic fault, the surface of the basement is displaced along the fault for more than 50 m in a vertical direction. BELOUSOV *et al.* (1994) described tectonic structures in the north-east direction. These tectonic structures were detected in the central part of Romashkino and Novoelohovskoje oil fields with the help of an analysis of geomorphologic data. Analysis of the map presented in Figure 5 shows that these Altunino–Shunaksky depression structures and faults of the north-west direction define the location of grouped events. Let us point out that lines connecting epicenters of the successively grouped events have mainly longitudinal and north-west directions.

It was found, that starting with 5, an increase of the coefficients of proportionality between the size of an earthquake fault zone and a zone of earthquake influence does not change a shape of grouped events influence zones. It leads only to an increase in the number of the grouping events. This allows estimation of the radius of influence zone as equal to earthquake fault lengths multiplied by five.

Figure 6 shows variations in time of the number of grouped events taking place in a period 1 month. Coefficient of proportionality between fault length and the size of

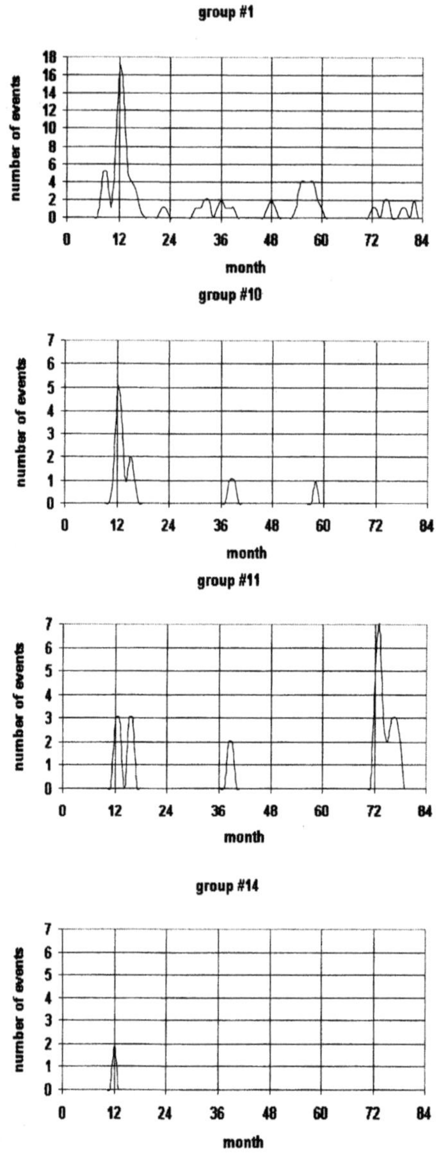

Figure 6
The distributions in time of spatial groups of events for the period 1986–1992. The number of groups corresponds to the consequence of "first in the group" events appearing.

the influence zone is taken here to be equal to 8 (to increase the statistical significance). Group number in Figure 6 corresponds to a succession of groups. Figure 7 shows seismic activity variation for the same period. The activity was calculated as a sum of cubic roots from energies of seismic events that took place in a 1-month period and was divided by average seismic energy of the studied group for the studied period. One can see several types of variations in the rate of grouped events appearance. For

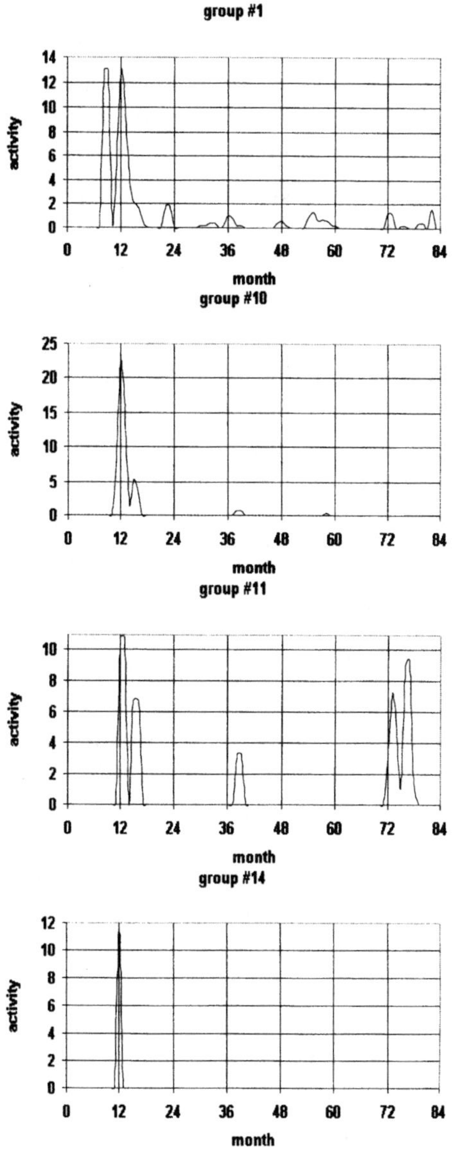

Figure 7
The change in seismic activity of spatial groups of events for the period 1986–1992.

example, for the first group an abrupt increase of grouped events appearance rate is observed, which is preceded by a less significant peak of the number of events, equal to five events per month. Thereafter that the rate of grouped events appearance decreases. During the entire period of observations, events belonging to this spatial group continue to appear; they form temporal groups of two events per month. The peak amplitude for the seismic activity of group 1 corresponds to the first peak observed in the number of events vs. the monthly plot of this group. Then seismic activity peaks decrease to a certain value, which is almost constant. This means that in this case events with extensive seismic energy work as a group forming factor. Another case is seen with group #10. There are no preliminary occurrences of grouped events. The rate of grouped event manifestations is maximal at the moment of appearance of the first event of the group. Later a decrease of events appearance rate takes place up to the maximum value of one event per month. After a considerable peak, the seismic activity continues to decrease during the entire observation period. One more type of spatial grouped events is represented by group #14. This is the group of events with the single peak for the number of events as well as for the seismic activity. Another type of group development is represented by group #11. Peaks of nearly equal amplitude obtained with a small number of events (3–4 events) per month are observed in 1986–87 and 1989; seismic activity decreases, and in 1992 an abrupt increase in the rate of events appearance and their energies takes place.

An analysis of the spatial groups distribution in time allows detection of the next characteristic features. More than a half of 41 group forming events took place in the period of seismic activity increase at the end of 1986–beginning of 1987. One or two groups took place during the period of 1988–1990. Seismic activity increase at the end of 1991 again is followed by the increase of the number of group forming seismic events. The duration of the period of existence of the group caused by the strong earthquakes at the end of 1986 is remarkable, because one may think that strong events and their aftershocks should have decreased the stress level in the epicentral zone and the seismic activity in this area should have diminished. We can suggest that such a long activation of the influence zone of the first considered events is caused by the influence of Romashkino hydrocarbon deposit exploitation on the seismic regime.

Figure 8 shows the dependence of the number of seismic events in groups and of the seismic activity in groups on the successive number of group forming events between 1986 and 1992. One can see that both distributions look similar. Four peaks with different amplitudes can be observed. These peaks correspond to the increase of the rate of groups of seismic events appearance, as well as to an increase of the seismic activity. From #1 to #36 group three peaks take place, with the amplitude decreasing from 110 to 28 events per group. Then a sharp peak is observed with the amplitude of 102 events in the group. In seismic activity distribution one can also see four peaks; the first and the fourth peaks are equal to 44,000 and 25,000 $J^{1/3}$, respectively. For the purpose of comparison, a normalized distribution of the seismic activity in time for all the events of 1986–1992 catalog is shown. One can see that the plot of the seismic activity of grouped events versus group numbers has the same character as the time distribution of seismic activity of all events.

The dependencies of the number of events in the group and of the total seismic activity in the group on the successive number of the group have the same character. This can be considered as the consequence of Gutenberg–Richter relation. It is less expected that distribution of seismic activity of all the events in time has the same

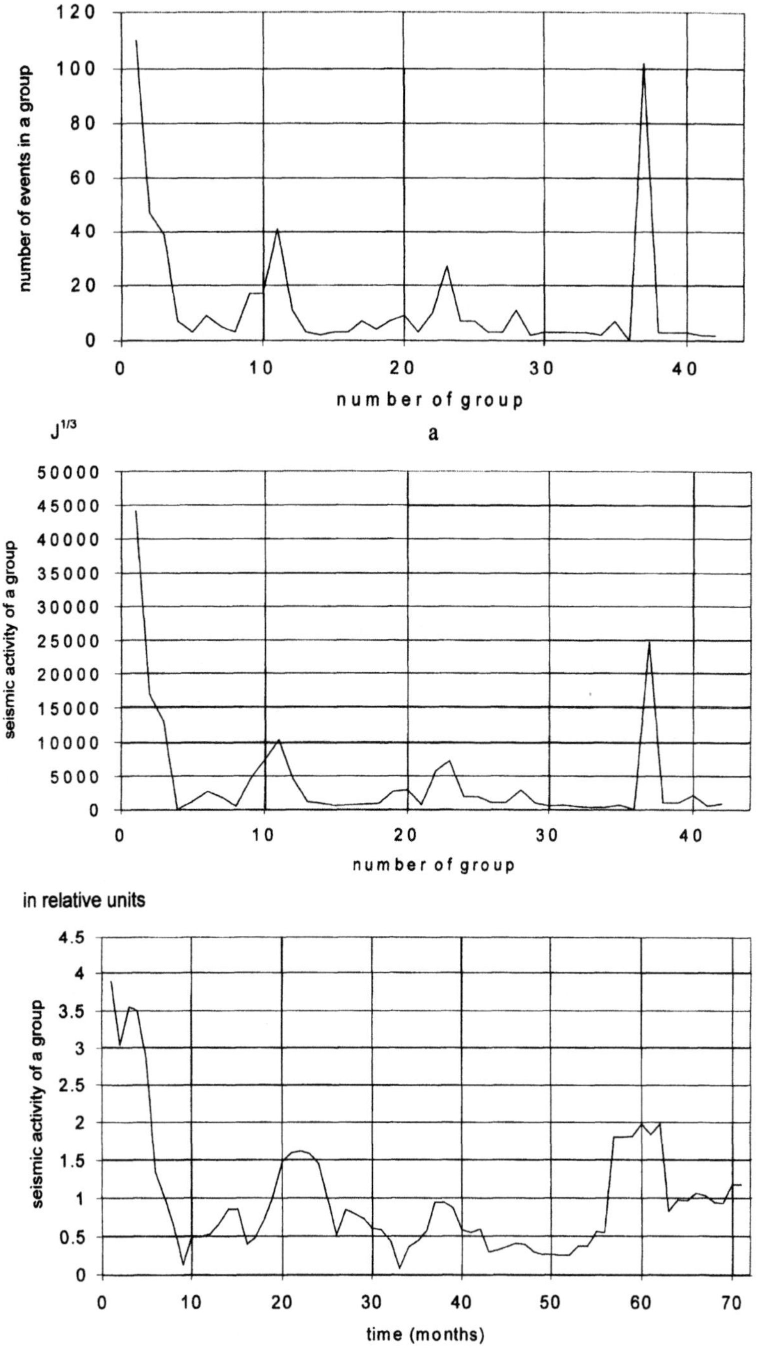

Figure 8
The number of events in group (a) and the seismic activity (b) versus successive numbers of groups; the seismic activity versus time (c) (1986–1992).

character as the distribution of seismic activity of grouped seismic events on the group numbers.

Let us consider in detail dependencies of the seismic activity and the number of events in time of group forming event occurrences. Every point in these plots corresponds to the number of events (or to the seismic activity) during the entire period of observations in a group with a certain number. A comparison of these plots with the plot of total seismic activity variation in time shows that the value of seismic activity of grouped events in the particular time interval defines the total number and energy of grouped events, not only in the considered time interval but also in the future. In other words, integral characteristics of a seismic regime (the number and the total energy of seismic events in the region which size is defined by the energy of grouped events) depends on the value of seismic activity at the moment of the first event occurrence. Consquently, it was found, that for the purpose of evaluating seismic activity variation in time, it is enough to know hypocenter coordinates and depths of the seismic events, energies and sequence (not the exact time) of their occurrences.

Figure 9a indicates distributions of the number of events in groups: histogram of the number of events in the groups and probability density function. These plots were constructed using data of 41 groups. The minimum number of events in a group equals 2, the maximum 110. Probability density function has a maximum at point 3 which means that the most often appearing number of events in a group equals 3. Distribution of the total seismic activity in groups during 1986–1992 is shown in Figure 9b. Minimum seismic activity is equal to 7 $J^{1/3}$ and the maximum to 44096 $J^{1/3}$. Density of seismic activity distribution has a maximum in the point 2000 $J^{1/3}$.

The same study was carried out for the period of 1991–1995. The number of groups for this period is 14. The minimal number of events in the group is 2 events, the maximal –15 events. The same way as during the period 1986–1992, groups consisting of three events are the most frequent. Seismic activity for this period changes from 1711 $J^{1/3}$ to 11022 $J^{1/3}$. Density of seismic activity distribution for the period 1991–1995 has a character similar to the density for the previous period. The maximum of distribution density is in the point 3000 $J^{1/3}$. The fact that the most frequent number of events in the group is equal to 3, can be explained in the following way. As a result of a new fault appearance or sliding along the existing one during the earthquake, a maximal increase in the stress takes place at the end points of the fault. These stress increase zones can become the zones of the next two earthquakes.

Pairs of Seismic Events

The presence of a large number of events following each other within a period less than 24 hours (about 60 events with $M < 1$, which is about 50% of the total number of events of that magnitude) is a characteristic feature of the Romashkino oil field seismicity. This fact supports the grouping of seismic events. One can surmise that

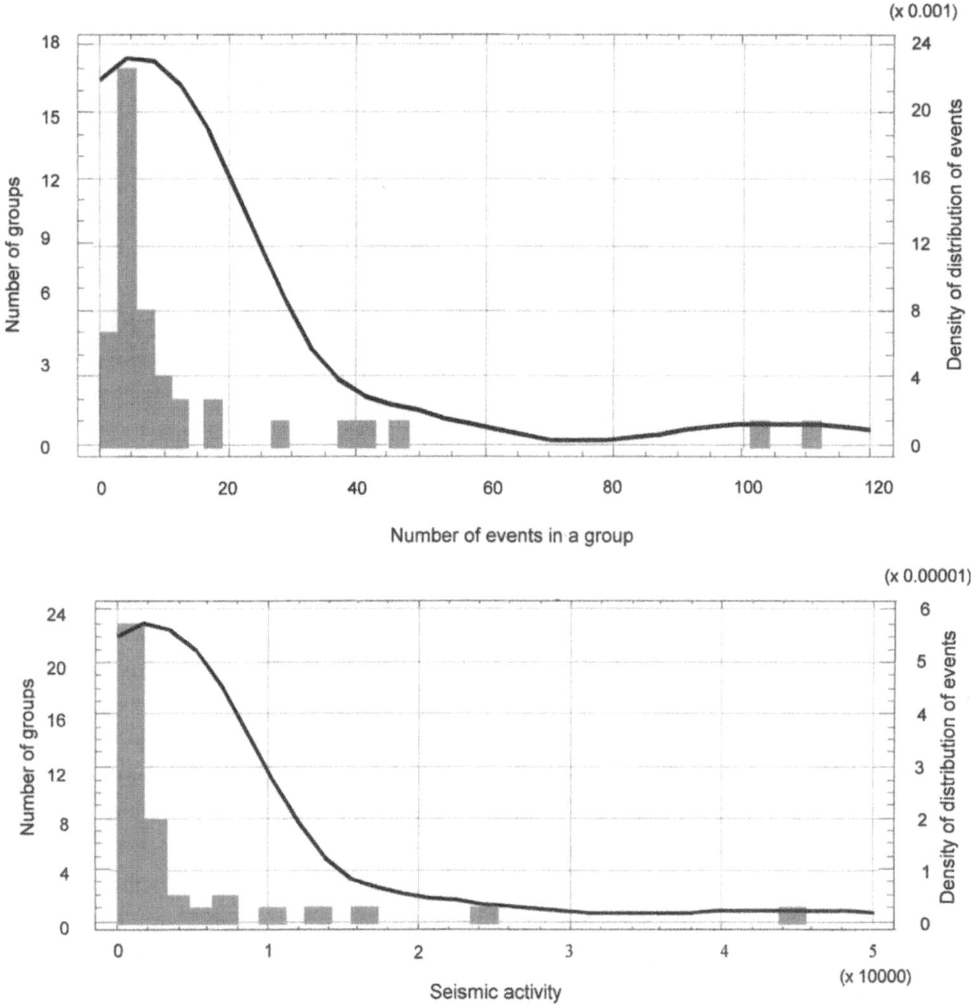

Figure 9
The distributions of the number of seismic events in groups (a) and seismic activities of the groups (b) in 1986–1992. Densities of distributions are shown by solid lines.

events grouped in time are somehow connected in space. For example, research works on laboratory study of seismic signal forming during a crack growth (TURUNTAEV, 1985) indicated, that for certain conditions of crack development, a seismic impulse appears when the crack exits the blocks boundaries. In this case epicenters of pairs of events could define surface projections of the points of stop-phase of movements along faults.

In Figure 10 epicenters of pairs of events are connected by lines. The following features were also shown in Figure 11: areas of seismic activity increase (contours,

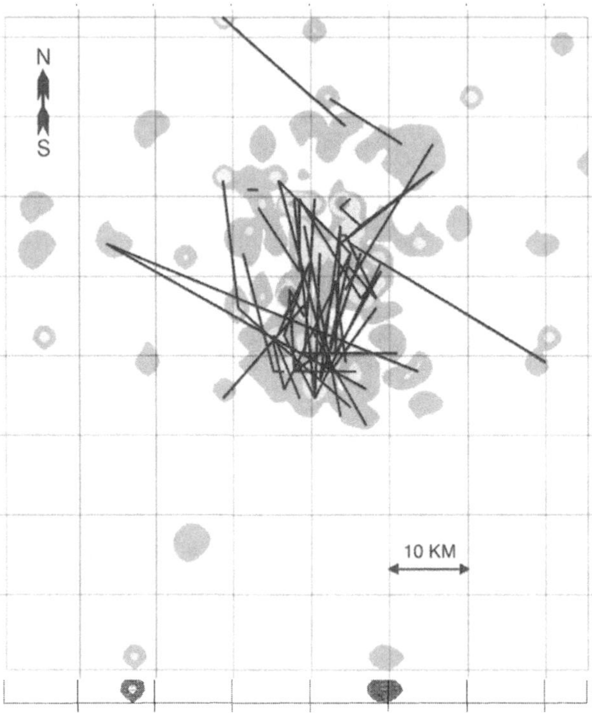

Figure 10
Lines connecting pairs of seismic events in the region of the Romashkino oil field. Contours of seismic activity are shown.

characterizing release of seismic energy per area unit), tectonic faults, Altunino–Shunaksky depression structure (dashed lines). One can see in Figure 10 that some of the lines are parallel to each other. Lines tend to group themselves, the main directions of which can also be observed. A high number of lines, connecting pairs of events, is related to the longitudinal Altunino–Shunaksky depression structure, dividing the deposit into two parts. Figure 11 shows the distribution of the directions of lines, connecting epicenters of pairs of events. One can see that the distribution has a clear maximum, corresponding to a longitudinal direction.

The above observation allows the notation, that pairs of events appear in the active fault zone, related with the Altunino–Shunaksky depression structure. Direction of the lines, connecting pairs of event epicenters, corresponds to the model of the regional stress field (BELOUSOV *et al.*, 1994).

Correlation Between Seismic Activity and the Process of Deposit Exploitation

Data on the monthly volumes of extraction and injection in the four most seismically active areas of the Romashkino oil field (Almetjevskaja, Severoalmetjevskaja, Minibaevskaja and Berezovskaja) for the period 1981 to 1992 were

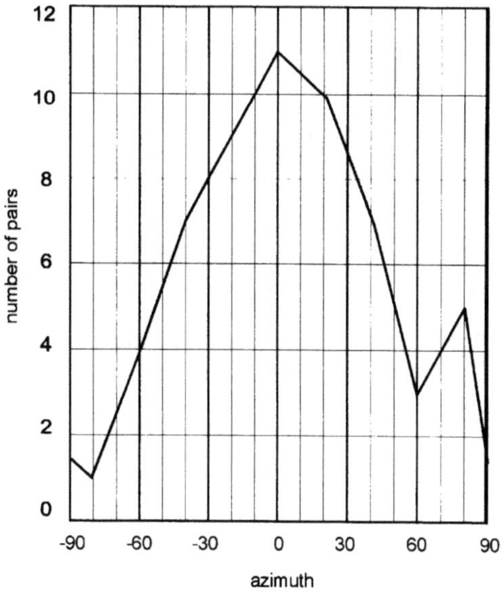

Figure 11
Azimuth distribution of lines connecting pairs of seismic events.

compared to the seismicity. Using these data, a pseudo-catalog of the extracted and injected volumes, and imbalance (a difference between the values of injected and extracted fluids) was made: monthly value was assigned a date (middle of a month), time (middle of a day), coordinates (approximately center of the considered area), a depth (1 km). The above described procedure for seismic activity calculation was applied to this pseudo-catalog and raw values were averaged over six months moving windows; these are shown in the bottom part of Figure 12. All presented values were normalized on their arithmetic averages for the considered period of time. The plots of the changes of monthly injection and extraction volumes (in normalized values) and imbalance between injections and extractions on the studied areas are shown together in Figure 12a; the same values smoothed by moving the 6-months window are shown in Figure 12b. Figure 13 displays the changes of the ratio of the volumes of extracted fluid to the volumes of injected water ("injection effectiveness") for the four areas separately. Change of seismic activity (since 1986) in the region of the Romashkino oil field in general is also shown. A certain inverse correlation between the oscillations of seismic activity and effectiveness of injecting wells can be observed.

To some extent, the characteristics of the temporal variations of injection effectiveness are related to the conversion in 1986 to a new method of fluid injections in cycles: some of the injection wells are switched off from time to time (nonstationary injection). During winter, due to the danger of freezing, all the wells are switched on, therefore the injected volumes in summer were decreased in comparison with the winter (which causes a seasonal component in effectiveness

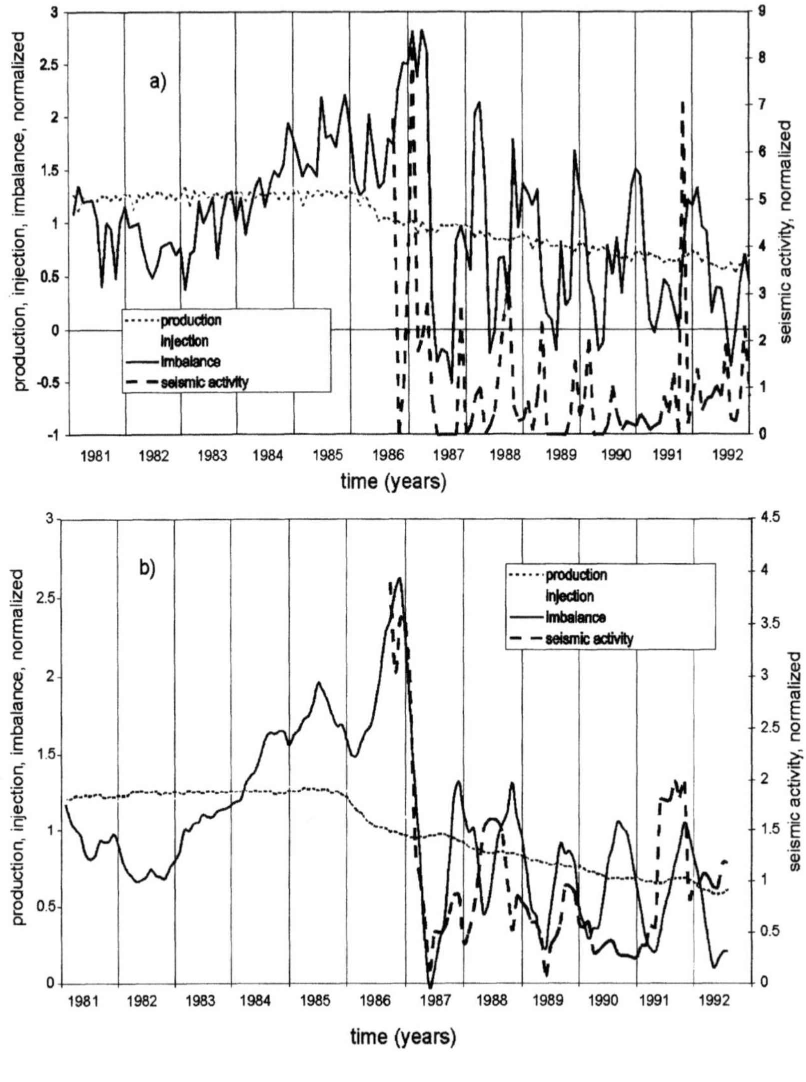

Figure 12
Injection, production volumes and imbalance in four areas of the Romashkino oil-field in comparison with variations of seismic activity: a) – monthly values; b) – smoothed by 6-month moving window.

oscillations). At the same time it is impossible to consider all the variations of the studied injection effectiveness as if caused by technology reasons.

In Figure 14 comparisons of the seismic activity changes in the Almetjevskaja, Severoalmetjevskaja, Minibaevskaja and Berezovskaja areas with the extracted and injected fluid volumes ratios are shown. For these plots the areas were approximated to rectangles.

In Figure 15 the seismic activity in the Almetjevskaja area is compared with changes in production and injection. An increase of injection volumes, which took

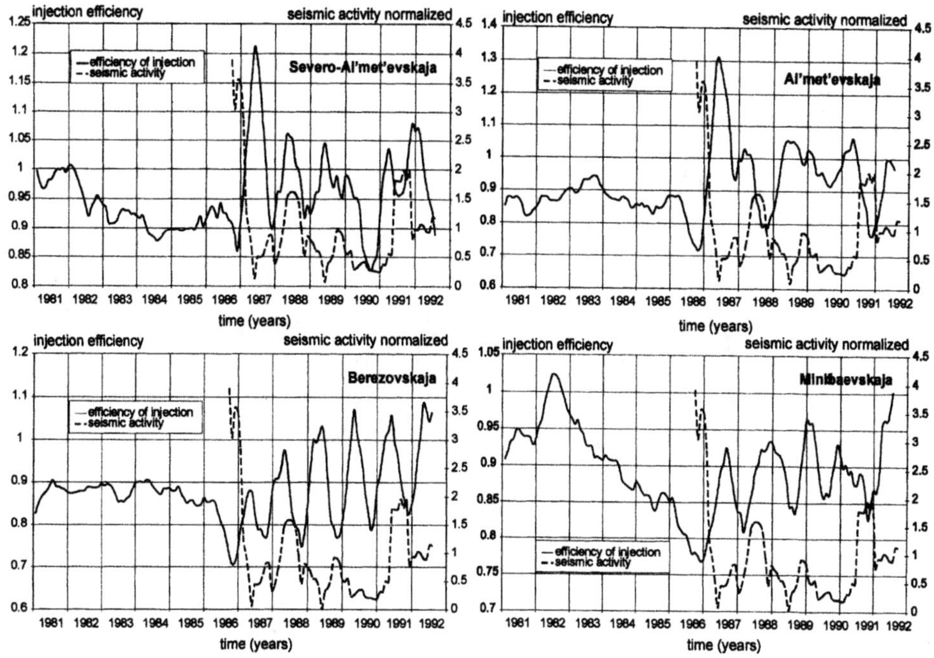

Figure 13

Variations of production to injection ratio for different areas of the Romashkino oil-field a) – Severo-Al'met'evskaja; b) – Al'met'evskaja; c) – Berezovskaja; d) – Minibaevskaja. Change of seismic activity is shown for the entire Romashkino region.

place some time before the two biggest increases in seismicity at the beginning and at the end of the studied time period on Almetjevskaja area, attracts attention. It is remarkable that production decreases during the periods in which the seismic activity increases, even when the injection increases (it can be seen for the first peak of the seismic activity). In due time, even weaker increases of the seismic activity were always accompanied with a decrease of total fluid production in the Almetjevskaja area.

It is interesting that, for example, in the Minibaevskaja area during an increase of seismic activity in 1991–92, both injection and production increase but an injection efficiency decreases at the same time. Regression analysis combined with a one-way analysis of variance reveals a statistically significant relation between the seismic activity variations in the four studied areas and production and injection regimes in the areas. Confidence level of the relationship is 99%.

Figure 16 shows cross-correlation coefficients of correlation between seismic activity on the four most seismically active areas of the Romashkino oil field and characteristics describing exploitation processes, such as extracted and injected fluid volume, imbalance (a difference between injected and extracted volume), injections effectiveness (a ratio of the extracted and injected volume). During the studied period a monotonous decrease of the injected fluids as well as the decrease of produced oil

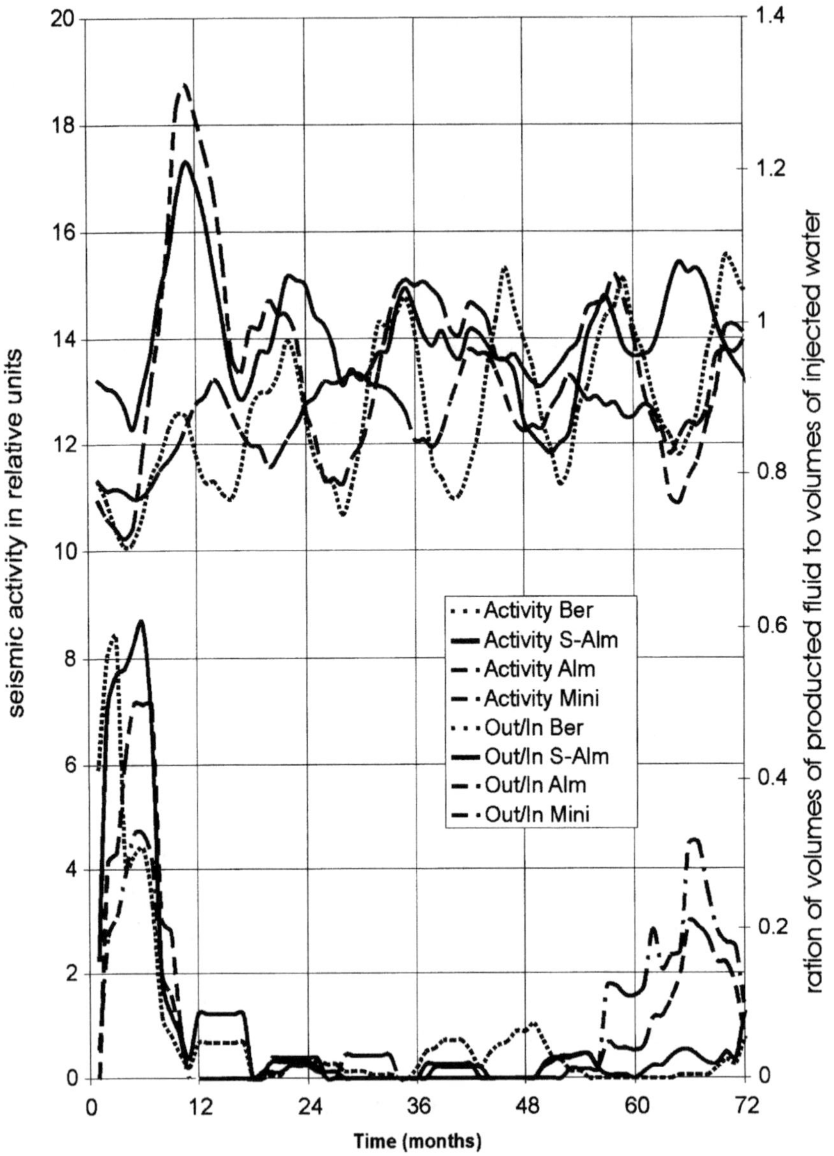

Figure 14
A comparison of the seismic activity changes in the Almetjevskaja, Severoalmetjevskaja, Minibaevskaja and Berezovskaja areas with the extracted to injected fluids volume ratio.

took place, which was caused by economical reasons. That is why Figure 16 also contains data on the correlation between seismic activity and the above values detrended (a linear trend was removed from the values).

It is remarkable that a correlation between seismic activity and exploitation parameters is high; and it is observed not only for a correlation between seismic activity and exploitation of the same area (for example a correlation between

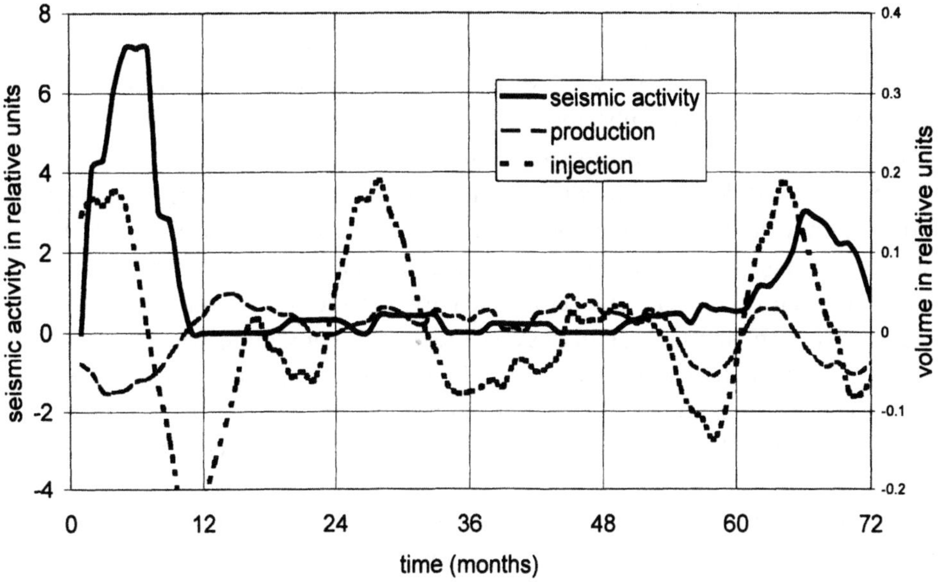

Figure 15
A comparison of the seismic activity changes in the Almetjevskaja area with the production and injection.

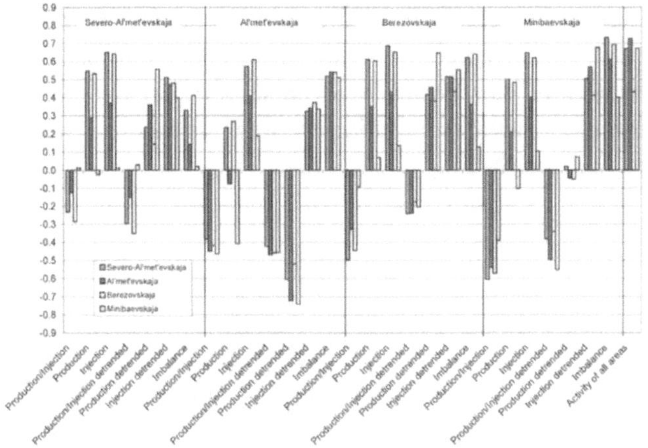

Figure 16
Correlation between the seismic activity and the parameters of exploitation.

seismicity in the Almetjevskaja area and injection in the Almetjevskaja area), but in several cases for correlation between seismic activity and exploitation parameters of different areas. Sometimes it can be explained by an overlap of rectangles approximating the areas of the Romashkino oil field, however the overlap took place for neighboring areas only. The highest absolute values of the correlation are observed for a correlation between the seismic activity in the Almetjevskaja area and detrended time series of production in the same area, between the seismic activity in the Minibaevskaja area (which is near Almetjevskaja) and detrended production in the Almetjevskaja area, between seismic activity in the Severoalmetjevskaja and Berezovskaja areas and imbalance in the Minibaevskaja area. Absolute values of these correlation coefficients are greater than 0.7.

We note that detrended production in the Almetjevskaja area only has a negative correlation with seismic activity on all the studied areas but its correlation coefficient is high enough. Detrended production in the Severoalmetjevskaja and Berezovskaja areas correlates positively with seismic activity, but the correlation coefficient is less than 0.5 (except for the Minibaevskaja area). The correlation between detrended production and detrended injection and imbalance is always positive and high enough. The correlation between detrended production and the injection efficiency is always negative, but the coefficient of the correlation reaches 0.5 in the Minibaevskaja area only.

A change of correlation coefficients (between seismic activity and detrended production, injection, imbalance and injection efficiency), due to a shift of data series in time relative to each other, was studied. The plot is shown in Figure 17. Positive lags correspond to positive shifts of seismic activity series in time relative to others data series. The most interesting are the plots for the Minibaevskaja and Almetjevskaja areas. Considering these plots one can see that changes in exploitation parameters precede changes in seismic activity. Maximal correlation is observed when lags are positive and equal to 1–2 months. Correlation coefficients reach 0.8 for the Minibaevskaja area (correlation between seismic activity and injection) and 0.7 for the Almetjevskaja area (correlation between seismic activity and imbalance, seismic activity and production).

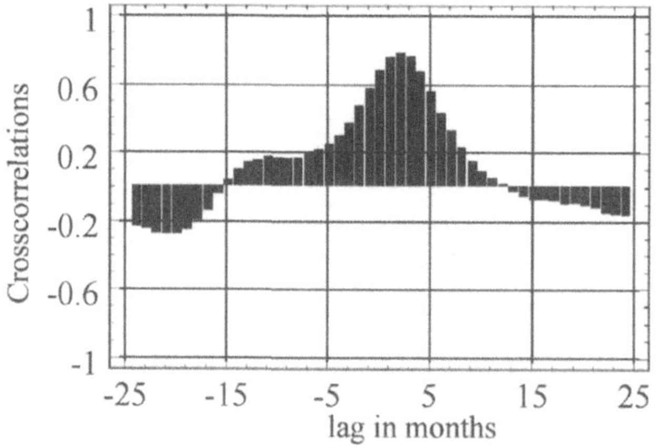

Figure 17
A change of correlation coefficients between the seismic activity and detrended injection due to a shift of data series in time relative to each other for the Minibaevslaya area.

Conclusion

A relation between seismic deformation processes activity in the region of a hydrocarbon field and parameters of hydrocarbon recovery was studied. It was shown that an analysis of data on temporal and spatial characteristic features of seismic activity with the help of modern methods allow collection of useful information on deformation processes taking place in reservoir and surrounding rocks. The information is extremely valuable not only for a well-timed forecast of a hazardous increase of seismic activity, but also for evaluation of temporal and spatial variations of fluid dynamics' characteristics of reservoir, of reservoir reaction on recovery and secondary recovery methods, for detection of fluid migration and so on. It was shown that the efficiency of water injection for oil recovery in the Romashkino oil field decreased during an increase of seismic activity; it increased when seismic activity became low. Grouping in time seismic events helps us to detect zones of active modern fracturing, which are zones of high permeability.

Furthermore, it was demonstrated that phase portrait of the Romashkino oil field seismicity resembles a phase portrait which can be obtained as a result of the Duffing equation solution. This result allows us to suggest that oscillations under action of periodical outer force can be considered as a mathematical model of the seismic processes in the studied region. An unstable injection (a cycling one) used in the Romashkino oil field seismic active areas can be regarded as the outer force.

It is very important to install a local permanent seismic network in advance. This study shows that to estimate the values of temporal and spatial parameters of seismic and deformation processes in the region of hydrocarbon fields, it is necessary to record the data during 1–1.5 years. In order to detect a dangerous seismic activity increase early enough, it is essential to maintain continuous seismic monitoring. The problem of oil company responsibility regarding earthquakes in the region of a hydrocarbon field can be resolved only if there are enough prolonged series of seismic activity recordings there. At the same time an oil company which installs a seismic network will have an additional source of direct data on deformation processes taking place in the hydrocarbon field during its development, and will have a high potential to improve oil recovery.

Acknowledgements

The published results were received based on data obtained by the local seismic network of Stock Joint Company "Tatneft". The authors would like to thank I.A. Iskhakov, head of seismic crew TNGF, and K.M. Mirzoev, chief of Tatarstan seismic survey, who provided us with the catalogue of seismic events and with the produced and injected fluid volume data. We would also like to thank Stephane Nechtschein and an anonymous reviewer for their elaborate comments and suggestions which assisted us in improving significantly the original manuscript. The support from "Tatneft" and the Russian Foundation for Basic Research (RFBR project # 98-05-64547) is gratefully acknowledged. The study is part of a cooperative project between Schlumberger and Institute for Dynamics of Geospheres, Russian Academy of Sciences.

References

BELOUSOV, T. P., MUHAMEDIEV, Sh. A., TURUNTAEV, S. B., YUNGA, S. L., ISKHAKOV, I. A., and KARAKIN, A. V., *Active faults, stresses state and seismicity of south-east Tatarstan.* In *Seismicity and Seismic Zones of Northern Part of Eurasia. Part 2* (ed. Ulomov, Moscow, Nauka, 1994), pp. 90–108. (in Russian).

Directions for Data Acquisition and Data Processing at Seismic Stations of the USSR United Seismic Network (Moscow, Nauka, 1981) (in Russian).

Geological Conditions of Oil and Gas Accumulation and Location (ed. Bakirov A. A., Moscow, Nedra, 1982) (in Russian).

GRASSO, J.-R. (1992), *Mechanics of Seismic Instabilities Induced by the Recovery of Hydrocarbons*, Pure appl. geophys., *139*, 3/4, 507–534.

GAUPTA, H. and RASTOGI, B., *Dam Earthquakes* (Moscow, Mir, 1979) (in Russian).

GORBUNOVA, I. V. and KAL'MET'EVA, Z. A., *Experimental Characteristics of Seismic Waves Generation in Weak Earthquakes Zones* (Frunze, Ilim, 1988) (in Russian).

GUTENBERG, B. and RICHTER, C. F. (1956), *Earthquake Magnitude, Intensity, Energy and Acceleration* (second paper). *Bull. Seismol. Soc. Am. 46*, 105–145.

HAKEN, H., *Advanced Synergetics.* (Berlin, Heidelberg, Springer-Verlag, 1983).

ISKHAKOV, I. A., *A Study of Seismic Regime in the South-east of Tatarstan Aimed to Reveal the Nature of Seismicity in the Region. A report of OMP 50/88* (Bugulma, 1988) (in Russian).

ISKHAKOV, I. A., SERGEEV, N. S., and BULGAKOV, V. YU., *A Study of Relation between Seismicity and Oil Fields Development. A Report of OMP 50/81* (Bugulma, 1992) (in Russian).

KASAHARA, Earthquakes mechanics (Cambridge University Press, 1981).

MILNE, W. G. and BERRY, M. J. (1976), *Induced Seismicity in Canada*, Eng. Geol. *10*, 219–226.

MIRZOEV, K. M., NEGMATULLAEV, S. H., and DASTURY, T. Yu., *An influence of mechanical vibrations on characteristic features of seismic energy release in the region of Nurec dam.* In *Seismological Investigations in the Regions of Large Dam Constructions in Tadjikistan* (Dushanbe, Donish, 1987), pp. 101–119 (in Russian).

MUSLIMOV, R. H., *An Influence of Geological Structures Distinguish Features on an Efficiency of Romashkino Oil Field Development* (Kazan, KSU, 1979) (in Russian).

MUSLIMOV, R. H., SHAVALIEV, A. M., HISAMOV, R. B., and YUSUPOV, I. G., *Geology, Development and Exploitation of Romashkino Oil Field* (in 2 volumes) (Moscow, VNIIOENG, 1995) (in Russian).

NICHOLSON C. and WESSON, R. L., *Earthquake Hazard Associated with Deep Well Injection – A Report to the U.S. Environmental Protection Agency* (U.S. Geological Survey Bulletin, 1991).

NIKOLAEV, N. I. (1977), *On the state of study of induced earthquakes, related with industrial activity.* In *An Influence of Industrial Activity on the Seismic Regime* (Moscow, Nauka, 1977), pp. 8–21 (in Russian).

PASECHNIK, I. P. (1977), *Earthquakes induced by underground nuclear explosions.* In *An Influence of Industrial Activity on the Seismic Regime* (Moscow, Nauka), pp. 142–152 (in Russian).

PONOMARYOV, V. S. and TEJTELBAUM, U. M., *Dynamics interactions between earthquakes focuses.* In *Regional Investigations of Seismic Regime* (Kishinev, Shtinitsa, 1974), pp. 79–92 (in Russian).

PRIGOGINE, I., *From Being to Becoming: Time and Complexity in the Physical Sciences* (W. H. Freeman and Company, San Francisco, 1980).

RODIONOV, V. N., SIZOV, I. A., and KOCHARYAN, G. G., *On the modeling of natural objects in geomechanics.* In *Discrete Properties of Geophysical Medium* (Moscow, Nauka, 1989), pp. 14–18 (in Russian).

RODIONOV, V. N., SIZOV, I. A., and TSVETKOV, V. M., *Foundations of Geomechanics* (Moscow, Nedra, 1986) (in Russian).

SADOVSKY *et al.* (1987), *Characteristic Dimensions of Rocks and Hierarchical Properties of Seismicity.* Physics of the Earth *2*, 3–15 (in Russian)

SADOVSKY, M. A., KOCHARYAN, G. G., and RODIONOV, V. N. (1988), *On the Mechanics of Block Rock Massif*, Report of Academy of Sciences of USSR *302*, 2, 193–197 (in Russian).

SADOVSKY, M. A. and PISARENKO, V. F., *Seismic Process in Block Media* (Moscow, Nauka, 1991) (in Russian).

SIMPSON, D. W. (1986), *Triggered Earthquakes*, Ann. Rev. Earth and Planet Sciences *14*, 21–42.

TURUNTAEV, S. B., *A Study of Various Model Wave Sources in Application to Seismology, Ph.D. Thesis* (Moscow, IPKON, 1985) (in Russian).

TURUNTAEV, S. B. (1993), *An Application of the Weak Seismicity Catalog for Detection of Active Faults in Rock Massif. The Monitoring of Rock Massif State during Long-time Exploitation of Large-scale Underground Works* (Appatity) (in Russian).

TURUNTAEV, S. B. (1994), *Development of Methods for Seismically Hazardous Blocks Detection and for Revealing of Tendencies of Rockbursts Hazard Increase, Methodological Principles of Control of Rock Massif State and Seismic Events Forecasting* (Moscow) (in Russian).

(Received March 31, 2000, revised/accepted December 22, 2000)

Pure appl. geophys. 159 (2002) 449–471
0033–4553/02/030449–23 $ 1.50 + 0.20/0

Pure and Applied Geophysics

Application of Kalman Filtering Techniques for Microseismic Event Detection

ERICK BAZIW[1] and IAIN WEIR-JONES[1]

Abstract—Microseismic monitoring systems are generally installed in areas of induced seismicity caused by human activity. Induced seismicity results from changes in the state of stress which may occur as a result of excavation within the rock mass in mining (i.e., rockbursts), and changes in hydrostatic pressures and rock temperatures (e.g., during fluid injection or extraction) in oil exploitation, dam construction or fluid disposal. Microseismic monitoring systems determine event locations and important source parameters such as attenuation, seismic moment, source radius, static stress drop, peak particle velocity and seismic energy. An essential part of the operation of a microseismic monitoring system is the reliable detection of microseismic events. In the absence of reliable, automated picking techniques, operators rely upon manual picking. This is time-consuming, costly and, in the presence of background noise, very prone to error. The techniques described in this paper not only permit the reliable identification of events in cluttered signal environments they have also enabled the authors to develop reliable automated event picking procedures. This opens the way to use microseismic monitoring as a cost-effective production/operations procedure. It has been the experience of the authors that in certain noisy environments, the seismic monitoring system may trigger on and subsequently acquire substantial quantities of erroneous data, due to the high energy content of the ambient noise. Digital filtering techniques need to be applied on the microseismic data so that the ambient noise is removed and event detection simplified. The monitoring of seismic acoustic emissions is a continuous, real-time process and it is desirable to implement digital filters which can also be designed in the time domain and in real-time such as the Kalman Filter. This paper presents a real-time Kalman Filter which removes the statistically describable background noise from the recorded seismic traces.

Key words: Microseismic monitoring, Kalman Filter, state-space, automated event detection, ambient noise.

1. Introduction

Microseismic monitoring systems are used in mining rock mechanics, civil construction, petroleum extraction, and heavy structural engineering. Extreme examples of energy release can cause violent rockbursts which result in fatalities and injuries among underground personnel and damage to mine structures (e.g., drifts, stopes, shafts, etc.). Microseismic systems are capable of detecting rock failures in the

[1] Weir-Jones Engineering Consultants Ltd., 2040 West 10th Ave., Vancouver, B.C. Canada, V6J 2B3. E-mail: wjgroup@weir-jones.com

vicinity of underground excavations caused by the sudden release of strain energy resulting from the redistribution of stresses around openings.

Various hydrocarbon production sites also benefit from seismic monitoring systems during certain phases of production. Primary or secondary extraction or the injection of material into the reservoir to enhance production can cause significant stress changes. These stress changes can result in failures of the overlying strata and the migration of hydrocarbons to aquifers or to the ground surface. Thus microseismic monitoring can be used to satisfy environmental concerns, meet regulatory requirements and assess the development of induced fracturing within the reservoir. In addition, the microseismic monitoring systems have been successful in identifying and locating casing failures due to steam stimulation in oil sands (TALEBI *et al.*, 1998).

During filling of hydroelectric or large irrigation reservoirs, changes in regional loading and pore pressures cause significant stress variations within the surrounding rock mass. These can induce a wide range of micro- and macroseismic events, some of which are capable of causing damage to adjacent structures or to the dam itself. Microseismic monitoring can locate and characterize these potentially hazardous induced events.

Irrespective of the cause of microseismic events, their reliable detection and placement on a common time base is of critical importance. This is because the arrival times of wavelets at various detector packages within a three-dimensional array provide the basis for the calculation of the location of the microseismic event. Imprecision or uncertainty in arrival time determination reduces the precision of the source location operation. In many microseismic monitoring situations – for example where there is interest in the behaviour of specific geological features or where events must be related to specific structures in a mining or hydrocarbon extraction environment – the accurate determination of event arrival times is the primary rationale for the installation of the system. Hence, techniques such as the one described in this paper open the way for the more widespread use of microseismic monitoring in a number of industrial applications.

The environmental and seismic models used in this paper are designed to fit into a Kalman Filter (KF) formulation. The Kalman Filter is an optimal (in a least-squares sense) recursive filter which is based on state-space, time-domain formulation of physical problems, thereby avoiding difficulties commonly associated with frequency domain filters (BAZIW, 1993), and which is ideal for real-time applications. The KF requires that the physical problem be modelled by a set of first-order differential equations which, with initial conditions, uniquely define the system behaviour. The filter utilizes a knowledge of system and measurement errors and statistical information about the initial conditions (BAZIW, 1988). Section 2 presents the KF microseismic event detection formulation.

The performance of the KF presented in this paper was first evaluated by generating synthetic source wavelets which have correlated earth noise added.

Section 3 presents results from the seismic data simulation and processed real data obtained from a heavy oil extraction facility in Northern Canada.

2. Kalman Filter Formulation

The Kalman Filter is an optimal (least-squares) recursive filter which is based on state-space, time-domain formulations of physical problems. Application of this filter requires that the physical problem be modified by a set of first-order differential equations which, with initial conditions, uniquely define the system behaviour. The filter utilizes knowledge of system and measurement dynamics, assumed statistics of system noises and measurement errors and statistical information regarding the initial conditions. Figure 1 illustrates the essential relation between the system, the measurements and the Kalman Filter.

Figure 1 indicates the scope of information the KF takes into account. As can be seen, the statistics of the measurement and state errors are essential components of the filter. The *a priori* information provides for optimal use of any number, combination and sequence of external measurements. The KF can be applied to problems with linear time-varying systems and with non-stationary system and measurement statistics. Problems with nonlinearities can be handled by linearizing the system and measurement equations. The Kalman Filter is readily applied to estimation, smoothing and prediction.

From the published technical literature, it seems that geophysical data processing has been primarily carried out by the application of frequency domain and steady-state filters such as the Weiner Filter (KANASEWICH, 1981). These filters put many restrictions on the structure of the problem. In general, one assumes non-time-varying system equations (where the coefficients of the system differential equation are constant), and statistically stationary error processes (GELB, 1978).

If the geophysical problem can be structured into one susceptible to the application of the Kalman Filter, it is possible that better state and system

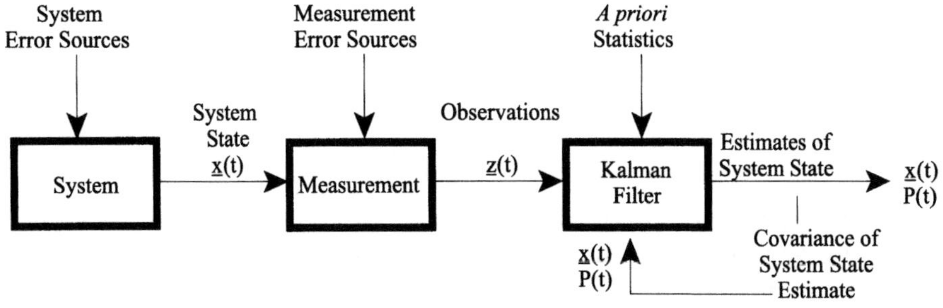

Figure 1
Block diagram of system, measurement, and Kalman Filter.

parameters would be obtained as compared to those derived from use of the frequency domain and steady-state filters. In addition, the KF is ideally suited for real-time optimal data filtering in situations where each new measurement provides for an updated state estimate. This feature of the KF is particularly useful for microseismic event detection. Another advantage of a Kalman Filter is that it allows one to use "state noise" to compensate for errors in the mathematical model. The use of state noise causes the filter to apply less weight to measurements made in the distant past, and to apply more weight to state vector estimates based on more recent measurement data.

BAYLESS and BRIGHAM (1970) were among the first researchers to fit geophysical problems into the Kalman Filter formulation. In addition, MENDEL (1983, 1995) has carried out extensive work in fitting geophysical problems into state-space representations.

2.1 Standard Kalman Filter Governing Equations

The KF is a method for estimating a state vector $\underline{x}$ from measurement $\underline{z}$. The state vector may be corrupted by a noise vector $\underline{w}$ and the measurement vector is corrupted by a noise vector $\underline{v}$. The KF filter is applicable for systems that can be described by a first-order differential equation in $\underline{x}$ and a linear (matrix) equation in $\underline{z}$. The KF can be described in both continuous and discrete form. The continuous state and measurement equations are given by

$$\dot{\underline{x}}(t) = F(t)\underline{x}(t) + G(t)\underline{w}(t) \tag{1}$$

$$\underline{z}(t) = H(t)\underline{x}(t) + \underline{v}(t) \tag{2}$$

where $\underline{x}$ is an n-vector, $\underline{w}$ is a p-vector, and $\underline{z}$ and $\underline{v}$ are m-vectors.

The random (vector) processes $\underline{w}$ and $\underline{v}$ are assumed to be zero mean, white noise processes, so that

$$\begin{gathered} E[\underline{w}(t)] = \underline{0} \quad \text{and} \quad E[\underline{v}(t)] = \underline{0}, \\ E\left[\underline{w}(t)\underline{w}(\tau)^T\right] = Q(t)\delta(t-\tau), \\ E\left[\underline{v}(t)\underline{v}(\tau)^T\right] = R(t)\delta(t-\tau). \end{gathered} \tag{3}$$

It is further assumed that $\underline{w}$ and $\underline{v}$ are statistically independent of each other, so that

$$E\left[\underline{w}(t)\underline{v}(\tau)^T\right] = \underline{0} \tag{4}$$

In the above equations the superscript T denotes the vector transpose, the E denotes the expected value operation and $\delta(t-\tau)$ is the Dirac delta function.

We use the notation $\hat{\underline{x}}(t'|t)$ to denote an estimate of the state at a time $t', t' \geq t_0$, based on measurements $\underline{z}(\tau)$ on the interval $t_0 \leq \tau \leq t$. It is convenient to make the following definitions:

Definition:

1. If $t' \geq t, \hat{\underline{x}}(t'|t)$ is a *predicted estimate.*
2. If $t' = t, \hat{\underline{x}}(t'|t)$ is a *filtered estimate.*
3. If $t' < t, \hat{\underline{x}}(t'|t)$ is a *smoothed estimate.*

The *estimation error* is defined as

$$\tilde{\underline{x}}(t'|t) \equiv \underline{x}(t') - \hat{\underline{x}}(t'|t) \tag{5a}$$

and the *performance index* of the estimation is given by

$$I[\tilde{\underline{x}}(t'|t)] = E\left[\tilde{\underline{x}}(t'|t)\tilde{\underline{x}}(t'|t)^T\right]. \tag{5b}$$

The estimation problem can now be stated as follows:

Estimation Problem: Given the system defined by eqs. (1) and (2) with statistics defined by (3) and (4), and measurements z(τ) over the interval $t_0 \leq \tau \leq t$, determine an estimate of $\hat{\underline{x}}(t'|t)$ such that $I[\hat{\underline{x}}(t'|t)]$ is minimized.

The solution to this problem is specified by the following equations for the filtered estimate:

State Estimation equation

$$\dot{\hat{\underline{x}}}(t) = F(t)\hat{\underline{x}}(t) + K(t)[\underline{z}(t) - H(t)\hat{\underline{x}}(t)], \tag{6}$$

with

$$\hat{\underline{x}}(t_0) = \hat{\underline{x}}_0$$

given

Kalman Gain matrix

$$K(t) = P(t)H(t)^T R(t)^{-1} \tag{7}$$

Estimation Error Covariance (Matrix Ricatti) equation

$$\dot{P}(t) = F(t)P(t) + P(t)F(t)^T + G(t)Q(t)G(t)^T - K(t)R(t)K(t)^T, \tag{8}$$

with

$$P(t_0) = P_0$$

given.

The error covariance matrix is a positive semi-definite symmetric matrix. A block diagram of the system, measurement and filtered estimates is shown in Figure 1.

Using the notation of GELB (1978), the corresponding discrete state and measurement equations are given by

$$\underline{x}_k = \Phi_{k-1}\underline{x}_{k-1} + \underline{w}_{k-1}, \quad \underline{w}_k \approx N(\underline{0}, Q_k) \tag{9}$$

$$\underline{z}_k = H_k\underline{x}_k + \underline{v}_k, \quad \underline{v}_k \approx N(\underline{0}, R_k). \tag{10}$$

In eqs. (9) and (10), symbol N denotes a normal distribution with mean $\underline{0}$ and variance Q_k and R_k, respectively. The discrete Kalman Filter estimation equations are outlined as follows:
State Estimate Extrapolation:

$$\hat{\underline{x}}_k(-) = \Phi_{k-1}\hat{\underline{x}}_{k-1}(+) \tag{11}$$

Error Covariance Extrapolation:

$$P_k(-) = \Phi_{k-1}P_{k-1}(+)\Phi_{k-1}^T + Q_{k-1} \tag{12}$$

State Estimate Update:

$$\hat{\underline{x}}_k(+) = \hat{\underline{x}}_k(-) + K_k[\underline{z}_k - H_k\hat{\underline{x}}_k(-)] \tag{13}$$

Error Covariance Update:

$$P_k(+) = [I - K_kH_k]P_k(-). \tag{14}$$

I is the identity matrix in eq. (14).
Kalman Gain Matrix:

$$K_k = P_k(-)H_k^T\left[H_kP_k(-)H_k^T + R_k\right]^{-1} \tag{15}$$

Initial Conditions:

$$E[\underline{x}_0] = \hat{\underline{x}}_0, \quad E\left[(\underline{x}_0 - \hat{\underline{x}}_0)(\underline{x}_0 - \hat{\underline{x}}_0)^T\right] = P_0. \tag{16}$$

The computational sequence for the discrete KF is outlined as follows:
A. At time index $k = 0$, specify initial conditions $\hat{\underline{x}}_0$, and P_0, and compute Φ_0 and Q_0.
B. At time index $k = 1$, compute $\hat{\underline{x}}_1(-), P_1(-), H_1, R_1$, and the gain matrix K_1
C. Using the measurement $\underline{z}_1$ at time index $k = 1$, the best estimate of the state at $k = 1$ is given by

$$\hat{x}_1(+) = \hat{x}_1(-) + K_1[z_1 - H_1\hat{x}_1(-)]$$

D. Update the error covariance matrix $P_1(+)$
E. At time index $k = 2$, a new measurement $\underline{z}_2$ is obtained and the computational cycle is repeated.

2.2 Microseismic Event Detection Kalman Filter

As was stated previously, the most important factor when applying the KF is that the physical problem must be modelled/approximated by a set of first-order differential equations. In addition, the statistical properties of the system and measurements need to be modelled such that they can be completely specified by first- and second-order statistics (i.e., expected values and covariances).

The microseismic seismic time series is typically comprised of ambient noise and seismic *P* waves and *S* waves. The first step in defining the Microseismic Event Detection Kalman Filter (MEDKF) is to provide a mathematical description of the ambient background noise.

2.2.1 MEDKF ambient noise model

The authors calculated the autocorrelation and power spectrum of a large number of microseismic ambient noise time series. From these results, it was possible to identify a mathematical model which sufficiently fits the seismic noise processes. Figure 2 illustrates the autocorrelation function and power spectral density of

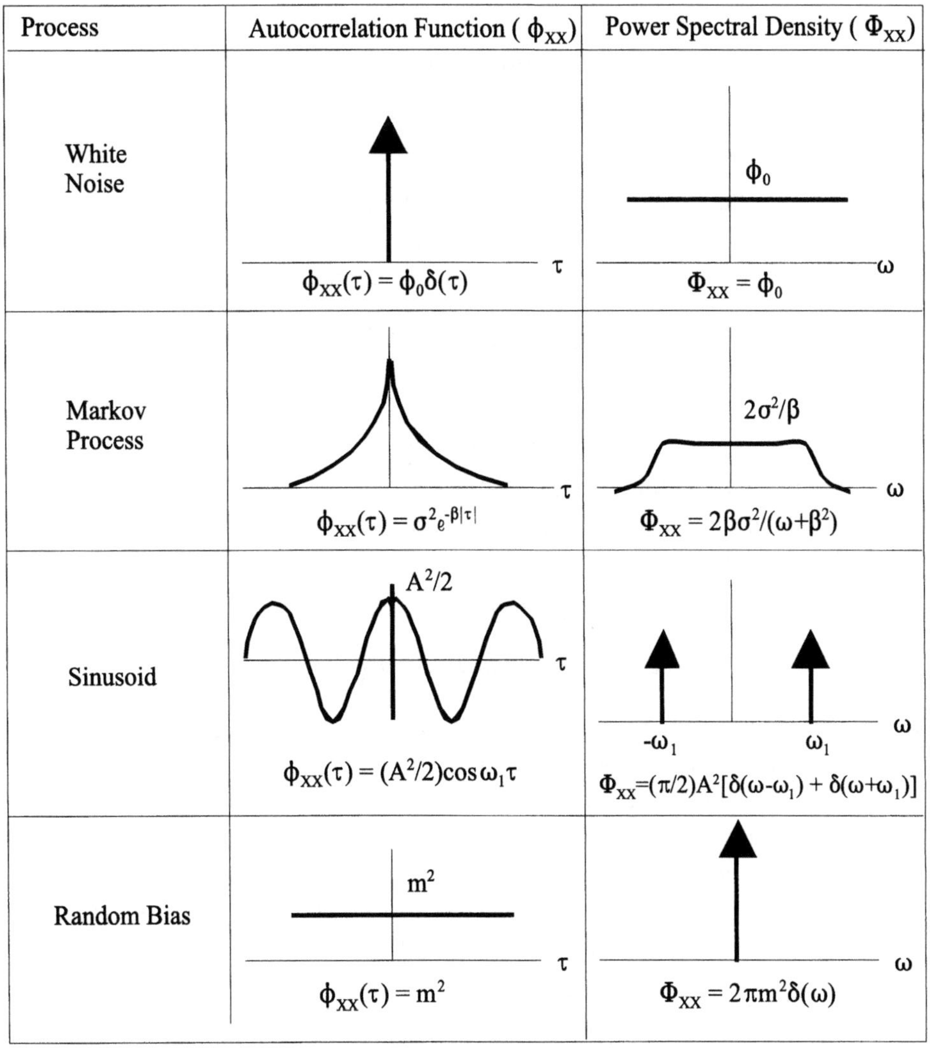

Figure 2
Descriptions of Common Random Processes (GELB, 1978).

common random processes (GELB, 1978). A Gauss-Markov process can be used to describe many physical phenomena (BAZIW, 1994) and is a good candidate to model the microseismic environmental noise.

The Gauss-Markov Process has a relatively simple mathematical description. As in the case of all stationary Gaussian processes, specification of the process autocorrelation completely defines the process. The variance, σ^2, and time constant, T_c (i.e., $\beta = 1/T_c$), define the first-order Gauss-Markov process. These parameters can be derived from the seismic time series by windowing on the noise portion of the trace and calculating the autocorrelation of the ambient noises. A Gauss-Markov process, $n(t)$, has an autocorrelation function defined as follows

$$\phi_{nn}(\tau) = \sigma^2 e^{-\beta|\tau|} \ . \tag{17}$$

The power spectrum for this autocorrelation function is given by

$$\Phi_{nn}(s) = \frac{2\sigma^2\beta}{-s^2 + \beta^2} \ . \tag{18}$$

In eq. (18), s is the Laplacian variable. From basic statistical properties, $n(t)$ can be assumed to be generated by passing Gaussian white noise through a linear system transfer function $\sqrt{(2\sigma^2\beta)/(s+\beta)}$; thus for the continuous system we have

$$\dot{n}(t) = -\beta n(t) + \sqrt{2\sigma^2\beta . w(t)} \tag{19}$$

with

$$E[w(t)w(\tau)] = \delta(t - \tau) \ .$$

To obtain the discrete form for eq. (19), we assume a sampling interval Δ and then solve eq. (19) over this interval. Since β is a constant, the discrete model for the Gauss-Markov process can be written as

$$\begin{aligned} &n_{k+1} = a_w n_k + b_w w_k \\ &where, a_w = e^{-\beta\Delta} \ and \ b_w = \sigma\sqrt{(1 - e^{-2\beta\Delta})} \end{aligned} \tag{20}$$

In eq. (20), w_k is a zero-mean, timewise-uncorrelated, unit-variance sequence with a Gaussian probability distribution function. n_k is therefore a zero-mean, exponentially-correlated random variable whose standard deviation is σ. The constant a_w can range in values from -1 to $+1$. For a stable variable, we restrict a_w to values between 0 and $+1$. For $a_w \rightarrow 0, n(t)$ changes rapidly and tends to be uncorrelated from sample to sample. For $a_w \rightarrow 1$, the behaviour of $n(t)$ becomes more sluggish and it tends to change little from sample to sample (LEAR, 1985).

2.2.2 MEDKF seismic wavelet model

The next step in defining the MEDKF system model is fitting the wavelet into a linear mathematical model. A seismic wavelet is typically modelled as an exponentially decaying cyclic waveform (SHERIFF and GELDART, 1982) as follows:

$$A(t) = A_0 e^{-h(t-t_0)} \sin[\omega(t-t_0)], \quad t \geq t_0 \ . \tag{21}$$

In eq. (21) $A_0 \equiv$ initial amplitude, $h \equiv$ damping factor, and $\omega \equiv$ dominant angular frequency (i.e., $\omega = 2\pi f$). The shot noise characteristics of the seismic wavelet make it challenging to model it in the time domain without introducing nonlinearities in the describing equations. We know that the wavelet will take on the form defined in eq. (21), but we do not know at what time (i.e., t_0) it will occur.

To simplify the mathematics and keep the KF in a linear form, we have modelled the seismic wavelet as a periodic process with a random walk amplitude. This process is defined as follows:

$$x_1(t) = x_2(t) \sin[\omega(t-t_0)], \quad t \geq t_0 \ . \tag{22}$$

In eq. (22), $x_1(t)$ is an approximation to the seismic wavelet defined by eq. (21), and $x_2(t)$ is the random walk process approximating A_0 in eq. (21). The random walk process facilitates the provision of some flexibility to the MEDKF when determining the arrival time and initial amplitude (i.e., t_0 and A_0 in eq. (21)) of the seismic wavelet understudy. The random walk process results when uncorrelated signals are integrated. The process is defined as its derivative being driven by white noise as follows:

$$\dot{x}_2(t) = w(t), \quad \text{where} \quad E[w(t)w(\tau)] = q(t)\delta(t-\tau) \ . \tag{23}$$

By using a random walk process to define the seismic wavelets amplitude, we are able to account for arrival time variations and maintain linearity in the KF formulation. The linear continuous differential equation (i.e., eq. (1)) defining the seismic wavelets is outlined as follows:

$$\dot{x}_1(t) = \omega \, x_2(t) \cos(\omega t) \tag{24a}$$

$$\dot{x}_2(t) = w(t), \quad \text{where } E[w(t)w(\tau)] = q(t)\delta(t-\tau) \ . \tag{24b}$$

In eq. (24a) it is assumed that $t_0 = 0$ with respect to eq. (22). In addition, it is also assumed that $x_2(t)$ is a constant when differentiating eq. (22) to obtain eq. (24a) (i.e., $F(t)\underline{x}(t)$ term in eq. (1)). The noise component of the system equation (i.e., $G(t)\underline{w}(t)$ in eq. (1)) handles the random walk term $\dot{x}_2(t) = w(t)$.

2.2.3 MEDKF governing equations

Substituting eqs. (19) and (24) into eq. (1) results in the continuous MEDKF matrix system equation defined as follows:

$$\begin{bmatrix} \dot{x}_1(t) \\ \dot{x}_2(t) \\ \dot{x}_3(t) \end{bmatrix} = \underbrace{\begin{bmatrix} 0 & \omega\cos(\omega t) & 0 \\ 0 & 0 & 0 \\ 0 & 0 & -\beta \end{bmatrix}}_{=F(t)} \begin{bmatrix} x_1(t) \\ x_2(t) \\ x_3(t) \end{bmatrix} + \underbrace{\begin{bmatrix} 0 & 0 \\ q(t) & 0 \\ 0 & \sqrt{2\sigma^2\beta} \end{bmatrix}}_{=G(t)} \begin{bmatrix} w_1(t) \\ w_2(t) \end{bmatrix} \ . \tag{25}$$

In eq. (25), $E[w_1(t)w_1(\tau)] = \delta(t-\tau)$ and $E[w_2(t)w_2(\tau)] = \delta(t-\tau)$. Both $w_1(t)$ and $w_2(t)$ are Gaussian white noise processes with mean zero and unity variance.

The discrete form of eq. (25) is given as

$$\begin{bmatrix} x_{1(k)} \\ x_{2(k)} \\ x_{3(k)} \end{bmatrix} = \underbrace{\begin{bmatrix} 1 & \Delta\omega\cos[\omega\Delta(k-1)] & 0 \\ 0 & 1 & 0 \\ 0 & 0 & e^{-\beta\Delta} \end{bmatrix}}_{=\Phi_{k-1}} \begin{bmatrix} x_{1(k-1)} \\ x_{2(k-1)} \\ x_{3(k-1)} \end{bmatrix} + \underbrace{\begin{bmatrix} 0 & 0 \\ q_{(k-1)} & 0 \\ 0 & \sigma\sqrt{1-e^{-2\beta\Delta}} \end{bmatrix}}_{=Q_{k-1}} \begin{bmatrix} w_{1(k-1)} \\ w_{2(k-1)} \end{bmatrix} . \tag{26}$$

In eq. (26) Δ is the sampling rate, $q(k-1) = q(t)\Delta$, and $w_{1(k-1)}$ and $w_{2(k-1)}$ are zero mean, unity variance, Gaussian white noise processes.

There is only one scalar measurement available for the MEDKF, which is a combination of both the seismic wavelet understudy and the ambient noise (i.e., $z_k = x_{1(k)} + x_{3(k)}$). This results in the following measurement matrix:

$$H_k = [1 \quad 0 \quad 1] \; . \tag{27}$$

Given a sequence of measurements $z_0, z_1, \ldots, z_k$, the *observability condition* defines our ability to determine $x_0, x_1, \ldots, x_k$ from the measurements. An observability check was carried out on the previously defined MEDKF formulation and it was found that the MEDKF was completely observable.

3. Results

3.1 Data Simulation

Prior to implementing the previously outlined KF formulation on real data, extensive testing was carried out on synthetic data. This section presents a portion of the test bed results. The first step in the simulation was to outline the source wavelets defined by eq. (21). The *P* wave was modelled with a frequency of 200 Hz, initial amplitude of 160 mm/s^2 and damping factor of 79/s. The *S* wave was modelled with a frequency of 70 Hz, initial amplitude of 200 mm/s^2 and damping factor of 50/s. The sampling rate was set at 0.05 ms and a total sampling time of 1 s was specified. Figure 3 illustrates the source wavelets generated with previous parameters specified.

The next step in data simulation was to specify different Gauss-Markov ambient noise processes. Figure 4 illustrates simulation results when the wavelets shown in Figure 3 have added ambient Gauss-Markov background noise superimposed. Plots a to e in Figure 4 have the following Gauss-Markov parameters specified:

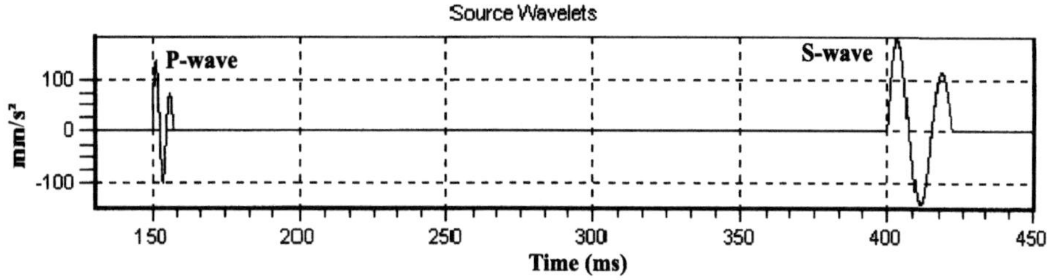

Figure 3
Simulating *P* and *S* wavelets by implementing eq. (21).

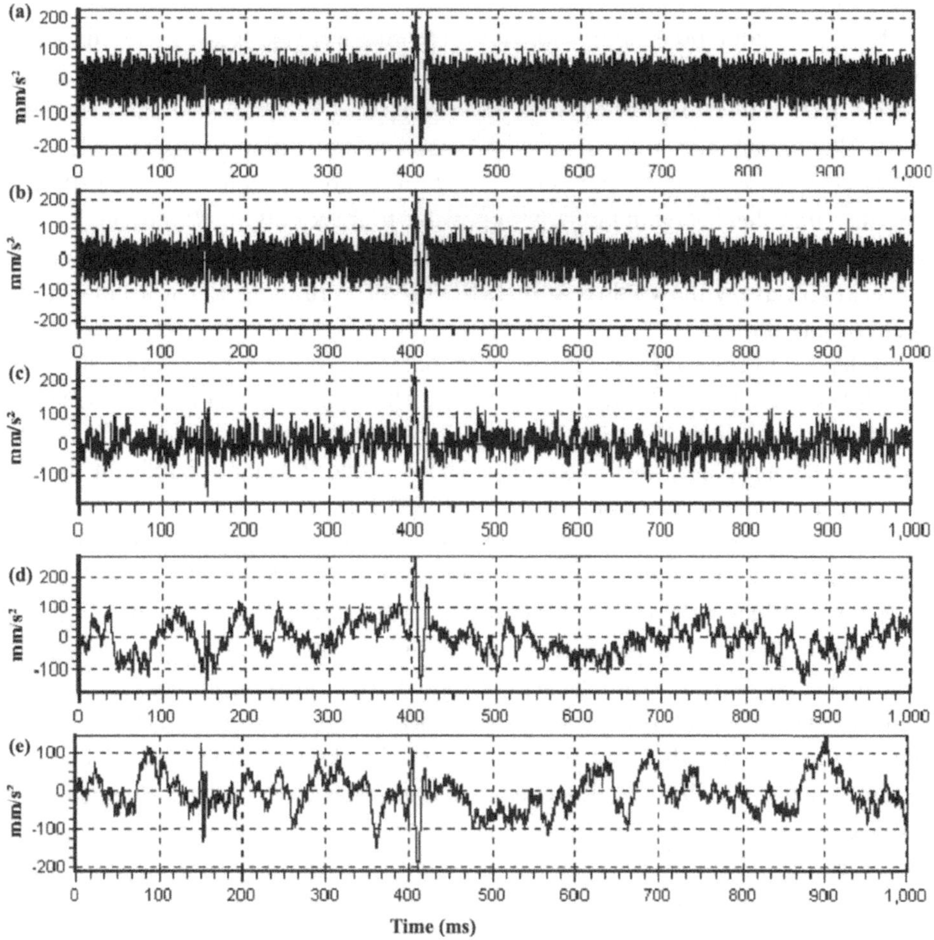

Figure 4
Superposition of wavelets illustrated in Figure 3 with varying ambient background noise.

Table 1

Gauss-Markov simulation parameters

Plot	Time constant (ms)	Variance (mm^2/s^4)
A	0.0001	1000
B	0.1	1000
C	1	1000
D	10	2000
E	20	2000

Figure 5 illustrates the autocorrelations of the Gauss-Markov noise processes illustrated in Figure 4(a) to (e). Figure 5 also illustrates the best fit curves where the ambient noise *Time Constant* and *Variance* parameters were identical to those specified above.

In these simulations and subsequent results from processed real data, the MEDKF filter was specified as a three-state filter as described by eq. (26). The *P*-wave MEDKF filter parameters were first specified, results obtained and then the *S*-wave filter parameters were set and subsequent filter estimates derived. Had the MEDKF been set to a five-state filter (i.e., to incorporate both *P* wave and *S* wave into the same filter), it would have been possible to track both the *P* wave and *S* wave simultaneously. Alternatively, it would have been possible to track both the *P* wave and *S* wave with two three-state MEDKF filters running in parallel in software multitasking threads.

Referring to eqs. (16) and (26), the filter parameters set for the *P*-wave filter were $E[x_1(0)] = 0, E[x_2(0)] = 0$, and $E[x_3(0)] = z_0$, where z_0 is the data measurement at $t_0 = 0, \omega = 2\pi f$ where $f = 200$ Hz, and $P_0 = 0$ except for elements $P_0[1,1] = 80$, $P_0[2,2] = 80$, and $P_0[3,3] = 0$. The values set for x_0 and P_0 had no significant effect on the MEDKFs performance because it was found that the error covariance and filter gain values reached steady responses rapidly. The random walk Q value was set to 0.3, where $Q = q(t)*\Delta$ and $\sqrt{q(t)} \approx 77.5\,\text{mm/s}^2$. The MEDKF responded robustly to variable Q values specified. In general, we have a more sluggish (low variability) response for lower Q values as compared to higher values.

The Gauss-Markov parameters of σ^2 and T_c (i.e., $1/\beta$) were derived adaptively and automatically from the recorded time series. The *S*-wave initial parameters were identical to those for the *P* wave except for the $f = 70\,\text{Hz}$ for the *S* wave. Since it is assumed that the ambient noise defined the time series error, the measurement error variance, $R(t)$, was specified at a fraction of the Gauss-Markov variance (i.e., $R(t) = \sigma^2(t)/1000$).

From our preliminary investigations, it was found that the best event detection state to track was the seismic wavelets amplitude (i.e., state x_2). Figure 6 illustrates the MEDKF results after tracking the *P* wavelet. As illustrated, we have obtained a dramatic improvement in the signal to noise ratio when comparing these results to the initial seismic time series outlined in Figure 4. Figure 7 shows the MEDKF results after tracking the S wavelet. As with the P wavelet MEDKF results, an impressive improvement was recorded in the signal-to-noise ratio after the implementation of the MEDKF.

As mentioned in Section 2, the error covariance (Martix Ricatti) equation is a fundamental part of the KF formulation. The error covariance matrix provides for the determination of the Kalman gain and it quantifies the filter performance.

The continuous error covariance matrix for the MEDKF is defined as follows

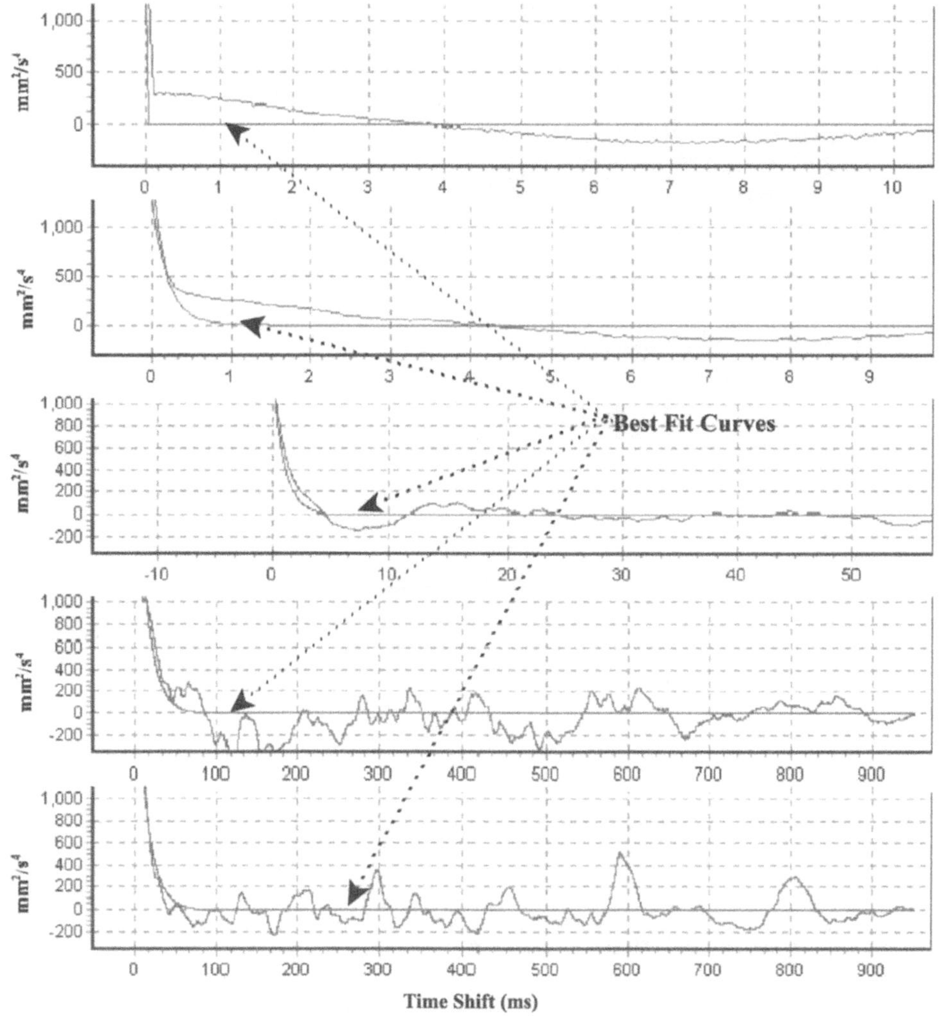

Figure 5
Applying a Gauss-Markov best fit to the simulated ambient noise autocorrelation.

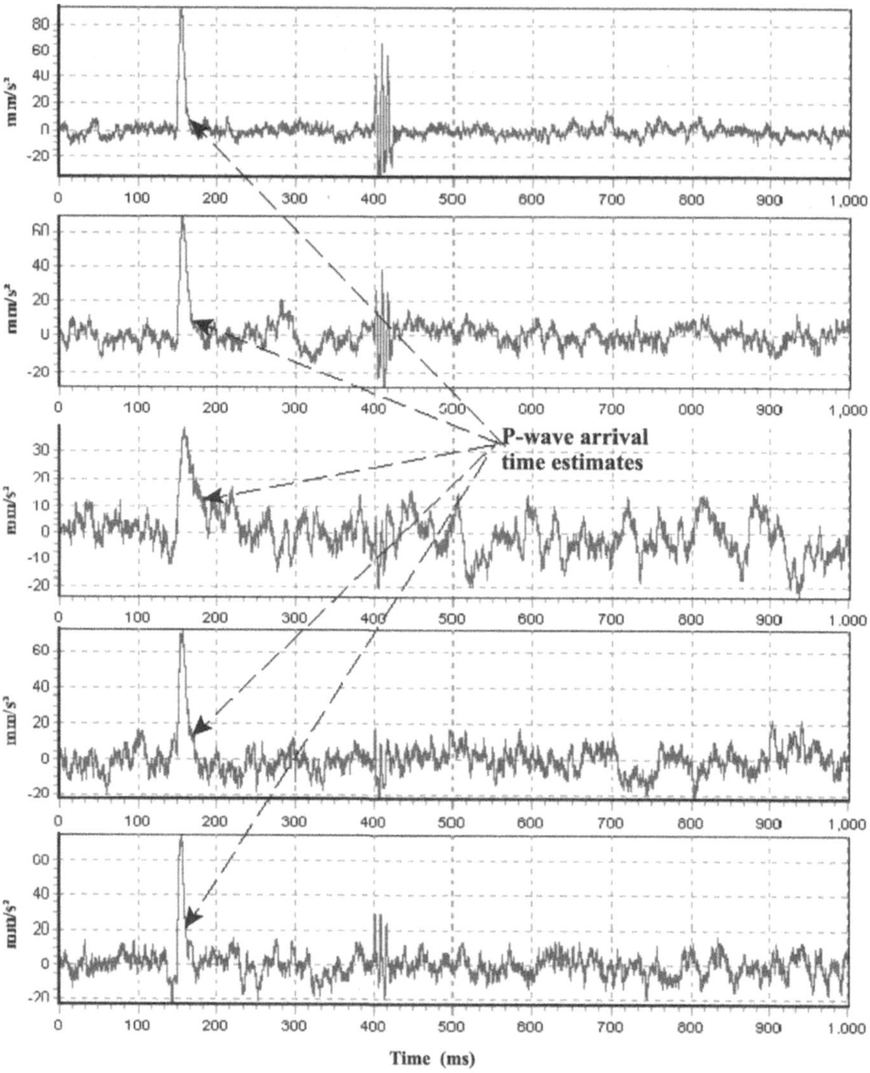

Figure 6
Kalman Filter results for the P wave. The seismic amplitudes (i.e., X_2) illustrated show excellent signal-to-noise ratio for the P-wave event detection.

$$\dot{P}(t) = \begin{bmatrix} 2P_{12}(t)\omega\cos(\omega t) & \bullet & \bullet \\ P_{22}(t)\omega\cos(\omega t) & Q & \bullet \\ P_{32}(t)\omega\cos(\omega t) - \beta P_{31}(t) & -\beta P_{32}(t) & -2\beta P_{33}(t) + \sigma^2(1 - e^{-(2\beta\Delta)}) \end{bmatrix} - B(t) \tag{28a}$$

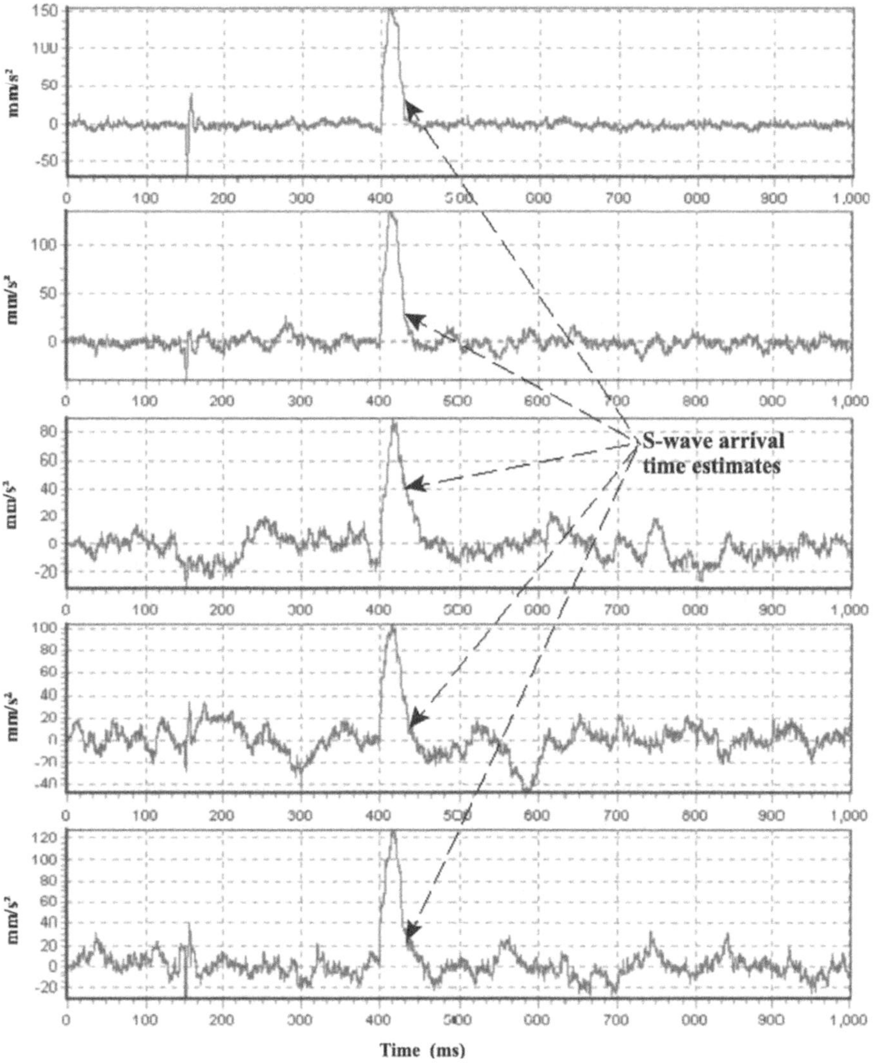

Figure 7
Kalman Filter results for the S wave. The seismic amplitudes (i.e., X_2) illustrated show excellent signal-to-noise ratio for the S-wave event detection.

where

$$B(t) = \left[\begin{array}{ccc} (P_{11}(t)+P_{13}(t))^2 & \bullet & \bullet \\ (P_{21}(t)+P_{23}(t))(P_{11}(t)+P_{31}(t)) & (P_{21}(t)+P_{23}(t))^2 & \bullet \\ (P_{31}(t)+P_{33}(t))(P_{11}(t)+P_{31}(t)) & (P_{31}(t)+P_{33}(t))(P_{12}(t)+P_{32}(t)) & (P_{31}(t)+P_{33}(t))^2 \end{array}\right] \Big/ R(t) \; . \tag{28b}$$

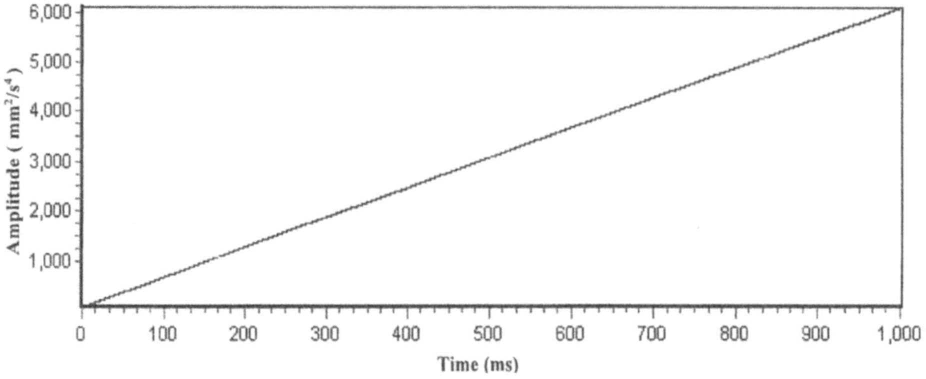

Figure 8
$P_{11}(k)$ response for $R(t) \Rightarrow \infty$ and MEDKF initial conditions.

For simplicity, the off diagonal terms in eqs. (28a) and (28b) have been omitted due to the symmetry of the error covariance matrix. The continuous error covariance for the amplitude estimate (i.e., state $x_2(t)$) is defined as

$$\dot{P}_{11}(t) = Q - (P_{21}(t) + P_{23}(t))^2 / R(t) \ . \tag{29}$$

We see from eq. (29), that if there is a high measurement (i.e., $R(t) \Rightarrow \infty)P_{11}(t)$ would grow linearly at a rate specified by the random walk variance Q. Figure 8 illustrates the discrete $P_{11}(k)$ value with a very high $R(t)$ specification and our initially specified Q value of 0.3. The fact that accurate measurements are utilized (i.e., low $R(t)$), allows the MEDKF to constrain our error covariance estimates and bring them to steady-state responses.

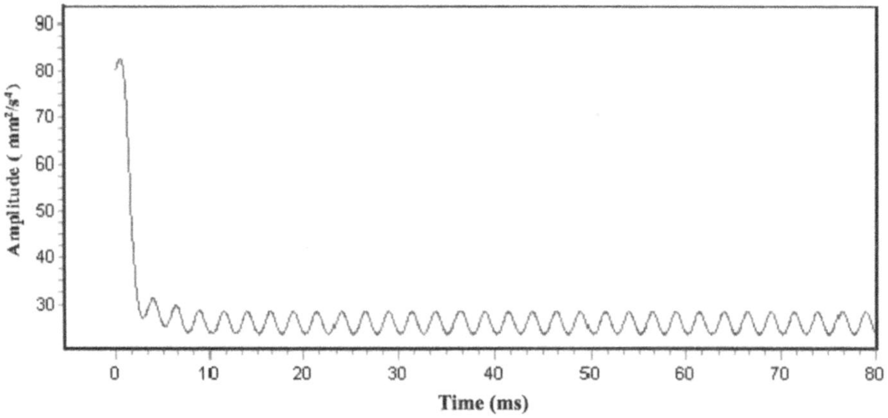

Figure 9
$P_{11}(k)$ response for $R(t) = \sigma^2(t)/1000$ and MEDKF initial conditions.

In general terms, the error covariance estimate $P_{11}(t)$ is directly proportional to Q, σ, and β, and inversely proportional to $R(t)$. In addition, because term $P_{21}(t)$ is a function of $\cos^2(\omega t)$, we would expect $P_{11}(t)$ to have components of oscillation at $2^*\omega$ where $\omega = 2\pi f$. Figure 9 illustrates the $P_{11}(k)$ responses for the P-wave estimate for test case A. As is shown, the error covariance estimate, $P_{11}(k)$, behaves as expected where it reaches a steady response rapidly from the initially specified $P_{11}(0)$. $P_{11}(k)$ then oscillates at the steady response rate of $2^*\omega$ or $f = 400$ Hz. It is also important to bear in mind that P_{11} reflects uncertainty of both the A_0 and t_0 parameters in eq. (21).

3.2 Processing Real Data

The data presented in this paper were obtained from some of the microseismic installations Weir-Jones Engineering Consultants has commissioned at heavy oil extraction operations in Northern Alberta. Earlier work in this region has been described by other authors (TALEBI *et al.*, 1998).

The Cold Lake area is located in Northern Alberta, Canada. Extensive, heavy oil reserves underlie the area. The oil sands occur in the Mannville Group of Lower Cretaceous age and range from 305 m to 610 m below surface. The primary reservoir is the Clearwater formation at an average depth of 457 m. The sand in the Clearwater is laterally and vertically continuous and pay thickness ranges from 15 m to 49 m. The reservoir has 11 billion m^3 of bitumen in place.

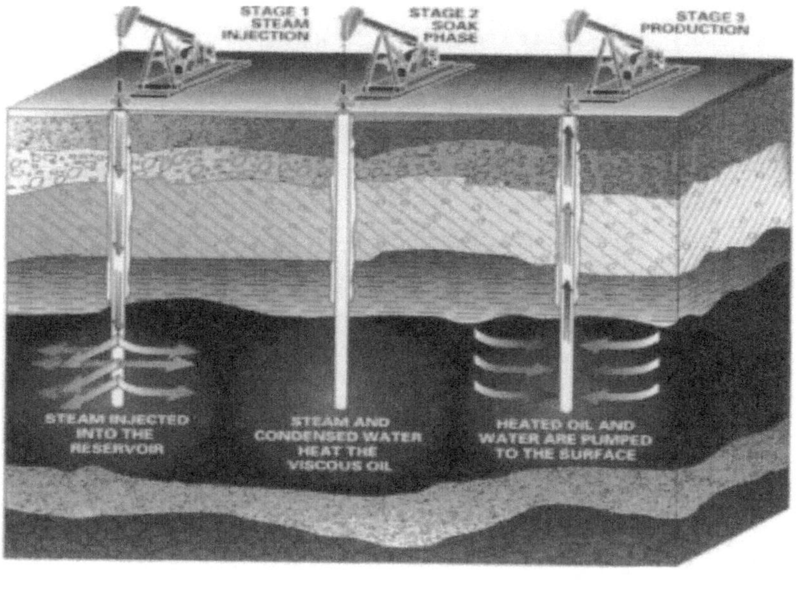

Figure 10
Typical steam injection mechanism in Northern Alberta.

The Clearwater sand is an unconsolidated clean sand with porosities between 30% and 35% and an average bitumen saturation of 70%. The absolute permeability ranges from $0.5 \times 10^{-12}\,m^2$ to $2 \times 10^{-12}\,m^2$ and viscosity reaches 150,000 cp. Initial reservoir conditions are 3 °C and 3 MPa. Figure 10 illustrates a typical steam injection mechanism in which there is pancake stratigraphy which is defined as follows:

Formation top	Depth (m)
Surface	0
Colorado Shales	149
Grand Rapids	319
Clearwater	432
McMurray	485

Overlying the Clearwater is the Grand Rapids formation, which consists of sands and shales. Above this is the Colorado Group of Upper Cretaceous age. It consists of marine shales that are impervious and separate the oil sands from the glacial tills near surface. Since most of the oil sands are immobile, additional heat and pressure are required to recover the bitumen. Three main processes used are cyclic steam stimulation, steam drive and steam assisted gravity drainage. A detailed technical description of the microseismic monitoring equipment and geophone array installation was presented by TALEBI (1998).

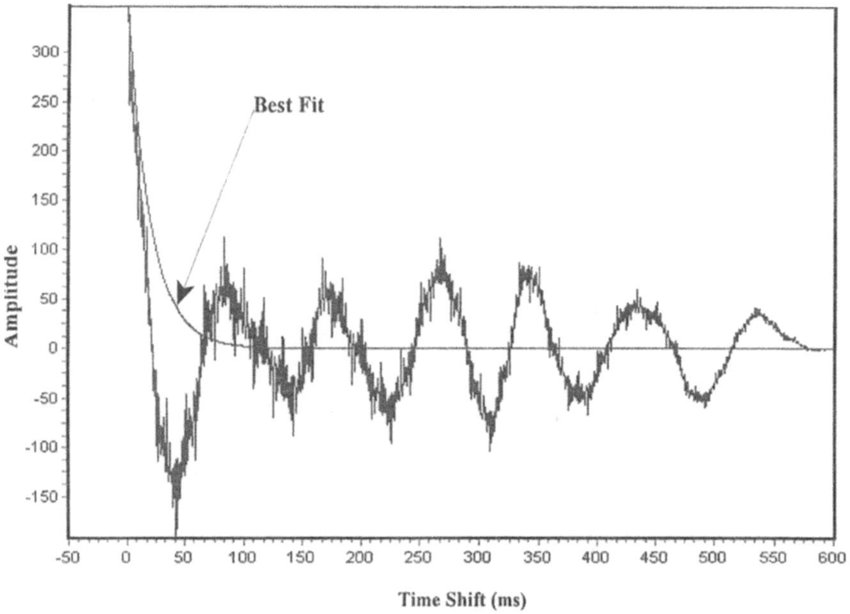

Figure 11
Autocorrelation calculated from the seismic time series presented in Figure 12 (a). The best fit parameters are $\sigma^2 = 350\,mm^2/s^2$ and $T_c = 20\,ms$.

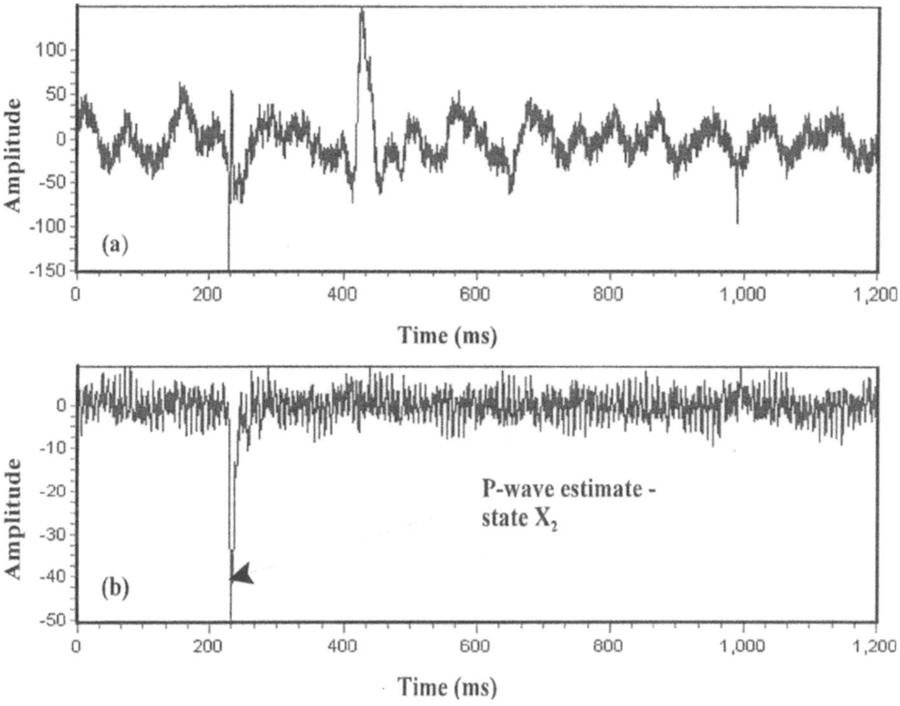

Figure 12
Input seismic trace (a), and the result based on the implementation of the Kalman Filter (b), as was outlined in section 2 ($f = 185\,\mathrm{Hz}$ for the P wave).

The data analysed in this paper were obtained using proprietary geophone arrays specifically developed by the authors for these applications. The geophone arrays consist of multiple sensing elements arranged on orthogonal axes. The arrays have a natural frequency of 10 Hz, are damped at 70% of critical, and have a bandwidth from 8 to 900 Hz.

Figure 11 illustrates the autocorrelation function of the seismic trace presented in Figure 12(a). This autocorrelation was calculated by windowing on the portion of data 600 ms and 1200 ms. The exponentially correlated process illustrated in Figure 11 was fitted with the parameters $\sigma^2 = 350\,\mathrm{mm}^2/\mathrm{s}^2$, and $T_c = 20$ ms. Figure 12(b) illustrates the results of implementing the MEDKF on the seismic trace presented in Figure 12(a) with the parameters specified by fitting the autocorrelation presented in Figure 11 and the dominant frequency set to 18.5 Hz for the P-wave response. The remaining MEDKF initial filter parameters were set in an identical manner to those specified in the previously outlined simulated results. As is shown in Figure 12(b), there has been a dramatic improvement in the signal to noise increase after implementation of the MEDKF.

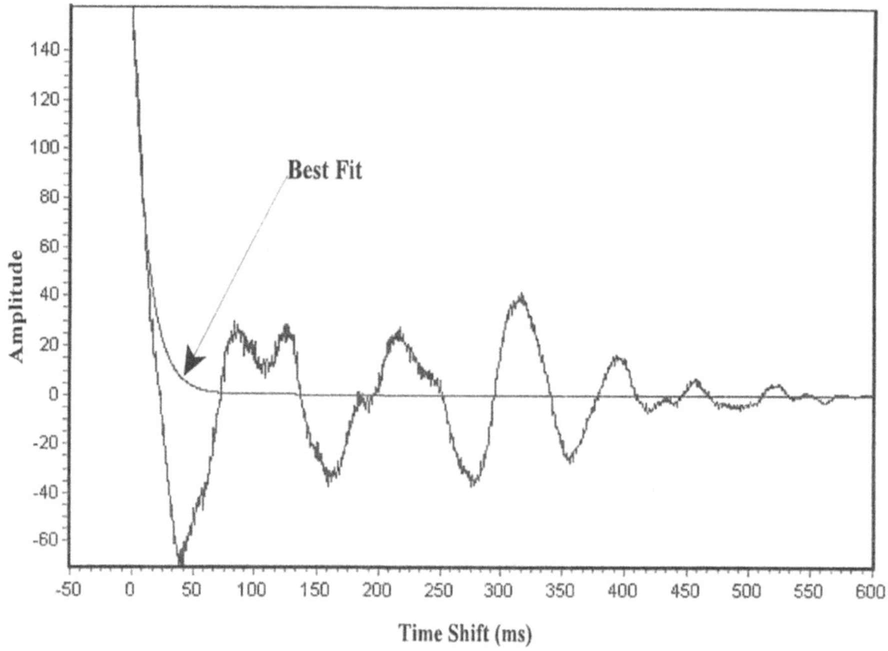

Figure 13
Autocorrelation calculated from the seismic time series presented in Figure 14(a). The best fit parameters are $\sigma^2 = 157\,\mathrm{mm}^2/\mathrm{s}^2$ and $T_c = 13\,\mathrm{ms}$.

The next set of data illustrates the MEDKF's abilitty to identify the *P*-wave and *S*-wave responses from the time series in real time. Figure 13 shows the autocorrelation function of the seismic trace presented in Figure 14(a). This autocorrelation was again calculated by windowing on the portion of data between 600 ms and 1200 ms. The Gauss-Markov noise process illustrated in Figure 13 was fitted with the parameters $\sigma^2 = 157\,\mathrm{mm}^2/\mathrm{s}^2$, $T_c = 20\,\mathrm{ms}$. In this analysis the frequency was set to 185 Hz for the *P*-wave response and a frequency of 30 Hz for the *S*-wave response. Figure 14(b) and 14(c) illustrate the results after filtering. Similarly to Figure 12(b), we again obtain excellent results where the *P*-wave and *S*-wave arrivals are clearly identified.

Figure 15(a) illustrates a time series where it is very difficult to identify the *P* wavelet visually, let alone automatically. In this test case the dominant frequency was set to 170 Hz for the *P*-wave response. The exponentially correlated noise process illustrated was fitted with the parameters $\sigma^2 = 7.7\,\mathrm{mm}^2/\mathrm{s}^2$, and $T_c = 20\,\mathrm{m\,sec}$. Figure 15(b) illustrates the filtered results where the *P*-wavelet response has been more clearly identified.

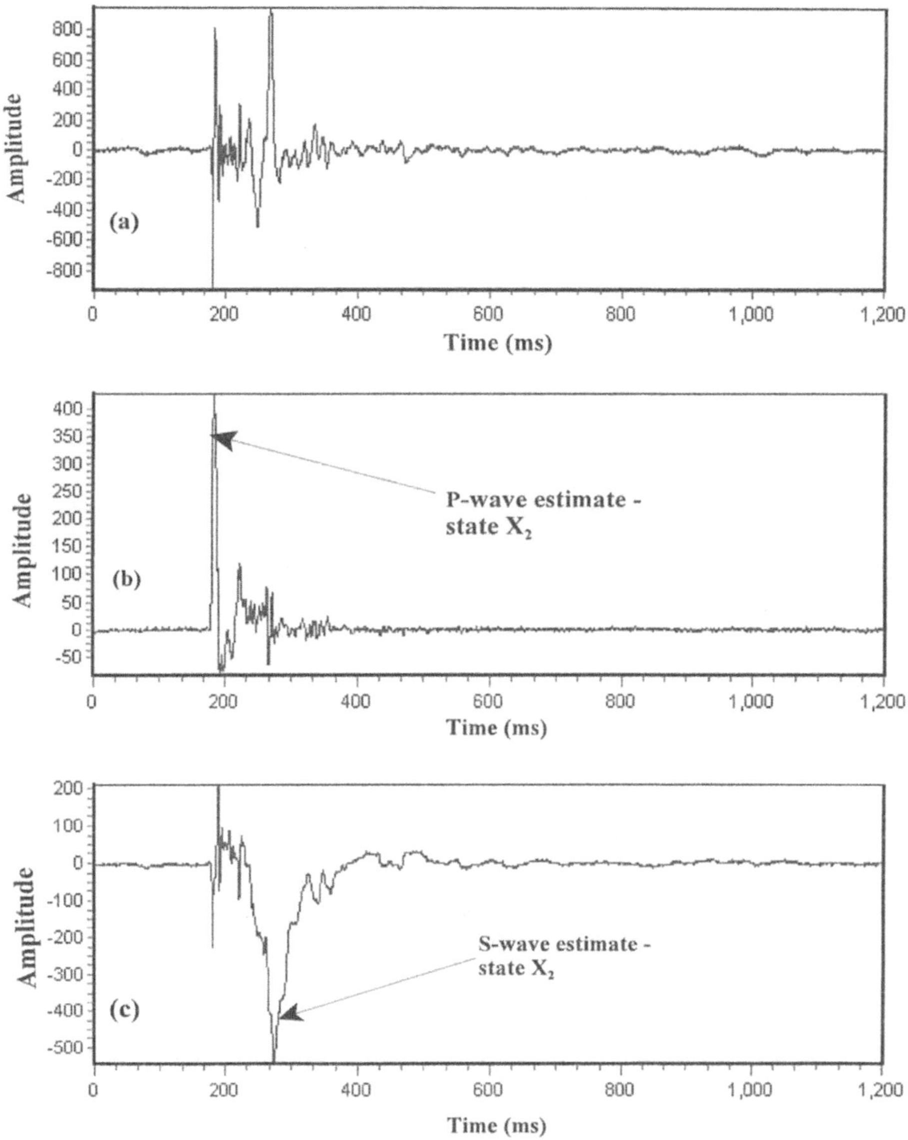

Figure 14
Input seismic trace (a); Kalman Filter *P*-wave arrival time estimation results ($f = 185\,\mathrm{Hz}$) (b); Kalman Filter *S*-wave arrival time estimation results ($f = 30\,\mathrm{Hz}$) (c).

4. Conclusions

In microseismic monitoring programs it is important to have reliable event triggering and source parameter characterization. Poorly defined passive microseis-

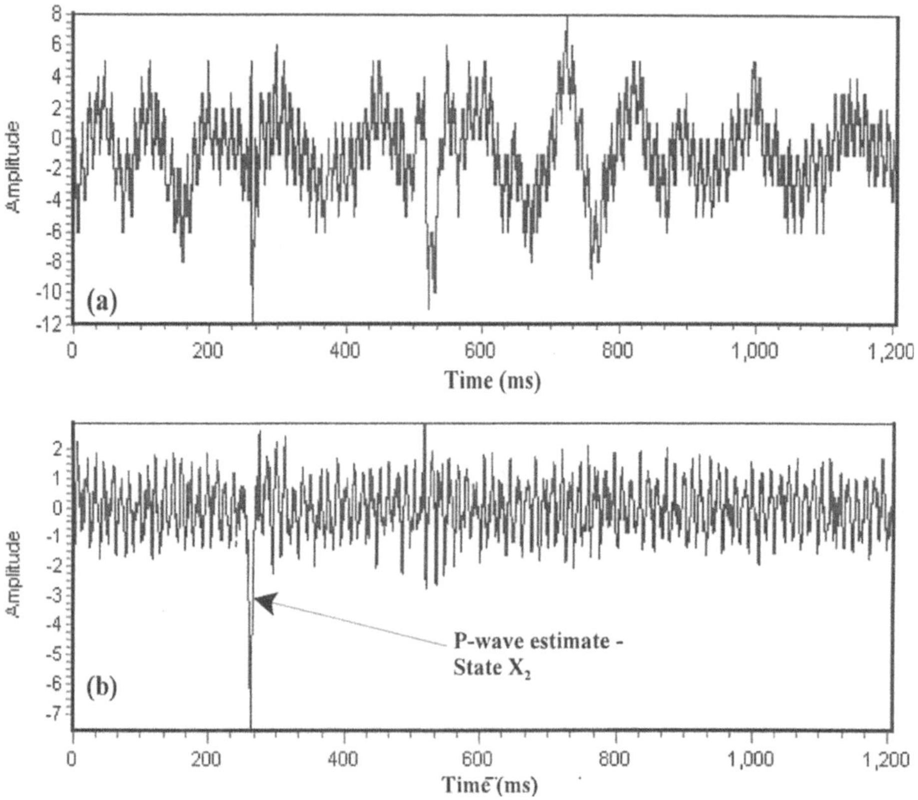

Figure 15
Input seismic trace (a), and Kalman Filter *P*-wave arrival time estimation results ($f = 170\,\mathrm{Hz}$, $\sigma^2 = 7.7\,\mathrm{mm}^2/\mathrm{s}^2$, and $T_c = 20\,\mathrm{ms}$).

mic triggering algorithms tend to acquire substantial amounts of data which demand extensive post-processing manipulation; they also increase the work and reduce the efficiency of the personnel operating the systems. This paper presented a real-time Kalman Filter which modelled the background noise as a Gauss-Markov process and the seismic wavelet as a periodic process with a random walk initial amplitude. The filter allows for online processing and circumvents many of the problems associated with frequency domain filters.

Both simulated signals and seismic data acquired from passive microseismic installations in Northern Alberta were analysed and the results evaluated in terms of the improvement in signal-to-noise ratio. The results are very encouraging; there is a substantial improvement in signal-to-noise ratio after implementation of the Kalman Filter. For example, for the simulated data presented, there is an approximate 12-old increase in the signal-to-noise ratio as a result of the post-processing. For the processed real data, the signal-to-noise ratio is enhanced by approximately six times (i.e., Fig. 12).

The authors' ultimate objective is to provide a system which offers microseismic system operators a reliable and cost-effective procedure which can process microseismic data in real-time and provide accurate event locations with a reduction in the amount of system operator intervention.

Acknowledgements

The authors would like to thank Robert Rasmusens for his software engineering assistance.

References

Bayless, J. W. and Brigham, E. D. (1970), *Application of the Kalman Filter to Continuous Signal Restoration*, Geophysics *35*, 2–23.

Baziw, E. J. (1993), *Digital Filtering Techniques for Interpreting Seismic Cone Data*, ASCE. J. Geotech. Engin. *119*(6), 998–1018.

Baziw, E. J. (1988), *Applications of Digital Filtering Techniques for Reducing and Analysing In-Situ Seismic Time Series*, MASc Thesis, Dept. of Civil Engineering, University of British Columbia, Vancouver, BC.

Baziw, E. J. (1994), *Dead Reckoning/Target Tracking in a ECDIS Environment*, Institute of Navigation's National Technical Meeting, San Diego California, January 26, 747–756.

Gelb, A., *Applied Optimal Estimation*, (4th edition, MIT Press, Cambridge, Mass. 1978).

Kanasewich, E. R., *Time Sequence Analysis* (3rd edition, The University of Alberta Press 1981), 230 pp.

Lear, W. M., *Kalman Filtering Techniques* (National Aeronautics and Space Administration Publication, Mission Planning and Analysis Division, No. JSC-20688 1985), pp. 172–176, 321–335.

Mendel, J. M., *Optimal Seismic Deconvolution An Estimation-Based Approach*, (1st Edition, Academic Press, 1983).

Mendel, J. M., *Lessons in Estimation Theory for Signal Processing, Communications, and Control* (Prentice Hall PTR, Englewood Cliffs, NJ, 1995).

Sheriff, R. E. and Geldart, L. P., *Exploration Seismology* Volume 1 (2nd edition, Cambridge University Press, Cambridge, UK 1982), 55 pp.

Talebi, S., Ge, M., Rochon, P., and Mottahed, P. (1994), *Analysis of Induced Seismicity in a Hard-rock Mine in the Sudbury Basin*, Ontario, Canada, Proceedings of the 1st North American Rock Mechanics Symposium, the University of Texas at Austin, June 1–3, 937–944.

Talebi, S., Nechtschein, S., and Boone, T. J. (1998), *Seismicity and Casing Failures Due to Steam Stimulation in Oil Sands*, Pure appl. geophys. *153*, 219–233.

(Received July 20, 2000, revised January 10, 2001, accepted January 11, 2001)

Pure appl. geophys. 159 (2002) 473–487
0033–4553/02/030473–15 $ 1.50 + 0.20/0

Pure and Applied Geophysics

Near-surface Microearthquakes at The Geysers Geothermal Field, California

JAMES T. RUTLEDGE,[1] MITCHEL A. STARK,[2] THOMAS D. FAIRBANKS,[3] and TIMOTHY D. ANDERSON[4]

Abstract—A 150-m length, 6-level, three-component, vertical geophone array was cemented into the 67- to 219-m depth interval (220 to 720 ft) of Unocal's well GDCF 63-29 during a plug and abandonment operation on April 7, 1998. Casing deformation has been observed in wells of the study area including the GDCF 63-29 well. An objective of the study was to determine if shallow deformation at The Geysers is manifested seismically. Near-surface microearthquake activity was monitored for a period of one year; during the latter four months, monitoring was supplemented with four surface stations to help constrain locations of shallow seismicity. Event locations occurring within about 750 m of the array bottom have been determined for the 10-week period January 6 to March 16, 1999. These events are distinct from surface-monitored seismicity at The Geysers in that they occur predominantly above the producing reservoir, at depths ranging from about 220 to 1000 m (600 to −180 m elevation). The shallow events tend to be episodic, with relatively quiescent periods of up to three weeks occurring between swarms. Event locations show a northeast-striking trend, similar to seismicity trends mapped deeper in the reservoir, and parallel to the strike of a major surface lineaments observed over the productive field. However, clear fault or fracture planes are not resolved from the hypocenters. Composite fault-plane solutions suggest oblique reverse faulting in the overburden. The shallowest seismicity terminates near the base of a serpentine unit, a contact which is the locus of most of the well casing deformations logged in the area, suggesting that reservoir contraction is accommodated along numerous discrete faults below the serpentine, but as continuous plastic deformation in the serpentine. It is hypothesized that the resulting strain discontinuity at the base of the serpentine explains the prevalence of wellbore deformation there. The shallow, above-reservoir microseismicity is strongly correlated in time with the injection and the deeper injection-induced seismic activity occurring in the reservoir immediately below. This suggests that deep injection-induced events trigger shallower events, by a remote triggering mechanism which has been observed at a larger scale at The Geysers and elsewhere.

Key words: Induced seismicity, microearthquake, wellbore deformation, geothermal.

Introduction

Since fractures usually dominate the contribution to permeability in geothermal reservoirs, the ability to map them far from boreholes has direct applications to

[1] Nambe Geophysical, Inc., Route 1 Box 104F, Santa Fe, NM 87501, USA. E-mail: jrutledge@lanl.gov
[2] Calpine Corporation, 10350 Socrates Mine Road, Middletown, CA 95461, USA.
[3] Nambe Geophysical, Inc., Route 1 Box 104F, Santa Fe, NM 87501, USA.
[4] Sonoma County Water Agency, P.O. Box 11628, Santa Rosa, CA 95406, USA.

reservoir development and management. It is well known that the gross flow paths affected by hydraulic fracturing can be mapped using the microearthquakes induced during the treatment. BARTON *et al.*, (1995) have shown correlations of high permeability along fractures that are oriented such that resolved shear stress is high. If this is generally true, it would imply that any reservoir stress changes that even weakly promote failure on critically stressed fractures could result in seismicity that reveals important or potentially important reservoir flow paths. Stress changes can also be induced outside the reservoir, where no pore-fluid content changes need occur, due to reservoir volume changes accompanying pressure and temperature drawdown (e.g., SEGALL, 1989; SEGALL and FITZGERALD, 1998). Microearthquakes induced above the reservoir could be used for monitoring and characterizing deformation in the overburden.

Seismic station coverage of The Geysers region was greatly enhanced in the mid-1970s by the U.S. Geological Survey, leading to early recognition and characterization of production-induced seismicity (MARKS *et al.*, 1978; MAJER and MCEVILLY, 1979; DENLINGER and BUFE, 1982; EBERHART-PHILLIPS and OPPENHEIMER, 1984; OPPENHEIMER, 1986). Hypocentral resolution was improved when steamfield operators, such as the Unocal-NEC-Thermal partnership (U-N-T), set up field wide array coverage (STARK, 1990), and was further refined with denser arrays covering specific areas within the field (KIRKPATRICK *et al.*, 1995; ROMERO *et al.*, 1995; JULIAN *et al.*, 1993). Detection thresholds for these arrays ranged as low as $M = -1$. Data collected on the high-resolution arrays have been used to further understand the reservoir by: 1) associating hypocenter patterns with fluid movement (STARK, 1990; KIRKPATRICK *et al.*, 1995), 2) associating seismic velocity anomalies with specific lithologies, reservoir processes, and saturation levels (ROMERO *et al.*, 1995; KIRKPATRICK *et al.*, 1997), and 3) inferring reservoir fracture behavior or mechanics by associating extraction/injection operations with seismic source characteristics (JULIAN *et al.*, 1993; KIRKPATRICK *et al.*, 1995; ROSS *et al.*, 1999).

Downhole seismic receivers deployed as deep as 712 m (~280 m above the reservoir) at The Geysers lowered the threshold of detection to $M = -3$, or 2 to 3 magnitude units below the limits of surface monitoring (ALBRIGHT *et al.*, 1998). Placing seismic sensors downhole also results in a concomitant increase in event detection rate for a given volume of rock and, in general, greatly improves the resolution of mapping active reservoir fractures or faults (e.g., PHILLIPS *et al.*, 1997; RUTLEDGE *et al.*, 1998a, b; GAUCHER *et al.*, 1998; PHILLIPS, 2000).

In this paper, we present microearthquake mapping results from monitoring with a borehole geophone array that was cemented within the upper 220 m (720 ft) of Unocal's well GDCF 63-29 in the southeast Geysers (Figs. 1 and 2). Approximately 0.5 m of subsidence has occurred in the immediate study area between 1977 and 1996 and has been attributed to deeper, production-induced reservoir contraction (MOSSOP and SEGALL, 1997). Near-surface strain has also been observed as casing

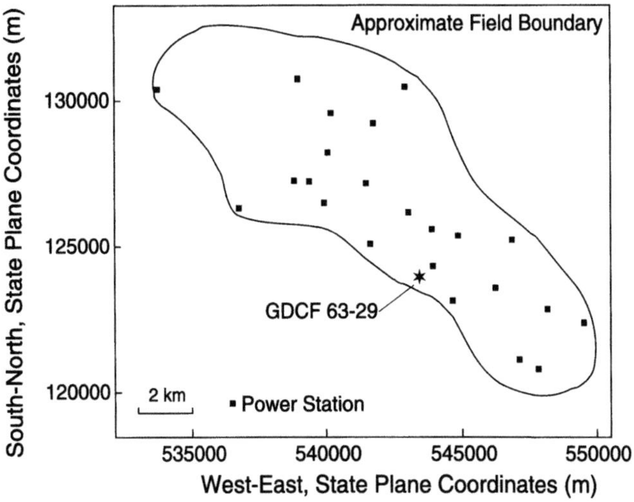

Figure 1
Location of the monitor well GDCF 63-29.

deformation in this area of the field, occurring over short depth intervals where the wells intersect the upper and lower contacts of a serpentine unit dipping to the northeast (Fig. 3). In fact, GDCF 63-29 was plugged and abandoned because of the severity of wellbore collapse concentrated at the base of the serpentine, at 244 m (800 ft) depth. Monitoring shallow microseismicity could potentially provide answers to questions regarding the nature of deformation affecting borehole integrity, such as: Is the deformation episodic? Are the collapse zones associated with shallow fault zones intersecting the wellbores? and can intersection with active, shallow faults or deformation zones be avoided?

Data

The downhole array installed in GDCF 63-29 is 152 m (500 ft) long with six three-component sondes spaced at 30 m (100 ft) intervals (Fig. 2). Deployment took place on April 7, 1998 during plug and abandonment of the well; all but the middle sondes S3 and S4 continued to operate with at least one functional geophone channel after one year of continuous monitoring. The temperature at the bottom of the array is 146 °C. Details of the deployment and array specifications are given in ALBRIGHT *et al.* (1998).

The array's short length and placement about 750 m above the reservoir makes it best suited for detecting and locating small microearthquakes associated with near-surface deformation. The downhole array was supplemented with four three-component, surface-deployed geophones surrounding the GDCF 63-29 monitor well in late 1998 and early 1999 (Fig. 2). It is possible to uniquely determine source

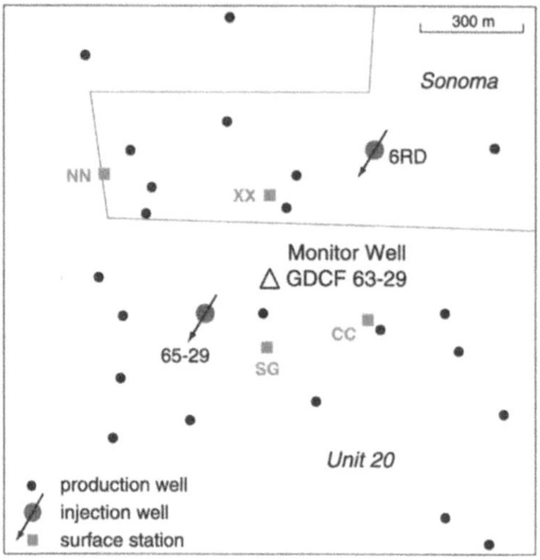

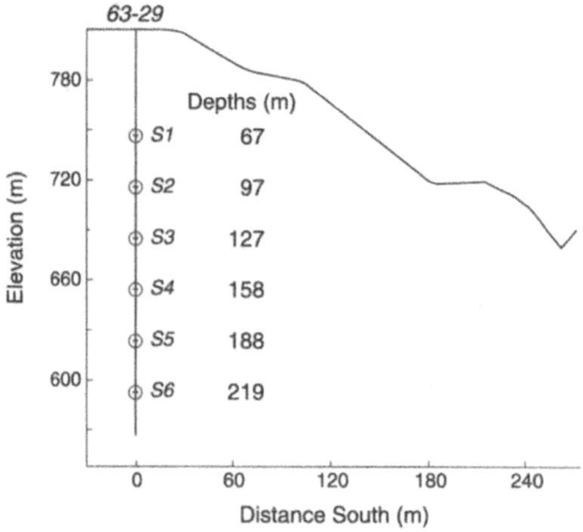

Figure 2
Map view of the monitor well GDCF 63-29, surface stations deployed to supplement the downhole array and the production and injection wells within 762 m (2500 ft) of the monitor well (top). Sonoma and Unit 20 refer to production units. Production and injection wells are mapped at their weighted midpoint of steam entries recorded while drilling. The lower diagram shows the downhole geophone configuration in GDCF-63-29. Topographic profile south of well is also shown. All depths are with respect to ground level at wellhead (811 m elevation). Station spacing is 30.48 m (100 ft).

locations from a single vertical array if good azimuthal data can be obtained from the horizontal component first-arrival particle-motion trajectories (e.g., RUTLEDGE *et al.*, 1998a). All events detected on the borehole array occur beneath the bottom sonde, most with steep travel paths from source to receiver. This results in low signal-to-noise ratio first arrivals on the horizontal components and hence, unreliable

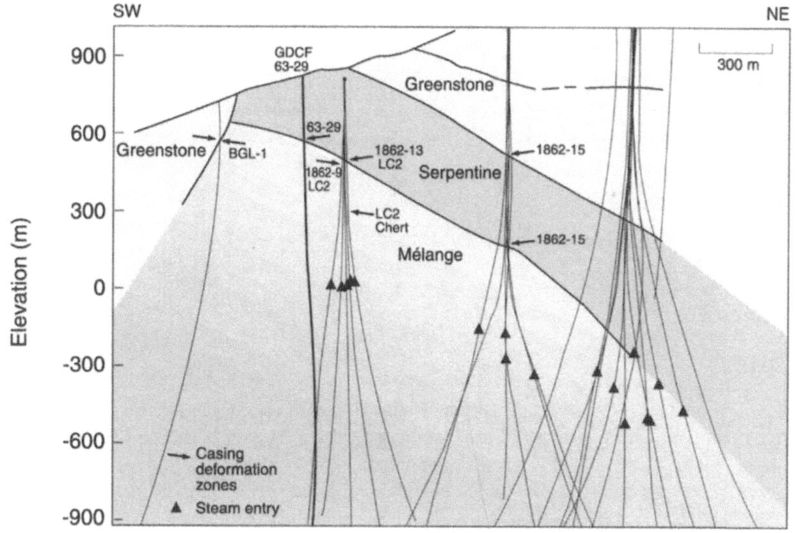

Figure 3
Geologic profile showing the well casing deformation zones association with upper and lower contacts of a serpentine unit. Profile is oriented N30E (viewing N60W). Steam entries delineate approximate top of reservoir.

azimuthal data. Adding the surface stations allowed locations to be determined using only the *P*- and *S*-arrival time data.

The six downhole receiver levels and three of the surface stations were equipped with OYO GS-20DM geophones (28 Hz downhole, 14 Hz at surface); the fourth surface station was a Mark Products 1-Hz, L4-3C geophone. Using existing telemetry lines in the field, surface station output were sent directly to the same PC-based data acquisition system used for the downhole receivers. Data were sampled at a 0.2 msec interval per channel. Downhole signal bandwidth above noise extends from about 20 to 400 Hz.

All data collected from April 7, 1998 to March 16, 1999, have been screened to select only those events occurring within about 750 m of the array (based on the criteria *S-P* arrival time difference ≤150 msec at the deepest station (S6, Fig. 2)). Details of event occurrence over an 88-day period in 1998 show that the shallowest seismicity is episodic. Within about 500 m of the array, relatively quiescent periods of up to three weeks are observed between swarms of events (Fig. 4). In this paper we present the locations of near-array events detected on the borehole receivers and at least three of our local surface stations for the period January 6 to March 16, 1999. A total of 535 events was detected within ~750 m of the deepest station for this 70-day period. An example of a high-quality, near-array event recorded on downhole sonde S5 is shown in Figure 5.

The downhole array also detects events occurring field wide, in common with the U-N-T surface array. However, the shallow, local events analyzed are unique to our

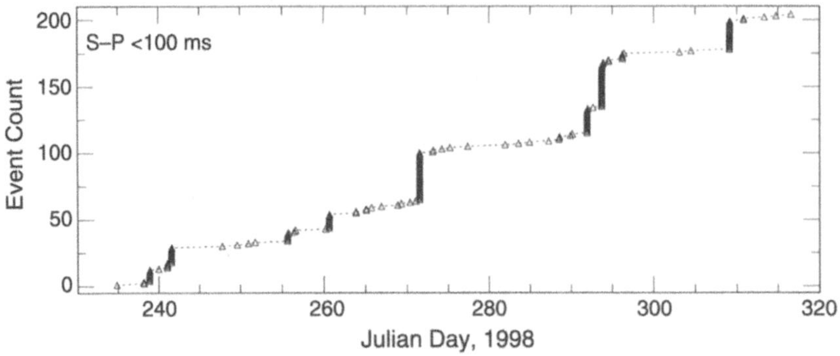

Figure 4
Cumulative event count from August 18 to November 13, 1998 for events within ~500 m of the array (*S*-*P*-arrival time difference < 100 msec).

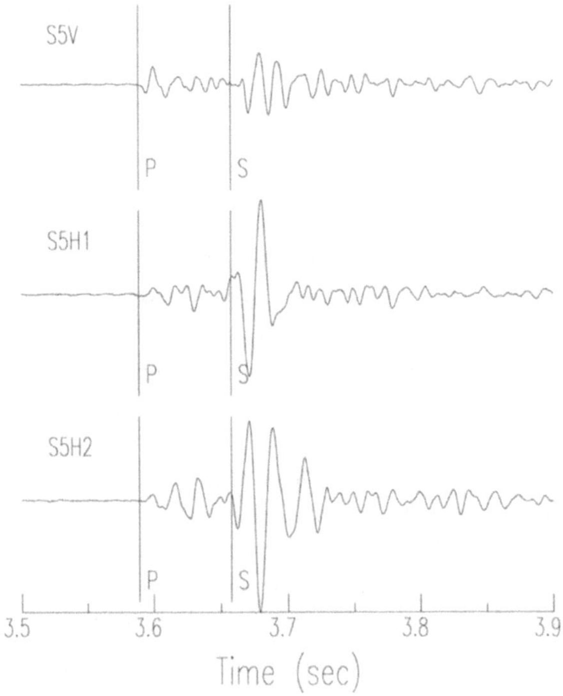

Figure 5
Example of a high-quality, shallow microearthquake recorded on downhole sonde S5.

downhole array because they are generally too small for common detection over the wider station spacing (≥1 km) of the U-N-T surface array.

Determining Source Locations

Seismic velocities were determined using 1) surface calibration shots, 2) observed travel-time differences across the downhole array, and 3) a velocity grid search to find the best fit between observed and computed arrival times. *P*- and *S*-wave velocities (V_p and V_s, respectively) were first computed for a shallow serpentine unit from the surface to 244 m depth using the surface shots and microearthquakes with travel paths within 10° of the vertical array. With these shallow velocities fixed, we then did a grid search for an underlying half-space's V_p and V_s values that minimized the difference between observed and computed arrival times in a least-squares sense. The grid search resolved V_s very well but could not constrain V_p. Our top-layer V_p value is within 2% of the velocity model of KIRKPATRICK *et al.* (1997) for the upper kilometer of the southeast Geysers. Therefore, we kept V_p constant across the 244 m serpentine contact. The final velocity model is summarized in Table 1. Trial runs to compare the resultant source locations and residual misfits with the model interface at 244 m oriented horizontal or dipping 30° to the NE (as in Fig. 3) showed little difference. The results shown are for the flat-layered model.

Of the 535 events, 304 with at least six arrival times identified over the full length of the downhole array, and three or more arrival time picks on the surface stations, were considered for mapping. 297 location solutions converged, of which 248 had root-mean-squares travel-time residuals less than 5 msec and location errors less than ±60 m (Fig. 6). *P*- and *S*-station corrections were applied, based on the median travel-time residuals determined from two initial location runs. Arrival-time errors were estimated from the standard deviations of the travel-time residuals, and ranged from 2 to 5 msec. The location error ellipses displayed in Figures 6, 7 and 8 only reflect the arrival-time data errors, data distribution and array geometry; velocity model uncertainties are not considered. Median principle error axes for the 248 events displayed is ±24 m.

Microearthquake Maps

The map view (Fig. 6) shows a gross northeast striking trend, similar to seismicity trends mapped deeper within the reservoir (e.g., ROMERO *et al.*, 1995)

Table 1

Summary of seismic velocities used

Top of layer		V_p (km/sec)	V_s (km/sec)
Depth (m)	Elevation (m)		
0	811	3.81	1.89
244	567	3.81	2.16

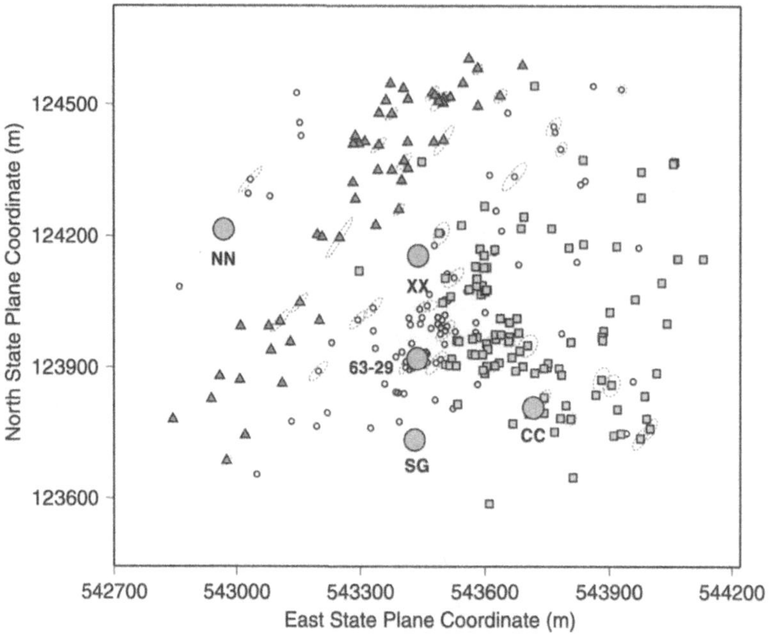

Figure 6
Map view of 248 microearthquakes detected on the downhole array in 63-29 and surrounding surface stations. The monitor well and surface stations are shown as larger shaded circles. These hypocenters had rms travel-time residuals ≤5 msec and location errors less than 60 m. Every fifth error ellipsoid projection is shown as dashed ellipses to show representative location errors. Two sub-clusters of locations are displayed as shaded square and triangle symbols and were identified based on the depth view shown in Figure 7. All other hypocenters are shown as open circles.

and parallel to the strike of a major surface lineament observed over the productive field (NIELSON and NASH, 1997). A depth view projecting the locations onto a plane orthogonal to the map trend is shown in Figure 7. Figure 8 depicts a depth projection along the same profile direction as Figure 3. The shallowest steam entries in this area occur at about −180 m elevation. The events mapped are distinct from most surface-monitored seismicity in that they lie above the producing reservoir. Preliminary, downhole monitoring at this site indicated that the shallow seismicity magnitudes ranged from about −3 to 0 (ALBRIGHT *et al.*, 1998). Two major clusters are identified in the strike-direction depth view (Fig. 7). The shaded square-symbol cluster forms a low-angle feature dipping approximately 30° southeast. The same symbols are used in the map view to aid in visualizing the three-dimensional distribution of hypocenters. The depth view looking along strike of the serpentine interface (Fig. 8) reveals that the shallowest seismicity terminates near the base of the serpentine, the boundary known to accommodate much of the near-surface strain, as manifest from the well casing deformations in the area (Fig. 3).

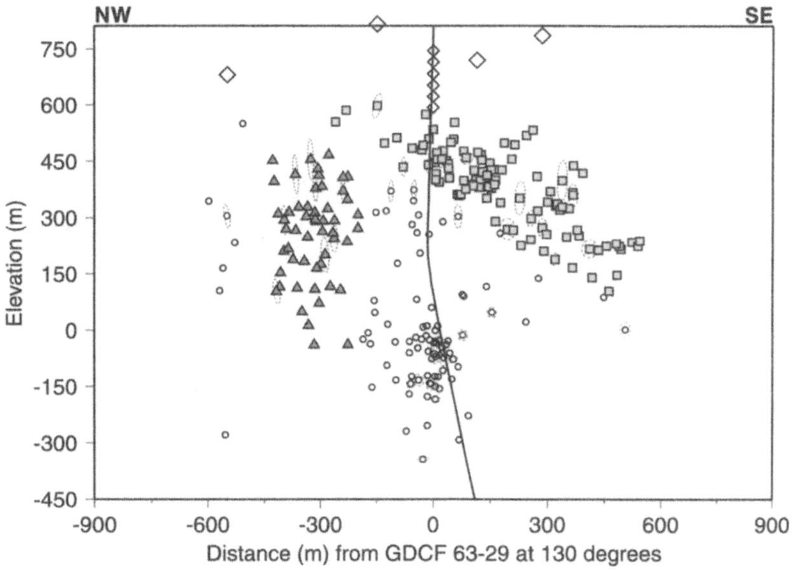

Figure 7
Depth locations of events projected onto a plane oriented 130° azimuth (viewing at N40E). Well GDCF 63-29 trace is shown. The open diamond symbols are the downhole and surface stations used. Two hypocenter groups are identified based on this projection and are distinguished by the shaded square and triangle symbols. All other hypocenters are shown as open circles. The map view of these same groups are shown in Figure 6 for a three-dimensional visual aid. Top of reservoir is at approximately −180 m. Every fifth error ellipsoid projection is shown as dashed ellipses to point out representative location errors.

Injection and Near-surface Seismicity

Seismicity and injection activity at The Geysers typically exhibit strong temporal and spatial correlations (STARK, 1990; BEALL *et al.*, 1999). Figure 9 shows this to be true in this small study area of The Geysers, for both the overburden (that is, above the shallowest steam entries) and the deeper, reservoir microseismicity. Figure 10 illustrates the difference in depth distribution for our near-array events and the U-N-T surface-network-detected events for the same epicentral area. The shallow, near-array events are not energetic enough for common detection over the larger station spacing (≥1 km) of the U-N-T surface array, therefore, the overburden seismicity is a subset of the total downhole-detected event population unique to our local array. The gross rate change in overburden seismicity tracks right along with changes in the deeper, reservoir seismicity. Both cumulative-event count curves increase in slope within about 1 week of the increased injection rate, starting at approximately day 380 (Fig. 9). We also looked for correlations between production and the shallow seismicity, using the production wells shown in Figure 2. No clear temporal correlation between production and event rate was evident.

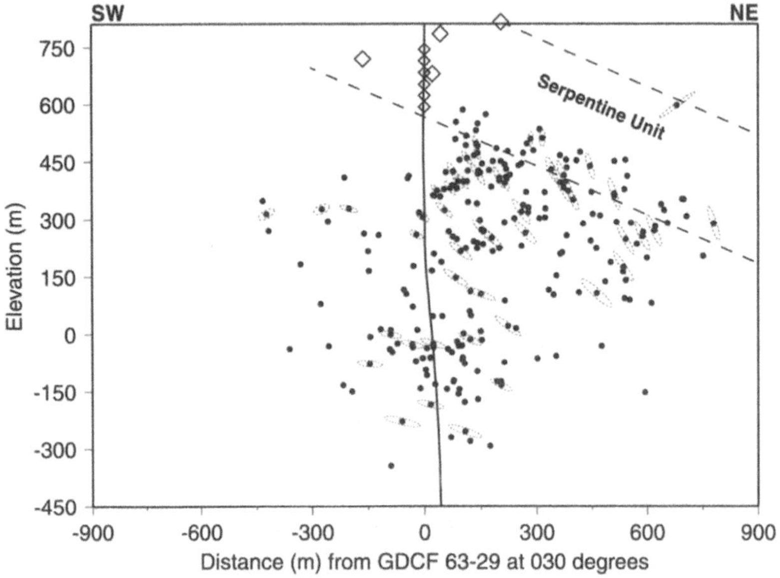

Figure 8
Depth locations of events projected onto a plane oriented N30E (viewing at N60W). Well GDCF 63-29 trace is shown. The open diamond symbols are the downhole and surface stations used. The dashed lines mark the approximate top and bottom contacts of a serpentine unit (as seen in Fig. 3). Top of reservoir is approximately at −180 m elevation. Every fifth error ellipsoid projection is shown as dashed ellipses to show representative location errors.

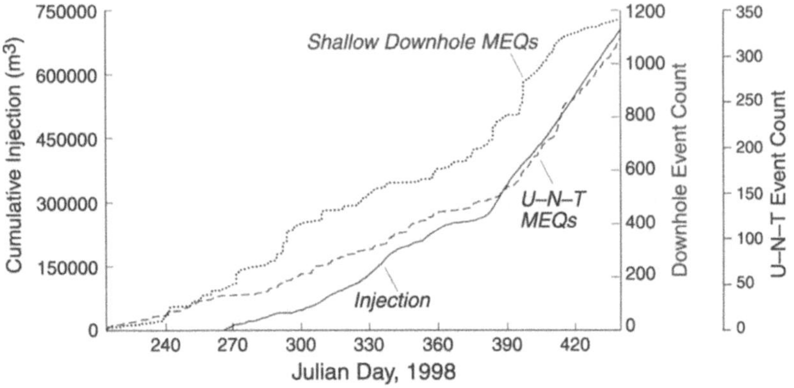

Figure 9
Cumulative injection volume in the study area and the cumulative event counts for the downhole detected events within about 750 m of the array and the U-N-T surface-array events with epicenters within 750 m of the monitor-well wellhead. Figure 2 shows the injection wells and the common epicenter area. Figure 10 shows the depth distribution of the two distinct populations of events; the downhole events considered for analysis (*S* - *P*-arrival time ≤150 msec) occur predominantly within the overburden, whereas the larger, wider-distributed surface array detects the deeper, higher-energy reservoir events. Injection in the area was ongoing prior to day 266, however complete records were not available. Both the reservoir and overburden microseismicity increase within about one week of the increased injection rate at day 380.

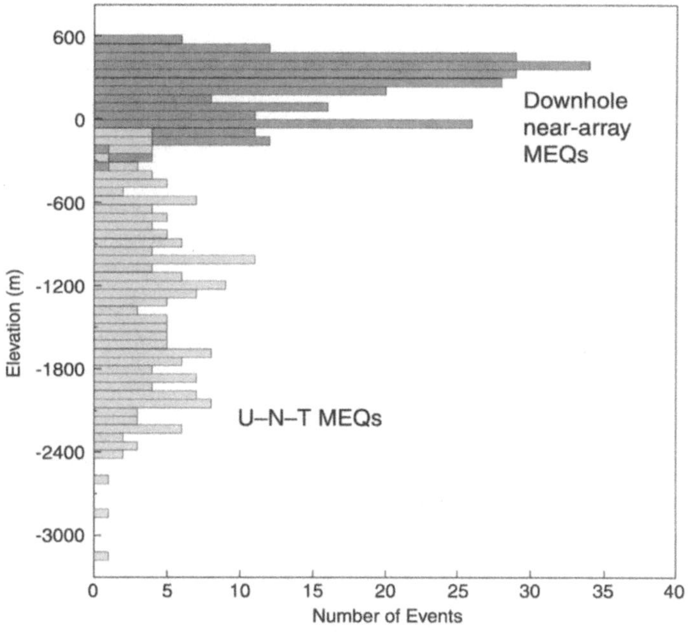

Figure 10
Depth distribution of microearthquakes detected on the U-N-T surface array at The Geysers and the shallow, downhole array in well 63-29. These are events with epicenters within the map area of Figure 2, over the common time interval January 6 to March 16, 1999. The near-array events consider for analysis from the downhole array (S - P-arrival time $\leq$150 msec) are predominantly within the overburden, whereas the larger, wider-distributed surface array detects only the deeper, higher-energy reservoir events. Top of reservoir for this area is at approximately −180 m elevation.

Focal Mechanisms

Composite fault-plane solutions were computed (Fig. 11) for the square- and triangle-symbol hypocenter groups (Figs. 6 and 7) using the computer program FPFIT (Reasenberg and Oppenheimer, 1985). Convergence to a single solution was achieved for both cases and both indicate a component of reverse faulting (triangle group predominantly so). The square-symbol group solution is better constrained and has fewer discrepant first-motion data.

Discussion

For depths between 1 and 5 km at The Geysers, inversion of focal mechanisms for stress tensors indicates a normal faulting stress regime that is transitional to strike-slip (i.e., the maximum horizontal stress is approximately equal to lithostatic load) (Oppenheimer, 1986). For the upper 1 km, focal mechanisms over large areas

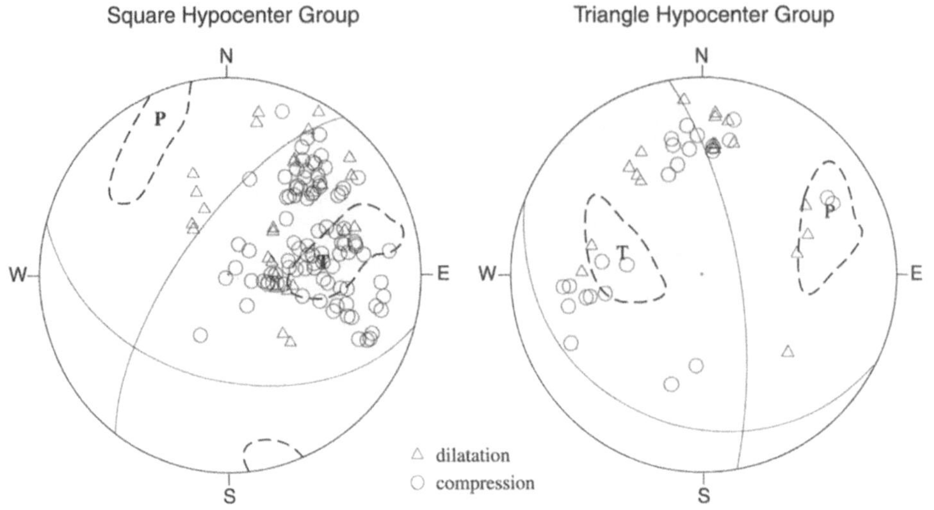

Figure 11
Composite fault-plane solutions as equal-area, lower-hemisphere projections for square- and triangle-symbol hypocenter groups shown in Figures 6 and 7. Compressive quadrants are shaded. Dashed boundaries around *P* and *T* axes show the uncertainty within the 90% confidence limit. Only first-motion data with signal-to-noise ratios ≥10 were used. The square-group solution has 112 first-motion data with 15% discrepant; the triangle-group solution has 43 first-motion data with 28% discrepant.

of the field include reverse faulting as well as normal and strike slip failures. As a result the stress regime is very poorly resolved at shallow depths. The stress regime could very well transition to reverse faulting at the near surface as the lithostatic load decreases. OPPENHEIMER (1986) also noted that the rugged topography at The Geysers could result in rapid lateral variations in the stress field and may in part explain the highly variable focal mechanisms observed at shallow depths over larger areas of the field. Reverse faulting at shallow depths will further be promoted as a result of reservoir contraction accompanying pressure and temperature drawdown (SEGALL and FITZGERALD, 1998).

The first motion data suggest a reverse faulting stress regime locally above the reservoir (<750 m depth). However, the lack of clear planar distribution of the grouped hypocenters leaves no corroborative evidence that the nodal planes are reasonable and leaves in question whether the compositing of first motion data is justified. The northeast striking nodal plane for the better-constrained, square-symbol-group solution (N37E) closely corresponds to the gross trends of seismicity seen in map view, and could indicate that failure predominantly occurs along similarly-oriented, steeply-dipping fractures (Fig. 11). The difference in *P*-axes orientations between the two solutions would indicate that the focal mechanisms do not represent planes of maximum resolved shear and/or that the principle stress axes orientations in the overburden varies over short distances (~600 m).

Serpentinites are known to be mechanically weak, particularly at the temperatures encountered for this study (< 150 °C) (MOORE *et al.*, 1997). The termination of seismicity near the base of the serpentine (Fig. 8) and the concentration of wellbore deformation at that interface (Fig. 3) suggest that strain transitions from brittle to more ductile deformation there. Volumetric contraction, accumulated on numerous discrete small faults below the serpentine, could be accommodated in the form of continuous plastic deformation in the overlying serpentine. This would create a strain discontinuity across the basal contact of the serpentine, subjecting the well casings to stress buildup and eventually failure. Repeated caliper logs from several wells of the area show spotty patterns of deformation rates and magnitudes. The caliper logs are not oriented, so no direction or sense of motion is obtained from these periodic "strain" measurements. Oriented caliper logs might indicate if a uniform sense of motion occurs at inter-well scales and may provide insight into the kinematics of strain concentration and transition at the serpentine boundaries.

The short-term correlation of shallow seismicity with injection is impossible to explain in terms of pore pressure diffusion or local reservoir volume reduction. Rather, the shallow seismicity may be triggered by the more energetic reservoir seismicity via dynamic (e.g., STARK and DAVIS, 1996) and/or quasi-static stress transfer.

Conclusions

Our ability to detect and locate the shallow seismicity can be attributed to placing geophones downhole and deploying surface stations close enough to enable detection of these low-magnitude events on multiple sensors. Composite fault-plane solutions suggest oblique reverse faulting above the reservoir. Seismicity is nearly absent above the base of a serpentine unit. This lithologic contact is known to accommodate much of the near-surface strain, as evident from the well casing deformations in the area. The concentration of casing deformation at the contact may be due to strain discontinuity between a plastic deformation regime in the serpentine and a brittle deformation regime below. The shallow, above-reservoir microseismicity is strongly correlated in time with the reservoir injection and reservoir injection-induced seismic activity occurring just below. This correlation is best explained as a small-scale example of remote triggering, as previously observed on a larger scale at The Geysers and elsewhere.

Acknowledgments

Unocal Corporation and Calpine Corporation granted access to the well and field facilities and provided personnel to assist in the work. We thank Steve Whitt and

Mike Woodall for their assistance in the field work. Baljit Singh, with assistance from Jim Thomson, planned and supervised the well abandonment and array installation operations. Joe Beall provided valuable discussions on data analysis and interpretation. This work was funded by the U.S. Department of Energy's Office of Geothermal Technologies.

References

ALBRIGHT, J. N., RUTLEDGE, J. T., FAIRBANKS, T. D., THOMSON, J. C., and STEVENSON, M. A. (1998), *Vertical Arrays for Fracture Mapping in Geothermal Systems*, Geotherm. Resources Counc. Trans. *22*, 459–463.

BARTON, C. A., ZOBACK, M. D., and MOOS, D. (1995), *Fluid Flow Along Potentially Active Faults in Crystalline Rock*, Geology *23*, 683–686.

BEALL, J. J., STARK, M. A., SMITH, J. L., and KIRKPATRICK, A. (1999), *Microearthquakes in the Southeast Geysers Before and After SEGEP Injection*, Geotherm. Resources Counc. Trans. *23*, 253–257.

DENLINGER, R. P. and BUFE, C. G. (1982), *Reservoir Conditions Related to Induced Seismicity at The Geysers Steam Reservoir, Northern California*, Bull. Seismol. Soc. Am. *72*, 1317–1327.

EBERHART-PHILLIPS, D. and OPPENHEIMER, D. H. (1984), *Induced Seismicity in The Geysers Geothermal Area, California*, J. Geophys. Res. *89*, 1191–1207.

GAUCHER, E., CORNET, F. H., and BERNARD, P. (1998), *Induced Seismicity Analysis for Structure Identification and Stress Field Determination*, SPE Paper 47324, Soc. Petro. Eng./Int. Soc. Rock Mech., Proc. Eurock 98, Trondheim, Norway.

JULIAN, B. R., MILLER, A. D., and FOULGER, G. R. (1993), *Non-shear Focal Mechanisms of Earthquakes at The Geysers, California, and Hengill, Iceland, Geothermal Areas*, Geotherm. Resources Counc. Trans. *17*, 123–128.

KIRKPATRICK, A., PETERSON, J. E., and MAJER, E. L. (1995), *Microearthquake Monitoring at the Southeast Geysers Using a High-Resolution Digital Array*, Proc. 20th Workshop on Geotherm. Reservoir Engineering, Stanford University, Stanford, California, 79–89.

KIRKPATRICK, A., PETERSON, J. E., and MAJER, E. L. (1997), *Three-dimensional Compressional- and Shear-wave Seismic Velocity Models for the Southeast Geysers Geothermal Field, California*, Proc. 22nd Workshop on Geotherm. Reservoir Engineering, Stanford University, Stanford, California, 399–410.

MAJER, E. L. and MCEVILLY, T. V. (1979), *Seismological Investigations at The Geysers Geothermal Field*, Geophysics *44*, 246–269.

MARKS, S. M., LUDWIN, R. S., LOUIS, K. B., and BUFE, C. G. (1978), *Seismic Monitoring at The Geysers Geothermal Field, California*, U.S. Geol. Surv., Open-File Rept. 78–798.

MOORE, D. E., LOCKNER, D. A., MA, S., SUMMERS, R., and BYERLEE, J. D. (1997), *Strengths of Serpentinite Gouges at Elevated Temperatures*, J. Geophys. Res. *102*, 14,787–14,801.

MOSSOP, A. and SEGALL, P. (1997), *Subsidence at The Geysers Geothermal Field, N. California from Comparison of GPS and Leveling Surveys*, Geophys. Res. Lett. *24*, 1839–1842.

NIELSON, D. L. and NASH, G. D. (1997), *Structural Fabric of The Geysers*, Geotherm. Resources Counc. Trans. *21*, 643–649.

OPPENHEIMER, D. H. (1986), *Extensional Tectonics at The Geysers Geothermal Area, California*, J. Geophys. Res. *91*, 11,463–11,476.

PHILLIPS, W. S., HOUSE L. S., and FEHLER, M. C. (1997), *Detailed Joint Structure in a Geothermal Reservoir from Studies of Induced Microearthquake Clusters*, J. Geophys. Res. *102*, 11,745–11,763.

PHILLIPS, W. S. (2000), *Precise Microearthquake Locations and Fluid Flow in the Geothermal Reservoir at Soultz-sous-Forets, France*, Bull. Seismol. Soc. Am. *90*, 212–228.

REASENBERG, P. and OPPENHEIMER, D. (1985), *FPFIT, FPPLOT and FPPAGE: Fortran Computer Programs for Calculating and Displaying Earthquake Fault-plane Solutions*, U.S. Geological Survey, Open-File Report 85–0739.

ROMERO Jr., A. E., MCEVILLY, T. V., MAJER, E. L., VASCO, D. (1995), *Characterization of the Geothermal System Beneath the Northwest Geysers Steam Field, California, from Seismicity and Velocity Patterns*, Geothermics *24*, 471–487.

ROSS, A., FOULGER, G. R., and JULIAN, B. R. (1999), *Source Processes of Industrially-induced Earthquakes at The Geysers Geothermal Area, California*, Geophysics *64*, 1877–1889.

RUTLEDGE, J. T., PHILLIPS, W. S., and SCHUESSLER, B. K. (1998a), *Reservoir Characterization Using Oil-production Induced Microseismicity, Clinton County, Kentucky*, Tectonophysics, *289*, 129–152.

RUTLEDGE, J. T., PHILLIPS, W. S., HOUSE, L. S., and ZINNO, R. J. (1998b), *Microseismic Mapping of a Cotton Valley Hydraulic Fracture Using Decimated Downhole Arrays*, 68th Ann. Mtg., Soc. of Explor. Geophys., Expanded Abstracts, 338–341.

SEGALL, P. (1989), *Earthquakes Triggered by Fluid Extraction*, Geology *17*, 942–946.

SEGALL, P. and FITZGERALD, S. (1998), *A Note on Induced Stress Changes in Hydrocarbon and Geothermal Reservoirs*, Tectonophysics *289*, 117–128.

STARK, M. A. (1990), *Imaging Injected Water in The Geysers Reservoir Using Microearthquake Data*, Geotherm. Resources Counc. Trans. *14*, 1697–1704.

STARK, M. A. and DAVIS, S. D. (1996), *Remotely Triggered Microearthquakes at The Geysers Geothermal Field, California*, Geophys. Res. Lett. *23*, 945–948.

(Received May 2, 2000, revised/accepted October 31, 2000)

Pure appl. geophys. 159 (2002) 489–516
0033–4553/02/030489–28 $ 1.50 + 0.20/0

Pure and Applied Geophysics

Determination of Stress State from Focal Mechanisms of Microseismic Events Induced During Hydraulic Injection at the Hijiori Hot Dry Rock Site

SHUNJI SASAKI[1] and HIDESHI KAIEDA[1]

Abstract—The stress state at the Hijiori hot dry rock site was estimated based on the inversion from focal mechanisms of microseismic events induced during hydraulic injection experiments. The best fit stress model obtained by inverting 58 focal mechanisms of seismic events simultaneously indicates that the maximum principal stress σ_1 is vertical, while the minimum principal stress σ_3 is horizontal and trends north-south. The average misfit between the stress model and all the data is 6.8°. The inversion results show that the average misfit is small enough to satisfy the assumption of homogeneity in the focal mechanism data and that the 95% confidence regions of σ_1 and σ_3 are well constrained, i.e., they do not overlap, suggesting that the inversion results are acceptable. The stress estimates obtained by the focal mechanism inversion essentially agree with other stress estimates previously obtained. It is therefore concluded that the focal mechanism inversion method provides a useful tool for estimating the stress state. The hypocentral distributions of microseismic events associated with the hydraulic fracturing experiments are distributed around the plane that spreads to almost east–west from the injection wells and declines to the north at a high angle. The vertical orientation and east–west strike of the seismic events are essentially coplanar with the caldera ring-fault structure in the southern portion of the Hijiori Caldera. This indicates that tensile fractures of intact rock were not being created, but pre-existing fractures were being re-opened and developed in the direction of the maximum horizontal principal stress, although microseismic events were caused by shear failures.

Key words: Induced seismicity, microearthquakes, focal mechanism, stress inversion, hot dry rock, hydraulic fracturing.

1. *Introduction*

Since 1984, the New Energy and Industrial Technology Development Organization (NEDO) has been conducting a research program for a hot dry rock (HDR) geothermal energy system at Hijiori, Yamagata Prefecture, Japan (KURIYAGAWA and TENMA, 1999). The extraction of geothermal energy from HDR requires the creation of an artificial geothermal reservoir (fracture) between an injection well and a production well. This geothermal reservoir, which acts as a heat exchange interface, is created by hydraulic fracturing. Since the fractures created by hydraulic fracturing

[1] Central Research Institute of Electric Power Industry, 1646 Abiko, Abiko, Chiba 270-1194, Japan. E-mail: s-sasaki@criepi.denken.or.jp

at depth are controlled by the stress state and the permeability, the stress states at HDR test sites have been extensively estimated by using various methods such as the hydraulic fracturing method (e.g., CORNET and JULIEN, 1989), laboratory tests by the Acoustic Emission method (e.g., SHIN *et al.*, 2000) and the Differential Strain Curve Analysis method (DSCA) (e.g., OIKAWA *et al.*, 1993).

On the other hand, to obtain information pertaining to the shape and propagation of hydraulically created fractures and regarding the direction of major fluid flow away from the injection well, monitoring and analysis of microseismic events associated with hydraulic fracturing experiments have been carried out by various researchers. Results of monitoring of hydraulic fracturing experiments have been reported (e.g., JUPE *et al.*, 1992; CORNET *et al.*, 1997; PHILLIPS *et al.*, 1997; SASAKI, 1998). The analysis of microseismic locations provides information on the fracture growth, the main flow direction, and the drilling target of a new well, while the focal mechanisms of the microseismic events provide other information on the stress state (e.g., CORNET and JULIEN, 1989; CORNET and YIN, 1995; TEZUKA and NIITSUMA, 2000).

In assessing stress states from a focal mechanism data set, two problems arise. First, it is often assumed that the averaged P, B, and T axes from a number of focal mechanisms coincide with the maximum, intermediate and minimum principal stress directions. However, it is widely recognized that these axes from a single focal mechanism may vary significantly from the principal stress directions (MCKENZIE, 1969). In other words, these axes represent the moment tensor and not the stress. The second problem is the ambiguity as to which of the two nodal planes is the actual fault plane. A focal mechanism represents two possible planes and slip directions due to the interchangeability of the fault planes and auxiliary nodal planes. Unfortunately, the constraint the focal mechanism places on the stress field is different for each of the possible fault planes. To invert focal mechanisms for the stress state, therefore, we must know which nodal plane is the true fault plane. Normally, selecting the preferred nodal plane from each focal mechanism is done on the basis of local geology and tectonics. However, there is no objective means by which to make this selection. Several inverse methods have been proposed to determine the stress field which is most consistent with a set of focal mechanism data (e.g., VASSEUR *et al.*, 1983; GEPHART and FORSYTH, 1984; MICHAEL, 1987; RIVERA and CISTERNAS, 1990). Among them, GEPHART and FORSYTH (1984) have developed an inversion method that attempts to select the correct fault planes while inverting the stress state.

Between 1988 and 1995, NEDO conducted a number of hydraulic fracturing and circulation experiments at the Hijiori HDR site and successfully created two reservoirs: a shallow reservoir at a depth of 1800 m and a deep reservoir at 2200 m. In this paper attention focuses on estimating of the stress state at the Hijiori HDR site by inverting the focal mechanisms of microseismic events observed during these experiments. After a brief description of the Hijiori HDR site, microseismic activities associated with the experiments are presented. Next, the stress state at the Hijiori

HDR site is estimated by applying the stress inversion method of GEPHART and FORSYTH (1984) to the focal mechanism data set of microseismic events. Finally, the relation between the induced microseismicity, the stress state and the hydraulically induced fractures at the Hijiori HDR site is discussed.

2. *Hijiori HDR Site*

The location of the Hijiori HDR site in Yamagata Prefecture is shown in Figure 1, along with the locations of four wells (SKG–2, HDR–1, HDR–2 and HDR–3) used as injection and production wells in the experiments.

The topography is characterized by a volcanic depression called the Hijiori caldera, whose diameter is about 1.5 to 2 km (Fig. 1). The HDR site is located at the southern rim of the Hijiori Caldera where four wells have been drilled to between 1800 m and 2300 m. According to the cores recovered from the wells, the order of geological stratigraphy from depth to the surface is as follows: Pre-Tertiary

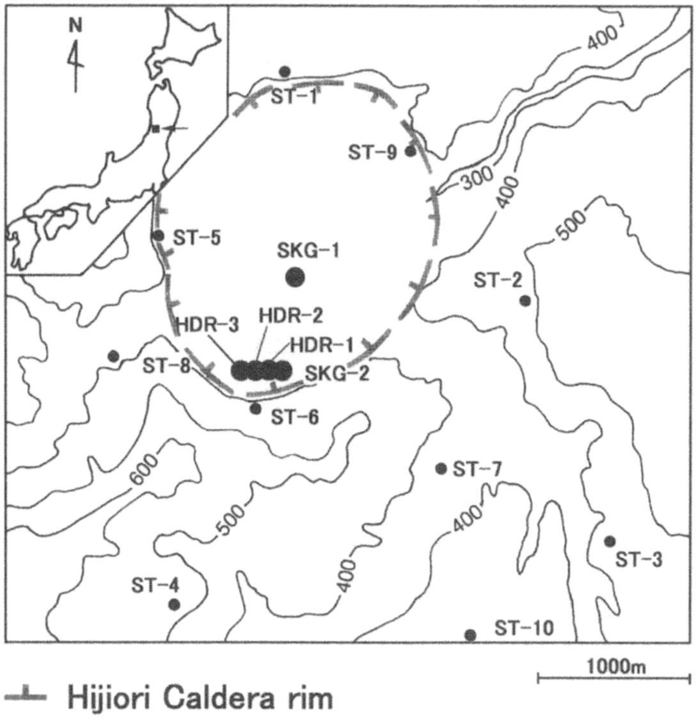

Figure 1
Locations of SKG–2, HDR–1, HDR–2, and HDR–3 wells at the Hijiori HDR site. ST–1 to ST–10 (dots) are borehole seismic stations. Information from SKG–1 and SKG–2 was used to construct the velocity structure model for region.

granodiorite which forms the bedrock, the Neocene formation, ejecta of the Quaternary period and alluvium. The depths from the heads of each well SKG–2, HDR–1, HDR–2 and HDR–3 to the granodiorite range from 1450 m to 1486 m.

The experiment at the Hijiori HDR site began in 1984, using the existing geothermal exploration well SKG–2 which has a depth of 1802 m with casing to a depth of 1788 m. The second well, HDR–1, was drilled in 1987 to a depth of 1805 m with casing to a depth of 1513 m. HDR–1 was extended to a depth of 2205 m in 1989. In the same year, the third well, HDR–2, was drilled to a depth of 1910 m with casing to a depth of 1510 m. In 1990, the fourth well, HDR–3, was drilled to a depth of 1907 m with casing to a depth of 1516 m. HDR–3 was deepened to 2303 m in 1993 and HDR–2 to 2302 m in 1994. Shown in Figure 2 are the trajectories of SKG–2, HDR–1, HDR–2 and HDR–3.

The temperatures at the bottom of SKG–2 (depth: 1802 m) and HDR–1 (depth: 2205 m) are 253 and 271 °C, respectively.

3. Outline of Experiments

NEDO conducted a number of hydraulic fracturing and circulation experiments at the Hijiori HDR site between 1988 and 1995, and succeeded in creating two

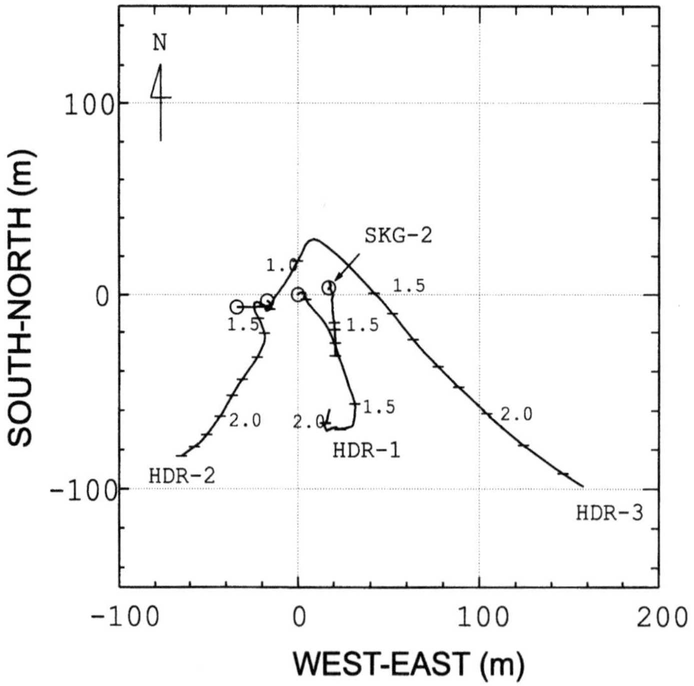

Figure 2
Drilling trajectories of SKG–2, HDR–1, HDR–2, and HDR–3 wells (plan view).

reservoirs: a shallow reservoir at a depth of 1800 m, and a deep reservoir at 2200 m as shown in Figure 3. The outline of these experiments is shown in Table 1. Since the experiments at the Hijiori HDR site are summarized by KURIYAGAWA and TENMA (1999), a brief description of the main experiments conducted by NEDO at the Hijiori HDR site is given here.

3.1 Hydraulic Fracturing Experiment in 1988

A hydraulic fracturing experiment was carried out using SKG–2 as an injection well, and a shallow reservoir was created at a depth of 1800 m. In the experiment, a total of about 2000 m^3 of water was injected from SKG–2 at four stages of flow rate,

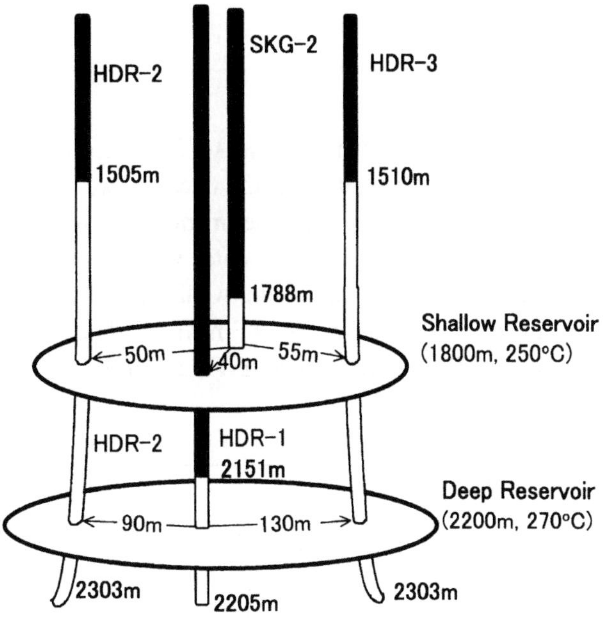

Figure 3
Schematic representation of locations of SKG–2, HDR–1, HDR–2, and HDR–3 wells, and the shallow and deep reservoirs at the Hijiori HDR site.

Table 1

Outline of experiments at the Hijiori HDR site

	Type of experiment	Injection well	Injection interval (m)	Maximum of flow rate (m^3/min)	Total injection volume (m^3)	Production well
1988	hydraulic fracturing	SKG–2	1788–1802	6	1,961	
1989	circulation	SKG–2	1788–1802	2	45,921	HDR–1,–2
1992	hydraulic fracturing	HDR–1	2151–2205	4	2,115	
1995	circulation	HDR–1	2151–2205	2	51,500	HDR–2, –3

i.e., 1, 2, 4 and 6 m^3/min. The changes of wellhead pressure and injection flow rate at SKG–2 with time are shown in Figure 4(a). As shown in this figure, the maximum wellhead pressure at SKG–2 was 16 MPa when the injection flow rate was 6 m^3/min.

3.2 Circulation Test in 1989

A 30-day circulation test was conducted by injecting water into SKG–2 and producing hot water and steam from HDR–1 and HDR–2. The injection flow rate was kept at 1 $\sim$ 2 m^3/min, and the wellhead pressure at SKG–2 was 6 MPa initially and decreased to 4.5 MPa. The total amount of water injected during this test was about 44,500 m^3, and hot water and steam of about 13,000 m^3 were recovered from HDR–1 and HDR–2. The changes of wellhead pressure and flow rate at SKG–2 are shown in Figure 4(b).

3.3 Hydraulic Fracturing in 1992

After HDR–1 was deepened from 1805 m to 2205 m, a hydraulic fracturing experiment was carried out using HDR–1 as the injection well, and a deep reservoir was created at a depth of 2200 m. In the experiment, the flow rate was changed in three steps, i.e., 1, 2 and 4 m^3/min, and the total volume of injection during the experiment reached 2115 m^3. The maximum wellhead pressure at HDR–1 was 25 MPa when the injection flow rate was 4.5 m^3/min. The changes of wellhead pressure and flow rate at HDR–1 are shown in Figure 4(c).

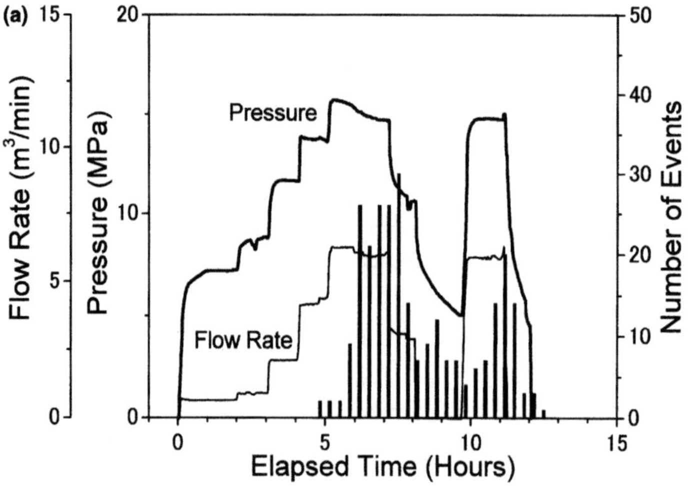

Figure 4(a)

Wellhead pressure, injection flow rate at SKG–2 and microseismic event rate per 20 minutes observed at ST–7 versus time for the 1988 hydraulic fracturing experiment. Water injection was started at 20:00, July 19, 1988. Microseismic events with amplitudes equal to or larger than 10 μkine were simultaneously counted by an event counter.

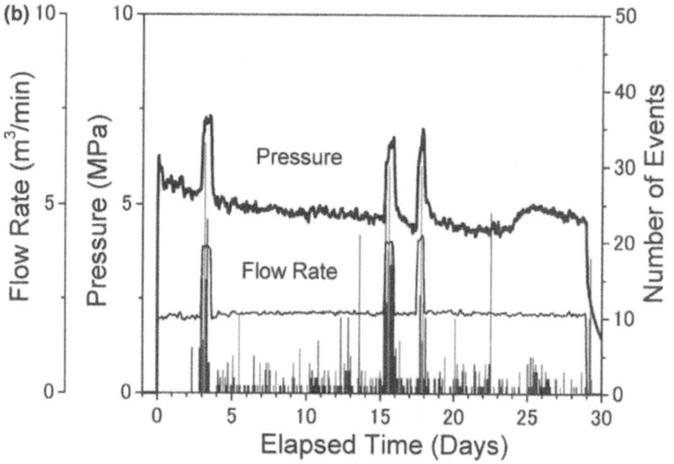

Figure 4(b)
Wellhead pressure, injection flow rate at SKG–2 and microseismic event rate per 2 hours observed at ST–7 versus time for the 1989 circulation test. Water injection was started at 16:00, October 19, 1989. Microseismic events with amplitude equal to or larger than 10 μkine were simultaneously counted by an event counter.

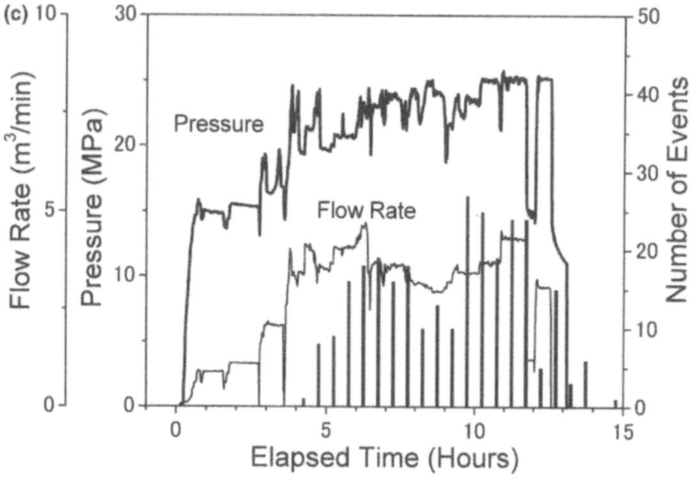

Figure 4(c)
Wellhead pressure and injection flow rate at HDR–1 during the 1992 hydraulic fracturing experiment and seismic event rate per 30 minutes observed at ST–7. Water injection was started at 21:09, July 21, 1992. Microseismic events with amplitudes equal to or larger than 10 μkine were simultaneously counted by an event counter.

3.4 Circulation Test in 1995

A 25-day circulation test was performed using HDR–1 as the injection well and deepened HDR–2 and HDR–3 as production wells. At the beginning of the test, water was injected into HDR–1 at the maximum flow rate of about 3.5 m^3/min. Then the flow rate was kept at around 1 and 2 m^3/min. The total amount of water injected during the experiment was about 51,500 m^3, and hot water and steam of 20,600 m^3 were recovered from HDR–2 and HDR–3. The maximum wellhead pressure at HDR–1 was 15 MPa when the injection flow rate was 3.5 m^3/min. The changes of wellhead pressure and flow rate at HDR–1 are shown in Figure 4(d).

4. Seismic Observation System

It is of importance to deploy a large number of seismometers surrounding the hydraulic fracturing well to obtain accurate locations of the microseismic events and to obtain a satisfactory azimuthal coverage for focal mechanism determination. A borehole seismic network was constructed, consisting of 10 stations (ST–1 ~ ST–10) deployed in a circle at a radius of 1.5 to 3 km around the injection and production wells. The locations of each seismic station are shown in Figure 1, together with those of SKG–2, HDR–1, HDR–2 and HDR–3 wells. To reduce noise and improve detection of high-frequency signals, seismometers were cemented in the boreholes at depths ranging from 50 m to 150 m. The seismometers used were three-component geophones with a natural frequency of 5 Hz. Output from the seismometers was amplified by 40 dB and transmitted over wire to a data acquisition room located within the HDR test site to keep time with accuracy. There, signals were amplified

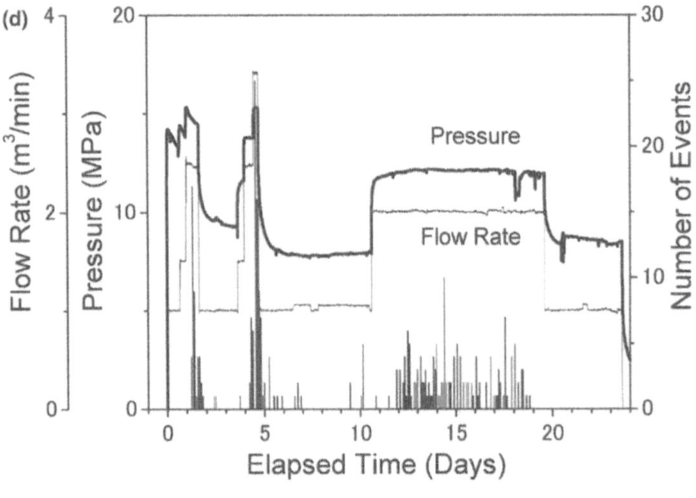

Figure 4(d)
Wellhead pressure and injection flow rate at HDR–1 during the 1995 circulation test and microseismic event rate for every 2 hours observed at ST–7. Water injection was started at 12:34, August 1995. Microseismic events with amplitude equal to or larger than 10μkine were simultaneously counted by an event counter.

again and recorded digitally on a triggered system. The digital recording system used consisted of an EWS and a 32-channel AD converter with a 16-bit resolution. The sampling rate was 1 kHz per channel and the total duration of the recorded signal was 1.8 sec. Recording was triggered by an automatic routine. The overall response was considered flat over the frequency range of 10–200 Hz, with a 90 dB dynamic range. This was backed up by continuous analogue recording on magnetic tape.

5. *Microseismic Activity, Pressure and Flow Rate*

The relations among the microseismic event rate, the amount of water injected into the injection wells and the wellhead pressure during the two hydraulic fracturing experiments and two circulation tests are shown in Figure 4(a) ~ Figure 4(d). In the 1988 hydraulic fracturing experiment, the first microseismic events were observed six hours following the start of pumping, after a total of about 900 m^3 of water had been injected (Figure 4(a)). The injected volume of water is almost equal to the stimulated volume ($= 1000$ m^3) during the 1986 hydraulic fracturing experiment. This indicates that the microseismicity during the 1988 hydraulic fracturing experiment did not start until the previously stimulated region created during the 1986 injection was re-inflated. As shown in Figure 4(a), the increases and decreases in seismic activity are significantly correlated to changes in the injection flow rate and to changes in the wellhead pressure. This tendency is similar to the relation between the injection pattern and the occurrence frequency of microseismic events associated with the 1992 hydraulic fracturing experiment and the 1989 and 1995 circulation tests (Figure 4(b) ~ Figure 4(d)). It appears that the seismicity rate strongly depends on the injection flow rate and on the wellhead pressure as well. However, the increases and decreases in seismic activity significantly lag the corresponding changes in the injection flow rate, but would probably be much better correlated to changes in the level of wellhead pressure. Thus it is more probable that a higher fluid pressure in the fracture is the cause of induced seismicity.

6. *Induced Microseismicity*

The microseismic events were located using a least-squares algorithm. All events with more than five *P*-wave arrivals and at least two *S*-wave arrivals were selected. The initial hypocenter was assumed to be at the bottom of the injection well and the initial origin time was determined from the first *P*-arrival time and *S*-*P*-time at ST-7 whose travel time was the first among all stations for most of the events. The *P* velocity structure model was developed using data from well logging in SKG–1 and SKG–2. The *S*-wave velocities were calculated by assuming Poisson's ratio to be 0.25. The velocity model used for location is shown in Table 2. Since the *S*-wave

Table 2

Velocity model used for location

Thickness (km)	Depth (km)	V_p (km/sec)	V_s (km/sec)
0.11	0.11	2.10	1.24
0.14	0.25	2.50	1.48
0.20	0.45	3.10	1.83
0.20	0.65	3.60	2.13
0.20	0.85	4.00	2.36
0.61	1.46	4.70	2.78
0.54	2.00	5.20	3.07
∞	∞	5.40	3.19

velocity was rather arbitrarily assumed and it was difficult to pick *S* waves reliably, *S*-wave arrival times were assigned a weight of 0.2, while *P*-wave arrivals were given full weight in the location procedure.

The seismic stations are located at the edge of or outside the Hijiori Caldera. This makes it difficult to locate microseismic events using a uniform velocity model since there are very probably lateral variations in geological structure across the seismic network. There are also large variations in station altitude between 214 m to 453 m above sea level. Station corrections were used in the location algorithm to account for the variations in altitude and structure. These were determined from a series of explosive calibration shots conducted at the bottom of SKG–2 in 1988, 1989 and HDR–1 in 1992. The quantity of explosives used in each blasting was 1 kg. The average of the *P*-wave travel-time residuals at each station observed from the blasts was used as *P*-wave station correction. *S*-wave station correction was obtained from *P*-wave station correction by assuming Poisson's ratio to be 0.25. When no station corrections were used, the blasting position obtained by calculation deviated from the actual position by approximately 140 m in both horizontal and vertical directions. On the other hand, the blasting position determined by including station corrections in the location algorithm was within 10 m of the known location. The hypocenters of microseismic events located in proximity to the injection wells are thus considered reliable to within 10 m. The station corrections to hypocenter locations of the microseismic events induced during the 1988 hydraulic fracturing experiment are compared in Figure 5. The hypocenters determined without station corrections are shown on the right and those determined with station corrections are on the left. The former hypocenters are diffused and are located west of the injection well, while the latter hypocenters are distributed linearly in the direction of N70°E from the injection well and delineate the fracture created by the experiment. The quality of the locations can also be seen from the RMS travel-time residual. The RMS travel-time residuals of the two location results are 35.3 ms and 5.5 ms, respectively. The reduction in the RMS residual using station corrections indicates

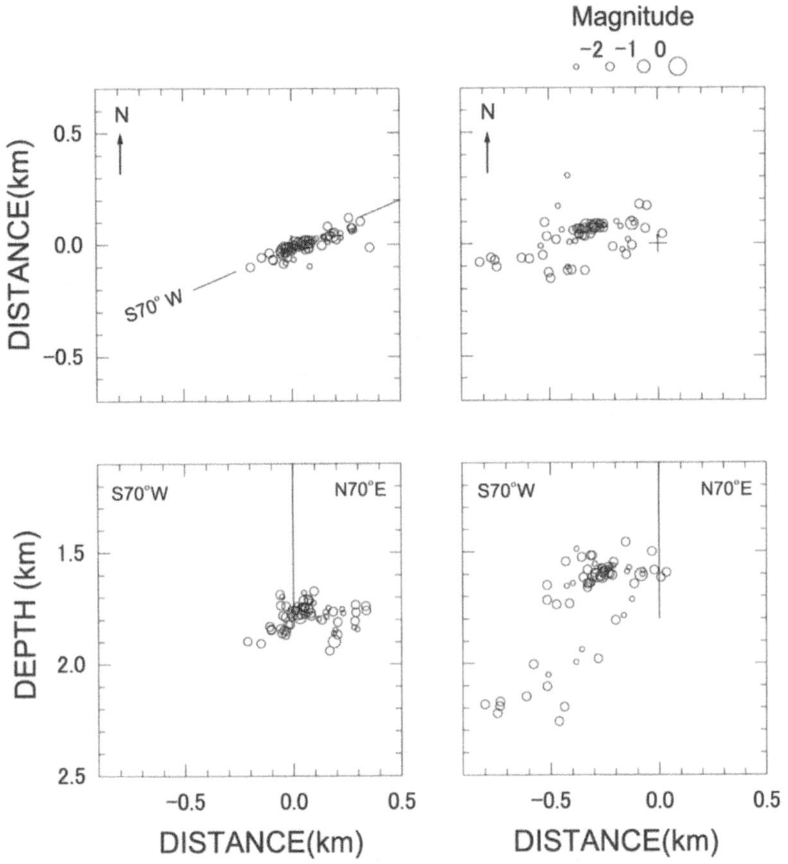

Figure 5
Comparison of hypocenters determined without station corrections (right) and those determined with station corrections (left) for the 1988 hydraulic fracturing experiment. Top: Epicentral distribution. Bottom: Vertical cross section projected onto a N70°E–S70°W plane where microseismic events are arranged linearly. Cross shown in the plan view indicates the well head of SKG–2. The trajectory of SKG–2 is shown in the section views. Symbols are proportional to size of microseismic event.

that the latter locations are an improvement over the former ones. These results indicate that the microseismic events can be located better with the station corrections.

The locations of the microseismic events induced during the four experiments are shown in Figure 6. Altogether, a total of 792 events was located by the borehole seismic network: 65 events during the 1988 hydraulic fracturing experiment, 270 during the 1989 circulation test, 127 during the 1992 hydraulic fracturing experiment and 330 during the 1995 circulation. The magnitudes of these events were determined from the maximum velocity amplitudes of the events, based on the Watanabe's method (WATANABE, 1971). The magnitudes range from −3.3 to 1.2 and average −2.2.

6.1 Microseismicity Induced in the 1988 Experiment

Figure 6(a) shows that the epicenters of microseismic events are distributed slightly south of SKG–2, and extend toward the N70°E direction from SKG–2, with a length of about 500 m. Most of the microseismic events which are located between 1700 and 1900 m below the surface are distributed vertically.

6.2 Microseismicity induced in the 1989 Experiment

The seismic cloud primarily lies northeast of SKG–2, in agreement with the hypocenter distribution of microseismic events induced in the 1988 hydraulic fracturing experiment (Fig. 6(b)). Near SKG–2, hypocenter depths ranged between

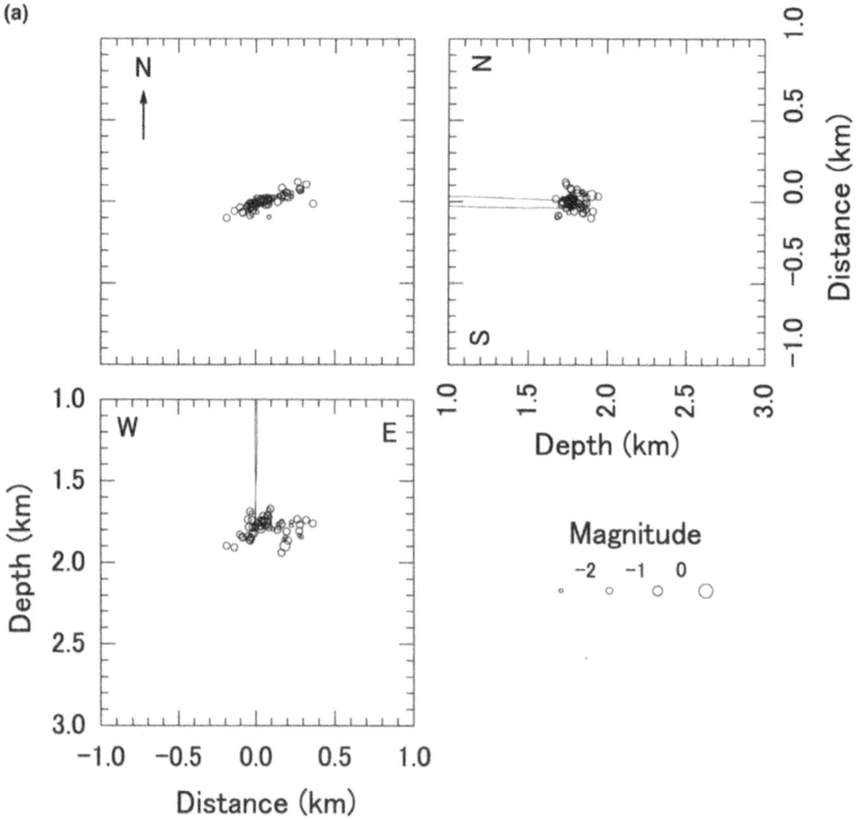

Figure 6(a)
Hypocenter distribution of the microseismic events induced during the 1988 hydraulic fracturing experiment. Upper left: Epicentral distribution. Upper right: Vertical cross section projected onto a south-north plane. Lower left: Vertical cross section projected onto an east–west direction. The trajectories of the injection and production wells are shown in the plan view and section views. Symbols are proportional to size of microseismic event.

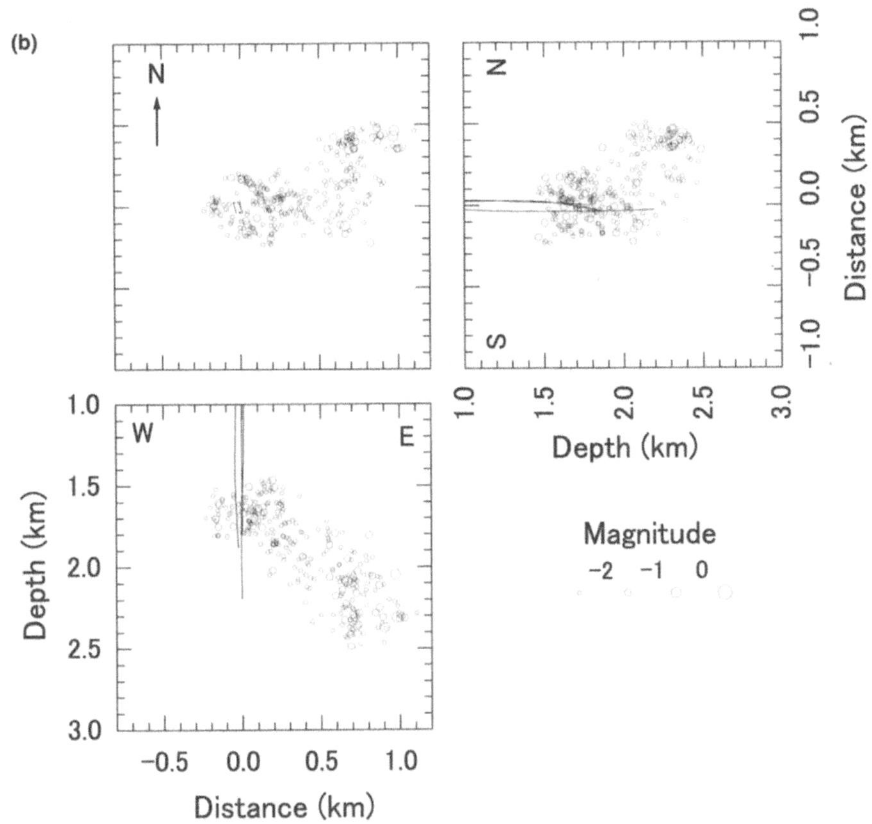

Figure 6(b)
Hypocenter distribution of the microseismic events induced during the 1989 circulation test. See Figure 6(a) caption.

1500 m and 1800 m below the surface, and were concentrated in a position shallower than the injection zone. However, the microseismic events tended to occur at a greater depth as they migrated away from SKG–2. The epicenters of microseismic events were densely distributed in the N70°E direction from SKG–2, similar to the trend observed in 1988. The hypocenter distribution of microseismic events induced by the circulation test occurred over a larger region compared to those that accompanied the 1988 hydraulic fracturing experiment.

6.3 Microseismicity Induced in the 1992 Experiment

Figure 6(c) shows that the hypocenters of microseismic events are distributed around a plane that spread east–west from HDR–1 and declined to the north at a high angle. The seismic cloud is distributed between 2,000 m and 2,500 m below the surface, and is clearly deeper than the 1988 seismic cloud.

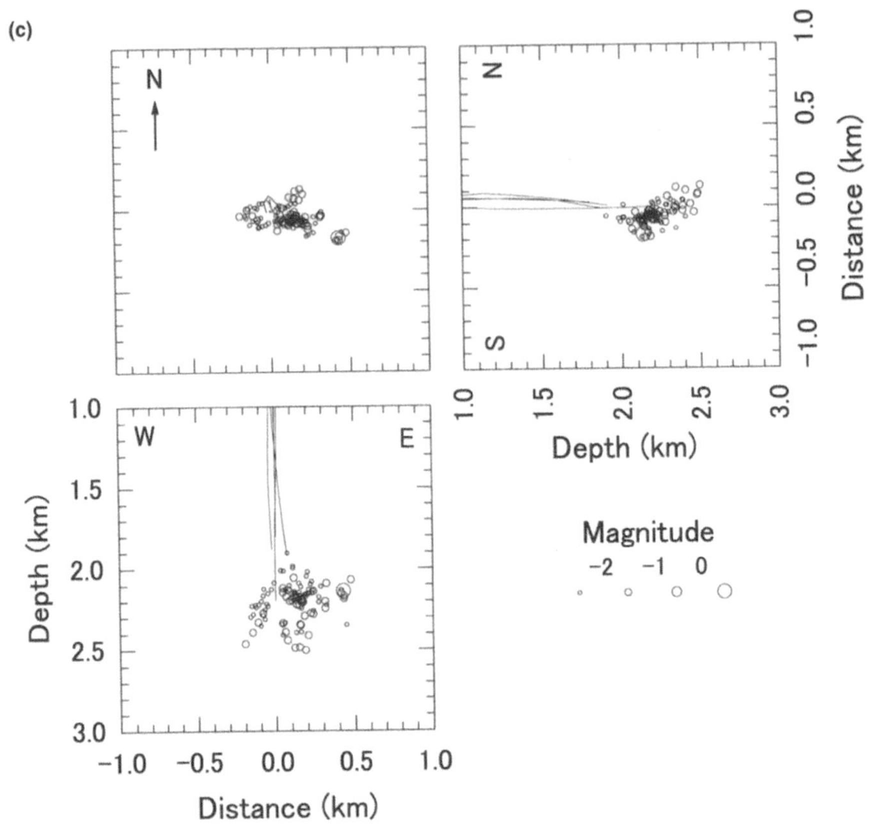

Figure 6(c)
Hypocenter distribution of the microseismic events induced during the 1992 hydraulic fracturing experiment. See Figure 6(a) caption.

6.4 Microseismicity Induced in the 1995 Experiment

The epicenters of microseismic events are distributed in the east–west direction from HDR–1 with a more diffused cloud of seismicity around the injection well (Figure 6(d)). There is a tendency that microseismic events occur outside the focal area of microseismic events associated with the 1992 hydraulic fracturing experiment and that the events whose magnitudes are larger than −1 occur at the margin of the active area. The locations of the microseismic events induced during the 1995 circulation test seem to bear a similar relationship to those of the 1992 hydraulic fracturing experiment, as the 1989 circulation test did to the 1988 hydraulic fracturing experiment. However, there is no deepening of the seismic cloud and it seems to be evenly spaced around the injection point.

In the case of the hydraulic fracturing experiments with a high injection pressure (25 MPa in the 1992 experiment), microseismic events were distributed around a

(d)

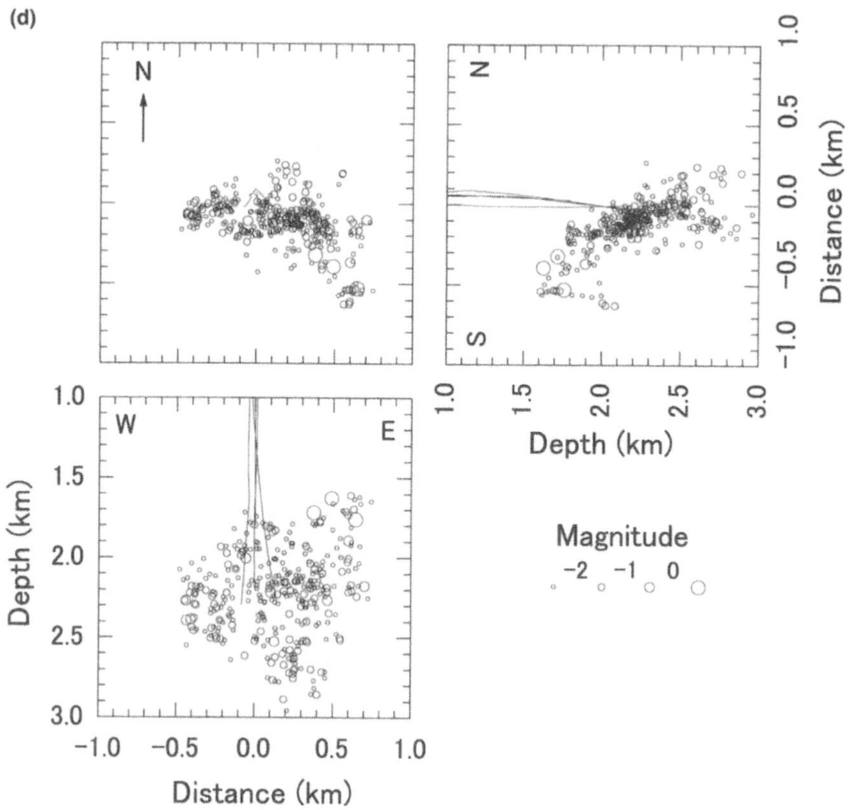

Figure 6(d)
Hypocenter distribution of the microseismic events induced during the 1995 circulation test. See Figure 6(a) caption.

straight line as shown in Figures 6(a) and 6(c). On the other hand, seismicities accompanying the circulation tests with a low injection pressure (4.5–6 MPa in the 1989 circulation test) were diffused and showed seismic clouds (Figs. 6(b), (d)). Assuming that an increased pore fluid pressure reduces the effective stresses and results in the occurrence of a shear failure, SASAKI (1998) estimated the permeability from the space-time distribution of microseismic events induced during the 1989 circulation test as 10^{-16} m^2, which is intermediate between the permeability of the core samples of granodiorite taken from the production well and that of fractured rocks obtained by an injection test between the injection and production wells. Therefore SASAKI (1998) concluded that the seismic cloud accompanying the circulation test is due to the permeation of water into pre-existing fractures which slip when the effective stress is reduced by the increased pore fluid pressure accompanying the hydraulic injection. Microseismic events occur and migrate along these pre-existing fractures. The diffusion of the microseismicity during a hydraulic fracturing experiment with a low injection pressure is also reported at the Ogachi HDR site (KAIEDA *et al.*, 1995) and it is a characteristic common to seismic activities associated with fluid injection; such as microearthquakes in geothermal areas (ITO AND SUGIHARA, 1988), reservoir associated earthquakes (GUPTA and RASTOGI, 1976; TALWANI and ACREE, 1984/85), earthquakes associated with volcanic activity (MIZOUE and YOKOTA, 1991; SUZUKI, 1992), and earthquakes induced by the injection of water into rock (OHTAKE, 1974; RALEIGH *et al.*, 1976).

7. *Focal Mechanism and Stress Field*

7.1 Focal Mechanism

Among the microseismic events induced during the hydraulic fracturing and circulation experiments shown in Figure 6, microseismic events with more than 8 initial *P*-wave data were selected and used in the focal mechanism analysis. Besides *P*-wave polarity data, the *P*- and *S*-wave amplitude ratios on the vertical component seismograms, with at least three seismograms, were also read to constrain the range of possible solutions (KISSLINGER, 1980). Best fit double–couple focal mechanisms were determined using the fault plane fit grid-search program FOCMEC of SNOKE *et al.* (1984). The program FOCMEC gives all acceptable solutions consistent with the data. Generally, the poorer the data, the larger the number of possible solutions. In order to select high quality focal mechanisms, the solutions with one plane not constrained to better than $\pm 15°$ were rejected. As a result, from 103 selected microseismic events, 45 were rejected because of poorly constrained fault plane solutions, resulting in a final data set of 58. This higher rejection rate is due to the lower station density. Examples of focal mechanism solutions are shown in Figure 7. The quality of focal mechanisms can be judged from Figure 8, which shows all possible solutions for each microseismic event: all satisfying the polarity data without polarity errors. The mechanism adopted in Figure 7 is chosen from the set of consistent mechanisms shown in Figure 8 by selecting the median nodal plane strike. Table 3 summarizes the focal mechanism results, listing strike, dip, and rake of both nodal planes, date, hypocenter location and magnitude. The number of focal mechanisms obtained in the 1988 hydraulic fracturing experiment, the 1989 circulation test, the 1992 hydraulic fracturing experiment and the 1995 circulation test were 9, 18, 7 and 24, respectively. The magnitudes of these events range from −2.1 to 0.6 and the average is −1.4. The distributions of the *P* and *T* axes of the focal mechanisms of microseismic events obtained are shown in Figure 9. The *P* and *T* axes obtained from the focal mechanisms of microseismic events induced during the 1988 hydraulic fracturing experiments are well concentrated close to vertical and nearly horizontal directions and tend to be nearly north–south, respectively. The *P* and *T* axes obtained from the focal mechanisms of microseismic events induced during the 1992 hydraulic fracturing experiments are also concentrated, inclining

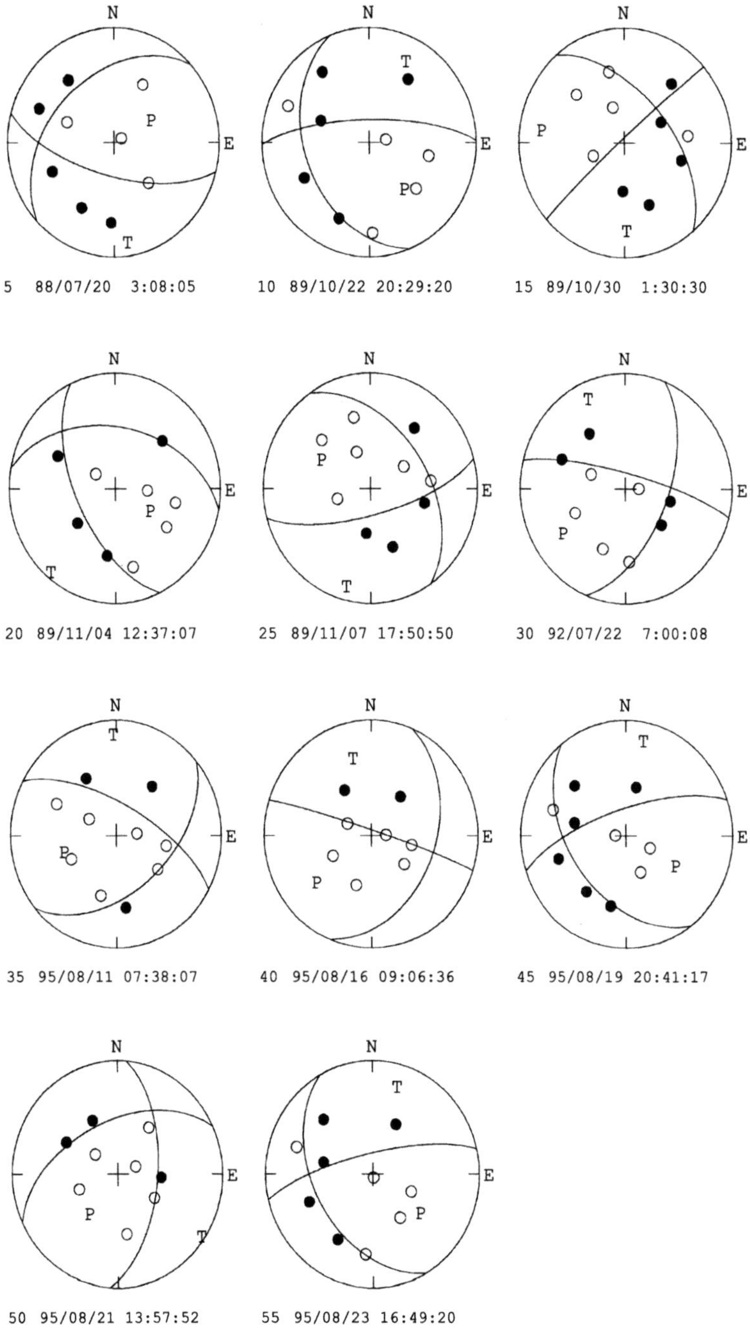

Figure 7

Examples of focal mechanism solutions obtained. Lower-hemisphere equal area projection is used. Solid and open circles are compressions and dilatations, respectively. Symbols *P*, *T* are the *P* and *T* axes. The label below each mechanism gives the identification number (Table 3) and origin time of the microseismic event.

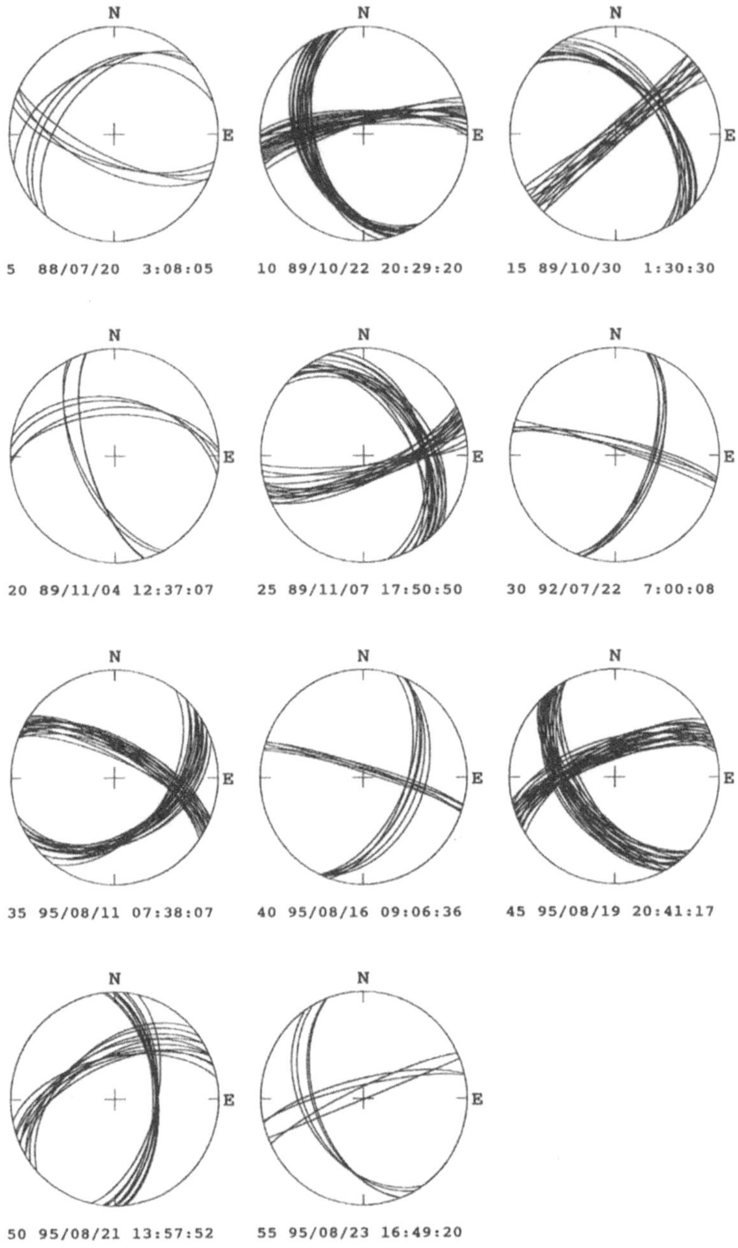

Figure 8

Lower-hemisphere projections of the focal sphere for fault planes of all focal mechanism solutions shown in Figure 7. The mechanism adopted in Figure 7 is chosen from the set of focal mechanisms shown in Figure 8 by selecting the median nodal plane strike.

Table 3

Focal mechanism solutions

No.	Date Yr/Mo/Da	Hr:Min	X km	Y km	Z km	M km	Nodal Plane 1 Strike deg.	Nodal Plane 1 Dip deg.	Nodal Plane 1 Rake deg.	Nodal Plane 2 Strike deg.	Nodal Plane 2 Dip deg.	Nodal Plane 2 Rake deg.
1	1988/07/20	2:08:36	−0.05	−0.02	1.78	−1.7	96	58	−78	255	34	−108
2	1988/07/20	2:29:03	−0.01	−0.01	1.82	−1.5	115	73	−58	231	36	−149
3	1988/07/20	2:55:15	0.04	0.00	1.76	−1.6	81	50	−97	271	40	−82
4	1988/07/20	3:01:28	0.08	0.00	1.72	−1.4	130	57	−66	270	41	−122
5	1988/07/20	3:08:05	−0.03	−0.04	1.85	−1.7	105	66	−57	227	40	−141
6	1988/07/20	6:43:48	0.00	−0.01	1.79	−1.8	111	69	−69	243	29	−134
7	1988/07/20	7:09:58	0.06	0.01	1.76	−0.5	139	61	−55	263	44	−136
8	1988/07/20	7:17:37	0.01	0.00	1.76	−1.4	124	68	−68	257	30	−133
9	1988/07/20	7:17:59	0.03	0.01	1.78	−1.4	130	56	−65	271	41	−121
10	1989/10/22	20:29:20	0.10	−0.07	1.64	−1.9	269	74	−127	158	40	−26
11	1989/10/22	23:50:17	0.38	−0.01	1.94	−2.0	337	48	−63	119	48	−117
12	1989/10/26	0:43:20	0.18	−0.22	1.51	−2.1	289	85	105	37	16	18
13	1989/10/27	4:38:06	0.13	0.04	1.63	−1.7	229	52	−109	79	42	−67
14	1989/10/28	18:29:13	0.27	0.05	1.73	−2.1	53	66	−147	308	61	−28
15	1989/10/30	1:30:30	0.72	0.35	2.16	−2.0	228	87	145	320	55	3
16	1989/11/01	5:57:25	0.56	0.35	1.86	−1.5	243	61	−137	128	53	−37
17	1989/11/04	8:05:06	0.20	0.04	1.85	−1.9	9	70	−23	107	69	−158
18	1989/11/04	9:54:21	0.04	−0.08	1.79	−1.7	113	90	−55	203	35	−180
19	1989/11/04	10:55:50	1.00	0.35	2.32	−1.7	107	70	−47	217	47	−152
20	1989/11/04	12:37:07	0.56	−0.15	2.06	−1.9	155	59	−55	280	46	−134
21	1989/11/04	12:40:54	0.64	−0.11	2.14	−1.9	277	87	35	185	55	177
22	1989/11/05	8:06:29	0.06	0.15	1.65	−2.0	297	56	−102	138	36	−73
23	1989/11/06	8:00:13	0.66	0.40	2.23	−1.7	61	64	−136	308	51	−34
24	1989/11/06	9:31:26	0.67	−0.17	1.95	−1.7	94	90	−140	4	50	0
25	1989/11/07	17:50:50	0.89	0.43	2.31	−1.2	75	71	−133	326	46	−27
26	1989/11/11	17:36:31	0.69	0.34	2.49	−1.3	81	85	−105	334	16	−18
27	1989/11/14	12:45:25	−0.06	0.18	1.75	−1.7	183	87	−40	275	50	−177
28	1992/07/22	6:41:25	0.11	−0.09	2.06	−1.7	287	80	−28	23	62	−169
29	1992/07/22	6:59:31	0.15	−0.05	2.23	−1.8	3	79	−136	263	47	−15
30	1992/07/22	7:00:08	0.13	−0.07	2.22	−1.6	284	79	−33	22	57	−167
31	1992/07/22	8:07:52	0.17	−0.08	2.19	−1.8	257	56	−108	108	38	−65
32	1992/07/22	9:50:26	0.43	−0.17	2.14	0.3	27	62	−131	268	48	−39
33	1992/07/22	10:24:52	0.31	−0.03	2.24	−1.5	25	67	−134	274	48	−31
34	1992/07/22	10:56:32	0.32	−0.06	2.29	−1.7	350	53	−65	131	44	−120
35	1995/08/11	7:38:07	0.31	−0.04	2.20	−1.7	299	67	−46	50	48	−149
36	1995/08/11	13:15:16	−0.08	−0.18	1.96	−1.1	120	77	−90	300	13	−90
37	1995/08/11	15:47:22	0.10	−0.12	2.22	−0.8	302	68	−35	46	58	−154
38	1995/08/13	15:34:25	0.19	−0.09	2.67	−1.1	170	56	−90	350	34	−90
39	1995/08/16	8:01:02	0.49	−0.39	1.63	0.6	261	75	−121	147	35	−28
40	1995/08/16	9:06:36	0.26	−0.08	2.70	−1.1	288	86	−50	22	22	−174
41	1995/08/19	5:04:57	0.42	−0.41	1.77	−1.1	286	62	−67	65	65	−126
42	1995/08/19	9:27:40	−0.40	−0.12	2.18	−1.4	269	74	−127	158	158	−26
43	1995/08/19	9:47:30	−0.40	−0.12	2.18	−1.4	257	74	−125	155	155	−14
44	1995/08/19	10:53:14	−0.44	−0.08	2.27	−1.0	251	71	−127	140	140	−26
45	1995/08/19	10:54:12	−0.38	−0.13	2.23	−1.4	252	52	−133	142	142	−27
46	1995/08/20	0:19:07	0.46	−0.18	2.21	−1.5	306	76	−32	45	45	−164

Table 3

Continued

No.	Date Yr/Mo/Da	Hr:Min	X km	Y km	Z km	M km	Nodal Plane 1 Strike deg.	Nodal Plane 1 Dip deg.	Nodal Plane 1 Rake deg.	Nodal Plane 2 Strike deg.	Nodal Plane 2 Dip deg.	Nodal Plane 2 Rake deg.
47	1995/08/20	8:12:34	−0.41	−0.08	2.22	−1.3	208	79	−136	108	108	−15
48	1995/08/20	19:36:44	−0.35	−0.04	2.19	−1.7	53	58	−48	173	173	−137
49	1995/08/20	23:14:12	−0.39	0.02	2.46	−1.2	170	67	−90	350	350	−90
50	1995/08/21	13:57:52	0.33	0.12	2.52	−1.6	5	58	−132	245	245	−43
51	1995/08/21	14:03:09	−0.40	0.00	2.43	−1.1	170	66	−90	350	350	−90
52	1995/08/21	22:15:50	0.60	−0.10	2.22	−1.3	172	78	−38	272	272	−165
53	1995/08/22	2:56:18	0.60	−0.61	2.01	−1.8	174	77	−27	270	270	−165
54	1995/08/23	16:48:07	0.22	−0.10	2.72	−1.4	170	57	−90	350	350	−90
55	1995/08/23	16:49:20	−0.17	−0.15	2.05	−1.1	257	75	−132	150	150	−22
56	1995/08/24	5:12:19	0.25	−0.10	2.73	−1.2	170	58	−90	350	350	−90
57	1995/08/24	12:16:46	0.42	−0.39	1.78	−1.6	265	57	−78	64	64	−107
58	1995/08/24	19:36:19	0.13	−0.04	2.53	−0.5	140	63	−90	320	320	−90

The origin is at wellhead of SKG–2; the X direction is horizontal and positive to the east, the Y direction is horizontal and positive to the north, the Z direction is vertical, positive downwards.

westward from vertical and horizontal and tend to be nearly north–south, respectively. The distributions of the *P* and *T* axes of the focal mechanisms for the hydraulic fracturing experiments indicate that the averaging of the *P* and *T* axes, which has classically been used to make approximate stress determinations, may be good indicators of regional maximum and minimum principal stress directions. On the other hand, the *P* and *T* axes obtained from the focal mechanism solutions of microseismic events associated with the circulation tests in 1989 and 1995 show rather diffused distributions, although these distributions appear to show similarity to those of the hydraulic fracturing experiments.

7.2 Stress Inversion Method

Several inverse methods have been proposed for determining the principal stress directions of the stress field which is most consistent with focal mechanism data (e.g., VASSEUR *et al.*, 1983; GEPHART and FORSYTH, 1984; MICHAEL, 1987; RIVERA and CISTERNAS, 1990). All of these methods are based on the following two assumptions: first, that the shear stress on the fault plane is parallel to the slip direction and, secondly, that the stress field is uniform throughout the volume in which the seismic events occurred. Such methods are capable of resolving four of the six independent components of the stress tensor. The four constrained values are commonly parameterized by three principal stress directions, i.e., the maximum σ_1, intermediate σ_2, and minimum σ_3 principal stress directions and by the stress ratio R, as defined

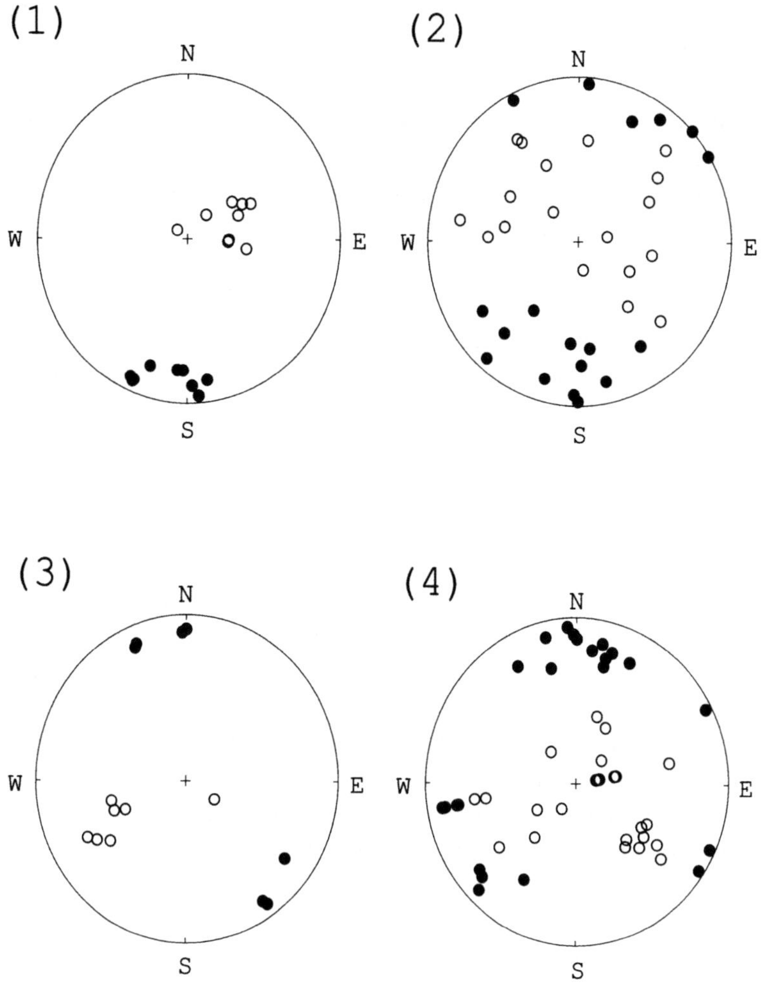

Figure 9
Lower-hemisphere projections of *P* and *T* axes of the focal mechanisms of microseismic events associated with (1) the 1988 hydraulic fracturing experiment, (2) the 1989 circulation test, (3) the 1992 hydraulic fracturing experiment, and (4) the 1995 circulation test, respectively. Open circles represent *P* axes and solid circles *T* axes.

below, which describes the relative magnitude of the principal stresses and hence constrains the shape of the stress ellipsoid.

$$R = \frac{\sigma_2 - \sigma_1}{\sigma_3 - \sigma_1} \ . \tag{1}$$

Assuming that $\sigma_1 \geqq \sigma_2 \geqq \sigma_3$, the stress ratio varies from 0 to 1. When σ_2 and σ_3 are nearly equivalent (uniaxial deviatoric compression), R approaches 1, and when σ_2 and σ_1 are nearly equivalent (biaxial deviatoric compression), R approaches 0. The two remaining stress parameters characterizing a uniform pressure component and a scaling factor cannot be determined from focal mechanism data.

One of the difficulties encountered in the stress inversion arises from the fact that a focal mechanism yields two nodal planes however no method exists generally for identifying which of these two planes is the actual fault. Indeed, when one of the principal stress directions is close to being perpendicular to the fault plane, a very slight change of fault plane orientation significantly alters the slip direction. GEPHART and FORSYTH (1984) have developed an inversion method that enables an objective selection of the preferred nodal plane from each focal mechanism. In the following, therefore, the stress inversion program FMSI of GEPHART (1990) was used to determine stress directions at the Hijiori HDR site from focal mechanisms. The program FMSI, which implements the method of GEPHART and FORSYTH (1984) inverts focal mechanism data to determine the best fitting values of four stress parameters without facing the problem of the ambiguity of nodal planes. The misfit measure between each considered stress model and the data is taken to be the average minimum angular rotation angle around any axis necessary to bring into agreement the slip direction of each focal mechanism and the slip direction predicted by the stress model. The best fitting stress model is found by minimizing the average of the individual misfits. The grid search was carried out according to the following steps. At the start, initial estimates of σ_1 and σ_3 axes are first selected. With initial σ_1 and σ_3 axes selected, all stress directions are fixed because the σ_2 axis is orthogonal to the σ_1 and σ_3 axes. Then, the misfit is calculated for a range of R. The process is progressively repeated using new trial values of σ_1 and σ_3 axes. Misfits for each sampled value of four model parameters (σ_1, σ_2, σ_3, and R) are tabulated, and misfit-based confidence limits over which solutions are not statistically different from the best solution are determined using one-norm statistics.

7.3 Results

Figure 10 shows the results obtained by inverting all 58 focal mechanisms. The left projection shows the distribution of the orientations of the P and T axes for the 58 focal mechanisms. The right projection shows 95% confidence limits for the orientation of principal stress directions. The best fit stress model is $\sigma_1 = (4, 82)$, $\sigma_2 = (270, 1)$, $\sigma_3 = (180, 8)$ (trend and plunge, respectively), and $R = 0.4$. The average misfit between this stress model and all the data is 6.8°.

Estimates of stress directions by inversion of focal mechanisms might be meaningless if they are poorly constrained or derived from heterogeneous data sets. LU *et al.* (1997) proposed two acceptance criteria for results of stress inversions. The first criterion for judging the quality of the inversion results is the limit of the 95% confidence regions of the orientation of the principal stress axes. A stress model is accepted if the 95% confidence regions of the maximum and minimum principal directions are well constrained, i.e., they do not overlap. The second criterion to estimate the degree of heterogeneity in a sample is the average misfit of the inversion. LU *et al.* (1997) accepted solutions with the average misfits of less than 6° as satisfying the assumption of homogeneity. This requirement is based on the

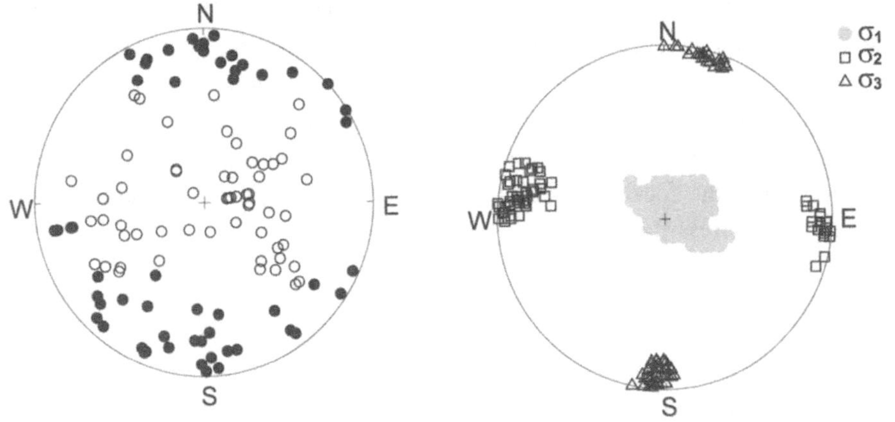

Figure 10
Equal area plots showing orientations of *P* and *T* axes and stress inversion results. (Left) orientations of *P* axes (open circles) and *T* axes (solid circles) from 58 focal mechanisms. (Right) orientations of σ_1 axes (solid circles), σ_2 axes (open squares), and σ_3 axes (open triangles) of the stress models within the 95% confidence limit of our analysis.

observation that synthetic data sets with errors of 15° (average of the uncertainties in the strike, dip, and rake) lead to the average misfit with less than 6° (WYSS *et al.*, 1992; GILLARD *et al.*, 1996) which is reasonable for our data set, since the average values are less than 15°. They also accepted solutions with the average misfits ranging from 6° to 7.5° if the principal stresses are resolvable at the 95% confidence level. Inversions for stress directions resulting in the average misfit of more than 7.5° were rejected because they violate the assumption of homogeneity.

As shown in Figure 10, the 95% confidence regions of the maximum and minimum principal stress directions are well constrained. The average misfit of the inversion is 6.8° and is less than the acceptance criterion (7.5°) of LU *et al.* (1997). Thus our inversion result is interpreted as acceptable on the basis of the criteria of LU *et al.* (1997).

The result suggests that the maximum principal stress beneath the Hijiori HDR site is vertical, the intermediate and minimum principal stresses are horizontal and tend to be east–west and north–south, respectively.

The direction of principal stresses is one of the key factors which control the direction of man-made fracture propagation. Therefore, the stress state at the Hijiori HDR site was extensively investigated using various methods such as a geological survey of Hijiori (NEDO, 1992), natural fractures observed by borehole televiewer (BHTV) in HDR–2 and HDR–3 (NEDO, 1992), mapping of shear fractures by the microscopic observation of core specimens (NEDO, 1992), the differential strain curve analysis (DSCA) of core specimens (OIKAWA *et al.*, 1993), the inversion of data from drilling induced fractures observed on BHTV images (OKABE *et al.*, 1995), and the focal mechanism analysis, supplemented with microseismic doublet analysis

(TEZUKA and NIITSUMA, 2000). These results, as summarized by YAMAGUCHI *et al.* (1996), showed that the maximum principal stress direction in the Hijiori area is nearly vertical, and the minimum is horizontal and close to north–south. The stress estimate obtained by inverting the focal mechanisms of microseismic events essentially agrees with other stress estimates previously obtained. It is therefore concluded that the focal mechanism inversion method provides a useful tool for estimating the stress state.

8. Discussion

According to the classic hydraulic fracturing concept, fractures due to hydraulic fracturing are created by a tensile failure and extend along a direction parallel to the maximum principle stress (HUBBERT and WILLIS, 1957). The hypocentral distributions of microseismic events associated with the 1988 and 1992 hydraulic fracturing experiments are distributed around the plane that spreads nearly east–west from the injection wells and declines to the north at a high angle. The strike of the induced seismicity is nearly parallel to the direction of the maximum horizontal principal stress. This suggests that tensile fractures of intact rock were newly-created by hydraulic fracturing. However, the vertical orientation and east–west strike of the seismic events are essentially coplanar with the caldera ring-fault structure in the southern portion of the Hijiori caldera. This indicates that tensile fractures of intact rock were not being created, but pre-existing fractures were being re-opened and developed in the direction of the maximum principal stress, although microseismic events were caused by shear failures. The wellhead pressures at the injection wells (SKG–2 and HDR–1) support the implication that intact rock was not hydraulically fractured. The pressure at the tip of the growing fracture must be at least equal to the minimum principal stress. The minimum principal stress estimate at a depth of 1800 m derived from the oriented cores taken from HDR–1 and HDR–2 is 33 MPa (OIKAWA *et al.*, 1993). However, the wellhead pressures in the 1988 and 1992 experiments were 16 MPa and 25 MPa, respectively, and did not exceed the minimum principal stress.

The most likely mechanism of the hydraulically induced fracture at the Hijiori HDR site may be explained using Hill's model (HILL, 1977) for volcanic earthquake swarms. HILL (1977) suggested, based on a model wherein pre-existing weak planes lie along the direction of the maximum principal stress, that shear fracture should occur with fracture connecting between the tips of the two neighboring weak planes, in addition to tensile fracture accompanying magma intrusion into the weak planes. The generic mechanism of the induced microseismic events and the earthquake swarms are similar. According to Hill's model, although the main hydraulic fractures themselves are caused by re-opening pre-existing fractures that align along the maximum principal stress direction and are almost aseismic, they can induce shear failure on pre-existing fractures that occur in rocks surrounding the main hydraulic fracture. Microseismic events caused by shear failure are probably due to high pore fluid pressures (PEARSON, 1981; FEHLER, 1989). As to the aseismic propagation of hydraulic fractures, SASAKI (1995) has shown that the seismic energy radiated from a tensile failure is small as compared with that of a shear failure, so that it is difficult to detect the seismic waves caused by tensile failures.

9. Conclusions

The stress state at the Hijiori hot dry rock site was investigated using the focal mechanisms of the microseismic events associated with hydraulic injection experiments. The inversion results, obtained by inverting the 58 focal mechanisms of seismic events simultaneously, show that the 95% confidence regions of the maximum and minimum principal directions obtained are well constrained and that the average misfit is 6.8° which is small enough to satisfy the acceptance criterion (7.5°) of LU *et al.* (1997), suggesting that the focal mechanism data used for the inversion are homogeneous and that the inversion results are meaningful. The stress results indicate that the maximum principal stress direction is vertical, the intermediate and minimum principal stress directions are horizontal and trend east–west and north–south. The stress estimates obtained by the inverting focal mechanisms of microseismic events essentially agree with other stress estimates previously obtained. It is therefore concluded that the focal mechanism inversion method provides a useful tool for estimating the stress state.

The strike of the microseismicity associated with the 1988 and 1992 hydraulic fracturing experiments is nearly parallel to the direction of the maximum horizontal principal stress. This suggests that tensile fractures of intact rock were newly created by hydraulic fracturing. However, the vertical orientation and east–west strike of the seismic events are essentially coplanar with the caldera ring-fault structure in the southern portion of the Hijiori Caldera. This indicates that tensile fractures of intact rock were not being created, but pre-existing fractures were being re-opened and developed in the direction of the maximum horizontal principal stress, although microseismic events were caused by shear failures.

The most likely mechanism of the hydraulically induced fracture at the Hijiori HDR site may be explained using Hill's model (HILL, 1977) for volcanic earthquake swarms. According to Hill's model, although hydraulic fractures themselves are almost aseismic and are not directly responsible for observable microseismic events, they can induce shear failure on pre-existing fractures that occur in rocks surrounding the main hydraulic fracture. Microseismic events caused by shear failure are probably due to high pore fluid pressures.

Acknowledgements

We would like to thank S. Sato, Yoshizawa Giken K.K. for his assistance. We would also like to thank A. Snoke, Virginia Polytechnic Institute and State University, J.-R. Grasso, Observatoire de Grenoble and an anonymous reviewer for their comments and suggestion to modify the manuscript. We are especially grateful to M. Fehler, Los Alamos National Laboratory, for providing the focal mechanism stress inversion code to us and for helpful discussions. This research was conducted under contract with the New Energy and Industrial Technology Development Organization (NEDO) from FY 1997 to FY 1998.

REFERENCES

CORNET, F. H. and JULIEN, PH. (1989), *Stress Determination from Hydraulic Test Data and Focal Mechanisms of Induced Seismicity*, Int. J. Rock Mechanics Min. Sci. and Geomech. Abs. *26*(3/4), 235–248.

CORNET, F. H. and YIN, J. (1995), *Analysis of Induced Seismicity for Stress Field Determination and Pore Pressure Mapping*, Pure appl. geophys. *145*(3/4), 677–700.

CORNET, F. H., HELM, J., POITRENAUD, H., and ETCHECOPAR, A. (1997), *Seismic and Aseismic Slips Induced by Large-Scale Fluid Injections*, Pure appl. geophys. *150*, 563–583.

FEHLER, M. (1989), *Stress Control of Seismicity Pattern Observed during Hydraulic Fracturing Experiments at the Fenton Hill Hot Dry Rock Geothermal Energy Site, New Mexico*, Int. J. Rock Mechanics Min. Sci. and Geomech. Abs., *26*(3/4), 211–219.

GEPHART, J. (1990), *FMSI: A FORTRAN Program for Inverting Fault/Slickenside and Earthquake Focal Mechanism Data to Obtain the Regional Stress Tensor*, Comput. Geosci. *16*(7), 953–989.

GEPHART, J. and FORSYTH, W. (1984), *An Improved Method for Determining the Regional Stress Tensor Using Earthquake Focal Mechanism Data: Application to the San Fernando Earthquake Sequence*, J. Geophys. Res. *89*(B11), 9305–9320.

GILLARD, D., WYSS, M., and OKUBO, P. (1996), *Type of Faulting and Orientation of Stress and Strain as a Function of Space and Time in Kilauea's South Flank, Hawaii*, J. Geophys. Res. *101*(B7), 16,025–16,042.

GUPTA, H. K. and RASTOGI, B. K., *Dams and Earthquakes* (Elsevier, Amsterdam 1976).

HILL, D. P. (1977), *A Model for Earthquake Swarm*, J. Geophys. Res. *82*, 1347–1352.

HUBBERT, M. K. and WILLIS, D. G. (1957), *Mechanics of Hydraulic Fracturing*, Petrol. Transact. Am. Soc. Min. Eng. *210*, 153–168.

ITO, H. and SUGIHARA, M. (1988), *Estimate of in situ Permeability and Fluid Flow Paths from Hypocenter Migration of Microearthquakes during the Build-up Tests of Geothermal Wells in the Takinoue Geothermal Area*, J. Geotherm. Res. Soc. Japan. *10*, 305–319 (in Japanese with English abstract).

JUPE, A. J., GREEN, A. S. P., and WALLROTH, T. (1992), *Induced Microseismicity and Reservoir Growth at the Fjallback Hot Dry Rocks Project, Sweden*, Int. J. Rock Mechanics Min. Sci. and Geomech. Abs. *29* (4), 343–354.

KAIEDA, H., FUJIMITSU,Y., YAMAMOTO, T., MIZUNAGA, H., USHIJIMA, K., and SASAKI, S. (1995), *AE and Mise-à-Lamasse Measurements during a 22-day Water Circulation Test at Ogachi HDR Site*, Japan, Proc. World Geotherm. Cong. 2695–2700.

KISSLINGER, C. (1980), *Evaluation of S-to-P Amplitude Ratios for Determining Focal Mechanisms from Regional Network Observations*, Bull. Seismol. Soc. Am. *70*(4), 999–1014.

KURIYAGAWA, M. and TENMA, N. (1999), *Development of Hot Dry Rock Technology at the Hijiori Test Site*, Geothermics. *28*, 627–636.

LU, Z., WYSS, M., and PULAN, H. (1997), *Details of Stress Directions in the Alaska Subduction Zone Form Fault Plane*, J. Geophys. Res. *102*(B3), 5385–5402.

McKenzie, D. (1969), *The Relationship between Fault Plane Solutions for Earthquakes and the Directions of Principal Stresses*, Bull. Seismol. Soc. Am. *59*(2), 591–601.

Michael, A. (1987), *Use of Focal Mechanisms to Determine Stress: A Control Study*, J. Geophys. Res. *92* (B1), 357–368.

Mizoue, M. and Yokota, T. (1991), *The Relation between Propagation of Tensile Crack and Migration of Hypocenters Associated with the 1987 Earthquake Swarm of off Izu Peninsula*, Earth Monthly *140*, 63–73 (in Japanese).

New Energy and Industrial Technology Development Organization (1992), *FY 1992 Summary of Hot Dry Rock Geothermal Power Project in Japan.*

Ohtake, M. (1974), *Seismic Activity Induced by Water Injection at Matsushiro, Japan*, J. Phys. Earth *22*, 163–173.

Oikawa, Y., Matsunaga, I., and Yamaguchi, T. (1993), *Differential Strain Curve Analysis to Estimate the Stress State of the Hijiori Hot Dry Rock Field*, Int. J. Rock Mechanics Min. Sci. and Geomech. Abs. *30*(7), 1023–1026.

Okabe, T., Shinohara, N., Takasugi, S., and Hayashi, K. (1995), *Stress Analysis of Induced Fractures by Inversion Method (Part 2) – Application to Field Data*, Proc. 92nd SEGJ Conference, pp. 11–15.

Pearson, C. (1981), *The Relationship between Microseismicity and High Pore Pressures during Hydraulic Stimulation Experiments in Low Permeability Granitic Rocks*, J. Geophys. Res. *86*, 7855–7864.

Phillips, W. S., House. S. L., and Fehler, C. M. (1997), *Detailed Joint Structure in a Geothermal Reservoir from Studies of Induced Microearthquake Clusters*, J. Geophys. Res. *102*(B6), 11,745–11,763.

Raleigh, C. B., Healy, J. H., and Bredehoeft, J. D. (1976), *An Experiment in Earthquake Control at Rangely, Colorado*, Science, *191*, 1230–1237.

Rivera, L. and Cisternas, A. (1990), *Stress Tensor and Fault Plane Solutions for a Population of Earthquakes*, Bull. Seismol. Soc. Am. *80*(3), 600–614.

Sasaki, S. (1995), *Study on Characteristics and Mechanism of Acoustic Emission Induced by Hydraulic Fracturing*, Ph.D. Thesis, Tohoku Univ., Sendai, Japan (in Japanese).

Sasaki, S. (1998), *Characteristics of Microseismic Events Induced during Hydraulic Fracturing Experiments at the Hijiori Hot Dry Rock Geothermal Energy Site, Yamagata, Japan*, Tectonophysics, *289*, 171–188.

Shin, K., Ito, H., and Oikawa, Y. (2000), *Stress State at the Ogachi Site*, Proc. World Geothermal Congress 2000, 1749–1754.

Snoke, J. A., Munsey, J. W., Teague, A. G., and Bollinger, G. A. (1984), *A Program for Focal Mechanism Determination by Combined Use of Polarity and SV-P Amplitude Ratio Data*, Earthquake Notes *55*(3), 15.

Suzuki, S. (1992), *Linear Distributions of the Earthquake-swarm Epicenter and the Volcanic Craters in the Northern Tokachi District, Hokkaido.* In *Physics of Earthquake Field in the Crust*, pp. 33–38 (in Japanese with English abstract).

Talwani, P. and Acree, S. (1984/85), *Pore-pressure Diffusion and the Mechanism of Reservoir-induced Seismicity*, Pure appl. geophys. *122*, 947–965.

Tezuka, K. and Niitsuma, H. (2000), *Stress Estimated Using Microseismic Clusters and its Relationship to the Fracture System of the Hijiori Hot Dry Rock Reservoir*, Eng. Geol. *56*, 47–62.

Vasseur, G., Etchecopar, A., and Philip, H. (1983), *Stress State Inferred from Multiple Focal Mechanisms*, Ann. Geophys. *1*, 291–297.

Watanabe, H. (1971), *Determination of Earthquake Magnitude at Regional Distance in and near Japan*, J. Seismol. Soc. Japan, *24*, 189–200 (in Japanese with English abstract).

Wyss, M., Liang, B., Tanigawa, W. R., and Wu, X. (1992), *Comparison of Orientations of Stress and Strain Tensors Based on Fault Plane Solutions in Kaoiki, Hawaii*, J. Geophys. Res. *97*(B11), 4769–4790.

YAMAGUCHI, T., OIKAWA, Y., MATSUNAGA, I., and SATO, Y. (1996), *Estimation of in situ Stress State at Hijiori Test Site*, Proc. 3rd International HDR Forum, 13–14.

(Received May 5, 2000, revised/accepted November 7, 2000)

Pure appl. geophys. 159 (2002) 517–541
0033–4553/02/030517–25 $ 1.50 + 0.20/0

Pure and Applied Geophysics

Detailed Fracture System of the Soultz-sous-Forêts HDR Field Evaluated Using Microseismic Multiplet Analysis

HIROKAZU MORIYA,[1] KATSUHISA NAKAZATO,[1]
HIROAKI NIITSUMA[1] and ROY BARIA[2]

Abstract—The reservoir structure of the Soultz HDR field has been investigated by examining induced microearthquake multiplets. Microseismic events with similar waveforms have been selected from microseismic data obtained during a 1993 hydraulic fracturing experiment. Precise relative arrival times and source locations have been determined by cross-spectrum analysis. The cross-spectrum analysis decreased the residual from 0.75 ms to 0.1 ms. The estimated orientations of the multiplet planes are consistent with fracture orientations detected in Soultz boreholes. A comparison between the stress field and the orientation of structural planes suggests that the structural planes were under a critical condition of frictional slip.

Key words: Multiplet, HDR, microearthquakes, induced seismicity.

1. *Introduction*

Mapping technologies for evaluating engineered subsurface reservoirs have made major advances in recent years, with doublet/multiplet analysis, collapsing method and clustering analysis having been developed (MORIYA *et al.*, 1994; JONES and STEWART, 1997; PHILLIPS *et al.*, 1997; PHILLIPS, 2000).

The doublet/multiplet analysis method, which determines the precise relative locations of similar seismic events by cross-spectrum analysis, is a feasible mapping technique to evaluate the distribution and orientation of fractures (POUPINET *et al.*, 1984; MORIYA *et al.*, 1994; TEZUKA and NIITSUMA, 1996, 1997; GAUCHER *et al.*, 1998). A multiplet is a group of seismic events with similar waveforms, but different origin times, and is considered to be the expression of stress release on the same fracture. The source locations provide information on the distribution of seismically activated fractures, and are related to the elevation of fluid pressure.

The multiplet approach was introduced for the evaluation of artificially created geothermal reservoirs at HDR sites worldwide. For example, FRÉCHET *et al.* (1989)

[1] Department of Geoscience and Technology, Graduate School of Engineering, Tohoku University, Sendai, 980-8579 Japan. E-mail: hirokazu@ni4.earth.tohoku.ac.jp
[2] SOCOMINE, Soultz-sous-Forêts, France.

detected microseismicities associated with a hydraulic fracturing experiment at *Le Mayate de Montagne*, France, and evaluated fluid migration by accurately relocating the source location of similar events. MORIYA *et al.* (1994) estimated the direction of fracture growth, associated with lost circulation during well drilling at Kakkonda geothermal field, (Japan), using a group of similar-induced microseismicities. These researchers showed that fluid flow in a fracture can be estimated using the source migration of an induced multiplet.

TEZUKA and NIITSUMA (1997) reported that approximately 90% of the located microearthquakes in the Hijiori Hot Dry Rock (HDR) field (Japan), were doublet/multiplet events. They relocated the source locations using doublet/multiplet analysis, and showed that the orientations of fractures inferred from the relative source locations correlated well with the observed orientation of permeable fractures intersecting Hijiori boreholes. In addition, TEZUKA and NIITSUMA suggested that the seismically delineated structural planes represent those fracture planes that were best oriented for shear slip. At the Ogachi Hot Dry Rock field (Japan), it has also been demonstrated that the orientation of seismically delineated structural planes is consistent with fractures observed using BHTV in boreholes (MORIYA *et al.*, 2000).

In the Soultz HDR field (France), the reservoir structure was evaluated using microearthquake patterns determined by precise mapping methods, including multiplet analysis, the collapsing method and clustering (JONES 1997; GAUCHER *et al.*, 1998; BARIA *et al.*, 1999; PHILLIPS, 2000; ROWE *et al.*, 2002). For example, the collapsing method emphasizes structures by a statistical procedure, and was applied to the entire seismic cloud in order to delineate the structural planes. GAUCHER *et al.* (1998) analyzed several multiplets near the fracturing well, and correlated the orientations and positions of the seismically delineated structural planes with those of borehole imaged fractures.

PHILLIPS (2000) precisely mapped several clusters in the seismic cloud, and identified fluid flow paths within fractures. He also discussed pore-pressures, needed to cause slip for the orientation of each delineated structure, and suggested that seismically delineated active fractures require little added pressure to fail (PHILLIPS, 2000).

The above researchers provide important details, which allow a better understanding of reservoir characteristics. The application of precise mapping techniques to the entire seismic cloud is essential for detailed evaluation of fluid flow paths in the fracture system. Application of the microseismic multiplet helps to clarify the fracture system, because the relative source locations on the multiplets may be used to precisely identify small structures that may not be recognized by other mapping techniques.

In this paper we analyze induced multiplets to estimate the detailed structure of the entire seismic cloud of the Soultz HDR field, with more than 10,000 microseismic events examined. Features of inferred structural planes, well logging and stress field data are discussed. The delineated structures are compared with the clusters estimated by PHILLIPS (2000), to understand the relationship between the multiplet analysis and clustering analysis. The characteristics of the fracture system are described by integrating the delineated structures and the stress field.

2. *Seismicity Associated with Hydraulic Fracturing Experiments Conducted in 1993*

Figure 1 shows the location of the Soultz-sous-Forêts HDR field. The European HDR project at Soultz-sous-Forêts was founded by France, Germany and the European Commission (EC) in 1987. The Soultz field is located in the Rhine Graben, which forms part of the Western European Rift System, with basement Soultz granite having intruded Devonian-Early Carboniferous rocks. The Soultz granitic

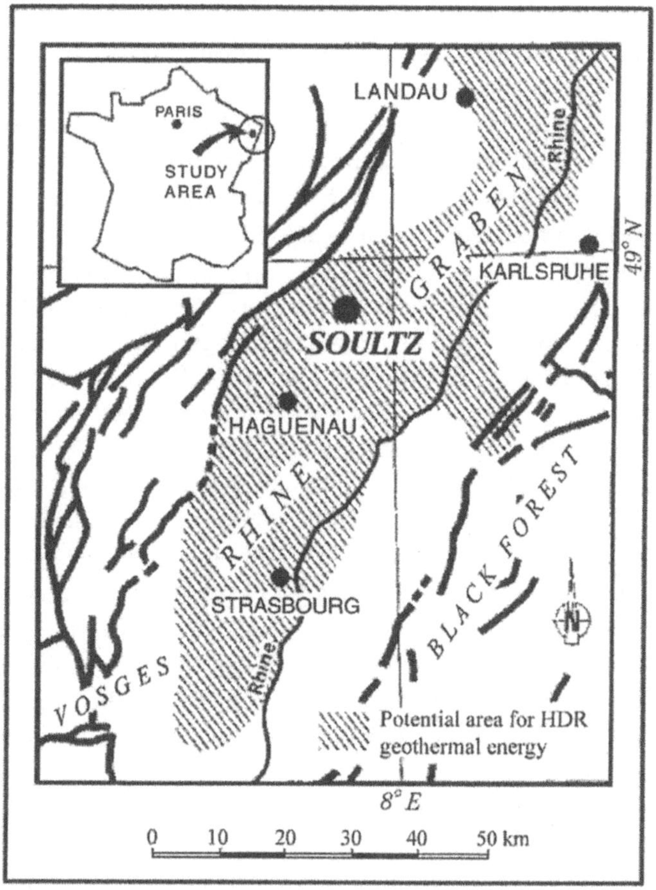

Figure 1
Location of the Soultz-sous-Forêts HDR field.

horst is an elevated structure in the Rhine Graben, but in the Soultz HDR field site area is covered by sedimentary layers up to 1400-m thick (BARIA *et al.*, 1999).

Well GPK-1 is located at the Soultz HDR field site. The well was drilled to 3590 m depth (open hole below 2850 m depth), and since 1987 it has been used for a number of detailed experiments. A major hydraulic fracturing experiment was undertaken at GPK-1 in September 1993. For twenty days, about 45,000 m^3 of fresh water was injected into granite 'reservoir rock' with 10 MPa maximum wellhead pressure, while the flow rate was increased to 50 l/s. Through the test, it was demonstrated that the fracture network in the basement rock was well developed, with enhanced permeability and a substantial increase in transmissivity (BARIA *et al.*, 1999).

The second deep well, GPK-2, was drilled in 1995 to a depth of 3876 m. This well is positioned around 450 m south of GPK-1, with a maximum downhole temperature of 168 °C at 3800 m. In 1995, GPK-2 was stimulated using brine, with a circulation test performed to confirm the hydraulic connectivity between GPK-1 and -2. In 1996, the open hole section of GPK-2 was again stimulated, with a total of 28,000 m^3 of fluid injected at a maximum wellhead pressure of 13 MPa, and flow rates of up to 78 l/s (GÉRARD *et al.*, 1997), to improve its injectivity. An inspection was undertaken of the borehole wall by BHTV (Borehole Televiewer) and FMI (Formation Micro Imager) after completion of the hydraulic fracturing experiment, and stress measurements related to hydraulic fracturing were also performed (GENTER and DEZAYES, 1993; BARIA *et al.*, 1995). Observations made in the wells suggest that there are two principal joint sets, (i) striking N10°E, dipping 65°W and (ii) striking N170°E, dipping 70°E, respectively. The maximum horizontal stress direction was estimated to be N170°E, with the magnitude of the minimum principal stress being relatively close to hydrostatic pressure (BARIA *et al.*, 1999). After these and other studies were conducted, well GPK-2 was extended to about 5000 m depth, with a hydraulic fracturing test carried out in 2000.

Figure 2 shows a plan view of the Soultz site. Induced microseismicities were detected using three downhole 4-component detectors and one hydrophone, installed in wells #4550, #4616, #4601 and EPS1. The 4-component detector consists of accelerometers mounted in a housing and set in sand at the bottom of the borehole. The three 4-component seismic detectors and hydrophone were set at depths of 1500 m, 1420 m, 1600 m and 2850 m, respectively. The downhole seismic detectors were installed in the same basement rock, and high quality signals were readily detected, since the influence of surface waves could be ignored and the transfer function from source to detectors is simple. The frequency band of the acquisition system was from 10 Hz to 1 kHz, and the signals were digitized by 5 kHz sampling frequency (BARIA *et al.*, 1999).

Using these detectors, source locations for more than 10,000 events have been determined (BARIA *et al.*, 1999). In this paper we analyze the induced seismicities from the 1993 hydraulic fracturing experiments. Figure 3(a–c) show source locations for all of the analyzed events, with the location of 10,182 events having been plotted.

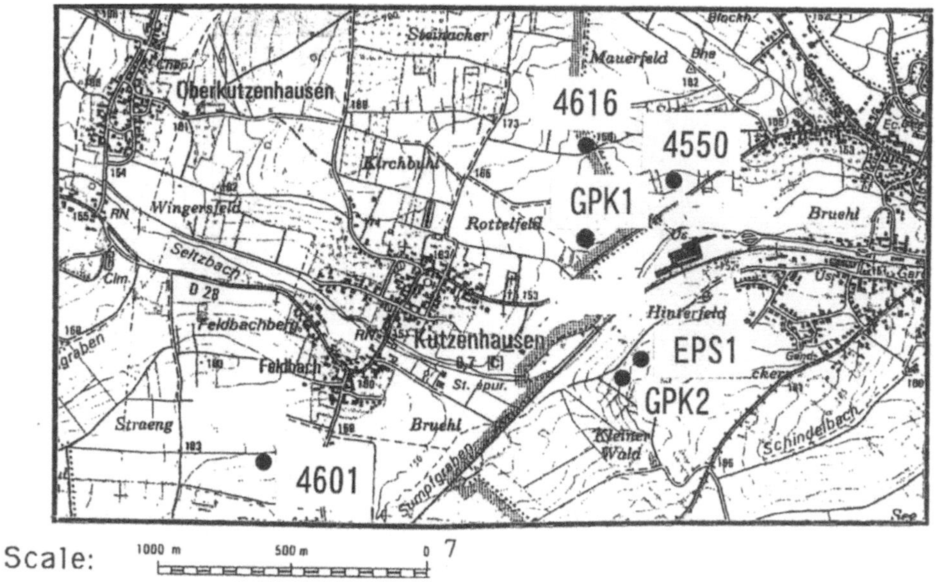

Figure 2
Map of the Soultz HDR field, and deployment of wells. GPK-1 is the fracturing well. The microseismic data used in this paper were obtained using downhole 4-component accelerometers in well 4550, 4616 and 4601, and a hydrophone in EPS-1.

As shown in this figure, the seismic cloud has approximate dimensions of 0.5 by 1.2 by 1.5 km, striking N30°W and dipping nearly vertically (PHILLIPS, 2000).

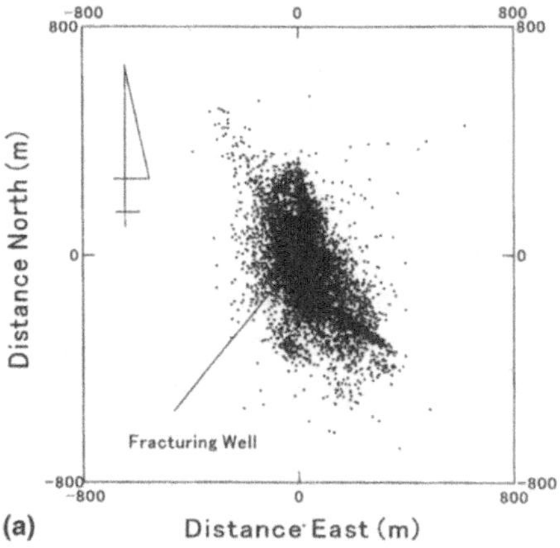

Figure 3
Source locations of induced seismicities, where a total of 10,182 events are plotted. The major outlet points in the GPK-1 are inferred to be the intervals of 2850–2900 m, 3090–3100 m, 3230–3240 m and 3320–3325 m. GPK-1 is an open hole below 2850 m.

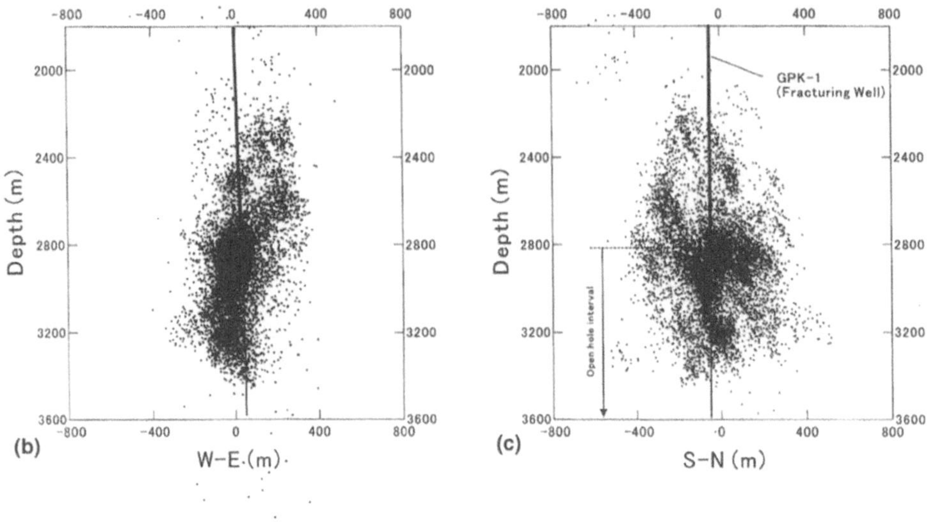

Figure 3b, c

3. *Detection of Multiplets*

The hydraulically induced microseismic multiplet is analyzed in this study. Firstly, it is necessary to define the 'multiplet', because various researchers in different fields have considered the multiplet differently. Here we regard similar microseismicities as a multiplet, on the basis of independently defined criterion. We define the multiplet as being a group of seismic events whose *P* and *S* waves onset, or where several periods of waves exhibit a high cross correlation. The onset of body waves is important for precise mapping, since they contain most information about source locations. The similarity of coda waves is excluded from the consideration, in that coda waves are mainly scattered waves whose probability of occurrence conforms to a random process, and their characteristics are strongly influenced by the wave propagation medium.

The search for a group of similar waveforms was performed by the following procedure: initially, the signal was transformed from 4-component signals to 3-component signals using a least-squares method (JONES and ASANUMA, 1997). We used the coherency function to objectively identify multiplet seismic events. The coherency function is the normalized cross spectrum and a measure of the similarity between two time series, as a function of frequency. The calculation of coherency was carried out using a horizontal component signal at the sensor in well #4550, because (i) the detector worked throughout the experiment, and (ii) because the signals at #4550 have a higher *S*/*N*, where the stability of the detector was evaluated using 4-component coherency analysis (JONES and ASANUMA, 1997). Another reason is that the seismic detector in well #4550 is located closer to the fracture point, and that the detected signal includes wider frequency components. It is important not to ‘miss’ the multiplet, through the search for similar seismic events, using the coherency function.

The coherency in *P*- and *S*-wave windows was calculated for the combination of 10,182 events, and there are ${}_{10182}C_2$ combinations of coherencies. The average

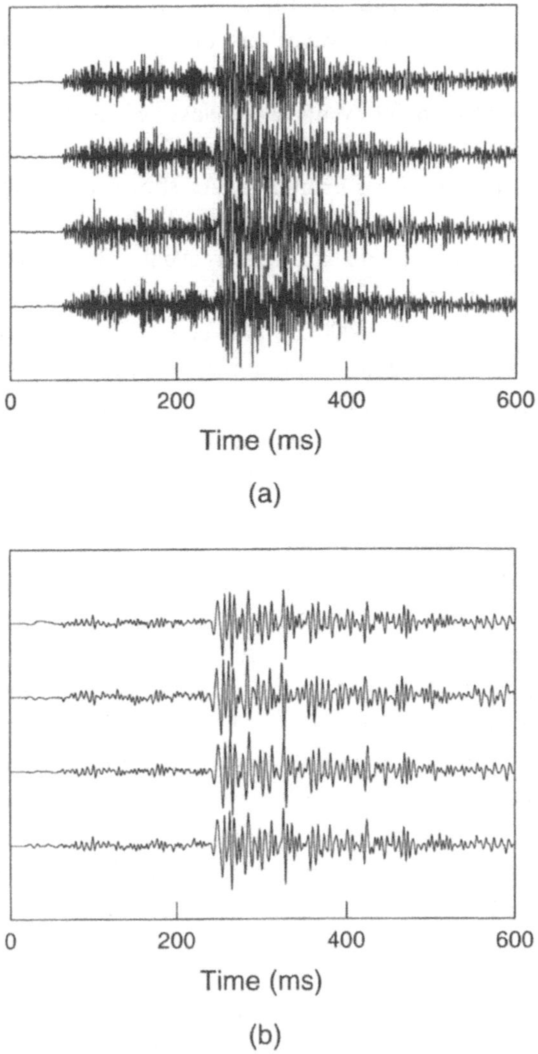

(b)

Figure 4
Examples of multiplet waveforms. (a) Raw waveforms, and (b) 150 Hz low-pass filtered waveforms. After filtering, the similarity of the waveforms is clear.

coherency between 100 Hz to 200 Hz was considered, and the group of seismic events with a coherency above 0.68 was regarded as the initial candidates for multiplets. Through visual observation, we identify the events as a multiplet if they have specific similarity at the *P*- and *S*-wave onset of waveforms. Figure 4 shows waveforms of multiplet events. Figure 4(a) depicts the raw waveforms and Figure 4(b) describes the waveforms after low-pass filtering at 150 Hz. After removing the high frequency component, the similarity of *P* and *S* waves, over their initial few periods, is clear. Spiky waves appear after the *P* and *S* waves, and they are most probably reflected waves from the top of granite, at a depth of 1417 m. It is also possible, however, that the 'spiky waves' are produced by borehole reverberation, because these waves appear periodically.

As a result, a total of 6039 events were classified as doublets and multiplets. Figure 5(a) shows a histogram of detected doublets and multiplets, where the frequency of events per cluster is examined. As shown in Figure 5, there are 600 doublet groups and 440 triplet groups, where 'doublet' means a pair of similar events and 'triplet' means a trio of similar events. The maximum number of multiplet events in a group is 35 events. The number of events per cluster is less than that detected by PHILLIPS (2000). The reason for the apparent discrepancy is the different criterion used to search for similar waveforms. In our experience, the size of each cluster is greater where we used a lower value of coherency as a criterion, and a lower frequency component for calculating the average value of coherency. We analyzed a total of 115 multiplets, with 1028 multiplet events, each consisting of more than five events. The histogram of the number of analyzed multiplets is shown in Figure 5(b).

Figure 6 shows the ratio of the events identified as multiplets and other events, during three-hour intervals. The figure indicates that multiplets occurred throughout the hydraulic fracturing experiment, and that the ratio of multiplet occurrence is greater at the initial stage of hydraulic fracturing, when flow rate and injection pressures are low. It is likely that the rock fractures have orientations which satisfy Coulomb's criterion for shear slip, thus a large population of similar fracture events is initiated, with the many fractures generating a group of similar seismic events.

To ascertain the time interval for the occurrence of similar events is important, since it may be related to the velocity of fluid flow into the fracture. Figure 7 shows a histogram of the time interval for the occurrence of events, which highlights the time difference between a previous and subsequent event. It is suggested that the time required for reoccurrence of a similar event is typically only a few minutes, as shown in Figure 7. This means that the multiplet will tend to occur at an adjacent location, and that a fracture having the same physical property and generating a similar seismic event, is seismically activated for only a short period of time. The time interval for occurrence of an event would be related to fluid velocity into the pre-existing fracture.

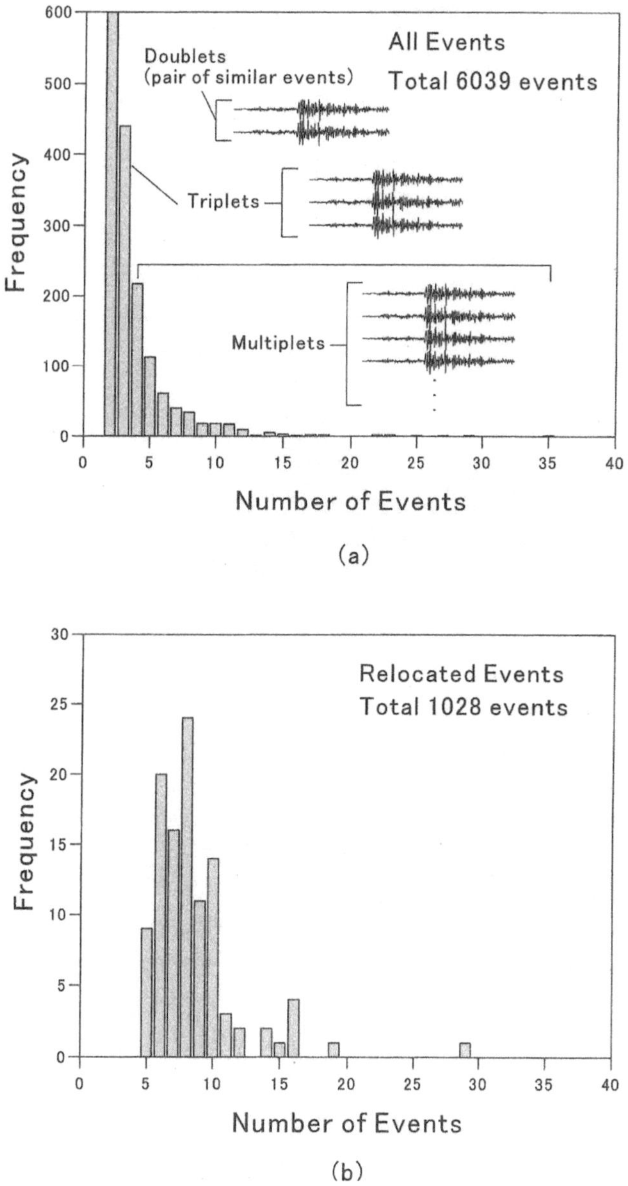

Figure 5
Histogram of (a) the number of similar events per cluster, and (b) analyzed multiplets. A total of 6039 events are classified, with multiplets consisting of more than five similar events, being used for relocation. Most of the detected groups consist of two similar seismic events.

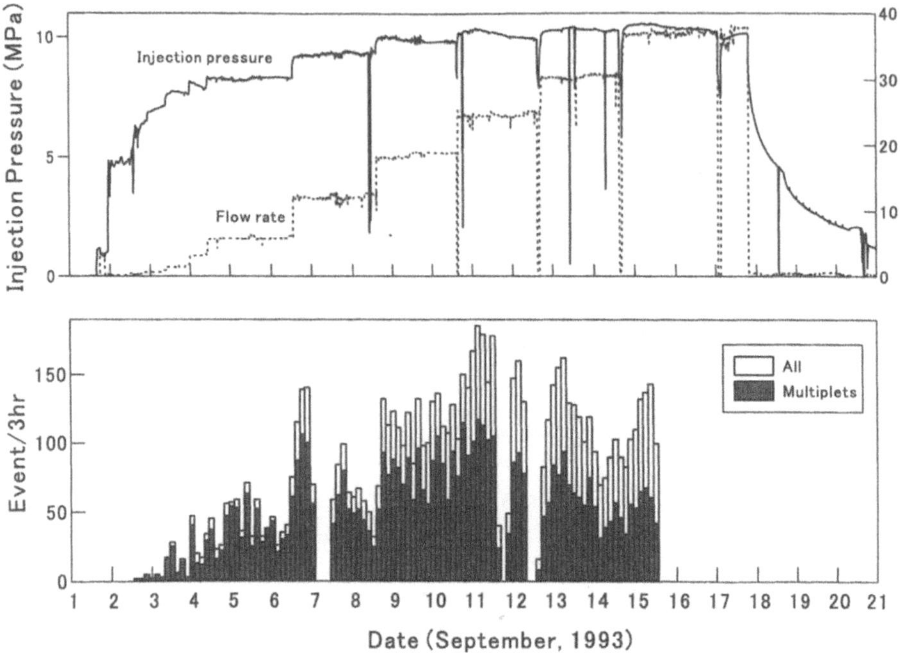

Figure 6
(a) History of injection pressure and flow rate, and (b) histogram of events identified as multiplets, and other analyzed events, at three-hour time intervals. The number of observed seismic event increase with flow rate and injection pressure. On the other hand, the population of multiplets decreases with flow rate and injection pressure.

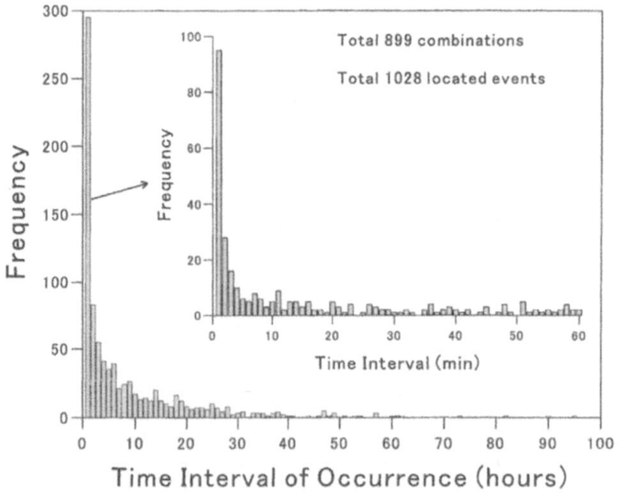

Figure 7
Histogram of time interval for event occurrence. The distribution of time interval for consecutive events was considered, with time intervals of 1 hour and 1 minute examined. Most of the similar events occur within 1 minute.

4. Relocation of Multiplets and Evaluation of Detailed Structure

4.1 Relative Source Location

In this paper we analyzed the multiplets that consist of more than five similar events, as shown in Figure 5(b), which incorporates a total of 115 multiplets. In order to estimate accurate, relative travel times of multiplet *P* and *S* waves, a moving window cross-spectrum analysis was applied to the waveforms. The relative travel-time delays for the *P* and *S* waves was estimated for four stations. The waves have dominant energy within 150 Hz to 200 Hz, consequently the time windows for cross-spectrum analysis using FFT algorithm were set to 51.2 ms, which corresponds to around 10 periods.

The cross spectra were calculated for 10 different time windows shifted by 0.4 ms steps on the *P*- and *S*-wave regions, and the spectra at each time window were averaged as a complex. Cross spectra were obtained for each time windows, and are generally complex numbers. Subsequently, the real and imaginary parts at shifting time windows were individually averaged to acquire the mean value of cross spectrum. After obtaining the mean value of the cross spectrum, the slope of the phase within a specified frequency band could be calculated to determine the relative travel-time difference of *P* and *S* waves (MORIYA *et al.*, 1994). The relative travel-time delays were calculated for all combinations of events in the multiplet, with all of the events assigned to a 'master' event. The iterative optimization procedure by TEZUKA and NIITSUMA (1996) was applied to the estimated time delays, in order to reduce the random error in time delay estimation. The relationship between source locations for each multiplet group was determined by the optimized travel-time delay, under a homogeneous velocity structure. The absolute position of each multiplet cluster was moved, so that the center of gravity for the source location is the same before and after relocation.

Figure 8(a–c) show the absolute locations of 115 multiplets, where a total of 1,028 events are plotted, as determined by JHD (Joint Hypocenter Determination). Figure 9(a–c) describe the relative source locations of multiplets shown in Figure 8. Figures 10(a–c) and 11(a–c) show the absolute and the relative source locations of multiplets near the fracturing well, where the thick line denotes the locus of the fracturing well.

The detection of wave arrival times on the cross-spectrum analysis resulted in a decrease of residuals from 1.89 ms to 0.1 ms. The value of 1.89 ms is the mean residual of 1028 analyzed events, prior to relocation. A residual of 0.1 ms corresponds to an uncertainty of location of about 1 m. As shown in Figures 10 and 11, the source locations for multiplets occur in small clusters.

4.2 Structural Planes Delineated from Multiplet Source Locations

Multiplets are considered to be events that occur on the same fracture plane, yet there is a possibility that the structural planes identified by the multiplet sources

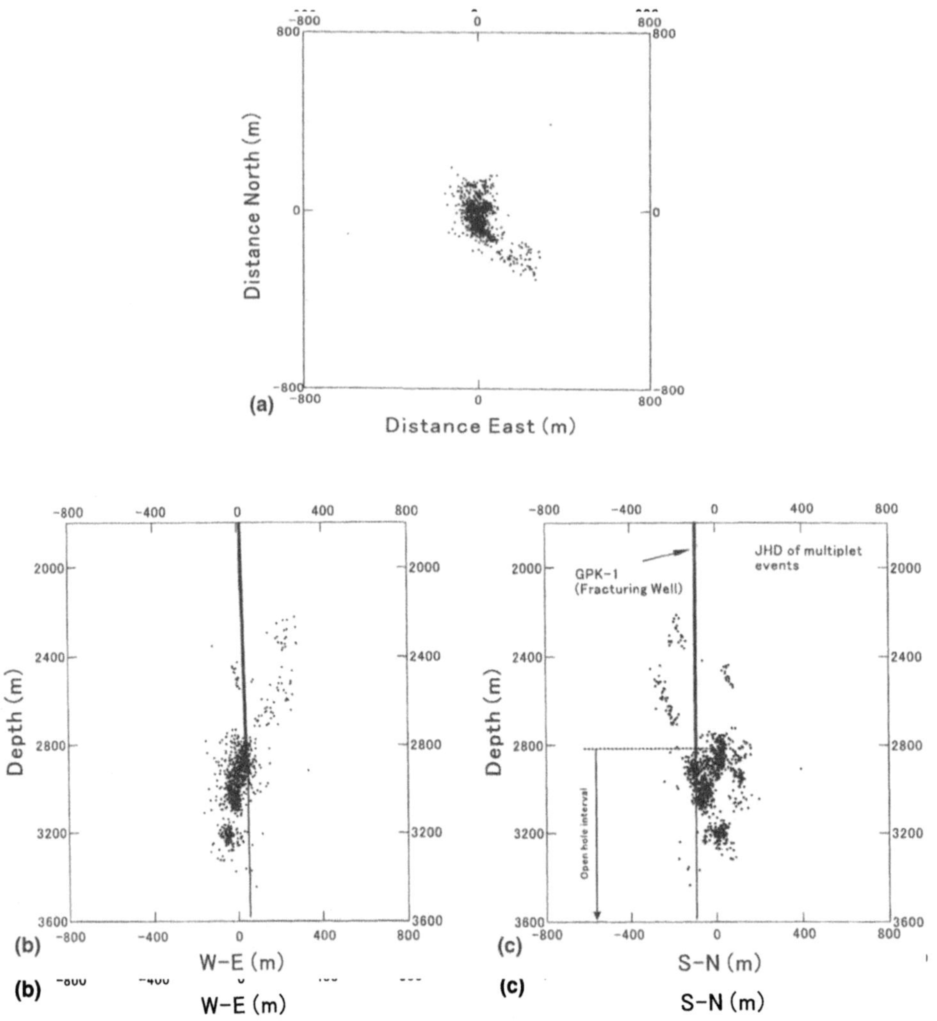

Figure 8

Absolute source locations of analyzed multiplets, with 115 groups of multiplets plotted. The multiplets tend to occur near the fracturing well.

represent either seismically activated structures and/or permeable fractures induced by hydraulic fracturing. Therefore, we introduced the principal component analysis to quantitatively evaluate the shape and source distribution of each multiplet cluster. By using the coordinates of respective source locations, the principal component analysis can quantify the three-dimensional configuration of the seismic cloud (i.e., whether the sources are distributed as a linear feature or as an ellipse). We calculated the first and second contribution ratios, as defined by eigenvalues of a 3×3 variance-covariance matrix. The variance-covariance matrix is defined using the source location coordinates of each cluster. The eigenvalues of the matrix represent the

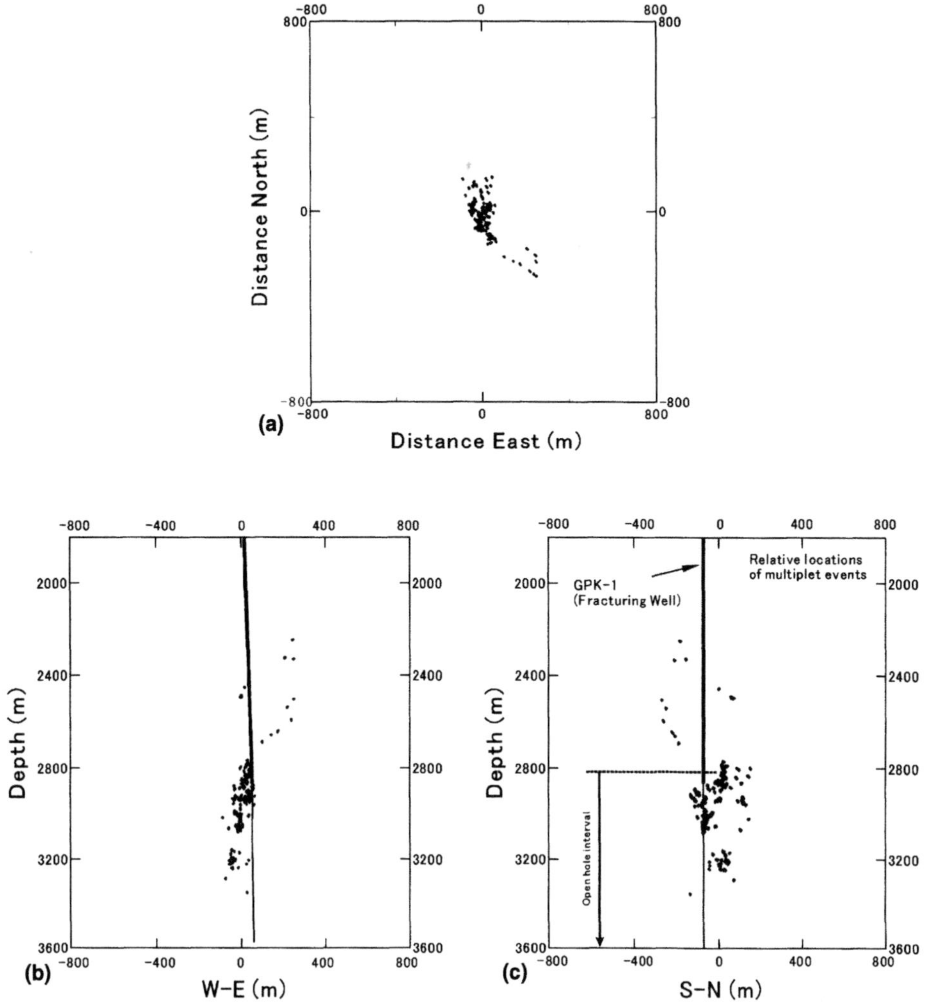

Figure 9
Relative source locations of 115 groups of analyzed multiplets, and 1028 events determined using the relative travel-time delay of P wave, as detected by the cross-spectrum analysis. The center of gravity of each cluster is the same before and after relocation.

variances to each principal axes, where the source distribution is approximated as an ellipsoid. Therefore, the contribution ratios, defined by the eigenvalues, indicate the configuration of source distribution.

Figure 12(a and b) are histograms showing the contribution ratios of multiplet clusters. The first contribution ratio represents the degree of linearity, while the first and second contribution ratio indicates the degree of flatness of plane. As shown in Figure 12(b), most multiplet groups have the first and second contribution ratios larger than 0.9. This result means that the relative sources of the multiplets form planar structures, rather than linear features.

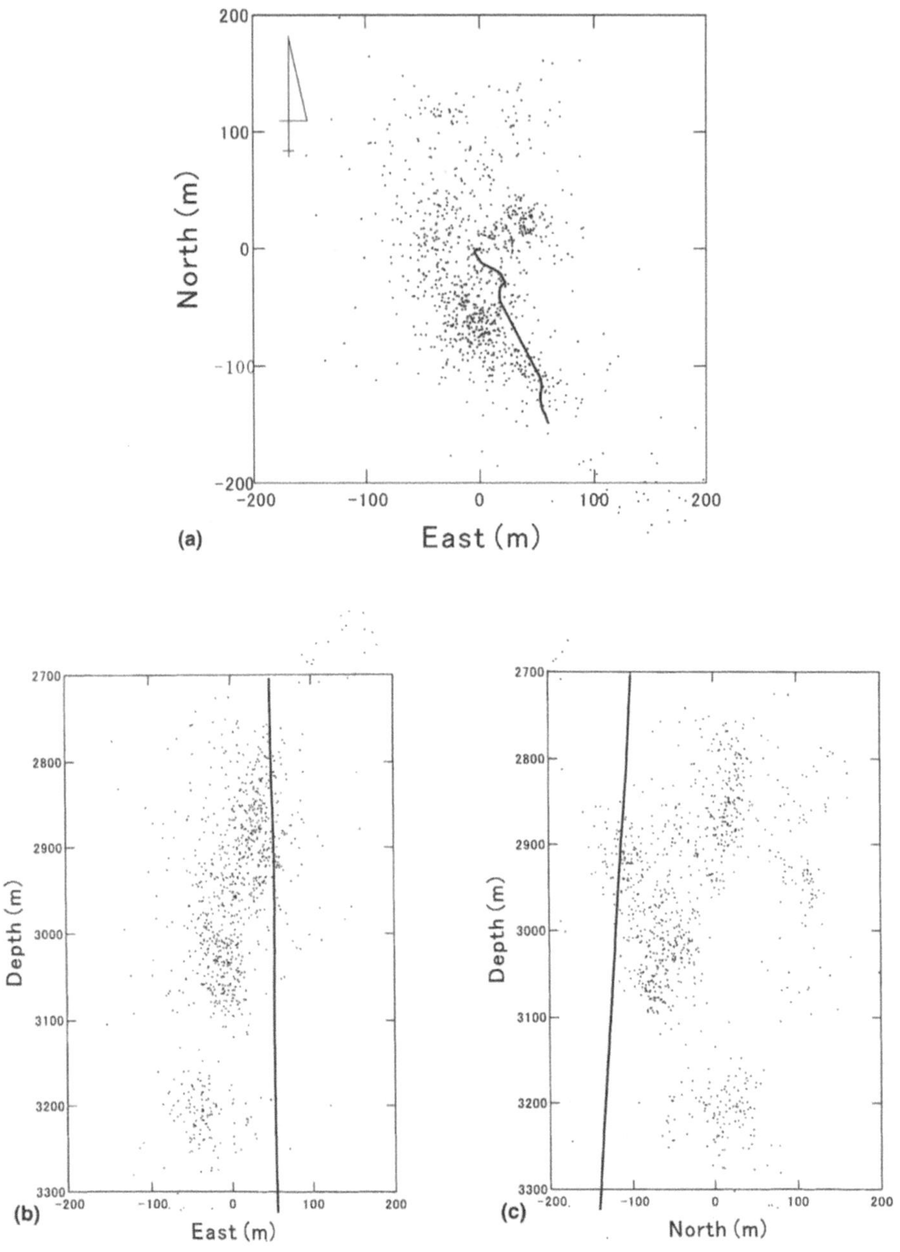

Figure 10
Absolute source locations of multiplets near the fracturing well. The thick line denotes the locus of the fracturing well.

Figure 13(a–c) display a side view of the multiplet clusters, the square root of first eigenvalues and the square root of third eigenvalues, respectively. The square root of the first eigenvalue indicates the standard deviation to direction of the maximum principal axis of the approximated ellipsoid, which ranges from 1.5 m to 7 m. On the other hand, the square root of the third eigenvalue represents the approximate thickness of the ellipsoid, which is about 1.5 m. The frequency range, to search for similar waveforms was 100 to 200 H and, with wavelengths from 45 m to 23 m. The size of each multiplet cluster is smaller that the wavelength.

Figure 14(a) shows the distribution of the poles to planes, determined for all of the analyzed multiplets. The strikes of the structural planes are approximately oriented NNW-SSE. The fracture distribution has been obtained from the analysis of core samples and borehole imaging logs from GPK-1 (GENTER and DEZAYES, 1993). BARIA *et al.* (1993) reported that the Soultz granite is pervasively fractured, with a mean joint spacing of about 3.2 joints/m. Figure 14(b) shows a Schmidt isodensity projection of poles to fracture, detected using BHTV (Borehole Televiewer), in the fracturing well (GPK-1) between 1376 m and 3600 m depth, where the thickness means the density of fractures. Figure 14(a and b) indicate that the structural planes identified from the source distributions of each multiplet were pre-existing fracture planes. Figure 14(c) shows the poles of multiplets distributed near the fracturing well, from 2687 m to 3349 m depth. Figure 14(d) shows contours of the poles of discontinuities, as detected by FMI (Formation Micro Imager), after GENTER and DEZAYES (1993). Although the depth interval is not identical, orientations of fractures obtained from the multiplets and the borehole imaging do agree. Therefore, we may infer that the multiplet planes indicate that the pre-existing fractures were activated by the increase of pore pressure.

Figure 15 shows the dip of structural planes, as a function of depth (i.e., the dip of multiplet planes are plotted at the corresponding depth of each multiplet cluster, and the dip of a vertical plane is 90 degrees). Our results indicate that the multiplet planes have various dips at depths of around 2900 m, 3200 m and 2500 m, respectively. On the other hand, the multiplet planes have dips of nearly 90 degrees at the other depths. It is reported that the major intervals of water outlets from the well are at 2850–2900 m, 3090–3100 m, 3230–3240 m and 3320–3325 m, with the percentage contribution of the top zone (2850–2900 m) being 65% when the injection pressure approached the magnitude of the minimum principal stress (BARIA *et al.*, 1995). Consequently, the plane dips are likely to be most variable near the outlet point (around 2900 m depth; within the interval described by A in Fig. 15), and progressively more uniform further away from the outlet point. This may be due to a larger pore pressure near the outlet point, which causes frictional shear slip on a wide range of orientations as predicted by Coulomb's law.

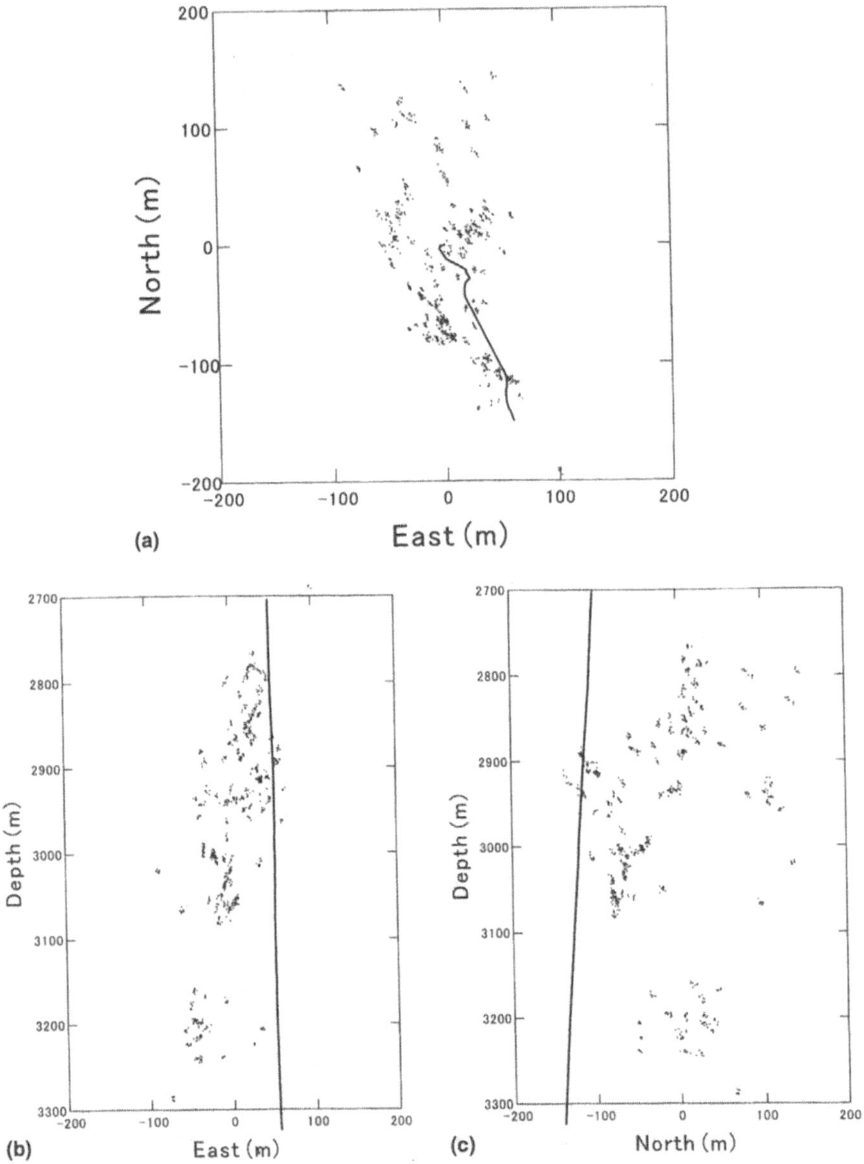

Figure 11

Relocated source distribution of multiplets near the fracturing well. The multiplet source locations plot together because the location accuracy is improved, and the residual is around 0.1 ms after the relocation.

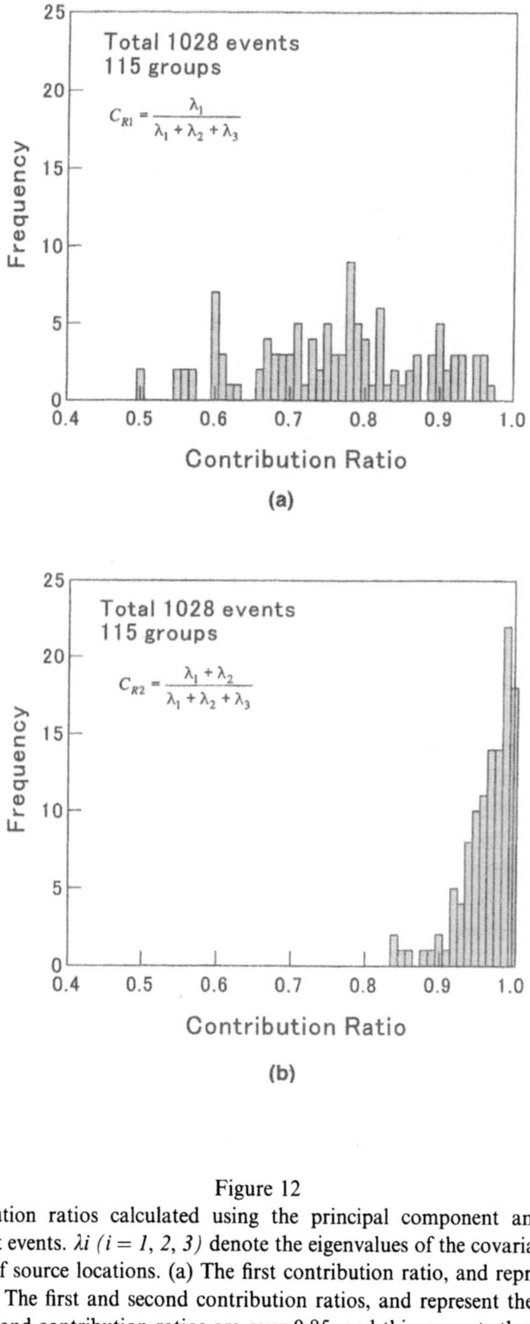

Figure 12
Histogram of contribution ratios calculated using the principal component analysis on the source coordinates of multiplet events. λi $(i = 1, 2, 3)$ denote the eigenvalues of the covariance matrix calculated using the coordinates of source locations. (a) The first contribution ratio, and represents the linearity of source distribution. (b) The first and second contribution ratios, and represent the degree of flatness of plane. The first and second contribution ratios are over 0.85, and this suggests that the structural planes defined by the source locations of multiplets tend to show the planar structures.

5. *Fracture System Interpreted from the Multiplets Data*

5.1 Orientation of the Structural Planes and the Critical Pore Pressure for Shear Slip

To understand the relationship between the seismically delineated structures and the stress field, it is important to discuss the dynamics and fluid flow in the fractures. In the Soultz field, PHILLIPS (2000) examined the fracture system and the stress condition at two different depths around 2675 m and 3200 m in well. He suggested that the critical pore pressure for these fractures is small (PHILLIPS, 2000).

We calculated the stress condition for all multiplet planes distributed in the hydraulic fracturing experiment seismic cloud. Figure 16 shows the critical pore pressure for frictional slip, according to Coulomb's law, using a representative stress field at the depth of 2900 m, where the coefficient of friction is 0.8. The dots in Figure 16 denote the poles of structural planes, which are projected onto a lower hemisphere. At a depth of 2900 m, the maximum (vertical direction) and intermediate (N170°E) principal stresses are close to each other, and pre-existing fractures which have NE poles are inferred to slip easily due to the increased pore

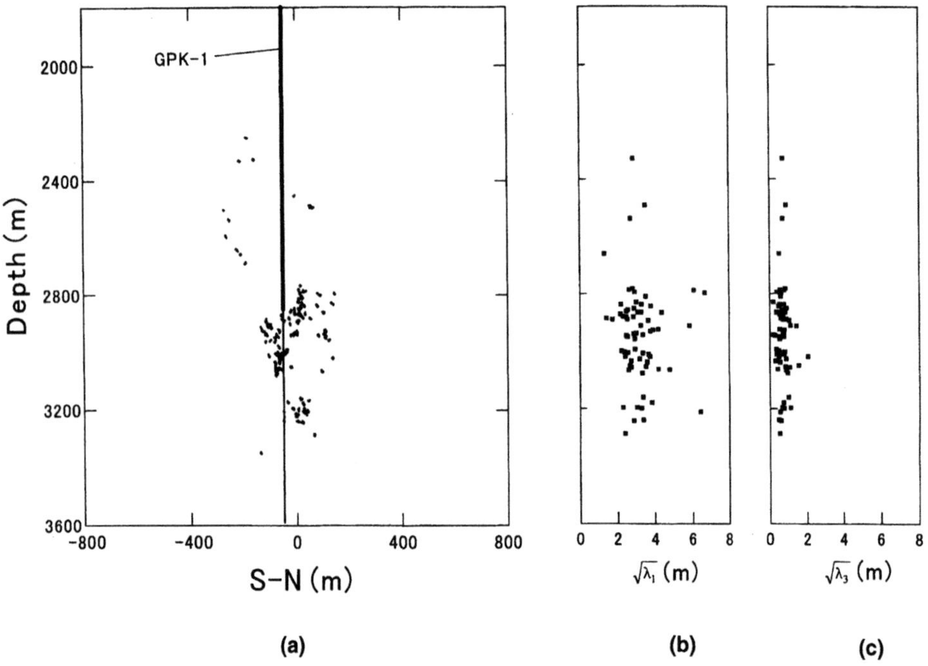

Figure 13

Side view of multiplet clusters, square root of first eigenvalues and square root of second eigenvalues. The square root of first eigenvalue, which approximately indicates the length of maximum principal axis of source ellipsoid, ranges from 1.5 m to 7 m. On the other hand, the square root of third eigenvalue, which approximately represents the thickness of the ellipsoid, is around 1.5 m.

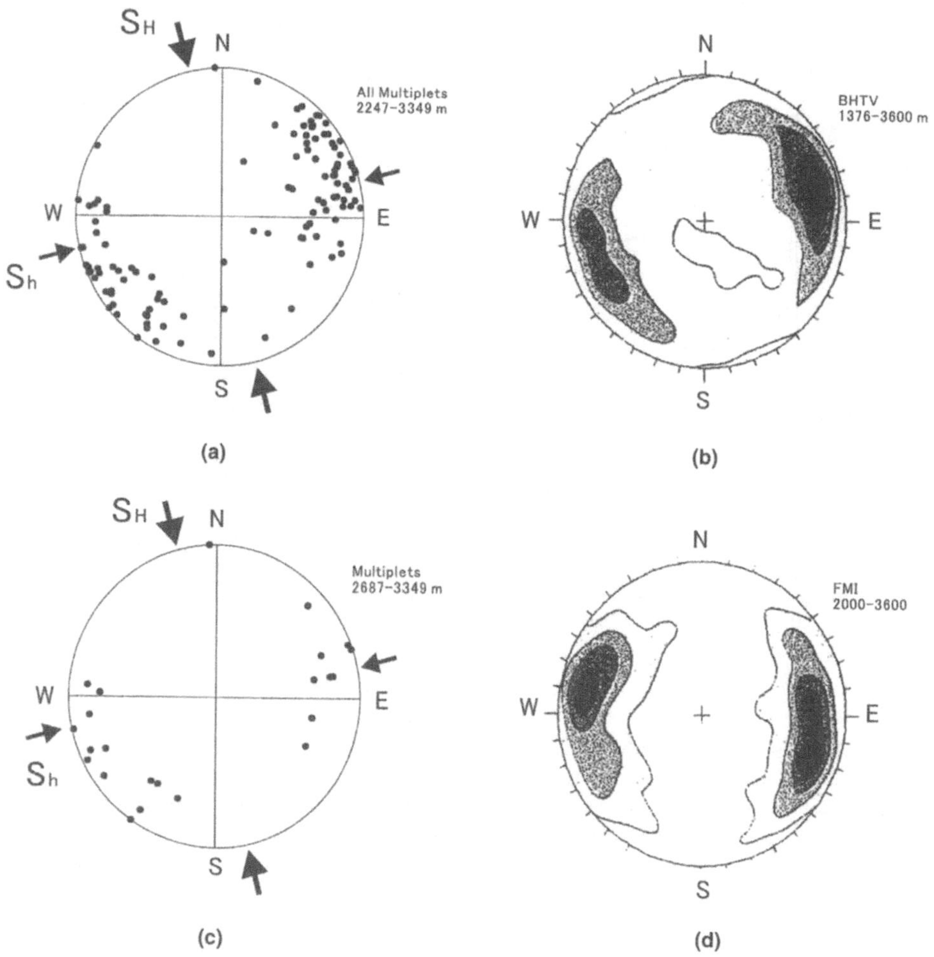

Figure 14

Poles to estimated structural planes and observed fractures projected on the lower hemisphere (GENTER and DEZAYES, 1993). (a) Structural planes delineated multiplet source distributions within the depth interval from 2247 m to 3349 m, (b) fracture distribution detected by BHTV survey within the depth interval from 1376 m to 3600 m, (c) poles of structural planes near the fracturing well within the depth interval from 2687 m to 3349 m and (d) fracture distribution detected by FMI within the depth interval from 2000 m to 3600 m. The thickness of the contour represents the density of poles.

pressure. The maximum wellhead pressure during the hydraulic fracturing experiment was about 10 MPa, at which the calculated pore pressure near the outlet point, at the depth interval between 2900–2950 m, was ideally around 40 MPa. Figure 16 indicates that nearly 80% of the structural planes can slip at pore pressures reaching 40 MPa, and that the structural planes delineated from

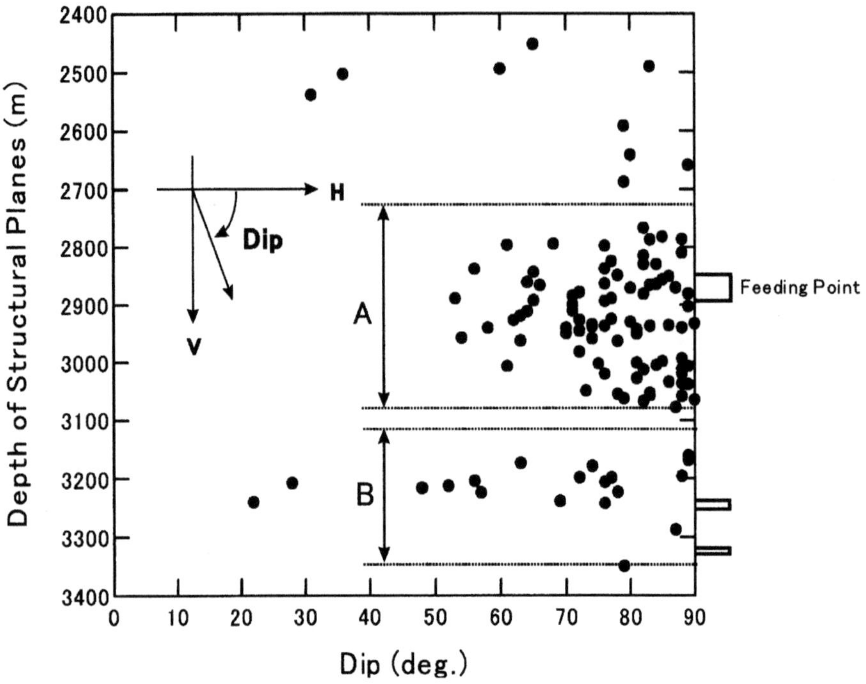

Figure 15
Dip of multiplet planes with depth. The outlet points exist at depth intervals of 2850–2900 m, 3090–3100 m, 3230–3240 m and 3320–3325 m, and the dip of plane has a wider range near the outlet points. It is suggested that the fractures with various dips cause shear slip near the outlet point because of higher pore pressure during the hydraulic fracturing test.

the relative source locations of multiplets are those fractures which slip most easily.

5.2 *Implication of the Structural Planes*

Figure 17 depicts the projection of structural planes to the horizontal plane, where the lines represent the strike of structural planes and dots denote source locations of the multiplets. It is reported that the maximum horizontal stress direction is N170°E within the depth interval from 2000 m to 3315 m in the well (BARIA *et al.*, 1999). The strikes of structural planes vary by roughly 20 degrees with respect to the maximum horizontal stress direction. As such, it is reasonable to infer that the fracture planes are oblique to the maximum stress direction. The fractures behave as a seismic source, associated with frictional shear slip that resulted from an increase in pore pressure.

We consider a fracture system with a mesh structure, in terms of multiplet planes, in which a strong directional permeability might be developed (HILL, 1977; SIBSON, 1996; TEZUKA and NIITSUMA, 2000). That is, the major fluid flow fractures, which

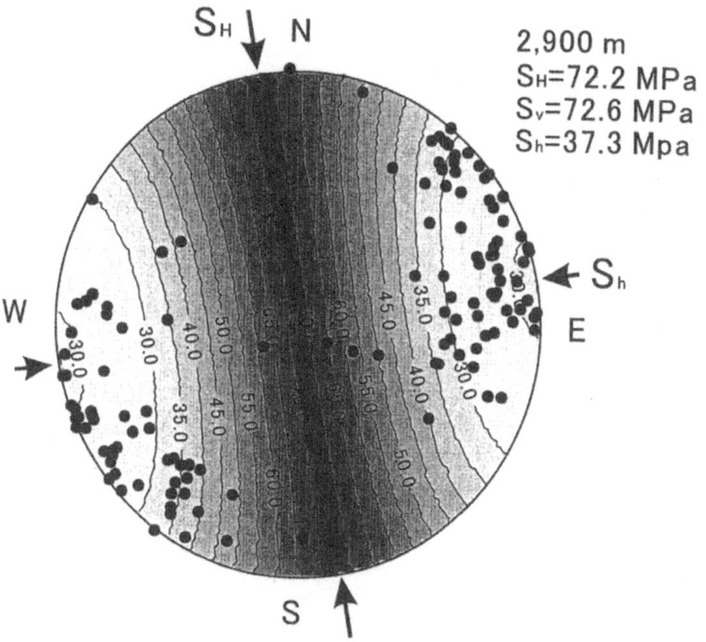

Figure 16
Critical pore pressure for frictional slip according to the Coulomb's law, where the stress field at the depth of 2900 m ($S_H = 72.2$ MPa, $S_v = 72.6$ MPa, $S_h = 37.3$ MPa) is considered as a representative stress field and the coefficient of friction is given as 0.8. The stress field determined by the hydraulic fracturing method is used to calculate the normal, shear and critical pore pressure on each multiplet plane. The dots denote the poles of the structural planes projected on the lower hemisphere. It is suggested that 80% of multiplet planes have the critical pore pressure below 30 MPa, which means that these multiplet planes cause shear slip due to the small increase of fluid pressure inside the fractures.

may be not seismically active, are distributed in a zone and oriented to the direction of maximum stress and approximately perpendicular to the direction of minimum stress. In addition, fractures that are oblique to the principal stress direction and induce shear slip, interlinked the major fluid flow fractures. This interpretation suggests that the fluid flow paths are mainly created by shear slippage along pre-existing fractures and interlinking of major fluid flow fractures, resulting from increased fluid pressure. Multiplet planes clearly represent mediation fractures, that induce shear slip and consequently play an important role in extending permeable fractures.

5.3 Characteristics of the Fracture System

The major flow point in the fracturing well was detected just below the casing shoe, and we examined critical pore pressure as a function of the distance from the outlet point to each multiplet cluster. Figure 18 shows the relationship between the

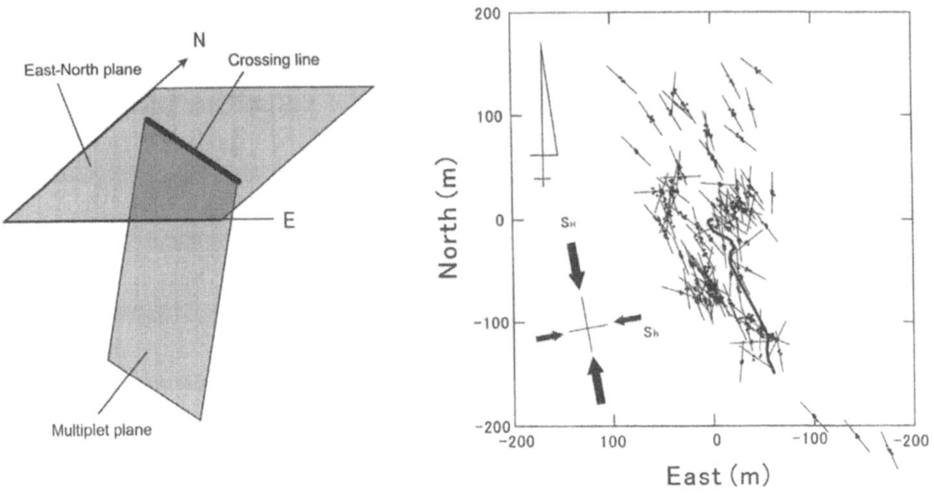

Figure 17
Projection of structural planes to the horizontal plane, where the lines represent the strike of the structural planes and the dots denote the source locations of multiplets. The strikes have angles to the maximum horizontal stress direction, and which represents that the multiplet planes show the fracture planes with shear slip due to the increase of fluid pressure. That is the multiplet planes would represent the mediation fractures that caused shear slip and play an important role in extending the permeable fractures, if we interpret the fracture system such that the fluid flow paths are mainly created by shear slippage along the pre-existing fractures, due to the elevation of fluid pressure according to Hill's mesh model (HILL, 1997).

distance from the outlet point to the centre of multiplet clusters, with pore pressure increase (ΔPp) for shear slip, for analyzed multiplets within the 2700 to 3100 m depth interval. The dots denote the increase in fluid pressure required to cause shear slip on each multiplet plane. We consider that ΔPp is the minimum increase of pore pressure required to induce shear slip in the fracture. The increase of fluid pressure at the outlet point is about 10 MPa, which corresponds to the pumping pressure, although ΔPp is less with distance due to pressure loss and diffusion. The critical pore-pore pressure at a distance of 250 m is estimated to be around 5 MPa. We interpret that the fluid pressure transmits away from the fracturing well, which clearly describes the characteristics of the fracture system at the Soultz site. That is, the fractures are seismically activated by a small increase of pore pressure and extend to more than several hundred meters.

6. Conclusions

We have analyzed the microseismic multiplets, induced during hydraulic fracturing experiments, to deduce the detailed structure and characteristics of the

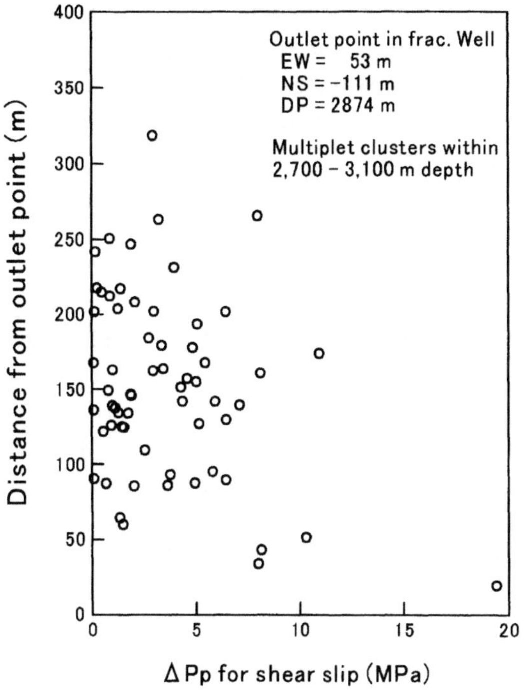

Figure 18
Relationship between the distance from the outlet point to the centroid of multiplet clusters and the pore-pressure increase (ΔP_p) for shear slip, where the multiplets within the interval from 2700 to 3100 m depths are analyzed. The dots denote the increase of fluid pressure needed to cause shear slip for each multiplet plane. The increase of fluid pressure at the outlet point was around 10 MPa, corresponding to the pumping pressure, and it decreases with the distance due to pressure loss and diffusion. From Figure 18, it is suggested that the fluid pressure transmits away from the fracturing well, and that the fractures were seismically activated by a slight increase of pore pressure and extend to more than several hundred meters.

Soultz-HDR, granite-hosted fracture system. We searched for multiplets from a larger seismic cloud using the coherency function, and indicated here that the multiplets tended to be observed at the early stages of the fracturing experiment. The use of the multiplets has enabled precise source locations to be determined, and demonstrated that relocated multiplet events occurred in small planar clusters on a scale of only a few meters.

A correlation between the orientation of multiplet planes, fracture distribution detected by borehole imagers and the stress field, indicates that the multiplet planes represent seismically activated pre-existing planes and that they play an important role interlinking fluid flow structures in a mesh fracture-reservoir system. The fluid pressure needed to cause shear slip, calculated using the Coulomb's law for the multiplet planes, indicates that the pumping pressure is easily transmitted to the fractures, and away from the fracturing well.

We suggest that our mapping technique, based on the identification of induced multiplets is sound, and allows for the recognition of individual fractures in a fractured reservoir. Furthermore, integrated interpretation of the stress field provides information on the characteristics of the fractured reservoir, as well as the overall structure of the fracture system.

Acknowledgement

We would particularly like to thank Dr. W. S. Phillips (Los Alamos National Laboratory), and also Dr. G. Bignall (Tohoku University) for their comments and suggestions. This work was carried out as a part of the MURPHY/MTC International Collaborative Projects, supported by NEDO (International Joint Research Grant) and Ministry of Education, Science, Sports and Culture (Grant-in-aid for International Collaborative Research). We also extend our gratitude to Socomine for providing the microseismic data and borehole imaging results from the European HDR site at Soultz.

References

Baria, R., Garnish, J., Baumgärtner, J., Gérard, A., and Reinhard, J. (1995), *Recent Developments in the European HDR Research Programme at Soultz-sous-Forêts (France)*, Proc. Word Geothermal Congress, Florence, Italy, 2631–2637.

Baria, R., Baumgärtner, J., and Gérard, A. (1999), *European HDR Research Programme at Soultz-sous-Forêts (France)* 1987–1996, Geothermics *28*, 655–669.

Fréchet, J., Martel, L., Nikolla, J., and Poupinet, G. (1989), *Application of the Cross-spectral Moving-window Technique (CSMWT) to the Seismic Monitoring of Forced Fluid Migration in a Rock Mass*, Int. J. Rock Mech. Min. Sci. and Geomech. Abstr. 26, 221–233.

Gaucher, E., Cornet, F. H., and Bernard, P. (1998), *Induced Seismicity Analysis for Structure Identification and Stress Field Determination*, Paper SPE 47324, Proc. SPE/ISRM, Trondheim, Norway.

Gérard, A., Baumgärtner, J., Baria, R., (1997), *An attempt towards a conceptual model derived from 1993–1996 hydraulic operations at Soultz*. In: Proceedings of NEDO International Symposium, Sendai, Japan, 2, pp. 329–341.

Genter, A., and Dezayes, C. (1993), *Fracture Evaluation in GPK1 Borehole by Using FMI Data*, Field Report, BRGM Orléans.

Genter, A., Castaing, C., Dezayes, C., Tenzer, H., Traineau, H., and Villemin, T. (1996), *Comparative Analysis of Direct (Core) and Indirect (Borehole Imaging Tools) Collection of Fracture Data in the Hot Dry Rock Soultz Reservoir (France)*, J. Geophys. Res. *102*, 15,419–15,431.

Hill, D. P. (1977), *A Model for Earthquake Swarms*, J. Geophys. Res. *82*(8), 1347–1352.

Jones, R., and Asanuma, H. (1997), *Analysis of the Four-component Sensor Configuration*, MTC/NEDO Internal Report, vol 2.

Jones, R., and Stewart, R. (1997), *A Method for Determining Significant Structures in a Cloud of Earthquakes*, J. Geophys. Res. *102*, 8245–8254.

Li, Y., Cheng, C. H., and Töksoz, N. (1998), *Seismic Imaging Geometry and Dynamic Growth Process of a Hydraulic Fracture Zone at Fenton Hill, New Mexico*, Geophysics, *63*, 120–131.

MORIYA, H., NAGANO, K., and NIITSUMA, H. (1994), *Precise Source Location of AE Doublets by Spectral Matrix Analysis of Triaxial Hodogram*, Geophysics *59*, 36–45.

MORIYA, H., NAKAZATO, K., and NIITSUMA, H. (2000), *Study of Microseismic Doublet/Multiplet for Evaluation of Fracture System in Soultz HDR Field*, Proc. World Geothermal Congress 2000, 3807–3812.

MORIYA, H., NIITSUMA, H., and KAIEDA, H. (2000), *Reevaluation of Reservoir Structure at Ogachi HDR Field by Precise Source Location of AE Multiplet*, Proc. World Geothermal Congress 2000, 3801–3806.

PHILLIPS, W. S., HOUSE, L., and FHELER, M. (1997), *Detailed Joint Structure in a Geothermal Reservoir from Studies of Induced Microearthquake Clusters*, J. Gephys. Res. *102*, 11,745–11,763.

PHILLIPS, W. S. (2000), *Precise Microearthquake Locations and Fluid Flow in the Geothermal Reservoir at Soultz-sous-Forêts, France*, Bull. Seismol. Soc. Am. *90*, 1, 212–228.

POUPINET, G., ELLSWORTH, W. L., and FRÉCHET, J. (1984), *Monitoring Velocity Variations in Crust Using Earthquake Doublets: An Application to the Calaveras Fault, California*, J. Geophys. Res. *89*, 5719–5731.

ROWE, C. A., ASTER, R. C., PHILLIPS, W. S., FEHLER, M. C., and BORCHERS, B. (2002), *Using Automated, High-Precision Repecking to Improve Delineation of Microseismic Structures at the Soultz Geothermal Reservoir*, Pure appl. geophys., submitted.

SIBSON, R. H. (1996), *Structural Permeability of Fluid-driven Fault-fracture Meshes*, J. Structural Geology *18*(8), 1031–1042.

TEZUKA, K. (1996), *MTC/NEDO Project Internal Report*, vol 1, 1996.

TEZUKA, K., and NIITSUMA, H. (1996), *Microseimic Doublet Analysis for Fracture Characterization in Hijiori HDR Test Site*, Proc. 3rd SEGJ/SEG International Symposium, 148–155.

TEZUKA, K., and NIITSUMA, H. (1997), *Integrated Interpretation of Microseismic Clusters and Fracture System in a Hot Dry Rock Artificial Reservoir*, SEG Expanded Abstracts, 657–660.

TEZUKA, K., and NIITSUMA, H. (2000), *Stress Estimated Using Microseismic Clusters and its Relationship to the Fracture System of the Hijiori Hot Dry Rock Reservoir*, Engin. Geology *56*, 47–62.

(Received May 5, 2000, revised/accepted December 22, 2000)

Pure appl. geophys. 159 (2002) 543–562
0033 – 4553/02/030543–20 $ 1.50 + 0.20/0

Pure and Applied Geophysics

Evaluation of Subsurface Structure at Soultz Hot Dry Rock Site by the AE Reflection Method in Time-frequency Domain

NOBUKAZU SOMA,[1] HIROAKI NIITSUMA[2] and ROY BARIA[3]

Abstract—Because the Soultz Hot Dry Rock (HDR) site, France, is to be expanded to a scientific pilot plant of greater depth, measurement of the deep area below the predeveloped artificial reservoir is gaining importance. In this paper, we present estimates of deep subsurface structure at the Soultz HDR site, obtained by a reflection method using acoustic emission (AE) signals, that is, induced seismicity, as the wave source. First, we briefly describe the AE reflection method in the time-frequency domain with wavelet transform. Then we show estimates of the subsurface structure by using 101 high-energy AE events observed in 1993. We compare and discuss the results obtained, using the AE reflection method in two wells with other independent borehole observations. Furthermore, we present the results of an investigation of the frequency dependence of reflectors identified by hodogram linearity as a possible means of further characterizing detected structures.

Key words: Reflected AE, linearity of three-dimensional hodogram, wavelet transform, time-frequency domain.

1. Introduction

The European Hot Dry Rock (HDR) programme is located at Soultz-sous-Forêts in northeastern France, some 50 km north of Strasbourg (Fig. 1). The site has been developed since 1987 with support from France, Germany, and the European Commission (EC) (BARIA *et al.*, 1995). In order to obtain geothermal energy from the deep crust, an artificial geothermal reservoir was created by hydraulic experiments. In the summer and autumn of 1997, a four-month long circulation test was conducted to demonstrate the possible future use of the HDR system as a renewable energy source. The test showed that the system had an acceptable production rate and negligible water losses, and required minimum input energy (BAUMGÄRTNER *et al.*, 1998). Since this successful test, development has proceeded to greater depths in order to reach temperatures high enough for practical use. One well, GPK2, was deepened from 3.5 km to 5.0 km in the spring of 1999.

[1] National Institute for Resources and Environment, 16-3 Onogawa, Tsukuba 305-0035, Japan. E-mail: n.soma@aist.go.jp

[2] Graduate School of Engineering, Tohoku University, Aoba01, Aoba-ku, Sendai 980-8579, Japan.

[3] SOCOMINE, Rute de Kutzenhausen BP39, F67250 Soultz-sous-Forêts, France.

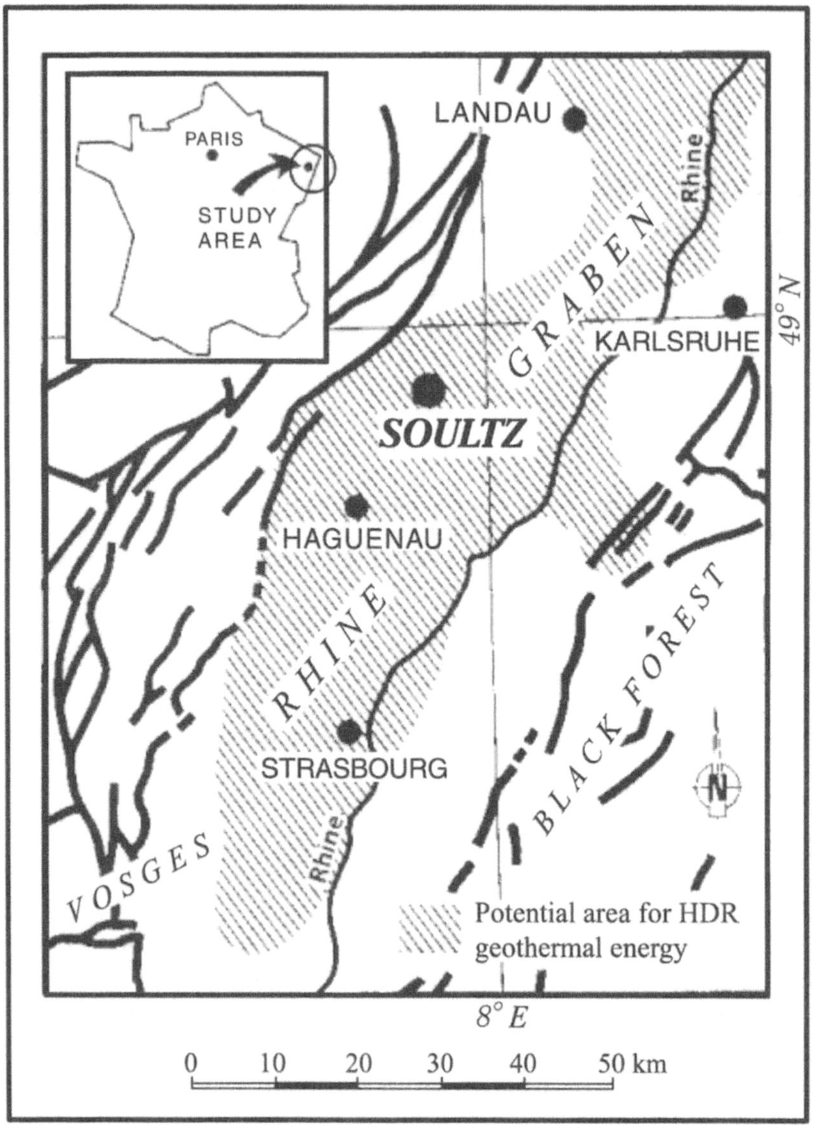

Figure 1
Location map of the Soultz HDR site.

In the Soultz project, induced seismicity, called acoustic emissions (AE), is commonly used to investigate the extension of the artificial reservoir or the exploitation flow path, or to model the connectivity between the injection and production well. However, induced seismicity, as the technique is generally employed, can measure conditions only in the area of the geothermal reservoir, even though it is important to understand conditions beneath or adjacent to the reservoir if the HDR system is to be expanded for practical use. Thus far, there has been no suitable technique for exploring those regions of a geothermal field.

Therefore, we developed a reflection technique in which natural or induced seismicity can be used as the wave source. We call it the "AE reflection method" (SOMA and NIITSUMA, 1997; SOMA *et al.*, 1997). This method has the following advantages: (i) the seismic energy of AE is greater than that of the more usual artificial wave sources, so it is possible to detect waves reflected from great depths; (ii) seismic sources are located in the deep rock mass as an inverse VSP (vertical seismic profile), thus there is less attenuation of the reflected waves compared to a survey from the surface; (iii) the horizontal distribution of seismic sources is relatively wide with no topographical restrictions; (iv) since the method requires only a passive measurement system, it is much less expensive and thus more practical for long-term, continuous observations; (v) if the proper source distribution is available, the method may be able to detect subvertical structures such as faults, which are a difficult target for a conventional technique used from the surface; (vi) use of a downhole multi-component detector allows observation of an enormous number of events, making three-dimensional processing possible even with only one detector.

The AE reflection method was initially developed for the time domain, and we have previously demonstrated its possibilities in geothermal fields (SOMA and NIITSUMA, 1997; SOMA *et al.*, 1997), although the resolution was not sharp enough for practical development. We have recently enhanced the method by developing a method of analysis in the time-frequency domain with wavelet transform (SOMA, 1998). We can expect the following advantages from time-frequency domain analysis with the AE reflection method, compared with time domain analysis: (1) higher resolution and accuracy for the detection of reflected waves, owing to the characteristics of wavelet transform; (2) less ambiguity because the method, unlike time domain analysis, does not require use of the sliding-time-window technique, which time domain analysis is forced to use; (3) robustness against noise because it is possible to choose the frequency component to avoid noise in the analysis; and (4) it offers various possibilities for the development of further signal processing techniques for imaging and characterizing subsurface structures.

In this study we estimated the deep subsurface structure in the Soultz HDR site using the AE reflection method in the time-frequency domain. We describe the Soultz site and the AE observation method, including how to detect reflected waves and image subsurface structure using the AE reflection method. Then we show the estimates of the deep subsurface structure in the Soultz HDR site obtained using waveforms recorded in 1993. We also make precise estimates of the structure along two boreholes and compare them with other well observations. Finally, we discuss the frequency dependence of the detected reflectors in an effort to characterize the deep subsurface structure.

2. Soultz HDR Site and AE Observation

The geology of the Soultz HDR site consists of shallow sedimentary layers and a deep crystalline rock mass consisting mainly of granite below a depth of approximately 1400 m. There are two geothermal wells (GPK1 and GPK2), four seismic wells, and one exploration well. The geothermal well GPK1 was drilled to a depth of 3.5 km and several hydraulic experiments were conducted. In 1993 massive hydraulic fracturing was performed at depths between 2800 and 3500 m in well GPK1, to create an artificial geothermal reservoir inside the basement granitic rock. Next, a second well, GPK2, was drilled in the vicinity of the artificial reservoir and a circulation system was successfully developed in 1995 (BAUMGÄRTNER *et al.*, 1996). After the success of a four-month circulation test in 1997, GPK2 was deepened to 5 km in the spring of 1999, in order to obtain higher temperatures. However, there is as yet sparse information available from the deepened part of GPK2.

During the massive hydraulic fracturing in 1993, about 20,000 AE events were observed with four-component downhole detectors (JONES and ASANUMA, 1997) deployed inside the basement rock and recorded as digital data sets with a 5-kHz sampling frequency. From the 1993 data sets we selected for analysis 101 high-energy events, which had been recorded in unusually long files. Unfortunately, we were able to use data from only one detector, designated E4550 (Fig. 1), because we could not collect enough usable waveform data from the other stations. We did not have reliable and precise hodograms from the other stations, probably due to minor electrical trouble during the experiment. The source location of the events was determined using all detectors (DYER *et al.*, 1994), since that procedure is less sensitive to hodogram accuracy. The source locations of the 101 events and the orientations of the cross sections later used for the subsurface structure estimates are shown in Figure 2. A typical AE waveform can be converted into a conventional three-component signal (see, for example, Fig. 3). Each waveform used contained about 3.0 s of data, which is longer than the usual AE measurement in this field, so that deeper subsurface structures could be seen. The dominant frequency of the waveforms was around 200 Hz. Many AE events recorded a high signal-to-noise amplitude ratio (S/N). We could identify the arrival of direct P and S waves relatively clearly because the site geology is mostly granite, which has low attenuation and high frequency signals. However, it was impossible to discriminate reflected waves from observed waveforms directly because they were covered by coda.

3. Outline of the AE Reflection Method in the Time-frequency Domain

The AE reflection method has two parts: distinguishing reflected waves from coda by the hodogram technique, and three-dimensional inversion for imaging with

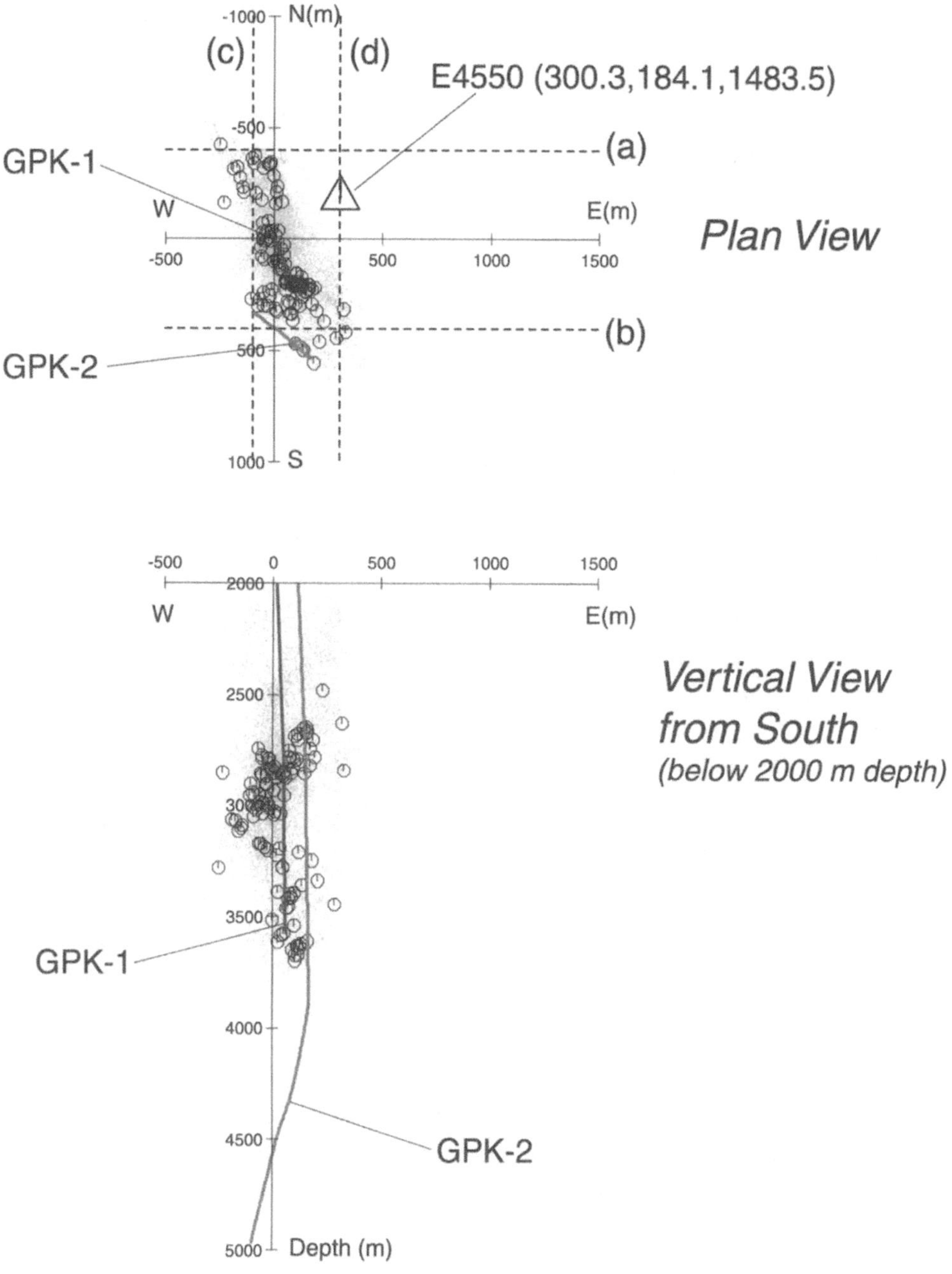

Figure 2
Source locations of the 101 AE events used for the analysis (circles). The triangle shows the detector at observation well E4550, and the gray cloud shows the distribution of all events observed in September, 1993. Trajectories of well GPK1 and well GPK2 which were deepened in 1999 are shown in the vertical section. Dotted lines in the plan view show the orientation of cross sections used for the estimates shown in Figure 5.

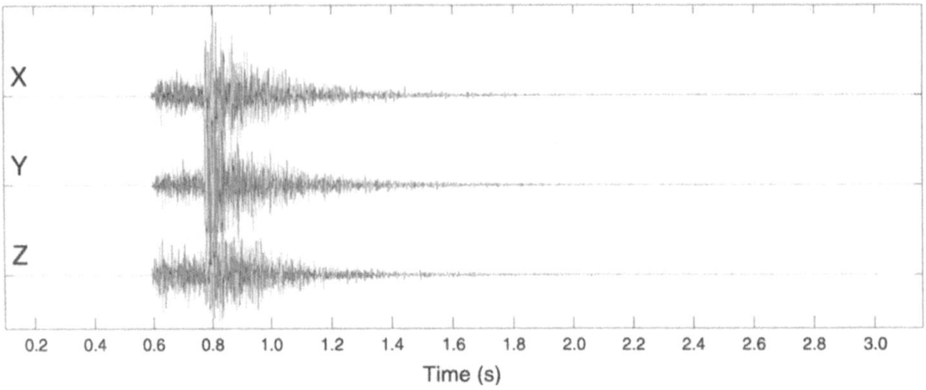

Figure 3
Example of a typical three-component AE waveform at the Soultz site.

a non-systematic source distribution (SOMA and NIITSUMA, 1997). We previously demonstrated in geothermal fields the potential of the method, which was initially developed for the time domain (SOMA and NIITSUMA, 1997; SOMA *et al.*, 1997). More recently we developed the time-frequency domain analysis method with wavelet transform in order to make this technique more practically useful (SOMA, 1998).

Because the reflected waves are covered by coda of the direct waves, conventional analysis based on wave energy or amplitude is useless for the analysis of reflected AE signals from a geothermal field. In the AE reflection method, the reflected waves are detected by investigating the shape of three-dimensional hodograms. The shape of the hodogram shows the trace of three-dimensional particle motion during the arrival of waves, and is known to indicate wave condition (NAGANO *et al.*, 1986). The shape of the hodogram is spherical when an incoherent signal such as random noise is detected, however it becomes relatively linear on the arrival of direct *P* or *S* waves, which are regarded as coherent signals (SOMA and NIITSUMA, 1997; SOMA *et al.*, 1997). In our study, reflected waves are regarded as coherent signals, since we assumed that the reflector has scale, and coda are normally regarded as incoherent signals (AKI and CHOUET, 1975). Therefore, using the AE reflection method we can distinguish reflected waves from coda by investigating the shapes of the hodograms.

For the quantitative evaluation of hodogram shape we used a covariance matrix method in the frequency domain, which is also called a spectral matrix (SAMSON, 1977). In this study we regarded the wavelet transform as a time-frequency signal representation (RIOUL and VETTERLI, 1991), which shows a smoothed time-varied spectral similar to a short-time Fourier transform (STFT) (HLAWATSCH and BOUDREAUX-BARTELS, 1992). Generally, a time-frequency signal representation allows us to compose a spectral matrix in the time-frequency domain (MORIYA and

NIITSUMA, 1996). Hence, we can define a spectral matrix $\mathbf{S}_{WT}(b, a)$ of the hodogram by using the wavelet transform as formula (1),

$$\mathbf{S}_{WT}(b,a) = \begin{pmatrix} W_{xx}(b,a) & W_{xy}(b,a) & W_{xz}(b,a) \\ W_{yx}(b,a) & W_{yy}(b,a) & W_{yz}(b,a) \\ W_{zx}(b,a) & W_{zy}(b,a) & W_{zz}(b,a) \end{pmatrix} , \tag{1}$$

where $W_{ij}(b, a) = W_i(b, a)W_j^*(b, a)$, a: frequency (scale), b: time (shift), $W_i(b, a)$: wavelet coefficient of i component, * : complex conjugate.

The spectral matrix is generally complex but always satisfies the conditions of a Hermitian matrix (MORIYA and NIITSUMA, 1996). We can treat this matrix as in the time-frequency domain since "scale" and "shift" in the wavelet transform correspond to "frequency" and "time," respectively. Then we can redefine Samson's global polarization coefficient, $Cp(b, a)$, (SAMSON, 1977) to be in the time-frequency domain, as follows:

$$C_p(b,a) = \frac{(\lambda_1 - \lambda_2)^2 + (\lambda_2 - \lambda_3)^2 + (\lambda_3 - \lambda_1)^2}{2(\lambda_1 + \lambda_2 + \lambda_3)^2} , \tag{2}$$

where $\lambda_i = \lambda_i(b, a)$ ($i = 1, 2, 3$; $\lambda_1 > \lambda_2 > \lambda_3$): eigenvalue of the matrix (1) for each time (shift: b) and frequency (scale: a).

By this parameter we can quantitatively evaluate the shape of the hodogram in the time-frequency domain; $Cp(b, a) = 1$ for an exactly linear hodogram and $Cp(b, a) = 0$ for a spherical hodogram. The arrival of coherent waves, such as reflected waves, should have a high Cp value. Hence, we can use a high value of the parameter $Cp(b, a)$ as an indicator of the arrival of reflected waves.

For imaging the subsurface structure, we established a three-dimensional inversion of the waveform which shows the time-frequency distribution of the linearity of the three-dimensional hodogram. Here we focused on only simple S-to-S-wave reflection because the energy of the S wave was dominant in our application and P-wave reflection may be concealed by the energy of the S wave.

The inversion concept is diagrammed in Figure 4. This inversion, which can generate a three-dimensional estimate from a small number of detectors, is based on a type of diffraction-stack migration as explained below. The inversion can create a deep subsurface image with data from only one detector if the source location has enough extension considering the objective area, although more detectors are better for accuracy and reliability.

[1] The linearity waveform $Cp(b, a)$, formula (2) of a three-dimensional hodogram in the time-frequency domain, is calculated. We can also obtain the polarization direction of the hodogram for each time and frequency component.

[2] We can assume the iso-delay ellipsoid for a delay $\Delta T(a_1)$ for scale "a_1", considering the detector, source, and path length. The virtual reflected points for $\Delta T(a_1)$ are located on the ellipsoid.

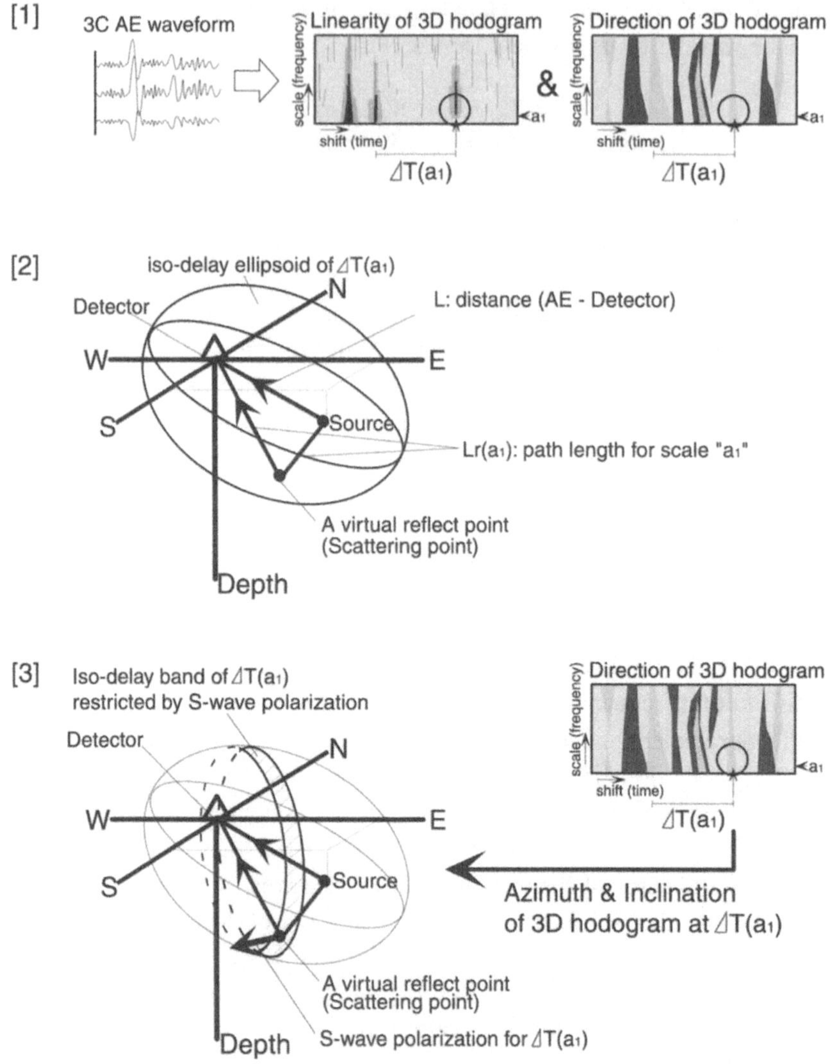

Figure 4
Diagram of the concept of three-dimensional inversion for the distribution of linearity of three-dimensional hodograms in the time-frequency domain.

[3] The strength of the linearity waveform for scale "a_1" is plotted on the ellipsoid.
[4] We repeat the process from [1] to [3] over all scales (frequency components) and all delays in the signal, and perform the same process for all of the usable AE events, after which the results are stacked.

In the inversion, we enhanced the resolution by restricting the virtual reflected points by examining the orthogonality between the propagation direction and *S*-wave polarization. The area in which the virtual reflected point occurs, where the strength of hodogram linearity is plotted, can be reduced if we select the area where *S*-wave motion is perpendicular to the propagation direction from the reflected point when we focus on a *S*-to-*S*-wave reflection. By this process, the virtual reflected point can be restricted to a band as in step [3] of Figure 4.

In addition, the effect of heterogeneous source distribution is compensated for by normalizing wave density for numerous nearby events for each source. The heterogeneity of source distribution causes the final image to have a dominant ellipsoidal artifact, which comes from a center of source distribution. Normalization can reduce the effect of the ellipsoidal artifact (SOMA and NIITSUMA, 1997). These two operations are effective when the inversion uses a natural distribution of induced seismicity.

The actual signal processing is done with time and frequency steps, since digitalized data sets are used. In this study, each time step was about 0.2 ms, corresponding to the sampling frequency during data acquisition, and we set seven frequency steps from 39.9 to 317.1 Hz for the wavelet analysis.

4. Three-dimensional Estimates at the Soultz HDR Site

We estimated the deep subsurface structure at the Soultz HDR site by the AE reflection method with the 101 high-energy events from 1993. The estimates obtained for the E-W and N-S cross sections are shown in Figure 5. The estimates were obtained from an E-W cross section 400 m north of the wellhead of injection well GPK1 (Fig. 5a), an E-W section 400 m south of GPK1 (Fig. 5b), a N-S section 100 m west of GPK1 (Fig. 5c), and a N-S section 300 m east of the well (Fig. 5d). The result obtained by time domain analysis from the same section as shown in Figure 5d was used for comparison with the time-frequency domain analysis results (Fig. 5e) (SOMA *et al.*, 1997). In the figure, darker colors indicate a higher value of stacked hodogram linearity, which is regarded as reflectivity in the analysis.

The time-frequency domain analysis showed a higher resolution and a better ability to detect reflectors than did the time domain analysis. A reflector at a depth of about 3000 to 3500 m was clearly seen using the time-frequency domain analysis (Fig. 5d), but was blurred in the time domain analysis because it was very close to the depth of the source distribution (Fig. 5e). The reflector seen at about 4000 m depth in the time domain analysis (Fig. 5e) was detected as two separate images at depths of 3950 m and 4200 m by the time-frequency domain analysis (Fig. 5d), suggesting that the resolution of the time-frequency domain analysis was higher. Deep reflectors below 4500 m depth were also detected more clearly than with the time domain analysis. The improved performance of the time-frequency domain analysis makes precise comparisons with other borehole observations possible.

However, the detected reflectors shown in Figure 5 appear curved because of the non-ideal and limited observation conditions. Since we were able to use only one

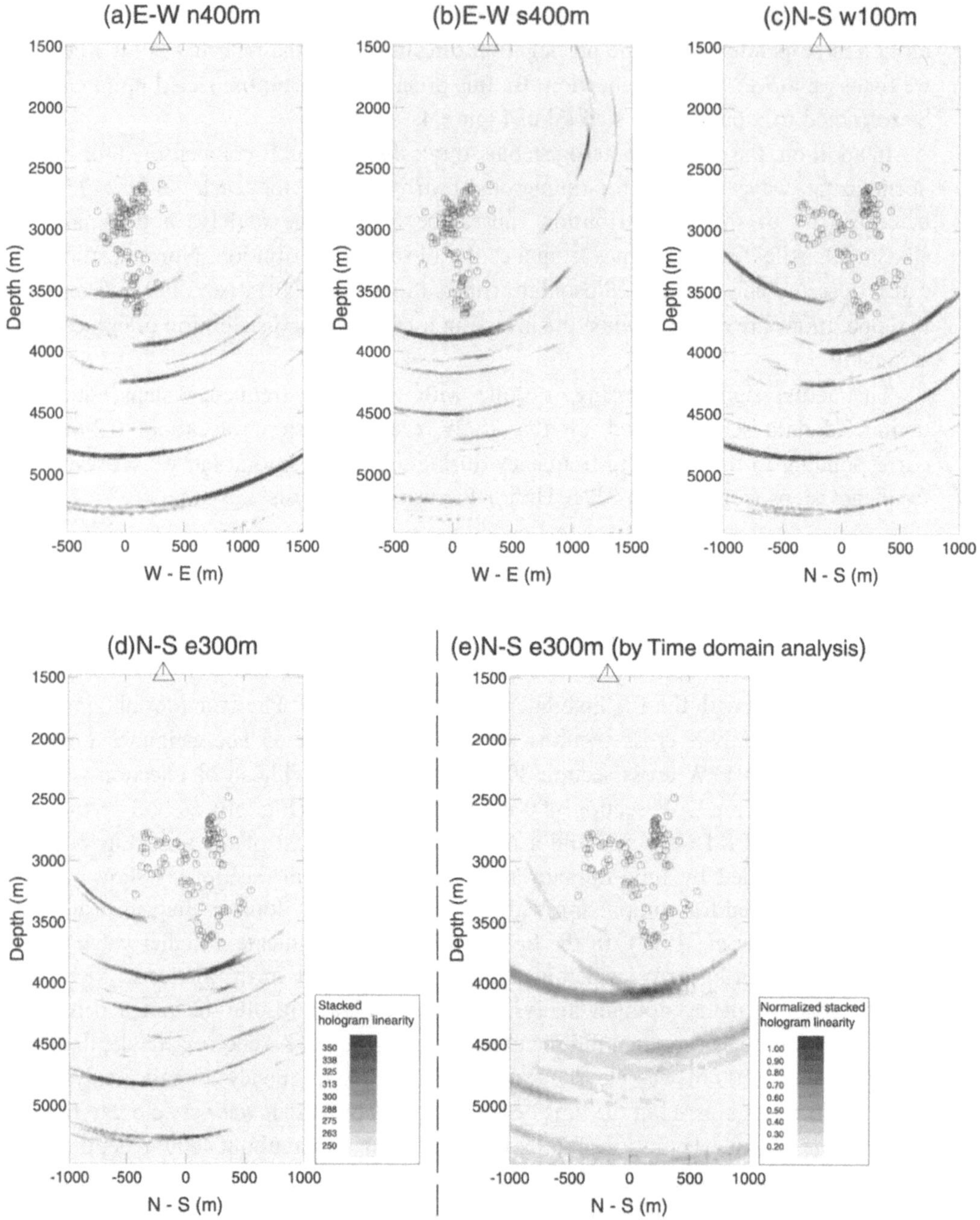

Figure 5

Cross sections of three-dimensional subsurface imaging by the AE reflection method in the time-frequency domain. (a): E-W cross section 400 m north from GPK1, (b): E-W section 400 m south, (c): N-S section 100 m west, (d): N-S section 300 m east, and (e): N-S section 300 m east analyzed in the time domain. The triangle and circles show the projections of the detector and sources, respectively.

detector, there was a strong geometrical effect. Therefore, we will not discuss the orientation or shape of the detected reflectors, but limit our discussion to the depth of the reflectors. We can roughly define the depth of each reflector using the point along the curved shape having the highest value, although the curved shape of the reflector affects the accuracy with which the depth can be determined. Thus reflectors were detected at depths of about 3500 m, 3950 m, 4200 m, 4500 m, 4800 m, and 5200 m.

The reflectors at 3950 or 4200 m depth may show the bottom of the artificially created reservoir because fracture conditions can be expected to change markedly there. To interpret the other reflectors, it is necessary to compare them carefully with other information collected independently.

5. Comparison with Other Borehole Data Sets

Since the AE reflection method generates three-dimensional volumetric results, we can estimate the subsurface structure wherever we choose. In order to make precise comparisons with borehole observations, we chose to make our estimates exactly along the trajectories of boreholes GPK1 and GPK2. Changes in hodogram linearity with depth (Fig. 6a), calculated from the AE reflection data and assumed to indicate reflectivity, were compared with several types of independent observations: the number of fractures per 10 m, based on formation micro-imager (FMI) log data (Fig. 6b), acoustic impedance log data (Fig. 6c), thermal flow meter log data (Fig. 6d), and helium log data (Fig. 6e). Hodogram linearity values at greater depths are larger because fewer events at shallow depths are usable if errors arising from the influence of direct waves are to be avoided. Therefore, data shown in the figure from depths shallower than 3100 m were considered unreliable and were disregarded.

Several major peaks in hodogram linearity indicated reflectivity at depths of 3100 m, 3200 m, 3330 m, 3375 m, 3500 m, and 3520 m (Fig. 6a) and agreed well with the fracture density data (Fig. 6b). The variations in the acoustic impedance log data also corresponded to the peaks obtained using the AE reflection method (Fig. 6c). These results suggest that the AE reflection method can detect fractures or fracture zones.

Thermal flow meter and helium log data are indicators of permeability. Major permeable zones in GPK1 occurred at depths of about 3192 m to 3293 m, 3370 m to 3406 m, and 3489 m to 3496 m (BARIA *et al.*, 1994). These zones were clearly detected in the AE reflection data. However, at present it is not possible to determine directly from AE reflection data whether or not a reflector indicates a permeable zone.

Hodogram linearity data were also collected in the deep portion of GPK2 (Fig. 7a), and compared with a geological section constructed from a synthetic petrographic log obtained from microscopic examination of drilling cuttings (GENTER *et al.*, 1999). The hodogram linearity was calculated from the AE reflection

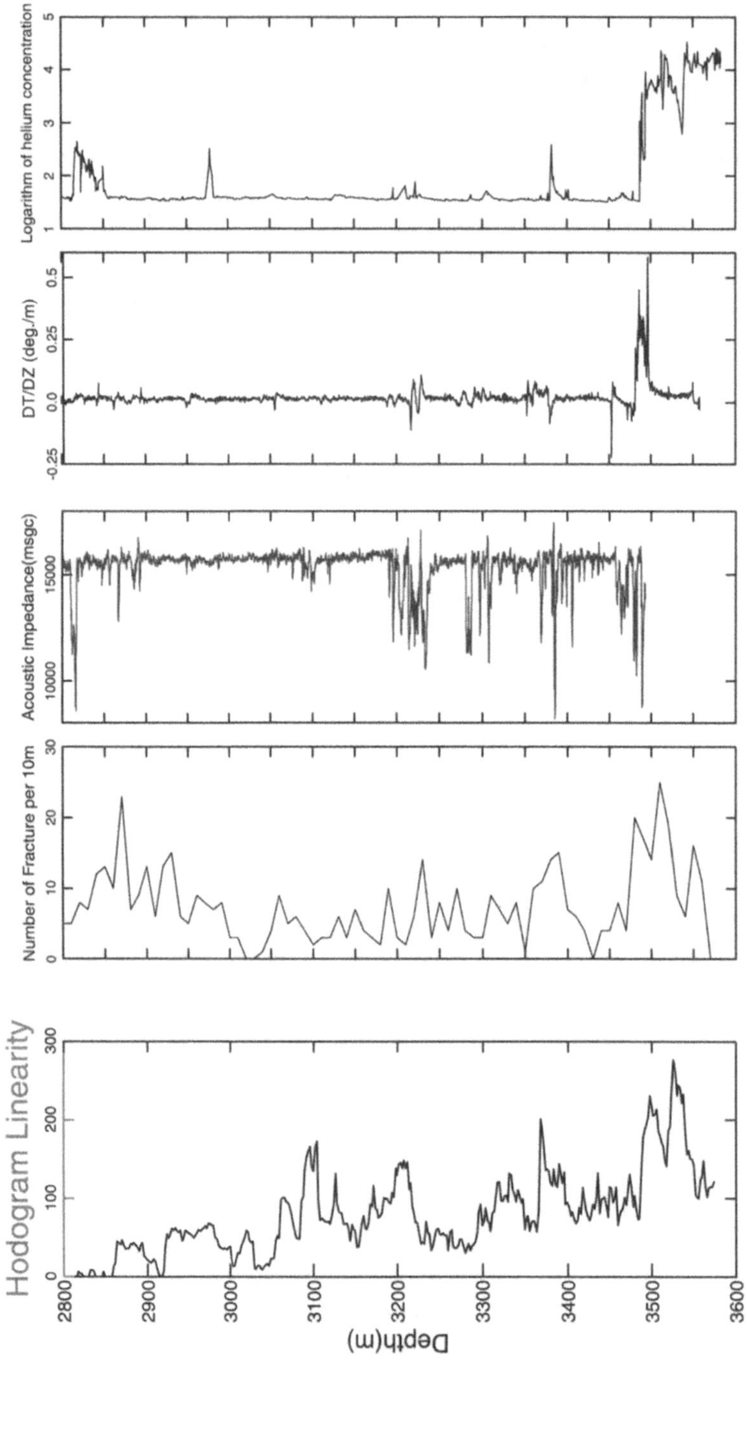

Figure 6

Comparison between the AE reflection method and other borehole observations in GPK1. (a): Estimation along GPK1 by the AE reflection method, (b): Number of fractures per 10 m based on FMI logging, (c): Acoustic impedance logging, (d): Thermal flow meter logging, and (e): Helium logging.

data before the actual drilling, and we adjusted the estimates to the actual well trajectory of the deepened borehole after well completion.

The AE reflection data showed probable reflectivity at depths of 3850 m, 4000 m, 4150 m, 4550 m, 4720 m, 4780 m, 4850 m, and 4950 m. Based on the relationships observed in GPK1, these results may indicate naturally existing structures such as fracture zones at great depth, since the data sets used were collected in 1993, before any hydraulic stimulation at these depths had been attempted.

The possible existence of many fracture zones has been reported, based on the microscopic examination of drilling cuttings obtained during GPK2 deepening (GENTER *et al.*, 1999). Major fracture zones composed of very highly altered granite were found at depths of about 4575 m and 4775 m. The most attractive permeable zone for the next phase of the project may be found at these depths because they will be helpful for developing a new deep artificial reservoir.

The AE reflection method clearly detected reflectors at depths of 4550 m and 4780 m, which correspond well to the two highly altered fracture zones found by examination of the drilling cuttings. Other peaks also appear to correspond to depths at which there was indication of the presence of other fracture zones. Therefore, there may exist many fracture zones at great depth at the Soultz HDR site. Accordingly, we can expect that deep HDR development at the Soultz site will be similar to development already completed at shallow depths.

6. *Frequency Dependence of Detected Reflectors*

In an attempt to characterize the detected structures, we examined the frequency dependence of hodogram linearity at each reflector depth. Since the AE reflection method is processed in the time-frequency domain, we can easily calculate frequency dependence. In this study we did not analyze the frequency dependence of wave amplitude but that of hodogram linearity, because the evaluation of wave amplitude covered by coda is too complicated. We investigated total hodogram linearity at each reflection point for each frequency component. Then we normalized it by the time resolution of each frequency component in a three-dimensional inversion.

The distribution of normalized hodogram linearity shows the frequency dependence of the estimate of subsurface structure (Fig. 8). Depth is shown on the vertical axis and the frequency component between 39.9 and 317.1 Hz is shown on the horizontal axis. The original hodogram linearity data are shown alongside. As before, data from depths less than 3100 m were disregarded.

Clear variations in frequency dependence among reflectors were observed in GPK1. For example, at depths of about 3100 m, 3200 m, 3520 m, the reflectors showed a tendency for lower frequency components to have more linear hodograms. On the other hand, the reflector at a depth of approximately 3400 m showed

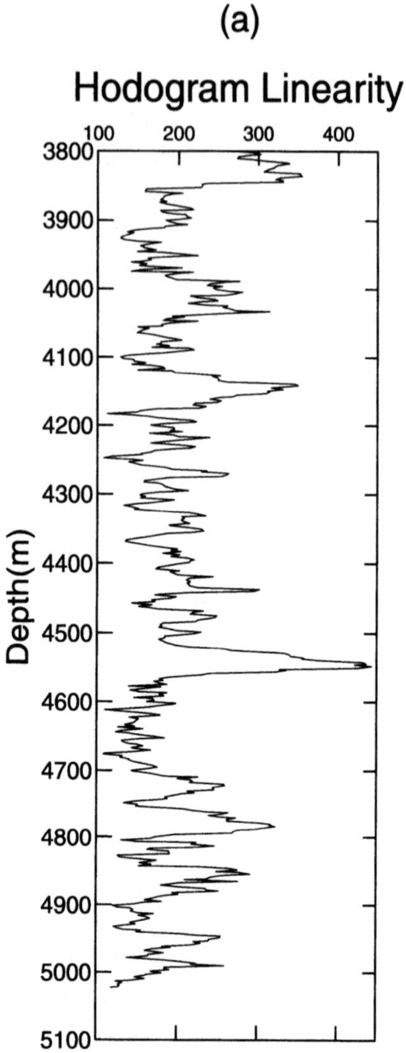

Figure 7
Comparison between the AE reflection method and drilling cuttings analysis in the deep well GPK2. (a): Estimation by the AE reflection method along the new well trajectory of GPK2 drilled in early 1999, (b): Synthetic log of GPK2 between 3880 and 5090 m depth according to analysis of the cuttings (GENTER *et al.*, 1999).

hodogram linearity in the middle- to high-frequency range. This kind of variation probably is related to the character of the structure since the nature of the reflection is a function of wavelength.

Some differences among the four reflectors were evaluated by a precise examination of borehole data sets. The fracture zones at about 3100 m, 3200 m, and 3520 m depth are known to have major flow, although major flow has not been found at 3400 m (EVANS, 2000). This difference may be shown by the frequency

(b)

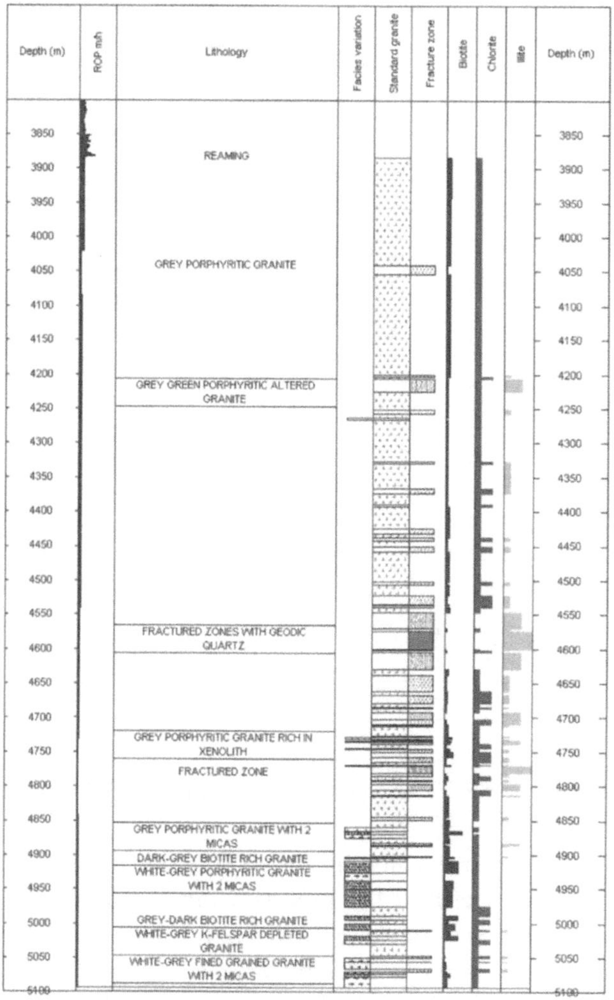

Figure 7b

dependence of hodogram linearity. The reflector with a higher frequency component may be a less permeable structure.

In the context of the Soultz HDR project, we assume that permeability is increased by the shearing and jacking of fractures. Thus, if the surface of a fracture is rough, the permeability of the fracture may be higher. Another possible source of high permeability is a multifracture zone. Furthermore, the extent of the fracture also affects permeability. These possible origins of permeability are also important factors affecting wave reflection.

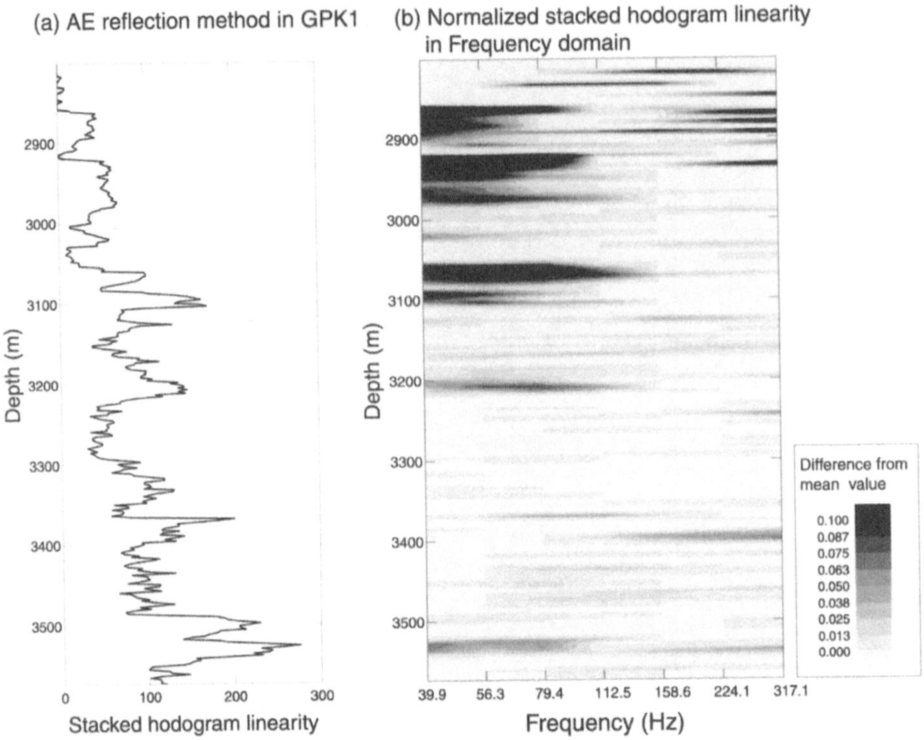

Figure 8
Estimate by the AE reflection method and frequency dependence of reflectors in GPK1.

Considering such permeability in the HDR system, the higher frequency component is greatly affected and is seen as a random nonlinear hodogram at a highly permeable reflector. The lower frequency component, conversely, can be reflected if the reflector is of considerable enough extent. Frequency dependence mainly reflects the shape of the surface and extent of the fracture, which relates to permeability in the context of the Soultz HDR site because the frequency dependence here does not relate to wave energy but to the shape of a three-dimensional hodogram with normalized wave energy.

However, the mechanism by which frequency affects the shape of the hodogram is not yet clear. Possible explanations for the frequency dependence include the physical properties of the material within the fracture and differences in the elastic properties of materials across the fracture. Further study is necessary to conclude the mechanism of frequency dependence of the shape of the three-dimensional hodogram.

Most of the reflectors in the deep part of GPK2 displayed middle to high frequency dependence (Fig. 9), except for those at 4100 m and 4500 m depth. This pattern implies that most of the fracture zones are less permeable than those in

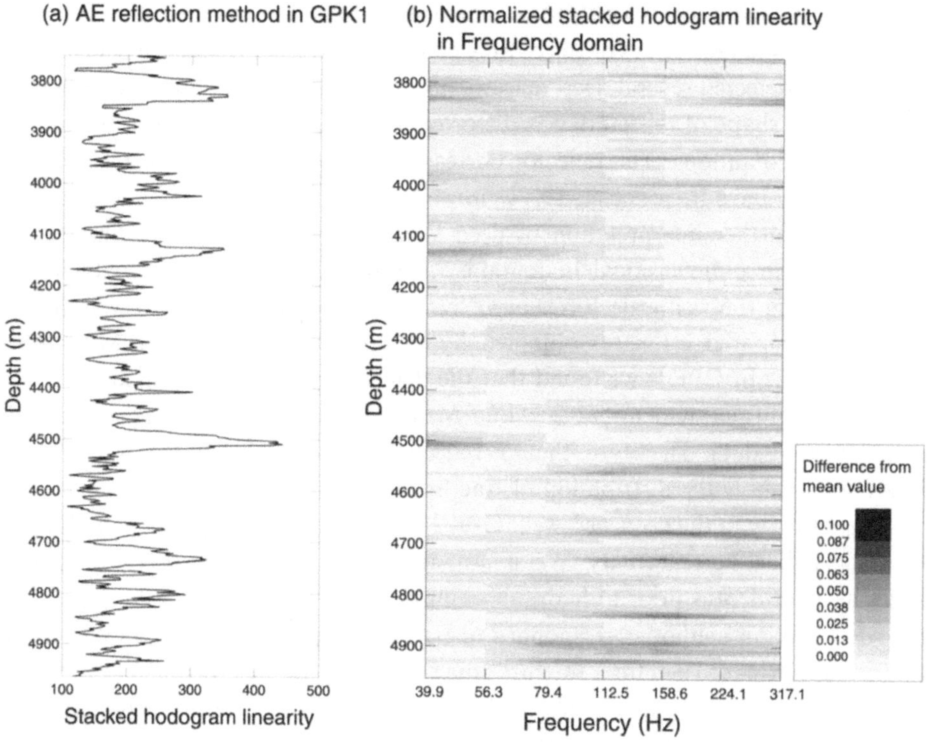

Figure 9
Estimate by the AE reflection method and frequency dependence of reflectors in the deepened part of GPK2.

GPK1 because the estimates were calculated under prestimulation conditions. The reflectors showing lower frequency dependence may be zones with relatively high permeability under predeveloped conditions. These may be promising zones for the next deep development project. Moreover, after the next large-scale fracturing, some reflectors with high frequency dependence may be expected to change to lower frequency dependence, if some natural fracture zone becomes activated as a dominant flow path.

Thus, this technique for analyzing the frequency dependence of reflectors is useful for characterizing deep subsurface structure in geothermal fields.

7. *Conclusion*

We estimated the deep subsurface structure at the Soultz HDR site by the AE reflection method in the time-frequency domain where natural or induced seismicity was used as a wave source. We compared data obtained using the AE reflection method in two boreholes with other independent borehole observations. We also examined the characteristics of the reflectors by an analysis of frequency dependence.

First we detected several deep reflectors on the general three-dimensional image, although the shape of reflectors was distorted due to artifacts related to the limited observation conditions. The reflector corresponding to the bottom of the artificial reservoir may have been detected as well.

For the further interpretation of deep subsurface structures, we estimated the linearity of three-dimensional hodograms, assumed to indicate reflectivity in our analysis, along the well trajectories of GPK1 and the deepened part of well GPK2. Then we compared estimates obtained by the AE reflection method with other borehole data. In GPK1, we found that the detected reflectors corresponded to the distribution of fracture density based on FMI log data as well as to peaks in acoustic impedance log data. Therefore, the AE reflection method can be used at the Soultz HDR site to show the distribution of fracture zones.

The estimate in the deepened well, GPK2, also showed many peaks in hodogram linearity, indicating reflectivity. We assume that these peaks correspond to fracture zones at great depths prior to hydraulic stimulation. The structures estimated using the AE reflection method corresponded to fracture zones reported from the analysis of drilling cuttings. Therefore, fracture zones may exist at great depths at the Soultz HDR site as well as at shallow depths. Consequently, there seems to be potential for successful deep development at the site.

We also investigated the frequency dependence of reflectors detected by hodogram linearity in GPK1 and GPK2. There is clear variation in the frequency component at each reflector position. Although the mechanism of frequency dependence is not yet clear, this study suggests the possible utility of the method to characterize deep subsurface structures. Lower frequency dependence appears to be associated with reflectors at depths of known major flow zones, and not to be associated with reflectors where low permeability has been found.

The AE reflection method can potentially provide us substantial information regarding subsurface structure in areas adjacent to or below the reservoir of a geothermal field, where it is difficult to employ conventional survey methods. Since the analyses are done using *S*-wave polarization, the method is more sensitive to the detection of thin structures such as fracture zones than normal *P*-wave amplitude analysis. With a better observation network, the AE reflection method is likely to provide a more reliable and precise three-dimensional image, which will be very useful for practical subsurface development.

Acknowledgments

This work was carried out as a part of the MTC Project and MURPHY Project, both of which are supported by the New Energy and Industrial Technology

Development Organization (NEDO), Japan (International Joint Research Grant). This work was also supported in part by the Japan Society for the Promotion of Science (Grant JSPS-RFTF 97P00901). We would also like to thank SOCOMINE for providing the data from the European HDR site at Soultz-sous-Forêts, which is supported mainly by the European Commission, BMBF (Germany), and ADEME (France).

References

AKI, K. and CHOUET, B. (1975), *Origin of Coda Waves: Source, Attenuation and Scattering*, J. Geophys. Res. *80*, 3322–3342.

BARIA, R., BAUMGÄRTNER, J., GÉRARD, A., and JUNG, R. (1994), *Scientific Results Associated with Large-scale Hydraulic Injection Tests (A Summary of the Scientific Programme Carried out in 1993; 11/1/94). The European Geothermal Energy Project at Soultz-sous-Forêts*, Socomine Internal report.

BARIA, R., GARNISH, J., BAUMGÄRTNER, J., GÉRARD, A., and JUNG, R. (1995), *Recent Developments in the European HDR Research Programme at Soultz-sous-Forêts (France)*, Proc. World Geothermal Congress, 1995, Florence, Italy, International Geothermal Association, vol. 4, pp. 2631–2637, ISBN 0-473-03123-X.

BAUMGÄRTNER, J., JUNG, R. GÉRARD, A., BARIA, R. and GARNISH, J. (1996), *The European HDR Project at Soultz-sous-Forêts: Stimulation of the Second Deep Well and First Circulation Experiments*, Proc. 21st Workshop Geothermal Reservoir Engineering, Stanford University, Stanford, California, SGP-TR-151, pp. 267–274.

BAUMGÄRTNER, J., GÉRARD, A., BARIA, R., JUNG, R., TRAN-VIET, T., GANDY, T., AQUILINA, L., and GARNISH, J. (1998), *Circulating the HDR Reservoir at Soultz: Maintaining Production and Injection Flow in Complete Balance – Initial Results of the 1997 Circulation Experiment*, Proc. 23rd Workshop Geothermal Reservoir Engineering, Stanford University, Stanford, California, SGP-TR-158.

DYER, B., JUPE, A., JONES, R. H., THOMAS, T., WILLIS-RICHARDS, J., and JAQUES, P. (1994), *Microseismic Results from the European HDR Geothermal Project, Soultz-sous-Forêts, Alsace, France – 1993*, CSM Associates Limited, Internal report.

EVANS, K. (2000), *The Effect of the 1993 Stimulations of well GPK1 at Soultz on the Surrounding Rock Mass: Evidence for the existence of a Connected Network of Permeable Fractures*, Proc. World Geothermal Congress, 2000, Japan (in press).

GENTER, A., HOMEIER, G., CHEVREMONT, Ph., TENZER, H. (1999), *Deepening of GPK-2 HDR Borehole, 3880–5090 m, (Soultz-sous-Forêts, France). Geological Monitoring*, BRGM Open File Report, R 40685, 40 pp.

HLAWATSCH, F. and BOUDREAUX-BARTELS, G. F., (1992), *Linear and Quadratic Time-frequency Signal Representations*, IEEE Signal Processing, April, 21–67.

JONES, R. and ASANUMA, H. (1997), *Analysis of the Four-component Sensor Configuration*, MTC/NEDO Internal Report, vol. 2.

MORIYA, H. and NIITSUMA, H. (1996), *Precise Detection of a P-wave in Low S/N Signal by Using Time-frequency Representations of a Triaxial Hodogram*, Geophysics *61* (5), 1453–1466.

NAGANO, K., NIITSUMA, H., and CHUBACHI, N., *A new automatic AE source location algorithm for downhole triaxial AE measurement*. In *Progress in Acoustic Emission III* (eds. Yamaguchi, K. *et al.*) (The Japanese Soc. for NDI 1986) pp. 396–406.

RIOUL, O. and VETTERLI, M. (1991), *Wavelets and Signal Processing, IEEE Signal Processing Magazine 8*, 14–38.

SAMSON, J. C. (1977), *Matrix and Stokes Velocity Representations of Detectors for Polarized Waveforms: Theory, with Some Applications to Teleseismic Waves*, Geophys. J. Roy. Astr. Soc. *51*, 583–603.

SOMA, N. (1998), *Evaluation of Deep Geothermal Reservoir Structure by Means of a Reflection Method Using Acoustic Emission*, Doctoral Thesis, Graduate School of Tohoku University (in Japanese).

SOMA, N. and NIITSUMA, H. (1997), *Identification of Structures within the Deep Geothermal Reservoir of the Kakkonda Field, Japan, by a Reflection Method Using Acoustic Emission as a Wave Source*, Geothermics *26*, 43–64.

SOMA, N., NIITSUMA, H., and BARIA, R. (1997), *Estimation of Deeper Structure at the Soultz Hot Dry Rock Field by Means of Reflection Method Using 3C AE as Wave Source*, Pure appl. geophys. *150*, 661–676.

(Received May 5, 2000, revised/accepted November 3, 2000)

Pure appl. geophys. 159 (2002) 563–596
0033–4553/02/030563–34 $ 1.50 + 0.20/0

Pure and Applied Geophysics

Using Automated, High-precision Repicking to Improve Delineation of Microseismic Structures at the Soultz Geothermal Reservoir

C. A. ROWE,*[1] R. C. ASTER,[1] W. S. PHILLIPS,[2] R. H. JONES,[3] B. BORCHERS[4] and M. C. FEHLER[2]

Abstract—An automatic, adaptive, correlation-based algorithm for adjusting phase picks in large digital seismic data sets provides significant improvement in resolution of microseismic structures using only a small fraction of the time and manpower which would be required to re-analyze waveforms manually or semi-automatically. We apply this technique to induced seismicity at the Soultz-sous-Forêts geothermal site, France. The method is first applied to a small, previously manually repicked subset of the catalogue so that we may compare our results to those obtained from painstaking, visual, cross-correlation-based techniques. Relative centroid-adjusted hypocenters show a decrease in median mislocation from 31 to 7 m for preliminary and automatically adjusted picks, respectively, compared to the manual results. Narrow, intersecting joint features not observed in the preliminary hypocenter cloud, but revealed through manual repicking, are also recovered using the automatic method. We then address a larger catalogue of ~7000 microearthquakes. After relocating the events using automatic repicks, the percentage of events clustering within 5 m of their nearest neighbor increases form 5 to 26% of the catalogue. Hypocenter relocations delineate narrow, linear features previously obscured within the seismic cloud, interpreted as faults or fractures which may correspond to fluid propagation paths, or to changes in stress as a result of elevated pore pressures. RMS travel-time residuals for the larger data set are reduced by only 0.2%; however, phase-pick biases in the preliminary catalogue have influenced both the velocity model and station correction calculations, which will affect location residuals. These pick biases are apparent on the adjusted, stacked waveforms and correcting them will be important prior to future velocity model refinements.

Key words: Cross correlation, phase repicking, automatic algorithm, precise relocation, microearthquakes.

Introduction

The Soultz-sous-Forêts geothermal field is located in the Alsace region of eastern France, on the western edge of the Rhine Graben (Fig. 1), about 50 km

---

[1] Department of Earth and Environmental Science, New Mexico Tech, Socorro, NM 87801, USA.
[2] EES-4, MS D443, Los Alamos National Laboratory, Los Alamos, NM 87545, USA.
[3] ABB Offshore Systems Ltd., Rosemanowes, Penryn, Cornwall T10 9DU, United Kingdom.
[4] Department of Mathematics, New Mexico Tech, Socorro, NM 87801, U.S.A.
* Corresponding Author: C. A. Rowe, Now at Department of Geology and Geophysics, University of Wisconsin,1215 W. Dayton St. Madison, WI 53706, USA. E-mail: char@geology.wisc.edu

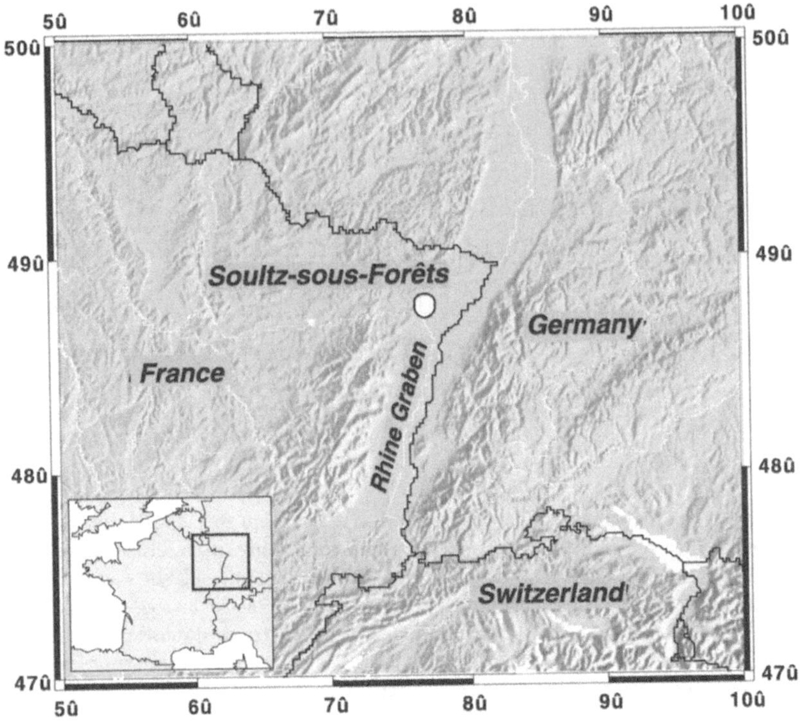

Figure 1
Location map for the Soultz-sous-Forêts reservoir (from PHILLIPS, 2000), situated on the western edge of the Rhine Graben in eastern France. This valley is an extensional rift exhibiting high geothermal heat flow.

north of Strasbourg (BARIA *et al.*, 1999). It is situated within the former Pechelbron oil field, in an area of high subsurface temperature gradients (BARIA *et al.*, 1999). The target geothermal reservoir resides within granitic rock between 2 to 4 km depth (BARIA *et al.*, 1999). This granite has been studied in detail using high-resolution borehole imaging and has been found to host two distinct populations of hydrothermally sealed fractures, striking N20°E and N30°W, respectively (GENTER and TRAINEAU, 1996). Stress estimates at Soultz indicate that the maximum horizontal stress within the reservoir is oriented roughly N10°W (JUNG, 1991; CORNET and JONES, 1994); hence many observed fractures strike subparallel to the maximum horizontal stress and are favorably oriented for hydraulic stimulation of the reservoir (BARIA *et al.*, 1999). Hydrofracturing at the reservoir in September and October, 1993 resulted in over 16,000 microearthquakes, recorded at three deep, down-hole multi-component seismometers and one down-hole hydrophone (Fig. 2). Numerous studies have been undertaken using Soultz microearthquakes to estimate growth of the newly fractured portion of the rockmass (JONES *et al.*, 1995), fracture orientations (e.g., GAUCHER *et al.*, 1998) or flowpaths (e.g., PHILLIPS, 2000; STARZEC *et al.*, 2000).

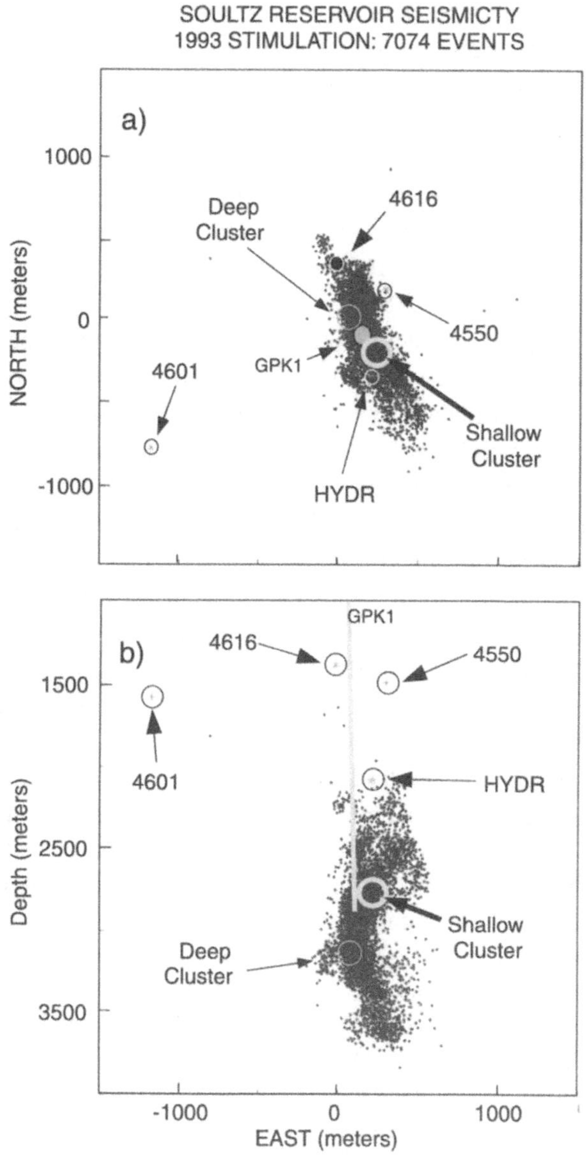

Figure 2

Soultz-sous-Forêts geothermal reservoir (after ASTER and ROWE, 2000). Shown are the four borehole sensors (multicomponent seismometers 4550, 4616 and 4601, and hydrophone HYDR) in both map view (a) and east–west cross section (b). The cloud of preliminary hypocenter locations is shown, along with the position of injection well GPK1. Gray circles indicate approximate locations for manually repicked clusters from PHILLIPS (2000).

Interpretation of seismogenic features depends critically upon the accuracy and interpretability of the source regions illuminated by the estimated hypocenters. In typical microearthquake studies at a local or regional scale, hypocenter mislocations of up to several kilometers may be tolerated, when only the identification of general seismic source regions and trends (and associated possible macroseismic effects) are the primary concern. Significant improvement in event locations for local and regional networks has been demonstrated through improved velocity modeling via joint hypocenter determination (JHD) (e.g., PUJOL, 1992) or local three-dimensional tomography (e.g., KISSLING, 1988; BLOCK, 1991; BLOCK *et al.*, 1994).

Earthquake location accuracy, however, depends not only upon the characterization of the seismic velocity structure through which seismic body waves have travelled, but also upon the ability to estimate accurately the arrival times of those waves. The most common human or computer picking approach is done one event at a time, for many recording stations. Slight to moderate picking inconsistencies are manifested ubiquitously at any particular station among even very similar events, resulting from variations in signal-to-noise ratio and other effects, as well as human subjectivity. Although one source of systematic mislocation can be successfully addressed through improved velocity modeling, fine details of the seismogenic structures are still commonly blurred by the intrinsic "fuzziness" of hypocenters arising from the remaining random pick errors and associated travel-time residuals (JONES and STEWART, 1997). The picking inconsistencies and resulting hypocentral scatter likewise affect other applications which rely upon the parametric data and associated error estimates. Such applications as reciprocal array processing (SPUDICH and BOSTWICK, 1987), spatial mapping of frequency-magnitude distribution (e.g., WIEMER and WYSS, 1997), and evolution of fractal dimension (e.g., NANJO *et al.*, 1998), as well as general understanding of spatio-temporal seismicity trends in many contexts, will benefit from correction of large numbers of picks. Certainly the reduction of pick scatter is crucial to mapping of very small-scale structures such as the fracture networks within a geothermal field (e.g., ROFF *et al.*, 1996; PHILLIPS *et al.*, 1997), repeating earthquakes (e.g., NADEAU *et al.*, 1995) or deep fault drilling (e.g., THURBER *et al.*, 2000), where precision to a scale of a few meters or less is desired.

A statistical approach to addressing hypocenter fuzziness caused by pick inconsistencies has been presented by JONES and STEWART (1997) and has been incorporated into JHD inversion by FEHLER *et al.* (2000). This technique, referred to as "collapsing," is based on assumptions of Gaussian pick errors and the observation that randomly perturbed points spread out in space. Hypocenters are permitted to move towards an event-based center of gravity, constrained by the dimensions of their error ellipsoids. Although collapsing can be prone to artifacts, synthetic tests as well as comparison to results from exhaustive, manual repicking of real data sets have demonstrated considerable success (e.g., JONES and STEWART, 1997; FEHLER *et al.*, 2000). This suggests that, in the absence of waveform data which might permit correction of the picks, collapsing, in combination with techniques for adjusting the velocity model and/or near-surface static corrections, may be a good approach to improve delineation of seismogenic features.

It is possible, however, to address picking inconsistencies directly when digital waveform data are available, using signal processing methods to adjust the picks for consistency among events exhibiting waveform similarity (e.g., DODGE *et al.*, 1995; SHEARER, 1997, 1998; RUBIN *et al.*, 1998, 1999; ASTER and ROWE, 2000; ROWE, 2000; ROWE *et al.*, 2001). Subsequent hypocenter relocation may then proceed using either single-event location, JHD, or simultaneous velocity and hypocenter inversion. Sharpening of the seismogenic source volume image has been demonstrated through relative relocation using waveform cross correlation techniques (e.g., FREMONT and MALONE, 1987; DEICHMANN and GARCIA-FERNANDEZ, 1992; GOT *et al.*, 1994; NADEAU *et al.*, 1995; GILLARD *et al.*, 1996; SHEARER, 1997, 1998; LEES, 1998; RUBIN *et al.*, 1998, 1999; WALDHAUSER *et al.*, 1999) or combinations of differential cross correlation and catalogue travel times (WALDHAUSER *et al.*, 1999). In these studies, typical errors of 1 km or more have been reduced to relative location errors of a few tens of meters or less, allowing previously undetected structures within the seismic source zone to be observed.

We have developed a new, automatic phase-repicking algorithm (ASTER and ROWE, 2000; ROWE, 2000; ROWE *et al.*, 2001) that addresses the correction of scatter which arises from inconsistent picking, by using estimated relative pick lags to calculate corrected picks for consistency among similar waveforms (e.g., DODGE *et al.*, 1995) without imposing constraints from preliminary catalogue parameters on our corrected picks. This method does not address mislocations which may arise from unmodeled velocity variations; however, coupling the improved picks with better velocity models and more sophisticated location methods should further improve the accuracy of reservoir imaging in future work. This study presents the results of applying our repicking algorithm to seismicity recorded at the Soultz reservoir during September and October, 1993.

Data

The stimulated volume of the Soultz reservoir occupies roughly 0.48 km^3 in a depth range of about 2100 to 3800 m. Median distances from microearthquakes to the nearest and most distant receiver were approximately 800 m and 2100 m, respectively. Data were digitized at a sampling rate of 5 KHz. The borehole network consisted of three, 4-component (redundant) sensors and one hydrophone. Four-component data were projected into three orthogonal components prior to our analysis.

Location of the Soultz microearthquakes from preliminary analyst picking resulted in the seismicity cloud shown in Figure 2, consisting of hypocenters of $\sim$7000 events which will be used in our analysis. This subset of microearthquakes was chosen from the larger catalogue based on the availability of seven phase picks from preliminary analysis. Unlike relocation efforts in large data sets recorded by large regional seismic networks (e.g., SHEARER, 1997, 1998; WALDHAUSER *et al.*, 1999), the Soultz data present significant additional location difficulties resulting from the sparse, four-station network. In the case of Soultz, all available phases are needed to adequately constrain hypocenters, and it was therefore not advisable to neglect arrivals which were noisy or nodal (PHILLIPS, 2000). Further, the sparse network geometry (Fig. 2) can cause asymmetric errors due to the location of the most distant station (4601), which provides much of the east-west location control.

Preliminary hypocenters determined using single event location (SEL) methods with initial arrival time picks and velocities determined by JUPE *et al.* (1994) are shown in Figure 2. These events were located with four *P*-wave and three *S*-wave arrivals (no *S* was available from the hydrophone channel, as *S* is generally unreadable there). A uniform *P*-wave velocity of 5.85 km/s and *S*-wave velocity of 3.34 km/s were used, consistent with data from a shot fired in well GPK1 at a depth of 3360 m. Individual station corrections determined from the calibration shot and applied for SEL hypocenters (JUPE *et al.*, 1994), and station corrections determined by a recent JHD inversion, are shown in Table 1.

Because detailed velocity investigation is beyond the scope of this study, we adopt the velocities and JHD station corrections determined by JUPE *et al.* (1994) and also used by PHILLIPS (2000) for the purpose of our comparative location analysis.

Previous Re-location Work at Soultz

GAUCHER *et al.* (1998) analyzed a small number of multiplets within the 1993 injection seismicity cloud. They obtained relative relocations defining structures near the borehole, which were correlated with fractures observed using acoustic imaging and well logs.

PHILLIPS (2000) relocated microearthquakes in two densely populated volumes within the seismic cloud by manually repicking relative arrival times of phases. The "shallow" and "deep" clusters, contained roughly 200 and 350 events, respectively. Approximate locations of these clusters are indicated by gray circles on Figure 2. Relocations from this manual repicking collapsed the diffuse clouds of seismicity into narrow, intersecting features which exhibited temporal seismicity patterns showing

Table 1

Station corrections (ms) for SEL and JHD Soultz microearthquake locations

Station	SEL *P*	SEL *S*	JHD *P*	JHD *S*
4550	0.0	0.0	−5.49	−8.25
4601	−17.0	−19.0	−16.4	−18.9
4616	−5.5	−7.0	0.5	1.32
HYDR	3.0	N/A	1.73	N/A

initial tight clusters or alignments of events which subsequently spread into adjacent areas. These early alignments in the sequence have been interpreted as joints or joint/fracture intersections that may represent fluid flow paths (PHILLIPS, 2000).

MORIYA *et al.* (2001) have addressed precise relative relocation of Soultz data through identification of highly similar earthquake doublets and multiplets. They have estimated relative hypocenter relocations using the cross-spectral method of POUPINET *et al.* (1984), and have observed numerous well-defined, narrow seismic lineations. MORIYA *et al.* (2001) further note that the high proportion of doublets and multiplets seen in the early period of the stimulation declines as time progresses. These observations are in good agreement with those of Phillips (2000) and Phillips *et al.* (2001), and we discuss them further, below.

Automatic Repicking Technique

We have developed a method which combines the most advantageous features of existing waveform cross-correlation-based phase adjustment methods (e.g., DODGE *et al.*, 1995; SHEARER, 1998; RUBIN *et al.*, 1998, 1999) in an automatic package which can be easily implemented even for very large (many thousands of events) data sets with a manageable amount of analyst interaction. Details of the signal processing, clustering and conjugate gradient techniques and our innovations are reported elsewhere (e.g., ASTER and ROWE, 2000; ROWE, 2000; ROWE *et al.*, 2001); however, we summarize our method below.

Signal Processing Tools

Pick adjustments are estimated by comparing station-common waveforms having preliminary picks, and estimating relative lags between these picks. Lag estimation proceeds in two steps: A coarse integer-sample correlation, and a subsequent fine correlation step which provides a subsample relative lag refinement.

Coarse (integer-sample) cross-correlation provides initial lag adjustments for each waveform pair. For multicomponent data, we apply polarization filtering, derived from eigenvalue decomposition of a joint signal covariance matrix, to improve P and S signal-to-noise levels prior to waveform comparison. Prior to correlation, we calculate the cross-spectrum and cross-coherency for each pair of signals. The cross-coherency is used as a filtering function to adaptively downweight incoherent frequency bands prior to cross-correlation. We then calculate a suite of cross-correlation functions from overlapping, narrow-band-filtered representations of the (polarization- and coherency-filtered) waveform pair. The integer-rounded, energy-weighted mean correlation maximum and its associated standard deviation are calculated from these functions (ASTER and ROWE, 2000; ROWE, 2000; ROWE *et al.*, 2001). Waveform pairs exhibiting sufficiently high integer correlation values (>0.8 for the Soultz catalogue) with low associated lag standard deviations (<2

samples) are aligned to the nearest sample, rewindowed, and passed to the subsample lag estimation algorithm.

A cross-spectral phase slope method (e.g., POUPINET *et al.*, 1984; ITO, 1995) is used to obtain subsample lag estimates following integer lag estimation. Because unbiased estimation of the cross-spectral phase slope is essential to obtaining accurate estimates of subsample lags and corresponding errors, we apply multitaper spectral estimation to this problem. A suite of eigenspectral measurements may be combined in a simple or weighted average to provide low spectral leakage, low variance spectral estimate and associated standard errors (e.g., THOMSON, 1982; PARK *et al.*, 1987; ASTER and ROWE, 2000; ROWE, 2000; ROWE *et al.*, 2001). Conjugate multiplication of corresponding eigenspectra for the waveform pair provides multiple linearly independent estimates of the cross-spectral phase, yielding a mean cross-spectral phase slope estimate with associated dimensionally meaningful standard deviation.

After integer and subsample lags and their respective standard deviation estimates are obtained for each waveform pair, we sum the lags to provide a total lag estimate; the standard deviation of the pick lag is the quadrature sum of the coarse and fine lag standard deviations (ASTER and ROWE, 2000; ROWE, 2000; ROWE *et al.*, 2001).

Solving for Consistent Pick Adjustments

After all possible waveform comparisons have been made, the resulting, overdetermined system of individual interevent lags may be solved to obtain consistent pick adjustments and associated uncertainties. We do this using an iterative conjugate gradient approach formulated to operate efficiently with a sparse matrix storage scheme (ASTER and ROWE, 2000; Rowe, 2000; Rowe *et al.*, 2001). The minimum L_1-norm residual solution is used, with a numerical modification for near-zero residuals to avoid gradient discontinuities, (e.g., ROWE, 2000; ROWE *et al.*, 2001). This provides improved robustness to outliers over L_2 methods (e.g., PARKER and MCNUTT, 1980; SHEARER, 1997). We further improve the solution by conservatively rejecting outliers in this massively overdetermined problem and successively resolving the system (ROWE and ASTER, 1999; ASTER and ROWE, 2000; ROWE, 2000; ROWE *et al.*, 2001). We calculate 1-sigma error bars via Monte Carlo perturbation of the data vector.

Event Association

We have found that solving the entire catalogue in one step does not provide optimal pick corrections because inconsistent waveform types are compared (ROWE, 2000; ROWE *et al.*, 2001). The choice of an appropriate event association method has proven to be crucial to improving the hypocenter locations. Because of the high degree of waveform variability within the small seismogenic volume, the Soultz catalogue has some corresponding difficulties which may not be encountered in other settings. Our event association and pick adjustment strategy is outlined below.

We have adopted an event clustering strategy based upon waveform similarity (e.g., ASTER and SCOTT, 1993; REVIERE-BARBIER and GRANT, 1993). We treat the maximum cross-correlation value for each event pair as a measure of distance in similarity space, so that the correlation matrix may be viewed as a spatial (dis-) similarity matrix.

We obtained good intra-cluster results when we selected one of the most variable phases and used it as our clustering basis. The S-wave arrivals for station 4550 exhibit tremendous variation in waveform, therefore we selected an extended (0.05 s) correlation window for that phase, filtered using the signal cross-coherency function for each waveform pair. Cross-correlation of these waveforms provided a similarity matrix for a highly variable parameter in the seismogram database. We then applied a dendrogram-based hierarchical pair-group classification clustering scheme (e.g., LUDWIG and REYNOLDS, 1988; CARR *et al.*, 1999; ROWE, 2000; ROWE *et al.*, 2001) to build similarity groups. We have adopted a cross-correlation peak value of 0.75 as a minimum similarity criterion for any cluster.

Intra-cluster Adjustments

Within each cluster we then cross-correlate all phases and solve the respective first-difference systems to determine optimal pick adjustments. Because of significant variation in initial pick error from event to event (adjustments range from a fraction of a sample to 50 samples or more), as well as variations in waveform signal-to-noise ratio, we find that correlation window length can strongly affect waveform alignment even within highly similar clusters. The best choice can vary from cluster to cluster, as well as from phase to phase within a cluster. We therefore calculate cross-correlations for a suite of window lengths for each phase, and choose the parameters which produce the best mean cross-correlation and lag standard deviation values for the phase being addressed (ROWE, 2000; ROWE *et al.*, 2001).

Once lags are estimated for all phases within a cluster, we solve for pick adjustments using our L_1-norm-minimizing conjugate gradient solver, imposing an *a postiori* zero mean constraint so that the mean of the adjusted picks corresponds to the preliminary pick mean. The corrected picks can provide very precise relative arrival times within a similarity cluster; however, inter-cluster relationships must still be addressed to achieve consistent inter-cluster picks, which may correspondingly correct the inter-cluster spatial relationships. Some means of correcting this bias without relying upon preliminary locations or traveltimes is desirable.

Inter-cluster Adjustments

We have found that a hierarchical correlation approach using stacked seismograms for each cluster provides good adjustments for inter-cluster relationships. For each phase, we calculate vertical waveform stacks within each of the clusters, where stacks are aligned on the adjusted picks (see Fig. 3). In this way, each cluster may be viewed as having a composite catalogue location represented by the hypocenter calculated from its waveform stacks. To adjust the stacked waveforms for each phase (hence correcting for inter-cluster pick biases and, ultimately, inter-cluster relative

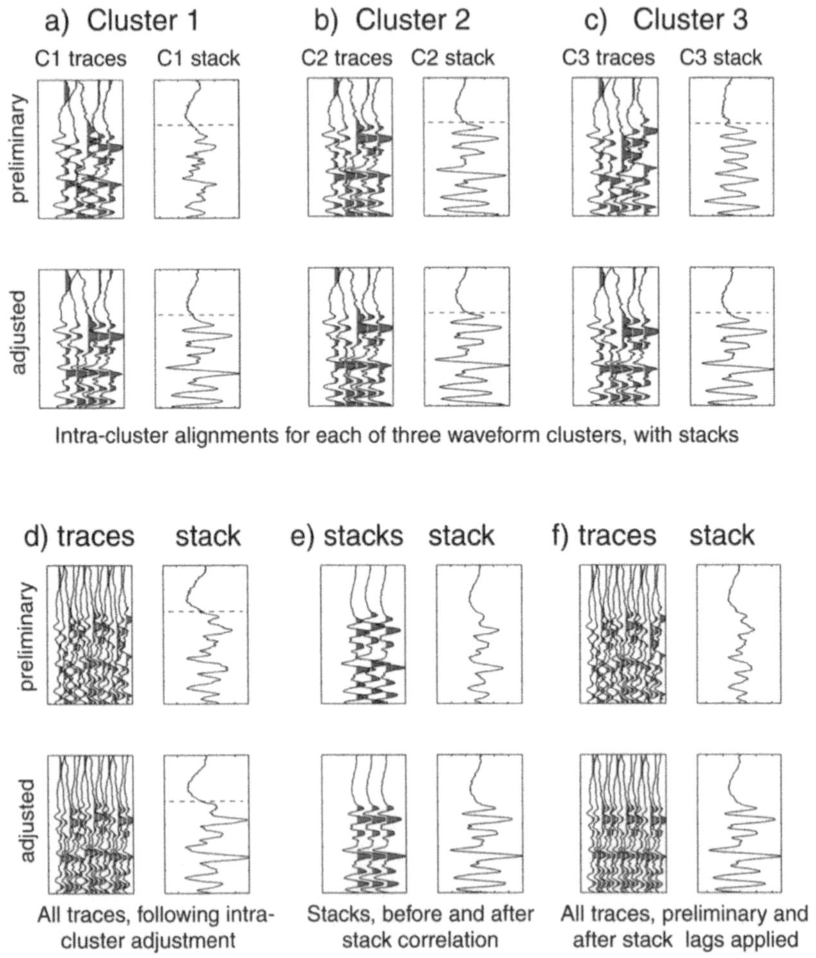

Figure 3
Hierarchical correlation, lag adjustment and stacking method. (a–c) illustrate three synthetic clusters of five events each. Upper panels show traces aligned on preliminary picks and associated waveform stack; lower panels show traces aligned on adjusted picks, with associated stack. Horizontal dashed lines indicate pick times on the stacked trace. (d) Clusters from a, b and c displayed together to show initial pick inconsistency (upper panel) and remaining biases among aligned clusters (lower panel). (e) Stacks from clusters of a, b and c showing stack mean pick misalignment (top) and aligned stacks (bottom). (f) the same three clusters showing initial misalignments (top) and final, corrected alignments (bottom) after both intra-cluster (a–c) and inter-cluster (e) lags have been applied to the individual traces.

centroid location), we apply the cross-correlation and conjugate gradient solution to the stacks. Each stack is assigned a relative pick lag which would serve to align stacks (mean picks) for every correlated cluster. We apply the stack lag to all corresponding phases for member events within each cluster, providing a final adjustment for inter-cluster relationships (Fig. 3). The tightly constrained intra-cluster relative relationships are thus preserved while the addition of the stack lag information adjusts inter-cluster centroids with respect to one another.

Final Pick Adjustments

There may still be a catalogue-wide absolute picking bias for some phases. For example, successively later picks for smaller events with lower signal-to-noise ratios – particularly for more distant stations whose arrivals are more emergent to begin with – may result in a net late pick bias. Addressing any such final absolute pick offset may be appropriately handled through adjusting to a master event, application of an autopicking routine to the second-order stacks, or application of consistent station corrections as derived through JHD or other joint location methods, although this latter approach does not serve to separate velocity delays from pick error delays. We discuss this further, below.

Automatic Repicking Test Case

The Soultz reservoir seismicity presents an excellent test case for development and application of the automated repicking algorithm, although it also presents some significant relocation issues not encountered with many larger seismic networks in tectonically active regions. The primary difficulty lies in the small number of recording stations, as discussed above. Generally, relocation studies using waveforms from large seismic networks (e.g., SHEARER, 1997, 1998; RUBIN, 1998, 1999) are able to neglect poor quality arrivals without compromising overall ability to obtain robust hypocenter locations. At the Soultz reservoir there are only four sites where downhole sensors were operating during the 1993 stimulation; hence, discarding any phase arrivals can destabilize the hypocenter solution process, particularly given the difficult network geometry (Fig. 2).

On the other hand, the sensors are situated in boreholes at depths between 1376 and 2075 m, resulting in very low background noise and good microearthquake signals. Three of the instruments are multicomponent sensors, which allow polarization filtering for improving signal correlation (particularly for *S* waves).

We first tested our method using the deep cluster events of PHILLIPS (2000) as a benchmark against which to compare our results (ROWE *et al.*, 1998, ROWE and ASTER, 1999; ASTER and ROWE, 2000; ROWE, 2000). The painstaking Phillips repicks and relocations were assumed to be the "correct" answer; step-wise comparison with these allowed us to gauge improvements in the development of our technique.

Initial correlation on the *S*-wave picks for station 4550 using a window length of 50 ms (about 5 cycles) provided a similarity matrix for clustering the 311 events within the Phillips deep cluster which were included in our catalogue. This resulted in eighty similarity groups of varying sizes, dominated by doublets (two-event clusters). Figure 4 shows a histogram of similarity cluster membership for this trial data set. Eighteen events were orphaned (unassociated) in the clustering process; the remaining 293 were addressed in the analysis.

Applying the automatic repicking algorithm to *P* and *S* phases for the deep cluster, we demonstrate the resulting waveform alignment in Figure 5. The

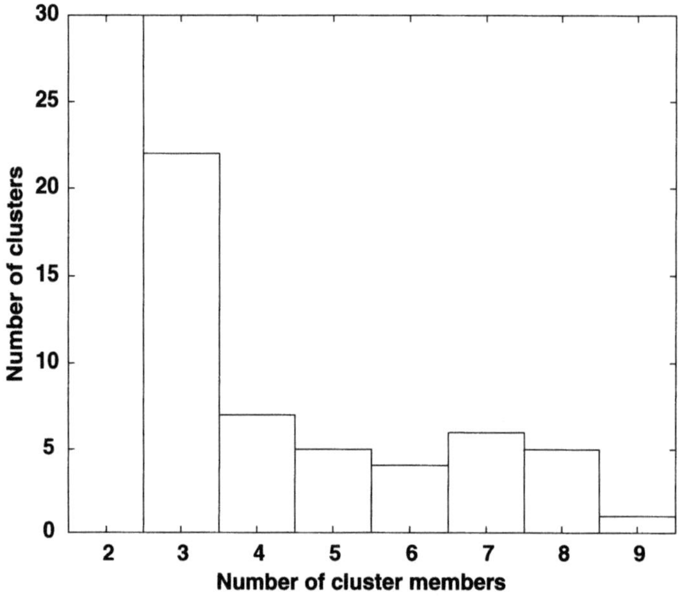

Figure 4
Earthquake cluster sizes for PHILLIPS (2000) deep cluster. Histogram shows the number of families for each category of doublet, triplet, quadruplet, and so forth. A total of eighty clusters were identified.

▶

Figure 5
Waveform alignment plots for events in the deep cluster (PHILLIPS, 2000). Waveforms are plotted as a gray-scale surface with white peaks and gray troughs. Each phase is depicted in six panels. Waveform alignments are on the left, stacks of the waveforms are on the right. Top panels show waveforms aligned on preliminary picks. Center panels show waveforms (and stack) aligned on the repicks arising from intra-cluster relative repicks. Lower panels show alignments and stacks of waveforms after inter-cluster adjustments have been made. (a) *P* waves for station 4550 are shown in the two left-hand columns. *S* waves for station 4550 are shown in the two right-hand columns. (b) *P* waves for station 4601 are shown in the left-hand columns; *S* waves are shown in the two right-hand columns. (c) *P* waves for station 4616 shown on the left-hand side, *S* waves for station 4616 shown on the right. (d) *P* waves for the hydrophone. No preliminary *S* was picked for the hydrophone, as this instrument was underdamped and the waveforms were often clipped.

uppermost panels for each phase illustrate alignment on preliminary analyst picks. Center panels represent alignments after intra-cluster adjustments have been made; the lowermost panels show alignment following correction for inter-cluster consistency. Stacked waveforms for each alignment plot are shown on the right, with the

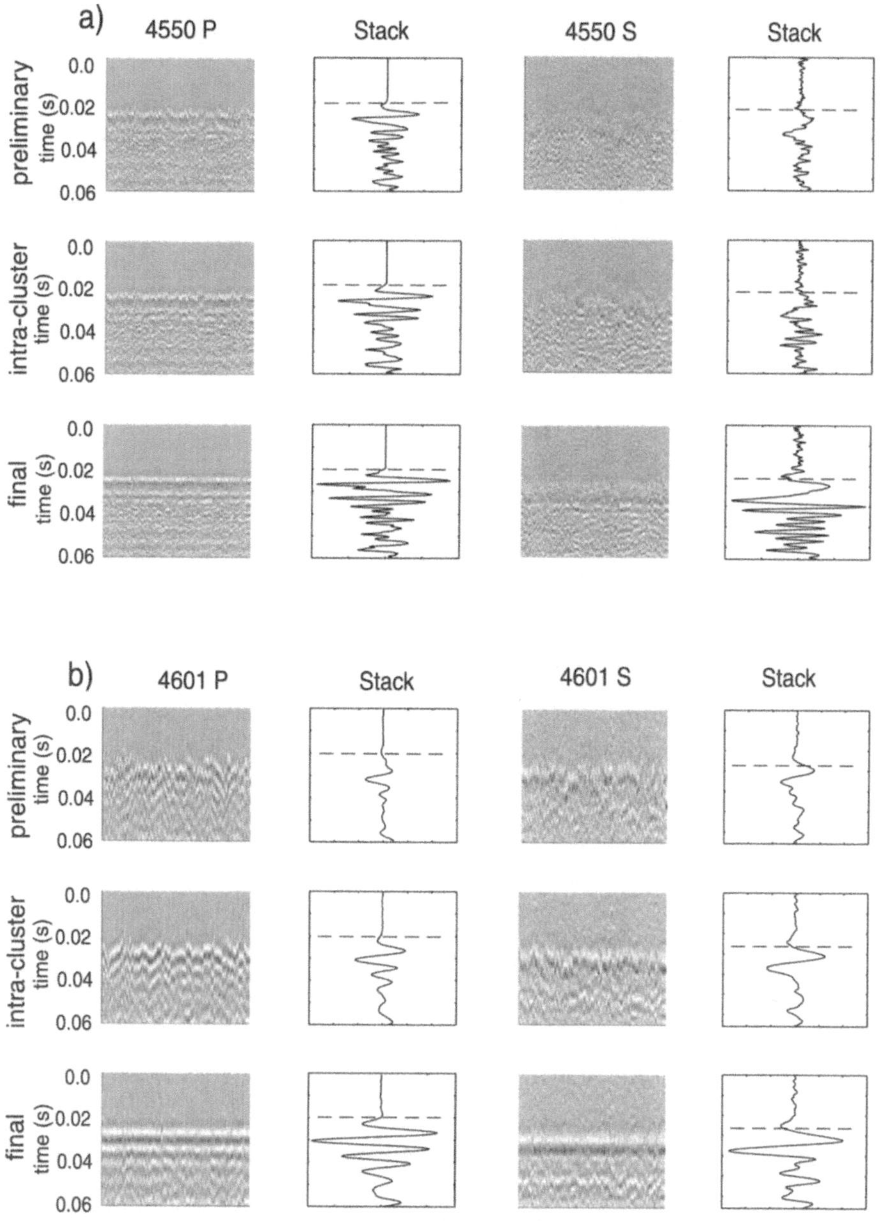

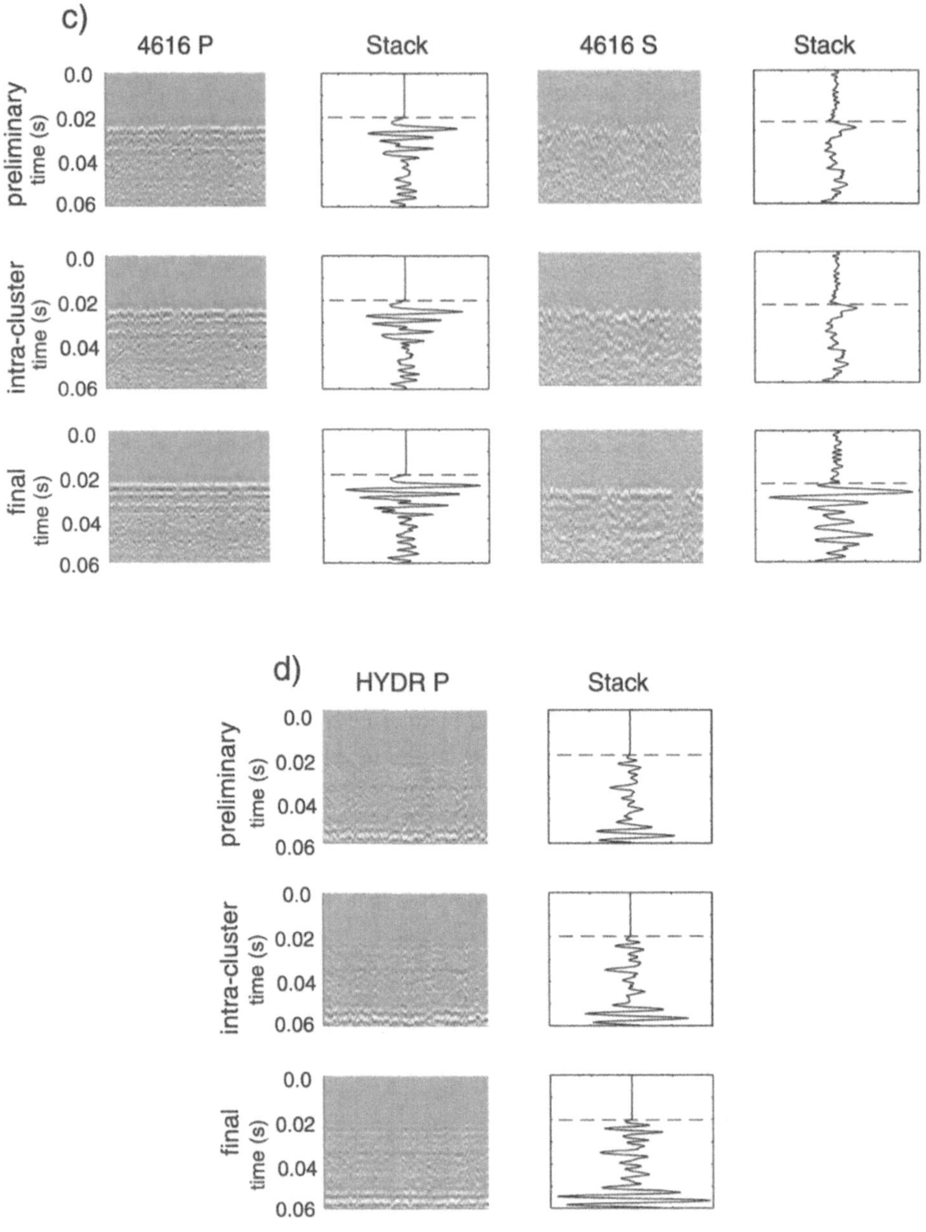

Figure 5c, d

aligned picks indicated by horizontal black bars. The increase in stack amplitude provides a qualitative measure of the repicking success.

Note that Figure 5 does not show the filtered and projected waveforms which were cross-correlated to obtain the pick lags. Because the filtering and optimal signal rotation may be different for every interevent cross-correlation, it is not possible to show these in a single figure. We show instead alignments of unfiltered traces. For *P*-wave alignments, we show the vertical component, and for *S*-wave alignments, we display the east-west (*X*) horizontal component. Improvement in waveform alignment is especially marked for *S* waves. For instance, at station 4616 the summed *S*-wave stack amplitude after repicking is nearly three times that of the preliminary stack.

After cross-correlating and solving for consistent pick adjustments, we relocated the deep cluster earthquakes using the same location program (single event, damped least-squares) and velocity parameters ($V_p = 5.85$ km/s, $V_s = 3.34$ km/s) determined by JUPE *et al.* (1994) and used by PHILLIPS (2000). Figure 6a, b illustrates hypocenter locations for preliminary picks and manual repicks, respectively. In Figure 6c we show hypocenters calculated using automatic intra-cluster repicks, to illustrate the increased degree of clustering obtained by this step. These hypocenters correspond to the waveform alignments in the center panels for each phase shown in Figure 5. Tight intra-cluster alignments result in increased clustering among hypocenters, but the large inter-cluster biases have not yet been addressed. Note that the structures in Figure 6b are still not clear in Figure 6c, although hypocenters exhibit significantly increased clustering over the preliminary locations of Figure 6a. Figure 6d shows the relocations following inter-cluster stack correlation and adjustment (corresponding to picks aligned in the lowermost panel for each phase in Fig. 5). The clearly defined, near-orthogonal intersecting joint features illuminated by the manual repicking in Figure 6b and discussed in detail in PHILLIPS (2000) and PHILLIPS *et al.* (2001) are also clearly distinguishable in Figure 6d. Median hypocenter mislocation, compared to relocations using the Phillips picks, is reduced from 31 to 7 m after we have adjusted centroid median biases.

Because Phillips correlated peaks and troughs with the visual cross-correlation, then applied station corrections to his picks to approximate first-break consistency for locating events, it is difficult to compare directly the relocation quality in terms of hypocenter RMS traveltime errors; however, we iteratively adjusted the JUPE *et al.* (1994) JHD station corrections to minimize mean traveltime residuals for automatic repicks and preliminary picks. The resulting RMS travel-times errors are 0.68 ms, 0.25 ms and 0.14 ms for preliminary, automatic and manual picks, respectively, with associated standard deviations of 0.62, 0.34 and 0.24 ms, respectively.

Application to the Larger Data Set

When we had completed testing of the automatic algorithm using the PHILLIPS (2000) deep cluster, we turned to the task of relocating the ~7,000 events in the Soultz catalogue that had seven preliminary phase picks.

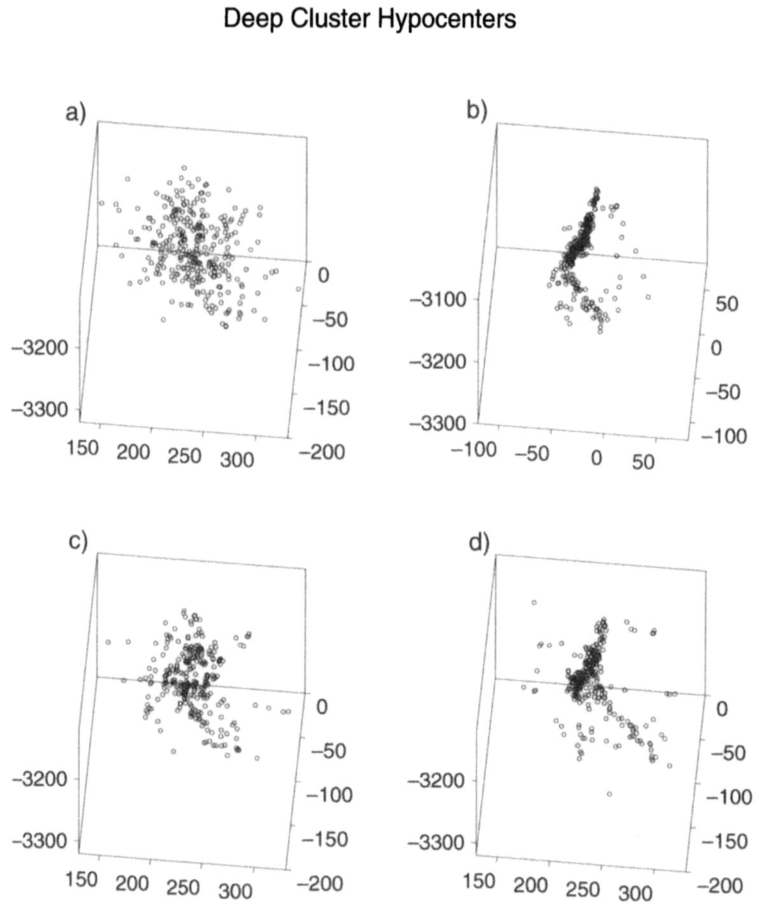

Figure 6
Hypocenter locations for Phillips' deep cluster. (a) Preliminary locations, based on initial, analyst picks. Note the random scatter in the distribution of seismicity. (b) Relocations based on PHILLIPS (2000) manual repicking. The diffuse cloud of hypocenters (a) has converged into two nearly orthogonal, intersecting joint features and a few scattered events. (c) Relocations based on intra-cluster pick adjustments (corresponding to waveform alignments in center panels of Figure 5). (d) Relocations after inter-cluster stack adjustments have been applied to individual traces (corresponds to alignments in lower panels of Figure 5).

Relocations

Processing the ~7000-event catalogue proceeded in the hierarchical correlation/clustering method as discussed for the deep cluster (above). Following cross-correlation on the *S* waves at station 4550, the algorithm segregated the catalogue into 2408 clusters, plus 661 unassociated (orphaned) events. Figure 7 shows the resulting cluster memberships. We show the number of clusters of each size (from 2 to 18 members) in Figure 7a, while plotting the cumulative proportion of the catalogue comprised of the different cluster sizes in Figure 7b. Picking inconsistencies

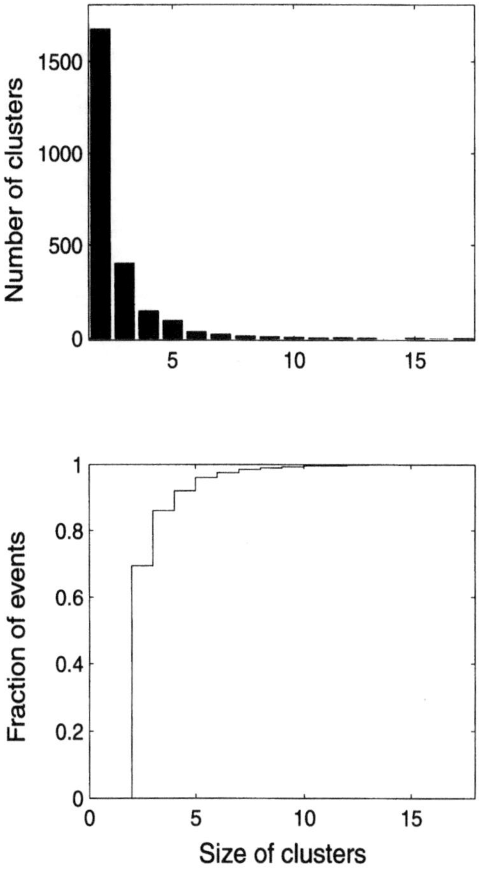

Figure 7
Cluster memberships for the ~7000-event data set. (a) Histogram shows the number of families for each category of doublet, triplet, quadruplet, and so forth. A total of eighty clusters were identified. (b) Cumulative fraction of the catalogue assigned to clusters of two, three, four and so forth. Note that doublets and triplets comprise over 87% of the data, indicating that adjusting relative pick lags among clusters is important to the final resolution of spatial relationships among earthquakes.

were resolved within similarity clusters with median lag adjustments ranging from 0.251 ms for station HYDR *P* picks to 1.6 ms for station 4601 *S* picks (Table 2). After adjusting each of the seven phase picks for consistency within each of the 2408 clusters, the aligned waveforms for each phase were stacked to provide a composite waveform (e.g., Fig. 3). These cluster stacks were then correlated with corresponding stacked waveforms for all other clusters so that inter-cluster picking inconsistencies might be resolved, and the final pick lag adjustments made. Inter-cluster stacked waveform adjustments are shown in Table 3. Maximum inter-cluster lag adjustments were 17.2 and 11.1 ms for *S* waves and *P* waves, respectively; both maxima were observed for station 4601, the most distant station from the seismogenic volume (Fig. 2).

Table 2

Intra-cluster relative pick adjustments and standard deviations

Station	Median *P* lag (ms)	*P* st.dev (ms)	Median *S* lag	*S* st.dev (ms)
4550	0.34	0.03	0.76	0.05
4601	1.1	0.06	1.6	0.06
4616	0.38	0.03	0.8	0.06
HYDR	0.25	0.03	N/A	N/A

Table 3

Inter-cluster stack pick adjustments and standard deviations

Station	Median *P* lag (ms)	*P* st.dev (ms)	Median *S* lag (ms)	*S* st.dev (ms)
4550	0.77	1.2	1.5	2.4
4601	1.8	2.7	3.0	4.7
4616	0.85	1.3	2.0	2.8
HYDR	0.55	0.96	N/A	N/A

Automatic repicking and relocation of our ~7000 event data set yielded significant improvement in illumination of interpretable seismogenic features. Median RMS traveltime error for repicked hypocenters decreased from 3.741 to 3.739 ms compared to original picks, using the station corrections derived from JHD (e.g., JUPE *et al.*, 1994). This moderate change of 0.2% indicates that overall location errors are far more heavily influenced by the oversimplified velocity model and the station correction parameters, which were derived to optimize fitting of the preliminary picks for a larger data set, than by moderate pick adjustments; however, waveform alignments and improved resolution of fracture and joint features indicate that the pick adjustments are providing improved hypocenter locations.

Figures 8–11 show the results of relocating our ~7000-event data set. We show map-view and three-dimensional projections for depth slices at 2500–2900 m (Fig. 8), 2900–3100 m (Fig. 9), 3100–3300 m (Fig. 10) and 3300–3700 m (Fig. 11). The three-dimensional projections have been chosen for optimal view angle to show the emerging structures within the seismicity cloud at each level. In each of Figures 8–11, we depict preliminary hypocenter locations in the left-hand panels, and relocated hypocenters in the right-hand panels. We have used the same SEL location algorithm (PHILLIPS, 2000) for both preliminary and repicked events to demonstrate the level of improvement which is possible solely from automatic repicking.

Relocated shallow events are shown in Figure 8. Figure 8a shows these events in map view. Black arrows (B) and (C) indicate view azimuths in Figure 8b, c. Shallow earthquakes exhibit roughly parallel lineations in map view, striking N40°W in the more southerly part of the reservoir, although the seismicity does not define these

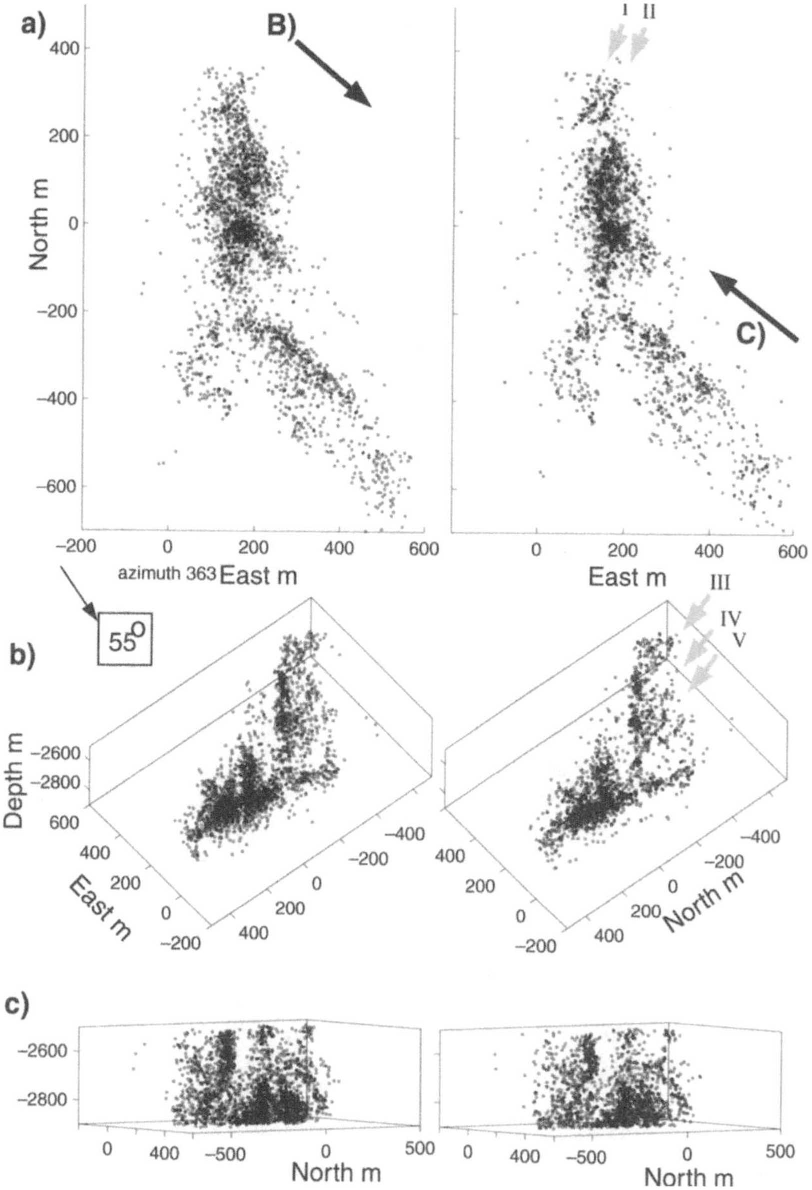

Figure 8
Preliminary (left) and relocated (right) hypocenters for shallow (2500 to 2900 m depth) Soultz seismicity. Details are discussed in the text. (a) Map view. Arrows (B) and (C) indicate the view directions for (b) and (c). (b) Steeply inclined view at 55° of shallow seismicity, looking towards S50°E. (c) Cross-section looking N40°W along strike of the segmented, southern portion of the shallow reservoir.

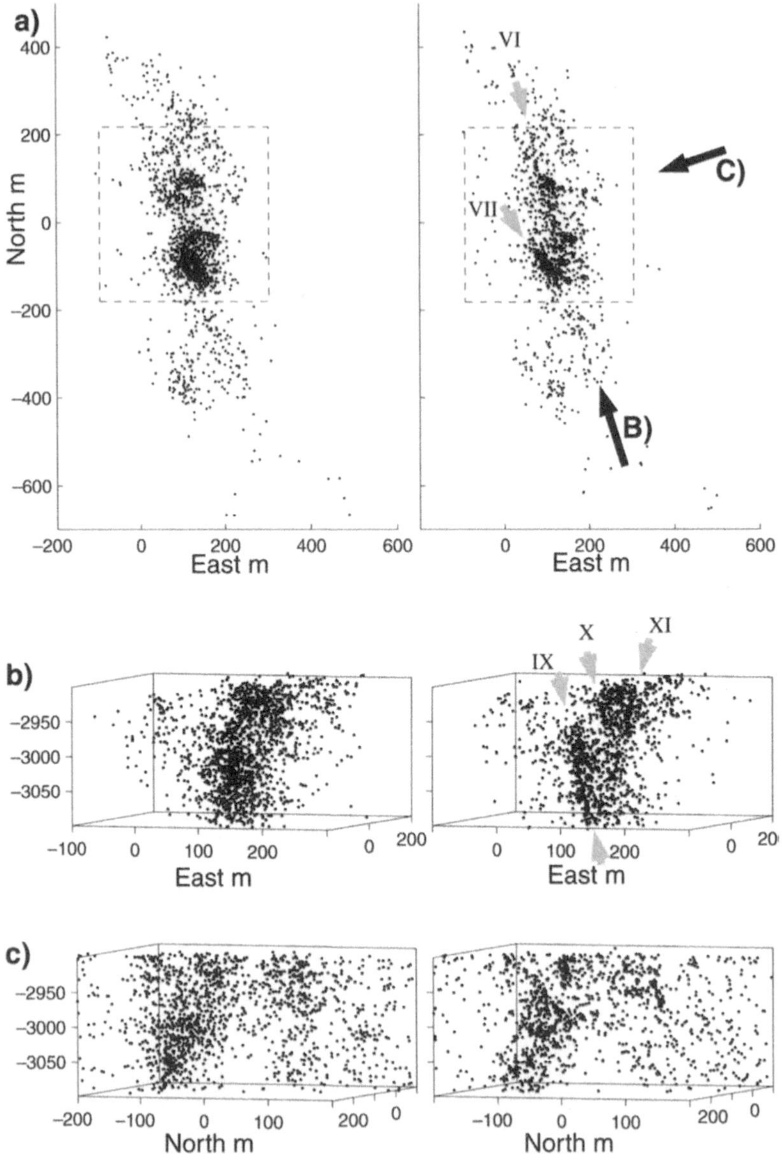

Figure 9
Intermediate depth seismicity (between 2900 and 3100 m). Details are discussed in the text. (a) Map view. Cross-section view angles in (b) and (c) are indicated by direction arrows (B) and (C). (b) Cross-sectional view looking N18°W, along strike of the clustered features seen in map view. (c) Broadside view looking S72°W.

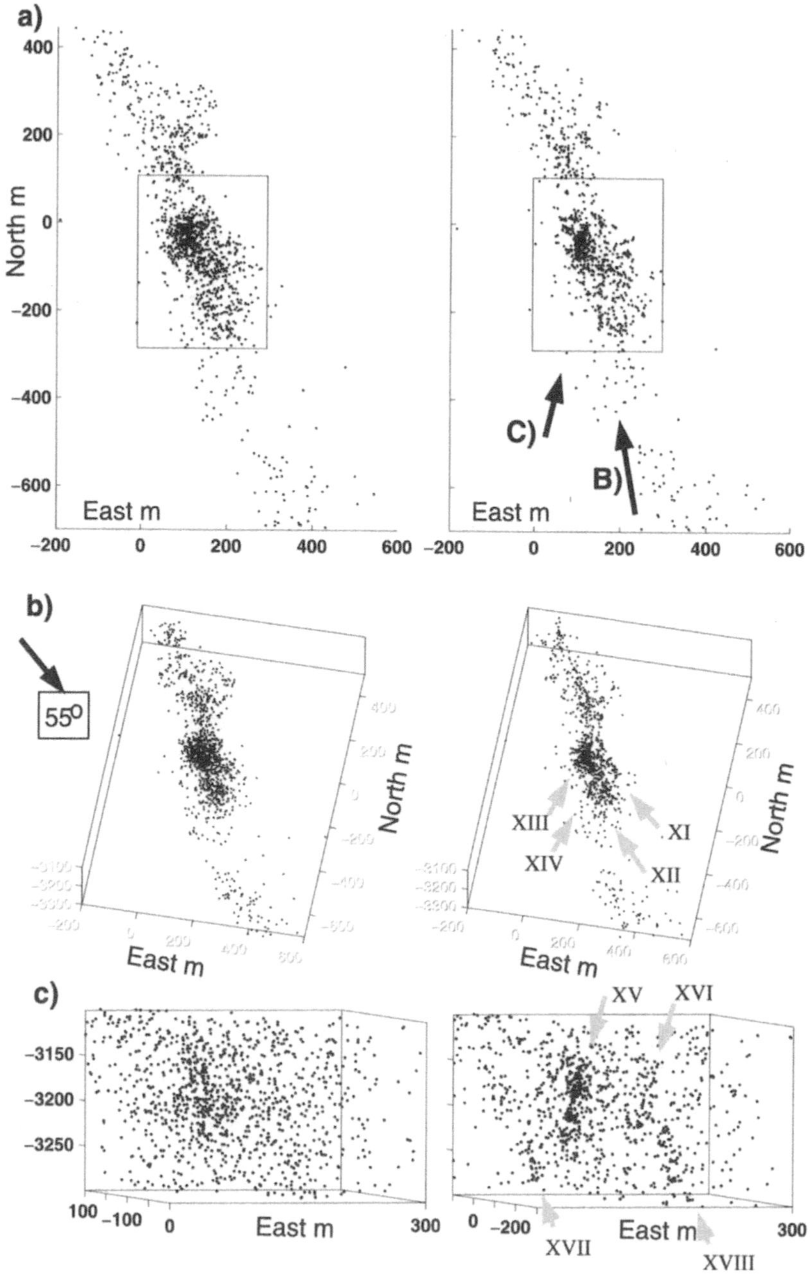

Figure 10

Deep seismicity between 3100 and 3300 m. Details discussed in the text. (a) Map view. Dashed box outlines the close-up elevation shown in c. View azimuths for (b) and (c) are indicated by arrows (B) and (C). (b) Elevated view at 55° inclination, azimuth N10°W, approximately coincident with Phillips' deep cluster intersection axis. (c) Cross-section looking N14°E, along strike of Phillips' "primary" plane.

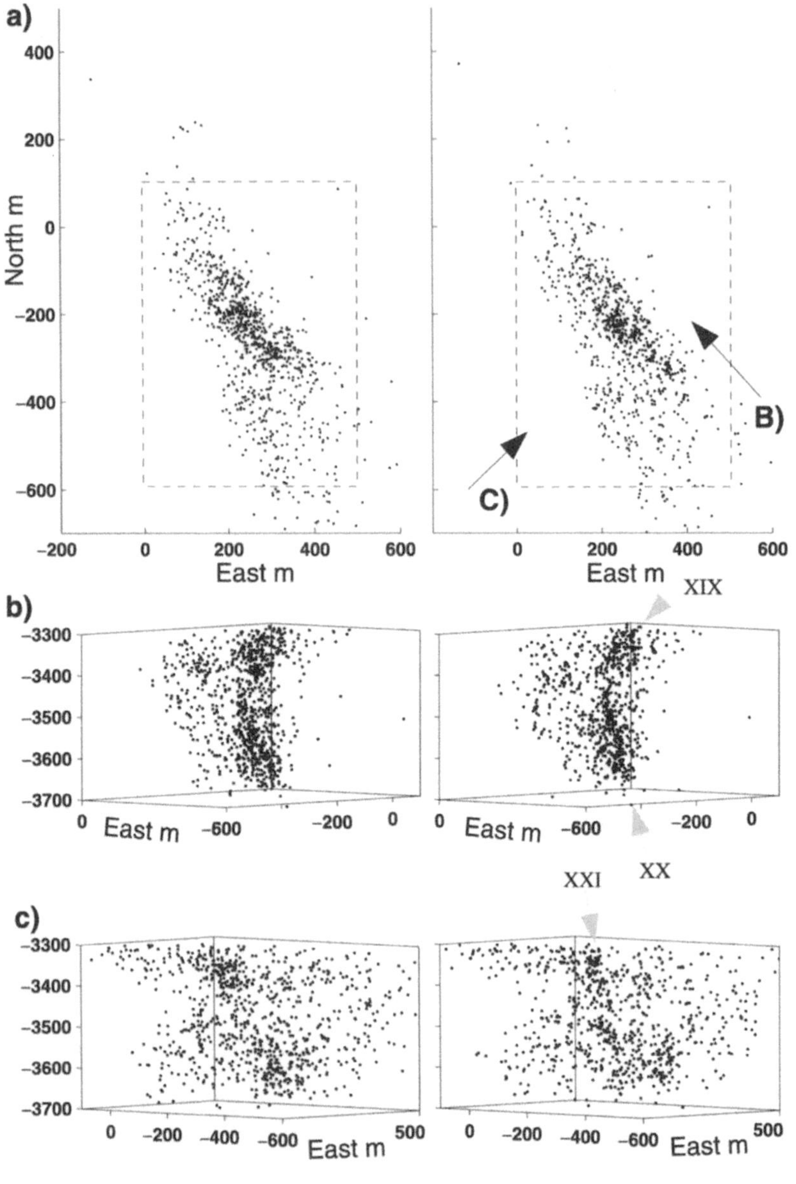

Figure 11
Deep seismicity (3300–3700 m) associated with 1993 injection. Details are discussed in the text. (a) Map view. Dashed box represents region shown in the elevations in (b) and (c), whose azimuths are indicated by arrows (B) and (C). (b) View looking N40°W along strike of the general trend in (a). (c) Broadside view, perpendicular to (b).

trends as clearly in the more northerly part. This orientation is similar to the orientation of one of the populations of hydrothermally sealed fractures identified by GENTER and TRAINEAU (1996). To the north, lineations suggest fractures or faults with an orientation of approximately N20°E (indicated by gray arrows I, II at the top of the figure). An elevated view looking S50°E with a 55° inclination (Fig. 8b) suggests that the N30°W seismicity trend dominating the southern part of the reservoir at these depths is made up of discrete, subparallel planes (indicated by gray arrows III, IV, V) steeply dipping towards the northeast. These structures are not discernible at all in the preliminary hypocenter cloud (8b, left-hand panel). In Figure 8c we show a cross-section looking towards N55°W. This shows the steeply eastward-dipping nature of the southern lineations, as well as a northerly dip for the isolated planes of seismicity seen in 10a at 200–300 m north.

Intermediate-depth events (between 2900 and 3100 m) are plotted in Figure 9. In map view the northwesterly-striking features may still be seen, although lineations are less laterally extensive. Two linear features are suggested, one striking N15°W (gray arrow VI) and one approximately N30°W (gray arrow VII). Seismicity did not expand as far to either side of the borehole at this level as it did at a shallower depth, suggesting a change in the fracture system and its connectedness at approximately 2900 m depth. The dashed boxes in Figure 11a indicate the area displayed in close-up cross-section in Figure 9b, c. Black arrows (B) and (C) indicate view azimuths for Figure 9b, c. In Figure 9b, looking N10°W, we see that after relocation the hypocenters define planar features which dip steeply to the north (gray arrows IX and X), but also a lineation suggesting an intersecting feature (gray arrow XI). Viewing these features broadside in Figure 9c (perpendicular to strike) illuminates small linear seismicity clusters within their planar surfaces.

Deeper (between 3100 and 3300 m) events are shown in Figure 10. Relocated hypocenters exhibit considerable improvement in clustering as seen in map view (Fig. 10a), particularly in the dense concentration of events at (100, −50). These events correspond to the depth of the PHILLIPS (2000) deep cluster, discussed earlier. Black arrows (B) and (C) indicate view azimuths in Figure 10b, c. In Figure 10b we see the inclined fracture ("secondary") plane (gray arrows XI and XII) of the Phillips deep cluster (see Fig. 6) as well as the northerly-trending ("primary") plane (gray arrows XIII and XIV) that it intersects. Relocated seismicity suggests additional lineations parallel to these major features which may indicate a series of subparallel seismogenic fractures activated by the reservoir stimulation. The cross-section in Figure 10c looks towards N14°E, along strike of the Phillips "primary" plane. The steeply westward-dipping lineation (gray arrow XV) is echoed by another, less well-defined plane some 100 m to the east (gray arrow XVI); two deeper lineations dipping steeply towards the east (gray arrows XVII and XVIII) are suggested in this view as well.

Figure 11 shows the deepest seismicity at Soultz associated with the 1993 injections, extending from 3300 to 3700 m depth. In map view (Fig. 11a) we see the diffuse cloud is resolved after relocation into a dense, northwesterly-trending feature that appears to be comprised of discrete patches of seismicity. Black arrows (B) and (C) indicate view azimuths for Figure 11b, c, respectively. In a cross-section looking N40°W (Fig. 11b), this concentrated zone of seismicity is revealed to be composed of two sets of roughly planar features, striking N40°W. The shallower features, extending to ~3500 m depth, appear to be subvertical lineations which dip towards the south (gray arrow XIX), whereas the deeper feature is a well-defined plane extending from ~3500 to ~3700 m depth, dipping steeply to the north (gray arrow XX). Viewed broadside (Fig. 11c), both shallow and deep planes consist of discrete seismicity patches which may correspond to distinct rupture zones on the seismogenic planes. Quasilinear edges (gray arrow XXI) to these zones suggest truncation by some linear or planar feature intersecting the seismic plane.

Discussion

The microearthquakes comprising this data set are the result of two different periods of reservoir stimulation, in September and October, 1993. Figure 12 illustrates preliminary hypocenter locations plotted as a function of time for the stimulation. In Figure 12a we show hourly event counts (for the 7000 microearthquakes we studied). Figure 12b, c, d illustrates the seismicity distribution through time in East–West (X), North–South (Y) and depth (Z) dimensions, respectively. Approximately 80% of the microearthquakes occurred during the September stimulation, during which ~25,000 m^3 of water were injected over an open-hole depth range between 2800 and 3500 m at rates of up to 36 l/s (CORNET and JONES, 1994) and overpressures of ~10 MPa (JUPE *et al.*, 1998). The bottom portion of the injection well GPK1 was filled with sand (depths > 3480 m) to prevent flow into a previously identified, deep fracture zone. Seismicity during this first stimulation began at a depth of ~2900 m, adjacent to the injection site, and expanded over time in an elongate cloud oriented roughly N-S at shallower depths (< 2900 m) and roughly NW–SE below this (CORNET and JONES, 1994).

Histograms of hypocentral depths of the seismicity resulting from the two stimulations are shown in Figure 13. The September stimulation, shown in Figure 13a, exhibits a strong peak in seismicity at a depth of ~2900 m. The October stimulation has the greatest number of its events occurring at ~3600 m depth (Fig. 13b). The time-varying hypocentral depth plot in Figure 12d shows this, and demonstrates that the removal of sand from the final 110 m of well GPK1 prior to the second stimulation causes the primary seismic activity to be generated by new slip within the deep fracture system, while the shallower portion remains comparatively quiet until overpressures are raised sufficiently to induce further failure on the previously destressed fracture system (JUPE *et al.*, 1998).

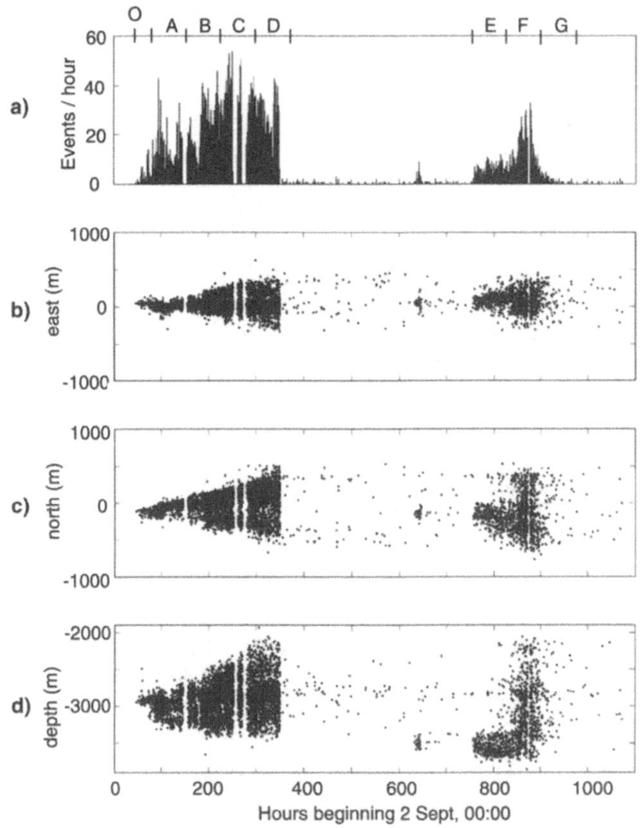

Figure 12
Spatial development of Soultz seismicity as a function of time, during the September and October 1993 injections over open-hole depths of 2800–3500 m in well GPK1 (Fig. 2). (a) Hourly event counts. Segments A–D and E–G represent 72-hour time intervals examined in Figures 14 and 15. (b) Longitude (meters east). (c) Latitude (meters north). (d) Hypocentral depth.

The greatest concentration of events within a volume might be expected to yield the greatest number of similar earthquakes. Intuitively we can look at the growing volume of seismicity and suspect that in the earlier hours of stimulation, when seismicity is concentrated within a small volume, many similar earthquakes should be observed. MORIYA *et al.* (2001) have shown through their analysis of doublets and multiplets that seismicity is dominated by doublets/multiplets during early stimulation, whereas the later activity has a significantly lower proportion of highly similar events. PHILLIPS (2000) noted a similar trend in his manual re-analysis, demonstrating that tight lineations of relocated hypocenters are formed prior to the spreading of seismicity into adjacent planar features. Phillips interprets the tight lineations of hypocenters as primary flowpaths for the injected fluid. Waveform similarity (i.e.,

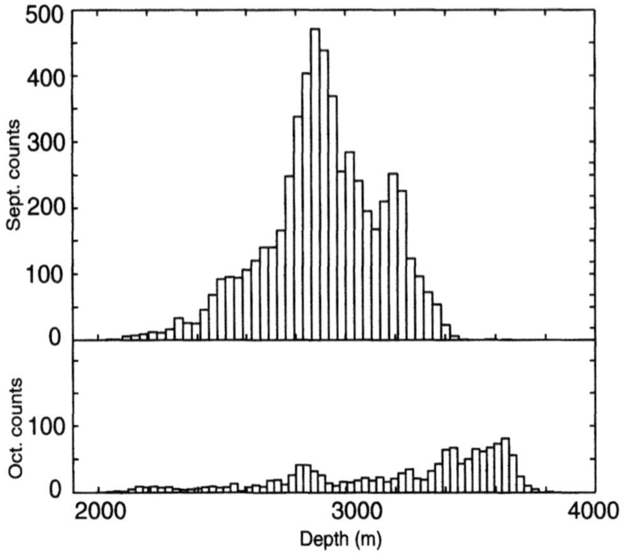

Figure 13
Histograms of hypocentral depths during the September and October, 1993 reservoir stimulations (Fig. 12) is dominated by earthquakes occurring at about 2900 m depth. After removal of the sand at the bottom of GPK1, the October stimulation (b) exhibits greatest activity at 3400–3600 m depths, with a secondary peak at the fracture zone near 2900 m depth.

membership in multiplets) is more likely to be observed within the early, tightly clustered events where source and path effects may be more consistent.

We expand upon these observations by calculating time-varying and spatially varying hypocenter density within the stimulated volume. The seismogenic volume in any time interval $d\tau$ may be estimated as the product of the maximum X, Y and Z dimensions during $d\tau$. These dimensions are determined to first order by estimating a three-dimensional envelope about the time-varying seismicity. MORIYA *et al.* (2001) report nearly 90% multiplet/doublet fraction during the first 48 hours of stimulation. Mean hypocenter density within this time period is $\sim 10^5$ events per km^3. Figure 14 shows event counts, approximate reservoir volume and estimated mean event density for the first five intervals (from Fig. 12) following the onset of stimulation. Event count per unit volume appears to decline in a roughly logarithmic fashion (Fig. 14c). The highest value of event density, $10^5/km^3$, corresponds to the MORIYA *et al.* (2001) estimated 90% doublet/multiplet fraction for events. The lowest density value, $\sim 1400/km^3$, corresponds to MORIYA *et al.*'s approximately 50% multiplet fraction.

Identification of planar and linear structures based on microseismicity may lead to identification of likely fluid flow directions and/or potential reservoir connectivity. FEHLER *et al.* (1987) developed an innovative method of identifying likely fracture planes along which earthquakes may have occurred. ROFF *et al.* (1996) further investigated linear and planar features through analysis of waveform amplitude

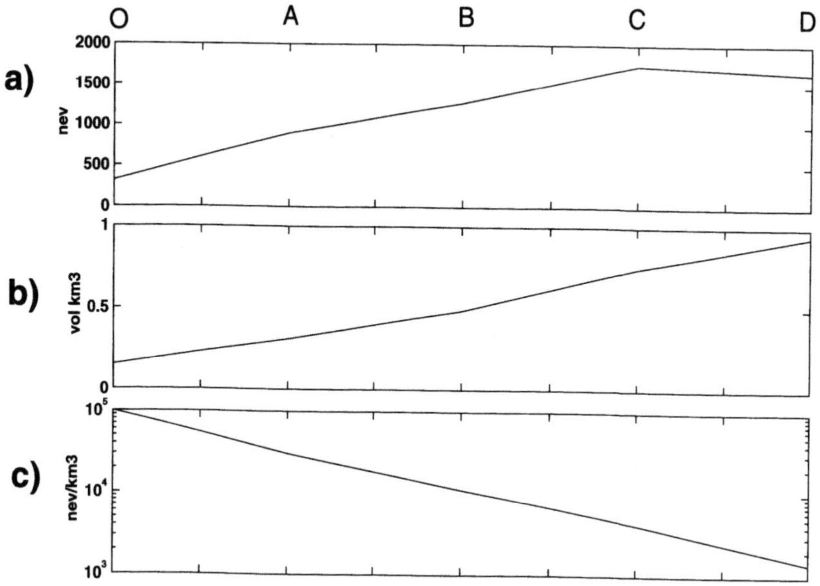

Figure 14
(a) Event counts from the first five 72-hour time intervals (A–D) for the September 1993 stimulation, as indicated in Figure 12a, beginning at point **O** (1623 hours, 2 Sept.). Vertical axis represents the number of earthquakes, horizontal axis indicates time interval (slight decrease in seismicity for the final data point is a reflection of the shorter time interval of active stimulation; reservoir deflation began during this interval). (b) Reservoir volume in each time segment. Vertical axis represents volume in km^3. The volume was approximated by estimating an envelope about the seismicity in the X, Y and Z directions as shown in Figure 14b–d, and calculating the mean volume for each time interval. (c) Event density estimate for each time interval. Vertical axis indicates estimated number of events per km^3, plotted on a logarithmic scale.

ratios, and PHILLIPS *et al.* (1997) investigated cluster structures via manual hypocenter relocations and slant stack amplitudes. STARZEC *et al.* (2000) found a correlation between the shapes of variograms of shear displacement and fracture orientation, suggesting a correlation of spatial variability of seismic slip and fracture complexity.

Tracking the highest concentrations of seismicity as a function of time may lead to the temporal mapping of fluid flow paths (e.g., PHILLIPS, 2000) or, alternatively, adjacent patches of failure on faults activated by pore pressure increases (CORNET and YIN, 1995); this may lead to better understanding of the dimensions and time development of important reservoir features. We illustrate in Figure 15 an east-west cross-section of the hypocenters in each of the seven 72-hour intervals A–G identified in Figure 12a. Upper panels depict original locations; lower panels show relocated hypocenters. Because we required the existence of seven phase arrivals to relocate events, and the period prior to time segment (a) (beginning at point O, indicated on Figure 12a) offered no hydrophone arrivals, we are unable to compare relocations for that period. We have calculated hypocentral density by counting neighbors within a sphere of 5 m radius centered on each hypocenter during each of the time intervals.

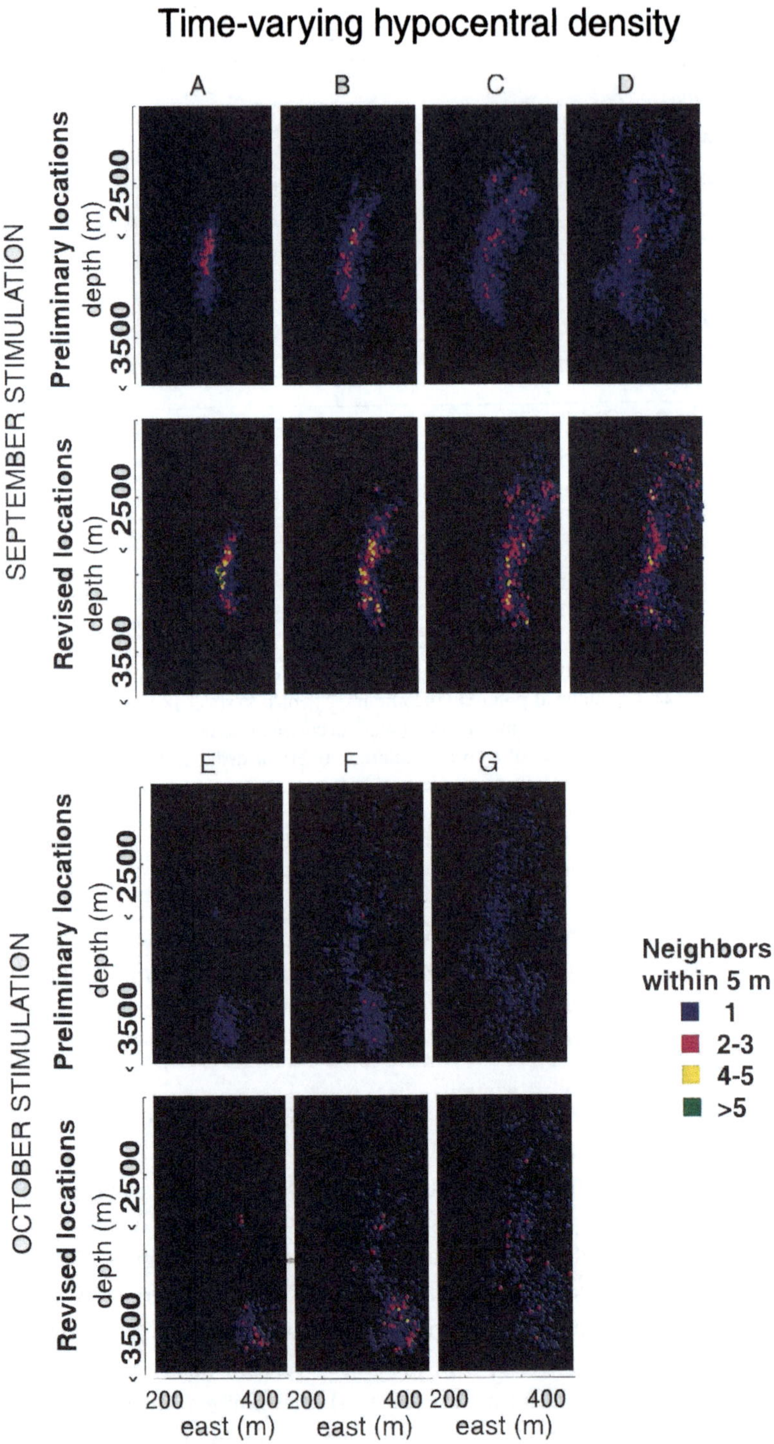
Time-varying hypocentral density
A
B
C
D
SEPTEMBER STIMULATION
Preliminary locations
Revised locations
depth (m)
2500
3500
E
F
G
OCTOBER STIMULATION
Neighbors within 5 m
1
2-3
4-5
>5
200
400
east (m)

Table 4

Percentages of singlet/multiplet events in 72-hour time segments

Time segment	a	b	c	d	e	f	g
% singlet preliminary	88	93	97	98	100	99	100
% multiplet preliminary	12	7	3	2	0	1	0
% singlet relocated	67	69	76	80	81	84	92
% multiplet relocated	33	31	24	20	19	16	8

Using the method of SNOKE (1987) to estimate source radius from corner frequencies, we have found that the median source radius for the relocated data subset to be approximately 3 m, so most events which fall within the 5 m search radius will have abutting or overlapping rupture patches. Upper panels of Figure 15 show results for the seven time intervals (A–G) using preliminary locations; lower panels illustrate relocated events. Hypocenters are color-coded to indicate spatial density, as indicated on the color key in Figure 15. Total numbers of earthquakes depicted are 6871 and 6217 for preliminary and relocated events, respectively. The discrepancy arises out of the number of events which were orphaned during the clustering process; these earthquakes were not repicked and so are not included in the relocations.

Overall, a clear increase in clustering after relocation is indicated in Figure 15. The tightest clustering remains nearest the injection point during the September stimulation (panels A–D), although the appearance of doublet events (magenta) on the edges of the developing reservoir margins suggests a concentration of activity at a migrating pressure front, although this may not necessarily reflect the locations of fluid paths (e.g., CORNET and YIN, 1995; CORNET *et al.*, 2000). A similar, albeit smaller, trend is apparent in panels (E) and (F), which correspond to the later, October, stimulation following removal of sand from the bottom of the borehole. Among events in the seven time periods (A–G) displayed in Figure 15, 74% of relocated events do not locate within 5 m of their nearest neighbor, whereas 95% of original hypocenters fail to cluster. Table 4 presents the singlet (events which have no neighbors within a 5 m radius) versus multiplet (two or more events within a 5 m radius) statistics by time interval, comparing percentage of singlet events for preliminary and relocated hypocenters during the seven time periods of Figure 15.

We note that relocating events solely through the use of adjusted phase picks may still not provide optimal relative and absolute hypocenter relationships, without

◀

Figure 15

Spatially varying hypocentral density as a function of time. Preliminary (top panels) and revised (bottom panels) depicting seismic development of the Soultz reservoir during each of the time intervals A–D, and E–G of Figure 14a. Color represents the number of neighbors within a 5 m radius. Intervals A–D correspond to the September stimulation, when the open hole extended from 2800 to 3400 m depth and the bottom of the well was sanded. Intervals E–G correspond to October injection following removal of the sand.

addressing inadequacies of the velocity model and station corrections. The Soultz velocity model and station corrections were derived from a single downhole shot (e.g., JUPE *et al.*, 1994) in combination with analysis of microseismicity using the preliminary phase picks. We observe systematic biases in these phase picks, which, if uncorrected, will contribute to the inaccuracy of preliminary velocity assumptions and subsequent analyses using those velocities. For example, pick times shown in Figure 5b demonstrate that the mean *S*-wave pick for station 4601 is clearly late, occurring more than a quarter cycle (~6 ms) after the *S*-wave arrival. Because we impose a zero mean constraint on our pick adjustments, this phase pick bias is an artifact of the preliminary pick distribution and has contributed to velocity and station correction estimates at Soultz. The bias may be accommodated in part by imposing an *a postiori* station correction, but this does not address the problem of separating actual pick bias from possible velocity heterogeneities. Removing this artifact from the catalogue using the waveform stacks is the subject of ongoing work; such a final processing step should help in future efforts to refine the very simple, uniform velocity model currently used for earthquake location at Soultz.

Summary

We have demonstrated an automatic phase-repicking algorithm to reduce the random component of earthquake location that arises from phase-picking inconsistency among similar earthquakes. We apply this technique to microearthquakes at the Soultz-sous-Forêts reservoir in the Rhine Graben of eastern France. The algorithm was tested using the results of careful, manual repicking on two isolated clusters of seismicity within the Soultz catalogue, then applied to the larger data set.

Results of hypocenter relocation on the test subset using the automatically adjusted picks demonstrate that the algorithm can recover detailed structures within the microseismicity cloud with a degree of improvement approaching that possible with careful, manual repicking. Median hypocenter discrepancies compared to PHILLIPS (2000) careful manual relocations improved from 31 to 7 m upon adjusting the preliminary phase picks with our automated algorithm. In the cases of distorted or clipped waveforms, such as with the Soultz hydrophone, human intervention may still be necessary if optimal adjustment of every phase for every earthquake is desired.

Application to a data set of ~7000 Soultz microearthquakes results in significant enhancement of seismogenic fractures and faults within the reservoir, consistent with detailed studies on smaller selected subsets. Event associations with nearest neighbors, estimated by counting the number of earthquakes within a 5 m radius, increases from 5 to 26% of the catalogue. These linear joint and fracture structures tend for the most part to be steeply dipping, with predominant northwest–southeast orientations, although northeasterly-striking fracture sets also comprise a significant portion of the resolved structure. The previously diffuse cloud also exhibits subparallel, *en echelon* linear features with a more northerly strike, among shallow

(2500–2900 m) events. Between 2900 and 3100 m we observe features which may represent intersecting faults or fractures, some dipping steeply to the east, others dipping steeply to the west. Overall, the image presented by relocated seismicity indicates that slip is occurring on fractures with varying orientations, whose characteristics change with depth. Evaluation of the relationships among these complex structures may yield better insights into reservoir flow behavior.

Correction of the phase arrival estimates to reduce random errors in the hypocenter location process can sharpen the results of subsequent applications including precise event location via SEL, JHD and/or collapsing techniques, as well as seismic tomography, reciprocal array processing, spatial *b* value or fractal dimension analysis and other imaging techniques which depend upon highly consistent phase picks. Additionally, the correlation, intra-cluster adjustment, stacking and subsequent stack correlation illuminate systematic biases in the preliminary phase picks which have contributed to inaccuracies in the initial velocity models. Identification and removal of these systematic data errors will allow for better modeling of the reservoir in the future.

Acknowledgements

The authors wish to thank D. Alde for help with data selection criteria. Preliminary phase picking for the full catalogue was undertaken by T. Thomas. H. Tobin and J. Schlue provided thorough and insightful reviews of the manuscript. We appreciate the helpful comments and insights of two anonymous reviewers. This work was supported by funding from the New Energy and Industrial Development Organization of Japan through the Laboratory of Hiroaki Niitsuma of Tohoku University, Sendai, Japan; support was also provided under the auspices of the U.S. Department of Energy by Sandia National Laboratory under Contract DE-AC04-94AL850000 and by the National Science Foundation Office of Polar Programs grant #9419267. Views expressed here are those of the authors and not necessarily those of the organizations for which they work.

References

Aster, R. C., and Scott, J. (1993), *Comprehensive Characterization of Waveform Similarity in Microearthquake Data Sets*, Bull. Seismol. Soc. Am. *83*, 1307–1314.

Aster, R. C., and Rowe, C. A. (2000), Automatic phase-pick refinement and similar event association in large seismic data sets. In *Advances in Seismic Event Location* (eds. Thurber, C., and Rabinowitz, N.) (Kluwer, Amsterdam, 1993) pp. 231–263.

Baria, R., Baumgärtner, J., Gerard, A., Jung, R., and Garnish, J. (1999), *European HDR Research Programme at Soultz-sous-Forêts (France) 1987–1996*, Geothermics *28*, 655–669.

Block, L. (1991), *Joint Hypocenter-Velocity Inversion of Local Earthquake Arrival Time Data in Two Geothermal Regions*, Ph.D. Dissertation, Massachusetts Institute of Technology, 448 pp.

BLOCK, L. V., CHENG, C. H., FEHLER, M. C., and PHILLIPS, W. S. (1994), *Seismic Imaging Using Microearthquakes Induced by Hydraulic Fracturing*, Geophysics *59*, 102–112.

CARR, D. B., YOUNG, C. J., HARRIS, J. M., ASTER, R. C. and ZHANG, X. (1999), *Cluster analysis for CTBT seismic monitoring*, Seismol. Res. Lett. *70*, 227–228.

CORNET, F. H., SHAPIRO, S. A., AUDIGRANCE, P., and ROYER, J. J. (2000), *Large-scale in situ Permeability Tensor of Rocks from Induced Microseismicity; Discussion and Reply*, Geophys. J. Int. *140*, 465–473.

CORNET, F. H., and JONES, R. (1994), *Field Evidence on the Orientation of Forced Water Flow with Respect to the Regional Principal Stress Directions*, Proc. 1st North Amer. Rock Mech. Symp., Austin, Texas, 61–69.

CORNET, F. H., and YIN, J. (1995), *Analysis of Induced Seismicity for Stress Field Determination and Pore Pressure Mapping*, Pure appl. geophys. *145*, 677–700.

DEICHMANN, N., and GARCIA-FERNANDEZ, M. (1992), *Rupture Geometry from High-precision Relative Hypocentre Locations of Microearthquake Clusters*, Geophys. J. Int. *110*, 501–517.

DODGE, D. A., BEROZA, G. C., and ELLSWORTH, W. L. (1995), *Foreshock Sequence of the 1992 Landers, California Earthquake and its Implications for Earthquake Nucleation*, J. Geophys. Res. *100*, 9865–9880.

FEHLER, M., HOUSE, L., and KAIEDA, H. (1987), *Determining Planes along which Earthquakes Occur: Method and Application to Earthquakes Accompanying Hydraulic Fracturing*, J. Geophys. Res. *92*, 9407–9414.

FEHLER, M., PHILLIPS, W. S., JONES, R., HOUSE, L., ASTER, R., and ROWE, C. (2000), *A Method for Improving Relative Earthquake Locations*, Bull. Seismol. Soc. Am. *90*, 775–780.

FREMONT, M.-J., and MALONE, S. D. (1987), *High Precision Relative Locations of Earthquakes at Mount St. Helens, Washington*, J. Geophys. Res. *92*, 10,223–10,236.

GAUCHER, E., CORNET, F. H., and BERNARD, P. (1998), *Induced Seismicity Analysis for Structure Identification and Stress Field Determination*, Proc. SPE-ISPM Rock Mechanics in Petroleum Engineering Conference *1*, 545–554.

GENTER, A., and TRAINEAU, H. (1996), *Analysis of Macroscopic Fractures in Granite in the HDR Geothermal well EPS-1, Soultz-sous-Forêts, France*, J. Volc. Geotherm. Res. *72*, 121–141.

GILLARD, D., RUBIN, A. M., and OKUBO, P. (1996), *Highly Concentrated Seismicity Caused by Deformation of Kilauea's Deep Magma System*, Nature *384*, 343–346.

GOLDSTEIN, P., DODGE, D., FIRPO, M., and RUPPERT, S. (1998), *What's new in SAC2000? Enhanced Processing and Database Access*, Seismol. Res. Lett. *69*, 202–205.

GOT, J.-L., FRECHÊT, J., and KLEIN, F. W. (1994), *Deep Fault Plane Geometry Inferred from Multiplet Relative Relocation Beneath the South Flank of Kilauea*, J. Geophys. Res. *99*, 15,375–15,386.

HARRIS, M. and YOUNG, C. (1997), *MatSeis: A Seismic GUI and Tool-box for MATLAB*, Seism. Res. Lett. *68*, 267–269.

ITO, A. (1995), *High Resolution Relative Hypocenters of Similar Earthquakes by Cross-spectral Analysis Method*, J. Phys. Earth *33*, 279–294.

JONES, R. H., and STEWART, R. C. (1997), *A Method for Determining Significant Structures in a Cloud of Earthquakes*, J. Geophys. Res. *102*, 8245–8254.

JONES, R. H., BEAUCE, A., JUPE, A., FABRIOL, H., and DYER, B. C. (1995), *Imaging Induced Microseismicity during the 1993 Injection Tests at Soultz-sous-Forêts, France*, Proc. World Geotherm. Cong., Florence, 2665–2669.

JUNG (1991), *Hydraulic Fracturing and Hydraulic Testing in the Granitic Section of Borehole GPK1, Soultz-sous-Forêts*, Geotherm. Sci. Technology *3*, 149–198.

JUPE, A. J., JONES, R. H., WILLIS-RICHARDS, J., DYER, B., NICHOLLS, J., and JACQUES, P. (1994), *Report on HDR Phase 4 – Soultz Experimental Programme 1993/1994, CSM Associates Ltd.*, IR02/12.

JUPE, A. J., JONES, R., DYER, B., and WILSON, S. (1998), *Monitoring and Management of Fractured Reservoirs Using Induced Microearthquake Activity*, Soc. Petroleum. Engineers contribution SPE 47315 in International Collaborative Project on Mapping of Engineered Geothermal Reservoirs, MTC Project Report.

KISSLING, E. (1988), *Geotomography with Local Earthquake Data*, Rev. Geophys. *26*, 659–698.

LANCE, G. N., and WILLIAMS, W. T. (1967), *A General Theory for Classificatory Sorting Strategies 1. Hierarchical Systems.*, Computer J. *10*, 271–276.

LEES, J. M. (1998), *Multiplet Analysis at Coso Geothermal*, Bull. Seism. Soc. Am. *88*, 1127–1143.

LUDWIG, J. A., and REYNOLDS, J. F. (1988), *Statistical Ecology, a Primer on Methods and Computing*, (John Wiley & Sons, New York) 337 pp.

MORIYA, H., NAKAZATO, K., and NIITSUMA, H. (2001), *Detailed Fracture System of Soultz-sous-Forêts HDR Field Evaluated by Using Microseismic Doublet/Multiplet*, Pure appl. geophys., submitted.

NADEAU, R., FOXALL, W., and MCEVILLY, T. (1995), *Clustering and Periodic Recurrence of Microearthquakes on the San Andreas Fault at Parkfield, California*, Science *267*, 503–507.

NANJO, K., HAGAHAMA, H., and SATOMURA, M. (1998), *Rates of Aftershock Decay and the Fractal Structure of Active Fault Systems*, Tectonophysics *287*, 173–186.

PARK, J., LINDBERG, C. R., and VERNON, F. L. III (1987), *Multitaper Spectral Analysis of High-frequency Seismograms*, J. Geophys. Res. *92*, 12,675–12,684.

PARKER, R. L., and MCNUTT, M. K. (1980), *Statistics for the One-norm Misfit Measure*, J. Geophys. Res. *85*, 4429–4430.

PHILLIPS, W. S., HOUSE, L. S., and FEHLER, M. C. (1997), *Detailed Joint Structure in a Geothermal Reservoir from Studies of Induced Microearthquake Clusters*, J. Geophys. Res. *102*, 11,745–11,763.

PHILLIPS, W. S. (2000), *Precise Microearthquake Locations and Fluid Flow in the Geothermal Reservoir at Soultz-sous-Forêts, France*, Bull. Seismol. Soc. Am. *90*, 212–228.

PHILLIPS, W. S., RUTLEDGE, J. T., HOUSE, L. S., and FEHLER, M. C. (2001), *Induced Microearthquake Patterns in Hydrocarbon and Geothermal Reservoirs: A Review*, Pure appl. geophys., submitted.

POUPINET, G., ELLSWORTH, W. L., and FRECHET, J. (1984), *Monitoring Velocity Variations in the Crust Using Earthquake Doublets: An Application to the Calaveras Fault, California*, J. Geophys. Res. *89*, 5719–5731.

PUJOL, J. (1992), *Joint Hypocentral Location in Media with Lateral Velocity Variations and Interpretation of the Station Corrections*, Phys. Earth Planet. Interiors *75*, 7–24.

REVIERE-BARBIER, F., and GRANT, L. T. (1993), *Identification and Location of Closely Spaced Mining Events*, Bull Seismol. Soc. Am. *3*, 1527–1546.

ROFF, A., PHILLIPS, W. S., and Brown, D. W. (1996), *Joint Structures Determined by Clustering Microearthquakes Using Waveform Amplitude Ratios*, Int. J. Rock Mech. Min. Sci. Geomech. Abstr. *33*, 627–639.

ROWE, C. A. (2000), *Correlation-Based Phase Pick Correction and Similar Earthquake Family Identification in Large Seismic Waveform Catalogues*, Ph.D. Dissertation., New Mexico Tech.

ROWE, C. A., ASTER, R. C., FEHLER, M. C., PHILLIPS, W. S., and Kyle, P. R. (1998), *Hypocenter Relocation in Tectonic and Volcanic Seismic Data Sets Using Automatic Phase Correlation: Application of Cross Correlation, Signal Envelope and Principal Components Methods to Improve Pick Consistency*, EOS *79*, F588.

ROWE, C. A., and ASTER, R. C. (1999). *Application of Automatic, Adaptive Filtering and Eigenspectral Techniques to Large Digital Waveform Catalogs for Improved Phase Pick Consistency and Uncertainty Estimates*, EOS *80*, F660.

ROWE, C. A., ASTER, R. C., BORCHERS, B., and YOUNG, C. J. (2001), *An Automatic, Adaptive Algorithm for Refining Phase Picks in Large Seismic Data Sets, Part I: Technique*, Bull. Seismol. Soc. Am., in review.

RUBIN, A. M., GILLARD, D., and GOT, J.-L. (1998), *A Reinterpretation of Seismicity Associated with the January 1983 Dike Intrusion at Kilauea Volcano, Hawaii*, J. Geophys. Res. *103*, 10,003–10,015.

RUBIN, A. M., GILLARD, D., and GOT J.-L. (1999), *Streaks of Microearthquakes Along Creeping Faults*, Nature *400*, 635–641.

SHEARER, P. M. (1997), *Improving Local Earthquake Locations Using the L_1 Norm and Waveform. Cross Correlation: Application to the Whittier Narrows, California, Aftershock Sequence*, J. Geophys. Res. *102*, 8269–8283.

SHEARER, P. M. (1998), *Evidence from a Cluster of Small Earthquakes for a Fault at 18 km Depth Beneath Oak Ridge, Southern California*, Bull. Seismol. Soc. Am. *88*, 1327–1336.

SNOKE, J. A. (1987), *A Stable Determination of (Brune) Stress Drops*, Bull. Seismol. Soc. Am. *77*, 530–538.

SPUDICH, P., and BOSTWICK, T. (1987), *Studies of the Seismic Coda Using an Earthquake Cluster as a Deeply Buried Seismograph Array*, J. Geophys. Res. *92*, 10,526–10,546.

STARZEC, P., FEHLER, M., BARIA, R., and NIITSUMA, H. (2000), *Spatial Correlation of Seismic Slip at the HDR-Soultz Geothermal Site: Qualitative Approach*, Bull. Seismol. Soc. Am. *90*, 1528–1534.

THOMSON, J. D. (1982), *Spectrum Estimation and Harmonic Analysis*, Proc. IEEE *70*, 1055–1096.

THURBER, C., ROECKER, S., MCEVILLY, T., HASLINGER, F., STEINBERG, D., and ELLSWORTH, W. (2000), *Earthscope at Parkfield, California: Setting the Stage for Drilling*, EOS *81*, S318.

WALDHAUSER, F. ELLSWORTH, W. L., and COLE, A. (1999), *Slip-parallel Seismic Lineations on the Northern Hayward Fault, California*, Geophys. Res. Lett. *26*, 3525–3528.

WIEMER, S. and WYSS, M. (1997), *Mapping the Frequency-magnitude Distribution in Asperities: An Improved Technique to Calculate Recurrence Times?* J. Geophys. Res. *102*, 15,115–15,128.

(Received October 30, 2000, revised/accepted February 2, 2001)

Pure appl. geophys. 159 (2002) 597–617
0033 – 4553/02/030597–21 \$ 1.50 + 0.20/0

Pure and Applied Geophysics

Reservoir-induced Seismicity in Brazil

MARCELO ASSUMPÇÃO,[1] VASILE MARZA,[2,3] LUCAS BARROS,[2] CRISTIANO CHIMPLIGANOND,[2] JOSÉ EDUARDO SOARES,[2] JURACI CARVALHO,[2] DANIEL CAIXETA,[2] ALEXANDRE AMORIM[2] and EDMAR CABRAL[2]

Abstract—A compilation of 16 cases of reservoir-induced seismicity in Brazil is presented with maximum magnitudes ranging from 1.6 M_L to 4.2 m_b. The compilation includes: location of the main epicentral area with respect to the reservoir (inside the lake, at the margin, or outside), predominant geology, and the temporal distribution of the main phase(s) of activity (initial or delayed in relation to impoundment). Data on the regional stress field for some reservoirs is also included. Four recent cases are discussed in more detail: Tucuruí, Nova Ponte, Miranda, and Serra da Mesa. A comparison with all other reservoirs deeper than 30 m and 50 m suggests that the hazard for induced-seismicity varies within Brazil: the NE part of the intracratonic Paraná basin has higher hazard as compared with the southern part of the same basin. No correlation of the induced hazard with variations in natural seismicity can be observed.

Key words: Intraplate seismicity, Brazil, induced earthquakes, seismic hazard.

Introduction

In the late 1960s and early 1970s, after the confirmation that impoundment of large reservoirs worldwide could cause moderate-sized earthquakes (magnitudes $6.5 > m_b > 5$) with damaging potential, studies of reservoir-induced seismicity attained great scientific and practical interest. It was soon realized that aseismic areas do not necessarily have lower potential for inducing earthquakes. In Brazil, studies of reservoir-induced seismicity (RIS) started around 1972 when an event of magnitude 3.7 caused intensity VI MM in a small reservoir (only 20 m deep) in the state of Minas Gerais. GUPTA (1992) described nine cases of RIS in Brazil, and GOMIDE (1999) analyzed 15 cases. Here we present an updated summary of 16 cases of RIS in Brazil, with magnitudes reaching 4.2 m_b (Tables 1 and 2), in a contribution to future statistical studies of the correlation with reservoir size, geology, tectonic stress field

[1] Inst. Astronômico e Geofísico, Universidade de São Paulo, Rua do Matão 1226, 05508-900, São Paulo, SP, Brazil. E-mail: marcelo@iag.usp.br
[2] Observatório Sismológico, Universidade de Brasília, Campus Universitário, 70910-900, Brasília, DF, Brazil. E-mail: obsis@unb.br
[3] Seismological Laboratory, National Institute (R & D) for Earth Physics, RO-76900 Bucharest-Magurele, Romania. E-mail: marza@unb.br

and impoundment history, such as done for example by BAECHER and KEENEY (1982) and FENG *et al.* (1995).

Besides depth and volume of the reservoir, other factors such as geology and stress regime have been investigated as possible parameters facilitating RIS hazard estimation. CASTLE *et al.* (1980) had suggested that both normal and strike-slip stresses have higher RIS hazard than compressional regimes. On the other hand, KNUEPFER *et al.* (1979) and BAECHER and KEENEY (1982) proposed that strike-slip stress regime have slightly higher RIS hazard than compressional or tensional stresses. They also showed that regions of sedimentary geology are about twice more prone to RIS than igneous or metamorphic regions. The present updated summary of Brazilian cases may contribute to the future resolution of such discrepancies and improve estimates of RIS hazard levels.

General Features of Reservoir-induced Seismicity

The studies and observations of RIS worldwide in the last three decades have shown that:

(a) Only a small percentage of reservoirs induce seismicity. In a recent worldwide compilation, GUPTA (1992) lists 68 accepted cases, including certain cases considered doubtful by other authors (such as CASTLE *et al.*, 1980). The existence of tens of thousands of reservoirs worldwide, which have caused no observable seismicity, indicates that the probability of a reservoir-inducing seismicity is low, and very special conditions are necessary for the occurrence of RIS. BAECHER and KEENEY (1982) estimate that reservoirs with height >100 feet have a probability around 10% of causing RIS (using only RIS larger than about 3 m_b) . In Brazil, of 17 reservoirs deeper than 90 m only two caused induced seismicity with $m_b \geq 3$, which is consistent with BAECHER and KEENEY's estimate. Under special conditions, however, (very large reservoirs, appropriate stress regime and geological conditions) RIS hazard can exceed 50% (BAECHER and KEENEY, 1982); i.e., more than half of the reservoirs will have RIS.

(b) It has long been recognized that the effect of the reservoir in the tectonic stress field (i.e., the weight of the water and the pore pressure at depths) is small compared to the stresses released by some of the largest induced earthquakes. This implies that reservoir induced seismicity can only occur if the area is already under near-critical tectonic stresses. In fact some authors prefer the term "triggered" seismicity, instead of "induced", to better convey the fact that high near-critical stresses are usually necessary for the occurrence of the earthquake activity (see TALWANI, 2000, for a discussion on this subject).

(c) Low levels of natural seismicity do not necessarily imply less hazard of induced seismicity. The identification of several cases of RIS in aseismic regions (e.g., SIMPSON, 1976; CASTLE *et al.*, 1980; GUPTA, 1992) shows that little correlation exists between natural seismicity and the likelihood that a reservoir will induce seismicity. More studies of this aspect are necessary because the known number of RIS may be underestimated in seismic regions where identification is more difficult (see for instance the debate related to the Oroville reservoir in California, as reported by GUPTA, 1992). In this paper we briefly show that RIS hazard within Brazil does not seem to correlate with variations of the regional level of intraplate seismicity.

(d) It is generally agreed that the maximum possible induced earthquake cannot exceed the maximum possible natural earthquake in the region (e.g., GUPTA, 1992). This is consistent with all reported empirical evidence to date, and also with the concept that near-critical pre-existing stresses are necessary for the occurrence of RIS. In Brazil the maximum induced magnitude was 4.2 m_b, in the SE seismic province, a region where natural earthquakes up to 5.1 m_b have been observed (BERROCAL *et al.*, 1983, 1984).

The mechanisms by which earthquakes are induced have been extensively studied (e.g., TALWANI and ACREE, 1984/85; SIMPSON *et al.*, 1988; SIMPSON and NARASIMHAN, 1990; TALWANI, 1997). The main aspects involve the perturbation of stresses and pore pressures at depths caused by the weight of the reservoir, and the diffusion of additional weight-induced pore pressures to hypocentral depths from the reservoir at the surface. Both theoretical (e.g., BELL and NUR, 1978; SIMPSON, 1986; ROELOFFS, 1988) and observational studies (e.g., HEALEY *et al.*, 1968; ZOBACK and HICKMAN, 1982) have shown that the controlling factors in these processes are the pre-existing tectonic stresses and pore pressures, the permeability of the rock masses and fracture systems, the strength of fault systems, and the relative orientation between the tectonic stresses and potential fault systems.

Because of the inhomogeneous properties of the rocks beneath a reservoir (such as permeability and fracture systems, as well as the local stress field) the induced seismicity can show complex temporal and spatial patterns in response to the impoundment history. For example, migration of activity from one area of the reservoir to another can be observed in same cases, such as in the Açu reservoir, in NE Brazil (FERREIRA *et al.*, 1995).

Proposals to classify the patterns of induced activity into two main components have been presented by SIMPSON *et al.* (1988) and TALWANI (1995, 1997). The two categories of SIMPSON *et al.* (1998) are "*rapid*" (when activity begins immediately following first filling, or major changes in water level, and dies out after a few years) and "*delayed*" (when the predominant seismicity, including the largest event, occurs several years after impoundment, i.e., after a number of annual cycles in the water fluctuation). TALWANI (1995, 1997) defines the two categories as "*initial*" (associated with initial impoundment or large lake level fluctuation; usually seismicity is widespread in the periphery of the lake) and "*protracted*" (after the effect of the initial seismicity has diminished; it persists for many years without significant decrease in frequency and magnitude; epicenters can be beneath the lake or in the surrounding areas). Although some reservoirs show only an initial (or rapid)

Table 1

Reservoir characteristics

#	Dam name (state)	Dam height (m)	Maximum water depth (m)	Reservoir volume (km^3)	Start of impoundment	Predominant geology	Regional stress; SH_{max} direction (reference)
1	Porto Colômbia & Volta Grande (MG/SP)	40 & 45	35 & 37	1.46 & 2.30	1973 April & 1973 September	Basalt	
2	Nova Ponte (MG)	142	132	12.8	1993 October	Basalt/gneiss	Compression; NNE-SSW? (b)
3	Cajuru (MG)	23	20	0.20	1954	Granite-gneiss	Strike-slip; E-W (b, c)
4	Capivara (PR/SP)	60	55	10.5	1976 January	basalt	
5	Tucuruí (PA)	106	90	45.8	1984 September	Metamorphic	
6	Balbina (AM)	42	35	17.5	1987 October	Sediments/ granite-gneiss	Possibly compression; NNW-SSE? (a, f)
7	Miranda (MG)	85	82	1.14	1997 August	Basalt/gneiss	Compression; NNE-SSW? (b)
8	Paraibuna-Paratinga (SP)	98 & 104	90	4.74	1974 & 1976	Granite-gneiss	strike-slip; NE-SW (g)
9	Jaguari (SP)	67	53	1.5	1969 December	Granite-gneiss	
10	Capivari-Cachoeira (PR)	61	58	0.18	1970 July	Gneiss	
11	Açu (RN)	31	31	2.4	1985	Granite-gneiss	Strike-slip; E-W (d, e)
12	Serra da Mesa (GO)	150	146	54.4	1996 October	Granite	Compression; NNW-SSE (a)
13	Marimbondo (MG/SP)	90	86?	6.15	1975	Basalt	
14	Sobradinho (BA)	43?	40	34.1	1977	Granite-gneiss/schist	
15	Emborcação (MG/GO)	158	154	17.5	1981 August	Gneiss	
16	Xingó (SE/AL)	140	110?	3.8	1994 June	Metamorphic	
				Doubtful cases			
17	Peti (MG)	43	42	0.042	1946	Gneiss	
18	Furnas (MG)	127	106	23	1963	Quartzite	Strike-slip; E-W (a)
19	Três Marias (MG)	70	60?	21	1962	Sediment	

References: (a) ASSUMPÇÃO (1992). (b) ASSUMPÇÃO *et al.* (1997). (c) ASSUMPÇÃO (1998b). (d) FERREIRA *et al.* (1995). (e) FERREIRA *et al.* (1998). (f) LIMA *et al.* (1997). (g) MENDIGUREN (1980).

response, many reservoirs exhibit a mixed behavior, with a later phase of activity following the initial response.

Despite the progress attained in explaining the mechanisms of RIS, it is not possible to predict the occurrence of induced seismicity of a future reservoir, because of the practical difficulties in accurately mapping, in a large volume of rock beneath the reservoir, key parameters such as *in situ* stresses, permeability of the rock masses and geometry of fracture systems. From a practical point of view, statistical studies of previous cases (e.g., CASTLE *et al.*, 1980; BAECHER and KEENEY, 1982; FENG *et al.*, 1995) can be useful for hazard evaluation of future reservoirs. In this respect, more complete compilations of past cases of RIS worldwide, including their temporal/spatial behavior, should contribute to better hazard evaluation of future reservoirs.

Reservoirs with Induced Activity in Brazil

Tables 1 and 2 list all cases of Brazilian reservoirs where induced activity has been reported. Besides the 16 confirmed cases, three other reservoirs with doubtful seismicity have been included for the sake of completeness. Table 1 supplies information concerning the reservoir size, main rock type in the reservoir area, and stress regime when available. Table 2 lists the main characteristics of the activity; the quality of the earthquake information varies considerably, depending on the number of seismic stations used in the studies. Locations of the reservoirs are shown in Figure 5, to be discussed later.

The magnitude adopted in this paper is the teleseismic m_b. For events recorded in the range of 200 to 1500 km by Brazilian stations, a regional scale (m_R) is used which is equivalent to the teleseismic m_b in the range 3.8 to 5.5 (ASSUMPÇÃO, 1983). The regional magnitude m_R reproduces the teleseismic m_b with a standard error of about 0.2 units (ASSUMPÇÃO, 1983, 1998a). In reservoirs where the events are very small, local magnitudes (M_L) or duration magnitudes (m_D) are used and tied to the regional m_R when possible. The uncertainties of the duration magnitudes are about 0.3 units.

The largest induced earthquake in Brazil was m_b 4.2, associated with two close-by reservoirs, Porto Colômbia and Volta Grande, which started filling in April and September 1973, respectively. Small events started to be felt by the local population near Porto Colômbia reservoir in November 1973. The largest event occurred in February 1974 with MM intensities VI–VII, causing damage to some peasant houses. Although no seismic station was installed near the reservoir, macroseismic information from the local population showed that the activity had died out by mid-1974, about one year after impoundment of the second reservoir (KNUEPFER *et al.*, 1979; BERROCAL *et al.*, 1984).

The temporal/spatial behavior of the seismicity in each reservoir is included in Table 2. We use the category "*initial*" for the cases where the main burst of activity followed impoundment in less than a year or so, with the largest event occurring

Table 2

Characteristics of the induced seismicity

#	Dam name (state)	Seismicity type	Largest events					Comments	Refs.
			Date	Magnit. (m_b)	I_o (MM)	ΔT (years)	Location		
1	Porto Colômbia & Volta Grande (MG/SP)	Initial	1974.02.24	4.2	VI–VII	~1	Margin?	Largest induced event in Brazil, no local stations	Be84, Kn79, Gu92
2	Nova Ponte (MG)	Initial,	1995.04.21	3.5	IV–V	1.5	Margin	Local telemetered network	As97, Ma99,
		Delayed	1998.05.22	4.0	VI	4.5	Outside	after new maximum water level	this paper
3	Cajuru (MG)	Delayed	1971.08.08	3.5	V–VI	17	Outside?	Local stations since 1975;	Be84, Vi97,
			1972.01.23	3.7	VI	18	Outside?	correlation with water level	Gu92
			1976.05.23	3.2	–	22	Outside	during 1978–1985	
4	Capivara (PR/SP)	Initial,	1976.01.25	<3	V–VI	0.1	Margin	No local stations during filling;	Mi9l, Gu92
		Delayed	1979.03.27	3.7	V–VI	~3	Margin	stations from 1978–1997;	As95
			1989.01.07	3.7	VI	13	Margin		
5	Tucuruí (PA)	Initial,	1985.11.02	3.2	–	~1	Inside	Local network	Ve92,
			1987.04.01	3.3	–	~2.5	Margin?		this paper
		Delayed	1998.03.02	3.6	IV–V	14	Inside	after large water changes?	
6	Balbina (AM)	Initial	1990.03.25	3.4	–	2.5	Outside	Local network	Ve91
7	Miranda (MG)	Initial	1998.04.07	2.4	felt	0.8	Margin?	Local stations,	this paper
		Delayed	2000.05.06	3.3	V–VI	2.7	Margin	macroseismic	
8	Paraibuna-Paraitinga (SP)	Initial	1977.11.16	3.0	IV	~1	Inside	Local stations since 1977	Ri89
9	Jaguari (SP)	Delayed	1985.12.17	3.0	V–VI	16	Margin	No felt events in the first years	MR95
10	Capivari-Cachoeira (PR)	Initial	1971.05.21	<3	VI	~1	–	macroseismic information	Be84, Gu92
11	Açu (RN)	Delayed	1994.08.26	3.0	IV?	9.5	Inside/ Margin	No felt events in the first years	Fe95

12	Serra da Mesa (GO)	Initial	1999.06.13	2.2	–	~3?	Margin	Local telemetered network	this paper
13	Marimbondo (MG/SP)	Initial	1978.07.25	2.0 M_L	not felt.	~3	Margin	Local station	Gu92
14	Sobradinho (BA)	Initial	1979.07.05	1.9 M_L	not felt.	~2	Inside	Local network	Gu92
15	Emborcação (MG/GO)	Initial	1982.05.20	1.6 M_L	not felt.	~1	Inside	Local network	VA86
16	Xingó (SE/AL)	Initial	1994.07.20	1.7	III–IV	~0.1	Margin	Local network	Be00
					Doubtful cases				
17	Peti (MG)	Initial?	1948.11.10	n.a.	IV	~2	–	Macroseismic information	Ve92b
18	Furnas (MG)	Initial?	1966.11.15	n.a.	IV–V	3.5	–	Macroseismic information	Be84
19	Três Marias (MG)	Delayed?	1969	n.a.	felt	~7	–	Macroseismic information	Ve92b

ΔT = time interval (years) since beginning of impoundment.

References: As95) ASSUMPÇÃO *et al.* (1995). As97) ASSUMPÇÃO *et al.* (1997). Be00) J. Berrocal (pers. comm). Be84) BERROCAL *et al.* (1984). Fe95) FERREIRA *et al.* (1995). Gu92) GUPTA (1992). Kn79) KNUEPFER *et al.* (1979). Ma99) MARZA *et al.* (1999a,b). Me80) MENDIGUREN (1980). MR95) MIOTO and RIBOTTA (1995). Mi91) MIOTO *et al.* (1991). Ri89) RIBOTTA (1989). VA86) VELOSO and ASSUMPÇÃO (1986). Ve91) VELOSO *et al.* (1991). Ve92a) VELOSO (1992a). Ve92b) VELOSO (1992b). (Vi97) VIOTTI *et al.* (1997).

within about two years or so. Our "initial" activity includes rapid response due to reservoir loading as well as diffusion to short distances and shallow depths. Most reservoirs seem to manifest only an initial response. Other reservoirs display a later seismicity well after the initial response. Some reservoirs show a pronounced decrease in the initial activity and a strong reactivation delayed by several years (such as Capivara and Tucuruí). In other reservoirs, the later activity seems to spread for decades with very slow decrease rate, sometimes correlating with water level fluctuations, which could be classified perhaps as "protracted" behavior (such as Cajuru, Paraibuna and Açu). However, for the sake of simplicity, we have used the term "*delayed*" for any significant activity (including the largest event, or others of similar magnitude) that occurred several years after impoundment.

In some reservoirs, earthquakes occurring several years after impoundment have been attributed to previous short-term changes in the water level. In Nurek and Aswan, for example, abrupt changes in water level over a few days impacted the rate of seismicity (SIMPSON and NEGMATULLAEV, 1981; SIMPSON *et al.*, 1990). At Koyna reservoir, earthquakes larger than 5 m_b seem to occur whenever the water level exceeds the previous maximum (TALWANI, 2000). In such cases, the inducing mechanism can include some elastic effect from the recent change in water load, typical of "rapid" or "initial" activity. We present below evidence of three cases in Brazil in which a large, delayed earthquake could be associated with recent water level changes. However, because most Brazilian reservoirs have not been thoroughly studied, we will maintain the classification of "delayed" activity even for the cases in which a short-term contribution from the water level change is possible.

In some cases the initial activity is not well known because instrumental monitoring started well after filling of the lake. Several reservoirs have shown both initial and delayed activity, such as Capivara and Tucuruí. It should be borne in mind, also, that activity is still continuing in many reservoirs, consequently this classification may change as the activity evolves. For instance, Miranda reservoir, impounded in 1997, showed a clear initial activity which rapidly decreased in the following years; only one or two seismographic stations were kept in operation to reduce costs. In May 2000 an event with m_R 3.3 occurred, causing MM intensities V–VI, the largest induced earthquake to date.

One of the best examples of delayed seismicity is the Cajuru reservoir. With a dam only 23 m tall, it was impounded in 1954. Small earthquakes started to be felt by the local population in 1970 with the largest event (3.7 m_R) in 1972 generating MM intensities VI. Instrumental monitoring started in 1975 and seismicity was still being recorded 20 years later in 1995 (VIOTTI *et al.*, 1997) with occasional events ranging from 2.5 to 3.0 m_R being felt by the local population (ASSUMPÇÃO *et al.*, 1997). Although no instrumental monitoring had been used in the first decade, no events were felt during this period. Valid evidence of a correlation between water level and the number of events seems to exist for the period 1978–1985 (VIOTTI *et al.*, 1997) which confirms the induced nature of the activity.

Paraibuna/Paraitinga reservoir caused induced seismicity soon after filling in 1976; the largest event, 3.0 m_R, occurred in November 1977 about one year later (MENDIGUREN, 1980; RIBOTTA, 1989). Twenty years later small events are still recorded by the local network and occassionally felt by the local population (L.C. Ribotta, personal communication). Paraibuna/Paraitinga has been classified as having only initial seismicity because the later activity, although being almost in steady-state for many years, did not show events with magnitudes comparable to the initial phase.

Recent Cases of RIS

We now briefly describe the main features of some cases of recent activity in Brazilian reservoirs, not published before.

Tucuruí (# 5 in Tables 1, 2)

Pre-impoundment monitoring for five years with a single station (*TUC1*) detected no local events. Initial seismicity occurred soon after impoundment of the reservoir in 1984 (Fig. 1a). A network of up to four analog stations showed that this initial seismicity was concentrated beneath the reservoir, south of the dam (Fig. 1b); the largest event in this initial period had magnitude 3.2 m_R. A second burst of seismicity followed the next year with some epicenters west of the dam and magnitudes reaching 3.3 m_R. Hypocentral depths are not reliably determined: the local data give depths in the range of 0 to 6 km with large uncertainties. Despite the large epicentral uncertainties (about ± 3 to 5 km) a general SW-NE trend can be recognized for this initial activity (Fig. 1b). In the reservoir area, a NS trending belt of metamorphic rocks, including a thrust fault, separates mainly Phanerozoic sedimentary rocks east of the lake (Barreiras sandstones) from Precambrian granitic/gneissic rocks on the west (Fig. 1; ELETRONORTE, 1977). Some NW-SE oriented faults were tentatively recognized on the western part of the lake (ELETRONORTE, 1977). However, as noted by VELOSO (1992a), no major tectonic feature can be correlated with the SW-NE trend of the epicentral distribution.

Seismicity decreased considerably the following years and the network was deactivated because of high operational costs. The largest event as yet occurred in March 1998 reaching a magnitude 3.6 m_b and epicenter near the deepest part of the lake (Fig. 1b). Only one analog local station was operating at the time, nonetheless the event was well recorded by the pIDC (prototype International Data Centre) which calculated an epicenter consistent with the average *S-P* interval of the aftershocks detected by the local station *TUC2* ($S\text{-}P = 2.9 \pm 0.1$). This largest event caused MM intensity of about IV–V near the dam, which is not inconsistent with the pIDC epicenter. No reliable information is available on the focal depth of this event.

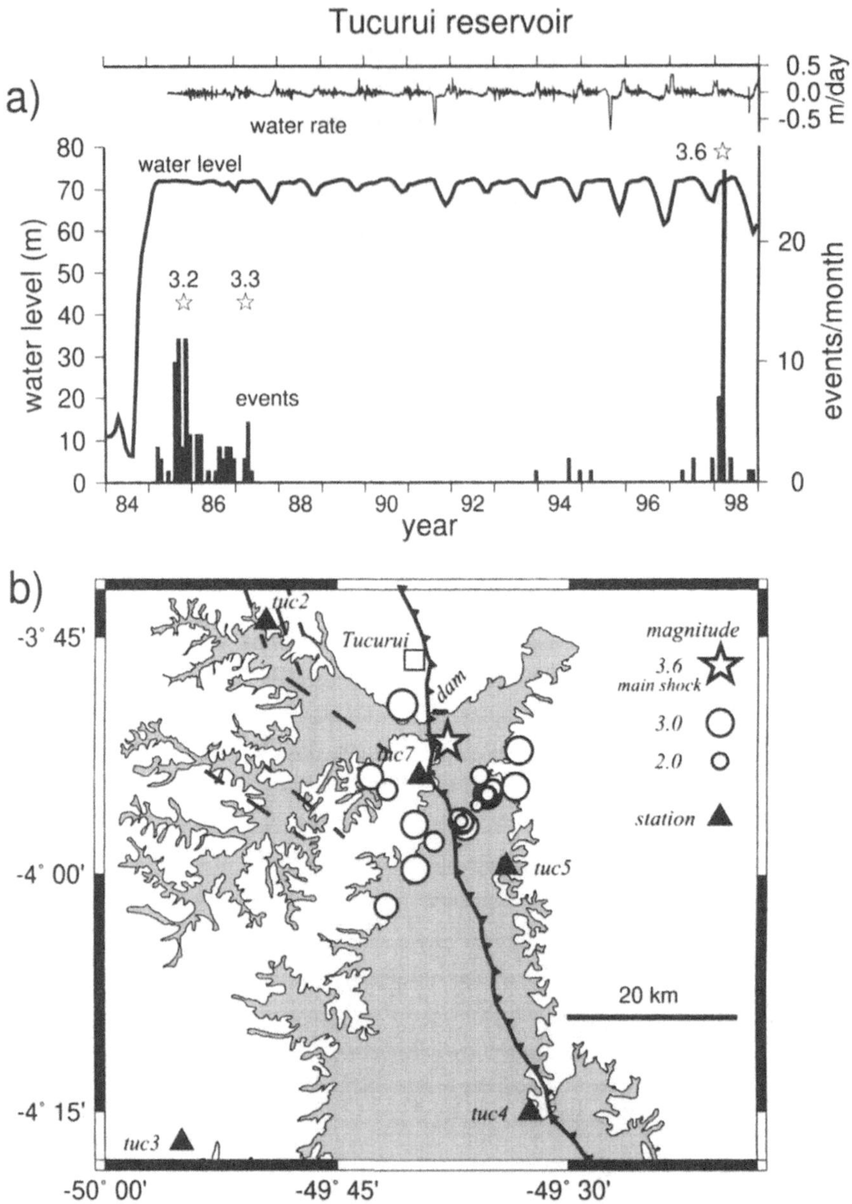

Figure 1

Seismicity in Tucuruí reservoir (#5 in Tables 1 and 2), state of Pará. (a) Temporal evolution: number of monthly events (histogram), monthly averaged water level (thick line), and rate of water level change (top trace); the stars denote the largest magnitudes. (b) Epicentral distribution: triangles denote seismographic stations, circles denote epicenters (errors in the range ± 3 to 5 km) with duration magnitudes above 1.6; the star is the main shock of March 1998 with magnitude 3.6 m_b. N-S oriented dented line is a reverse fault; dashed lines west of the dam are inferred faults of unknown type.

Tucuruí reservoir shows two main cycles of activity: one initial activity following impoundment and another phase in 1997–1998. It is possible that this last phase could have been caused by the large water variations in preceding years: a drop of ~10 m in 1996 and an abrupt increase in early 1997 (Fig. 1a). In mid-1997 the water reached its highest level of 72.5 m. In Tucuruí the water level is measured daily with centimeter accuracy. Figure 1a (top part) shows the rate of water level change in meters/day. No anomalously large short-term variation can be clearly related to the 1998 event. Also, the maximum water level in 1997 (72.5 m) does not seem to be significantly different from previous peaks of 72.4 m reached in 1990, 1992 and 1996. Therefore, because of the weak evidence for a causal relationship between the 1998 event and water level variations in the preceding years, we have preferred to classify the 1998 earthquake as delayed activity (Table 2). Unfortunately, no focal mechanisms are available for the Tucuruí events and hypocenters are not accurate enough to better understand the mechanism of induction.

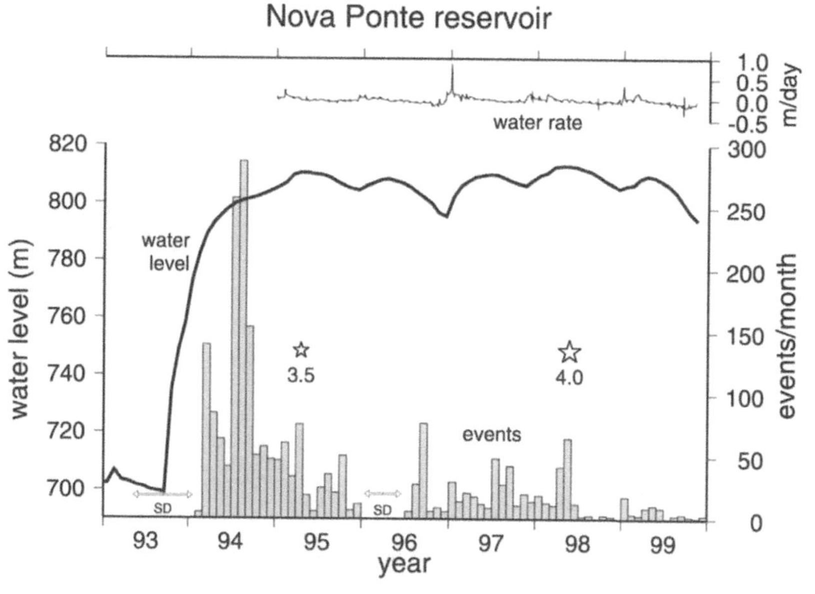

Figure 2
Seismicity in Nova Ponte reservoir (#2 in Tables 1 and 2), state of Minas Gerais. (a) Temporal evolution: number of monthly events (histogram), monthly averaged water level (thick line) referred to mean sea level, and rate of water level change (top trace); the stars denote the two largest events. The number and location of seismographic stations varied during the monitoring and may have affected the detection threshold; SD indicate the two periods without data when the stations were down. (b) Epicentral map for the period 1994 to 1999: circles are small events (magnitudes in the range 0.5 to 3) showing the two main areas of activity: the area in the south showed only initial seismicity which died out more rapidly than the main area near the lake; epicentral errors range from 0.3 to 1.5 km. The stars denote the epicenters of the two main events: the smaller star near the lake is the 3.5 m_b event of 1995, and the larger star further to the south is the 4.0 m_b event of 1998.

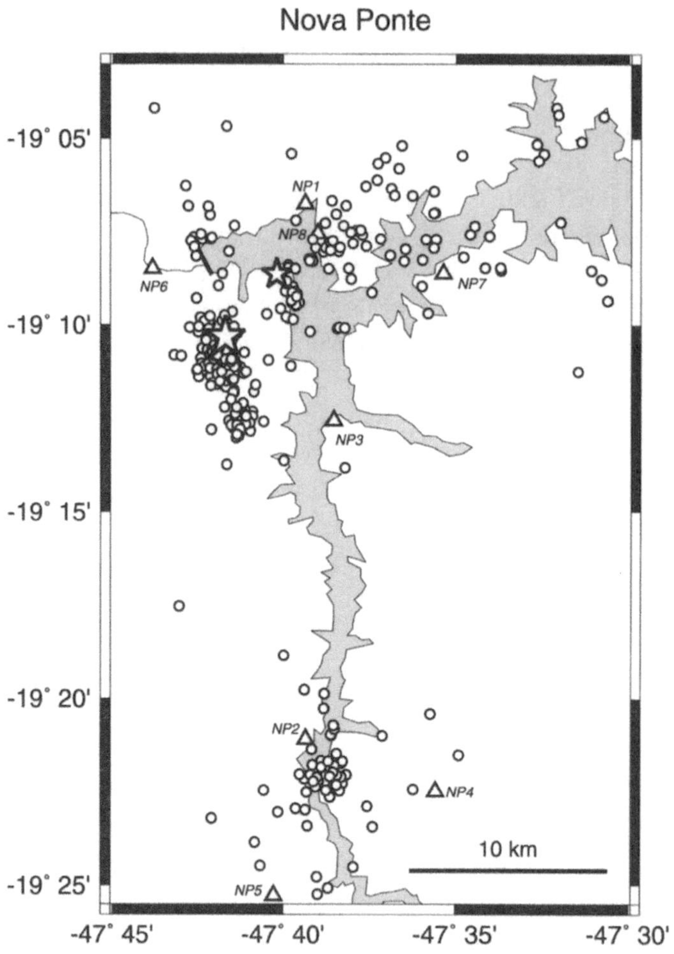

Figure 2b

Nova Ponte (# 2 in Tables 1, 2)

Pre-impoundment monitoring of the Nova Ponte reservoir was carried out with one station (*NP1*) for eight years. Unfortunately, this local station *NP1* experienced operational problems and malfunctioned during impoundment which started in October 1993 (Fig. 2a). Events started to be felt by the local population in January 1994. Five analog stations were deployed in February/March 1994, revealing seismic activity in two well separated areas: one small area about 25 km south of the dam in a narrow branch of the lake, and another, larger area at the margins of the deepest part of the reservoir (Fig. 2b). A four-station vertical-component telemetered network was installed in 1995 and more stations have since been added to the network. The number and efficiency of the seismic stations have been rather variable, so the number of monthly events shown in Figure 2a may be affected by uneven coverage.

Despite the uneven detectability of the network, a clear initial phase can be seen in Figure 2a with the largest event (3.5 m_R) occurring near the lake when the water level was about to reach its first maximum of 810.6 m, about 1.5 years after the start of impoundment. The initial activity far from the dam decreased rapidly in the first few years and attained a maximum magnitude of only about 2.0 m_D (VELOSO *et al.*, 1994). The activity near the dam has continued with a maximum magnitude of 4.0 m_b (MM intensities up to VI) occurring in May 1998, more than four years after impoundment. Compared with the initial activity, the main event of 1998 occurred further from the lake margin, which is probably due to the delayed response from pore pressure diffusion. GOMIDE (1999) used the increase of the active area near the lake to estimate the seismic hydraulic diffusivity (k_s) at Nova Ponte as 1.5 m^2/s, which is in the range (0.1 to 10 m^2/s) observed at other reservoirs (TALWANI and ACREE, 1984/85). Also, it is interesting to note that, although no abrupt change in the water level can be observed shortly before the main shock of May 1998 (top trace in Fig. 2a), it occurred three months after the water level had exceeded the previous maximum of 1995, and about 10 days after a new maximum of 813.1 m had been reached.

The 1998 main event in Nova Ponte (the second largest induced earthquake in Brazil; Table 2) seems to have occurred after two days of immediate foreshocks which were preceded by a quiescent period of 12 days, according to MÂRZA *et al.* (1999a).

Composite focal mechanisms have been presented for the Nova Ponte southern cluster (VELOSO *et al.*, 1994) and the northern cluster (ASSUMPÇÃO *et al.*, 1997), indicating predominance of reverse faulting and roughly NE-SW oriented *P* axes. Unfortunately, accuracy of hypocentral depths has been inadequate to define fault planes in the Nova Ponte area.

Miranda (# 11 in Tables 1, 2)

Seismic activity was observed at Miranda immediately following impoundment of the reservoir, started in August 1997, in a clear case of initial response (Fig. 3a). Although no local station had operated near Miranda dam, the network monitoring the Nova Ponte reservoir since 1992, about 30 to 50 km away, had detected no event near Miranda during its pre-impoundment period. One local station was installed during impoundment (*MIR1*) followed by other stations later. The largest event in the initial months had a duration magnitude 2.1 m_D. During 1998 the activity decreased with occasional small reactivation periods; the largest magnitude during the initial phase was 2.4 m_D. In 1999 and early 2000 the activity decreased considerably, and most stations were deactivated to reduce costs.

On May 06, 2000 a delayed large event occurred with magnitude 3.3 m_R and MM intensities V–VI. Unfortunately, the only two local stations were down and the records from the nearest stations (Nova Ponte reservoir, ~50 km away) do not allow an accurate epicentral determination for this main shock. In Figure 3b we show the macroseismic epicenter defined by a country house damaged with several wall cracks. Four portable local stations were deployed a few days after the event, but no aftershocks were detected, indicating that the Miranda main shock of 2000 occurred as an isolated single event. Figure 3a shows that the earthquake occurred during a *decrease* of the water level.

Epicentral accuracy is highly variable due to changing station configuration and equipment problems. In Figure 3b we show only the best epicenters, determined by at least three stations with both *P* and *S* arrivals read by cross-correlation of the waveforms. Although the absolute epicentral errors can reach up to 1 km, the *relative* location errors are very small, usually less than 0.1 or 0.2 km. This allowed the identification of at least three clusters of epicenters aligned in the SW-NE direction. The cluster nearest the dam aligns rather well with a SW-NE branch of the lake which probably defines a geological fault. Although no geological faults have been mapped yet in the Miranda area, a more regional mapping reflects a predominance of SW-NE oriented faults about 30 km NW of the dam (CEMIG, 1995).

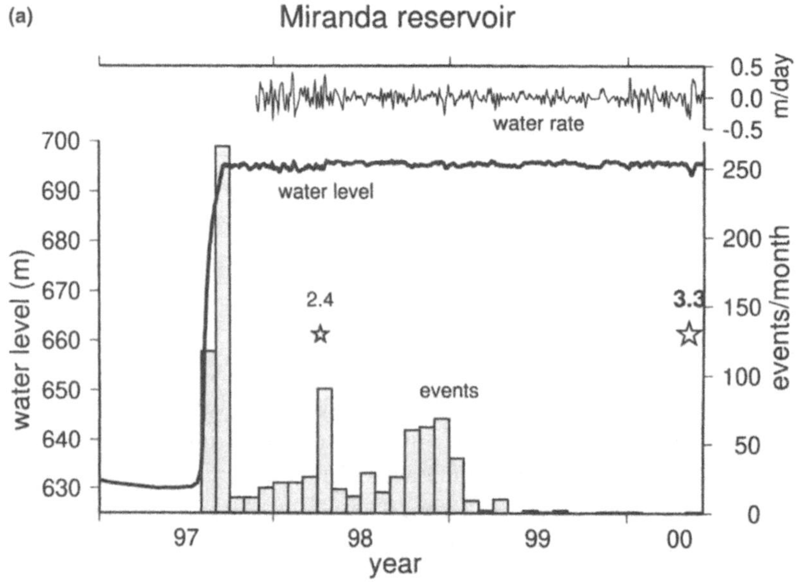

Figure 3
Seismicity in Miranda reservoir (#7 in Tables 1 and 2), state of Minas Gerais. (a) Temporal evolution: number of monthly events (histogram), daily water level (thick line), and rate of water level change (top trace); stars denote the largest event of the initial phase and the delayed, main event of May 2000 with magnitude 3.3 m_b. (b) Geology and epicenters: circles are well determined epicenters with three or more stations (relative locations accurate to within ± 0.2 km); note the three SW-NE oriented clusters; magnitudes in the range 1.0 to 2.4 m_D; the smaller star shows the largest event of April 1998 (2.4 m_D) in the initial phase; the larger star shows the macroseismic epicenter of the delayed, magnitude 3.3 m_R event of May 2000.

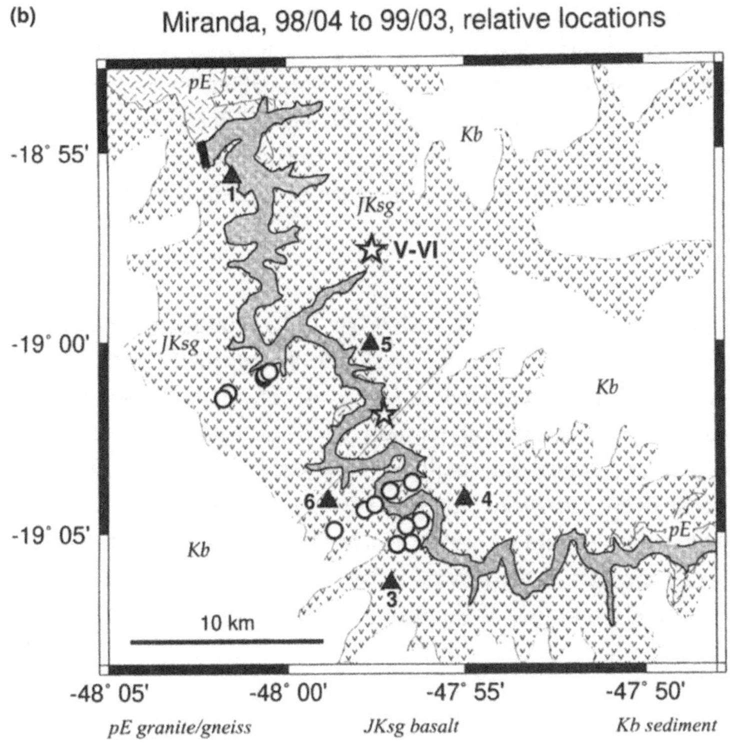

Figure 3b

The geology beneath the Miranda reservoir is characterized by Mesozoic flood basalts (Serra Geral Formation), no more than a few hundred meters thick, overlying Proterozoic gneisses. The seismic activity has occurred far from the deepest part of the reservoir, mainly on its margins, probably in SW-NE oriented fractures reactivated by the lake impoundment. The epicenters are all in basalt areas; the focal depths are usually around 1 km, however the accuracy is imprecise to define whether the activity is occurring in the basalt layers or in the basement. It is possible that fractures in the basalt provide conduits for pore pressure to diffuse to deeper levels, as suggested for Bhatsa, Koyna and Warna reservoirs in the Deccan traps of India (GUPTA, 1992; TALWANI, 2000; P. Talwani, personal communication).

Preliminary focal mechanisms indicate predominance of reverse faulting with nodal planes consistent with the SW-NE epicentral alignment. The focal mechanisms in the nearby Nova Ponte reservoir, together with preliminary mechanisms in Miranda, are consistent with regional compressional stresses roughly oriented NNE-SSW. In a compressional environment, the initial seismicity occurring immediately upon lake impoundment (Fig. 3a), with epicenters in the margins of the lake, suggests an inducing mechanism of elastic response to the water load for the initial phase, also called "rapid" response. The delayed activity, especially the main shock of May 2000, is probably due to the long-term effect of pore-pressure diffusion coupled, perhaps, with additional short-term elastic effects from a rapid decrease in the water level. Further studies, especially focal mechanism determinations, are necessary to better understand the occurrence of the main event of May 2000.

Serra da Mesa (# 12 in Tables 1, 2)

Filling of the reservoir started in October 1996. About one year later small local events (maximum magnitude 2.2 m_b) started to be recorded by the local four-station vertical-component telemetered network (Fig. 4). The activity is very sparse although the epicenters seem to occur preferentially in the margins of the lake.

The local events were too small to allow determination of focal mechanisms. However local stress measurements by hydraulic fracturing near the Serra da Mesa dam (CAPRONI and ARMELIN, 1990; ASSUMPÇÃO, 1992) indicate NW-SE oriented compressional stresses which would be consistent with the epicenters being preferentially at the margins of the lake (e.g., BELL and NUR, 1978).

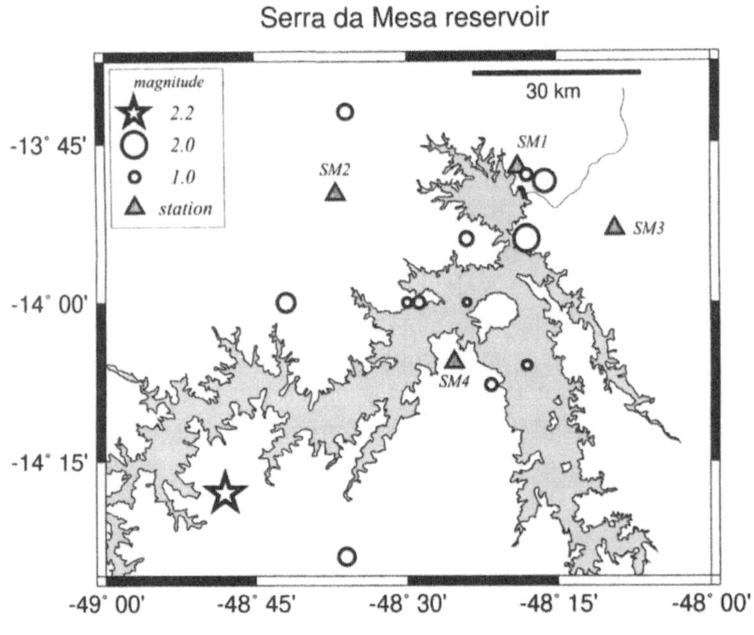

Figure 4
Seismicity in Serra da Mesa reservoir (#12 in Tables 1 and 2), state of Goiás. Epicentral errors are about ± 10 km. The star is the largest event with 2.2 m_R. Triangles are seismographic stations.

Discussion and Conclusions

Figure 5 shows the locations of all 16 confirmed cases of RIS in Brazil (Tables 1 and 2) compared with the background level of natural seismicity and the location of all other reservoirs deeper than 30 m (taken from CBGB, 1999). In Figure 5a, the size of the closed circle denotes the magnitude of the largest reservoir-induced event (note the different scale for magnitudes of the natural events). The background seismicity is not the complete catalog, but has been filtered, according to the following time-varying completeness thresholds: $m_b > 6.0$ after 1940, $m_b > 5.0$ after 1961, $m_b > 4.1$ after 1967, and $m_b > 3.2$ after 1980 (similar to the values used by ASSUMPÇÃO *et al.*, 1997). This filtered catalog better represents the spatial variations in natural seismicity levels in the continental area of Brazil.

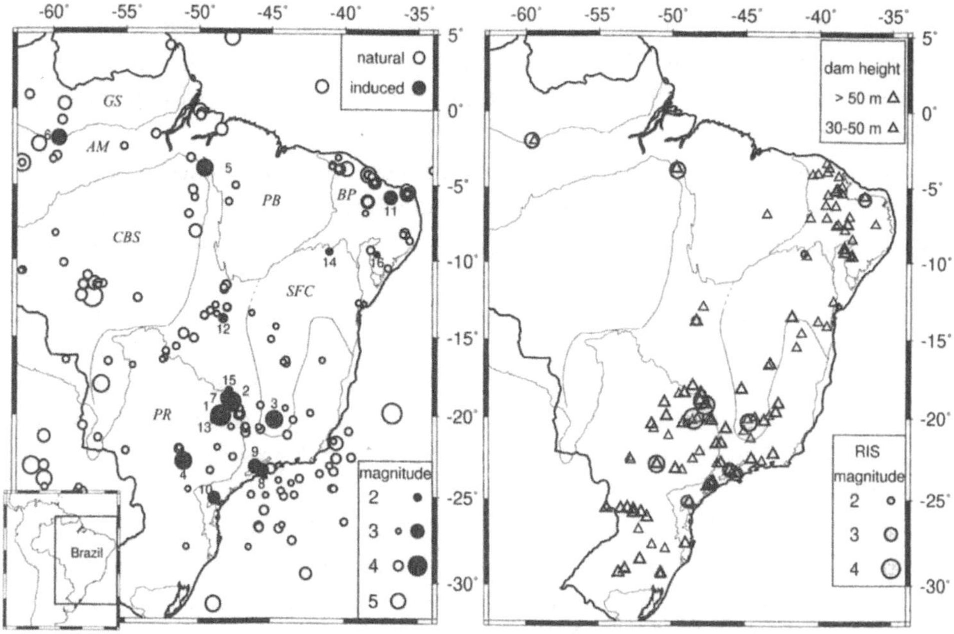

Figure 5

Location of the reservoirs with induced seismicity. Thin lines refer to the main geological provinces in Brazil. (a) Comparison with natural seismicity: note the different scales for the maximum induced event (closed circles) and for the natural seismicity (smaller open circles); numbers refer to Tables 1 and 2. *AM* = Amazon basin, *BP* = Borborema Province; *CBS* = Central Brazil shield, *GS* = Guyana shield, *PB* = Parnaíba basin, *PR* = Paraná basin, *SFC* = São Francisco craton, other unnamed areas are Upper Proterozoic fold belts and coastal marginal basin. (b) Comparison with other reservoirs where RIS has not been reported (triangles); smaller and larger triangles refer to reservoirs with depths ranging from 30 to 50 m, and larger than 50 m, respectively.

The distribution of dams in Figure 5b may give a rough idea of the RIS hazard in Brazil. In the central and northern parts of Brazil, very few reservoirs have been built and thus no conclusions can be drawn yet. We will then only discuss the data for the eastern part of the country. Although more detailed analyses are still necessary, some interesting observations can be made.

In the Borborema province (NE corner of Brazil, north of 8°S, and east of the Parnaíba basin) where natural seismicity is relatively high (Fig. 5a) only one reservoir has caused induced seismicity (Açu, # 11, 31 m deep) among approximately 20 others of similar height. Although only Açu was instrumentally monitored, many events were felt by the local population near its margins. No other reservoir in the Borborema province has had any confirmed reports of macroseismic effects. Although the majority of dams in that region are lower than 50 m, there is no indication that the higher natural seismicity makes NE Brazil more hazardous to RIS, compared with other regions of the country.

In the southern part of the Paraná basin (south of 25°S) no activity has been reported in 13 reservoirs, most of them deeper than 50 m (Fig. 5b). In this part of the country, only about five dams have been instrumentally monitored (only those finished after 1980), none of which showed any seismicity (i.e., either no events or magnitudes less than about 1). This area also has extremely low natural intraplate seismicity (Fig. 5a).

A concentration of six RIS cases, three of them with magnitudes around 4 m_b, can be observed in the northeastern part of the Paraná basin; an area with about 24 reservoirs, half of them deeper than 50 m (Fig. 5b). The natural seismicity of the NE part of the Paraná basin seems to be lower than other seismic areas in Brazil (Fig. 5a). The natural seismicity seems to occur near the NE margin of the Paraná basin, somewhat offset from the induced seismicity. In the NE part of the Paraná basin, besides the concentration of RIS cases, three cases of seismicity induced by artesian wells have been observed (BERROCAL *et al.*, 1984; YAMABE and HAMZA, 1996). No explanation for this apparently higher induced hazard has been found yet.

Despite the small number of cases, it seems that RIS hazard in Brazil is not uniform: some areas seem to be more prone to induced seismicity (such as the NE part of the Paraná basin) while the RIS hazard in other areas are much lower (such as the southern part of the Paraná basin). No correlation between RIS hazard and the level of natural seismicity can be observed with the present data.

Considerable emphasis has been given in recent decades to RIS in deep (more than ~100 m) reservoirs. Although deep reservoirs clearly have a decidedly higher potential for causing induced seismicity, Tables 1 and 2 show that shallower reservoirs can also cause seismicity, with significant engineering and social concern (MM intensities up to VI) and that the 100 m figure should not be used to reduce studies of RIS in smaller reservoirs.

Acknowledgements

We thank all Brazilian hydroeletric power companies (CEMIG, CESP, CHESF, COPEL, ELETRONORTE, ELETROSUL, FURNAS, ITAIPU) for allowing us to publish results of many years of seismic monitoring of their reservoirs. We also thank Luís Carlos Ribotta and Jesús Berrocal for fruitful discussions, and Marlon Pirchiner for his unpublished results of Miranda epicenters. We are grateful to Pradeep Talwani and David Simpson for their valuable comments which enhanced the paper. Work was supported by CNPq grants 30.0227/79, 30.0537/00-0 and FAPESP 97/36040-4.

References

Assumpção, M. (1983), *A Regional Magnitude Scale for Brazil*, Bull. Seismol. Soc. Am. *73*, 237–246.

Assumpção, M. (1992), *The Regional Intraplate Stress Field in South America*, J. Geophys. Res. *97*, 11.889–11.903.

Assumpção, M., Freire, M., and Ribotta, L. C. (1995), *Sismicidade induzida no reservatório de Capivara: resultados preliminares sobre localização de fraturas ativas*, Proc. IV Intern. Congr. Braz. Geophys. Soc., vol. *2*, 961–964.

Assumpção, M., Barbosa, J. R., Berrocal, J., Bassini, A., Veloso, J. A. V., Mârza, V., Huelsen, M., and Ribotta, L. C. (1997), *Seismicity Patterns and Focal Mechanisms in SE Brazil*, Rev. Bras. Geofísica *15*, 119–132.

Assumpção, M., *Sismotectónica y esfuerzos en Brasil*. Física de la Tierra, *10*, 149–166. In *Sismicidad y Sismotectónica de Centro y Sudamérica*, (eds. E. Buforn and A. Udías) (Univ. Complutense, Madrid. 1998a).

Assumpção, M. (1998b), *Focal Mechanisms of Small Earthquakes in SE Brazilian Shield: A Test of Stress Models of the South American Plate*, Geophys. J. Int. *133*, 490–498.

Baecher, G. B and Keeney, R. L. (1982), *Statistical Examination of Reservoir-induced Seismicity*, Bull. Seismol. Soc. Am. *72*, 553–569.

Bell, M. L. and Nur, A. (1978), *Strength Changes due to Reservoir-induced Pore Pressure and Stresses and Application to Lake Oroville*, J. Geophys. Res. *83*, 4469–4483.

Berrocal, J., Assumpção, M., Antezana, R., Dias Neto, C. M., França H., and Ortega, R. (1983), *Seismic Activity in Brazil in the Period 1560–1980*, Earthq. Predic. Res. *2*, 191–208.

Berrocal, J., Assumpção, M., Antezana, R., Dias Neto, C. M., Ortega, R., França, H., and Veloso, J. (1984), *Sismicidade do Brasil*. IAG-USP/CNEN, São Paulo, Brazil, 320 pp.

Caproni, N. and Armelin, J. L. (1990), *Instrumentação das escavações subterrâneas da UHE Serra da Mesa*. Simpósio sobre Instrumentação Geotécnica de Campo – SINGEO90, ABGE, São Paulo, vol. 1, 249–257.

Castle, R. O., Clark, M. M., Grantz, A., and Savage, J. C. (1980), *Tectonic State: Its Significance and Characterization in the Assessment of Seismic Effects Associated with Reservoir Impounding*, Engin. Geology *15*, 53–99.

CBGB (1999), *Registro Nacional de Barragens*, Brazilian Committee on Large Dams, CD-ROM.

CEMIG (1995), *Mapa de integração geológica das usinas de Nova Ponte, Miranda e Capim Branco, escala 1:100.000*, CEMIG-Companhia Energética de Minas Gerais.

ELETRONORTE (1977), *Usina Hidrelétrica de Tucuruí, Projeto Básico – Apêndice geológico – geotécnico e de materiais*. Consórcio Engevix-Themag, vol. IV.

Feng, Deyi, Xuejun Yu and Jinqping Gu (1995), *Assessment of Potential Strength of an Induced Earthquake by Using Fuzzy Multifactorial Evaluation*, Pure appl. geophys. *145*, 149–153.

Ferreira, J. M., Oliveira, R. T., Assumpção, M., Moreira, J. A. M., Pearce, R. G., and Takeya, M. (1995), *Correlation of Seismicity and Water Level in the Açu Reservoir – An Example from NE Brazil*, Bull. Seismol. Soc. Am. *85*(5), 1483–1489.

FERREIRA, J., OLIVEIRA, R. T., TAKEYA, M. K., and ASSUMPÇÃO, M. (1998), *Superposition of Local and Regional Stresses in NE Brazil: Evidence Form Focal Mechanisms around the Potiguar Marginal Basin*, Geophys. J. Int. *134*, 341–355.

GOMIDE, L. C. (1999), *Nature and History of Reservoir-induced Seismicity in Brazil*, M.Sc. Thesis, Univ. of South Carolina, 60 pp.

GUPTA, H. K. (1992), *Reservoir-induced Earthquakes*, Developments in Geotechnical Engineering *64*, Elsevier, 364 pp.

HEALEY, J. H., RUBEY, W., GRIGGS, W., and RAYLEIGH, C. B. (1968), *The Denver Earthquakes*, Science *161*, 1301–1310.

KNUEPFER, P. L., PACKER, D. R., and WITHERS, R. J. (1979). *An Analysis of Reported Cases of Reservoir-Induced Seismicity and the Potential for Future Occurrences*, Simpósio sobre Sismicidade Natural e Induzida, ABGE, São Paulo, Proceedings, 69–106.

LIMA, C., NASCIMENTO, E., and ASSUMPÇÃO, M., (1997), *Stress Orientations in Brazilian Sedimentary Basins from Breakout Analysis – Implications for Force Models in the South American Plate*, Geophys. J. Int. *130*(1), 112–124.

MÂRZA, V. I., BARROS, L. V., SOARES, J. E. P., CARVALHO, J. M., CHIMPLIGANOND, C. N., and CAIXETA, D. F. (1999a), *A Precursory Seismicity Pattern Associated to the Nova Ponte (MG) Reservoir-induced Main shock of 1998 May 22 ($m_R = 4.0$)*, Proc. 6th Int. Congr. Brazilian Geophys. Soc., Rio de Janeiro, Brazil, 4 pp (CD-ROM).

MÂRZA, V. I., BARROS, L. V., SOARES, J. E. P., CARVALHO, J. M., FONTENELLE, D. P., CHIMPLIGANOND, C. N., CAIXETA, D. F., GOMES, I. P., FURTADO, G. O., CARIM, A. L., SOUZA, I. F., CALIMAN, E. H., and BARROS, J. B. (1999b), *Some Aspects of Reservoir-induced Seismicity in Brazil*, Proc. 23rd Brazilian Congr. on Large Dams, Belo Horizonte, Brazil, vol. 1, 199–211 (in Portuguese with English abstract).

MENDIGUREN, J. A. (1980), *A Procedure to Resolve Areas of Different Source Mechanisms When Using the Method of Composite Nodal Plane Solution*, Bull. Seismol. Soc. Am. *70*, 985–998.

MIOTO, J. A., and RIBOTTA, L. C. (1995), *Atividade sísmica nas proximidades do reservatório da barragem de Jaguari, Rio Jaguari (SP)*. Proc. 4th Int. Congr. Brazilian Geophys. Soc., vol. II, pp. 983–986.

MIOTO, J. A, RIBOTTA, L. C., and VERDIANI, A. C. (1991), *Aspectos Geológico-Estruturais da Sismicidade Relacionada ao Reservatório de Capivara (SP/PR)*, Proc. 2nd International Congress of the Brazilian Geophys. Soc., vol. 1, 513–520.

RIBOTTA, L. C. (1989), *Magnitude Local e Análise Preliminar do Risco Sísmico para Atividade Sísmica Induzida no Reservatório de Parabuna/Paratinga*, Proc. 1st International Congr. Brazilian Geophys. Soc., vol. I, 384–391.

ROELOFFS, E. A. (1988), *Fault Stability Changes Induced Beneath a Reservoir with Cyclic Variation in Water Level*, J. Geophys. Res. *93*, 2107–2124.

SIMPSON, D. W. (1976), *Seismicity Changes Associated with Reservoir Loading*, Eng. Geology *10*, 123–150.

SIMPSON, D. W. and NEGMATULLAEV, S. K. (1981), *Induced Seismicity at Nurek Reservoir, Tadjikistan, USSR*. Bull. Seismol. Soc. Am. *71*, 1561–1586.

SIMPSON, D. W. (1986), *Triggered Earthquakes*, Ann. Rev. Earth Planet. Sci. *14*, 21–42.

SIMPSON, D. W., LEITH, W. S., and SCHOLZ, C. H. (1988), *Two Types of Reservoir-induced Seismicity*, Bull. Seismol. Soc. Am. *78*, 2025–2040.

SIMPSON, D. W., GHARIB, A. A., and KEBEASY, R. M. (1990), *Induced Seismicity and Changes in Water Level at Aswan Reservoir, Egypt*, Gerlands Beitr. Geophys. *99*, 191–204.

SIMPSON, D. W. and NARASIMHAM, T. N. (1990), *Inhomogeneities in Rock Properties and their Influence on Reservoir-induced Seismicity*, Gerlands Beitr. Geophys. *99*, 205–219.

TALWANI, P. and ACREE, S. (1984/85*)*, *Pore Pressure Diffusion and the Mechanism of Reservoir-induced Seismicity*, Pure appl. geophys. *122*, 947–965.

TALWANI, P. (1995), *Two Categories of Reservoir-induced Seismicity*, Proc. Int. Symp. on Reservoir-induced Seismicity (ISORIS'95), pp. 44–64.

TALWANI, P. (1997), *On the Nature of Reservoir-induced Seismicity*, Pure appl. geophys. *150*, 473–492.

TALWANI, P. (2000), *Seismogenic Properties of the Crust Inferred from Recent Studies of Reservoir-induced Seismicity – Application to Koyna*, Current Science, in press.

VELOSO, J. A. V., *Cases of RIS in the Brazilian Amazon Area.* In Proc. Tenth World Conference on Earthquake Engineering, 1992, Madrid, Spain, (A. A. Balkema, Rotterdam 19929), vol. 1, pp. 269–273.

VELOSO, J. A. V. (1992b), *Terremotos induzidos pelo homem*, Ciência Hoje *14*, 66–72.

VELOSO, J. A. V. and ASSUMPÇÃO, M. (1986), *Estudo da Sismicidade Associada ao Enchimento do Reservatório de Emborcação-CEMIG*, Proc. 34°Congr. Bras. Geol., vol. 6, pp. 2617–2627.

VELOSO, J. A. V., CARVALHO, J. M., FERNANDES, E. P., BLUM M. L. B., and ARAUJO, D. P. (1991), *Microearthquakes and the Balbina Lake, a Possible Case of Induced Seismicity in the Amazon Area*, Proc. 2nd International Congress of the Brazilian Geophys. Soc. (Salvador, Brazil), vol. II, pp. 508–512.

VELOSO, J. A. V., CARVALHO, J. M., HUELSEN, M. G., GOMIDE, L. C., and CHIMPLIGANOND, C. N. (1994), *Recent Seismic Activity in Nova Ponte Reservoir Area, Brazil.* Regional Seismological Assembly in South America, Brasília, Abstracts, 22.

VIOTTI, C. B., VELOSO, J. A. V., and GOMIDE, L. C. (1997), *Induced Seismicity at Cajuru Reservoir, Minas Gerais, Brazil*, Proc. 19th Int. Congr. on Large Dams, (Florence, Italy) pp. 1211–1225.

YAMABE, T. H. and HAMZA, V. (1996), *Geothermal Investigations in an Area of Induced Activity, NE São Paulo State, Brazil*, Tectonophys. *253*, 209–225.

ZOBACK, M. D. and HICKMAN, S. (1982), *Physical Mechanisms Controlling Induced Seismicity at Monticello Reservoir, South Carolina*, J. Geophys. Res. *87*, 6959–6974.

(Received May 9, 2000, revised/accepted November 13, 2000)

Note added in proof: In the Balbina reservoir (#6 in Table 1), an event with 3.7 m_R occurred on May 27, 2001, being felt near the lake. This characterizes another case of delayed activity (13.5 years delay) long after the initial activity shown in Table 2.